APPLIED GROUNDING & BONDING

Based on the 2023 *National Electrical Code*®

electrical training
IBEW - NECA ALLIANCE

Applied Grounding and Bonding is intended to be an educational resource for the user and contains procedures commonly practiced in industry and the trade. Specific procedures vary with each task and must be performed by a qualified person. For maximum safety, always refer to specific manufacturer recommendations, insurance regulations, specific job site and plant procedures, applicable federal, state, and local regulations, and any authority having jurisdiction. The *electrical training ALLIANCE* assumes no responsibility or liability in connection with this material or its use by any individual or organization.

Contents

Chapter 1 INTRODUCTION XIV

Grounding and Bonding for Safety 2
Earth in the Circuit 2
Grounding Concepts 3
Bonding Concepts 4
Minimizing Shock Hazards 5

Path for Current through the Body 6
Grounding and Bonding Concepts Together 8
Performance *Code* Language 9
Electrical Systems 10

Chapter 2 CIRCUIT BASICS AND OVERCURRENT PROTECTION 22

Circuit Fundamentals 24
Ohm's Law 24
Opposition to Current in Circuits 28
Current in Circuits (Normal and Fault Current)... 30
Overcurrent Protection Basics 31

Amperes Operate Overcurrent
    Protective Devices 33
Time and Current 35
Safety by System Design 41
Equipment Grounding Conductor Capacity 43

Chapter 3 USING THE *NATIONAL ELECTRICAL CODE* 48

Building a Solid Foundation 50
NEC Arrangement and Application 50
Enforcement and Approvals 52
Requirements, Exceptions, Alternatives,
    and Information 53
Permissive *Code* Language 54
Explanatory Information 54
Bracketed Text 54
Defined Terms 55
Code-Making Panel Responsibilities 55
NEC Article 100—Definitions of Grounding and
    Bonding Terms 56
Article 200—Use and Identification of
    Grounded Conductors 64

Article 250—Arrangement and Use 65
Part I—General 65
Table 250.66—Sizing AC Grounding Electrode
    Conductors 66
Table 250.102(C)(1)—Sizing the Grounded
    Conductor, Main Bonding Jumper, System
    Bonding Jumper, and Supply-Side Bonding
    Jumper for Alternating-Current Systems 67
Table 250.122—Sizing Equipment Grounding
    Conductors 68
Special Occupancies, Equipment,
    and Conditions 69
Chapter 8 71

Contents

Chapter 4 GROUNDING ELECTRODES AND THE GROUNDING ELECTRODE SYSTEM 74

Introduction 76
Grounding Electrode Defined...........................76
Purpose and Performance of Electrodes.......... 77
Grounding Electrode System Requirements...... 79

Establishing a Grounding Electrode System 79
Mandatory Grounding Electrodes 80
Types of Grounding Electrodes 81
Grounding Electrode Installation Requirements .. 85

Chapter 5 REQUIREMENTS FOR GROUNDED CONDUCTORS AT SERVICES... 98

Grounded Utility Supply Systems.....................100
First Line of Defense 101
Grounding Scheme for Services......................102
Grounded Conductor Routing
    and Connections 104
Main Bonding Jumpers in Service Equipment ..105
Dual-Fed Service Equipment..........................107
Minimizing Impedance in Service
    Grounded Conductors 107
Functions (Purposes) of the Grounded
    Service Conductor 109
Grounded Neutral Conductor 110

Requirements for Service Equipment
    (Listing)... 116
Grounded Conductor (Neutral) Disconnect
    Requirement for Services 119
Requirements for Services Supplied by
    Ungrounded Systems 120
Marking Equipment for
    Ungrounded Systems 121
Grounding of Service Raceways
    and Enclosures 122
Supply-Side Grounding at Other
    Than a Service 122

Chapter 6 GROUNDING ELECTRODE CONDUCTORS................................ 126

Grounding Electrode Conductor Basics...........128
Purpose of Grounding Electrode Conductors...128
Current in Grounding Electrode Conductors 130
Grounding Electrode Conductor Material.........130
Sizing Grounding Electrode Conductors 131
Using Table 250.66 133
Grounding Electrode Conductors
    For DC Systems.....................................134

Grounding Electrode Conductor Installation.....135
Effectiveness (Integrity) of the
    Grounding Path......................................144
Grounding Electrode Conductor
    Connection Locations.............................146
Grounding Electrode Conductor Connections ..148
Magnetic Field Concerns...............................150

Contents

Chapter 7   BONDING REQUIREMENTS 156

Definitions of Bonding Terms 158
Bonding Performance Criteria 158
Maintaining Continuity 160
Bonding Connections (Wire-Type Conductors) .. 163
Cleaning Coated Surfaces 163
Bonding Jumper
    and Bonding Conductor Length 164
Equipment Bonding Jumpers
    (Function and Purpose) 164
Sizing Requirements for the Supply Side
    and Load Side Applications 165

Service Bonding Rules (Supply Side) 166
Reducing Washers 168
Boxes With Concentric or
    Eccentric Knockouts 169
Sizing Supply-Side Bonding Jumpers
    (Wire Types) 169
Load Side Bonding Rules 170
Bonding Around Expansion Fittings and
    Loose-Joined Metal Raceways 173
Bonding Other Metal Parts 174

Chapter 8   EQUIPMENT GROUNDING CONDUCTORS 182

Equipment Grounding Conductor Defined 184
Equipment Grounding Conductor Basics 184
Equipment Grounding Conductor Installations .. 192
Equipment Grounding Conductor Connections .. 194

Equipment Grounding
    Conductor Identification 195
Equipment Grounding Conductor Sizing 197
Current in Equipment Grounding Conductors .. 202

Chapter 9   GROUNDING ELECTRICAL EQUIPMENT 206

Purpose of Grounding Equipment 208
General Grounding Rules for Equipment 208
Conductor Enclosure and Raceway Grounding
    Requirements 212
Methods of Grounding Equipment
    (Part VII of Article 250) 213
Connections of Equipment
    Grounding Conductors 216
Receptacle Grounding Connections 216

Receptacle Replacements 223
Grounding Appliances Using the
    Grounded Conductor 224
Auxiliary Grounding Electrode Requirements ... 225
Grounding Nonelectrical Equipment 227
Equipment Grounded by Secure
    Metal Supports 227
Use of the Grounded Conductor
    for Grounding 228

Contents

Chapter 10 ISOLATED/INSULATED GROUNDING CIRCUITS AND RECEPTACLES..........232

Electromagnetic Interference (EMI) in Grounding Circuits 234

Purpose of Isolated Grounding Circuits and Receptacles 236

Objectionable Currents in Grounding Paths 237

Power Quality System Grounding Analysis...... 238

Isolated Grounding Circuits 239

Use of Auxiliary Grounding Electrodes 244

Grounding and Bonding in Information Technology Centers 244

Signal Reference Structures (Grids) 246

Surge Protection 248

Chapter 11 GROUNDING AT SEPARATE BUILDINGS OR STRUCTURES 252

Definitions 254

Supplying Power to Separate Buildings or Structures... 254

Purpose of Grounding and Bonding at Separate Buildings or Structures........ 254

Grounding Electrode Requirement 255

Grounding Electrode Conductor 256

Feeder and Branch Circuit Requirements........ 258

Ungrounded Systems Supplying Separate Buildings or Structures.......................... 260

Supplied by a Separately Derived System 260

Building Disconnecting Means Requirements...262

Metal Water Pipe Bonding and Other Bonding............................... 263

Disconnecting Means Remote from Building or Structure 263

Buildings or Structures Supplied by Separately Derived Systems.............. 264

Buildings or Structures Supplied by an Ungrounded System....................265

Buildings or Structures Supplied by Generators..................................... 266

Chapter 12 GROUNDING ELECTRICAL SYSTEMS 272

Definitions 274

System Grounding...274

Methods of System Grounding 276

System Grounding Requirements 276

Mandatory System Grounding 277

Optional System Grounding...........................281

System Grounding Prohibition 285

Impedance Grounded Systems...................... 286

Ungrounded Systems (Concepts)...................287

Contents

Chapter 13 GROUNDING AND BONDING FOR SEPARATELY DERIVED SYSTEMS 290

Definitions 292
Determining a Separately Derived System 292
Grounding Requirements 293
Grounded Systems.....................................293
Grounding Electrodes.................................. 299
Bonding Water Piping and Building Steel 302

Outdoor Source 303
Ungrounded Systems 304
Generators and Transfer Equipment 305
Grounding DC Systems.................................311
Ungrounded DC Systems..............................313

Chapter 14 SPECIAL OCCUPANCIES AND CONDITIONS.............................318

Special Rules for Hazardous Locations 320
Special Rules for Health Care Facilities 324
Special Rules for Agricultural Installations 334

Mobile and Manufactured Home Grounding
    and Bonding Rules 338
Special Rules for Marinas, Boatyards,
    and Docking Facilities............................ 340

Chapter 15 GROUNDING FOR SPECIAL EQUIPMENT 344

Purpose of Grounding Equipment 346
Electric Signs and Outline Lighting Systems... 346
Electric Cranes and Elevators 351
Information Technology Equipment
    and Sensitive Electronic Equipment.........352

Grounding and Bonding Requirements for
    Swimming Pools and Similar Installations..356
Grounding Requirements
    for Solar PV Systems 369

Chapter 16 GROUNDING AND BONDING FOR COMMUNICATIONS SYSTEMS AND EQUIPMENT 376

Performance and Concepts 378
Definitions 378
Grounding and Bonding Performance 379
Connecting to a Grounding Electrode 380
Grounding Electrode Conductor Installation.....381
Intersystem Grounding and Bonding 381

Common Grounding and Bonding Rules
    for Communications Systems 384
Grounding and Bonding at Mobile Homes....... 389
Radio and Television Equipment
    and Antennas...................................... 390
Overvoltages and Lightning Events.................392

Contents

Chapter 17 GROUND-FAULT CIRCUIT INTERRUPTERS AND GROUND-FAULT PROTECTION OF EQUIPMENT 396

Ground-Fault Circuit Interrupters (GFCIs) 398

Ground-Fault Protection of Equipment (GFPE) ...406

Chapter 18 GROUNDING RULES FOR MEDIUM- AND HIGH-VOLTAGE SYSTEMS 420

Requirements for Grounding Systems 422

Grounding Methods for Systems
    Over 1,000 Volts 422

Solidly Grounded Systems 423

Grounding Service Supplied Alternating
    Current Systems 425

Impedance Grounding 426

Portable or Mobile Equipment Grounding 427

Grounding Equipment 428

Substation Grounding Requirements 430

Conductor Shielding and Stress Reduction..... 433

Grounding Through Surge Arresters 435

Engineered Grounding System Designs...........437

Common Grounding
    and Bonding Components...................... 440

Annexes 444

Annex A – Investigation and Testing of
    Footing-Type Grounding Electrodes
    for Electrical Installations 445

Annex B – Steel Conduit and EMT
    for Equipment Grounding 455

Annex C – Selecting Protective Devices -
    Short Circuit Current Calculations........... 467

Annex D – Chapter 9, Table 8
    Conductor Properties 479

Annex E – Lightning Protection Systems 481

Index 485

Acknowledgments

Technical Review

Johnston, Michael, NECA

electrical training ALLIANCE Staff Contributor

Palmer Hickman,
   Director of *Code* and Safety Training and
   Curriculum Development, technical editor

QR Codes

Consumer Product Safety Commission
Eaton
Electro Specialties, Inc.
National Electrical Contractors Association (NECA)
National Electrical Installation Standards (NEIS)
Steel Tube Institute
UL, LLC (UL Product IQ)
U.S. Department of Energy (DOE)

Content, Photograph, and Illustration Contributions

ABB
Atkins, Derrick
Cogburn Brothers, Inc.
Colgan, Rob, NECA
Cook, Donald R., Shelby County, AL
Dollard, Jr., James T., Philadelphia Electrical JATC
   and IBEW Local 98, Retired
Eaton
Edwards, Christopher
ERICO International Corporation
Fluke Corporation
Garvey, Tom
Harger Lightning and Grounding
Insulated Cable Engineers Association (ICEA)
Institute of Electrical
   and Electronics Engineers (IEEE)
Johnston, Michael, NECA
Maddox, Rick, Clark County, NV
McGovern, Bill, City of Plano, TX
Morse Electric Inc.
National Electrical Contractors Association (NECA)
National Fire Protection Association (NFPA)
Pass and Seymour/Legrand
Schneider Electric Square D Company
Siemens
Southwire Company, LLC
Steel Tube Institute
UL, LLC
Valley Electric Company VEC
Washington DC Joint Apprenticeship and Training
   Local 26 IBEW
Young Electric Sign Company (YESCO)

Features

Figures, including photographs and artwork, clearly illustrate concepts from the text.

Headers and **Sub-headers** organize information within the text.

Code **Excerpts** are "ripped" from the 2023 *National Electrical Code.*

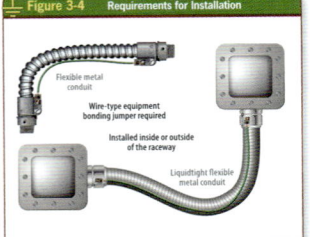

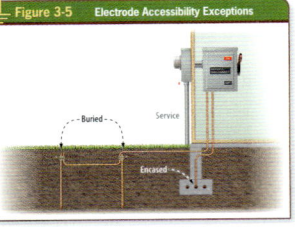

Fact boxes offer additional information related to Grounding and Bonding.

For additional information related to QR Codes, visit qr.njatcdb.org Item #1079

Quick Response Codes (QR Codes) create a link between the textbook and the Internet. They can be scanned using Smartphone applications to obtain additional information online. (To access the information without using a Smartphone, visit qr.njatcdb.org and enter the referenced Item #.)

Features

Clear, easy-to-read **Contents** pages in the front of the textbook and inside each Chapter enable the reader to quickly find important grounding and bonding content.

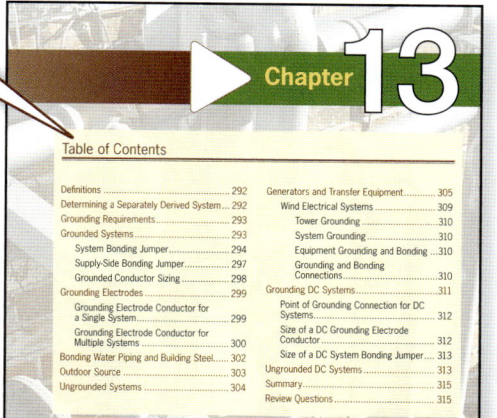

The **Introduction** and **Objectives** at the beginning of each chapter introduce readers to the concepts to be learned in each chapter.

At the end of each chapter, a concise chapter **Summary** and **Review Questions** reinforce the important concepts included in the text.

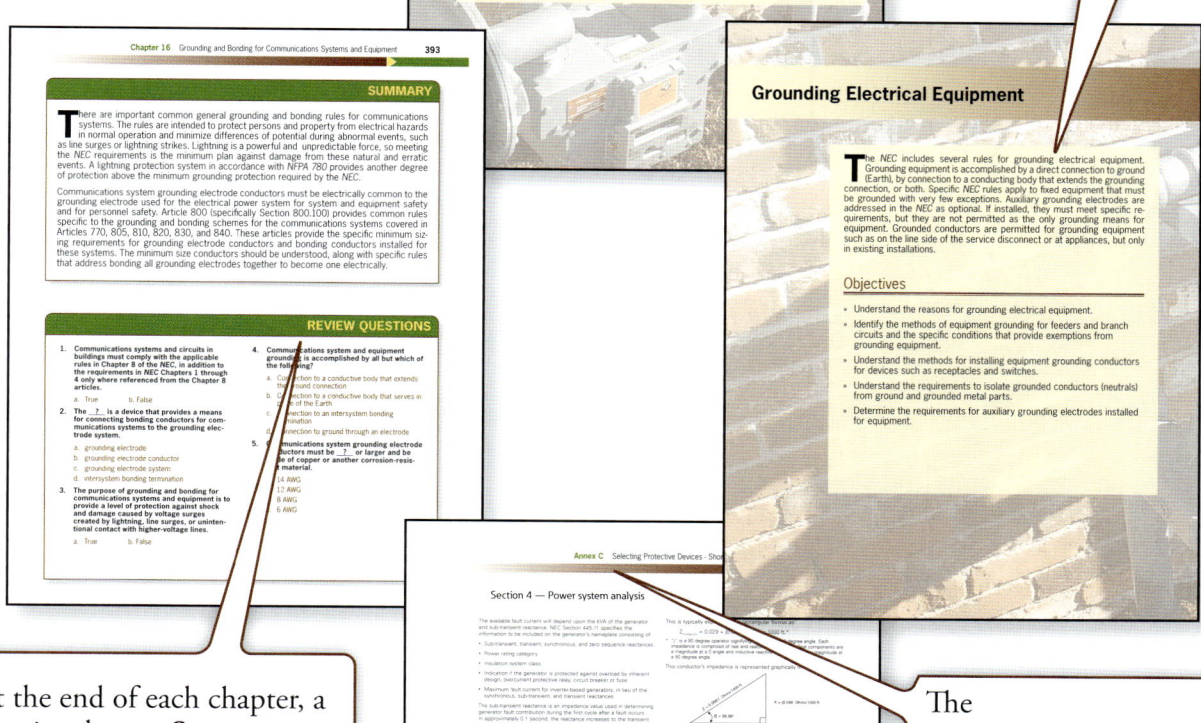

The **Annexes** include supplemental information related to grounding and bonding.

Introduction

Safe electrical circuits and systems depend on *Code*-compliant electrical grounding and bonding. Grounding and bonding are integral functions that must perform effectively to ensure safety for persons and property as required by the *National Electrical Code* (*NEC*). There are many myths and a lot of confusion related to grounding and bonding. This textbook aims to take the mystery out of this topic.

Grounding and bonding is not a complicated or difficult subject. This training material breaks down this subject and simplifies it for instructors and students. These concepts are much easier to learn once one achieves an understanding of how and why circuits and systems work the way they do. A good working knowledge about electrical circuits and current will allow the reader to understand how proper grounding and bonding results in circuits and systems that are essentially safe. It is equally important to develop a thorough understanding of the *NEC* and how to apply it to electrical installations.

There are many words and terms used in the electrical field—some slang, some defined in the *NEC*. The best approach to training students is to use defined terms for a common understanding of the rules. This training material places significant emphasis on the use and understanding of defined grounding and bonding terms. Using slang terms such as "ground wire" and "stingers" can result in the misapplication of *Code* requirements.

This material is built on the principles of establishing a solid foundation in the subject of grounding and bonding. Electrical Workers are trained in the classroom to understand theoretical operation of circuits and systems. Hands-on knowledge is gained in the field through the on-the-job segment of apprenticeship programs. The *electrical training ALLIANCE* Applied Grounding and Bonding curriculum is designed for an optimal teaching and learning experience with the instructor, veteran, and student in mind. Essential performance concepts are presented in a straightforward and concise fashion, making a challenging subject easy to master. Effective use of visual support, including actual images of products and photos from the field, assists workers and students of the *Code* in developing their knowledge of grounding and bonding while enhancing their abilities to accurately apply the *NEC* grounding requirements. The complex issues related to grounding and bonding systems are made simple through easy-to-read text and detailed explanations and diagrams that ensure an engaging learning experience. The author hopes that the information in this training program will be passed on to others entering this rewarding field in an industry that is second to none in skills, attitude, and knowledge.

The objective of this training material is to assist students of the electrical industry in developing and mastering the subjects of electrical grounding and bonding. This textbook starts by introducing basic performance concepts and graduates to more complex applications of grounding and bonding circuits used in electrical wiring systems. It is helpful to build knowledge of a subject in an order similar to the order of the tasks and phases of a construction project. The program is arranged in a fashion similar to that of a construction project. In other words, the starting point is in the ground (at the grounding electrodes), and the subjects are covered carefully and in detail through the entire electrical system, all the way to the final branch-circuit outlet and utilization equipment connections. Before getting into how to build the grounding and bonding circuits and systems, the textbook reviews various fundamentals to reinforce current electrical knowledge and elementary understanding of the *Code* to establish a good foundation for the more complex topics that follow.

About This Book

Applied Grounding and Bonding is an authoritative textbook on one of the most critical topics in the industry. It was originally developed and authored by Michael J. Johnston, NECA's executive director of codes and standards. Johnston is recognized as one of the leading experts in the field of grounding and bonding and is past chair and a longtime principal member of the *NEC* Correlating Committee. Prior to his service on the *NEC* Correlating Committee, Johnston most recently served as chairman and a longtime member of *NEC Code*-Making Panel 5, which is responsible for Article 250, Grounding and Bonding, in the *NEC*.

The textbook is uniquely designed to follow field installations the way a contractor and an Electrical Worker would install the grounding and bonding system in actual practice. Topics covered include "traditional" topics such as service, feeder, and branch circuit grounding and bonding, as well as more specialized and advanced topics such as grounding and bonding in information technology rooms, grounding and bonding in health care facilities, grounding and bonding in hazardous (classified) locations, and grounding and bonding requirements for medium- and high-voltage systems.

Original Contributing Developer

Michael J. Johnston is currently NECA's executive director of codes and standards. Prior to working with NECA, Johnston worked for the International Association of Electrical Inspectors as the director of education, codes, and standards. He also worked as an electrical inspector and electrical inspection field supervisor for the City of Phoenix, AZ, and achieved all IAEI and ICC electrical inspector certifications. Johnston achieved a BS in Business Management from the University of Phoenix. He served on *NEC Code*-Making Panel 5 in the 2002, 2005, 2008 cycles and chaired the *Code*-Making Panel 5 representing NECA during the 2011 *NEC* cycle. He is a past chair and former member of the *NEC* Correlating Committee and is a current member of the NFPA Standards Council. Among his responsibilities for managing the codes, standards, and safety functions for NECA, Johnston is secretary of the NECA Codes and Standards Committee. Johnston has been a consistent contributor to the *electrical training ALLIANCE* curriculum, authoring titles such as *Health Care Systems, Hazardous Locations, Applied Grounding and Bonding*, and *Significant Changes to the NEC*. Johnston is a member of the IBEW and has experience as an electrical Journeyman wireman, foreman, and project superintendent. He has achieved Journeyman and master electrician licenses in multiple states and is an active member of IAEI, ICC, SES NFPA, ASSP, the NFPA Electrical Section, Education Section, the UL Electrical Council, and the National Safety Council.

About the Subject Matter Expert Updating this Edition

Derrick Atkins is the training director at the Minneapolis Electrical JATC. He holds a Bachelor of Science degree in physics from the University of Minnesota and an associate of applied science degree in electrical construction and maintenance at Dunwoody College of Technology. He also holds a Class A Master Electrical license with the state of Minnesota.

Atkins joined the IBEW Local 292 as an apprentice electrician in 1996 and completed his apprenticeship in 2000. He has worked as a foreman and general foreman on a variety of industrial projects. In 2005, he taught full-time as an instructor for the Minneapolis Electrical JATC, and two years later worked as a project manager and estimator with his primary focus as design build of various industrial projects. He then returned to full-time teaching at the JATC and is now currently serving as the training director.

Introduction

Electrical Workers must be familiar with the basic grounding and bonding concepts and the related performance provisions contained in the *National Electrical Code®* *(NEC)*. Concepts such as what *ground* is, what *grounding* is, what *electrical bonding* is, and how and why these circuit elements work will help the Electrical Worker master knowledge in electrical grounding and bonding. Grounding and bonding are essential for safe electrical systems. The concepts are straightforward, yet many find them complex and challenging.

Objectives

» Recognize and understand key grounding and bonding terms.

» Understand the role of the Earth in the electrical grounding system.

» Understand the purpose of electrical grounding and bonding and fundamental grounding concepts for systems and equipment.

» Understand fundamentals related to bonding and connecting conductive parts and equipment to establish electrical continuity and conductivity.

» Understand grounding and bonding functions that perform simultaneously in electrical wiring systems.

» Understand performance requirements for electrical grounding and bonding and the application of the *NEC* rules.

» Recognize that if there is no connection to ground (the Earth), there is no grounding.

Chapter 1

Table of Contents

Grounding and Bonding for Safety 2

Earth in the Circuit.............................. 2

Grounding Concepts 3

Bonding Concepts 4

Minimizing Shock Hazards............................ 5

Path for Current through the Body 6

    Series Circuit through the Human Body 6

    Parallel Circuit through the Human Body 7

Grounding and Bonding Concepts Together 8

Performance *Code* Language 9

Electrical Systems.................................... 10

    Grounded Electrical Systems 10

    Ungrounded Electrical Systems.............. 15

Summary...................................... 20

Review Questions 20

GROUNDING AND BONDING FOR SAFETY

Section 90.2(A) of the *NEC,* titled *Practical Safeguarding,* contains the purpose of the *Code.* The *NEC* rules provide a practical safeguarding of people and property from hazards associated with the use of electricity (*NEC* 90.2). The *NEC* is not intended as a design specification or training manual for untrained people. Both grounding and bonding are necessary for safe electrical wiring installations. Electrical installations include safety circuits that accomplish grounding and bonding simultaneously. One function complements and supports the other. When non–current-carrying conductive parts of equipment are bonded together and then connected to the ground, bonding and grounding are accomplished, and enhanced electrical safety is the result.

EARTH IN THE CIRCUIT

From the beginning, the discovery of electricity brought many challenges and questions. Along with difficult debates about distribution of alternating current (AC) systems as compared with direct current (DC) systems, equally challenging decisions had to be made

about electrical grounding. Today, the use and installation of direct current (DC) systems is significantly more common in applications such as small- and large-scale energy storage systems. After much debate and experimentation early on, grounding electrical equipment and systems prevailed as the most popular and effective method of building safety into electrical systems, at least in the United States.

Many arguments have been made for and against electrical system grounding. For example, without grounding, the possibility of completing an electrical circuit through the human body is reduced; this is because the Earth is included in the electrical circuit in a grounded system, making shock hazards more likely. On the other hand, grounded electrical systems and equipment offer the advantage of circuit protection by allowing fuses or circuit breakers to operate during ground-fault conditions.

If electrical equipment or systems are grounded, the Earth is included in the circuit and connected to the electrical system. The terms *ground* and *grounded (grounding)* are defined in Article 100 of the *NEC* as follows:

Ground. The earth. (CMP-5)

Grounded (Grounding). Connected (connecting) to ground or to a conductive body that extends the ground connection. (CMP-5)

One of the most elementary concepts to remember is that throughout the *NEC, electrical grounding* refers to a connection to the Earth—to this planet. This is accomplished by using what is referred to in the *NEC* as a *grounding electrode.* **See Figure 1-1.**

Therefore, with the understanding that the Earth is part of any circuit or system that is grounded, the quality of the Earth as a conductor must be discussed.

The Earth is a large but somewhat poor conductor. The Earth should not be relied on to carry any current, be it normal current or fault current. **See Figure 1-2.**

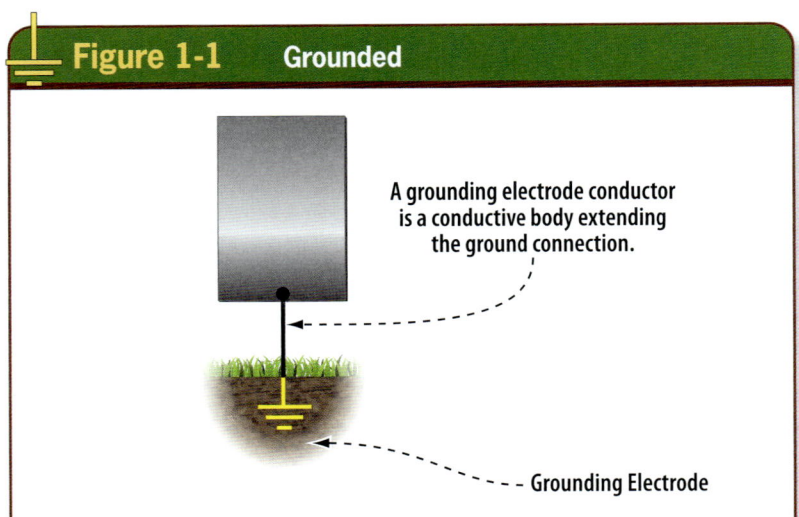

Figure 1-1 **Grounded**

A grounding electrode conductor is a conductive body extending the ground connection.

Grounding Electrode

Figure 1-1. *The term* grounded *means connected to the ground (the Earth) directly or through a conductive body that extends the grounding connection.*

Figure 1-2   The Earth as a Conductor

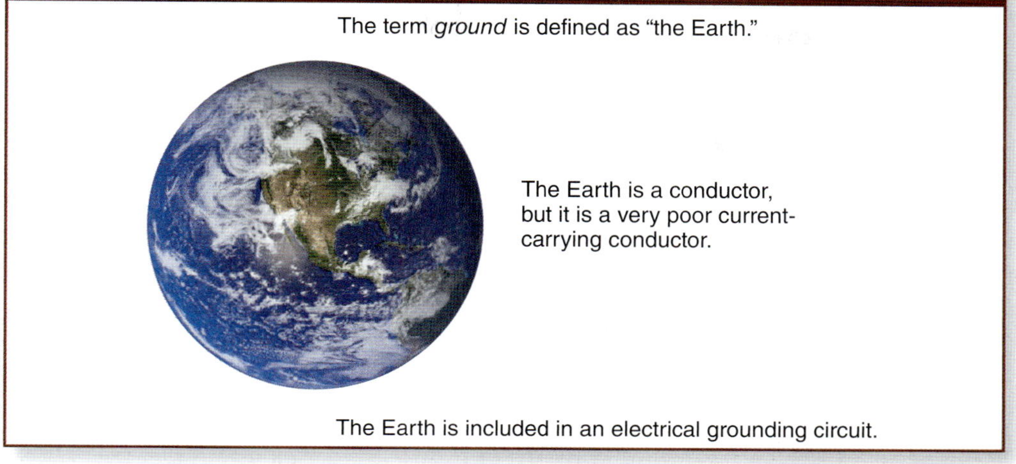

The term *ground* is defined as "the Earth."

The Earth is a conductor, but it is a very poor current-carrying conductor.

The Earth is included in an electrical grounding circuit.

Figure 1-2. When equipment and systems are connected to the Earth, it becomes a conductive path in the circuit, although not an effective path.

The Earth offers significant opposition to current in circuits, including grounding and bonding circuits. This opposition to current is known as either *resistance* or *impedance.* In DC circuits, the primary opposition to current is *resistance,* whereas in AC circuits, the opposition to current is referred to as *impedance.*

GROUNDING CONCEPTS

To understand electrical grounding concepts and performance, getting back to basics is necessary.

Grounding is a function of connecting to the Earth. Buildings and structures are supported by and frequently depend on a good connection to the Earth through a sufficient foundation or footing. **See Figure 1-3.**

Electrical systems are built on a similar foundation. The foundation of a grounded electrical system and grounded equipment is the *grounding electrode* or *grounding electrode system.* When an object qualifies as a grounding electrode, any connection to it renders a grounding function. Since the *NEC* defines the word ground simply as the Earth, if equipment or a system is grounded, it must be connected to this planet—to Earth.

For example, contrary to what many people think, the frame of an automobile does not qualify as grounded because there is no connection to the Earth. It is connected to the negative side of the vehicle battery, not to the ground. The same holds true for an aircraft. The frame of a jet airliner may be used as the reference for the 400-cycle power system onboard, but it is not grounded, as some people might claim. Ground is the Earth, by definition. Therefore, if there is no connection to the ground (the Earth), there is no grounding.

In order to accomplish grounding of equipment or systems, a connection to the planet Earth must be made. This connection is made through grounding

Figure 1-3   Foundations

Figure 1-3. Buildings are constructed on solid foundations. The grounding electrode system is the foundation of a safe electrical system.

Figure 1-4   Grounding and Bonding

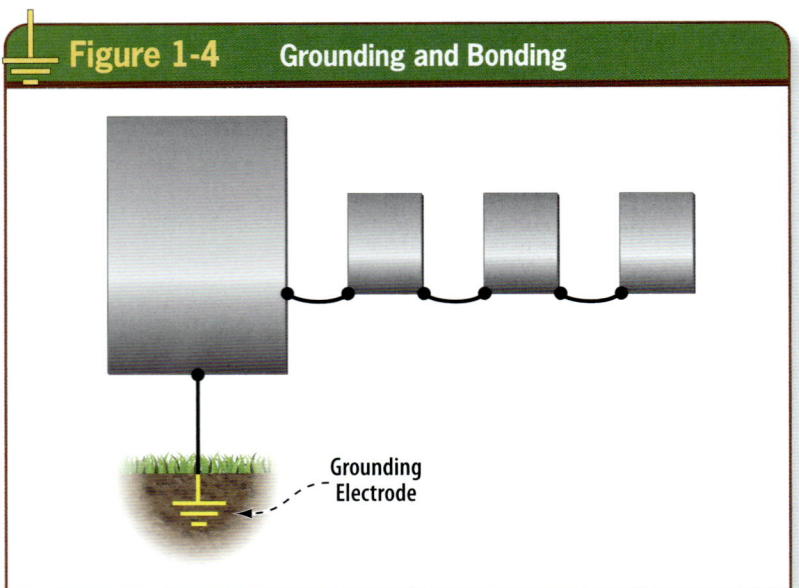

Grounding
Electrode

Figure 1-4. *Grounding is an ongoing process of connecting systems and equipment to the ground (Earth). Bonding is inherent to the grounding process.*

electrodes. Grounding is a function that results from conductive equipment, parts, or systems that are connected to ground (the Earth). The definitions of grounding and bonding terms in Article 100 provide a common means of communication when applying the *NEC* rules to installations and systems. **See Figure 1-4.**

BONDING CONCEPTS

When there is bonding, a connection is involved. The function of bonding is the process of connecting conductive objects together. When there is bonding, two or more conductive objects become one electrically. **See Figure 1-5.**

Bonded (Bonding). Connected to establish electrical continuity and conductivity. (CMP-5)

Bonding minimizes potential (voltage) differences between the conductive parts that are bonded together. Bonding is a function that results in the connection of conductive parts.

Therefore, if electrical equipment is grounded and connected to other equipment, bonding of all the connected equipment occurs and accomplishes grounding (the connection to the Earth).

Bonding is a function that results in non–current-carrying conductive parts being connected together electrically, establishing continuity between them and, in many instances, establishing adequate conductivity for current. Bonding safeguards against possible shock hazards and

Figure 1-5   Bonding

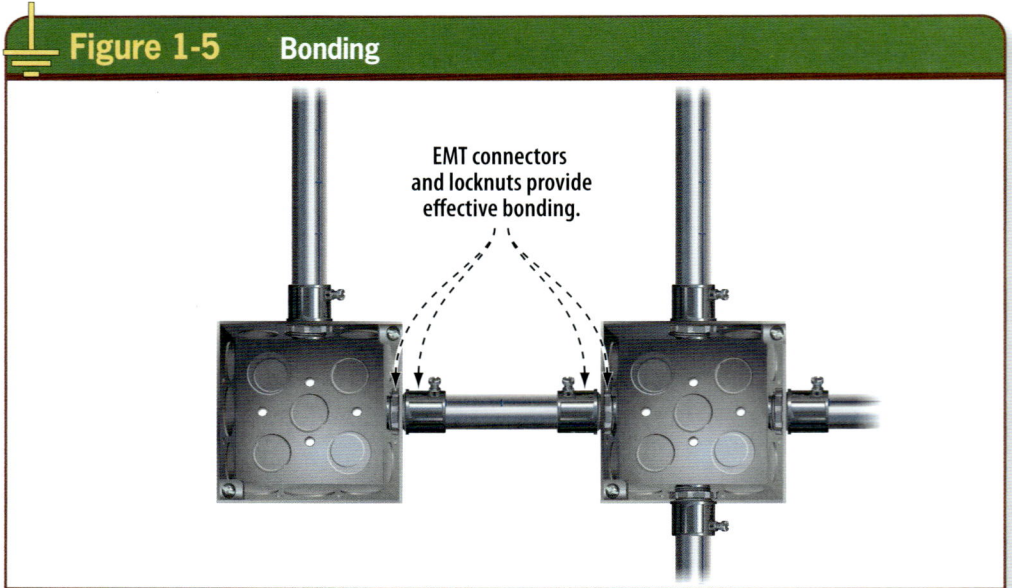

EMT connectors
and locknuts provide
effective bonding.

Figure 1-5. *Bonding is a process that establishes electrical continuity and conductivity between conductive parts.*

ensures effectiveness of a ground-fault current path for overcurrent protective device operation. Ineffective bonding can cause arcing or sparking at loose bonding connections under fault conditions, which can be a fire hazard.

MINIMIZING SHOCK HAZARDS

Grounding and bonding are the safety circuits installed with branch circuits and feeders of electrical systems. Electrical grounding and bonding are two protective processes that often occur simultaneously. These two processes not only provide essential protective functions for equipment and property, but also reduce shock hazards. The process of grounding a system or electrical equipment reduces potential differences between electrical systems, equipment, and the Earth. Bonding is the process of connecting conductive objects together so they become one electrically, thus minimizing potential differences between them while at the same time reducing possibilities of electric shock. **See Figure 1-6.**

Historically, grounding and bonding have been regarded as the primary methods of protecting systems, property, and personnel from electrical shock hazards and fire, while also offering significant advantages in system operation. Over the years, there have been many debates about whether greater protection is achieved by grounding systems and equipment or by isolating or otherwise not grounding electrical systems.

Contrary to the beliefs of many, grounding performs only some of the protective functions necessary for equipment and systems designed and installed according to the *NEC*. While grounding provides the degrees of protection described above, the *NEC* strongly emphasizes the critical role of an effective ground-fault current path in electrical safety systems, and how it relates to sure operation of overcurrent protective devices. This path for ground-fault current is essential in grounded systems.

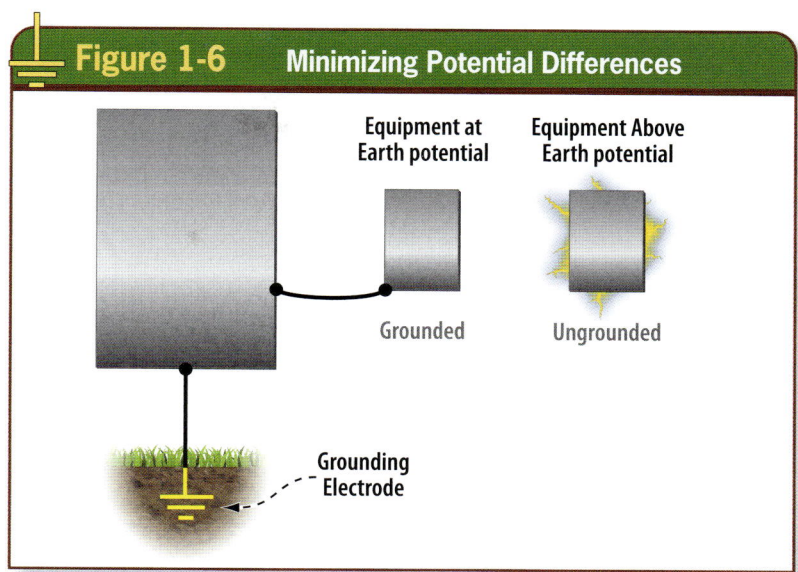

Figure 1-6. Grounding reduces potential differences to the ground in normal operation and during ground-fault events.

During a ground-fault event, current will return to the source through multiple paths, such as through equipment grounding conductors (EGCs), the Earth, conductive parts of equipment, a human body, or any combination of ground-fault return paths. **See Figure 1-7.**

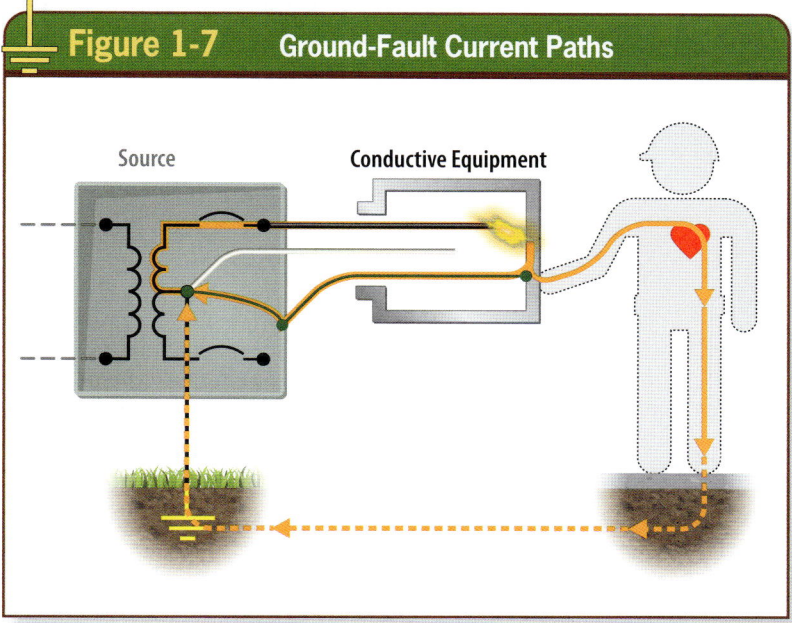

Figure 1-7. Ground-fault current tries to return to the source over any available conductive path.

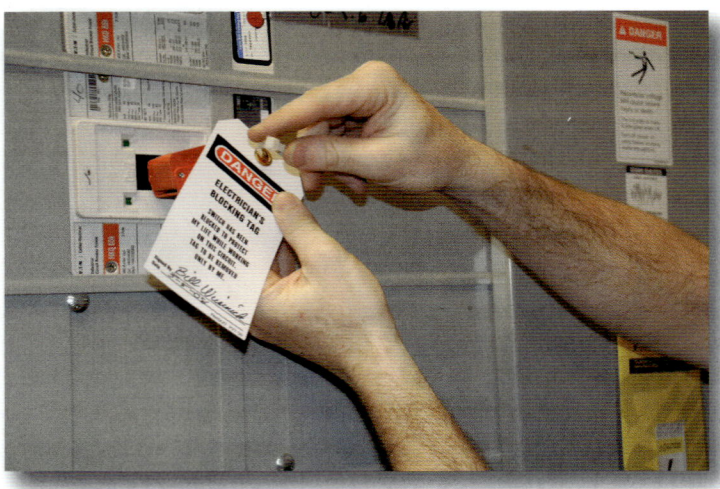

Protection from shock and other electrical hazards is achieved by establishing an electrically safe work condition in accordance with NFPA 70E: Standard for Electrical Safety in the Workplace.

PATH FOR CURRENT THROUGH THE BODY

When a person becomes a pathway for electrical current, the person will usually receive an electrical shock. Three conditions are necessary to result in an electrical shock:

1. There must be a contact point on the body for entry of the circuit current.

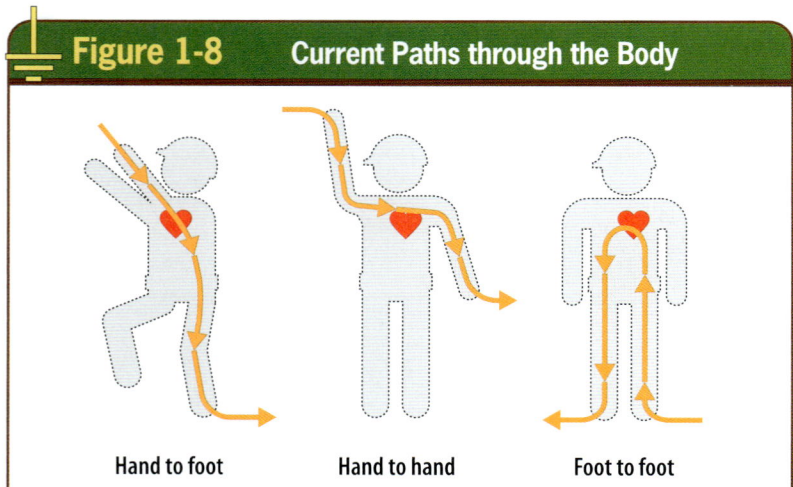

Figure 1-8   Current Paths through the Body

Hand to foot   Hand to hand   Foot to foot

Figure 1-8. A person in the electrical circuit becomes a conductive path for current. Shown here are some typical paths for current through the human body.

2. The current must exit through a second contact point.
3. There must be a voltage to force the current through the human body.

An electrical shock event can result in the person being seriously injured or even electrocuted. How serious the electrical shock injury is generally depends on three factors:

1. Amount of current (amperes).
2. Duration (time) the current is present.
3. Path the current takes through the body. **See Figure 1-8.**

Circuit frequency can also affect the severity of an electrical shock a person receives. The amount of damage an electrical shock will inflict on the human body is influenced by several other factors, such as moisture levels, temperatures, and the location on the body where the contact occurs.

Contact can occur in many ways. Current will take all paths available to return to its source. **See Figure 1-9.** People can become involved in electrical circuits or provide a pathway for current in a circuit through ground in one of two ways: they might be (1) in series contact with the electrical circuit, or (2) in parallel contact with the circuit.

For current to be present in a circuit, the circuit must be complete. In series contact, the person is the only current path to the ground and the EGC is not involved in the circuit.

Series Circuit through the Human Body

The severity of shock is usually more significant when the human body is in series with the complete electrical circuit.

When contact with energized parts occurs, and the body is also in contact with another conductive path to the source, all the current for the circuit can be present in series through the body.

The body is a conductor. The epidermis (skin) provides a protective covering and acts like insulation, except that it is conductive. Current will be present through the body if there is contact. As an example, electrical shock or electrocution can occur when series contact is made by inserting a conductive object into an energized receptacle while the body is also in contact with another grounded surface (such as a grounded metal baseboard heater) and is therefore a fault-current pathway to the source. This is one reason the *NEC* requires tamper-resistant receptacles in dwelling units and other occupancies where unsuspecting children are predominantly present. The current in a series circuit through the body is usually not enough to operate an overcurrent protective device, so severe shock or electrocution is possible. **See Figure 1-10.**

Parallel Circuit through the Human Body

A severe electrical shock or electrocution can also result when the human body is in parallel with other ground-fault current paths. In this situation, the fault current will divide between the body and all other conductive paths that are common to the source. The impedance in each path will provide opposition to current, and most of the current will exist on the paths with the lowest impedance.

Consider an insulation failure internal to an electrical product, such as a tool that is not double-insulated. The ungrounded (hot) conductor contact with the conductive parts of the tool is enough to cause the case to become energized, but the amount of current will be limited due to the loose connection. This can result in an insufficient amount of current to cause the circuit breaker or fuse to operate. If a person touches the energized part and another conductive path common to the source simultaneously, there are parallel paths for fault current; the body provides one of those fault current paths. Even though the current is not sufficient to cause the circuit breaker or fuse to operate, there can easily be sufficient current to cause a shock or electrocution. An EGC connection that is loose or ineffective can also limit the current.

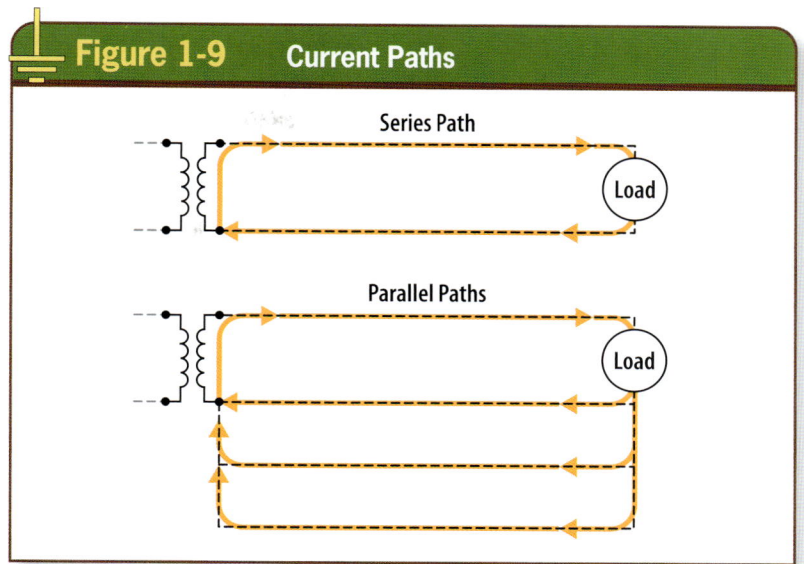

Figure 1-9   Current Paths

Series Path

Load

Parallel Paths

Load

Figure 1-9. Current will divide over all paths available when returning to the source.

In addition to missing, poor, or loose connections, circuit length and conductor size can contribute to impedance in equipment grounding conductors. Proper connections in the effective ground-fault current path(s) are critical. The effective ground-fault current path(s) must be of low impedance, have ample capacity for any fault current imposed, and effectively operate an overcurrent protective device if a ground fault occurs.

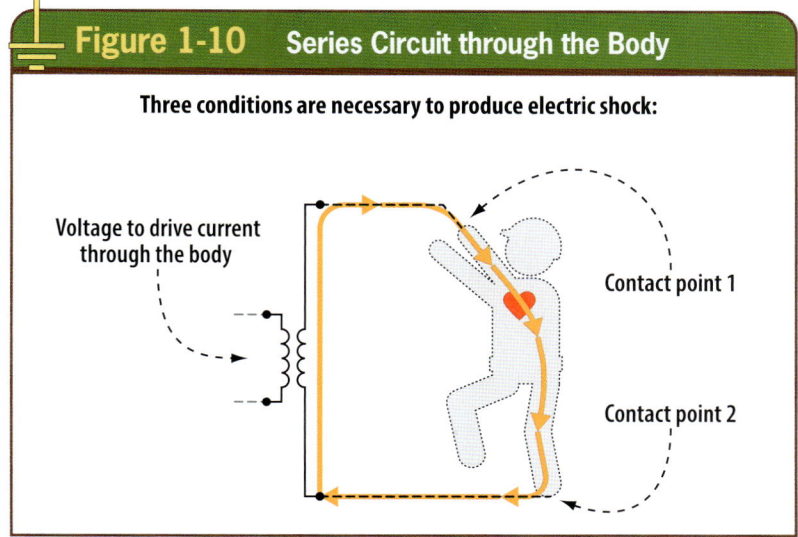

Figure 1-10   Series Circuit through the Body

Three conditions are necessary to produce electric shock:

Voltage to drive current through the body

Contact point 1

Contact point 2

Figure 1-10. Current in series with the body usually causes the most severe injuries and can often result in electrocution.

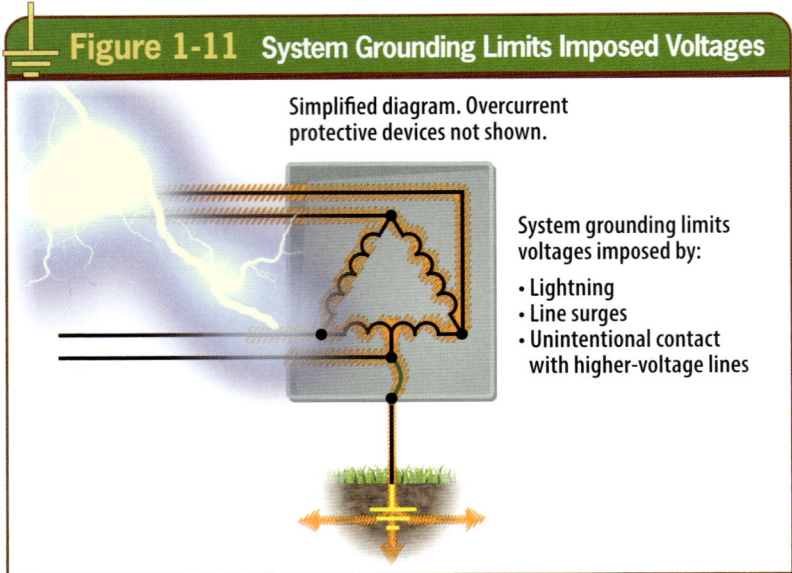

Figure 1-11 System Grounding Limits Imposed Voltages

Simplified diagram. Overcurrent protective devices not shown.

System grounding limits voltages imposed by:
- Lightning
- Line surges
- Unintentional contact with higher-voltage lines

Figure 1-11. System and equipment grounding can limit voltages imposed by lightning, line surges, and unintentional contact with higher-voltage lines.

For additional information, visit qr.njatcdb.org Item #2562

For additional information, visit qr.njatcdb.org Item #5331

For additional information, visit qr.njatcdb.org Item #2281

Research by Charles F. Dalziel contributed to the understanding of human "Let-Go" thresholds and ultimately to the development of the ground-fault circuit interrupter (GFCI). Let-Go thresholds for humans were established at four to six milliamperes, which is the trip level for Class A GFCI protection. More information about this topic is provided in Chapter 17 of this textbook. Expanded *NEC* requirements for ground-fault circuit interrupters have enhanced protection for people against electrical shock and electrocution. A Consumer Product Safety Commission (CPSC) paper entitled "Economic Considerations of GFCIs" provides information about estimated costs and direct benefits of expanding the use of ground-fault circuit interrupters in residential occupancies.

Fact

An effective grounding and bonding system depends on the integrity of many connections, which must be properly made and maintained.

GROUNDING AND BONDING CONCEPTS TOGETHER

Grounding and bonding circuits are safety circuits for electrical services, feeders, and branch circuits in today's wiring systems. Section 250.4 of the *NEC* provides performance language to assist users by describing what grounding and bonding must accomplish. Performance-based rules in the *NEC* make many other prescriptive requirements easier to apply and follow.

Electrical systems are grounded to limit potentials (voltages) imposed by lightning, line surges, and/or unintentional contact with higher-voltage lines. Grounding also stabilizes voltages during normal operation. Electrical equipment is grounded to establish the same or close to the same potential between the equipment and the ground (Earth). **See Figure 1-11.**

When electrically-conductive parts are effectively bonded together, electrical continuity and conductivity are established. A vitally important part of the grounding and bonding scheme is constructing an effective ground-fault current path. **See Figure 1-12.**

Care must be taken to ensure that good electrical connections are made for all the conductors, including, but not limited to, the service grounding scheme. Section 110.12 of the *NEC* is titled "Mechanical Execution of Work" and requires that electrical conductors and equipment be installed in a professional and skillful manner.

Any worthy training program must include a strong emphasis on good work practices and training people in their field in a way that results in a skilled and confident workforce. Good electrical installations generally look good and perform well.

The National Electrical Installation Standards (NEIS) are a library of professional quality and performance standards developed and maintained by the National Electrical Contractors Association (NECA).

PERFORMANCE *CODE* LANGUAGE

Part I of Article 250 provides general requirements for grounding and bonding. In the 1999 edition of the *NEC*, Article 250 was revised in a unique fashion to include text on what grounding and bonding must accomplish (performance language) in electrical wiring. Prior to that edition of the *NEC*, there were only fine-print notes (now informational notes) that explained grounding performance. Section 250.4 was a significant improvement in helping users understand the *NEC* requirements related to this subject. These general provisions tell users how grounding and bonding perform in grounded systems and ungrounded systems. In other words, these provisions explain what is anticipated when grounding and bonding meet the minimum requirements in the *NEC*.

The performance language must be analyzed closely to develop a strong understanding of what must happen electrically. This will simplify an Electrical Worker's understanding and improve accurate application of the other prescriptive requirements in Article 250 and throughout the *NEC*.

In its simplest form, grounding is made up of two concepts: system grounding and equipment grounding. **See Figure 1-13.**

Figure 1-12 Effective Bonding Connections

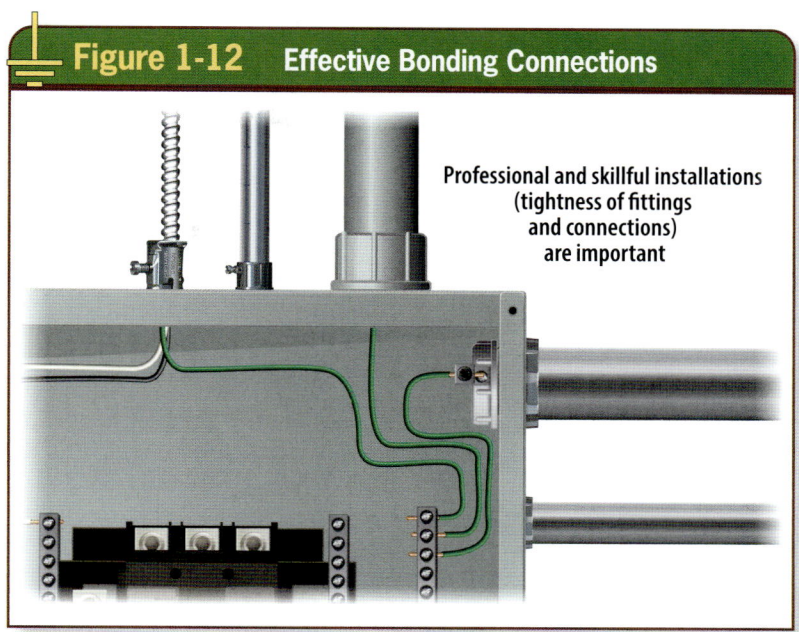

Professional and skillful installations (tightness of fittings and connections) are important

Figure 1-12. Bonding functions effectively establish electrical continuity and conductivity.

Electrical systems are either grounded or ungrounded. Grounded systems include one conductor of the system (source) that is intentionally connected to the ground. Grounded system conductors and EGCs are generally not permitted to be connected other than

Figure 1-13 System and Equipment Grounding

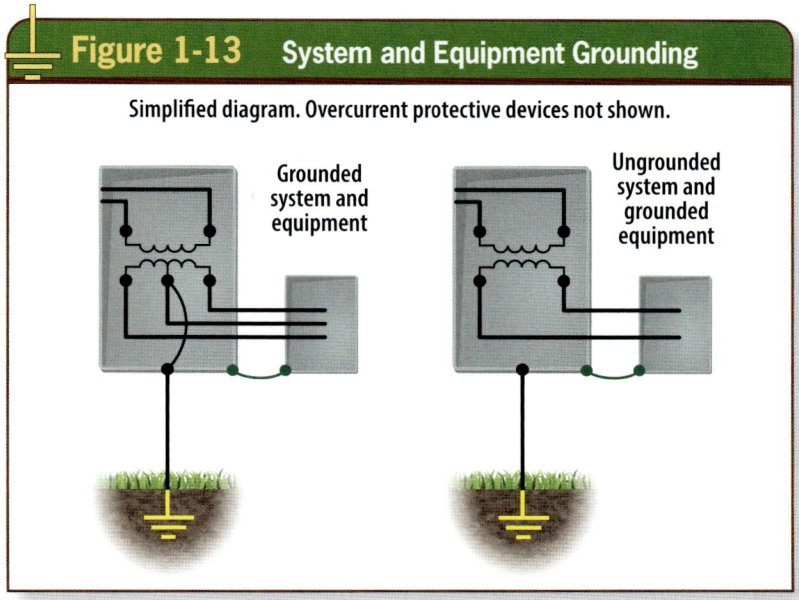

Simplified diagram. Overcurrent protective devices not shown.

Grounded system and equipment

Ungrounded system and grounded equipment

Figure 1-13. System grounding and equipment grounding are shown on the left side of the illustration, while only the equipment is grounded on the right.

at the point of grounding for the system. A clear differentiation and separation between these two conductors must be made.

In the simplest form, bonding is made up of two concepts: connecting for continuity and connecting for conductivity. Bonding must fulfill both functions in grounded systems and ungrounded systems. Bonding that is correctly done results in effective paths for ground-fault current and minimizes potential differences between conductive parts and equipment that are required to be bonded. **See Figure 1-14.**

It is important for workers to realize the performance aspects of grounding and bonding as they are constructing electrical systems. With a thorough understanding of what grounding and bonding are intended to accomplish for electrical systems and equipment, the *NEC* prescriptive requirements are much more understandable and easier to apply in design and installation.

The actual text from the *NEC* is provided, followed by commentary that provides further clarification and examples of what is intended by each of these performance provisions.

ELECTRICAL SYSTEMS

The *NEC* provides requirements for electrical systems (sources) that are grounded and for those that are ungrounded. Whether a system is grounded or not grounded, there are specific performance requirements that apply. The following provides a breakdown of the grounding and bonding performance requirements in the *NEC*.

Grounded Electrical Systems

Section 250.4(A) provides general performance requirements that apply to systems that are grounded. A grounded electrical system includes a conductor that is intentionally grounded. Systems can be grounded solidly or through an intentional resistance or impedance.

When an electrical system or source is grounded, a voltage-to-ground reference is established for the ungrounded conductors of the system. At the point of grounding, the grounded conductor of the system has the same potential as the ground (Earth). **See Figure 1-15.**

Grounded systems include one conductor that is intentionally connected to ground (Earth) and connected to

Figure 1-14   Accomplishing Effective Bonding

Figure 1-14. Bonding is accomplished by connecting conductive parts or materials together mechanically and electrically using suitable fittings, hardware, or wire-type bonding jumpers.

electrically-conductive parts enclosing the system circuit conductors, usually by a main bonding jumper at the service or by a system bonding jumper for a separately derived system. System grounding helps limit voltages above ground potential, such as those caused by surges, lightning, or contact with higher-voltage systems or lines.

When these types of events occur in systems that are grounded, because the system is connected to ground, the potentials on the grounded conductor and grounded equipment are kept equal to (or close to) the ground potential for the duration of the event. This reduces flashover possibilities and shock hazards through the grounding process.

250.4(A) Grounded Systems

(1) Electrical System Grounding. Electrical systems that are grounded shall be connected to earth in a manner that will limit the voltage imposed by lightning, line surges, or unintentional contact with higher-voltage lines and that will stabilize the voltage to earth during normal operation.

Two purposes are served by system grounding:
1. The system is connected to ground, limiting the voltage to ground, and
2. The voltage to ground is stabilized during normal operation

If a grounded system is compared to one that is ungrounded, the stabilization effect of grounding systems becomes clear. Ungrounded systems are more vulnerable to surges and other events that can cause the system output voltages to rise and fall. Systems that are solidly connected to ground are stabilized at a more constant voltage-to-ground output during normal load conditions. **See Figure 1-16.**

Ungrounded systems do not offer the same stability and are subject to rise and fall conditions because the system "floats" above ground and other conductive objects are not connected to it (other than through induction and capacitive coupling).

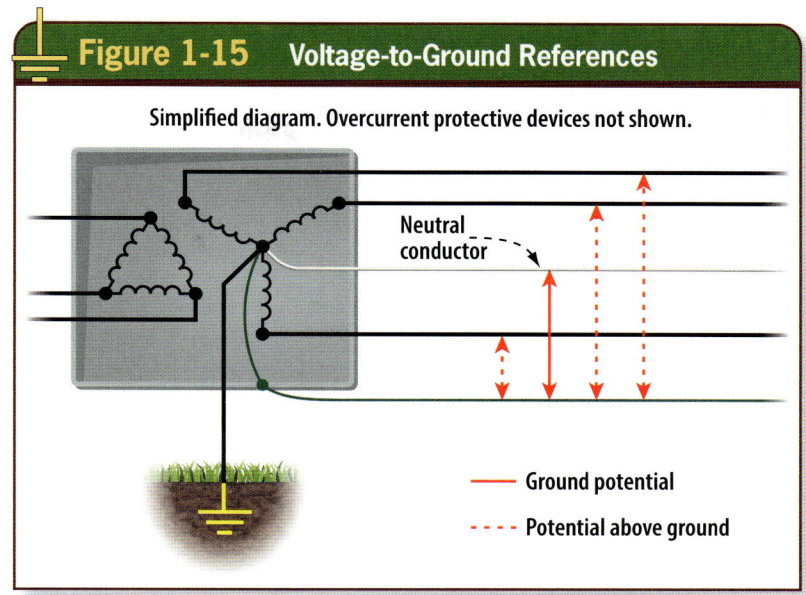

Figure 1-15. *System grounding creates a reference to the Earth, and one conductor becomes common to the Earth.*

An important consideration for limiting the imposed voltage is routing of bonding and grounding conductors in a manner that ensures that they are not longer than necessary to complete the connection without disturbing the permanent parts of the installation and ensures that unnecessary bends and loops

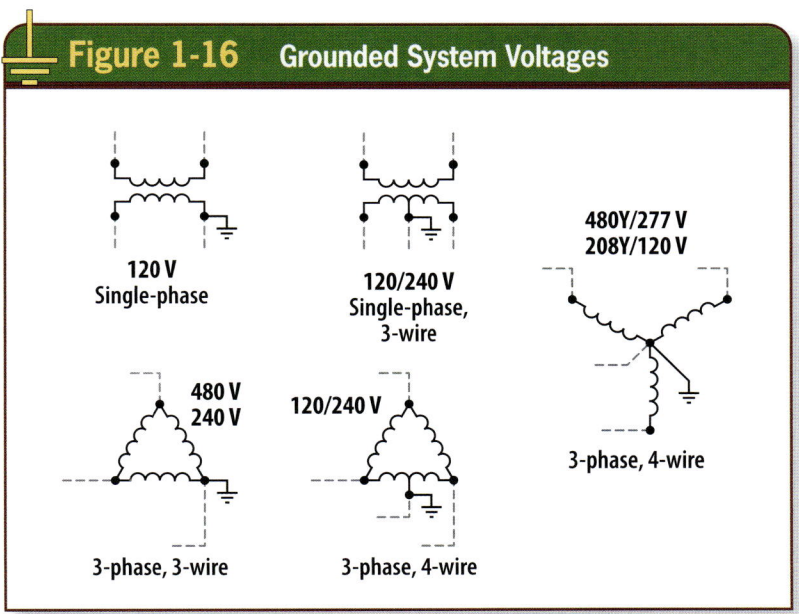

Figure 1-16. *Common system voltages are shown. The grounded conductor is a neutral in some cases and is a phase conductor in others.*

are avoided. Informational Note No. 2 follows 250.4(A)(1) and refers to *NFPA 780: Standard for The Installation of Lightning Protection Systems*, where grounding and bonding installation requirements specific to these systems can be found.

> **250.4(A) Grounded Systems**
>
> **(2) Grounding of Electrical Equipment.** Normally non–current-carrying conductive materials enclosing electrical conductors or equipment, or forming part of such equipment, shall be connected to earth to limit the voltage to ground on these materials.

Grounding of electrical equipment involves connecting equipment to ground (Earth), either directly or through a conductive body that extends the connection to ground. A conductive body extending the ground connection can be a bonding conductor, a grounding electrode conductor, or an EGC. **See Figure 1-17.**

If the *NEC* provides a requirement to ground equipment, it generally implies that there will be either a direct connection to the Earth or an indirect connection, usually through an EGC. The *NEC* refers to "normally non–current-carrying metal parts enclosing electrical conductors or equipment." This includes raceways, boxes, wireways, panelboards, switchboards, motor control centers, and so forth, all of which contain conductive parts that are not intended to carry any current during normal operation.

During ground-fault events, these conductive paths will experience a high amount of fault current for a short time until the circuit breaker or fuse protecting the affected circuit can clear the fault. EGCs are installed to connect equipment to the ground where required by the *NEC*. The grounded equipment is held at or close to ground (Earth) potential, because during normal operation and ground-fault conditions, grounding helps keep the potential from rising above ground until the overcurrent protective device operates and opens the circuit. This minimizes potential shock hazards.

Another performance function accomplished by equipment grounding is bonding. The function of bonding is inherent to the act of equipment grounding because bonding is the process of connecting to establish continuity.

The last performance characteristic of equipment grounding is to provide effective ground-fault current paths to facilitate overcurrent protective device operation. The term *ground fault* is used in the performance language of 250.4.

Figure 1-17 **Grounded Equipment**

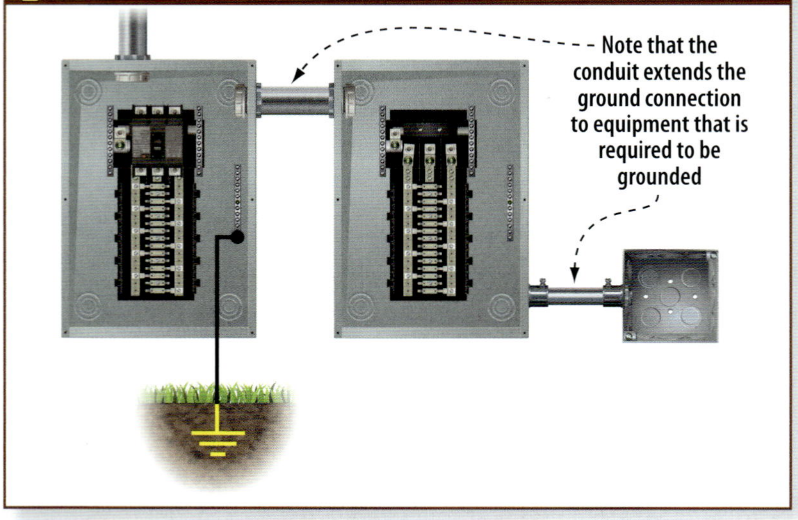

Note that the conduit extends the ground connection to equipment that is required to be grounded

Figure 1-17. *The phrase* grounded equipment *means the equipment is connected to the ground (Earth).*

> **Ground Fault.** An unintentional, electrically conductive connection between an ungrounded conductor of an electrical circuit and the normally non–current-carrying conductors, metal enclosures, metal raceways, metal equipment, or earth. (CMP-5)

Bonding is the process of connecting two or more things together. In the electrical world, the two or more things connected together are normally non–current-carrying conductive parts in the electrical installation. For example, when a length of metal conduit is attached to a metal box with two locknuts, the function of bonding is occurring. **See Figure 1-18.**

250.4(A) Grounded Systems

(3) Bonding of Electrical Equipment. Normally non–current-carrying conductive materials enclosing electrical conductors or equipment, or forming part of such equipment, shall be connected and to the electrical supply source in a manner that establishes an effective ground-fault current path.

Once this connection process is complete, the two conductive parts become one electrically and are considered bonded. Bonding ensures little or no potential differences between conductive parts, which results in electrical continuity between conductive parts. Bonding is also required in many *NEC* rules for safely conducting any imposed fault current. Sometimes bonding is required for shock protection only, such as when equipotential bonding grids are required for swimming pools or similar aquatic installations.

250.4(A) Grounded Systems

(4) Bonding of Electrically Conductive Materials and Other Equipment. Normally non–current-carrying electrically conductive materials that are likely to become energized shall be connected together and to the electrical supply source in a manner that establishes an effective ground-fault current path.

The *NEC* provides some bonding requirements for conductive materials foreign to the electrical equipment. Examples of such conductive materials within buildings or structures are structural metal frames and metal piping systems. **See Figure 1-19.**

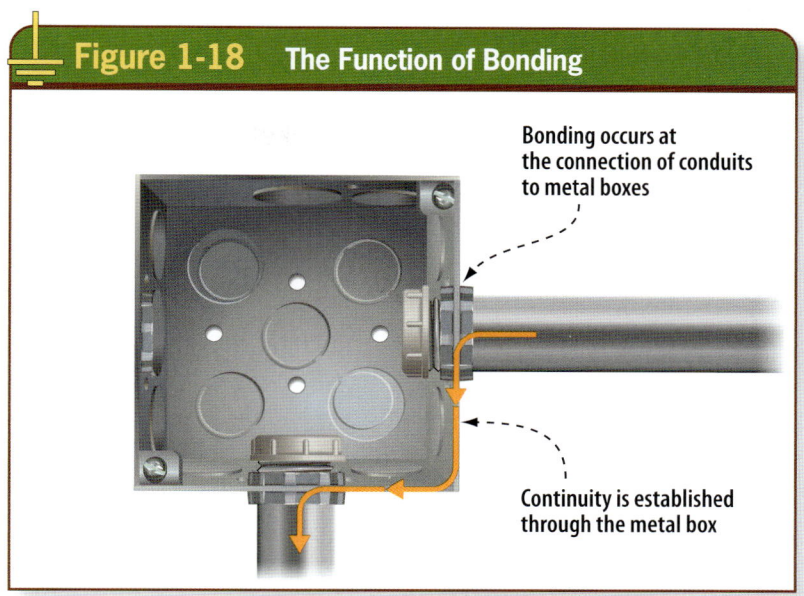

Figure 1-18 **The Function of Bonding**

Bonding occurs at the connection of conduits to metal boxes

Continuity is established through the metal box

Figure 1-18. Bonding establishes continuity and conductivity between conductive parts.

The bonding required by 250.4(A)(4) is intended to provide a direct path back to the source from materials that are "likely to become energized." The term "Energized, Likely to Become (Likely to Become Energized)" is new for the 2023 *NEC*. It describes the failure of electrical

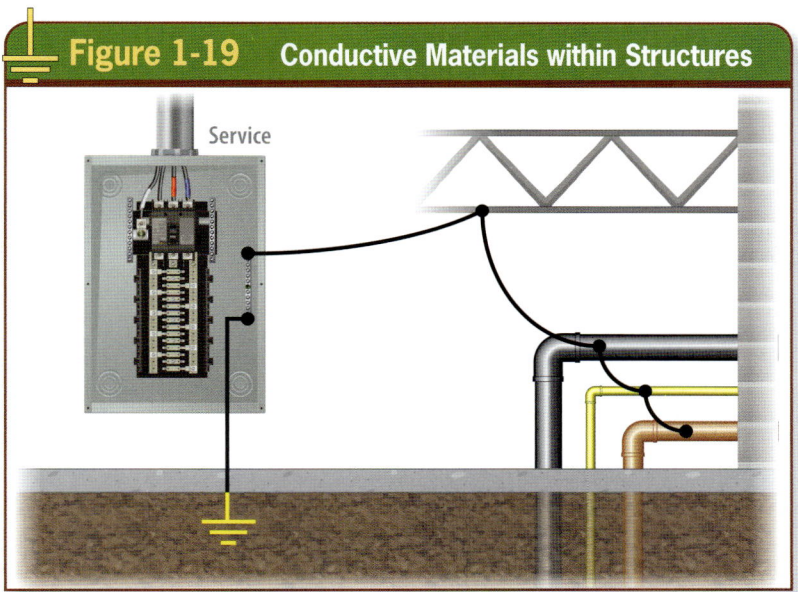

Figure 1-19 **Conductive Materials within Structures**

Service

Figure 1-19. Metal piping systems and structural steel can become energized during abnormal circumstances such as a ground fault. Bonding provides a deliberate ground-fault current path from such conductive materials.

insulation or electrical spacing that could cause conductive material to be energized. This requires careful judgement and application in the field.

> **Energized, Likely to Become. (Likely to Become Energized)** Conductive material that could become energized because of the failure of electrical insulation or electrical spacing. (CMP-5)

Bonding is a function that happens at each connection of conductive materials or equipment. For bonding to be effective and perform properly in ground-fault conditions, each fitting, bushing, connector, and coupling should be made up tight to keep impedance low if a fault occurs over the conduit, enclosures, or fittings. The definition of the term *bonded (bonding)* clarifies that it is a connection that establishes continuity and conductivity between conductive components.

> **250.4(A) Grounded Systems**
>
> **(5) Effective Ground-Fault Current Path.** Electrical equipment and wiring and other electrically conductive material likely to become energized shall be installed in a manner that creates a low-impedance circuit facilitating the operation of the overcurrent device or ground detector for impedance grounded systems. It shall be capable of safely carrying the maximum ground-fault current likely to be imposed on it from any point on the wiring system where a ground fault occurs to the electrical supply source. The earth shall not be considered as an effective ground-fault current path.

Figure 1-20   **Ground-Fault Current Paths**

Normal Current Path

Utility source   Service   Equipment

Ground Fault Current Path

Utility source   Service   Fault   Equipment

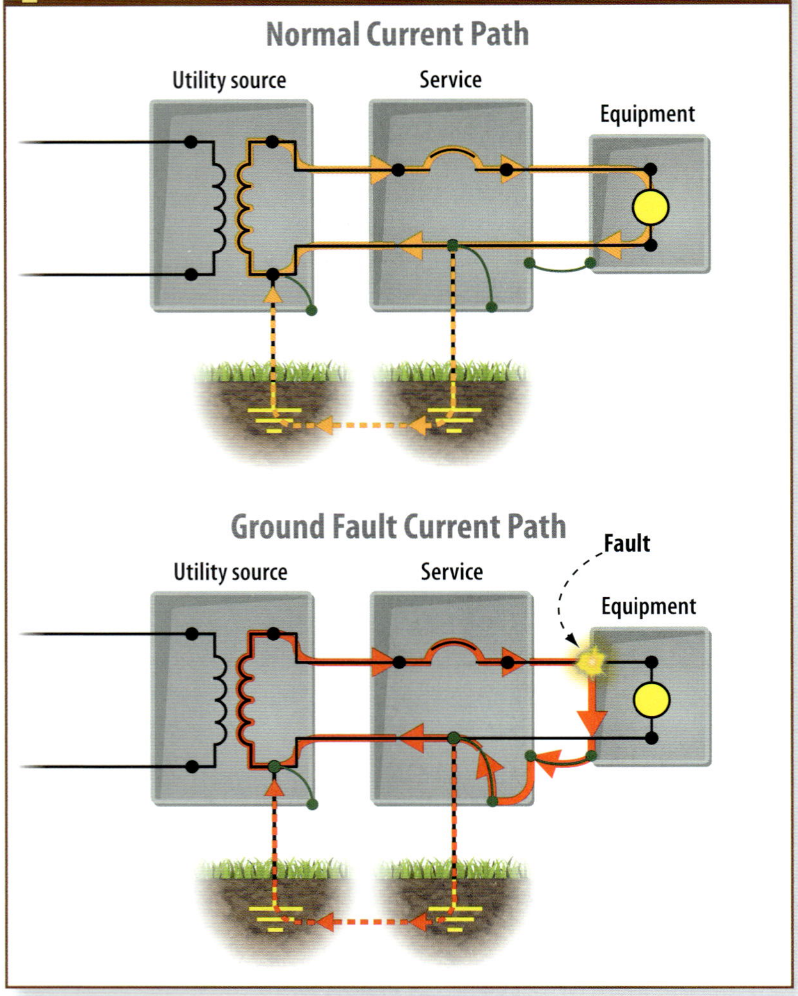

Figure 1-20. Effective ground-fault current paths facilitate overcurrent protective device operation.

The effectiveness of all bonding connections is directly related to professionalism and skillful mechanical execution of work. Loose fittings and poorly-installed systems that are not supported properly can affect bonding connections in metal wiring methods installed for services, feeders, and branch circuits. Professionalism and skill are essential in all aspects of electrical installations.

This performance requirement addresses installing electrical materials that form all or part of an effective ground-fault current path. An important aspect of this requirement is that it is related to the way the materials are installed. Because the effective ground-fault current path is intentionally installed, its effectiveness is measurable based on meeting minimum requirements of the *NEC* and being installed in a professional and skillful fashion. An effective ground-fault current path can be a wire-type, cable-type, or in the form of a conductive metal path such as a cable tray, conduit, tubing, wireway, or others that qualify for this use.

This path for fault current is relied on for fast, effective activation of overcurrent protective devices during a ground-fault event. Loose fittings, locknuts, or other fasteners can decrease the effectiveness of this path and increase impedance. Note that this path must meet specific criteria that are directly related to its performance. It must be of the lowest possible impedance. This will vary based on the different lengths, sizes, and other installation characteristics of branch circuits and feeders. This path must also be capable of carrying the maximum fault current likely to be imposed. In ground-fault conditions, the source will supply as much fault current as the system can deliver. **See Figure 1-20.**

These higher levels of current can cause significant damage to conductive paths or current-sensitive components of the circuit. Impedance is opposition to current in an AC circuit, so keeping the impedance as low as possible means that sufficient current will be present during ground-fault conditions, allowing fast operation of fuses or circuit breakers.

> **Ground-Fault Current Path, Effective. (Effective Ground-Fault Current Path).**
> An intentionally constructed, low impedance electrically conductive path designed and intended to carry current during ground-fault events from the point of a ground fault on a wiring system to the electrical supply source and that facilitates the operation of the overcurrent protective device or ground-fault detectors. (CMP-5)

The Earth is not permitted as an effective ground-fault current path. The restriction against using the Earth for this purpose is necessary because although the Earth is conductive, it offers variable levels of opposition to current. The Earth should not be

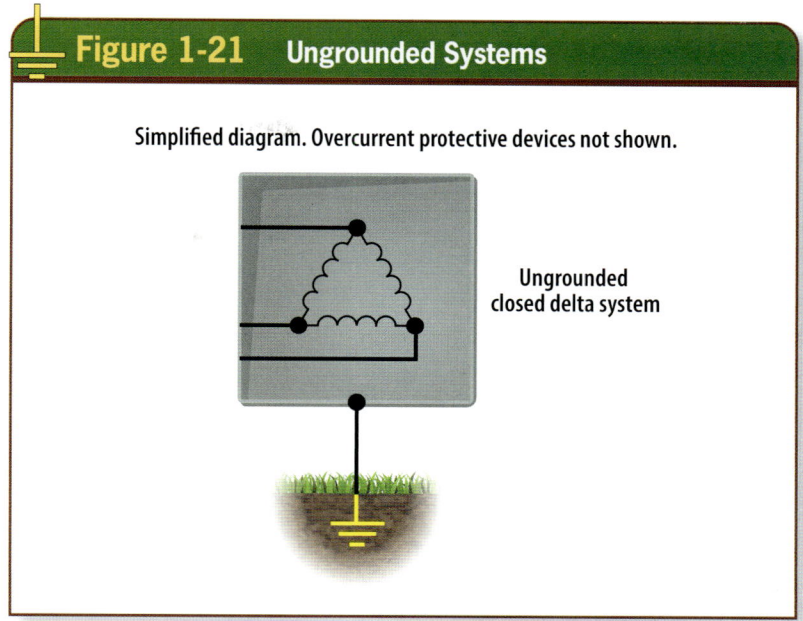

Figure 1-21. *Ungrounded electrical systems have no system connection to Earth from any of the circuit conductors of the system.*

depended on to conduct any steady-state current, so obviously the Earth can never be depended on to conduct fault current or to serve as an effective ground-fault current path.

Ungrounded Electrical Systems

Section 250.4(B) provides general performance requirements that apply to systems that are not grounded. Ungrounded systems are those that are not connected solidly or through any impedance to the Earth. Ungrounded systems do not include a conductor that is intentionally grounded. The ungrounded conductors operate without a solid voltage-to-ground reference. **See Figure 1-21.**

Voltage measurements to ground from the phase conductors of an ungrounded system are typically the result of distributed leakage capacitance and can vary depending on the circuit lengths, the size of conductors, and so forth. Although the system is ungrounded, the functions of grounding and bonding are still necessary for the equipment enclosing circuit conductors supplied from these systems. The following includes the *NEC* performance language for ungrounded systems and is followed by commentaries that break the provisions down into simple concepts.

250.4(B) Ungrounded Systems

(1) Grounding Electrical Equipment. Non–current-carrying conductive materials enclosing electrical conductors or equipment, or forming part of such equipment, shall be connected to earth in a manner that will limit the voltage imposed by lightning or unintentional contact with higher-voltage lines and limit the voltage to ground on these materials.

The performance language in 250.4(B)(1) is related to the grounding requirements for equipment installed for an ungrounded system. The non–current-carrying parts of raceways, equipment enclosures, and so forth must be grounded (connected to the Earth) in such a way as to limit overvoltage caused by lightning or unintentional contact with higher-voltage lines. Grounding equipment essentially places these conductive parts at or close to the same potential as that of the Earth. **See Figure 1-22.**

Therefore, if abnormal events should cause a voltage rise on these parts, the rise will be consistent with the rise in potential of the Earth. The objective is to keep the non–current-carrying conductive parts at the same potential as that of the Earth.

That is what grounding is all about. Grounding minimizes potential differences between the Earth and equipment in normal operation and during abnormal events such as ground faults.

If equipment is grounded, potential differences are reduced for the duration of time it takes overcurrent protective devices to operate. The equipment grounding requirements in 250.4(B)(1) can be compared with those in 250.4(A)(2) since the electrical performance expectations are essentially the same. An informational note follows 250.4(B)(1) and refers to *NFPA 780: Standard for The Installation of Lightning Protection Systems*, where grounding and bonding installation requirements specific to these systems can be found.

Figure 1-22   Equipment Required to be Grounded

Ungrounded system

Grounding is required for enclosures and raceways containing ungrounded system components & conductors

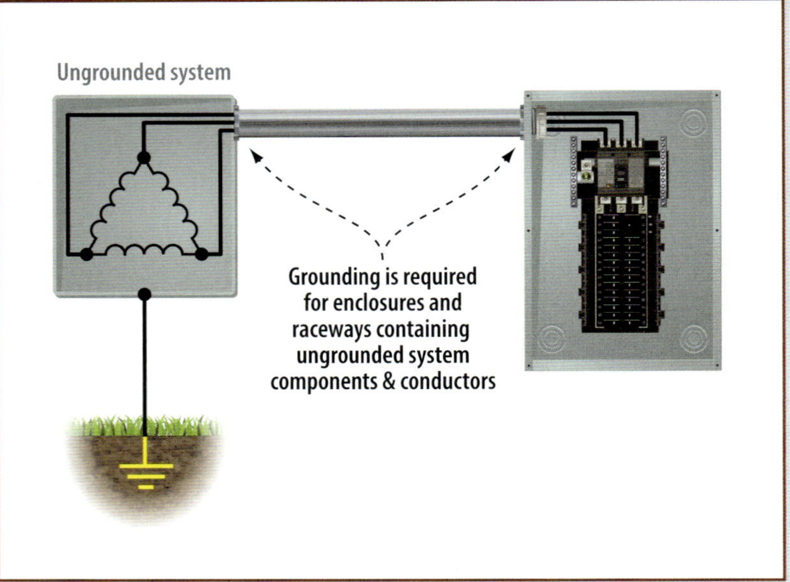

Figure 1-22. Raceways and electrical enclosures are examples of electrical equipment required to be grounded.

250.4(B) Ungrounded Systems

(2) Bonding of Electrical Equipment. Non–current-carrying conductive materials enclosing electrical conductors or equipment, or forming part of such equipment, shall be connected and to the supply system grounded equipment in a manner that creates a low-impedance path for ground-fault current that is capable of carrying the maximum fault current likely to be imposed on it.

If an electrical system is not grounded, there is no intentional connection between the system conductors and the Earth. When the supply conductors from ungrounded AC systems are installed in grounded metal raceways and enclosures, the effects of capacitance coupling are present, creating varying

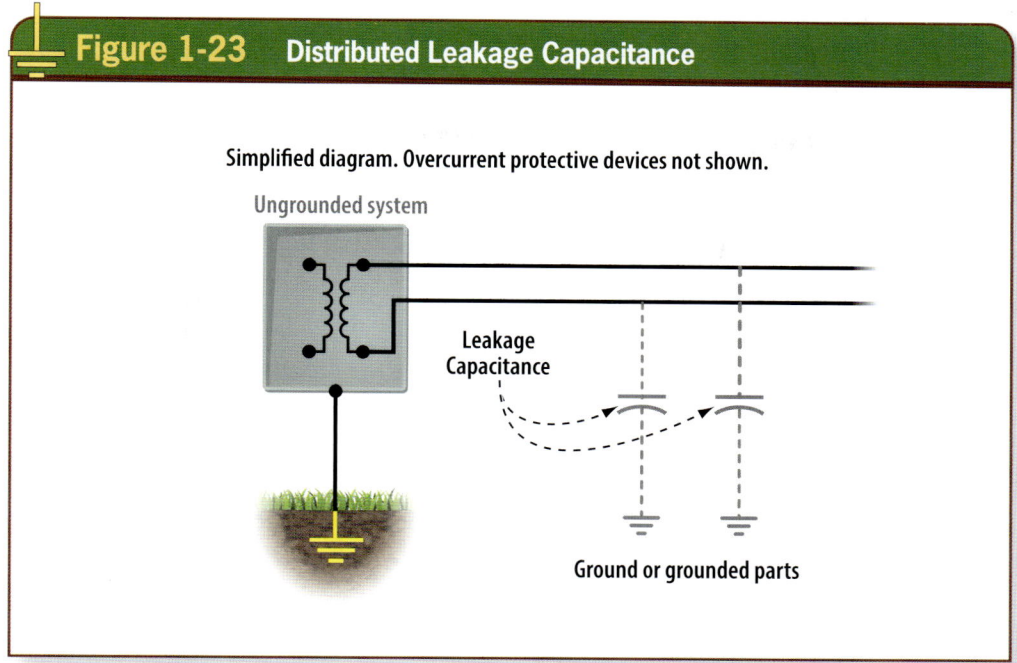

Figure 1-23   Distributed Leakage Capacitance

Simplified diagram. Overcurrent protective devices not shown.

Ungrounded system

Leakage Capacitance

Ground or grounded parts

Figure 1-23. Ungrounded AC systems (sources) produce distributed leakage capacitance (shown in dotted gray lines). No direct or solid connection to earth exists in ungrounded systems.

potential (voltage) differences between them. This potential difference can appear as voltage-to-ground on test instrument measurements, but this is the result of leakage capacitance. An example of an ungrounded AC system is a single-phase, 2-wire, 480-volt system. **See Figure 1-23.**

Although this capacitive coupling voltage is present, it has little or no effect in the event of a ground fault on the system. Instead, a ground fault from any of the ungrounded system conductors will often accidentally and ineffectively ground the system. The word *ineffectively* is used because these events are typically ground faults, which are often created by pinched wires or other similar failures of insulation. This differs greatly from the intentional system connection to the ground for solidly grounded electrical systems.

Thus, a first ground fault on an ungrounded system is usually an intermittent and unstable connection that will manifest through intermittent minor arcing and possibly eventually become

attached to the conductive raceway or enclosure. This condition should be identified and annunciated to qualified persons by a ground detection system. If a second phase-to-ground fault were to develop on a different phase than the first fault, a short circuit and ground-fault condition would result simultaneously. Heavy levels of fault current are then introduced into the system and will feed into the faults until an overcurrent protective device opens the circuit.

Bonding functions for metal equipment enclosures and metal raceways are important in overall safety. Connections to these conductive enclosures must be made up tightly and skillfully to perform effectively in ground-fault conditions. The bonded conductive parts and equipment must be able to maintain continuity and conductivity for any value of current imposed on them, including the highest value of fault current the system can deliver. Bonding is an essential function of electrical safety and is essential to the way the safety circuits operate.

250.4(B) Ungrounded Systems
(3) Bonding of Electrically Conductive Materials and Other Equipment. Electrically conductive materials that are likely to become energized shall be connected and to the supply system grounded equipment in a manner that creates a low-impedance path for ground-fault current that is capable of carrying the maximum fault current likely to be imposed on it.

This section deals with bonding of electrically-conductive materials that are likely to become energized. This is not necessarily limited to electrical materials; such conductive materials could include metal piping systems, metal building frames, or other electrically-conductive building components. Bonding of this material is necessary to put these parts at the same potential as the electrical grounding and bonding scheme of the service or system and to ensure a path for fault current to the system source if these parts are likely to become energized.

250.4(B) Ungrounded Systems
(4) Path for Fault Current. Electrical equipment, wiring, and other electrically conductive material likely to become energized shall be installed in a manner that creates a low-impedance circuit from any point on the wiring system to the electrical supply source to facilitate the operation of overcurrent devices should a second ground fault from a different phase occur on the wiring system. The earth shall not be considered as an effective fault-current path.

A low-impedance path for fault current is as important for ungrounded systems as it is for systems that are grounded. Bonding these conductive materials together and to the grounded equipment enclosures minimizes potential differences and provides a low-impedance path for any fault current likely to be imposed on them. Once again, failure to bond these conductive parts makes it possible for them to become energized, resulting in shock hazards and other possible fire hazards.

Ground-Fault Current Path. An electrically conductive path from the point of a ground fault on a wiring system through normally non–current-carrying conductors, grounded conductors, equipment, or the earth to the electrical supply source. (CMP-5)

The performance requirements in 250.4(B)(4) address the path for fault current that is necessary for wiring methods and equipment installed for ungrounded systems. Once again, a first phase-to-ground fault condition will ground the system, but not effectively, because these are usually loose, unintentional connections or contacts between any of the ungrounded phase conductors supplied by the system. The requirements for the fault current path in ungrounded systems are very similar to the performance requirements for the effective ground-fault current path in grounded systems. If this condition develops, the faulted phase conductor is forced to take on the same potential as that of the ground because all wiring methods and conductive parts of equipment are grounded.

Ground detection systems are required for ungrounded systems in accordance with 250.21(B). The *NEC* requires a ground detection system so that a fault indication is provided when a phase-to-ground fault event happens on an ungrounded system. This alerts qualified persons to this condition so that they can fix the problem before a second phase-to-ground event develops on another phase. **See Figure 1-24.**

Another important characteristic of a ground fault on ungrounded systems is that the path for fault current usually includes many conductive bodies other than those installed specifically as the

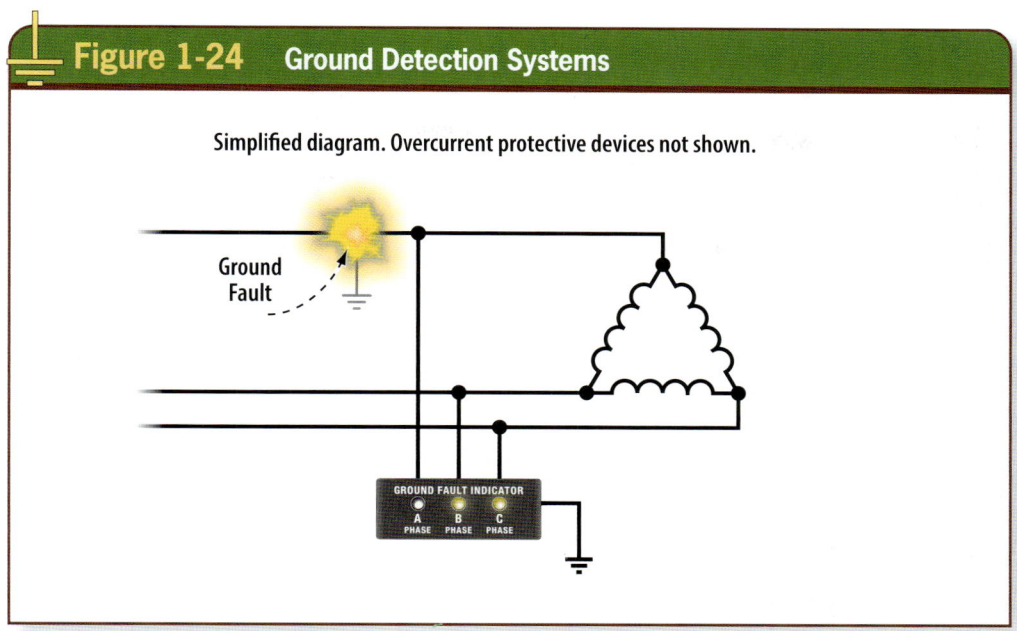

Figure 1-24   Ground Detection Systems

Simplified diagram. Overcurrent protective devices not shown.

Ground
Fault

GROUND FAULT INDICATOR
A   B   C
PHASE   PHASE   PHASE

Figure 1-24. *A ground detection system used in an ungrounded system provides annunciation of a first phase-to-ground fault condition on the system.*

ground-fault current path for the system. A ground fault on ungrounded systems, as on grounded systems, introduces current in multiple paths, such as combinations of EGCs, metal wiring methods, conductive electrical equipment enclosures, and other electrically conductive materials, including conductive piping, structural steel framing, metal ducts, cable shields, and even the ground (Earth) itself. This is the reason that the word *effective* is not used for the fault current path constructed for ungrounded systems.

In a first phase-to-ground fault, effective performance to facilitate overcurrent protective device operation is not the objective, as compared with the effective ground-fault current path constructed for grounded systems.

SUMMARY

Electrical grounding and bonding provides essential safety for people and property. The performance criteria of these important electrical functions have been analyzed and broken down into simple, yet thorough descriptions to provide ease of comprehension. The concepts of grounding and bonding are simple, but many have made them more complicated than necessary. Grounding simply involves connecting electrical systems and equipment to the Earth. Bonding simply means connecting equipment together electrically. It is the process of establishing and maintaining effective continuity and conductive connections between conductive parts. Grounding and bonding are actions that happen simultaneously in all electrical installations. The electrical safety system is heavily reliant upon effective grounding and bonding of equipment and systems.

REVIEW QUESTIONS

1. **Which of the following best defines the term** *ground*?
   a. An accidental or unintentional contact between an ungrounded conductor and the Earth or another grounded object
   b. The Earth
   c. The frame of an airplane
   d. The frame of an automobile

2. **__?__ means connected to establish electrical continuity and conductivity.**
   a. Bonded (bonding)
   b. Grounded (grounding)
   c. Guarded (isolated)
   d. Insulated

3. **Which of the following *NEC* terms is best defined as connected (connecting) to ground or to a conductive body that extends the ground connection?**
   a. Bonded
   b. Bonding
   c. Earthing (earthed)
   d. Grounded (grounding)

4. **__?__ is not considered an effective ground-fault current path.**
   a. Earth
   b. Equipment bonding jumpers
   c. Equipment grounding conductors
   d. Grounded conductors

5. **Which of the following is not a conductor in the grounding and bonding scheme for an ungrounded system?**
   a. Bonding conductors or jumper
   b. Equipment grounding conductor
   c. Grounding electrode conductor for grounding equipment enclosures and conductive parts
   d. System grounded conductor

6. **When a conductive object, equipment, or electrical system (source) is connected to ground, it is not always connected to the Earth.**
   a. True   b. False

7. **Which of the following characteristics are necessary for an effective ground-fault current path?**
   a. It is intentionally constructed.
   b. It must be an electrically-conductive path.
   c. It must have low impedance.
   d. All of the above.

8. **Which of the following terms is best defined as "not connected to ground or to a conductive body that extends the ground connection"?**
   a. Grounded
   b. Guarded (isolated)
   c. Impedance grounded
   d. Ungrounded

9. **Electrical systems are grounded (connected to the Earth) in a manner that will accomplish which of the following?**
   a. Limiting the voltage imposed by lightning
   b. Limiting voltages imposed by line surges or unintentional contact with higher-voltage lines
   c. Stabilizing the voltage to Earth during normal operation
   d. All of the above

10. **The Earth is in the electrical grounding circuit.**
   a. True   b. False

11. **Which of the following *NEC* terms means "connected to ground without inserting any resistor or impedance device"?**
   a. Grounded
   b. Impedance grounded
   c. Resistance grounded
   d. Solidly grounded

12. **___?___ refers to normally non–current-carrying conductive materials enclosing electrical conductors or equipment, or forming part of such equipment, that shall be connected to the Earth to limit the voltage to ground on these materials.**
   a. Bonding electrical equipment
   b. Electrical system grounding
   c. Grounding electrical equipment
   d. None of the above

13. **A ground-fault current path for a grounded system or an ungrounded system could include which of the following?**
   a. Equipment grounding conductors
   b. Metal piping systems
   c. The Earth itself
   d. All of the above

14. **Effective ground-fault current paths perform all but which of the following functions?**
   a. They are capable of safely carrying the maximum ground-fault current likely to be imposed on them from any point on the wiring system where a ground fault may occur to the electrical supply source.
   b. They facilitate the operation of fuses, circuit breakers, or ground detection systems for high-impedance grounded systems.
   c. They provide a low-impedance circuit.
   d. They require the Earth to perform its intended functions.

15. **The Earth must never be depended on to function as an effective ground-fault current path.**
   a. True   b. False

16. **A grounded electrical system always includes a conductor that is intentionally connected to ground.**
   a. True   b. False

17. **Grounding places a conductive object at or as close to ground (Earth) potential as possible.**
   a. True   b. False

18. **Bonding electrically-conductive equipment together establishes electrical continuity and conductivity to the electrical supply source in a manner that establishes an effective ground-fault current path.**
   a. True   b. False

19. **The process of grounding and bonding reduces potential differences between conductive parts and the Earth, thereby reducing shock hazards.**
   a. True   b. False

20. **Ground faults occur when electrical current in a circuit returns through which of the following paths?**
   a. Through a person to ground or through a combination of ground return paths
   b. Through conductive material other than the electrical system ground (metal, water pipes, and so forth)
   c. Through the equipment grounding conductor
   d. Any of the above

21. **For a person to receive an electrical shock, which of the following conditions must exist?**
   a. The current must exit through a second contact point on the body.
   b. There must be a contact point on the body for entry of the circuit current.
   c. There must be a voltage to force the current through the human body.
   d. All of the above.

22. **The severity of the electrical shock a human will receive is related to all of the following except ___?___.**
   a. the amount of current that passes through the body
   b. the length of time the current passes through the body
   c. the path the current takes through the body
   d. the type of overcurrent protective device installed in the system

Circuit Basics and Overcurrent Protection

Electrical circuits and systems operate normally if they are complete, intact, and without events such as a ground fault or short circuit. If an electrical circuit is not complete, no current is present in that circuit. This applies to both normal circuits that supply utilization equipment and the safety (grounding and bonding) circuits of electrical systems. As previously covered, electrical grounding and bonding should always be thought of as electrical safety circuitry. The operation of overcurrent protective devices is related to the effective ground-fault current path installed as part of the overall electrical safety system.

Objectives

» Understand the relationship between normal electrical circuits and grounding and bonding circuits in an electrical system.

» Apply circuit fundamentals to grounding and bonding and relate this relationship to overcurrent protective device operation.

» Understand voltage, current, resistance, and impedance in electrical circuits and how to apply Ohm's Law to electrical circuits.

» Know the importance of adequate equipment short-circuit current ratings and of proper selection of overcurrent protective devices (circuit breakers and fuses) with adequate interrupting ratings.

» Understand the importance of conductor insulation integrity, equipment grounding conductor capacity, and not exceeding the maximum withstand ratings of electrical conductors.

Chapter 2

Table of Contents

Circuit Fundamentals 24

Ohm's Law 24

    Voltage (Electromotive Force) 24

    Ohms (Resistance) 25

    Amperes (Current) 26

Opposition to Current in Circuits 28

Current in Circuits
(Normal and Fault Current) 30

Overcurrent Protection Basics 31

Amperes Operate Overcurrent
Protective Devices 33

Time and Current 34

Safety by System Design 41

    Protecting Conductor Insulation 41

    Protecting Wire-Type Equipment
    Grounding Conductors 43

Equipment Grounding Conductor Capacity 43

Summary 45

Review Questions 46

CIRCUIT FUNDAMENTALS

When installers are building electrical infrastructures on the premises, within buildings, and within structures, they install the circuit conductors using a variety of wiring methods recognized in the *NEC*. While wiring methods and materials are being installed, the safety circuits are also being installed. Electrical Workers must take great care not only with functional circuit wiring, but also when it comes to the grounding and bonding circuits associated with the system. **See Figure 2-1.**

Electrical Workers must have a good working knowledge of electrical theory to understand how and why electrical circuits function the way they do. This knowledge is necessary to assist in determining circuit capacity and conductor sizes for services, feeders, and branch circuits in addition to proper selection and application of overcurrent protection. The grounding and bonding system installed with ungrounded conductors for services, feeders, and branch circuits is an essential network of electrical circuits that must be complete and effective to operate overcurrent protection in the event of a ground fault on the system.

There is an important relationship between the grounding and bonding circuits and the overcurrent protection provided in electrical systems. Electrical circuits in their basic form include three components: voltage, resistance, and current. Ohm's Law and basic electrical theory are principles that apply to grounding and bonding circuits as well as the (normally) current-carrying conductors of the circuits.

OHM'S LAW

George Simon Ohm discovered the relationship between voltage, resistance, and current known as Ohm's Law. This relationship can be clearly illustrated in a simple pie chart. **See Figure 2-2.**

Voltage (Electromotive Force)

A volt is a unit of electrical pressure or force. In the electrical field, voltage is the electromotive force that pushes current through resistance. The voltage pushes the electrons of a circuit through the wires or other conductive components of the circuit. **See Figure 2-3.**

Voltage can be compared to water pressure. As the pressure in the water pipe increases, the flow increases. Likewise, as voltage increases, more current is

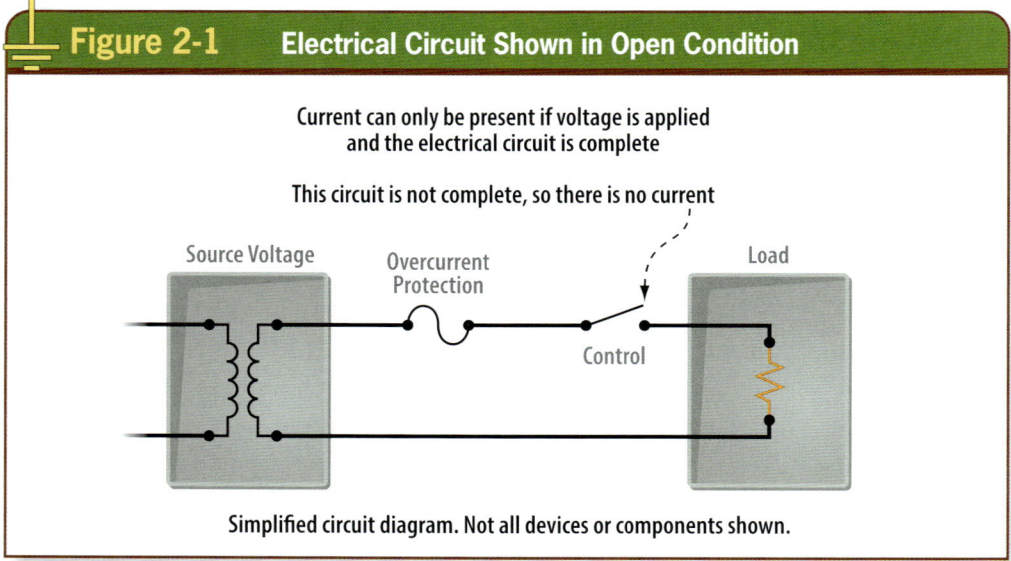

Figure 2-1 **Electrical Circuit Shown in Open Condition**

Current can only be present if voltage is applied and the electrical circuit is complete

This circuit is not complete, so there is no current

Source Voltage

Overcurrent Protection

Load

Control

Simplified circuit diagram. Not all devices or components shown.

Figure 2-1. A basic electrical circuit includes a source, circuit conductors, overcurrent protection, usually some method of control, and a load.

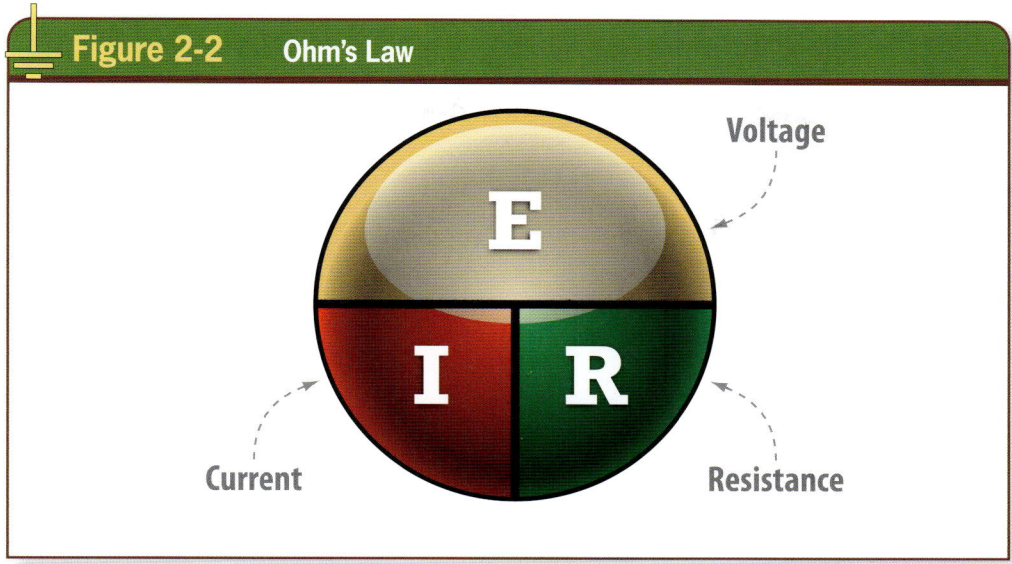

Figure 2-2. Basic Ohm's Law includes voltage (E), current (I) and resistance (R).

forced through the conductors of the circuit. Although voltage is the force that causes current in a circuit, voltage does not *flow* through a circuit—it is present in circuits. Because voltage in electrical circuits is the *electromotive* force in the circuit, it is expressed using the letter *E* in Ohm's Law. Note that voltage is measured in volts, abbreviated *V*.

Ohms (Resistance)

An ohm is the unit of electrical resistance in a circuit. More specifically, it is the amount of resistance that allows only one ampere of current flow when one volt is applied in the circuit. The symbol used to represent ohms is omega (Ω); note that this is the unit of measure. In Ohm's Law, the letter *R* is

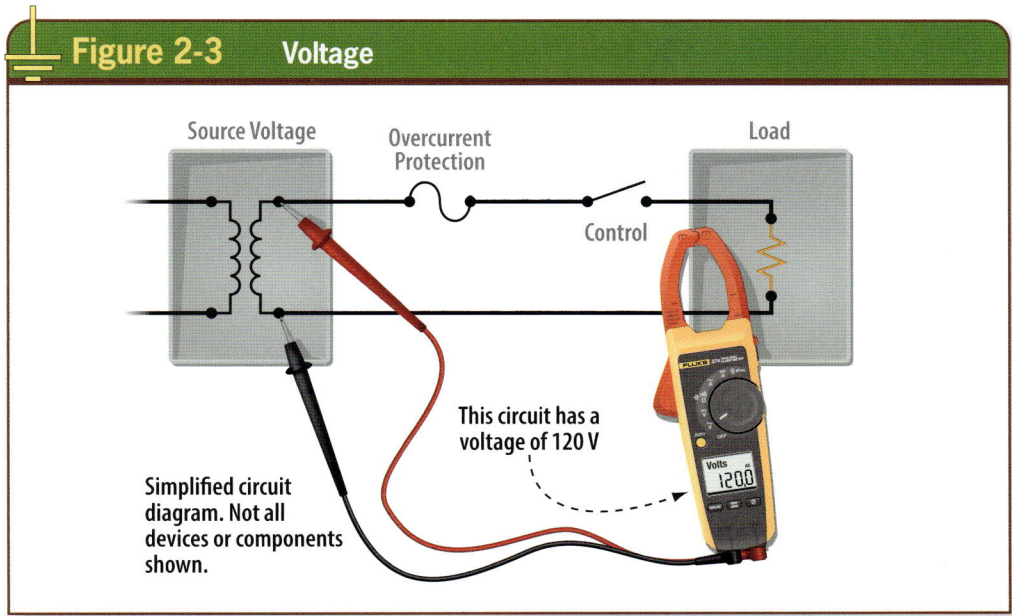

Figure 2-3. The voltage of an electrical circuit is the amount of pressure of electromotive (E) force in the circuit. This pressure is measured in volts.

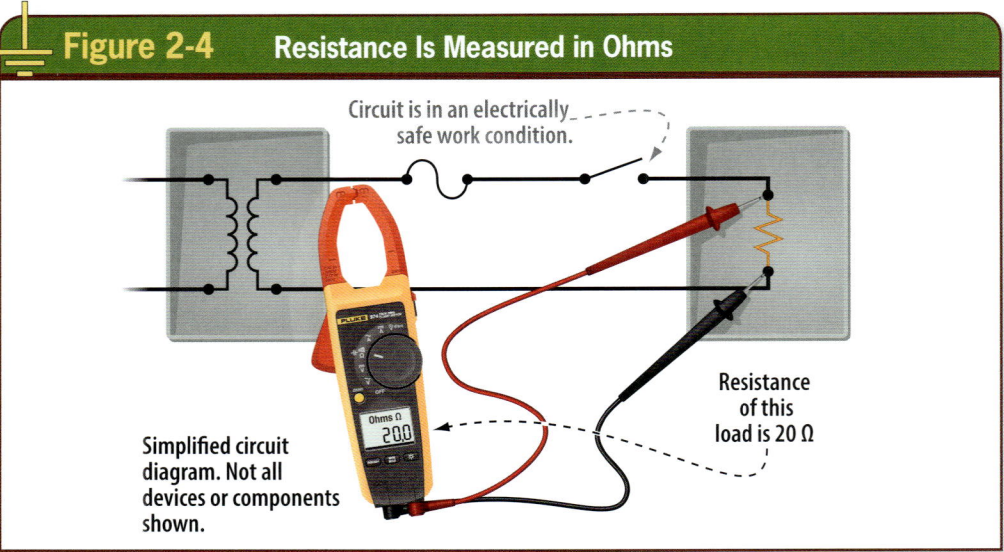

Figure 2-4 **Resistance Is Measured in Ohms**

Circuit is in an electrically safe work condition.

Simplified circuit diagram. Not all devices or components shown.

Resistance of this load is 20 Ω

Ohms Ω
20.0

Figure 2-4. An ohm is a unit of electrical resistance in the circuit. Resistance to current in a circuit rises as the ohms of the circuit increase.

used to denote resistance, which is measured in ohms. **See Figure 2-4.**

Resistance in an electrical circuit can also be compared to fluid piping systems. The larger the diameter of pipe, the less resistance to fluid flow. In electrical circuits, the larger the conductor (wire), the less resistance in the circuit. On the other hand, smaller conductors in a circuit offer greater resistance. (Note that this basic analogy does not account for any load resistance.)

It is important for those installing electrical equipment and systems and for other electrical professionals such as designers, engineers, and inspectors to apply these basic principles.

Amperes (Current)

The ampere or amp, abbreviated *A*, is the unit of measure used for the current in an electrical circuit. An ampere is defined as the flow of one coulomb per second. Because the amount of amperes in a circuit is referred to as the *intensity of current* in the circuit, the symbol *I* is used in electrical formulas using Ohm's Law. **See Figure 2-5.**

Ohm discovered this law of proportionality and concluded that it takes one volt to force one ampere through a resistance of one ohm. Another way of looking at this mathematical relationship is that the current in a direct current (DC) circuit is directly proportional to the voltage applied in a circuit, and current in a circuit is inversely proportional to the resistance in that circuit.

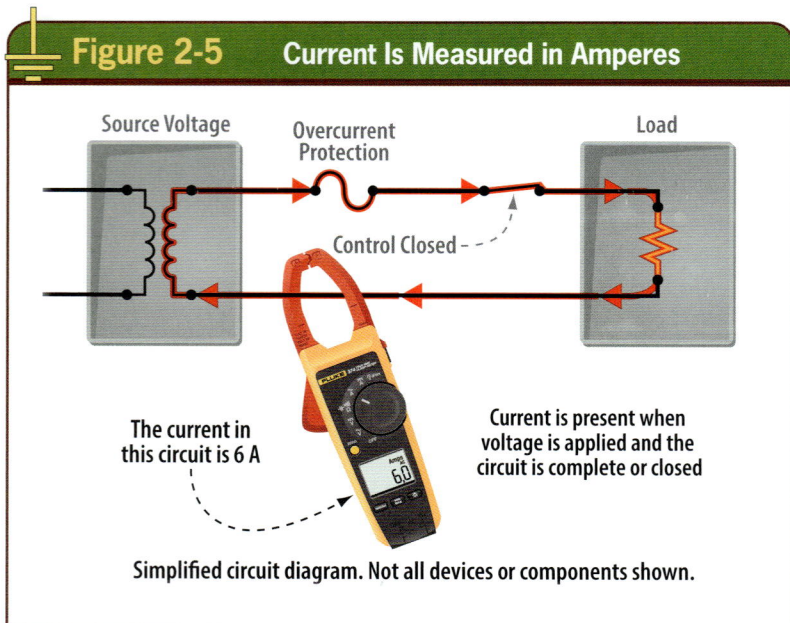

Figure 2-5 **Current Is Measured in Amperes**

Source Voltage
Overcurrent Protection
Load

Control Closed

The current in this circuit is 6 A

Current is present when voltage is applied and the circuit is complete or closed

Amps
6.0

Simplified circuit diagram. Not all devices or components shown.

Figure 2-5. An ampere is the measure used for current in an electrical circuit.

Based on the relationship reviewed earlier, the following are three simple forms of Ohm's Law:

$$E = I \times R$$

Voltage = Current x Resistance

$$I = \frac{E}{R}$$

Current = $\dfrac{\text{Voltage}}{\text{Resistance}}$

$$R = \frac{E}{I}$$

Resistance = $\dfrac{\text{Voltage}}{\text{Current}}$

These simple forms of Ohm's Law apply to both alternating current (AC) and DC circuits bearing only resistive loads. These three components of electrical circuits provide the basis for electrical calculations when any two of these values are known. **See Figure 2-6.** As an example, if the voltage of a circuit is 480 volts and the current of the circuit is 40 amperes, what is the resistance of the circuit? Because the problem is solving for resistance, the following formula is used:

$$R = \frac{E}{I}$$
$$= \frac{480 \text{ V}}{40 \text{ A}}$$
$$= 12 \ \Omega$$

Suppose that an electrical circuit has a voltage of 120 volts and a resistance of 40 ohms. How many amperes would be present in this circuit? Because the problem is solving for current (amperes), the following formula is used:

$$I = \frac{E}{R}$$
$$= \frac{120 \text{ V}}{40 \ \Omega}$$
$$= 3 \text{ A}$$

The last example solves for voltage of a circuit when current and resistance

are known. If the resistance of a circuit is 60 ohms and the current of the circuit is four amperes, Ohm's Law is applied as follows:

$$E = I \times R$$
$$= 4 \text{ A} \times 60 \ \Omega$$
$$= 240 \text{ V}$$

These formulas are valuable in determining current in electrical circuits when selecting sizes of circuit conductors and applying overcurrent protective devices (circuit breakers and fuses) in circuits.

Another valuable set of formulas is provided in Watt's Wheel. In this collection

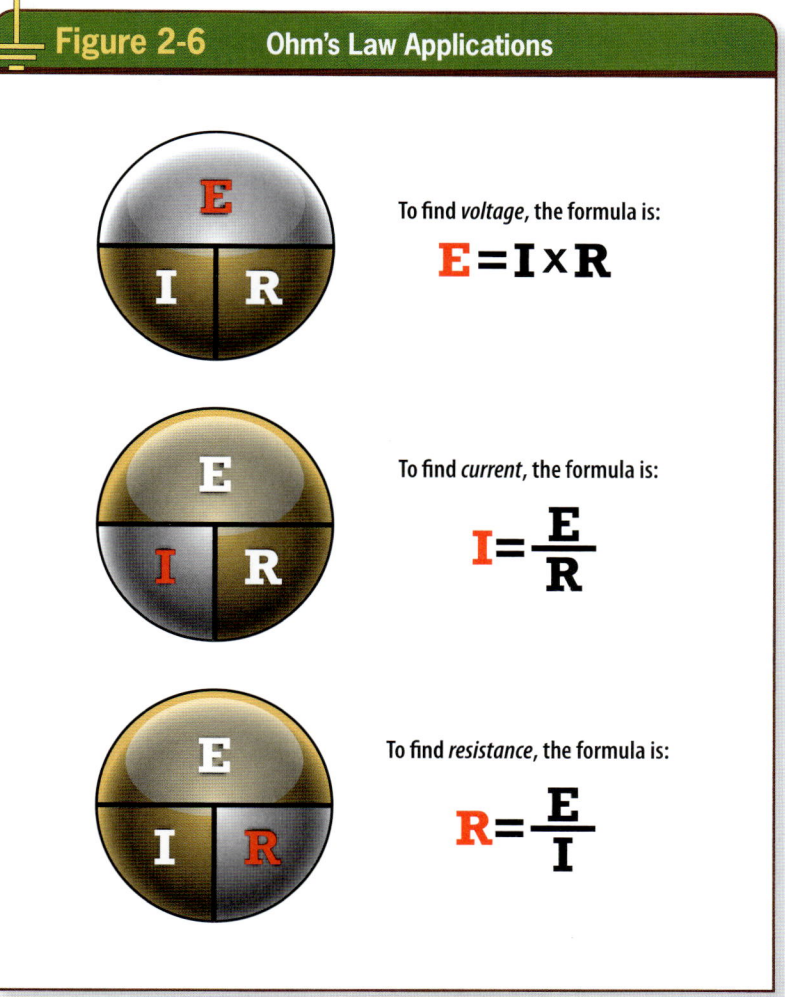

Figure 2-6   Ohm's Law Applications

To find *voltage*, the formula is:

$$\mathbf{E = I \times R}$$

To find *current*, the formula is:

$$\mathbf{I = \frac{E}{R}}$$

To find *resistance*, the formula is:

$$\mathbf{R = \frac{E}{I}}$$

Figure 2-6. *Finding values of voltage, current, and resistance in an electrical circuit is accomplished by inserting two of the known values.*

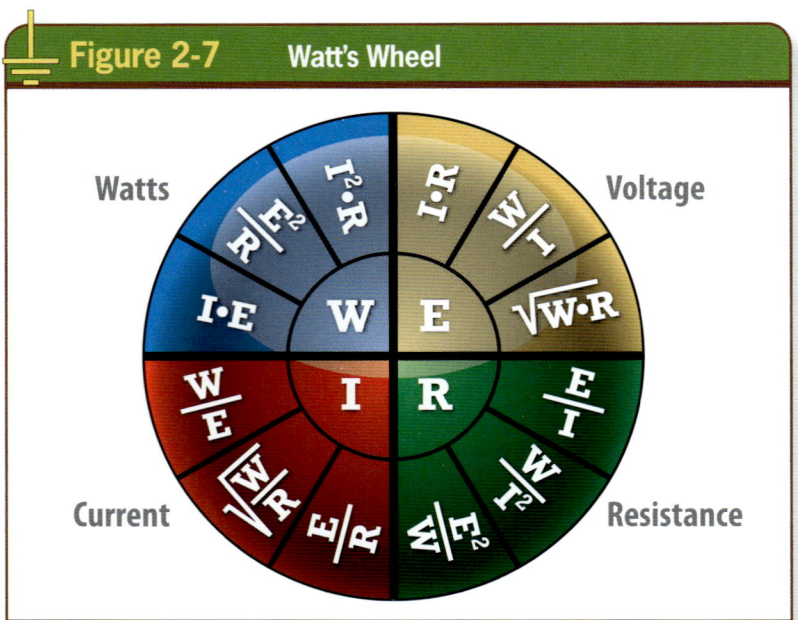

Figure 2-7   **Watt's Wheel**

Figure 2-7. Watt's Wheel is a formula chart for finding values of current, voltage, resistance, and power (watts) in electrical circuits.

of formulas, the watt (power of a circuit) is included in addition to volts, amperes, and resistance. Understanding how to apply Watt's Wheel can help solve for amperes, voltage, resistance, and wattage of a circuit. The symbol used for wattage in electrical formulas is *P*. Just as with Ohm's Law, if any two values of a circuit are known, the third can be calculated by using the appropriate formula from Watt's Wheel. **See Figure 2-7.**

Referring to the wheel, the three basic formulas for determining power (watts) in a circuit are as follows:

$W = I \times E$
Power = Current × Voltage

$W = I^2R$
Power = Current × Current × Resistance

$W = \dfrac{E^2}{R}$

Power = Voltage $\dfrac{\text{Voltage}}{\text{Resistance}}$

Using these formulas, for example, if the wattage and voltage are known,

current can be calculated. Consider an electric heater circuit, which normally has a heater nameplate that includes a wattage and voltage rating. If an electric heater is rated at 5,000 watts, and the circuit voltage applied is 277 volts, an equation solving for amperes (current) using Watt's Wheel looks like:

$$I = \frac{W}{E}$$
$$= \frac{5000 \text{ W}}{277 \text{ V}}$$
$$= 18.05 \text{ A}$$

Because it is possible to solve for current in a circuit with known wattage and voltage, correct circuit conductors and overcurrent protection can be designed and installed in electrical wiring systems.

OPPOSITION TO CURRENT IN CIRCUITS

Up to this point, the opposition to current in a DC circuit (resistance) has been used. In AC circuits, the total opposition to current is known as the *impedance* of a circuit. The impedance of an AC circuit includes the resistance, inductance, and capacitance. The letter *Z* is used to represent impedance in electrical formulas:

$$Z = \sqrt{R^2 + (X_L - X_C)^2}$$

Where

    Z = impedance

    R = resistance

    X_L = inductive reactance

    X_C = capacitive reactance

An AC circuit has different operating characteristics than a DC circuit primarily because the voltage and current are changing in amplitude and direction from the 0 point in the waveform. For example, in a 60-Hertz (cycle) circuit, the voltage and current change amplitude and direction 120 times per second. In a 400-Hertz (cycle) circuit, the voltage and current change amplitude and direction 800 times per second. As current moves through a circuit, a magnetic field is produced in a concentric form around each conductor. This field is present and constantly changing direction through each cycle of the waveform due to the change in current amplitude and direction. This magnetic field is the AC circuit component resulting from inductive reactance.

The other component present in most electrical circuits is capacitance. A capacitor, in its simplest form, is two conductors separated by insulation. From that basic description, it is easy to understand that when electrical circuits are installed, there is capacitance inherently built into the circuits.

Remember, the impedance of current in an AC circuit is the combination of resistance, inductance, and capacitance. Inductive reactance and capacitive reactance are not covered in this text, but there are excellent training resources available that provide detailed information about these subjects. Of most importance is that the total opposition to current in an AC circuit includes a resistive component or components as well as inductive and capacitive components.

The *NEC* uses the term *impedance* in relation to grounding and bonding in several ways. A common mention of impedance in the *NEC* is when a requirement refers to a low-impedance path for ground-fault current. This phrase refers to a path where the opposition to current is kept as low as possible so that the maximum amount of fault current operates the overcurrent protective device in fault conditions. The amount of impedance will vary from circuit to circuit based on the different physical characteristics of each

circuit, such as wire size, terminations, and circuit length. All these circuit characteristics contribute to the overall impedance in an AC circuit. Maintaining low impedance in electrical circuits is important.

The *NEC* includes two specific rules that help keep impedance low during normal operation through a process that results in a canceling effect. Sections 250.134(2) and 300.3(B) both require all conductors of a circuit to be installed together in the same raceway, cable, or trench. This requirement also includes the equipment grounding conductors (EGCs) because, in ground-fault or short-circuit conditions, the impedance of the circuit must be as low as possible to result in fast operation of the circuit breakers or fuses protecting the circuit. **See Figure 2-8.**

The importance of installing conductors in close proximity to each other was proven through test modeling and experiments performed by R. K. Kaufmann, an engineer at General Electric. Kaufmann discovered that when AC circuit conductors are not installed close together, inductive

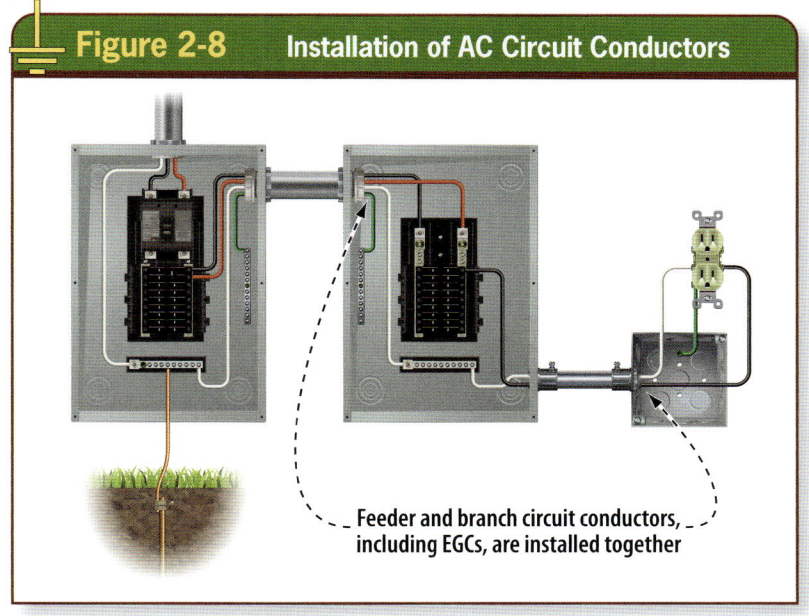

Figure 2-8 **Installation of AC Circuit Conductors**

Feeder and branch circuit conductors, including EGCs, are installed together

Figure 2-8. All conductors of AC circuits are required to be installed together in the same raceway, cable, or trench. This includes the ungrounded conductors, the grounded conductor, and the EGC of the circuit. [NEC 300.3(B)]

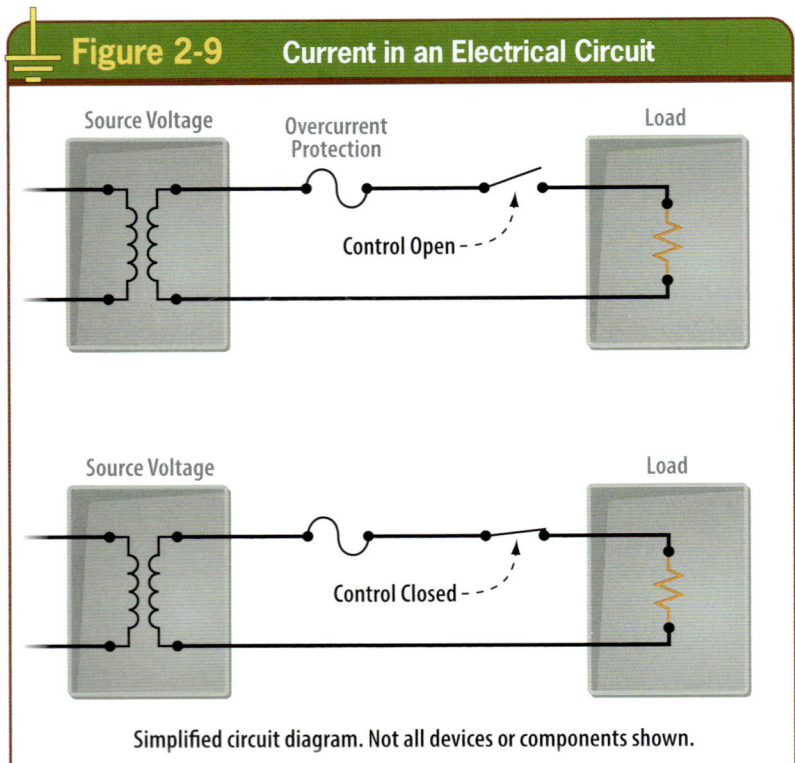

Figure 2-9 **Current in an Electrical Circuit**

Simplified circuit diagram. Not all devices or components shown.

Figure 2-9. Current is present in an electrical circuit only when a source voltage is applied, and the circuit is complete. This concept applies to the grounding and bonding circuits in addition to the normal current-carrying circuits.

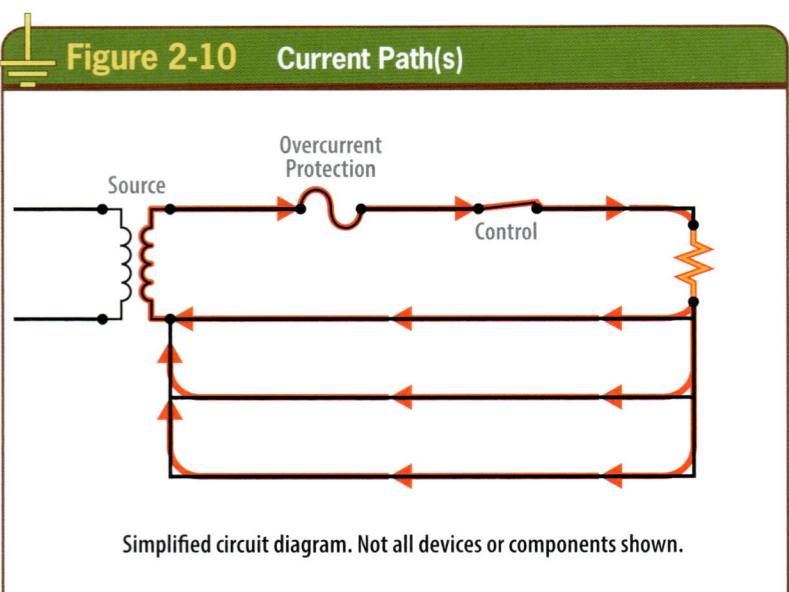

Figure 2-10 **Current Path(s)**

Simplified circuit diagram. Not all devices or components shown.

Figure 2-10. Current will divide over all paths (including the Earth) while returning to its source. The amount of current in each path depends on the amount of impedance in that path.

reactance increases the circuit impedance, thus having a negative effect on overcurrent protective device operation. By keeping conductors close together, the effects of inductive reactance and circuit impedance are minimized. Also, a canceling effect happens in circuits because the supply and return side of the circuit are close together; the magnetic fields moving in opposite directions cancel each other. If the conductors were to be separated, inductive reactance would increase, which would increase the impedance in the circuit. This is the reason the *NEC* requires conductors of an AC circuit to be routed together. This subject is explained in more detail in Chapter 8, which covers requirements and installation methods for equipment grounding conductors (EGCs).

CURRENT IN CIRCUITS (NORMAL AND FAULT CURRENT)

For current to be present in a circuit, the circuit must be complete. If a circuit is not complete, there is no current. This is true not only for the normal circuit wiring, but also for the grounding and bonding circuits. Without complete grounding and bonding paths in the circuit wiring, the protective system is compromised. Current will always return to its source. The source could be a generator, transformer, photovoltaic array, or battery. **See Figure 2-9.**

When a circuit is interrupted, the current returning to the source is also interrupted. Another important point here is that all current follows this principle. In other words, normal current will always return to its source and fault current will also return to its source. This concept is important to understand because it helps clarify the principles of operation for overcurrent protective devices such as circuit breakers and fuses. The path that current will take in a circuit is related to how many paths are available that are electrically connected to the source. Current will take any and all paths available to return to its source. As for the "path of least resistance" concept, current will divide over all parallel paths in the circuit. The amount of current in each separate conductive path is related to the impedance (opposition to current) in each of the paths involved. **See Figure 2-10.**

The paths with lower impedance will carry more current than the paths with greater impedance. Lower impedance in a circuit results in higher current in the circuit. With higher impedance in a circuit, the current is lower and encounters more opposition. This has a direct effect on how overcurrent protective devices operate and is the principle on which the requirement to install EGCs is based. Providing a reliable low-impedance path for ground-fault current by installing an EGC is essential for safe electrical systems. Remember, the Earth offers significant opposition to current and is never to be considered an effective path, even though some current can return to the source through the Earth.

> **Fault Current.** The current delivered at a point on the system during a short-circuit condition. (CMP-10)
>
> **Fault Current, Available (Available Fault Current).** The largest amount of current capable of being delivered at a point on the system during a short-circuit condition. (CMP-10)

OVERCURRENT PROTECTION BASICS

A key factor in protecting electrical equipment and people from electrical hazards is the proper application of overcurrent protective devices such as circuit breakers and fuses. Section 110.10 of the *NEC* requires the total circuit impedance, the overcurrent protective devices, the component short-circuit current ratings, and other circuit characteristics to be selected and coordinated so that the circuit protective devices can effectively respond and operate in a fault event. Furthermore,

Figure 2-11. *Equipment and overcurrent protective devices must have adequate normal current ratings, short-circuit current ratings, and interrupting current ratings.*

they must do so without causing extensive damage to the electrical equipment or conductors. Overcurrent protective devices such as circuit breakers and fuses must be selected and circuits designed to ensure that the short-circuit current rating of any system component is not exceeded should a short circuit or high-level ground fault occur. **See Figure 2-11.**

Electrical system components include the wire, bus structures, switching equipment, contactors, starters, overcurrent protective and disconnection devices, industrial control panels, and distribution equipment, all of which can have limited short-circuit current ratings. The short-circuit current rating is the amount of short-circuit current a component is rated to withstand without producing an explosion, fire, or shock hazard. If a short-circuit or ground-fault event exceeds a component's short-circuit current ratings, the component could be severely damaged or destroyed. Providing overcurrent protective devices with sufficient interrupting ratings solely in accordance with *NEC* 110.9 does not necessarily ensure adequate short-circuit protection for all components in the system.

The interrupting rating ensures only that the overcurrent protective device can safely open the circuit if applied within its rating. In cases where the available fault current exceeds the short-circuit current rating of an electrical component, the overcurrent protective device selected might not be able to limit the let-through energy to within the rating of that electrical component. The available fault current (sometimes referred to as *short-circuit current*) of a system is the maximum current that a system can deliver at any point on a wiring system. **See Annex C.** The level of available fault current in a system is typically highest at the source. The amount of available fault current can usually be obtained from the serving utility company, or it can be calculated conservatively by using an infinite bus value. This value must be known to properly calculate the amount of fault current at any point on the electrical system from the source to the furthest outlet. Since all AC circuits introduce impedance in the circuit, the amount of fault current will be reduced by the impedance of the circuit as circuits get longer and conductors get smaller.

Figure 2-12 **Ground Fault vs Short Circuit Condition**

Ground fault condition

Source

Single line-to-ground fault event

Grounded system

Load

Short circuit condition

Source

Line-to-ground fault event

Ungrounded system

Load

Line-to-ground fault event

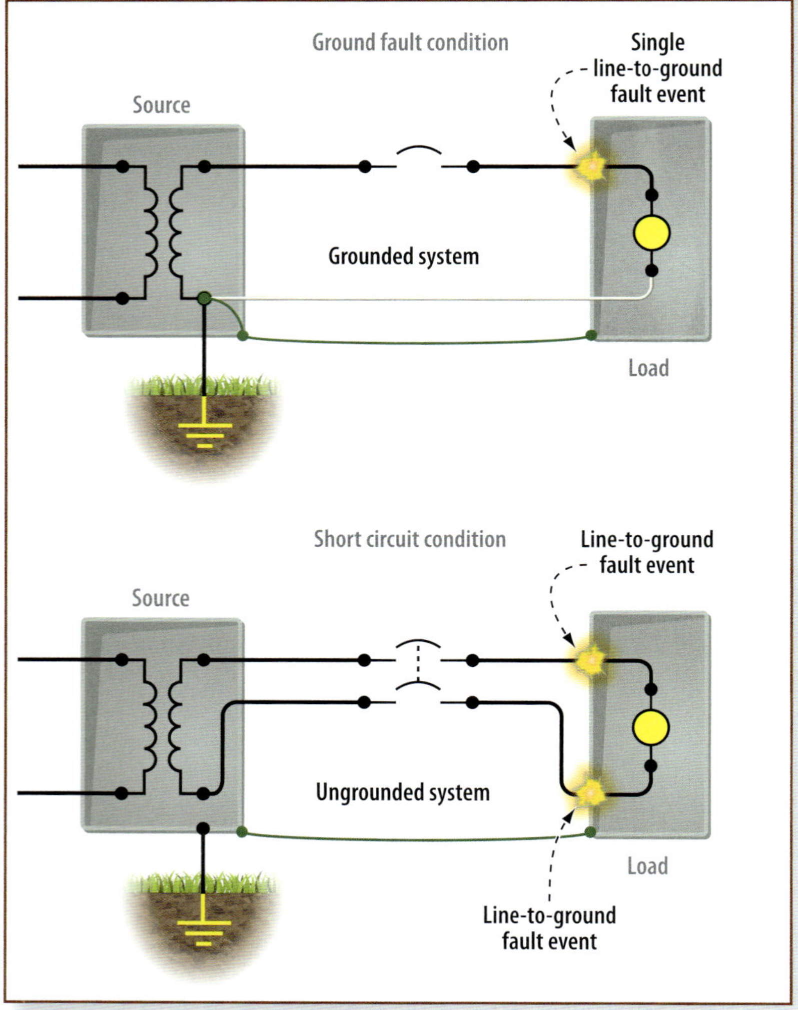

Figure 2-12. *The comparison of a ground-fault condition and a short-circuit condition shows both types of abnormal events that cause activation of overcurrent protective devices.*

Ground Fault. An unintentional, electrically conductive connection between an ungrounded conductor of an electrical circuit and the normally non–current-carrying conductors, metal enclosures, metal raceways, metal equipment, or earth. (CMP-5)

Interrupting Rating. The highest current at rated voltage that a device is identified to interrupt under standard test conditions. (CMP-10)

Short Circuit. An abnormal connection (including an arc) of relatively low impedance, whether made accidentally or intentionally, between two or more points of different potential. (CMP-10)

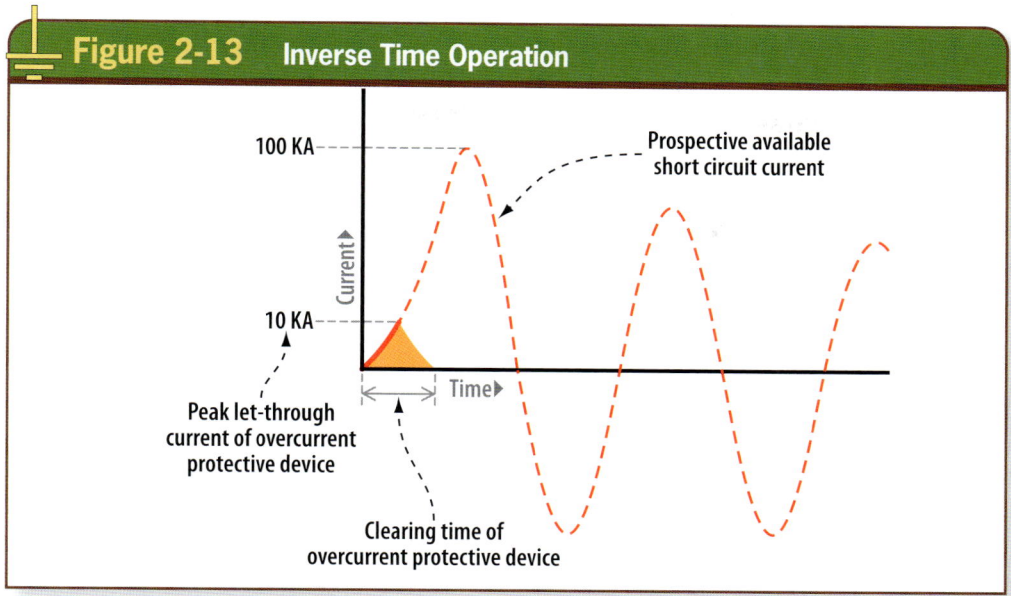

Figure 2-13   Inverse Time Operation

Current▶

100 KA - - - - - - - - - - - - - -

Prospective available
short circuit current

10 KA - - - - - - - - - -

Peak let-through
current of overcurrent
protective device

Time▶

Clearing time of
overcurrent protective device

Figure 2-13. *High values of fault current force quick overcurrent device operation.*

The effective paths for ground-fault current are protective safety circuits that facilitate overcurrent protective device operation during abnormal events such as ground faults and short circuits; these safety circuits are essential to protecting equipment and conductors of the system. The following definitions are helpful to understand the path for short-circuit and ground-fault current during these types of events in an electrical system.

The term *short-circuit* is now defined in the *NEC* as an abnormal connection between two or more points of different potential. It typically is an accidental connection between ungrounded conductors of a circuit or between any of the ungrounded conductors of the circuit and the grounded parts of equipment or the EGC. A ground fault is considered one type of short circuit. An overload (low-level overcurrent condition) is not considered a ground fault or short circuit. **See Figure 2-12.**

AMPERES OPERATE OVERCURRENT PROTECTIVE DEVICES

A simple way to look at how circuit breakers and fuses function is to consider how they respond to current. Normal current in a circuit will not cause a circuit breaker or fuse to operate. *Normal current* is load current that is below the rating of the circuit breaker or fuse protecting the circuit. During normal operation, overcurrent protective devices installed in open air will carry current up to their rating indefinitely without operating. Because overcurrent protective devices are installed within enclosures, the *NEC* requires overcurrent protective devices to be sized at 125% of the continuous load, unless the assembly and overcurrent protective device are listed for 100% operation of its rating. This accomplishes the 80% limitation of continuous loads imposed by product standards.

Overcurrent protective devices also have a fusing or a tripping point. The amount of current that will operate an overcurrent protective device is inversely proportional to time. In other words, more amperes should operate overcurrent protective devices quicker. The higher the level of current is, the faster the overcurrent device will react to the event. This operating characteristic of overcurrent protective devices is often referred to as inverse time operation. **See Figure 2-13.** Many overcurrent protective devices provide current-limiting characteristics that can enhance equipment protection.

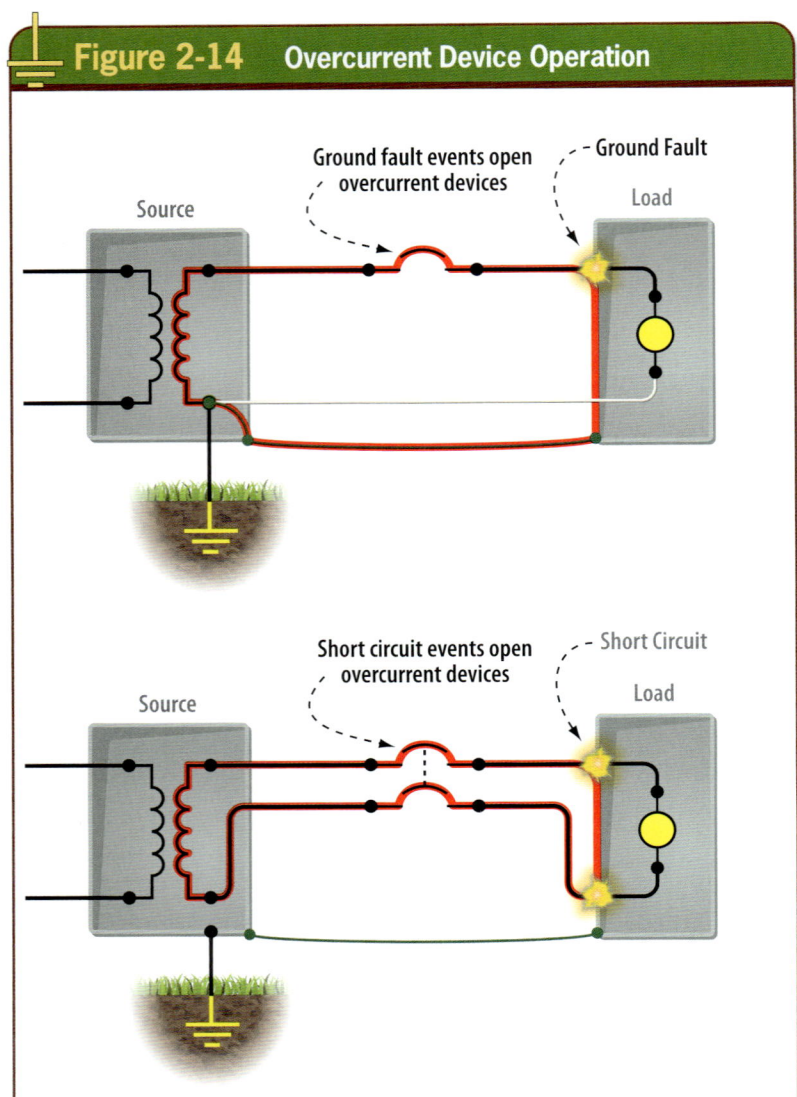

Figure 2-14   Overcurrent Device Operation

Ground fault events open overcurrent devices — Ground Fault

Source   Load

Short circuit events open overcurrent devices — Short Circuit

Source   Load

Figure 2-14. Overcurrent protective devices operate quickly to clear short-circuit and ground-fault current in circuits. Overload conditions cause much slower operation of overcurrent protective devices.

In a short-circuit event, there is little impedance or opposition to the maximum amount of current the electrical system is capable of supplying into the fault. A short-circuit condition is where a fault happens between two or more ungrounded circuit conductors or any ungrounded conductor and the grounded conductor (usually a neutral). These unintentional events typically cause an overcurrent protective device to operate quickly because of the rapid increase of fault current through the device. **See Figure 2-14.**

In a ground-fault event, the fault is between any ungrounded circuit conductor and the Earth, EGC, enclosing metal raceway, or other grounded metal equipment. The greatest amount of damage often occurs in the first half-cycle of time during a short circuit or ground fault. Excessive heat because of the high current can cause rapid deterioration of insulation or annealing of the conductor; it can even cause conductors to melt or vaporize in some cases. An effective ground-fault current path is integral to all electrical designs and installations to ensure fast operation of overcurrent protective devices during ground-fault or short-circuit conditions. This is imperative for safety and is a requirement of the *NEC* as provided in 110.10. **See Figure 2-15.**

The *NEC* indicates that the overcurrent protective device must function quickly enough to prevent extensive damage from occurring. Although *extensive damage* is not defined in the *NEC*, it can be surmised that the equipment should be able to safely handle the fault and clear without resulting in a fire or shock hazard.

Circuit breakers and fuses are related to overall equipment short-circuit current ratings. The *NEC* requires that circuit breakers and fuses in all electrical systems be applied within their ratings. The short-circuit current rating of equipment is the maximum amount of fault current that a component or assembly can safely withstand. Failure to ensure that the available fault current is within the short-circuit current ratings of equipment and the interrupting rating of overcurrent protective devices can result in severe damage or total component destruction in ground-fault or short-circuit conditions.

Fact

Although *extensive damage* is not defined in the *Code*, the equipment should be able to safely handle the fault and clear without resulting in a fire or shock hazard.

TIME AND CURRENT

Circuit breakers and fuses provide overcurrent protection for conductors and equipment. The general requirements for sizing overcurrent protective devices are provided in 240.4. Two concerns in electrical equipment and conductor protection designs are *how much current* the equipment and conductors can handle and for *how long*. Manufacturers of circuit breakers and fuses provide important data related to the operating time and current characteristics of these devices. Each overcurrent protective device has unique time and current graphs that are based on device type and size. This literature is usually in the form of a table or graph known as a *time-current curve*. These graphs indicate how long it takes the device to open or clear the overcurrent condition under various levels of overload and short-circuit current. **See Figure 2-16.**

Time-current curves for circuit breakers and fuses are simple to read. The time is shown on the graph's vertical axis, and the amount of current is provided along the horizontal axis. For circuit breakers, the vertical portion of the curve in the graph represents the instantaneous trip point. The region to the left of the curve is the overload region, and the region to the right is the short circuit (or instantaneous) region. The width of the time-current curve indicates the manufacturing tolerance; that is, the circuit breaker clears within that band. The ranges shown in the graph indicate the manufacturer's test values or proven tolerances of the device. Of course, this range varies based on the circuit breaker type and rating. The time for a circuit breaker to function is on the right or left side of the graph, and the multiples of current levels are shown at the bottom of the chart. For lower-rated circuit breakers, the instantaneous trip range is reached sooner than for larger devices. For example, using the circuit breaker chart, consider the operating characteristics of a 20-ampere circuit breaker. **See Figure 2-17.**

Now compare these operating characteristics to a 100-ampere circuit breaker.

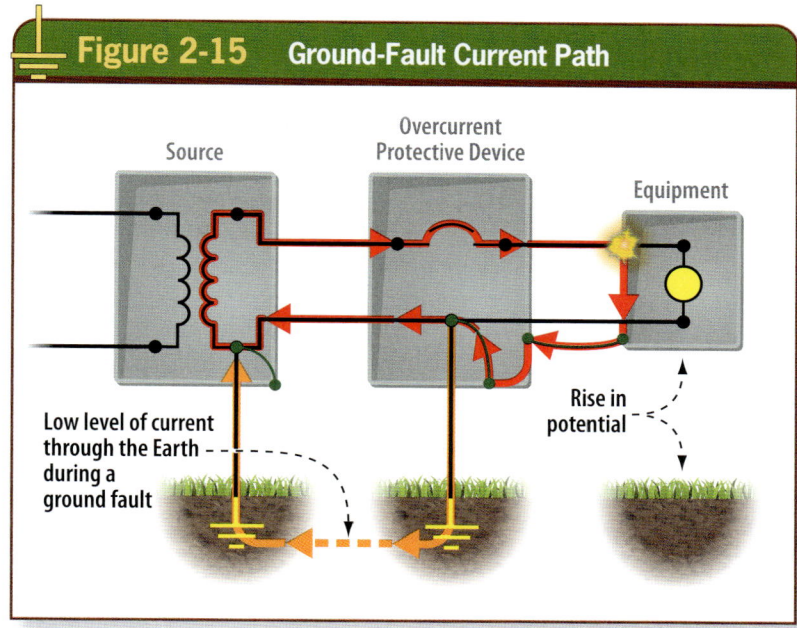

Figure 2-15. *Effective ground-fault current paths ensure fast overcurrent protective device operation.*

The amount of time (on the left side of the chart) needed to react to a short circuit or ground fault in a circuit is greater for the larger overcurrent protective device. **See Figure 2-18.**

The operating characteristics of fuses differ slightly from those of circuit breakers. Circuit breakers require a mechanical action to open, whereas fuses open because of melting of the element contained within the fuse. A time-current curve for fuses is a bit different in appearance in the short-circuit region due to the current-limiting ability of the fuse, but the information conveyed is the same. In a fuse time-current curve, the time is also shown on the vertical axis and the amount of current along the horizontal axis. For example, comparing the time-current curves for a 20-ampere and a 50-ampere Class RK1 fuse reveals that a current of 200 amperes will open the 20-ampere fuse in 1.7 seconds and the 50-ampere fuse in 55 seconds. **See Figure 2-19.**

The fuse time-current curves show various Class RK1 fuse ampere ratings that help determine the clearing times based on the levels of current. **See Figure 2-20.**

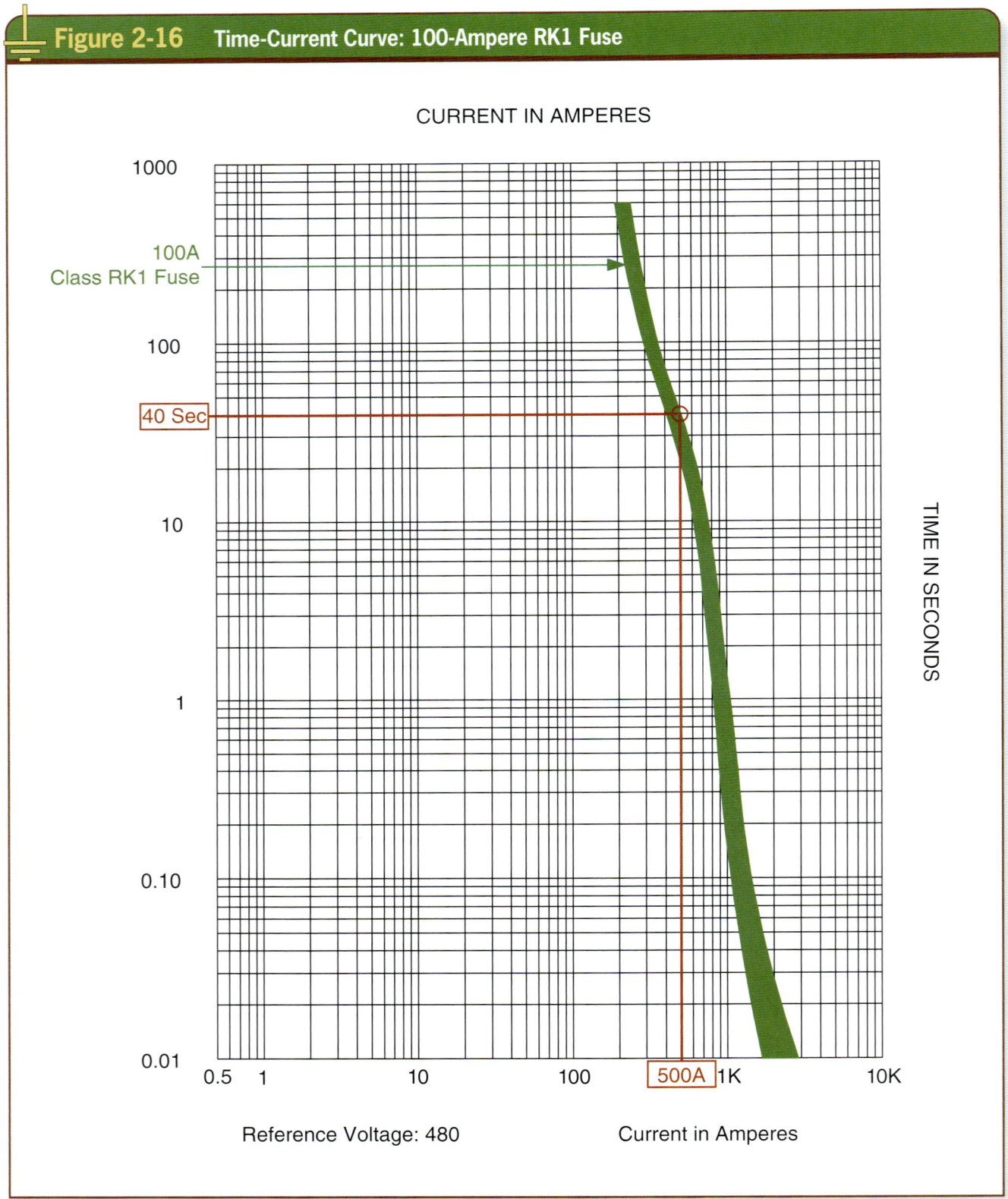

Figure 2-16   Time-Current Curve: 100-Ampere RK1 Fuse

CURRENT IN AMPERES

100A Class RK1 Fuse

40 Sec

TIME IN SECONDS

Reference Voltage: 480   Current in Amperes

500A

Figure 2-16. *The time-current curve for a 100-ampere RK1 fuse shows operating characteristics (in time and current) of this fuse in a graphic form.*

Figure 2-17   Time-Current Curve: 20-Ampere Circuit Breaker

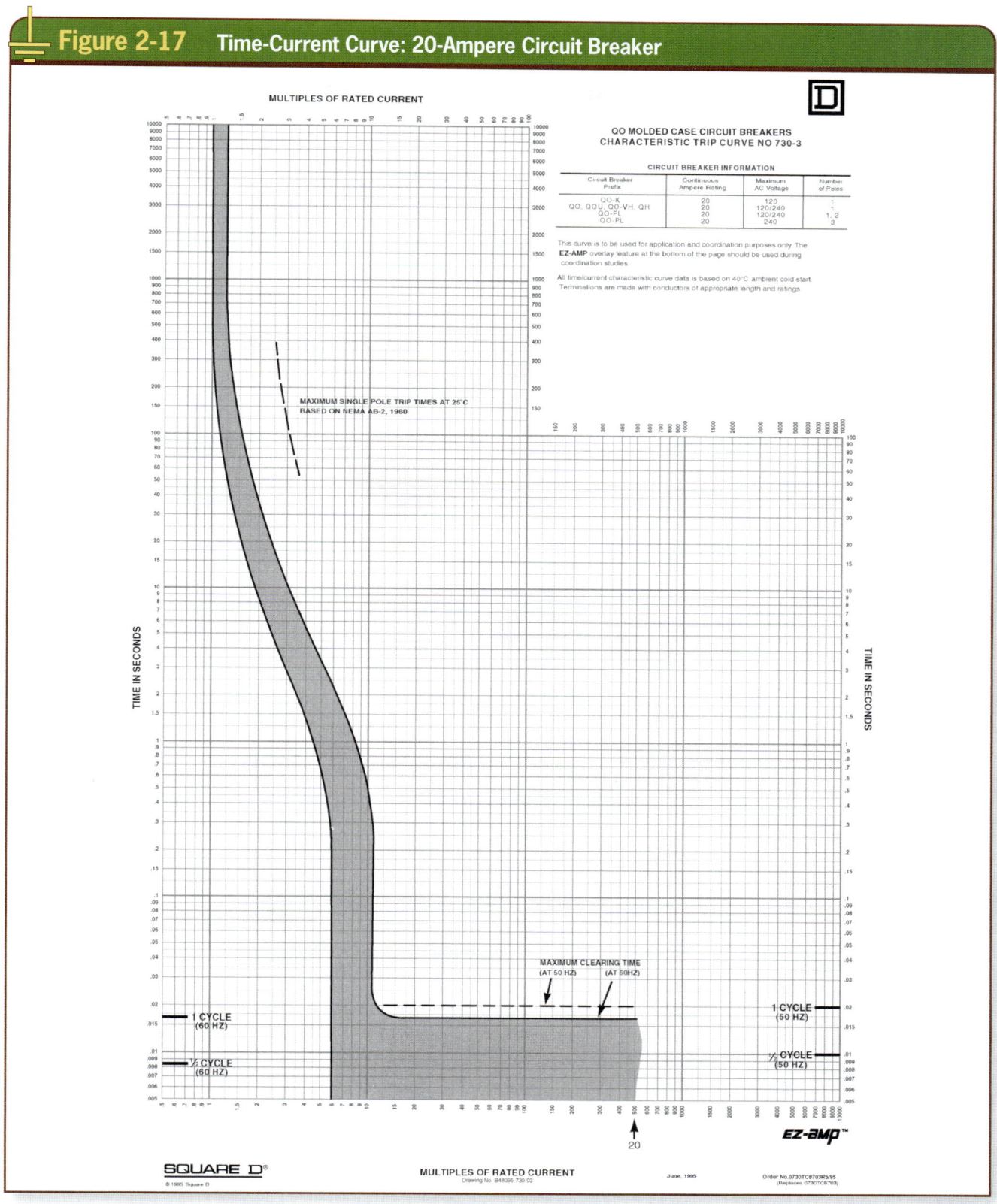

Figure 2-17. *The time-current curve for a 20-ampere circuit breaker shows operating characteristics (in time and current) of this circuit breaker in a graphic form.*

Courtesy of Schneider Electric Square D Company

Figure 2-18   Time-Current Curve: 100-Ampere Circuit Breaker

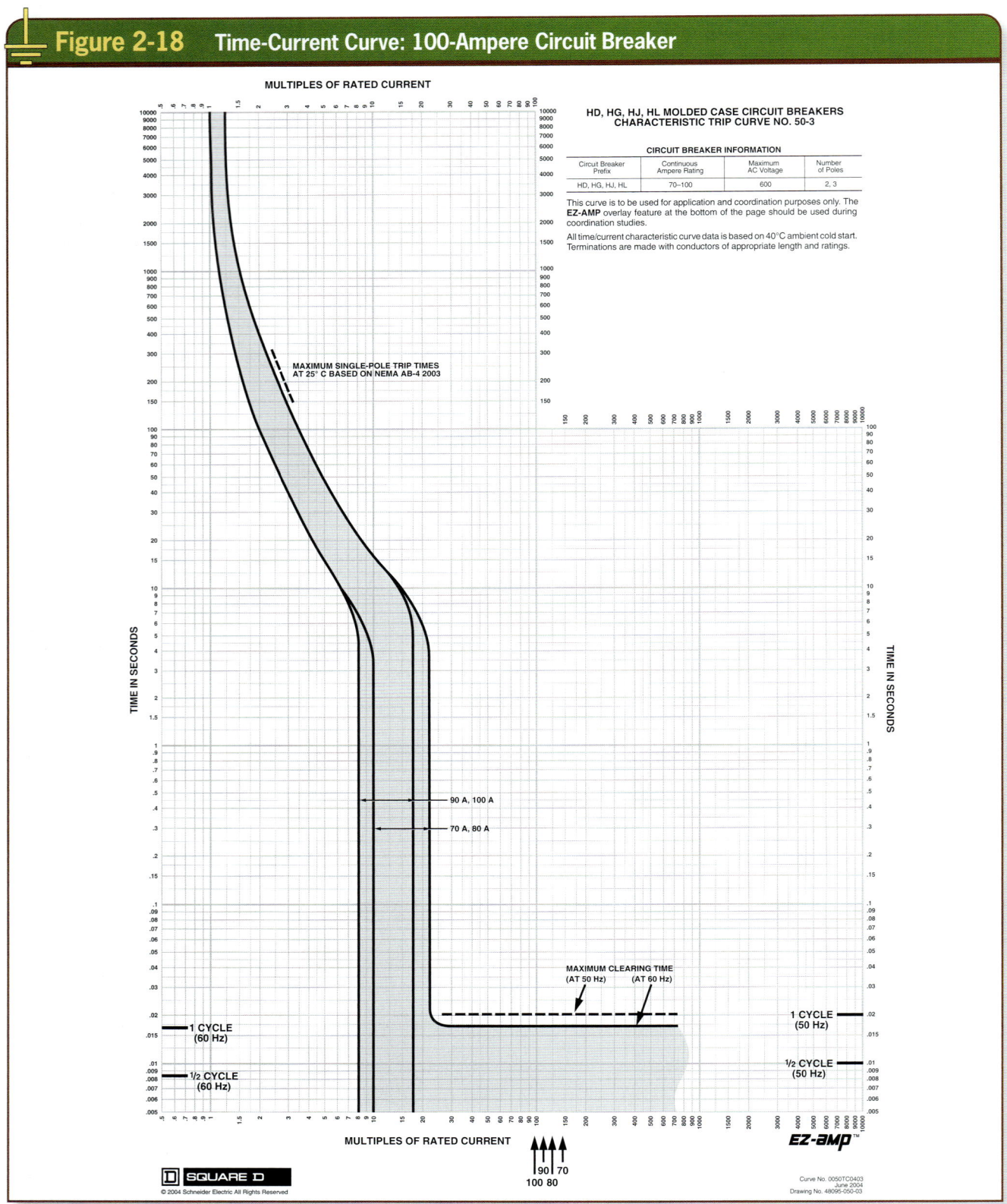

Figure 2-18. *The time-current curve for a 100-ampere circuit breaker shows operating characteristics (in time and current) in a graphic form.*

Figure 2-19 Comparison of Fuse Time-Current Curves

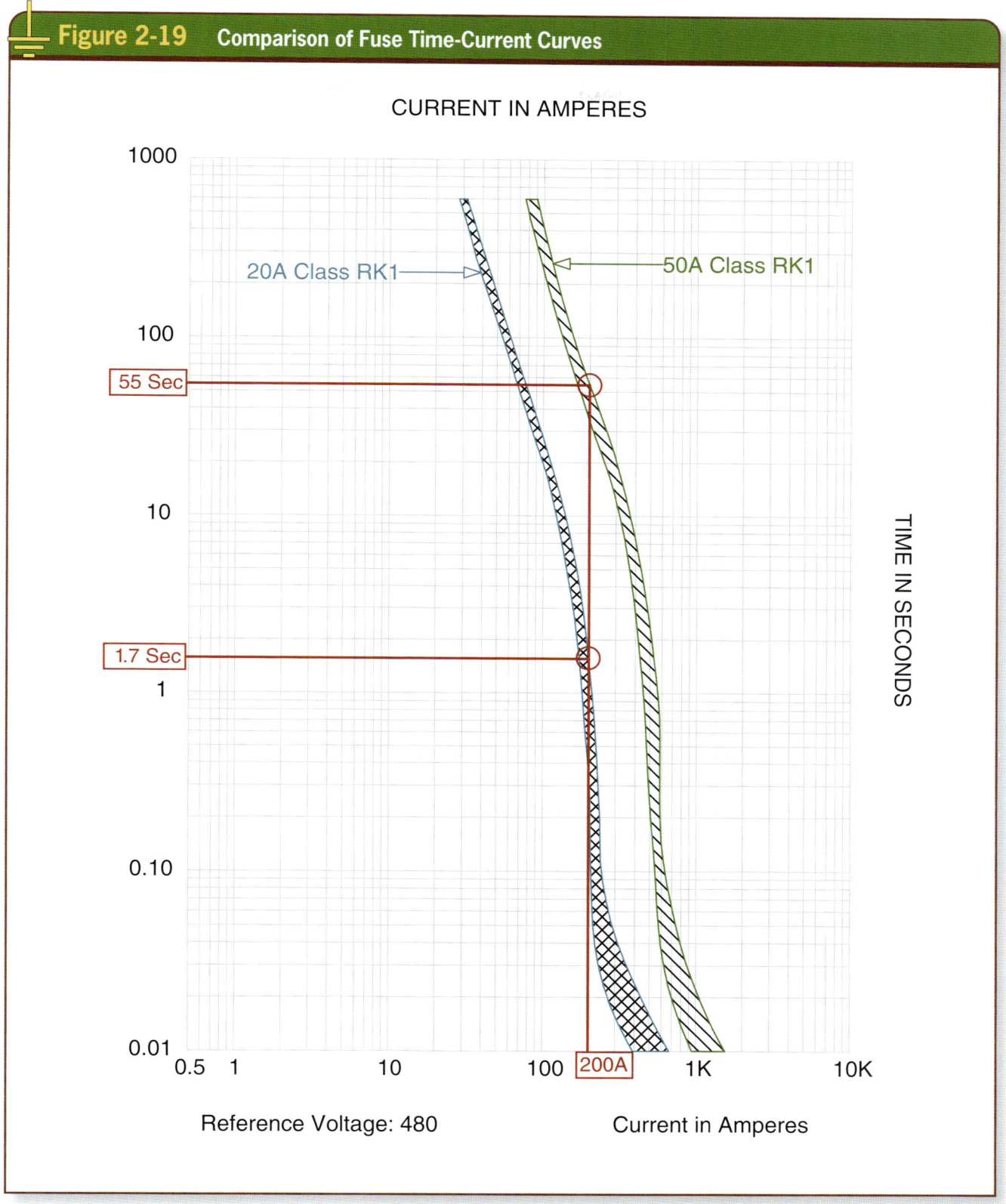

Figure 2-19. *The time-current curve chart for a 20-ampere and a 50-ampere RK1 fuse shows operating characteristics (in time and current) of these fuses in a graphic form.*

Courtesy of Eaton

Figure 2-20   **Fuse Time-Current Curve Comparisons**

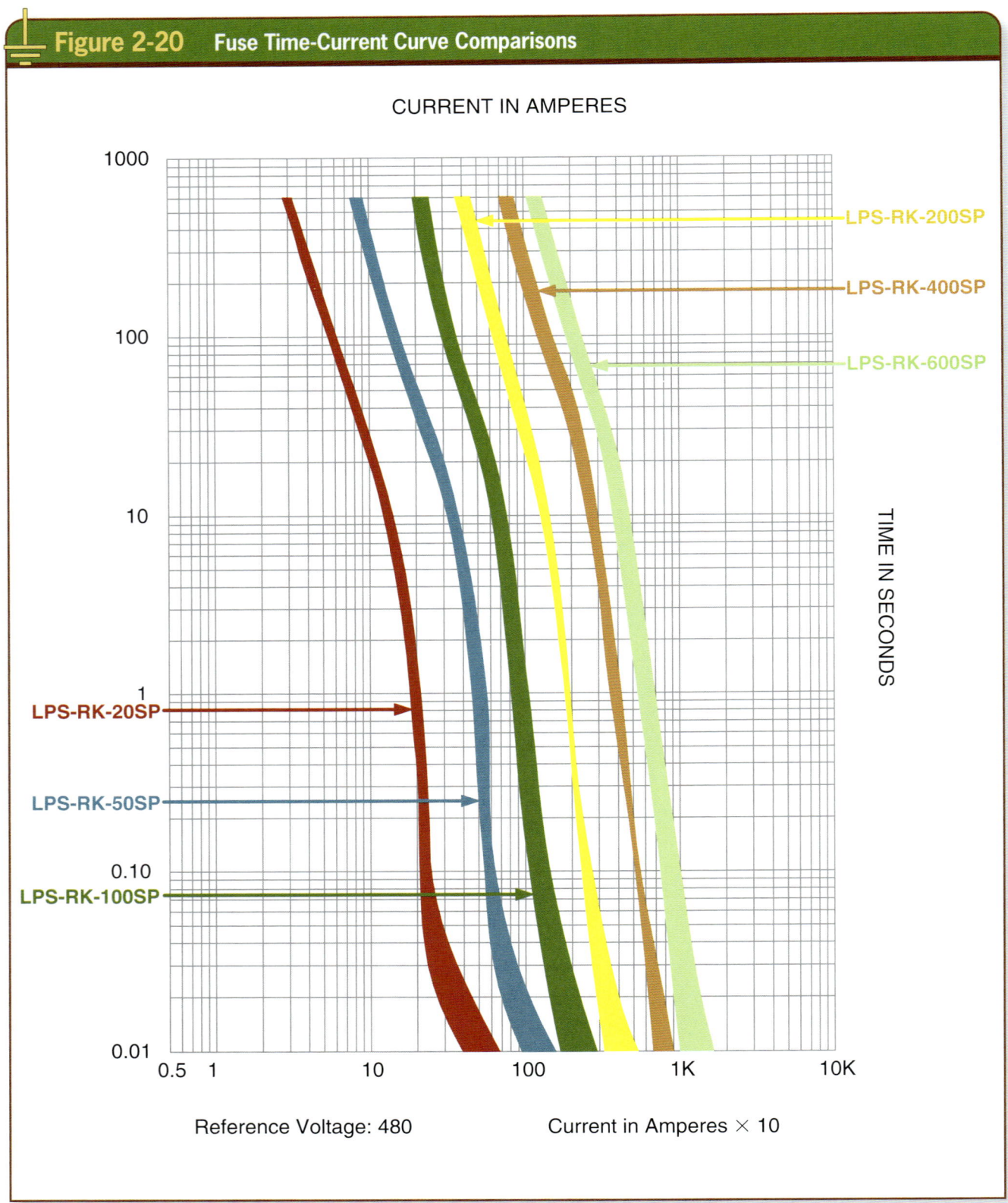

Figure 2-20. *The fuse time-current curve chart for various RK1 fuses shows operating characteristics (in time and current) of these fuses in a graphic form.*

Courtesy of Eaton

Understanding the information in overcurrent protective device time-current curves is helpful in determining whether conductors and equipment are properly protected.

SAFETY BY SYSTEM DESIGN

Overcurrent protection is required to protect conductors and equipment from extensive damage. *NEC* Section 110.10 provides general language that addresses this issue. Another method of achieving additional safety in electrical systems is to reduce let-through current so that incident energy (see *NFPA 70E*) is less during a short-circuit or ground-fault event. Current-limiting overcurrent protective devices can help provide such protection.

> **Incident Energy.** The amount of thermal energy impressed on a surface, a certain distance from the source, generated during an electrical arc event. Incident energy is typically expressed in calories per centimeter squared (cal/cm²).
>
> **Current-Limiting Overcurrent Protective Device.** A device that, when interrupting currents in its current-limiting range, reduces the current flowing in the faulted circuit to a magnitude less than that obtainable in the same circuit if the device were replaced with a solid conductor having comparable impedance.

Protecting Conductor Insulation

Power distribution systems continue to increase in capacity to handle demand. Where the transformer kilovolt-ampere rating (kVA) is high and the transformer impedance is low, the amount of available fault current is often exceedingly high. Ground faults and short circuits on systems with high levels of available fault current can cause serious damage to conductor insulation or to the conductor itself. The *NEC* requires conductors and completed wiring installations to be free from short circuits, ground faults, or any ground connections other than those required or permitted by specific *NEC* provisions.

Conductor insulation prevents the flow of electricity between points of different potential in an electrical system. Failure of the insulation system is one of the most common causes of problems in electrical installations in both high-voltage and low-voltage systems. Insulation tests on new or existing installations can determine the quality or condition of the insulation of conductors and equipment before they are energized. Good work practices include verifying circuit integrity before energizing any electrical circuits. This is usually accomplished using a suitable instrument that verifies conductor insulation dielectric integrity. **See Figure 2-21.**

Common causes of insulation failures are heat, moisture, dirt, and physical damage occurring during and after conductor installation. Insulation can also fail due to chemical effects, exposure to sunlight, and excessive voltage stresses. Insulation integrity must be maintained during overcurrent conditions as well. Conductor damage can include annealing and melting or vaporizing of the metal. Conductors that are subjected to too much current for too long can become annealed, which softens the properties of the conductor

TechTip!

Insulation tests on new or existing installations can determine the quality or condition of the insulation of conductors and equipment before it is energized.

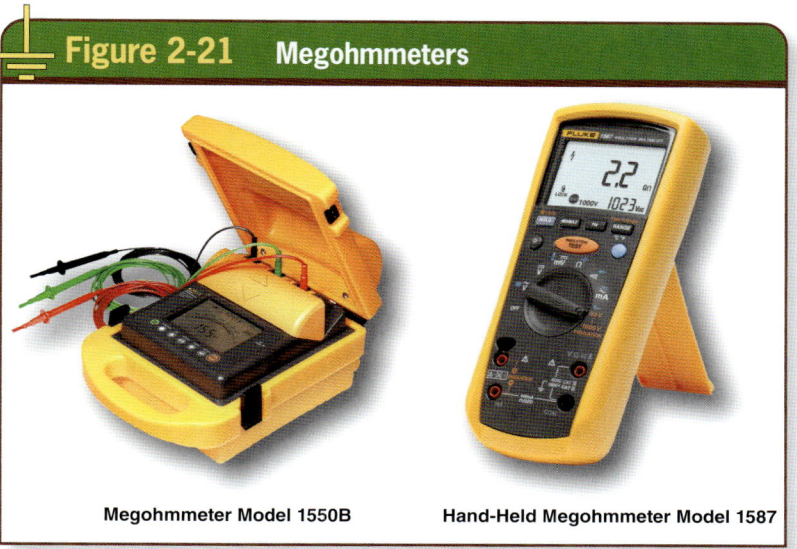

Figure 2-21   **Megohmmeters**

Megohmmeter Model 1550B   Hand-Held Megohmmeter Model 1587

Figure 2-21. Megohmmeters are used to determine conductor insulation integrity prior to energizing electrical circuits.

Courtesy of Fluke Corporation

metal. Annealing of conductors can result in loose connections, which can then in turn result in thermal conditions that damage the components or result in a poor conducting path. Melting or vaporizing of conductors can result in a serious fire hazard or arc flash hazard. In addition, an annealed or opened (due to melting or vaporizing) EGC can result in a shock hazard because there is a poor ground-fault return path or no path at all. People have been killed due to inadequate ground-fault return paths when a phase conductor energized normally non–current-carrying parts.

Properly sized conductors can carry current continuously and can withstand overcurrent conditions for some amount of time. The *NEC* provides ampacity tables in Article 310 and Article 315 for the permissible current values (normal load current). These tables provide current values that conductors can carry indefinitely. (Note that there are conductor ampacity correction and adjustment factors that may be applicable for certain conditions.) During a ground-fault or short-circuit event, the time a conductor can withstand these higher levels of current is significantly less. The higher the overcurrent protective device rating is, the less time a conductor can safely withstand the current.

A good set of guidelines addressing conductor insulation abilities has been established through engineering research by the Insulated Cable Engineers Association (ICEA). The ICEA developed this information to prevent damage to conductor insulation. The ICEA created a chart that shows that currents present for the times indicated in the graph produce maximum safe operating temperatures for each conductor size. The short-circuit current, conductor cross-sectional area, and overcurrent protective devices in the circuit should be applied in a manner that does not exceed the maximum short-circuit times in the ICEA chart. The informational note to *NEC* 240.4 references ICEA P-32-382-2018 for information on the maximum allowable short-circuit currents for insulated copper and aluminum conductors. **See Figure 2-22.**

Short-circuit protection is especially important related to EGCs because they are sized smaller than ungrounded conductors as indicated in Table 250.122. Wire-type EGCs must be selected and applied in system designs in a manner that does not leave them vulnerable to high current levels for periods that exceed their withstand ratings. For insulated EGCs, the following simple formula can be helpful

Figure 2-22   ICEA Chart

Short-Circuit Current Withstand Chart for Copper Cables with Thermoplastic Insulation
Allowable Short-Circuit Currents for Insulated Copper Conductors*

CONDUCTOR: COPPER
INSULATION: THERMOPLASTIC
CURVES BASED ON FORMULA

$$\left[\frac{I}{A}\right]^2 t = .0297 \log \left[\frac{T_2 + 234}{T_1 + 234}\right]$$

WHERE:

I = SHORT-CIRCUIT CURRENT-AMPERES
A = CONDUCTOR AREA-CIRCULAR MILS
t = TIME OF SHORT-CIRCUIT-SECONDS
T_1 = MAXIMUM OPERATING
        TEMPERATURE-75°C
T_2 = MAXIMUM SHORT-CIRCUIT
        TEMPERATURE-150°C

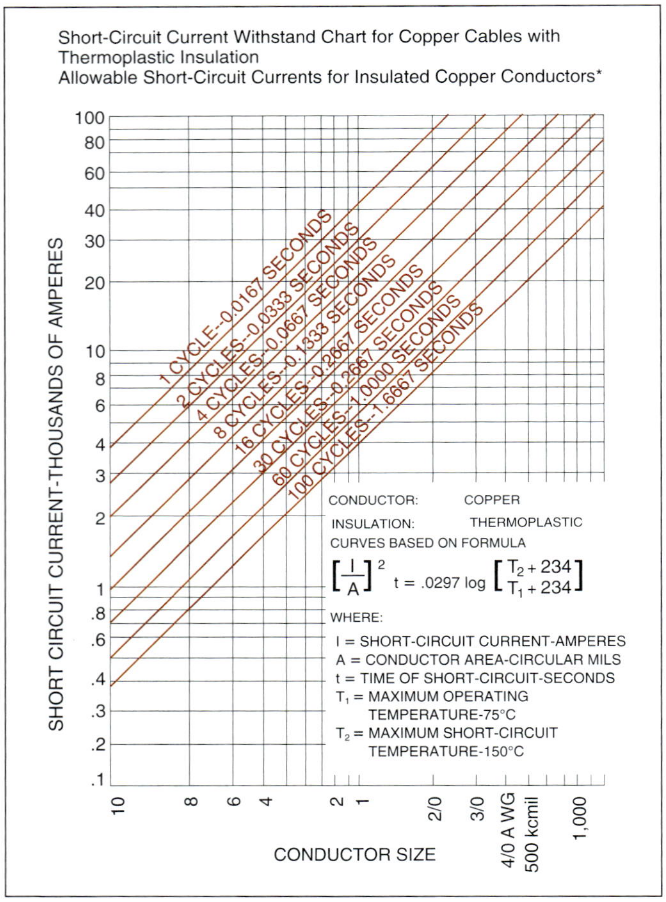

Figure 2-22. *The* Short-Circuit Current Withstand Chart for Copper Cables with Thermoplastic Insulation *provides information about the maximum current handling capabilities for various sizes of conductors over a duration of time.*

Printed with permission from the Insulated Cable Engineers Association (ICEA).

to verify the amount of current a conductor can safely withstand: for every 42.25 circular mils of area, an insulated conductor can safely carry one ampere for five seconds.

The measure of heat energy developed in a circuit during a short-circuit or ground-fault event is characterized in formulas as I^2t. This formula is simply the current (I) squared and then multiplied by the time (t), in seconds. Note that the type of thermal insulation on a conductor influences the true value of current it can safely withstand; it is not the same for all conductor insulation types. When using this method of calculating conductor withstand values, the insulation type and conductor size should be provided in the formula.

Protecting Wire-Type Equipment Grounding Conductors

It is essential that the integrity of conductors be protected to have safe electrical systems. This requirement applies to the ungrounded circuit conductors, the grounded conductors, and the protective EGCs of the wire type. Improper sizing of wire-type EGCs and/or use of improper types of overcurrent protective devices can result in their annealing, melting, or vaporizing before the circuit overcurrent protective device can clear the fault. EGCs are, in some installations, much smaller than the associated circuit conductors, usually about 25% of the size of the circuit conductor. (However, note that the value 25% mentioned here is only a rule of thumb, and one should always refer to the *NEC* to determine the minimum size wire-type equipment grounding conductor(s).)

Selecting the proper overcurrent protective device is an essential step in providing fast clearing times that will protect EGCs within the limits of the conductor withstand capabilities. Consideration must always be given to the entire system of protection, including the size of wire-type EGCs, their withstand ratings, the magnitude of ground-fault currents, and the

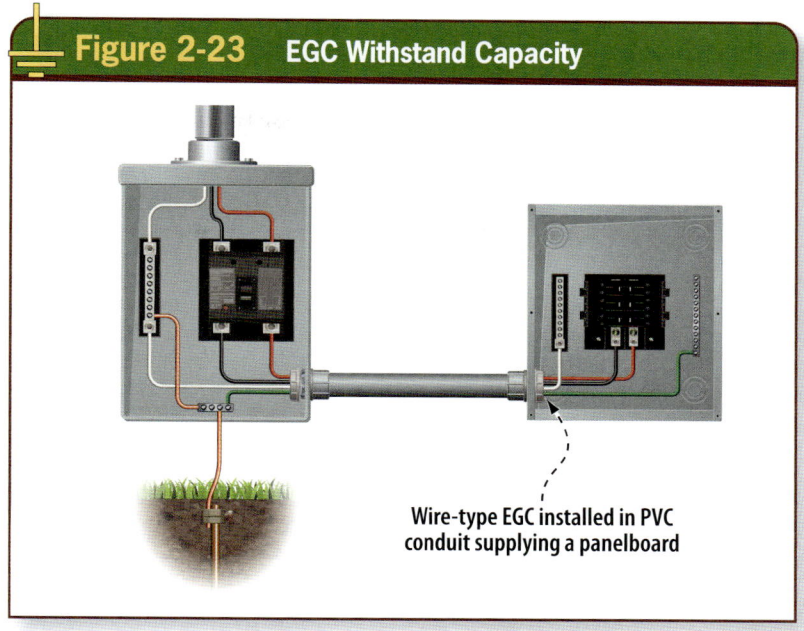

Figure 2-23 **EGC Withstand Capacity**

Wire-type EGC installed in PVC conduit supplying a panelboard

Figure 2-23. The minimum sizes required for wire-type EGCs must be able to effectively perform during ground-fault conditions. They must be able to withstand the fault current for the time it takes the overcurrent protective device to open.

operating characteristics of the overcurrent protective devices. Protective devices that do not operate fast enough might leave EGCs vulnerable to severe damage during a ground-fault event. The solutions are to either select faster overcurrent protective devices or increase the size of the EGC.

EQUIPMENT GROUNDING CONDUCTOR CAPACITY

NEC Table 250.122 provides minimum sizes for wire-type EGCs. Safety concerns are apparent when minimum sizes for EGCs are analyzed. Even though an EGC meets the size required by Table 250.122, other factors must be considered in determining full compliance with 110.10. Sections 250.4(A)(5) and 250.4(B)(4) provide performance criteria for EGCs. A note at the bottom of Table 250.122 refers to those sections. **See Figure 2-23.**

This note is a mandatory requirement, as compared with advisory notes provided elsewhere throughout the *NEC*. Wire-type EGC withstand capacity is the

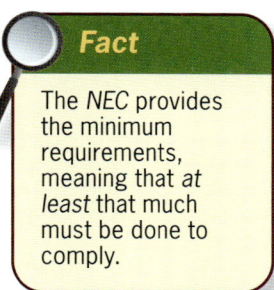

Fact

The *NEC* provides the minimum requirements, meaning that *at least* that much must be done to comply.

concern here. The issue of protecting EGCs was recognized and specifically addressed by Eustace C. Soares in the 1960s. Soares explored and provided engineering solutions to conductor withstand rating issues in his book titled *Grounding Electrical Distribution Systems for Safety*.

The values in Table 250.122 are *minimum* values and their validity and capacity often must be verified. This is an essential step in good engineering practices related to designing power systems. In some cases, the minimum sizes in Table 250.122 must be increased because of the available fault current and the current-limiting abilities of the overcurrent protective devices selected in the design. The best engineering practices include performing calculations of available fault current levels at various points on the electrical system. The circuit overcurrent protective device must be able to clear a short-circuit or ground-fault event without the EGCs being damaged. The EGC must not have insulation damage, and the conductor must not be annealed, melted, or vaporized after a fault occurs. Good design principles include using the circuit breaker or fuse manufacturer's literature to obtain operating characteristics of the overcurrent protective devices installed in the system. The let-through energy levels for these devices should be compared with the withstand ratings of the EGCs. Whenever the conductor withstand ratings are exceeded, the size of the EGC must be increased to ensure adequate capacity to perform as an effective ground-fault current path. The key word in this term is *effective*, which is measurable as it relates to size, capacity, and other physical and installation characteristics of the current path. The path for current must be effective during normal operation and during abnormal events such as ground faults.

SUMMARY

For current to be present in a circuit, the circuit must be complete. Current, be it normal or fault current, will always try to return to its source. Amperes operate overcurrent protective devices. Protection of conductors and equipment requires good engineering and design practices. For an effective ground-fault current path to provide the intended protective function, electrical circuit characteristics must be carefully analyzed and applied in systems. The total circuit impedance, overcurrent protective devices, component short-circuit current ratings, and other circuit characteristics must be selected and coordinated so that the circuit overcurrent protective devices can effectively respond and operate during a ground-fault or short-circuit event, and they must do so without causing extensive damage to the electrical equipment or conductors, including the equipment grounding conductors.

REVIEW QUESTIONS

1. A unit of electrical pressure in an electrical circuit is the __?__ of the circuit.

   a. amperage
   b. reactance
   c. resistance
   d. voltage

2. Resistance or opposition to current in a DC circuit is expressed in __?__.

   a. amperes (A)
   b. ohms (Ω)
   c. volts (V)
   d. watts (W)

3. For current to be present in any electrical circuit, the circuit must be __?__.

   a. closed or complete
   b. incomplete
   c. open
   d. switched

4. Using Ohm's Law, if a circuit has an applied voltage of 480 volts and a resistance of 40 ohms, how much current will be present in the circuit?

   a. 0.083 A
   b. 12 A
   c. 24 A
   d. 192 A

5. The total opposition to current in an AC circuit is known as the __?__ of the circuit.

   a. current
   b. impedance
   c. inductance
   d. resistance

6. The unit of electrical power is expressed in __?__.

   a. amperes
   b. resistance
   c. volts
   d. watts

7. If a circuit has an applied voltage of 120 volts and the current in the circuit is 30 amperes, how many ohms of resistance are present in the circuit?

   a. 4 Ω
   b. 40 Ω
   c. 90 Ω
   d. 3,600 Ω

8. An electric heating element has a resistance of 16 ohms and is connected to a voltage of 240 volts. How much current is present in the circuit?

   a. 0.066 A
   b. 7.5 A
   c. 15 A
   d. 19.20 A

9. If a five-kilowatt heater is connected to a 480-volt, single-phase circuit, how much current will the heater draw?

   a. 0.096 A
   b. 10.4 A
   c. 20.8 A
   d. 96 A

10. In a 60-Hertz (cycle) circuit, the voltage and current change amplitude and direction __?__ times per second.

    a. 60
    b. 120
    c. 240
    d. 400

11. The impedance of an AC circuit includes which of the following electrical characteristics?

    a. Capacitive reactance
    b. Inductive reactance
    c. Resistance
    d. All of the above

12. Keeping the opposition to current as low as possible in electrical bonding connections is necessary so that a maximum amount of fault current causes rapid operation of over-current protective devices in fault conditions.

    a. True   b. False

13. Generally, the *NEC* requires all conductors of AC circuits, including the equipment ground-ing conductors, to be run together in the same raceway, cable, or trench.

    a. True   b. False

14. The current in an electrical circuit will take which of the following paths to return to its source?

    a. All paths available, including the Earth
    b. Only the Earth (ground)
    c. The path of least resistance
    d. The path of most resistance

REVIEW QUESTIONS

15. Electrical equipment and overcurrent protective devices (such as circuit breakers and fuses) must be selected to ensure that the short-circuit current rating of any system component is not exceeded should a short-circuit or high-level ground-fault event occur.

    a. True   b. False

16. An unintentional, electrically-conducting connection between an ungrounded conductor of an electrical circuit and the normally non–current-carrying conductors, metal enclosures, metal raceways, metal equipment, or Earth best defines which of the following?

    a. A ground fault
    b. A short circuit
    c. An overcurrent condition
    d. An overload condition

17. The term __?__ refers to the highest current at rated voltage that a device is identified to interrupt under standard test conditions.

    a. interrupting rating
    b. short-circuit current rating
    c. voltage rating
    d. withstand rating

18. If a system can deliver 22,000 amperes of fault current, what is the minimum short-circuit current rating required for the equipment, including any overcurrent protective device(s) installed in the equipment?

    a. 5,000 A
    b. 10,000 A
    c. 22,000 A
    d. 42,000 A

19. The continuous load (current) on a circuit breaker is generally not permitted to exceed what percentage of the circuit breaker rating?

    a. 80%
    b. 100%
    c. 125%
    d. 150%

20. The two main concerns in electrical equipment and conductor protection designs are how much current the equipment and conductors can handle and for how long.

    a. True   b. False

21. The *NEC* does not require conductors and completed wiring installations to be free from short circuits, ground faults, or any ground connections other than as required or permitted by specific *NEC* provisions.

    a. True   b. False

22. Testing a wire for insulation integrity is typically accomplished by use of which of the following?

    a. A ground resistance tester
    b. A megohmmeter
    c. A voltage tester
    d. A wattage meter

23. The Insulated Cable Engineers Association (ICEA) demonstrated that for every 42.25 circular mils of area, an insulated conductor can safely carry one ampere for __?__.

    a. 1 second
    b. 5 seconds
    c. 10 seconds
    d. 60 seconds

24. All equipment grounding conductors must provide an effective ground-fault current path to facilitate overcurrent protective device operation.

    a. True   b. False

Using the *National Electrical Code*

The purpose of the *National Electrical Code* (*NEC*) is the practical safe-guarding of persons and property from hazards that arise from the use of electricity. The *NEC* provides the minimum safety requirements, meaning that it is the *least* that must be done for compliance. An important part of electrical construction is proper application of *Code* rules to installations and systems. It is important to understand all general *NEC* requirements and to know when and how these rules are modified due to special occupancies, special or unique conditions, and special equipment.

Objectives

» Understand the purpose of the *NEC* and the arrangement of Article 250.

» Understand key defined grounding and bonding terms in Article 100.

» Determine key requirements in the *NEC* that relate specifically to electrical grounding and bonding.

» Distinguish performance requirements from prescriptive requirements in the *NEC*.

» Understand how Chapters 5 through 7 of the *NEC* can modify or amend the requirements in Chapters 1 through 7 of the *NEC*.

» Understand the *NEC* tables related to grounding and bonding conductor sizing.

Chapter 3

Table of Contents

Building a Solid Foundation 50

NEC Arrangement and Application 50

Enforcement and Approvals 52

Requirements, Exceptions, Alternatives,
and Information 53

Permissive *Code* Language 54

Explanatory Information 54

Bracketed Text 54

Defined Terms 55

Code-Making Panel Responsibilities 55

NEC Article 100—Definitions of Grounding
and Bonding Terms 56

Article 200—Use and Identification
of Grounded Conductors 64

Article 250—Arrangement and Use 64

Part I—General 65

Table 250.66—Sizing AC Grounding
Electrode Conductors 66

Table 250.102(C)(1)—Sizing the Grounded
Conductor, Main Bonding Jumper, System
Bonding Jumper, and Supply-Side Bonding
Jumper for Alternating-Current Systems 67

Table 250.122—Sizing Equipment
Grounding Conductors 68

Special Occupancies, Equipment,
and Conditions 69

Example Modification from Article 517
in Chapter 5 69

Example Modification from Article 600
in Chapter 6 70

Example Modification from Article 770
in Chapter 7 71

Chapter 8 71

Summary 72

Review Questions 72

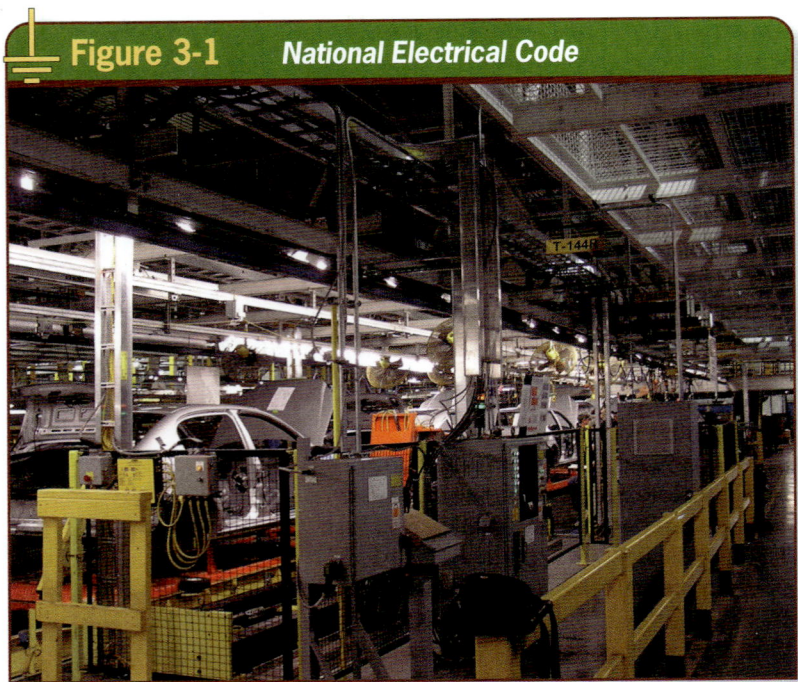

Figure 3-1. *The purpose of the* NEC *is to protect persons and property from hazards of electricity use.*

Courtesy of Valley Electric Company VEC

BUILDING A SOLID FOUNDATION

From the time the project is being designed and drawings are being created until the final receptacle is installed on the project, the requirements in the *NEC* apply. **See Figure 3-1.**

Buildings and structures must have a solid foundation on which they are built. Footings and foundations must have enough structural integrity to support all loads. Likewise, the grounding electrode system for a building serves as the foundation of the electrical system supplying the structure. Since the footing or foundation is part of the grounding electrode system, the concrete-encased electrode is one of the first electrodes installed.

Other grounding electrodes, such as metal water pipe electrodes and metal in-ground support structures, are often inherent to buildings. These electrodes are formed or installed as construction progresses. **See Figure 3-2.**

NEC ARRANGEMENT AND APPLICATION

The *NEC* is an installation code containing requirements that apply to electrical wiring and equipment. It consists of an introduction and nine chapters. Each chapter in the *NEC* contains articles that are further broken down into rules in the form of parts, sections, subdivisions, and lists. Also included are exceptions to rules, informational notes, and rules in tabular form. Article 90 serves as the introduction and provides users with essential information about the structure of the *NEC* and how the rules apply to electrical installations.

The very first provision in the *NEC*, Section 90.1, indicates that its purpose is the practical safeguarding of persons and property against hazards arising from the use of electricity. Section 90.2 provides a scope of the *Code*, clarifying what is covered and what is not covered by the *NEC* rules. Section 90.3 is an important provision that guides users in the correct application of the stated rules and how they must be applied. **See Figure 3-3.**

Chapters 1 through 4 of the *NEC* have general application, meaning these rules apply to all electrical installations. Chapters 5, 6, and 7 include rules for special occupancies, special equipment, and other special conditions. The provisions in Chapters 5, 6, and 7 modify or amend the requirements in Chapters 1 through 7, as there are also requirements in Chapters 5, 6, and 7 that modify requirements in other articles within Chapters 5, 6, and 7.

Chapter 8 is not subject to the general requirements of the other chapters except where the other rules are referenced from within Chapter 8. Chapter 9 of the *NEC* includes tables that are used in applying the other requirements of the *Code*. The Annexes in the back of the *NEC* contain information that is not mandatory but is often valuable in helping users understand how various requirements apply to installations and systems.

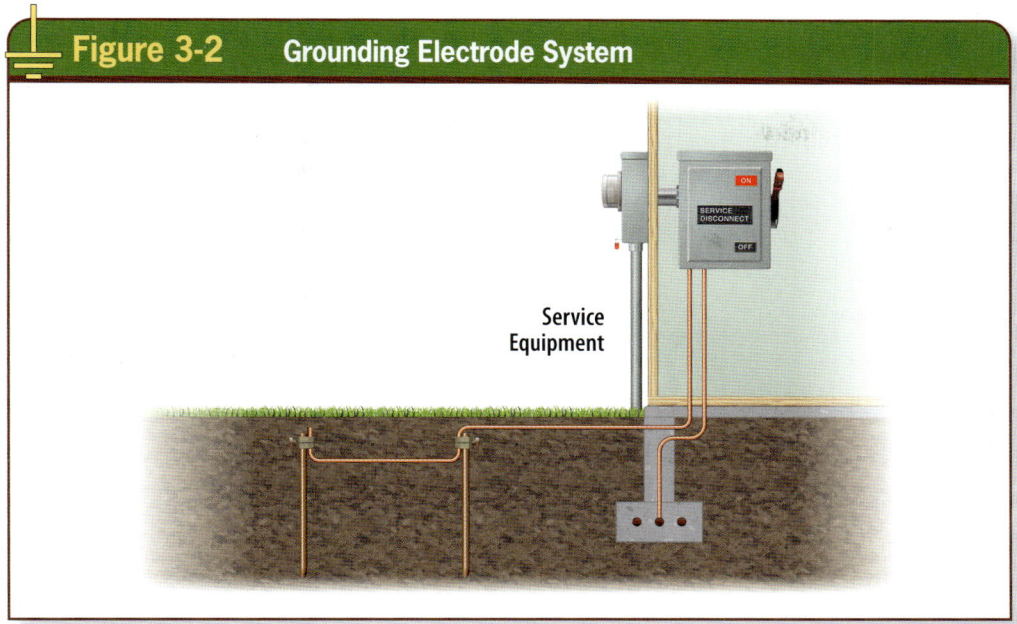

Figure 3-2 | **Grounding Electrode System**

Service
Equipment

Figure 3-2. *The foundation of the electrical system is the grounding electrode system. Grounding electrodes suitable for use in the grounding electrode system are provided in 250.52(A).*

There are many grounding and bonding requirements contained in Chapter 5 of the *NEC* that are more restrictive than the general rules contained in Article 250. For example, 250.118(6) recognizes listed liquid-tight flexible metal conduit (LFMC) as an acceptable equipment grounding conductor (EGC) if all the conditions indicated in 250.118(6)(a) through (e) are met. What this means is that listed liquidtight flexible metal conduit can be installed and used as an EGC without installing a wire-type EGC with the short run of listed liquidtight flexible metal conduit. *NEC* Chapter 5 includes requirements for special occupancies and often modifies the general requirements of the *Code* to be more restrictive. For example, Article 501 covers installations in Class I locations where explosion hazards exist, and 501.30(B) specifically restricts liquidtight flexible metal conduit and flexible metal conduit from being used as the sole bonding means. Equipment bonding jumpers are required for these installations. The result is a requirement to install a wire-type equipment bonding jumper in accordance with 250.102, which allows the bonding

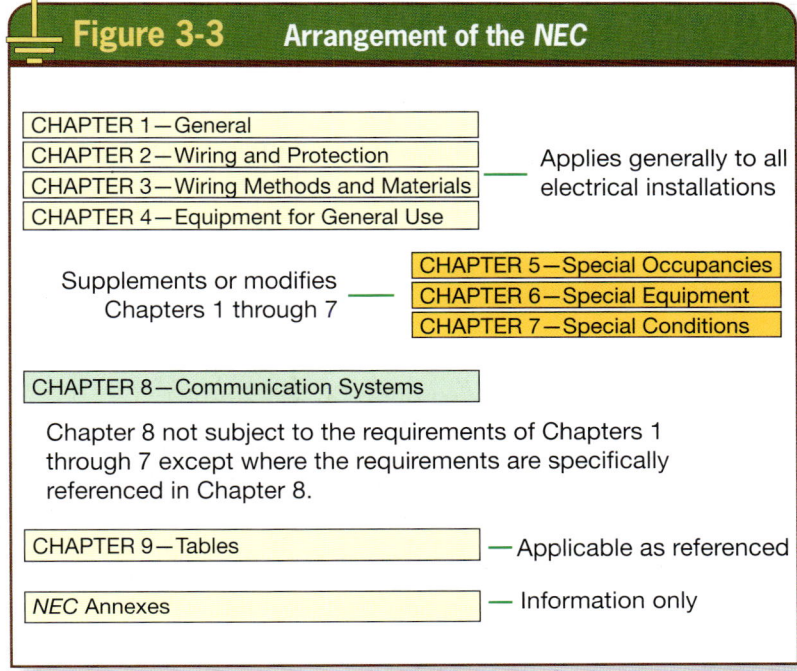

Figure 3-3   **Arrangement of the *NEC***

| CHAPTER 1—General |
| CHAPTER 2—Wiring and Protection |
| CHAPTER 3—Wiring Methods and Materials |
| CHAPTER 4—Equipment for General Use |

Applies generally to all electrical installations

Supplements or modifies
Chapters 1 through 7

| CHAPTER 5—Special Occupancies |
| CHAPTER 6—Special Equipment |
| CHAPTER 7—Special Conditions |

| CHAPTER 8—Communication Systems |

Chapter 8 not subject to the requirements of Chapters 1 through 7 except where the requirements are specifically referenced in Chapter 8.

| CHAPTER 9—Tables | — Applicable as referenced

| *NEC* Annexes | — Information only

Figure 3-3. *Section 90.3 provides the arrangement and application requirements for the NEC.*

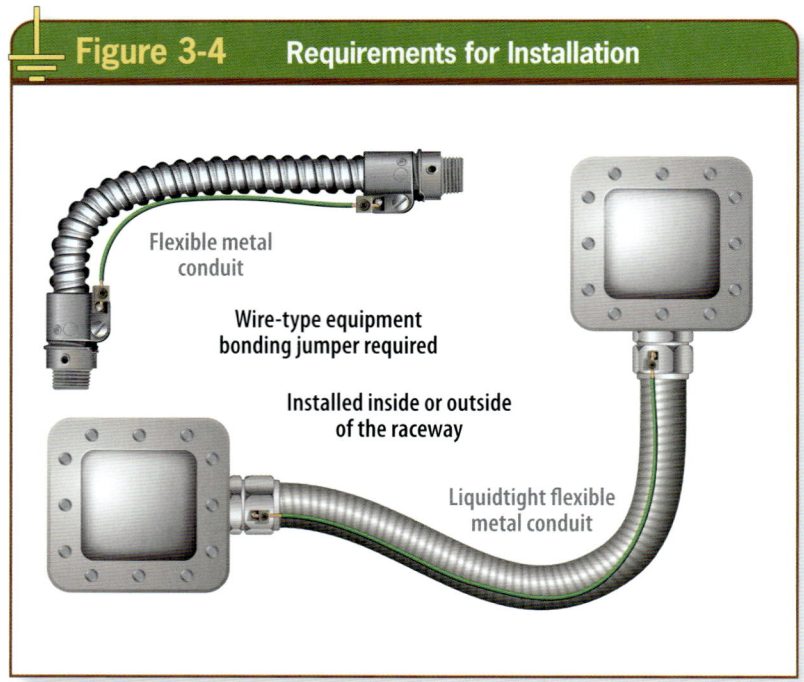

Figure 3-4 **Requirements for Installation**

Flexible metal conduit

Wire-type equipment bonding jumper required

Installed inside or outside of the raceway

Liquidtight flexible metal conduit

Figure 3-4. Section 501.30(B) modifies the general requirements in 250.118.

Fact

Section 501.30(B) specifically restricts flexible metal conduit and liquidtight flexible metal conduit from being used solely as an EGC and bonding means.

jumper to be placed either inside the conduit or outside the conduit. **See Figure 3-4.**

The reason for this requirement is to ensure that a length of liquidtight flexible metal conduit is not depended on for carrying ground-fault current in a Class I, Division 2 hazardous location. Both the conduit and equipment bonding jumper are used, meaning the liquidtight flexible metal conduit does not function as the sole effective path for ground-fault current, as that could result in arcing or sparks in the classified location during ground-fault conditions. This could become an ignition source in an explosive atmosphere where fumes or vapors such as gasoline are present.

Special consideration is necessary for grounding and bonding of exposed non–current-carrying metal parts of equipment, such as metal exteriors of motors, fixed or portable lamps, and luminaires, enclosures, and raceways. These parts must have an effective mechanical and electrical connection to reduce the possibility of arcs or sparks caused by ineffective or poor grounding and bonding methods.

ENFORCEMENT AND APPROVALS

Governing bodies have the authority to enforce the provisions of the *NEC* when it is adopted into law by a particular jurisdiction (90.4). The authority having jurisdiction (AHJ) is defined in Article 100 as a person or organization that is responsible for enforcing the *Code* and issuing approvals of installations and equipment covered by the rules of the *NEC*. The AHJ is also responsible for interpreting the requirements of the *NEC* and granting special permission as provided in some of the requirements.

It is important that workers establish a good working relationship with inspection authorities. This requires effective communication about how the rules will be applied to any given aspect of the installation. Being proactive is always the best approach when unsure of how a particular *NEC* requirement will be applied to an installation or system. Inspection authorities typically require electrical installations to be inspected and approved before they are concealed by building finishes. This requirement can vary among jurisdictions, so the inspection and approval procedures required in the jurisdiction where the project is being built should always be verified. Inspection jurisdictions often use listing and product certification as a basis for issuing approvals. Many electrical grounding and bonding products are listed to applicable safety standards. *UL 467* is the standard for Grounding and Bonding Equipment. Category KDER in the online UL Product iQ is a valuable resource for additional information about listed grounding and bonding equipment.

For additional information, visit qr.njatcdb.org Item #5333

REQUIREMENTS, EXCEPTIONS, ALTERNATIVES, AND INFORMATION

To apply the requirements in the *NEC*, the difference between mandatory requirements and permissive and informational provisions must be understood. A misunderstanding of how exceptions apply can lead to misapplication of mandatory rules. The mandatory requirements in the *Code* are characterized by the use of the terms *shall* or *shall not*.

> **250.68(A) Accessibility.** All mechanical elements used to terminate a grounding electrode conductor or bonding jumper to a grounding electrode shall be accessible.

This is a general requirement that must be followed. If there were no exceptions following this requirement, all connections to grounding electrodes would be required to be accessible. Knowing how impractical this would be, it makes sense that exceptions to the rule are included.

Exceptions to requirements modify only the rule they immediately follow, unless stated differently within the exception. That is, the exception may indicate that it applies to more than the rule it follows.

> **250.68(A) Accessibility.** All mechanical elements used to terminate a grounding electrode conductor or bonding jumper to a grounding electrode shall be accessible.
>
> *Exception No. 1: An encased or buried connection to a concrete-encased, driven, or buried grounding electrode shall not be required to be accessible.*

An exception to a rule can relax the requirement in that rule. For example, Exception No. 1 after 250.68(A) relaxes the requirement for the accessibility of

direct buried connections or connections that are encased in concrete. Because these types of connections are very unlikely to be disturbed or affected once buried or once concrete is poured, this exception provides a practical exemption to the general requirement. For the exception to apply, however, the connection means must be suitable for these locations. The means of connection are required to be evaluated and listed by a qualified electrical testing laboratory to indicate this suitability. **See Figure 3-5.**

There are also mandatory exceptions that use the terms *shall* or *shall not*. An example of a mandatory exception follows 230.95. This exception indicates that ground-fault protection of equipment shall not apply to a service disconnect for a continuous industrial process where a non-orderly shutdown introduces additional or increased hazards. In these cases, the *Code* recognizes a condition that introduces greater risks to persons or property.

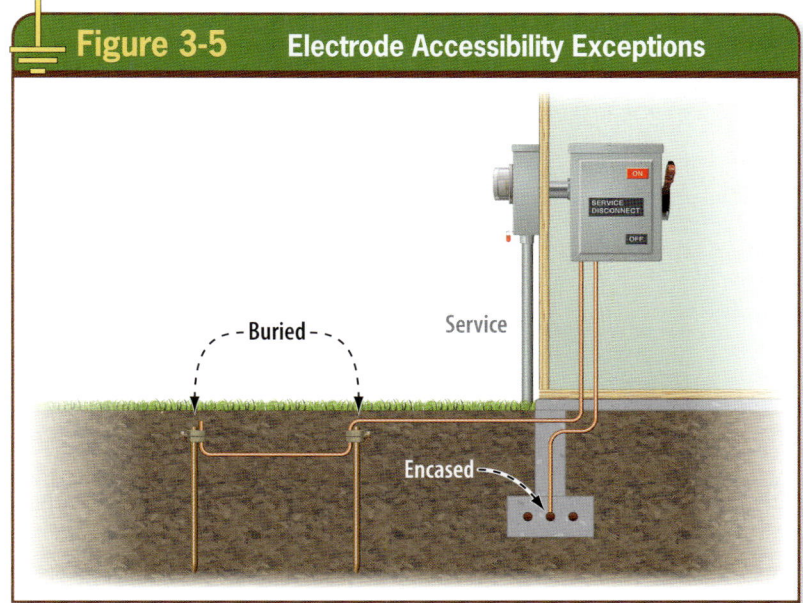

Figure 3-5. Exceptions modify only the NEC rule they follow, unless stated differently within the exception. Exception No. 1 to 250.68(A) relaxes an accessibility requirement for grounding electrode conductor connections. In an installation where the connections are direct buried and/or encased in concrete, Exception No. 1 applies.

Figure 3-6   Authority Having Jurisdiction (AHJ)

Figure 3-6. *Permissive requirements often include AHJ acceptability. The authority having jurisdiction interprets and enforces the NEC requirements, issues approvals, and can grant special permission.*

Courtesy of NECA

Figure 3-7   Notes to Tables

Table 250.122 Minimum Size Equipment Grounding Conductors for Grounding Raceway and Equipment (in part)

Rating or Setting of Automatic Overcurrent Device in Circuit Ahead of Equipment, Conduit, etc., Not Exceeding (Amperes)	Size (AWG or kcmil)	
	Copper	Aluminum or Copper-Clad Aluminum*
1000	2/0	4/0
1200	3/0	250
1600	4/0	350
2000	250	400
2500	350	600
3000	400	600
4000	500	750
5000	700	1250
6000	800	1250

Note: Where necessary to comply with 250.4(A)(5) or 250.4(B)(4), the equipment grounding conductor shall be sized larger than given in this table.

*See installation restrictions in 250.120.

Figure 3-7. *The note following Table 250.122 is a mandatory requirement. [Reproduction of Table 250.122 and Notes (in part)]*

PERMISSIVE *CODE* LANGUAGE

Permissive provisions in the *NEC* are characterized by the use of phrases such as *shall be permitted* or *shall not be required*. Permissive rules are options or alternative methods of achieving equivalent safety; they are not requirements. A closer review of permissive terms is essential because permissive rules are often misinterpreted. The qualifying feature of permissive rules involves determination of the AHJ, as indicated in 90.4.

The AHJ has the final approving authority and responsibility. When in doubt about a permissive requirement or provision in the *Code*, consult the AHJ. **See Figure 3-6.**

EXPLANATORY INFORMATION

Explanatory material is provided in the form of informational notes. These informational notes are advisory only and are not enforceable as requirements. Just as for exceptions to requirements, the *NEC Style Manual* requires that informational notes be positioned immediately following the rule or exception to which they apply. Often, informational notes clarify the requirement that precedes the note. Examples of informational notes are references to other standards, references to related sections of the *Code*, or information related to a rule.

Informational notes should not be confused with notes following tables, which are applicable requirements. A good example of this is found in the notes following Table 300.5 and Table 1 in Chapter 9. Another good example of a note to an *NEC* table is found following Table 250.122. **See Figure 3-7.**

BRACKETED TEXT

Brackets containing section references to another NFPA document serve as a guide to indicate the source of the extracted text. Standards referred to in the brackets are for reference only and are not intended to make the referenced standard applicable as a requirement.

Many examples of bracketed information can be found in Articles 500 and 517, but bracketed information is not limited to just those articles.

DEFINED TERMS

Defined terms help clarify the meaning of a rule in which the term appears. To fully understand how the *NEC* applies to grounding and bonding systems, a clear method or language of communication must be established.

Article 100 contains several defined terms used in the rules. The *NEC Style Manual* now requires all definitions to be in Article 100. Article 100 has been modified for the 2023 *NEC* to remove all parts, as it is not permitted to be subdivided. Definitions are arranged in alphabetical order, and definitions with subparts are listed with the base term first, followed by a comma and a modifying descriptor, for ease of searching. When the definition is listed in subpart form, the defined term is required to appear after the term in parentheses as it would appear in the *NEC*. A definition that only applies in one article has the article number in parentheses following the definition. At the end of all definitions, the *Code*-Making Panel (CMP) responsible for the definition is shown in parentheses. **See Figure 3-8.**

> **Ground-Fault Current Path, Effective. (Effective Ground-Fault Current Path)** An intentionally constructed, low-impedance electrically conductive path designed and intended to carry current during ground-fault events from the point of a ground fault on a wiring system to the electrical supply source and that facilitates the operation of the overcurrent protective device or ground-fault detectors. (CMP-5)

It is important to develop an understanding of defined grounding and bonding terms and promote appropriate use of these terms. *Code*-Making Panel 5 has been assigned this responsibility, and this panel now oversees

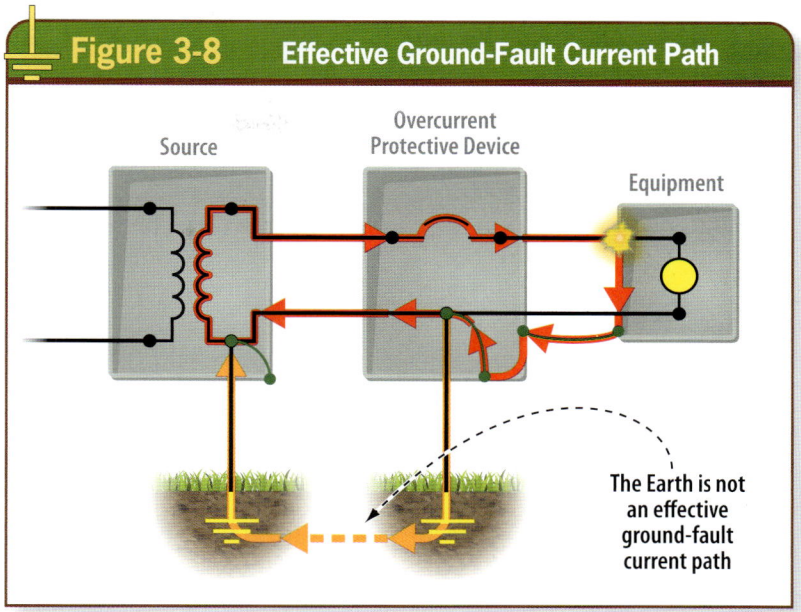

Figure 3-8. *The term* effective ground-fault current path *is defined in Article 100.*

Public Input and Public Comments for new grounding and bonding terms or revisions to existing terms; this group is also responsible for creating and revising grounding and bonding requirements every cycle.

Another important aspect of defined terms is that where a defined grounding or bonding word or term is used within a rule, it should be used in a manner consistent with how the term is defined. Using terms as they are defined in the *NEC* promotes effective and accurate application of requirements.

CODE-MAKING PANEL RESPONSIBILITIES

There are several *NEC* technical "subcommittees" that are referred to as *Code*-Making Panels. *Code*-Making Panel 5 is responsible for definitions of terms related to grounding and bonding and for Articles 200 and 250. During the 2008 *NEC* development cycle, the *NEC* Technical Correlating Committee (TCC) assigned a specific Task Group to work on revising grounding and bonding terms and to verify appropriate use of these terms throughout the *Code* to help clarify their meaning in the rules where they appeared. The result of that work was increased clarity of the requirements and how they apply to installations and systems.

The following examples demonstrate the value of this task group's efforts:

- *Example 1:* A grounding conductor is required to be grounded to the equipment grounding terminal bar of a panelboard.
- *Example 2:* The EGC is required to be connected to the equipment grounding terminal bar of the panelboard.

The first example is incorrect and and confusing wording. The correct wording is used in the second example. The difference is how the terms *grounding* and *grounded* are used. Article 100 defines these terms as "connected (connecting) to ground or to a conductive body that extends the ground connection." Therefore, the first example does not make technical sense, even though many people communicate it that way. It is important to use the language the *NEC* uses with specific respect to the meaning

of defined terms. This promotes a clear and consistent understanding of the requirements and which rules apply to various portions of the electrical system. Slang terms and trade jargon have different meanings to different people and often cause confusion. For example, some people refer to a bonding jumper as a *bond*. The term *bonding jumper* is defined in the *NEC*, whereas the term *bond* is not; therefore, *bond* should not be used.

NEC ARTICLE 100— DEFINITIONS OF GROUNDING AND BONDING TERMS

Article 100 of Chapter 1 in the *NEC* provides the definitions of common terms related to grounding and bonding. Within the *NEC*, the definitions are usually presented in the simplest form. The rule in which the term is used often indicates what is intended to be accomplished and clarifies the intended meaning of the rule with the support of its definition. Some definitions from the *NEC* are included here and followed by comments to help describe the implied meaning of each term.

> **Bonded (Bonding).** Connected to establish electrical continuity and conductivity. (CMP-5)

The term *bonded (bonding)* means connecting conductive objects together. Once this is accomplished, the potential differences between these objects are minimized and they become electrically one. What the bonding is intended to accomplish depends on the rule in which the term is used. For example, bonding at service equipment establishes electrical continuity and conductivity and ensures a path for fault current likely to be imposed, while the bonding required by 680.26 (equipotential bonding for swimming pools and similar installations) is concerned with establishing

Figure 3-9. Bonding Jumper Installation

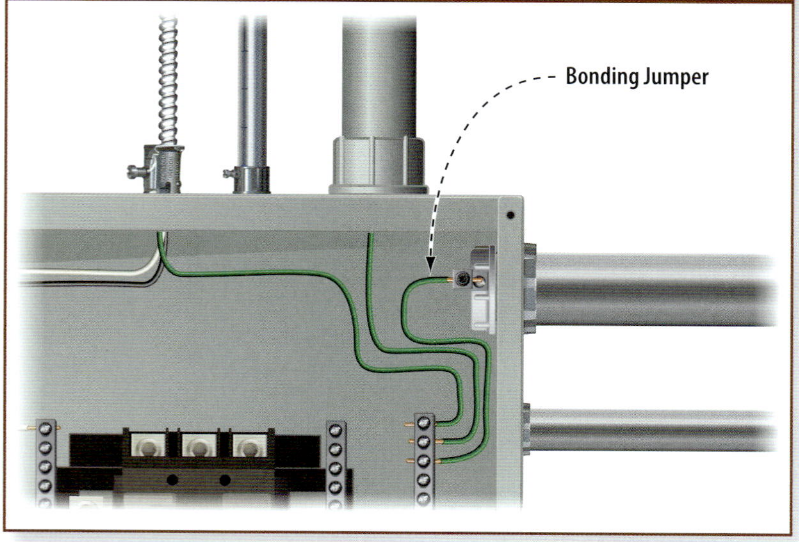

Figure 3-9. Bonding jumpers are reliable conductors that establish the required electrical continuity between metal parts required to be electrically connected.

an equipotential bonding grid that has little or nothing to do with fault current or facilitating operation of overcurrent protective devices. Therefore, the term is defined generally in Article 100, and what is intended to be accomplished by the bonding must be specifically addressed in the rule where the term is used.

> **Bonding Conductor (Bonding Jumper).** A conductor that ensures the required electrical conductivity between metal parts that are required to be electrically connected. (CMP-5)

Bonding conductors or jumpers are typically short conductors that are installed to establish required conductivity between conductive parts. A bonding jumper can serve to complete a conductive path to the Earth for grounding functions, can serve as a path for fault current, or can serve both purposes. **See Figure 3-9.**

An example of a bonding jumper is a conductor installed from a bonding/grounding bushing that is used to provide a conducting connection between a raceway and an enclosure.

> **Bonding Jumper, Equipment (Equipment Bonding Jumper).** The connection between two or more portions of the equipment grounding conductor. (CMP-5)

An equipment bonding jumper is usually a short conductor that establishes a connection between two or more portions of an EGC. For example, if an expansion fitting is used across a structural steel expansion joint in a building, an equipment bonding jumper is often installed to ensure bonding across the special fitting that is designed to expand and contract as the building moves. **See Figure 3-10.**

If rigid metal conduit (RMC) is used for the circuit crossing the expansion joint, the integrity of the EGC path must be assured by installing an equipment bonding jumper. This jumper joins the two lengths of RMC that serve as an EGC for the contained circuit.

> **Bonding Jumper, Main (Main Bonding Jumper).** The connection between the grounded circuit conductor and the equipment grounding conductor, or the supply-side bonding jumper, or both, at the service. (CMP-5)

The main bonding jumper is a conductive path between the grounded conductor bus (usually a neutral bus) and the enclosure and EGC or supply-side bonding jumper (or both) at the service equipment. Main bonding jumpers are installed at the service only, usually within the service

Figure 3-10   Equipment Bonding Jumper Installation

Figure 3-10. Equipment bonding jumpers are required to be installed with conduit expansion fittings. Listed expansion fittings should always be installed in accordance with the manufacturer's installation instructions.

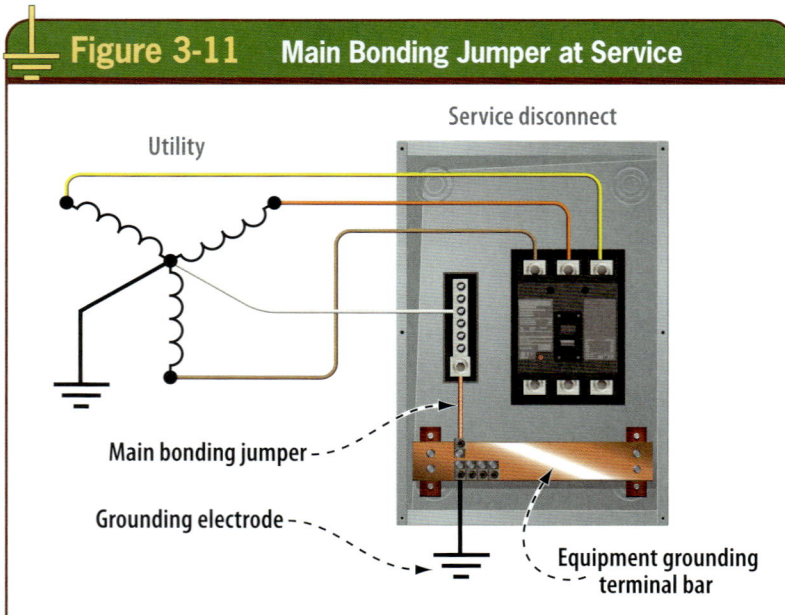

Figure 3-11   Main Bonding Jumper at Service

Service disconnect

Utility

Main bonding jumper

Grounding electrode

Equipment grounding terminal bar

Figure 3-11. The main bonding jumper connects the grounded conductor to the EGC(s) at the service.

disconnecting means enclosure. **See Figure 3-11.**

The main bonding jumper material can be copper, aluminum, copper-clad aluminum, or other corrosion-resistant material and can be a wire, bus, screw, or similar suitable conductor.

Figure 3-12   Ground is Earth by Definition

Figure 3-12. Grounding requires a connection to the planet Earth.

Ground. The earth. (CMP-5)

The term *ground* simply means the Earth. When this term is used in the *NEC*, the planet Earth is what is implied. **See Figure 3-12.**

Because the term *ground* is defined as the Earth, the connection implied in this definition is a connection to the Earth, or to a conductive body that extends the Earth connection, rather than a conductive body serving in place of the Earth. The Earth is a conductive body, but it is generally considered a poor conductor, especially when compared with typical conductor materials such as aluminum and copper. The resistivity of soil varies regionally and seasonally, and thus has differing conductive properties. Earth that is normally moist and not subject to large temperature differentials can have a very low resistivity, whereas sandy or rocky earth often offers higher levels of resistivity. The changes in seasons and fluctuations of moisture levels change the Earth's conductivity characteristics. The *Code* specifically prohibits the Earth from being used as an effective path for fault current because it is not suitable. It is a serious error to depend on the Earth as the sole return path for ground-fault current. Although the Earth is in the grounding circuit, it performs functions other than facilitating overcurrent protective device operation.

Ground Fault. An unintentional, electrically conductive connection between an ungrounded conductor of an electrical circuit and the normally non–current-carrying conductors, metal enclosures, metal raceways, metal equipment, or earth. (CMP-5)

A ground fault is an event that is unintentional and usually causes a circuit breaker or fuse to operate. This type of event is oftentimes caused by a failure or short-circuit condition when the supply system is under severe stress. The amount of current depends on the impedance of the circuit and the fault current available in the system. A ground-fault event can produce an arcing condition that can cause severe injury or death. This is another reason that the definition indicates that these events are unintentional.

> **Grounded (Grounding).** Connected (connecting) to ground or to a conductive body that extends the ground connection. (CMP-5)

The term *grounded (grounding)* clearly describes what is intended to be accomplished through the process of grounding. Since the term *ground* is defined as the Earth, a connection to the planet is implied.

This is typically accomplished through one or more grounding electrodes, such as a concrete-encased electrode or metal water pipe that is in contact with the earth. Conductive bodies that extend the ground connection can be a grounding electrode conductor or EGC. **See Figure 3-13.**

When a system or conductive object is solidly grounded, it is connected to the Earth by a conductive path (normally a grounding electrode conductor) that offers little or no resistance or impedance in the path to Earth. **See Figure 3-14.**

For example, a transformer or generator producing a system secondary or output voltage is required to be solidly grounded in accordance with 250.30(A).

An example of grounding that is not solid is an impedance-grounded system. The system is still connected to ground (Earth), but only through an impedance device or resistor.

> **Grounded Conductor.** A system or circuit conductor that is intentionally grounded. (CMP-5)

Systems that are grounded include one conductor that is intentionally connected to ground, either solidly or through an impedance device. A grounded conductor is one connected

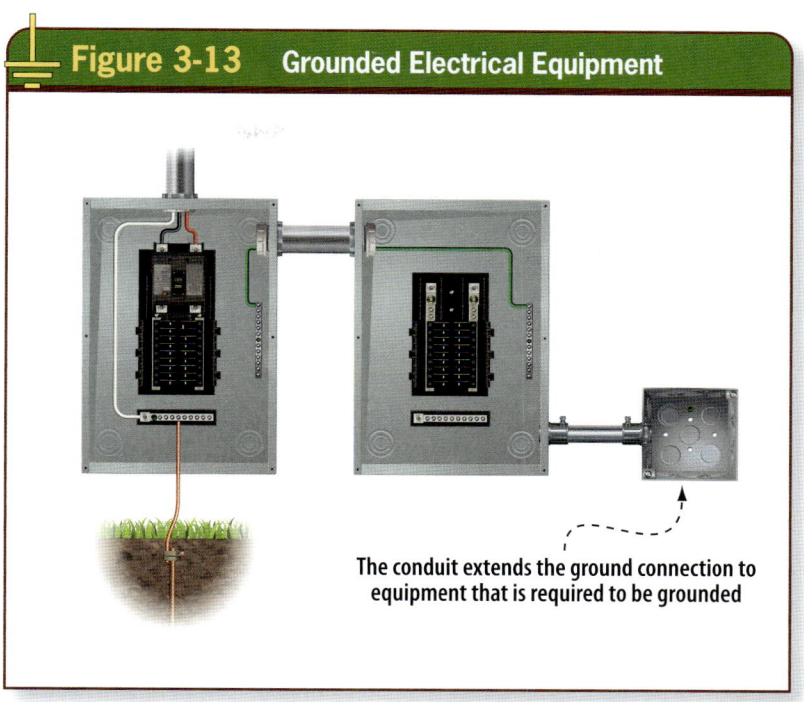

Figure 3-13. Grounding electrical equipment means ensuring that the equipment is grounded by connection to a conductive body that extends the ground connection.

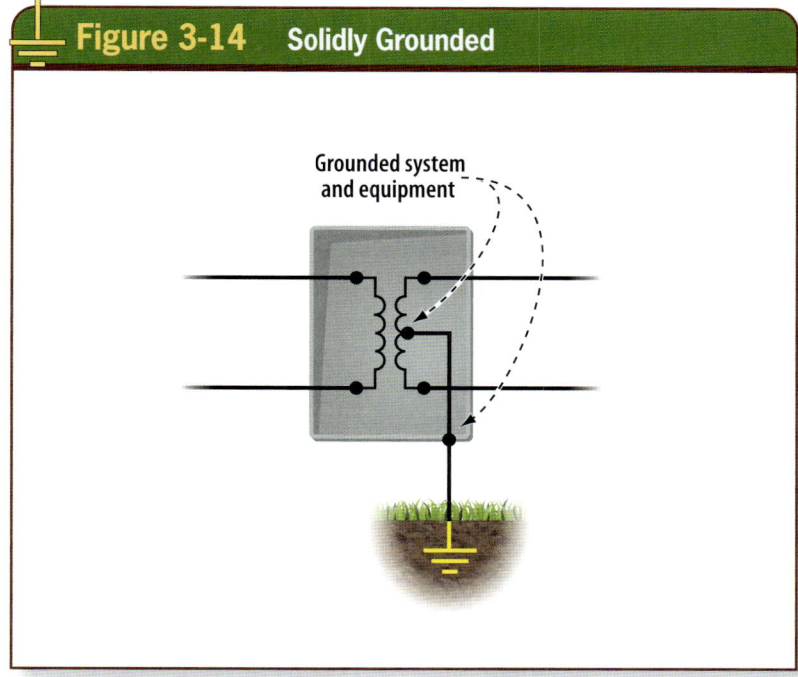

Figure 3-14. A solid connection to ground has no intentional impedance in the path. The system and equipment are at or close to Earth potential.

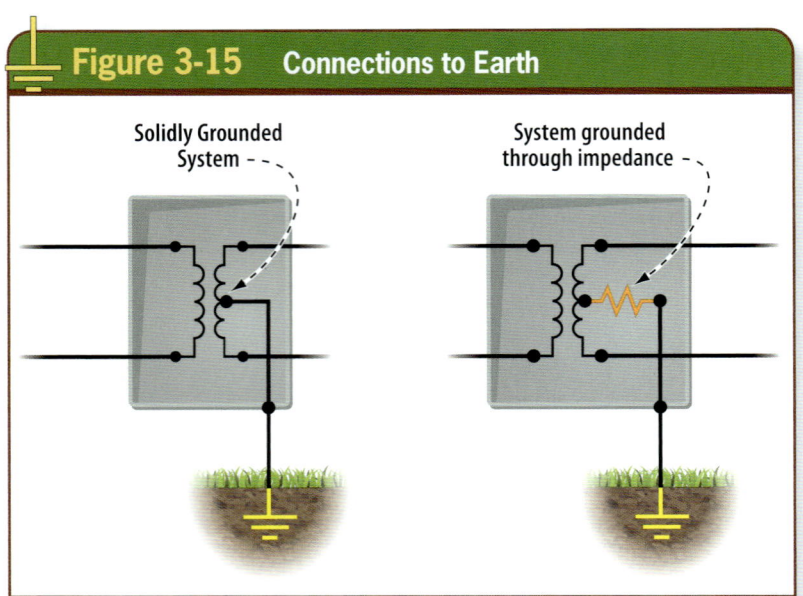

Figure 3-15 **Connections to Earth**

Solidly Grounded System - - -

System grounded through impedance - -

Figure 3-15. System grounded conductors are connected to the Earth either solidly (left) or through an impedance device (right).

a 3-phase, 3-wire delta system is grounded, creating a corner-grounded delta system. All the rules for grounded conductors apply regardless of whether the grounded conductor is a system neutral or a phase conductor. **See Figure 3-16.**

An EGC performs multiple functions through a single conductive path. The EGC serves as a conductive path to the ground (Earth), performs bonding functions, and facilitates overcurrent protective device operation when functioning as an effective ground-fault current path during ground-fault events. The types of EGCs are listed in 250.118. An EGC can be in the form of a wire-type conductor, a metal raceway, a metal cable tray, or another electrically conductive path recognized in 250.118. EGC installations must meet the provisions in 250.120. **See Figure 3-17.**

to the ground (Earth), usually by a grounding electrode conductor. **See Figure 3-15.**

Although many grounded conductors are system-neutral conductors, not all grounded conductors are neutrals. Sometimes a phase conductor of

Grounding Electrode. A conducting object through which a direct connection to earth is established. (CMP-5)

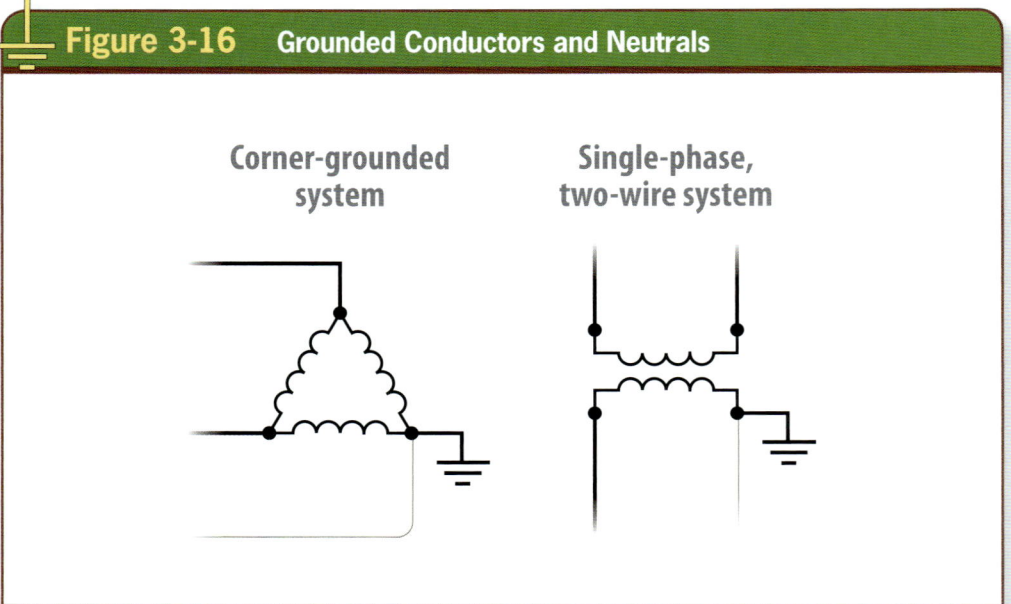

Figure 3-16 **Grounded Conductors and Neutrals**

Corner-grounded system

Single-phase, two-wire system

Figure 3-16. Most neutrals are grounded conductors, but not all grounded conductors are neutrals. Rules for grounded conductors, such as in 240.22 and 200.6, apply to all grounded conductors, not just neutrals.

Grounding electrodes are conductive objects in contact with the ground (Earth). For an object to qualify as a grounding electrode, there must be an Earth connection, the direct contact with the Earth must be maintained, and the object must be conductive. Electrodes identified in 250.52(A) are acceptable as grounding electrodes. Grounding electrodes have little or no effect on the operation of overcurrent protective devices.

A grounding symbol is often included on construction drawings. It is generally intended to mean a grounding electrode or grounding electrode system, not just a single ground rod. **See Figure 3-18.**

> **Grounding Electrode Conductor (GEC).** A conductor used to connect the system grounded conductor or the equipment to a grounding electrode or to a point on the grounding electrode system. (CMP-5)

Grounding electrode conductors are conductive bodies that extend the ground connection. They maintain potential differences at a minimum between the conductive objects and the

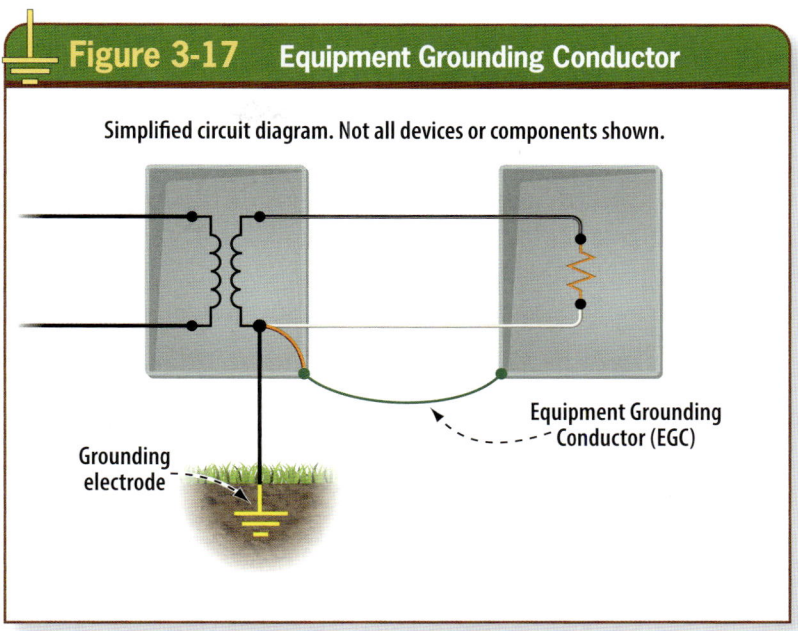

Figure 3-17   Equipment Grounding Conductor

Simplified circuit diagram. Not all devices or components shown.

Grounding electrode

Equipment Grounding Conductor (EGC)

Figure 3-17. *EGCs perform three essential functions: they (1) ensure the equipment is grounded, (2) establish bonding, and (3) provide an effective ground-fault current path.*

Earth electrode to which they are connected. Grounding electrode conductors are not intended to facilitate operation of overcurrent protective devices, but they do perform grounding and bonding functions. The grounding

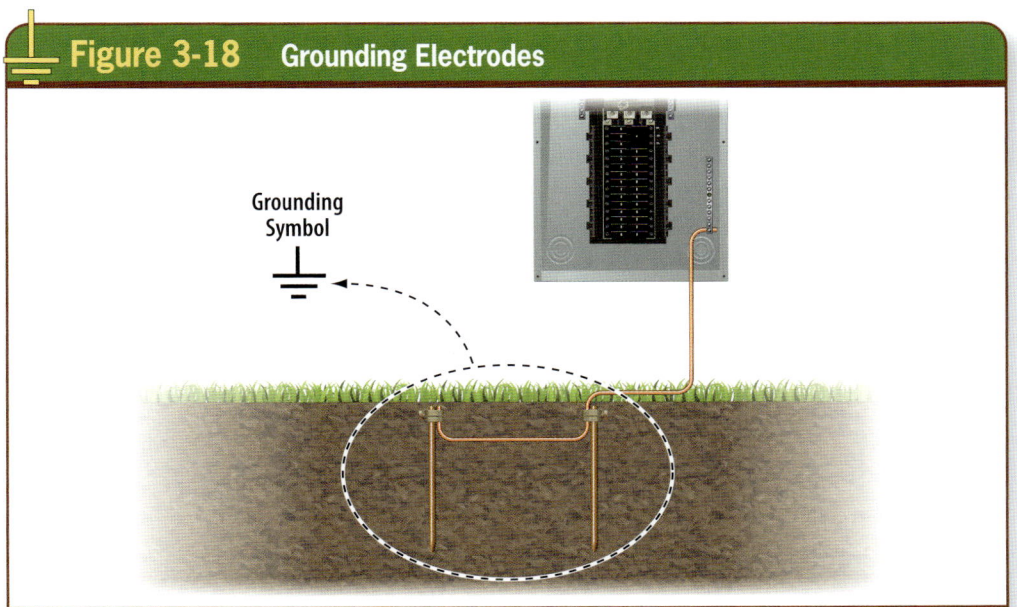

Figure 3-18   Grounding Electrodes

Grounding Symbol

Figure 3-18. *Grounding electrodes are conductive objects that establish a direct Earth connection. Note that the grounding symbol applies to all the grounding electrodes specified in 250.52(A).*

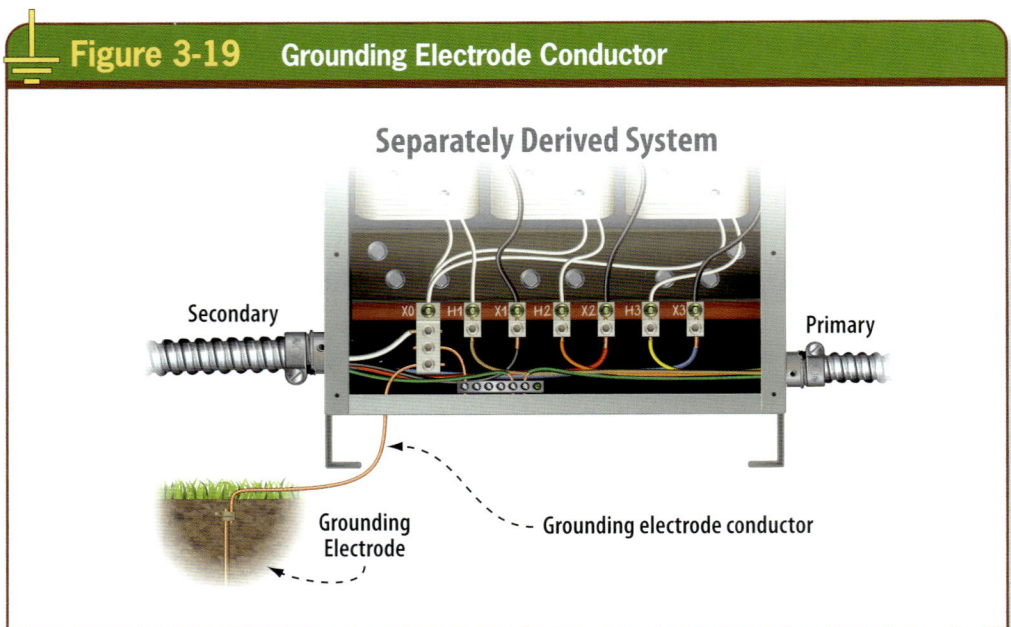

Figure 3-19. Grounding Electrode Conductor

Separately Derived System

Secondary   X0 H1 X1 H2 X2 H3 X3   Primary

Grounding Electrode

Grounding electrode conductor

Figure 3-19. Grounding electrode conductors are conductive bodies that extend the ground connection.

electrode conductor performs bonding between the Earth, grounded conductors, and other conductive parts required to be grounded. **See Figure 3-19.**

Bonding Jumper, System (System Bonding Jumper). The connection between the grounded circuit conductor, and the supply-side bonding jumper, or the equipment grounding conductor, or both, at a separately derived system. (CMP-5)

The system bonding jumper is located at a separately derived system and functions in similar fashion to that of the main bonding jumper. The term was defined to distinguish it from a main bonding jumper, which, by definition, is located only at service equipment.

The system bonding jumper is also addressed in 250.28 regarding sizing and physical characteristics. A system bonding jumper is required to be made of copper, aluminum, copper-clad aluminum, or another corrosion-resistant material and can be a wire, bus, screw, or similar suitable conductor.

The term *effective ground-fault current path* is used in the performance

Ground-Fault Current Path, Effective (Effective Ground-Fault Current Path). An intentionally constructed, low-impedance electrically conductive path designed and intended to carry current during ground-fault events from the point of a ground fault on a wiring system to the electrical supply source and that facilitates the operation of the overcurrent protective device or ground-fault detectors. (CMP-5)

provisions of 250.4(A) in Article 250; this term is also used in other *NEC* articles. The purpose of an effective ground-fault current path is to carry heavy levels of fault current during ground-fault conditions until the overcurrent protective device opens the offending circuit.

Effective ground-fault current paths are built in the field by electrical installers, so their effectiveness can be evaluated as part of engineering design or inspection. An example of an effective ground-fault current path is RMC installed for a branch circuit. Since RMC is recognized as an EGC in 250.118(2), the integrity of the conduit and its ability to serve in this fashion is related to adequate and effective support, tightness of fittings, and proper sizing. Therefore, the definition of the term indicates that this path is intentionally constructed. The effective ground-fault current path is usually installed simultaneously and in conjunction with conductors for feeders and branch circuits. Professionalism and skillful execution of work are important in constructing an effective ground-fault current path. **See Figure 3-20.**

> **Ground-Fault Current Path.** An electrically conductive path from the point of a ground fault on a wiring system through normally non–current-carrying conductors, grounded conductors, equipment, or the earth to the electrical supply source. (CMP-5)
>
> Informational Note: Examples of ground-fault current paths are any combination of equipment grounding conductors, metallic raceways, metallic cable sheaths, electrical equipment, and any other electrically conductive material such as metal, water, and gas piping; steel framing members; stucco mesh; metal ducting; reinforcing steel; shields of communications cables; grounded conductors; and the earth itself.

The term *ground-fault current path* is defined in Article 100. Ground-fault current paths will be any and all conductive paths that fault current will take as it is returning to its source. Ground-fault current paths can consist of combinations of EGCs, grounded conductors, metal raceways, metal cable sheaths, electrical equipment, and any other electrically-conductive material such as metal water and gas piping, steel framing members, stucco mesh, metal ducting, reinforcing steel, shields of communications cables, and the Earth itself. The difference between ground-fault current paths and effective ground-fault current paths is that the effective ground-fault current paths are intentionally constructed for fault-current–carrying purposes, whereas the

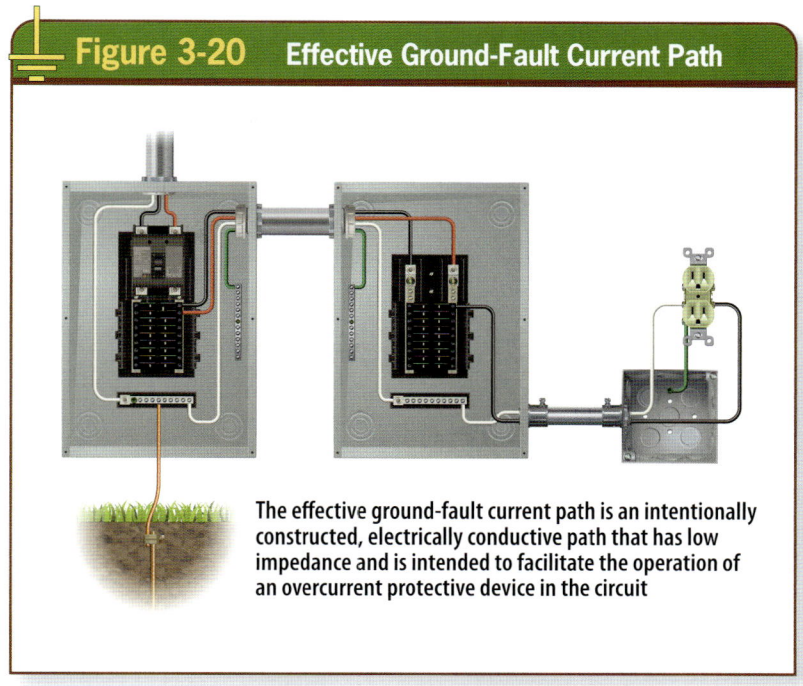

Figure 3-20   Effective Ground-Fault Current Path

The effective ground-fault current path is an intentionally constructed, electrically conductive path that has low impedance and is intended to facilitate the operation of an overcurrent protective device in the circuit

Figure 3-20. *Effective ground-fault current paths are constructed in the field by electrical installers.*

ground-fault current path includes any conductive path between the source and the point of the ground fault on a system.

> **Bonding Jumper, Supply-Side (Supply-Side Bonding Jumper).** A conductor installed on the supply side of a service or within a service equipment enclosure(s), or for a separately derived system, which ensures the required electrical conductivity between metal parts required to be electrically connected. (CMP-5)

The term *supply-side bonding jumper* is also defined in Article 100. A supply-side bonding jumper is typically installed on the line or supply side of an overcurrent protective device at the service equipment or for a separately derived system. If these supply-side bonding jumpers are of the wire type, they must be sized according to the requirements in 250.102(C) and Table 250.102(C)(1), which base the minimum size on the circular mil area of the largest ungrounded service-entrance conductor or largest ungrounded conductor of a separately derived system. Supply-side bonding jumpers need to be a bit more robust than load-side bonding jumpers and must have the capacity to carry fault current for the duration of time it takes an overcurrent protective device to open and clear the event. This overcurrent protective device could be on the supply side of the utility transformer or on the primary side of a transformer installed as a separately derived system.

ARTICLE 200 — USE AND IDENTIFICATION OF GROUNDED CONDUCTORS

Article 200 provides information regarding the use and identification of grounded conductors. In general, grounded conductors are identified using the colors white or gray. Conductors sized 4 AWG (American Wire Gauge) or larger are permitted to be identified by a continuous white or gray insulation color or by three continuous white or gray stripes along the entire length of a conductor insulation that is any color other than green;

alternatively, a distinctive white or gray marking can be used at the conductor terminations. **See Figure 3-21.**

The marking is typically applied in the field and must encircle the conductor or insulation. The *Code* does not specify how the marking should be affixed to the conductor. Many Electrical Workers use a strip of vinyl marking tape close to where the conductor terminates. A good practice is to use enough marking tape to ensure that the conductor is readily identified within the enclosure where it is occupying enclosure space with ungrounded conductors and EGCs. This helps differentiate them from one another. Article 200 also provides rules related to the terminal identification on electrical devices, such as the rules in 200.9 and 200.10. These rules indicate that the terminal on a device intended for the grounded conductor must be white or silver in color to readily distinguish the grounded conductor terminal from the terminals for ungrounded conductors, which are typically gold or brass in color. The screw or terminal for a grounded conductor is often identified with a silver or chrome color.

Figure 3-21   Identifying Grounded Conductors

White insulation

Gray insulation

White or gray marking at terminations

Three continuous white or gray stripes along the entire length of the conductor

Figure 3-21. Identification requirements for grounded conductors are provided in NEC 200.6 and 200.7.

ARTICLE 250— ARRANGEMENT AND USE

The scope of Article 250 does not specify different voltages, meaning that the rules in the article apply to installations at all voltage levels. Article 250 consists of 10 parts that are identified by the Roman numerals I through X. Each part addresses specific requirements for grounding and bonding in an electrical installation. Figure 250.1 is a schematic of the article and can assist users in quickly and accurately finding the appropriate rules.

PART I—GENERAL

Part I includes rules related to the system, circuit, and equipment grounding. Here the determinations can be made as to whether a system, equipment, or circuit is required to be grounded, is permitted to be grounded, or is not permitted to be grounded.

Section 250.1 gives the scope of Article 250 and provides information about the general requirements for grounding and bonding and specific rules as provided in 250.1(1) through (6). These sections address the following throughout the article:

- Systems, circuits, and equipment that are required, permitted, or not permitted to be grounded, and which circuit conductor must be grounded
- Location of grounding conductor connections to systems, sizes and types of grounding and bonding conductors, and the methods of grounding and bonding
- Conditions that permit guards, isolation, or insulation as alternatives to grounding

Below is a description of each part of Article 250:

Part I—General
Part II—System Grounding
Part III—Grounding Electrode System and Grounding Electrode Conductor
Part IV—Enclosure, Raceway, and Service Cable Connections
Part V—Bonding
Part VI—Equipment Grounding and Equipment Grounding Conductors
Part VII—Methods of Equipment Grounding Conductor Connections
Part VIII—Direct-Current Systems
Part IX—Instruments, Meters, and Relays
Part X—Grounding of Systems and Circuits of Over 1000 Volts

Part I of Article 250 includes all general provisions that apply in addition to the requirements in the subsequent parts of the article. Figure 250.1, which immediately follows 250.1, serves as a road map or blueprint of Article 250. A close review of Figure 250.1 reveals that it presents the topics in an order similar to the order of the various steps of installation in constructing grounding and bonding systems. An important feature is that bonding functions are common to all parts of the article. The box on the right of the figure contains "Part V Bonding," as bonding is inherent to all other parts. This training material is crafted to parallel how projects are built, and the specific aspects of grounding and bonding are covered in detail in these steps. **See Figure 3-22.**

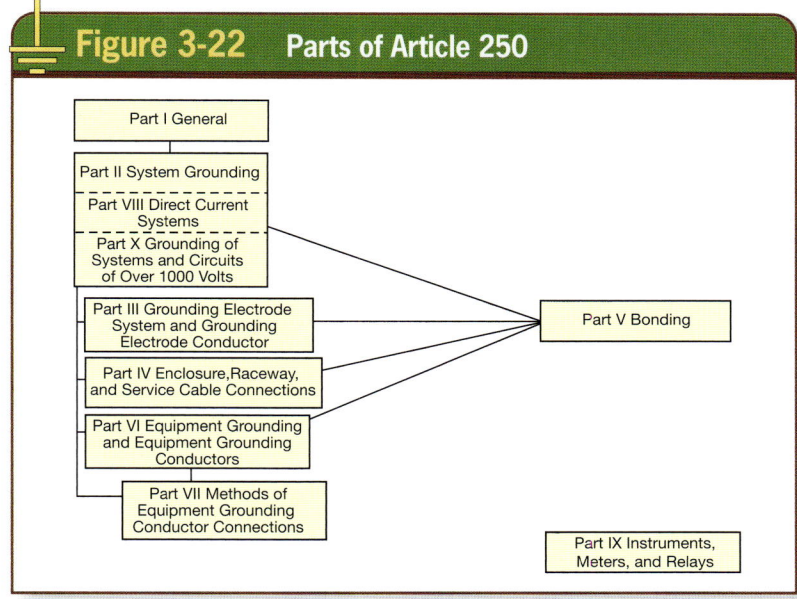

Figure 3-22. Parts of Article 250

Figure 3-22. The order of Article 250, as illustrated by Figure 250.1 from the NEC, parallels the order of many construction projects. [Reproduction of Figure 250.1]

Part I of Article 250 provides general requirements for grounding and bonding and includes some important performance language. Parts II, VIII, and X contain the requirements for system grounding, which includes systems that supply services and separately derived systems installed on the premises. Part II of Article 250 is used to make the determination as to whether these systems are required to be grounded and, if so, which rules apply.

Once this is known, the rules that follow apply to constructing the appropriate grounding and bonding circuits. Part III of the article is used to determine the requirements for the grounding electrode system and includes rules for installing grounding electrodes and grounding electrode conductors. When a building is being constructed from the ground up, grounding electrodes are often formed as part of the construction process.

Examples include the foundation or footing, a metal underground water piping system supplying the building, metal in-ground support structures that qualify as grounding electrodes, and so forth. This is one of the first tasks Electrical Workers perform on a construction site. Part IV of the article is all about enclosure, raceway, and serviced cable grounding.

TABLE 250.66—SIZING AC GROUNDING ELECTRODE CONDUCTORS

Table 250.66 is titled "Grounding Electrode Conductor for Alternating-Current Systems." This table is also used for sizing bonding jumpers between grounding electrodes that form the grounding electrode system for a building or structure. The maximum size required is 3/0 AWG copper or 250 kcmil aluminum or copper-clad aluminum conductor.

Figure 3-23 Table 250.66

Table 250.66 Grounding Electrode Conductor for Alternating-Current Systems (in part without notes)

Size of Largest Ungrounded Service-Entrance Conductor or Equivalent Area for Parallel Conductors (AWG/kcmil)		Size of Grounding Electrode Conductor (AWG/kcmil)	
Copper	Aluminum or Copper-Clad Aluminum	Copper	Aluminum or Copper-Clad Aluminum
2 or smaller	1/0 or smaller	8	6
1 or 1/0	2/0 or 3/0	6	4
2/0 or 3/0	4/0 or 250	4	2
Over 3/0 through 350	Over 250 through 500	2	1/0
Over 350 through 600	Over 500 through 900	1/0	3/0
Over 600 through 1100	Over 900 through 1750	2/0	4/0
Over 1100	Over 1750	3/0	250

Figure 3-23. *Table 250.66 is used for sizing grounding electrode conductors. [Reproduction of Table 250.66 (in part)]*

1. Table 250.66 is applied using sizes of the largest ungrounded service-entrance conductor or largest ungrounded derived conductor of a separately derived system.
2. This table is not based on overcurrent protective device ratings, in contrast to Table 250.122. Sizing grounding electrode conductors requires the use of Table 250.66 and Section 250.66 to determine the minimum sizes. Grounding electrode conductors need not be larger than the sizes in that table. **See Figure 3-23.**

When an engineering design specifies grounding electrode conductors larger than those provided in Table 250.66, the design specification usually takes precedence. The best practice is to install electrical equipment and conductors in accordance with engineering designs and in full compliance with applicable national and local codes.

TABLE 250.102(C)(1)— SIZING THE GROUNDED CONDUCTOR, MAIN BONDING JUMPER, SYSTEM BONDING JUMPER, AND SUPPLY-SIDE BONDING JUMPER FOR ALTERNATING-CURRENT SYSTEMS

Table 250.102(C)(1) is used for sizing equipment bonding jumpers (supply side), grounded conductors (minimum), or main bonding jumpers/system bonding jumpers; the minimum sizes in the table apply. There are three important notes following this table that are requirements. Table 250.102(C)(1) includes minimum copper conductor sizes and aluminum or copper-clad aluminum sizes. The correct column must be used when determining the minimum size conductor required for an installation. There is one difference: if the size of the largest ungrounded service conductor or largest derived system supply conductor exceeds 1,100 kcmil copper or 1,750 kcmil aluminum, then 12.5% must be used (as stated in Note 1). Selecting the

Figure 3-24   Table 250.102(C)(1)

Table 250.102(C)(1) Grounded Conductor, Main Bonding Jumper, System Bonding Jumper, and Supply-Side Bonding Jumper for Alternating Current Systems (in part without all notes)

Size of Largest Ungrounded Conductor or Equivalent Area for Parallel Conductors (AWG/kcmil)		Size of Grounded Conductor or Bonding Jumper* (AWG/kcmil)	
Copper	Aluminum or Copper-Clad Aluminum	Copper	Aluminum or Copper-Clad Aluminum
2 or smaller	1/0 or smaller	8	6
1 or 1/0	2/0 or 3/0	6	4
2/0 or 3/0	4/0 or 250	4	2
Over 3/0 through 350	Over 250 through 500	2	1/0
Over 350 through 600	Over 500 through 900	1/0	3/0
Over 600 through 1100	Over 900 through 1750	2/0	4/0
Over 1100	Over 1750	See Notes 1 and 2	

Notes (in part):
1. If the circular mil area of ungrounded supply conductors that are connected in parallel is larger than 1100 kcmil copper or 1750 kcmil aluminum, the grounded conductor or bonding jumper shall have an area not less than 12 1/2 percent of the area of the largest ungrounded supply conductor or…

Figure 3-24. Table 250.102(C)(1) is used for determining minimum sizes required for grounded conductors, main bonding jumpers, system bonding jumpers, and supply-side bonding jumpers for AC systems. [Reproduction of Table 250.102(C)(1) (in part)]

incorrect size can compromise safety. The grounded conductor or bonding jumper is not required to be larger than the largest ungrounded conductor or set of ungrounded conductors in parallel. For the purposes of this table, the term *bonding jumper* refers to main bonding jumpers, system bonding jumpers, and supply-side bonding jumpers. **See Figure 3-24.**

For installations where the ungrounded supply conductors and bonding jumpers are of different materials (copper, aluminum, or copper-clad aluminum), the minimum size of the grounded conductor or bonding jumper is based on the assumed use of ungrounded supply conductors of the same conductor material as the grounded conductor or bonding jumper and must have an ampacity equivalent to that of the installed ungrounded supply conductors.

If multiple sets of service conductors are installed as permitted in 230.40 Exception No. 2, or if multiple sets of ungrounded supply conductors are installed for a separately derived system, the equivalent size of the largest ungrounded supply conductor(s) shall be determined by the largest sum of the areas of the corresponding conductors of each set.

Where no service conductors are present, the supply conductor size shall be determined by the equivalent size of the largest service conductor required for the load to be served.

TABLE 250.122—SIZING EQUIPMENT GROUNDING CONDUCTORS

Table 250.122 is titled "Minimum Size Equipment Grounding Conductors for Grounding Raceway and Equipment." The key feature of this table is that it allows efficient determination of minimum sizes of wire-type EGCs. Several factors, such as voltage drop and high available fault currents, can affect the minimum sizes required for wire-type EGCs. **See Figure 3-25.**

The note at the bottom of Table 250.122 provides users with a valuable reminder that designs must meet the minimum sizes in the table, but in some cases, the size needs to be larger to ensure that an effective path for ground-fault current is achieved. Using this table is simple and is related to the size of a circuit breaker or fuse protecting the circuit.

For example, if a 1,000-ampere feeder is installed in polyvinyl chloride (PVC) conduit and requires a wire-type EGC, the EGC must not be smaller

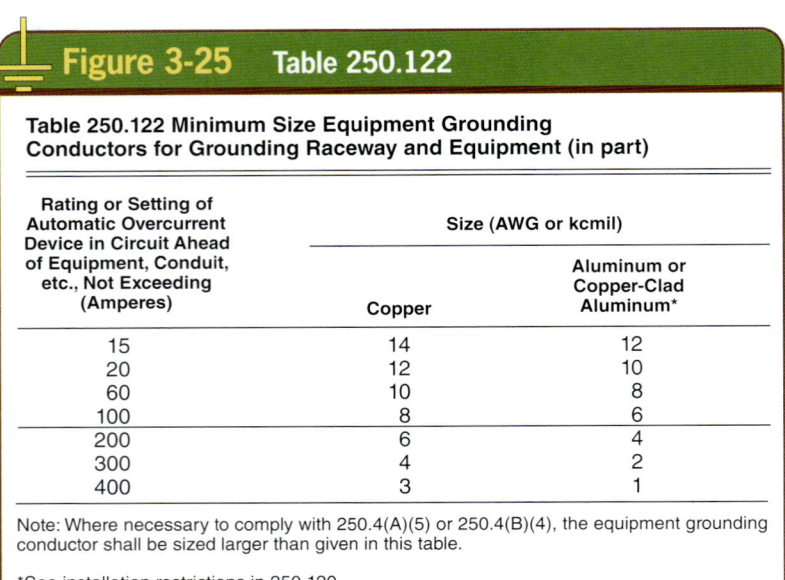

Figure 3-25   Table 250.122

Table 250.122 Minimum Size Equipment Grounding Conductors for Grounding Raceway and Equipment (in part)

Rating or Setting of Automatic Overcurrent Device in Circuit Ahead of Equipment, Conduit, etc., Not Exceeding (Amperes)	Size (AWG or kcmil)	
	Copper	Aluminum or Copper-Clad Aluminum*
15	14	12
20	12	10
60	10	8
100	8	6
200	6	4
300	4	2
400	3	1

Note: Where necessary to comply with 250.4(A)(5) or 250.4(B)(4), the equipment grounding conductor shall be sized larger than given in this table.

*See installation restrictions in 250.120.

Figure 3-25. *Table 250.122 (in part) is used for sizing wire-type equipment grounding conductors. [Reproduction of Table 250.122 (in part)]*

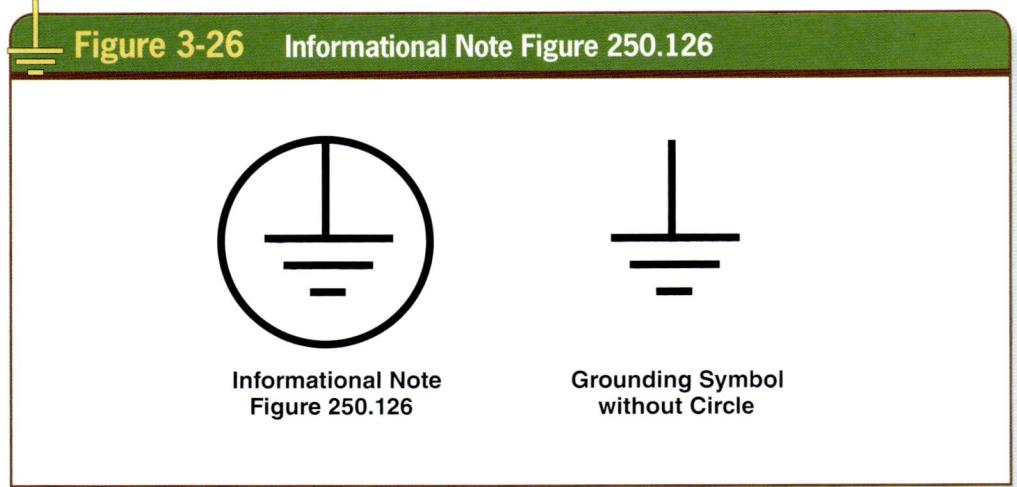

Figure 3-26   Informational Note Figure 250.126

Informational Note Figure 250.126

Grounding Symbol without Circle

Figure 3-26. *Informational Note Figure 250.126 provides an example of a symbol used to identify the grounding termination point for an EGC.*

than 2/0 AWG copper or 4/0 AWG aluminum or copper-clad aluminum, based on the values in the table. The minimum-size copper EGC for an 80-ampere branch circuit is 8 AWG. The minimum-size aluminum EGC for a 400-ampere feeder is 1 AWG.

Table 250.122 includes copper conductor sizes and aluminum or copper-clad aluminum sizes. The correct column must be used when determining the minimum size required for an installation. Selecting the incorrect size can compromise safety and the integrity of an effective ground-fault current path.

The *NEC* includes a grounding symbol in 250.126 and 406.10. These figures are in the form of informational notes because it is understood that there are other methods of identifying equipment grounding terminals on devices such as receptacles and switches. **See Figure 3-26.**

Part X of the article provides the rules addressing specific requirements that apply to systems greater than 1,000 volts. The general requirements in preceding parts of the article apply, unless specifically addressed within Part X. A good sample application of this concept is sizing EGCs for medium-voltage feeders. Section 250.190 indicates that EGCs that are not an integral part of a cable assembly must not be smaller than 6 AWG copper or 4 AWG aluminum or copper-clad aluminum. Other than those conditions, the EGCs are sized using Table 250.122. The purpose of medium- and high-voltage cable tape shields and concentric stranded shields that are part of the assemblies (315.44 Informational Note) is to address voltage stresses to insulation. The point here is that EGCs must have sufficient capacity as required in 250.122 and 250.4.

SPECIAL OCCUPANCIES, EQUIPMENT, AND CONDITIONS

Section 90.3 indicates that Chapters 5, 6, and 7 of the *NEC* modify or supplement the requirements in Chapters 1 through 7. Often the grounding and bonding rules in these later chapters are more restrictive than the general requirements. All general rules in Chapters 1 through 4 apply *in addition to* any modifications required by Chapters 5, 6, or 7.

Example Modification from Article 517 in Chapter 5

Sections 517.13(A) and (B) address branch circuits serving patient care spaces and require two separate EGC paths in the form of (1) a suitable metal raceway or cable armor and (2) a contained insulated copper EGC. This provides multiple EGC paths to ensure there is effective grounding and bonding as well as overcurrent protective device operation in the event of a ground fault on circuits serving these locations. **See Figure 3-27.**

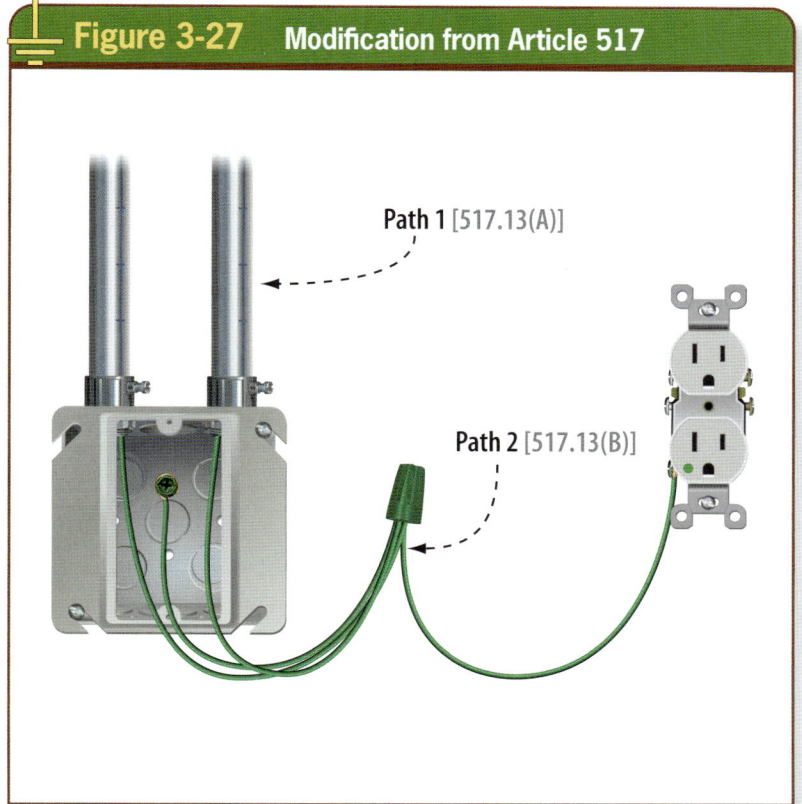

Figure 3-27 Modification from Article 517

Path 1 [517.13(A)]

Path 2 [517.13(B)]

Figure 3-27. Branch circuits serving patient care spaces require two EGC paths.

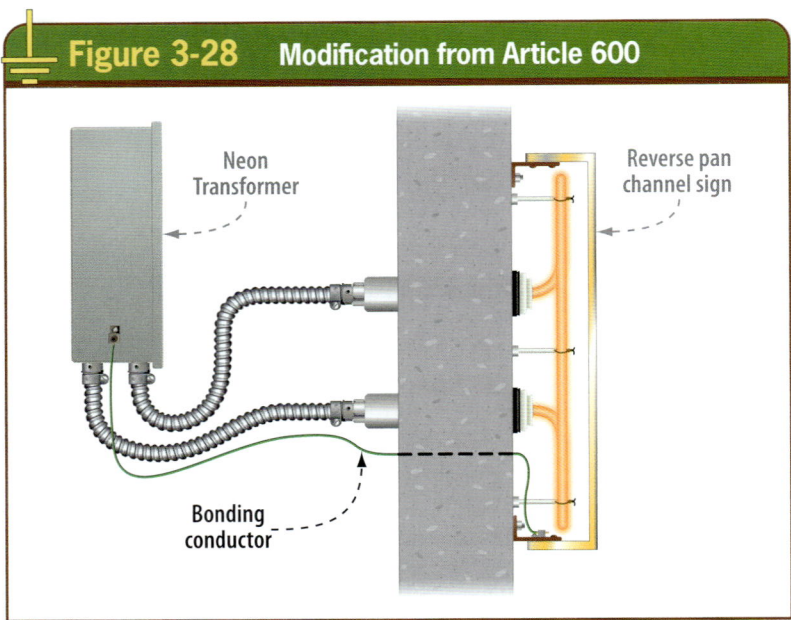

Figure 3-28   Modification from Article 600

Neon Transformer

Reverse pan channel sign

Bonding conductor

Figure 3-28. A 14 AWG copper conductor is required for bonding metal parts associated with a neon outline lighting system's high-voltage secondary circuit.

This is a modification of the general equipment grounding requirements contained in Part VI of Article 250. These modifications are more restrictive than the general rules and affect branch circuits in health care facilities.

Example Modification from Article 600 in Chapter 6

Section 600.7(B)(7) permits a bonding conductor greater than or equal to 14 AWG copper for neon transformers that are supplied by 20- or 30-ampere branch circuits. Normally these bonding conductors would be based on the rating of the circuit breaker or fuse protecting the branch circuit supplying the equipment, which would be 12 AWG or 10 AWG copper, respectively. **See Figure 3-28.**

Because a transformer is supplying these signs and outline lighting systems, the secondary circuits supplying the luminous tube(s) are high voltage. This results in secondary currents that are only in the milliamp range.

In addition to the low current on the secondary side, these circuits for sign and outline lighting systems are required to be equipped with transformers that provide secondary-circuit ground-fault protection. The 14 AWG bonding conductor is required to keep conductive parts on the secondary side of such systems at or as close to ground (Earth) potential as possible.

Example Modification from Article 770 in Chapter 7

Section 770.100 requires non–current-carrying conductive members of optical fiber cables to be grounded by the methods specified in 770.100(A) through (D). This modifies the rules in Article 250.

CHAPTER 8

Chapter 8 covers communications system installations and equipment. This chapter is not subject to the requirements in Chapters 1 through 7, except where those rules are referenced from within Chapter 8. The grounding and bonding rules for these limited-energy systems and equipment provide different safety benefits. Although the grounding and bonding concepts are the same, what is intended to be accomplished differs for limited-energy and communications systems. These grounding and bonding schemes are not intended to facilitate overcurrent protective device operation. The requirements deal more with connecting equipment and cable shields to the ground (Earth) and bonding to minimize potential differences. The grounding and bonding rules in Chapter 8 provide safety from events that can cause a voltage rise on these systems.

For example, Section 800.100(A)(3) indicates that the bonding conductor or grounding electrode conductor for communications systems and equipment shall not be smaller than a 14 AWG copper conductor. This section also requires that the bonding conductor or grounding electrode conductor shall have a current-carrying capacity not less than the grounded metal cable sheath member of the communications cable, the protected conductor of the communications cable, or the outer sheath of the coaxial cable, as applicable. The bonding conductor or grounding electrode conductor shall not be required to exceed 6 AWG.

This is a modification to the normal use of Table 250.66 for sizing grounding electrode conductors for electrical services or separately derived systems. Article 800 provides the specific sizing and installation requirements for grounding electrode conductors for communications systems. Chapter 16 of this textbook provides more specific information related to grounding and bonding for communications systems and equipment.

SUMMARY

A thorough working knowledge of the *NEC* is essential to fully understand electrical grounding and bonding and how the rules apply to electrical installations and systems. It is equally important to understand defined terms related to grounding and bonding and the overall arrangement of Article 250. Grounding and bonding terms are all defined in Article 100. The *NEC* serves as the basis for the information in this training material and should be used in conjunction with this textbook. Proper use and application of the *NEC* is essential not only for students of the *Code* but for seasoned professionals as well.

REVIEW QUESTIONS

1. The purpose of the *NEC* is to provide for __?__ of persons and property from hazards that arise from the use of electricity.
   a. complete protection
   b. exclusion
   c. isolation
   d. practical safeguarding

2. Which of the following can be compared with the foundation or footing of a building or structure?
   a. Equipment grounding system
   b. Grounding electrode system
   c. Overcurrent protective device system
   d. Signal reference grid structure

3. Which of the *NEC* chapters can modify the general requirements in the first four chapters?
   a. Chapters 5, 6, and 7
   b. The *NEC* Index
   c. Chapter 8
   d. Chapter 9

4. Chapter 8 is not subject to the requirements provided elsewhere in the *NEC* except where specifically referenced from within Chapter 8.
   a. True   b. False

5. The __?__ has the responsibility for making interpretations and enforcing *NEC* requirements.
   a. contractor
   b. engineer
   c. inspector or authority having jurisdiction
   d. owner

6. Exceptions to rules in the *NEC* modify only the rule they directly follow, unless stated otherwise in the exception.
   a. True   b. False

7. Mandatory requirements in the *NEC* are characterized by the use of which of the following terms?
   a. *Can*
   b. *May*
   c. *Shall* or *shall not*
   d. All of the above

8. Informational notes that follow rules in the *NEC* are enforceable as requirements as determined by the authority having jurisdiction.
   a. True   b. False

9. Permissive provisions in the *NEC* are characterized by the use of which of the following phrases?
   a. *May be required*
   b. *Shall be required*
   c. *Shall not*
   d. *Shall not be required* or *shall be permitted*

REVIEW QUESTIONS

10. The *NEC Style Manual* requires all definitions to be located in Article 100.
   a. True b. False

11. If any definitions of terms related to grounding and bonding were used only in Article 250, the Article would be indicated in parentheses immediately following the definition.
   a. True b. False

12. *A conducting object through which a direct connection to the Earth is established* best defines which of the following?
   a. Equipment grounding conductor
   b. Grounding electrode
   c. Main bonding jumper
   d. The Earth

13. Article 200 of the *NEC* provides rules for which of the following in the electrical system?
   a. Bonding jumpers
   b. Equipment grounding conductors
   c. Grounded conductors
   d. Grounding electrodes

14. Article 250 in the *NEC* provides the requirements for grounding and bonding and is made up of how many parts?
   a. 2
   b. 6
   c. 8
   d. 10

15. Part X of Article 250 provides requirements for __?__.
   a. direct-current systems
   b. enclosure, raceway, and service cable connections
   c. grounding of systems and circuits of over 1,000 volts
   d. methods of equipment grounding

16. Which of the following provides a graphic road map or blueprint of the arrangement of Article 250 in the *NEC*?
   a. Figure 250.1
   b. Figure 250.122
   c. Table 250.1
   d. Informational Note Figure 250.126

17. Which *NEC* table provides information about sizing wire-type equipment grounding conductors?
   a. Table 8, Chapter 9
   b. Table 250.1
   c. Table 250.66
   d. Table 250.122

18. *NEC* Table 250.66 is used for sizing grounding electrode conductors and bonding jumpers used to form the grounding electrode system.
   a. True b. False

19. Which of the following is covered by Chapter 5 of the *NEC*?
   a. Agricultural facilities
   b. Hazardous (classified) locations
   c. Health care facilities
   d. All of the above

20. Chapter 6 of the *NEC* provides rules that address special equipment such as elevators, cranes, and electric signs and outline lighting systems.
   a. True b. False

21. Article 250 includes conditions that permit guards, isolation, or insulation as alternatives to grounding.
   a. True b. False

22. Notes that follow a Table in the *NEC* are applicable as requirements.
   a. True b. False

23. Informational Notes following a Table in the *NEC* are applicable as requirements.
   a. True b. False

Grounding Electrodes and the Grounding Electrode System

When a construction project begins, one of the first steps is to install a footing and foundation system to support the building or structure. Because footings effectively connect the building to the Earth and provide a foundation of support, they often provide excellent grounding electrodes for electrical services and for systems installed later during construction. The *NEC* requires that all grounding electrodes present at each building or structure served be connected to form a grounding electrode system.

Objectives

» Understand what constitutes a grounding electrode by definition and understand the purpose served by grounding electrodes.

» Determine the requirements for a grounding electrode system on buildings or structures supplied by electrical services or feeders.

» Identify grounding electrodes that are inherent to the construction of a building or structure.

» Identify the types of grounding electrodes acceptable in accordance with the *NEC*.

» Determine proper installation requirements for grounding electrodes that must be installed.

» Understand the role of a lightning protection system in relation to a grounding electrode system for the power service or system of a building.

Chapter 4

Table of Contents

Introduction 76

Grounding Electrode Defined 76

Purpose and Performance of Electrodes 77

Grounding Electrode System Requirements ... 79

Establishing a Grounding Electrode System ... 79

Mandatory Grounding Electrodes 80

Types of Grounding Electrodes 81

    Metal Underground
Water Pipe Electrodes 81

    Metal In-Ground Support Structures 82

    Concrete-Encased Electrodes 83

    Ground Ring Electrodes 84

    Rod and Pipe Electrodes 85

    Plate Electrodes 85

    Other Local Metal Underground
Systems or Structures 85

Grounding Electrode
Installation Requirements 85

    Rod and Pipe Electrode Installation 86

    Soil Resistivity and Ground
Resistance Testing 87

    Plate Electrode Installation 89

    Electrode Spacing Requirements 90

    Water Pipe Electrode Supplement
and Bonding Jumpers 90

    Auxiliary Grounding Electrodes 91

    Items Not Permitted as
Grounding Electrodes 92

    Grounding Electrodes and
Utility Services 93

    Bonding of Lightning Protection
Systems to Service Electrode Systems ... 94

Summary 95

Review Questions 96

INTRODUCTION

Footings can be very simple and small or very complex and massive depending on the size of the building or structure. Footings and foundations serve as the means of connecting or attaching the building to the Earth. Engineering and designs for building footings and foundations are extremely important, as the weight of the structure must be adequately supported for the life of the building.

The structure must have a solid foundation on which it can be built. Since the footing, foundation, and various underground systems are installed during the early phases of construction, grounding electrodes are often concurrently established because they are inherent to the construction process; they are part of the building or structure.

Electrical Workers are usually on the job during the installation of footings and foundations. During this stage of the project, Electrical Workers add conduit sleeves and underground raceways and equipment for services, feeders, and branch circuits installed in the building slabs on grade.

The electrical service and grounding electrode system provide the foundation for the electrical wiring system in the building or structure served. **See Figure 4-1.**

Existing buildings already have footings and foundations, so the grounding electrodes should already be connected to existing services, separately derived systems, and other equipment. The grounding electrode system must be used when remodeling existing buildings, equipment, and electrical systems.

GROUNDING ELECTRODE DEFINED

The term *grounding electrode*, as defined in Article 100, refers to conductive objects and describes the primary function of a grounding electrode. An important feature of this definition is that the electrode is in direct contact with the Earth, making a connection. This means that the electrode is acceptable and performing as a grounding electrode without a wire being connected to it. **See Figure 4-2.**

The term *grounding electrode* is defined in the *NEC* as follows:

> **Grounding Electrode.** A conducting object through which a direct connection to earth is established. (CMP-5)

It is clear from this definition that a grounding electrode not only makes the connection to ground, but it also maintains this connection. A ground rod is not a grounding electrode until it is driven into the Earth to complete the connection. A metal water pipe is not a grounding electrode unless at least 10 feet of the piping is in direct contact with the Earth. Note that no depth is required for the water pipe, but historically, the water pipe electrodes were buried at least 24 inches deep or deeper because of local water service company installation regulations.

It is important to understand that the *NEC* does not distinguish between underground water piping that is outside a building perimeter or inside the

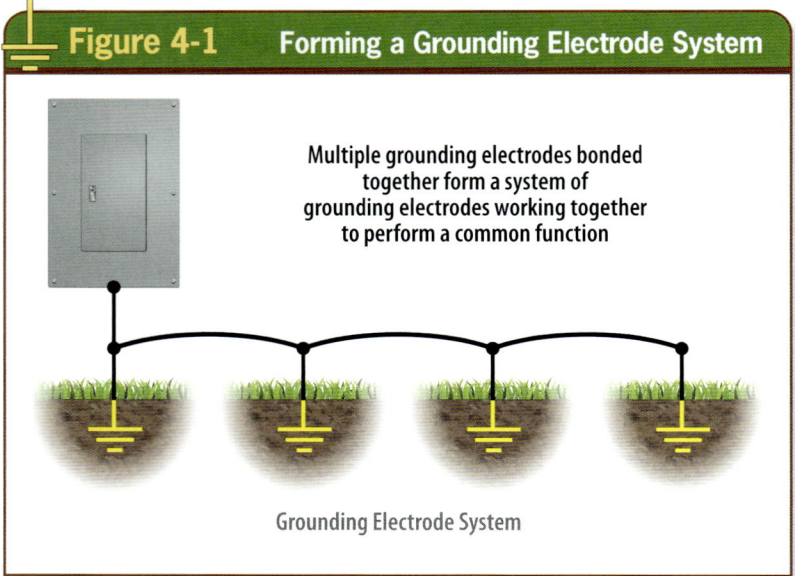

Figure 4-1   **Forming a Grounding Electrode System**

Multiple grounding electrodes bonded together form a system of grounding electrodes working together to perform a common function

Grounding Electrode System

Figure 4-1. A grounding electrode system is required at each building or structure, with each grounding electrode connected to form a grounding electrode system.

perimeter. The same concept of direct connection to the Earth is found for all grounding electrodes covered by 250.52(A). Foundations and footings are usually in direct contact with the Earth and are acceptable as grounding electrodes.

If a vapor barrier or any other isolation exists between a concrete footing and the Earth, it prevents a direct connection, interfering with the effectiveness of the footing and its ability to perform as an electrode. When the direct connection between an electrode and the Earth is impaired in this fashion, the footing or foundation does not meet the definition of a grounding electrode. In these cases, the footing cannot be depended on as an electrode because the direct connection to Earth is nullified. To determine the electrodes that must be used to form the grounding electrode system, it is important to be familiar with construction specifications and local requirements determining if such vapor barriers are a building code requirement.

Another example of concrete that is isolated from the Earth is a foundation wall coated with asphalt sealants or coatings used as vapor and moisture barriers. This affects the footing's ability to establish and maintain a direct connection to the Earth and the ability to perform as an electrode.

The concept is simple: a grounding electrode is a direct connection to the Earth through a conductive material.

PURPOSE AND PERFORMANCE OF ELECTRODES

The purpose of a grounding electrode is to function as the connection between grounded electrical systems and equipment and the Earth. The basic performance of grounding electrodes is provided in the definition of this term. **See Figure 4-3.**

The *NEC* does not provide details about the effectiveness or resistance of a grounding electrode or grounding electrode system, other than for a single rod, pipe, or plate electrode. The *NEC* also

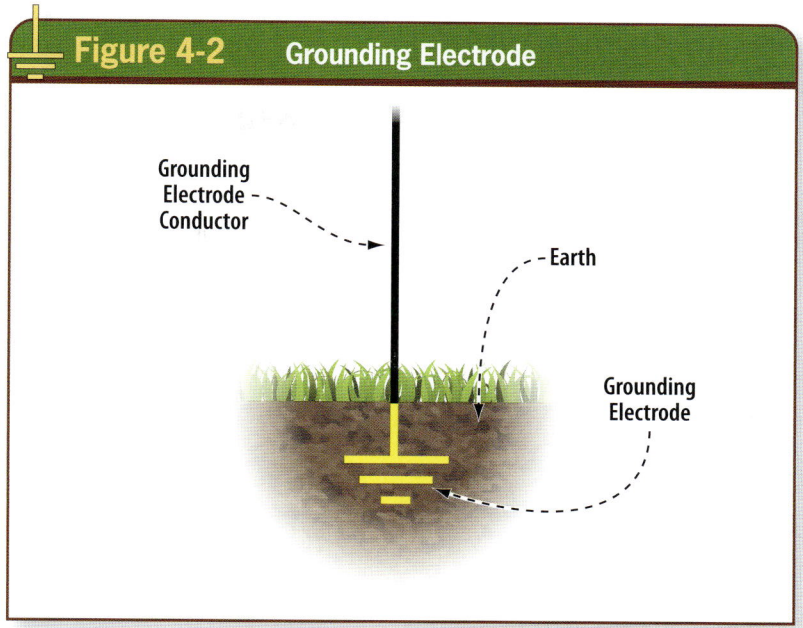

Figure 4-2. A grounding electrode is a conductive object through which a direct connection to Earth is established. A list of grounding electrodes is provided in 250.52(A).

does not address long-life or end-of-life expectancies of grounding electrodes.

Grounding electrode performance is essential if electrical services and systems are energized.

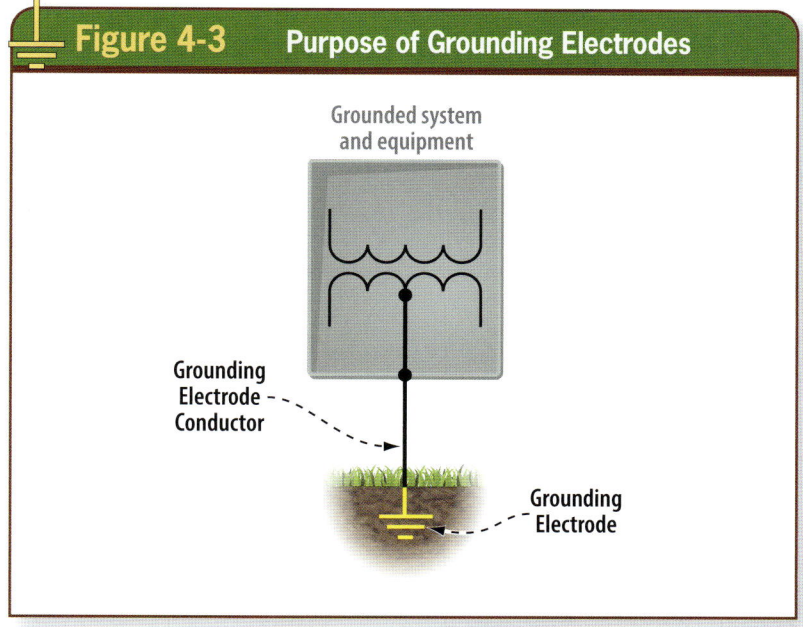

Figure 4-3. Grounding electrodes connect systems and equipment to ground.

For additional
information, visit
qr.njatcdb.org
Item #5335

Fact

The resistance
between the Earth
and a grounding
electrode varies
depending on geo-
graphical location,
seasonal condi-
tions, mineral con-
tent of the soil, and
other influencing
factors, such as
ambient tempera-
tures and Earth
temperatures.

The National Fire Protection Association (NFPA) conducted a national electrical grounding research project that analyzed the effects of soil and environmental conditions on grounding electrode integrity and performance. Several test sites were constructed across the country to account for various soil conditions. The study focused on the effectiveness of the various grounding electrodes addressed in the *NEC*. The study included periodic resistivity testing for the duration of the study, and the test results were recorded. Some grounding electrodes in the study performed better than others and maintained more consistent resistance measurements over time. This study was conducted using electrodes that were not connected to electrical power sources, so the influence of connected voltages, line surges, and other events were not included in the study as normally would be expected for grounding electrodes connected to energized electrical systems. The test was static only and measured periodic resistivity values that could indicate deterioration effects of grounding electrodes. As the study concluded, some of the test sites were removed, and the electrodes were removed and examined.

The findings revealed that some electrodes weathered well physically and electrically, while others suffered severe deterioration, indicating a shorter life expectancy. The type of soil conditions in a geographical area appeared to have a significant influence on the life and effectiveness of the various types of electrodes. This demonstrated that certain types of grounding electrodes seem to be more suited for some locations than others.

The Earth is a conductor, but its conductivity is variable and affected by the influence of nature. The connection to the Earth is usually better when the ground is moist or wet and tends to be less effective in dry or rocky soil conditions. It stands to reason that the resistance between the Earth and a grounding electrode varies depending on geographical location, seasonal conditions, mineral content of the soil, and other influencing factors, such as ambient temperatures and Earth temperatures.

If more than a single grounding electrode is installed and used, it is treated by the *NEC* as a grounding electrode system. Multiple grounding electrodes that form a system benefit from multiple Earth connection characteristics and locations for the same building or structure. **See Figure 4-4.**

It is reasonable to conclude that a system of multiple grounding electrodes should perform better than a single grounding electrode due to the multiple connections to the Earth. This is among the reasons why all grounding electrodes at each building or structure served must be interconnected to perform as a system of electrodes for the life of the building or structure.

Remember that the grounding electrode establishes and maintains a connection to the Earth. It has little to no effect in facilitating overcurrent protective device operation. The Earth is not suitable as an effective ground-fault current path. Many have insisted that the Earth is necessary in the safety system to cause overcurrent protective devices to open in ground-fault conditions, but it should be understood that the Earth is

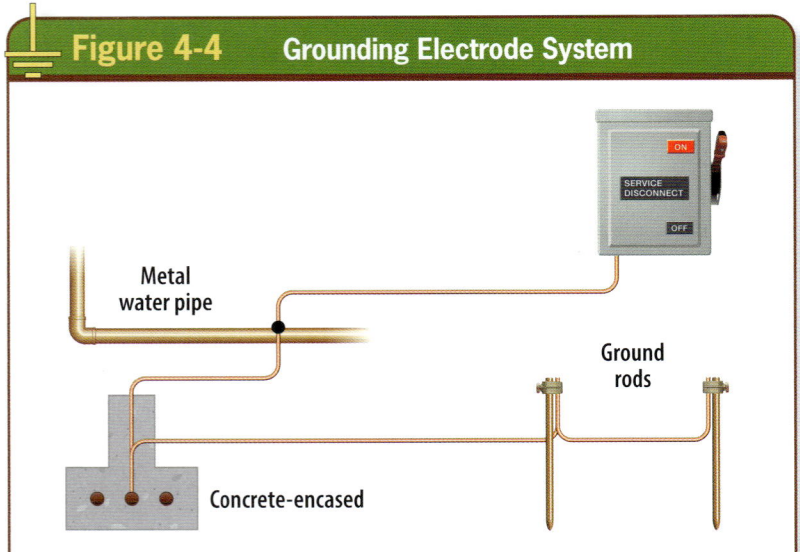

Figure 4-4   **Grounding Electrode System**

Metal
water pipe

Ground
rods

Concrete-encased

Figure 4-4. Multiple electrodes create a grounding electrode system.

not to be used as a path for normal current or as an effective ground-fault current path to carry any level of fault current necessary to cause circuit breakers or fuses to operate.

GROUNDING ELECTRODE SYSTEM REQUIREMENTS

The *NEC* provides rules indicating which electrodes must be used for grounding electrode systems. These requirements are not optional. Grounding electrodes and the grounding electrode system are covered in Part III of Article 250. The first sentence of 250.50 outlines the general requirement that all electrodes present at a building or structure be used to form a grounding electrode system. This section addresses each building or structure served. **See Figure 4-5.**

If a building or structure has no electrical system, there is no *NEC* requirement for a grounding electrode. A network of grounding electrodes for a lightning protection system may be installed on a structure that is not supplied by an electrical power service or system, such as a barn.

The grounding electrodes addressed in 250.52(A) include those that are inherent to building construction and electrodes that must be installed. Examples of grounding electrodes that Electrical Workers install are ground rods, plates, ground rings, and chemically enhanced electrodes. On the other hand, footings, metal in-ground support structures, underground water pipes, and local underground conductive systems or structures are examples of grounding electrodes that are inherent to many types of building construction. Section 250.50 is very clear: if there are any grounding electrodes present, they must all be used (bonded together) to form the grounding electrode system.

ESTABLISHING A GROUNDING ELECTRODE SYSTEM

A *system* is an assembly or combination of entities that form together in a complete and functional arrangement. From a simple point of view, a grounding electrode system includes multiple conductive objects (grounding electrodes) connected to form a collective connection to the Earth through several contact points.

There is a bonding requirement in 250.50 regarding an important aspect

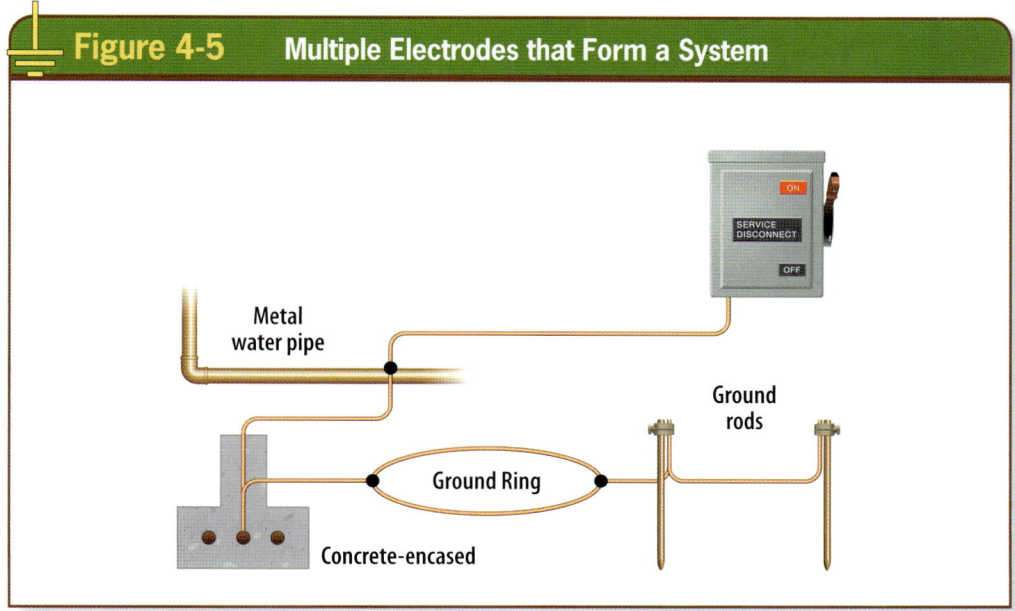

Figure 4-5 **Multiple Electrodes that Form a System**

Metal water pipe

Ground rods

Ground Ring

Concrete-encased

Figure 4-5. All grounding electrodes at a building or structure must be bonded together to form a system of grounding electrodes.

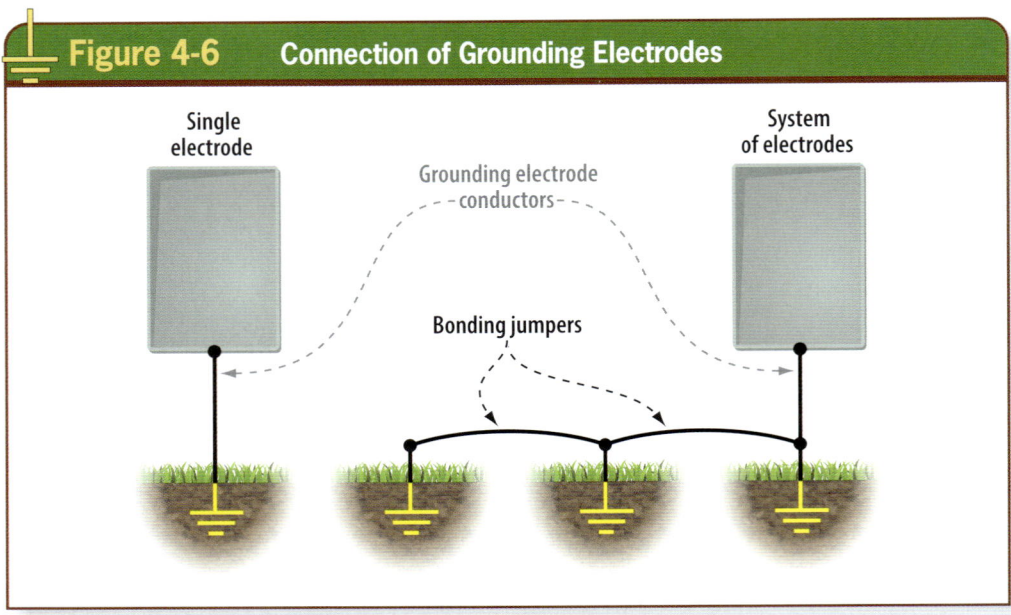

Figure 4-6   **Connection of Grounding Electrodes**

Single electrode

System of electrodes

Grounding electrode conductors

Bonding jumpers

Figure 4-6. *Bonding jumpers connect grounding electrodes together to form a grounding electrode system.*

of how the grounding electrode system should be developed on a project. As defined, bonding can be accomplished by a direct connection or by other connections that establish continuity and conductivity. Section 250.53(C) addresses bonding jumpers that are often used to connect grounding electrodes together to form a grounding electrode system. The bonding jumpers used to form a grounding electrode system must meet the installation requirements in 250.64. These bonding jumpers must be sized using 250.66, and the connections must meet the rules for grounding electrode conductor connections in 250.70. **See Figure 4-6.**

The bonding jumper size depends on its location in the system and the type(s) of grounding electrodes it is bonding together. For example, if 500-kcmil copper ungrounded service conductors are installed at the service and the first electrode connected is an underground metal water pipe electrode, a 1/0 AWG copper grounding electrode conductor is required in accordance with Table 250.66. If there is a jumper installed between a ground rod and the water pipe, the bonding jumper

is required to be sized no smaller than 6 AWG copper or 4 AWG aluminum or copper-clad aluminum.

Two factors govern the sizing for grounding electrode conductors and bonding jumpers used to form a grounding electrode system. First, the size of the largest ungrounded supply conductor is used for sizing the grounding electrode conductor and the bonding jumpers. Second, the type of grounding electrodes between which the bonding jumper is installed drives the minimum bonding jumper size. See Table 250.66 and review the alternative grounding electrode conductor sizing requirements addressed in 250.66(A), (B), and (C).

MANDATORY GROUNDING ELECTRODES

Grounding electrodes are often installed during building construction. The *NEC* offers no option among these grounding electrodes regarding selectivity for convenience or any other reason—if the electrodes are present, they must be used in the system of electrodes. If there are no grounding electrodes inherent to the construction of a building and the building is supplied with electrical power, a grounding electrode must be installed.

The *NEC* does not require that more than one grounding electrode be installed to form a system of installed electrodes where none are present; it only requires that at least one grounding electrode be installed. To that end, there are only a few choices for installing a grounding electrode, such as rod, pipe, or plate electrodes; ground rings; or other listed electrodes. There are various grounding electrode products that are evaluated and listed to UL Standard 467. An example of a listed grounding electrode is a listed chemical electrode assembly. As with any listed product, it is important to follow the manufacturer's installation instructions in accordance with 110.3(B).

TYPES OF GROUNDING ELECTRODES

The types of acceptable electrodes recognized in the *NEC* are provided in 250.52(A). This section is intended to provide the details and descriptions of each grounding electrode recognized for use by the *NEC*. The list of grounding electrodes in 250.52(A) provides the requirements for installers about the electrodes that must be used for the entire grounding electrode system. Specific installation requirements for grounding electrodes are provided in 250.53.

Metal Underground Water Pipe Electrodes

The water pipe electrode is described in 250.52(A)(1). This electrode has a long history in the *NEC* and is one of the first grounding electrodes to be required for use.

A metal water pipe electrode must have a minimum of 10 feet of metal underground water piping in direct contact with the Earth. This requirement does not mention depth, only that direct contact must exist. In the past, the water pipe electrode was the water service piping system routed to a building

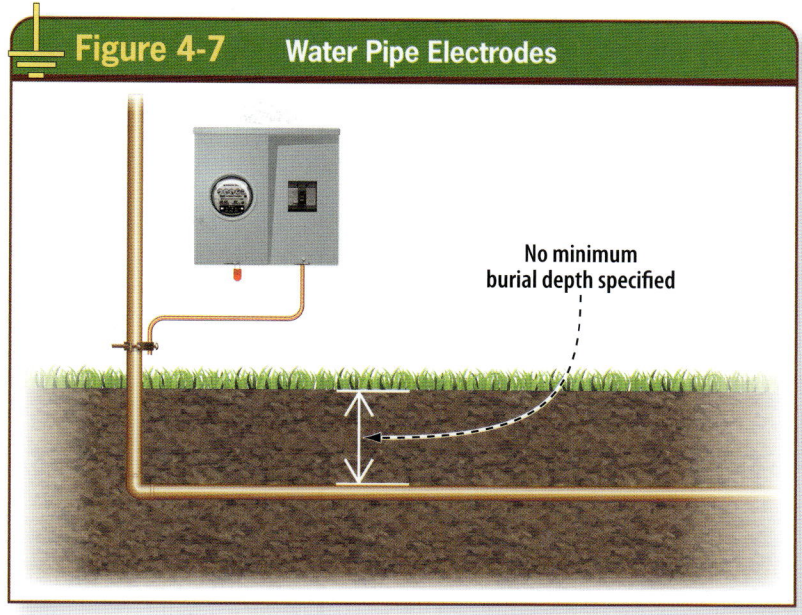

Figure 4-7. *A water pipe can serve as a grounding electrode if a minimum of 10 feet of conductive metal piping is in direct contact with the Earth.*

or structure, which was required by the water service provider to be buried at a minimum depth. That depth was typically based on the frost line and on ensuring protection from physical damage. Metal water pipe electrodes can be present outside the building perimeter or inside the footprint of the structure. **See Figure 4-7.**

The *NEC* prohibits the use of an above-ground metal interior water piping system that extends more than five feet from the point of entrance into the building as a connection point for grounding electrode conductors and bonding jumpers. This limits the length of water piping that can be used as a grounding electrode conductor or conductive path to the electrode. An exception relaxes this restriction for industrial, commercial, and institutional buildings or structures in which conditions of maintenance and supervision ensure qualified people service the installation. If these conditions are met, the interior metal water piping located more than five feet from the point of entry is permitted if it is exposed for the entire length.

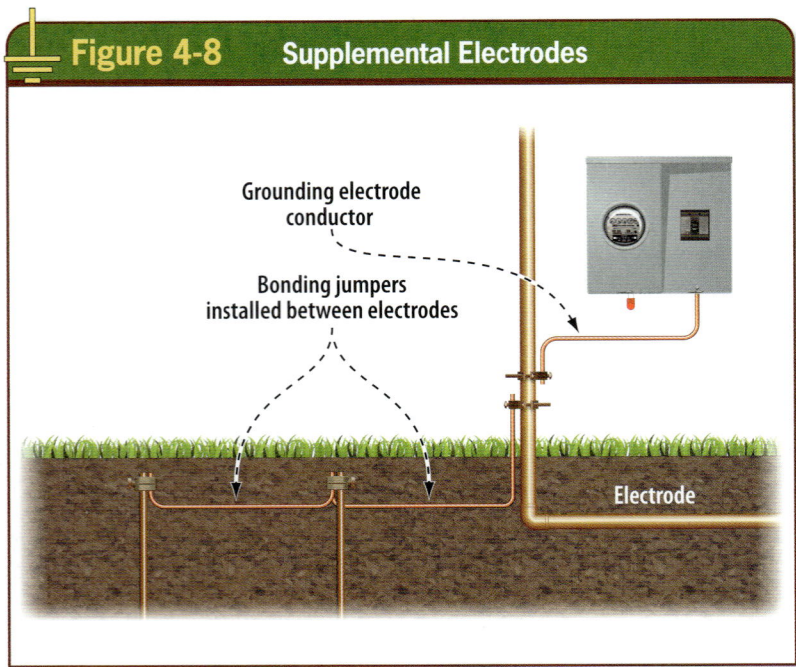

Figure 4-8   **Supplemental Electrodes**

Grounding electrode conductor

Bonding jumpers installed between electrodes

Electrode

Figure 4-8. Water pipe electrodes must be supplemented by at least one additional electrode of a type specified in 250.53(A)(2).

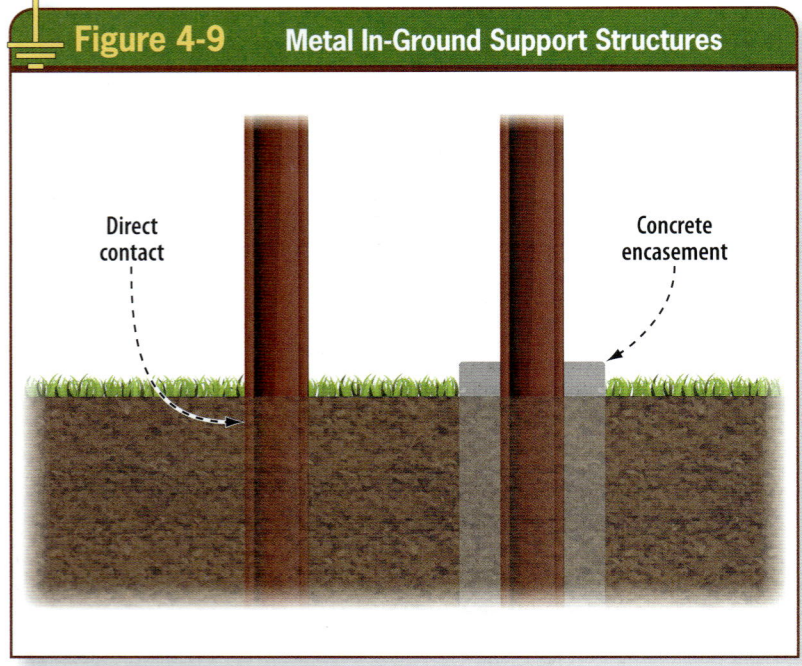

Figure 4-9   **Metal In-Ground Support Structures**

Direct contact

Concrete encasement

Figure 4-9. According to 250.52(A)(2), the metal in-ground support structure must be in direct contact with the Earth for 10 feet or more, with or without concrete encasement.

In recent years, nonmetallic types of piping have replaced many metal water piping systems. Replacements using nonmetallic piping or fittings can also interrupt the electrical continuity of the underground metal water piping. Once this evolution was recognized, the *NEC* included a requirement for any metal water pipe electrode to be supplemented by another grounding electrode. The idea is to have a backup in case the original water pipe electrode is removed or replaced with a nonmetallic pipe, thus eliminating the water pipe as a grounding electrode for electrical services and systems at the building or structure served. **See Figure 4-8.**

Metal In-Ground Support Structures

Many buildings or structures are built with metal in-ground support structures (structural steel) as a part of the foundation. A metal building frame is often a good grounding electrode. Section 250.52(A)(2) clearly indicates that a metal in-ground support structure must meet specific criteria to be acceptable as a grounding electrode. The metal frame (one or more metal in-ground support structures) of the building or structure must be directly connected to the Earth such that it is in contact with the Earth for 10 feet or more vertically, with or without concrete encasement. **See Figure 4-9.**

In instances where there are multiple metal in-ground supports for a building or structure served, then only one is required to be bonded to the grounding electrode system. Examples of metal in-ground support structures are pilings, casings, and other structural metal in direct contact with the Earth. Structural metal frame electrodes are reliable because they are unlikely to be disturbed for the life of a building.

The definition of *grounding electrode* clarifies that the metal building frame or metal in-ground support structure must be either in direct contact with the Earth or be connected electrically through a concrete-encased electrode. Under these two conditions, the metal building frames provide a conductive path to the Earth.

Section 250.68(C) addresses metal building frames and metal water piping systems that are not electrodes but are conductive paths to grounding electrodes that perform in similar fashion to grounding electrode conductors. Remember that an electrode, by definition, is a conductive object in direct connection with the Earth. Any conductive object (for example, structural steel or water piping) above the Earth can be a conductive path to the electrode.

Concrete-Encased Electrodes

Concrete-encased electrodes are described in 250.52(A)(3). Section 250.50 requires at least one concrete-encased electrode, as described in 250.52(A)(3), be included in the grounding electrode systems for buildings or structures. This rule applies to all buildings and structures with a foundation or footing having 20 feet or more of ½-inch or larger electrically-conductive rebar/reinforcing steel or 20 feet of a minimum 4 AWG bare copper conductor encased in concrete to form the electrode. If a concrete-encased electrode is not present at the building or structure supplied, it is not required that wire be installed in a concrete footing to form one, but it is an option. **See Figure 4-10.**

The *NEC* describes the concrete-encased electrode as bare, zinc galvanized, or another electrically-conductive coated rebar of not less than ½ inch in diameter installed in one continuous 20-foot length or in multiple pieces connected together by the usual steel tie wires, exothermic welding, welding, or other effective means to create a 20-foot or greater length. A concrete-encased electrode can also be constructed using 20 feet or more of bare copper conductor not smaller than 4 AWG.

Note that the 20 feet of conductive rebar or bare conductor used in a concrete-encased electrode only establishes the connection to the concrete. The combination of the concrete and the conductive component serve as the

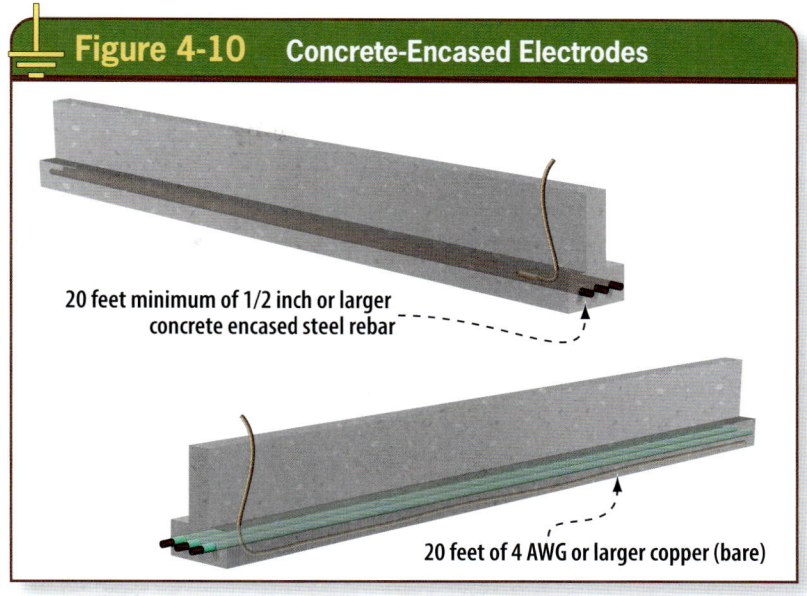

Figure 4-10   Concrete-Encased Electrodes

20 feet minimum of 1/2 inch or larger concrete encased steel rebar

20 feet of 4 AWG or larger copper (bare)

Figure 4-10. Concrete-encased electrodes must be included in the grounding electrode system for buildings or structures with concrete footings or foundations that provide not less than 20 feet of rebar (reinforcing steel) or bare copper wire.

grounding electrode, as clarified in the definition of the term.

Section 250.50 mandates the use of all grounding electrodes to form the grounding electrode system. This includes concrete-encased electrodes present at the building or structure. An exception to 250.50 relaxes this mandatory requirement for existing buildings and structures in which reaching the concrete-encased electrode could damage the structural integrity of the building or otherwise disturb the existing construction.

Because the installation of the footings and foundation is one of the first elements of a construction project and usually has been completed by the time the electric service is installed, this rule necessitates an awareness and coordinated effort on the part of designers and the construction trades to ensure that the concrete-encased electrode is incorporated into the grounding electrode system during the placement of rebar and concrete footings.

The concrete-encased electrode has proven that it offers optimal performance and longevity. The footing or foundation of any building will typically be there for as long as the building is.

Because all the rebar in the bottom perimeter of the building footing is usually tied together with tie wires, the electrode acts like a ground ring, but with much more surface area in the connection to the Earth. The footing is present around the bottom of the building perimeter, which means there is ground (Earth) contact from concrete-encased electrodes. Concrete retains moisture and is continuously absorbing moisture through the bottom of the footing. This keeps the connection between the footing and the Earth effective. The footing of a building is also typically the largest grounding electrode in each building or structure served.

The findings of Herbert G. Ufer in the 1940s and 1950s proved the effectiveness of concrete-encased grounding electrodes. Ufer was a vice president and engineer at Underwriters Laboratories who assisted the U.S. military with ground-resistance problems at installations in Arizona. The military required low-resistance (five ohms or less) to ground connections for lightning protection systems installed at its ammunition and pyrotechnic storage sites at the Navajo Ordnance Depot in Flagstaff and Davis-Monthan Air Force Base in Tucson. Ufer developed the initial design for a concrete-encased grounding electrode that consisted of ½-inch reinforcing bars 20 feet in length, placed within and near the bottom of two-foot-deep concrete footings for the ammunition storage buildings. Test readings over a 20-year period revealed steady resistance values of two to five ohms, which satisfied the specifications of the U.S. government at that time. This work eventually resulted in what we know of today as the concrete-encased grounding electrode in the *NEC*. The slang term for this electrode is *UFER* after the engineer who created it. More details about the research and findings of Herbert G. Ufer are provided in an article he wrote in October 1964. **See Annex A.**

Ground Ring Electrodes

Ground ring electrodes are described in 250.52(A)(4). Ground rings are electrodes that must be installed by Electrical Workers, as they are not inherent in the construction of a building. An interesting feature of the ground ring requirement is that it must encircle the entire building or structure and be not less than 20 feet in length. This can result in an extensive amount of work for installers when a ground ring electrode is specified as part of the grounding electrode system. **See Figure 4-11.**

A ring electrode must not be smaller than 2 AWG copper, though it can be larger. Aluminum or copper-clad aluminum is not permitted as a ground ring, in accordance with the restrictions in 250.52(B), because aluminum is vulnerable to corrosion and deterioration when in contact with the soil. An electrode that deteriorates and is susceptible to corrosion will obviously be ineffective.

The ring electrode is required to be buried not less than two and a half feet (30 inches) below grade level. This involves trenching and installing the bare conductor in contact with Earth.

The ground ring should not be confused with a counterpoise system installed for lightning protection systems. Grounding electrodes for building electrical services and systems cannot be used for lightning protection systems.

Figure 4-11   Ground Ring Electrodes

Figure 4-11. Ground ring electrodes must encircle the entire building or structure, be a minimum 2 AWG copper, and be buried a minimum of 30 inches below grade.

Ground ring installation requirements in the *NEC* specify a minimum depth, but do not provide a minimum distance from the exterior walls of a building. Good design practices specify installing it as close as practical to the building while remaining outside the dripline from the roof overhang and, when practical, below the permanent moisture level of the Earth, which can vary depending on geographical area. The depth of the building footing is usually a good depth to strive for when installing a ground ring, with the minimum depth at not less than two and a half feet.

Rod and Pipe Electrodes

Rod and pipe electrodes are described in 250.52(A)(5) and are required to be at least eight feet in length. Rod and pipe electrodes must consist of the following materials:

- Pipe or conduit used as grounding electrodes must not be smaller than metric designator 21 (trade size ¾). If the electrode is made of steel, it must have an outer surface that is galvanized or otherwise metal coated for corrosion protection. As covered in this section, a *pipe* is not the metal underground water piping system previously discussed.
- Stainless steel grounding electrodes and copper- or zinc-coated steel electrodes must be at least 15.87 mm (⅝ inch) in diameter, unless they are listed by a qualified electrical testing laboratory.

Plate Electrodes

Plate electrodes are described in 250.52(A)(7). A plate electrode must have not less than two square feet of surface contact between the plate and the soil. A one-foot square plate with two sides in contact with the Earth accomplishes this. Plate electrodes made of bare or conductively-coated iron or steel must be at least a quarter inch in thickness. Electrodes of nonferrous metal such as copper or brass must be at least 0.06 inch in thickness. **See Figure 4-12.**

Other Local Metal Underground Systems or Structures

The *NEC* recognizes local metal underground systems or structures as grounding electrodes in 250.52(A)(8). Examples of these types of electrodes are metal piping systems, conductive underground tanks, and underground metal well casings that are not bonded to a metal water pipe.

The underground structure or system mentioned in this provision must meet the criteria of what constitutes a grounding electrode. It must be a conductive object that is in direct contact with the Earth through an established and maintained connection. Remember that any coatings that encapsulate such structures can negate or nullify the effectiveness of such structures as grounding electrodes. The decision to use these types of electrodes should include careful analysis and investigation into the long-term effectiveness of this grounding electrode.

GROUNDING ELECTRODE INSTALLATION REQUIREMENTS

Section 250.53 provides grounding electrode installation rules that apply to those electrodes that are installed by

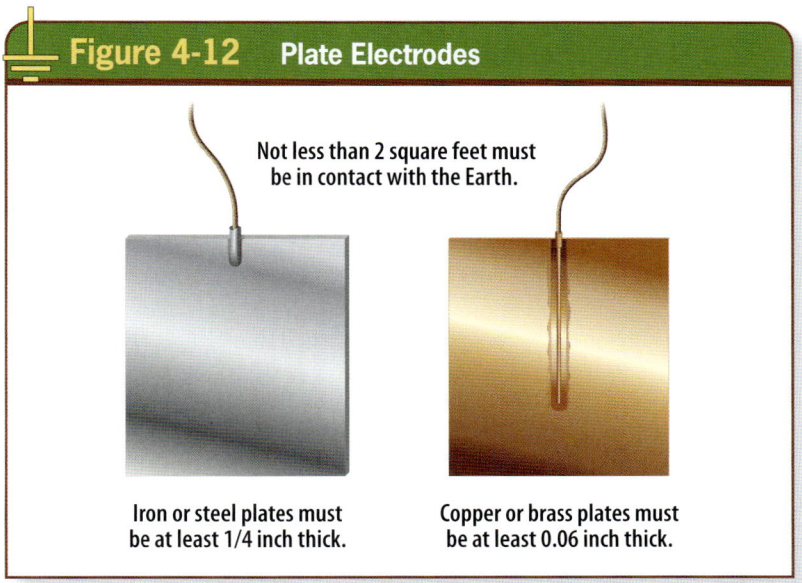

Figure 4-12   Plate Electrodes

Not less than 2 square feet must be in contact with the Earth.

Iron or steel plates must be at least 1/4 inch thick.

Copper or brass plates must be at least 0.06 inch thick.

Figure 4-12. *Plate electrodes must be buried a minimum of 30 inches deep.*

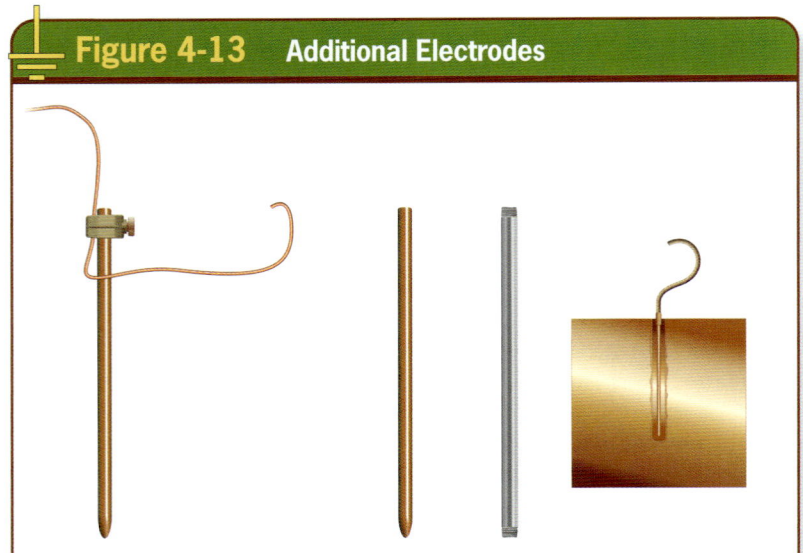

Figure 4-13   Additional Electrodes

Figure 4-13. *A single rod, pipe, or plate that does not meet the 25-ohm provisions must be supplemented by an additional electrode. The additional electrode can be any electrode specified in 250.52(A)(2) through (A)(8).*

Fact

Rod and pipe electrodes must be embedded below permanent moisture level when possible. This provides the most effective performance of the electrode.

Electrical Workers and are not typically inherent in construction, as is the case with metal in-ground support structures, underground metal water piping systems, and concrete-encased electrodes. The grounding electrodes addressed in this rule include rod, pipe, and plate electrodes, as well as ground rings. Rod, pipe, and plate electrodes must be spaced at least six feet from any other electrode. A requirement to supplement a metal underground water pipe with an additional electrode is also included in this section.

Rod and Pipe Electrode Installation

Rod, pipe, and plate electrodes are covered in 250.52(A)(5) and (A)(7). These are all grounding electrodes that must be installed if necessary. If installers choose to use rod, pipe, or plate electrodes because there are no electrodes present for use, a single electrode is required to be supplemented by an additional electrode of a type specified in 250.52(A)(2) through (A)(8). The supplemental electrode is permitted to be connected to the rod, pipe, or plate electrode; the grounding electrode conductor; the grounded service-entrance conductor; a nonflexible grounded service raceway; or any grounded service enclosure. **See Figure 4-13.**

There is an exception that permits using a single rod, pipe, or plate grounding electrode when the single electrode has a resistance to Earth of 25 ohms or less. If a single electrode meets the 25-ohm requirement, there is no need for any additional electrodes to be installed. The resistivity of a single grounding electrode such as a rod, pipe, or plate is determined by specialized Earth electrode testing methods and test instruments, although the *NEC* does not directly state that testing is required.

The *NEC* has been improved to require the two installed electrodes as a first choice rather than the method of installing the second electrode when the 25-ohm resistance is exceeded on the first. The requirement for installing two electrodes reflects how installers typically handle this requirement in the field.

Rod and pipe electrodes also must be embedded below permanent moisture level if possible. This provides the most effective performance of the electrode. Frost line depths can vary geographically and can present connection problems between the Earth and the electrode. Where the Earth freezes, the connection between the Earth and electrode is affected. Rod, pipe, or plate electrodes must also be clean and free of nonconductive coatings such as paint, enamel, and other substances that affect the connection between the Earth and the electrode.

Rod and pipe electrodes must be driven into the Earth to establish the best contact. When a grounding rod or pipe electrode is installed, it is required to be in direct contact with the Earth for a distance not less than eight feet. Sometimes workers have difficulty driving an 8-foot or 10-foot ground rod or pipe electrode, but the solution is not to cut the exposed portion of the rod off. This results in less than what is required by the *NEC* minimums for electrode contact with the Earth.

The *NEC* recognizes this challenge and offers alternatives that can be used when the electrode cannot be driven because of rock bottom. The first step in this hierarchy of alternatives is to install the electrode at an angle not more than 45 degrees from vertical. **See Figure 4-14.** Often, this will solve the installation problems but, if not, the *NEC* offers, as a third and last resort, the option of laying the rod or pipe electrode in a trench not less than 30 inches deep. Installers should be aware that laying the rod or pipe electrode in a trench is only a last resort as indicated by the hierarchy arrangement of 250.53(A)(4).

Soil Resistivity and Ground Resistance Testing

Some conditions and installations require low grounding electrode resistance. The overall resistance is the connection between the electrode and the soil itself. The overall resistivity has a lot to do with the connection to the Earth.

Resistivity values differ from location to location. In some areas of the world, the soil is rich with mineral content and stays relatively moist throughout the year, keeping the resistivity values low. This provides for low resistance values between the grounding electrode and the surrounding Earth. In other areas, the soil may be sandy or very dry and rocky. These conditions can present challenges in establishing and maintaining good, effective contact between the electrode and the Earth. **See Figure 4-15.**

There are some solutions that can help improve soil resistivity values. For example, it may help to increase the size or length of the grounding electrode(s). Special grounding rod couplings are available for lengthening rod-type electrodes. Rod or pipe electrodes can also be installed deeper into the Earth, which tends to lower the resistance of the grounding electrode connection to the Earth. Installing multiple grounding electrodes is another effective method of lowering the resistance in the grounding electrode connection to Earth. If multiple grounding electrodes are used to form a grounding electrode system, the resistance values in the connection to

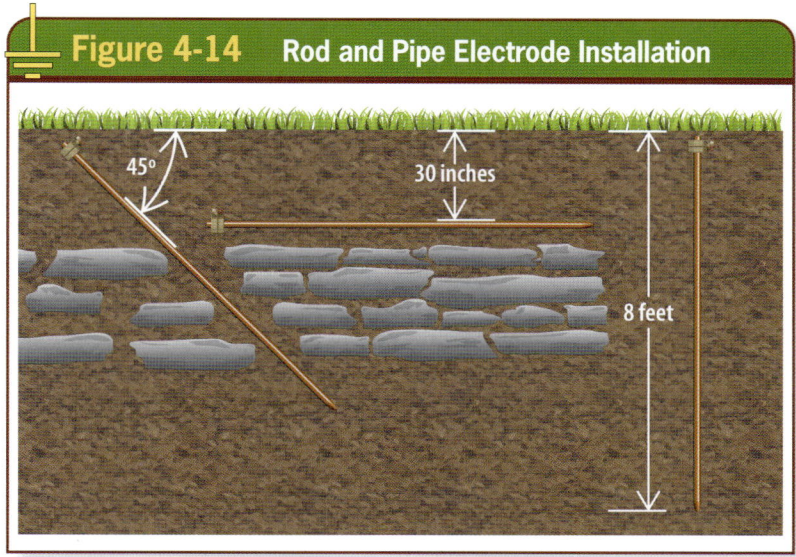

Figure 4-14   **Rod and Pipe Electrode Installation**

45°   30 inches   8 feet

Figure 4-14. *Rod and pipe installation must meet* Code *requirements or use an alternative method of installation.*

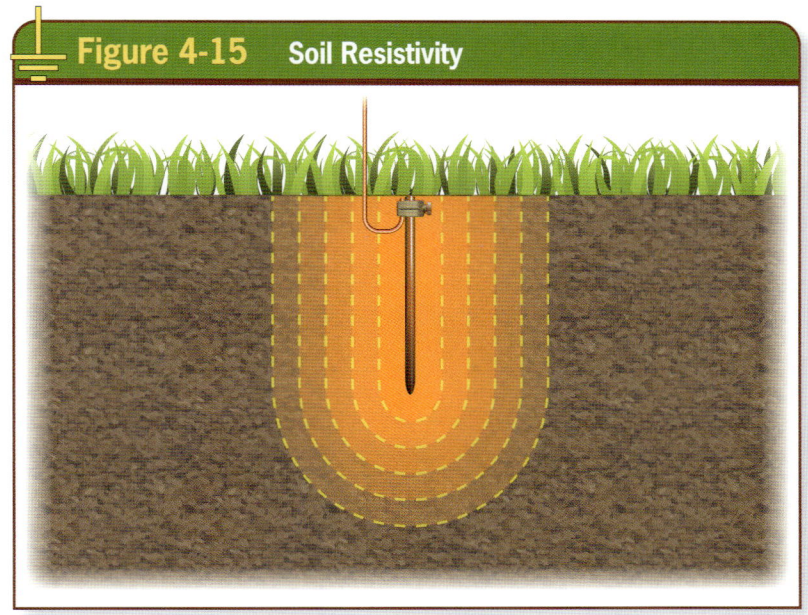

Figure 4-15   **Soil Resistivity**

Figure 4-15. *Resistance in the connection to Earth is a variable. Soil resistivity values differ from location to location—dry and rocky soils tend to have greater soil resistivity, while moist and mineral-rich soils typically have less resistivity.*

Earth are typically lower because of the benefits of a system of electrodes working together to maintain effective contact with the Earth.

Another effective method for lowering the Earth resistance is to treat the soil with suitable chemicals. Special listed

Figure 4-16   Chemically Treated Soil

Figure 4-16. Soil can be chemically treated to improve resistivity values.

Courtesy of Harger Lightning and Grounding

Figure 4-17   Ground Rod Resistance Meter

Figure 4-17. A clamp-on style electrode resistance meter measures the resistance between a grounding rod and the Earth by transmitting a signal through a grounding electrode conductor and using the receiver of the instrument to establish the resistance of the grounding electrode conductor or system.

Courtesy of Fluke Corporation

chemical electrodes are available for this use. **See Figure 4-16.** Using chemicals to treat the Earth for resistivity improvement can cause increased effects of corrosion on other metal in the Earth in close proximity. Chemically-treated soil can also be a concern for the environment. Check local environmental regulations in any locations where these types of electrodes are specified in the design or installed by choice.

Earth resistance testing applies the basic principles of Ohm's law. Ohm's law is the relationship between voltage, current, and resistance in a basic circuit. Applying voltage and measuring the resultant current can determine a resistance value. Ground resistance can be measured using a separate voltage supply, a voltmeter, and an ammeter. There are also instruments and equipment available that provide electrode-testing functions in a single unit.

One common method of Earth resistance testing is the three-point method, sometimes called the "fall-of-potential" or three-terminal grounding resistance testing method. This method uses two test electrodes in addition to the electrode being tested. The testing is required to be performed according to the test instrument manufacturer's instructions. The basic procedure is to ensure that the electrode being tested is not connected to the building electrode system. One test electrode should be installed approximately 100 feet from the electrode under test. The second test electrode should be installed approximately 62 feet from the electrode under test. This test method is also commonly referred to as the 62% test method. The two test electrodes should be in line with the electrode under test. The test leads of the resistance tester need to be connected to the grounding electrode being tested and the other temporary testing electrodes. The Earth resistance reading should be measured and recorded. Then, the test electrode in the middle should be moved one way or the other in 10-foot increments until the resistivity values are basically the same. This means the plateau area of the test has been determined. If the plateau reading cannot be determined, the test rod 100 feet from the electrode being tested needs to be installed at a greater distance. The testing procedure should then be repeated.

There are also clamp-on instruments that can provide ground resistance readings. These instruments incorporate a transmitter and receiver and are designed to measure circuit current, resistance, and leakage current. A clamp-on ground resistance meter measures the resistance between a grounding rod and the Earth by transmitting a signal through a grounding electrode conductor and using the receiver of the instrument to establish the resistance of the grounding electrode conductor or system. It can be used to determine soil resistivity. **See Figure 4-17.** Soil resistivity is measured to determine the type of soil conditions, the corrosive effects in the soil, and the best type of grounding electrode system to install. A four-point testing procedure can also be used to measure soil resistivity. For more comprehensive information on Earth testing, refer to the *Test Instruments and Applications* textbook used in the *electrical training ALLIANCE* curriculum.

Plate Electrode Installation

When workers install plate electrodes, the installation must provide a depth of not less than two and a half feet. The connection between the plate and the grounding electrode conductor will be buried in the soil, which drives the requirement that the connection means be listed as suitable for direct burial applications or be exothermically welded. The connection means also must be compatible with the plate metal and the grounding electrode conductor. Some plate electrodes are available with a lead already connected. **See Figure 4-18.** The exothermic welding process is also a popular means of connecting grounding electrode conductors to buried ground rods. **See Figure 4-19.**

Electrode Spacing Requirements

If rod, pipe, or plate electrodes are installed, each electrode must be spaced

Figure 4-18   Plate Electrodes with Connection Leads

Figure 4-18. A plate with an integral connection lead is suitable for connecting to a grounding electrode conductor.

Courtesy of ABB

Figure 4-19   Exothermic Welding

Figure 4-19. Exothermic welding connections can also be used to connect grounding electrode conductors to ground rod electrodes.

Courtesy of ERICO International Corp.

a minimum of six feet from another electrode. This requirement includes those ground terminals (grounding electrodes) for lightning protection systems.

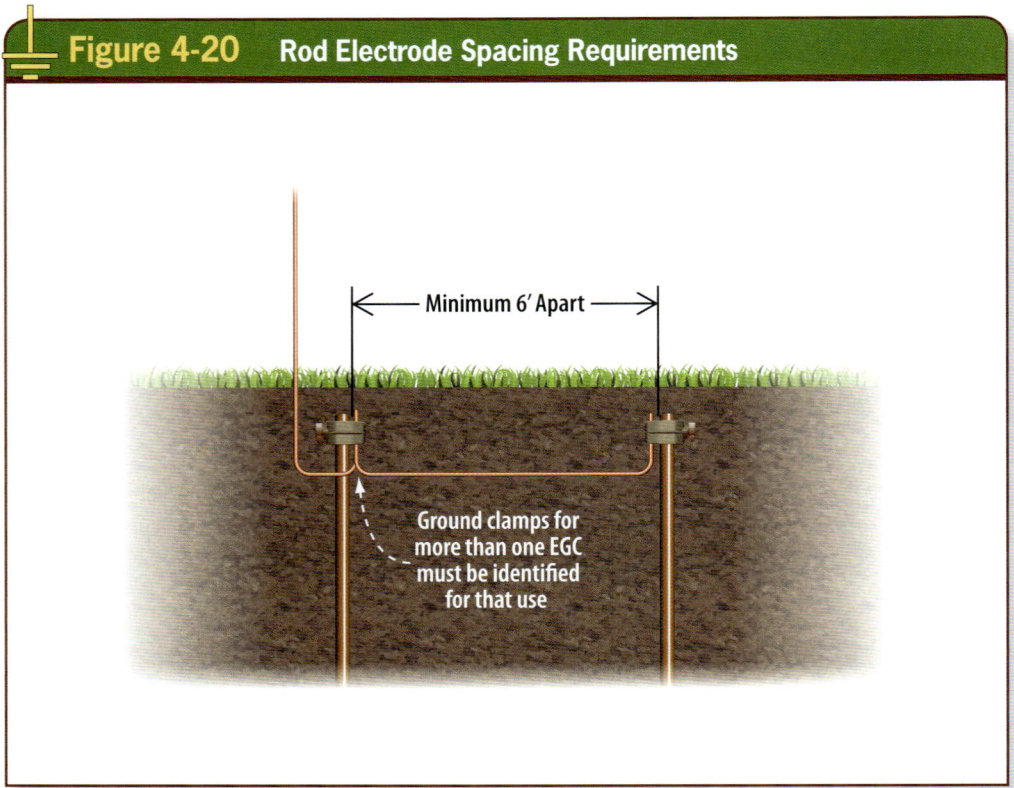

Figure 4-20   Rod Electrode Spacing Requirements

Minimum 6′ Apart

Ground clamps for more than one EGC must be identified for that use

Figure 4-20. *The NEC sets grounding electrode spacing requirements. The paralleling efficiency of rod electrodes is typically increased by spacing them not less than twice the length of the longest rod installed.*

Fact

When two or more grounding electrodes are bonded together, such as two ground rods that are installed six feet apart, they are considered to be a single grounding electrode system made up of two grounding electrodes.

The purpose of the spacing requirement is to reduce the effects of overlapping the spheres of influence associated with each electrode. The *NEC* requires a minimum spacing distance of not less than six feet, but it indicates that the paralleling efficiency of rods is increased by spacing them twice the length of the longest rod. **See Figure 4-20.**

Ideally, the spacing distance between these installed electrodes should be not less than twice the depth of the electrode, which for eight-foot rod- and pipe-type electrodes is 16 feet.

Water Pipe Electrode Supplement and Bonding Jumpers

Section 250.53(D)(2) indicates that metal underground water pipe electrodes be supplemented by an additional electrode of a type specified in 250.52(A)(2) through (A)(8). A common practice in the field is to install a ground rod to serve as this supplement, even though any of the electrodes specified in 250.52(A)(2) through (8) could be used. When the supplemental electrode is a rod, pipe, or plate type, the installation must meet the provisions in 250.53(A), meaning two such electrodes must be installed, unless a single electrode has a resistance of 25 ohms or less. **See Figure 4-21.**

The supplemental electrode is permitted to be connected to the grounding electrode conductor, the grounded service-entrance conductor, the nonflexible grounded service raceway, or any grounded service equipment enclosure.

When a metal water pipe electrode is used, the continuity of the grounding path or bonding connection to the piping must not depend on water meters,

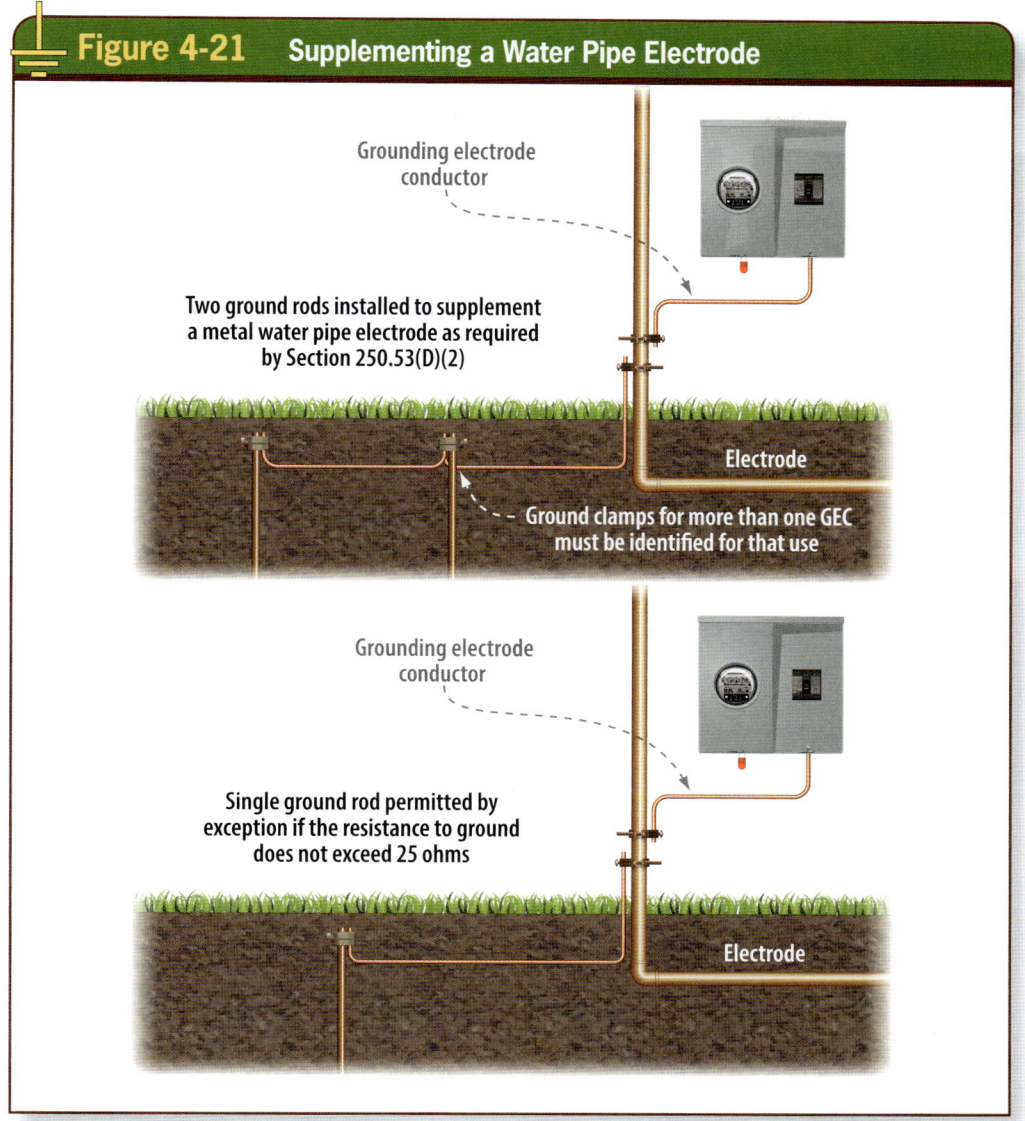

Figure 4-21 Supplementing a Water Pipe Electrode

Grounding electrode conductor

Two ground rods installed to supplement a metal water pipe electrode as required by Section 250.53(D)(2)

Electrode

Ground clamps for more than one GEC must be identified for that use

Grounding electrode conductor

Single ground rod permitted by exception if the resistance to ground does not exceed 25 ohms

Electrode

Figure 4-21. A water pipe electrode may be supplemented by two rod electrodes, or, if the resistance to ground for a single rod does not exceed 25 ohms, a single rod electrode.

filters, or similar equipment. Bonding jumpers are required to be installed around such equipment. The bonding jumper is also required to be installed with enough length to allow for servicing the equipment in the water line without removing the jumper and disrupting the grounding electrode system.

Auxiliary Grounding Electrodes

Auxiliary electrodes are those that are installed by choice and not to meet a requirement in the *NEC*. Auxiliary grounding electrodes are sometimes specified by electrical equipment manufacturers, electrical designers, and facility owners. The auxiliary grounding electrode is a connection to the Earth that is usually in close proximity to the equipment it supplements, although the *NEC* does not specify a distance from the auxiliary electrode to the equipment.

An important aspect of the auxiliary grounding electrode installation is that it is connected to equipment that is also connected to an equipment grounding conductor (EGC). A good example of auxiliary electrodes is the installation of grounding electrodes at lighting pole

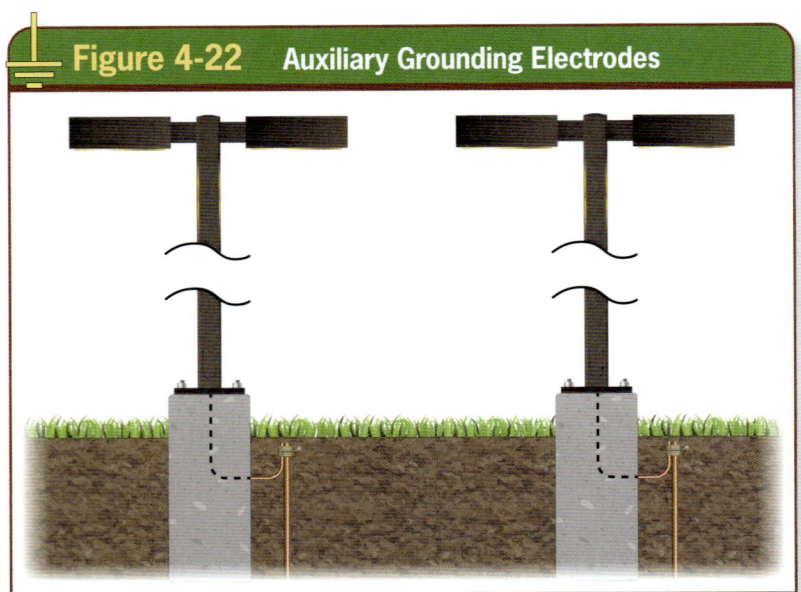

Figure 4-22 **Auxiliary Grounding Electrodes**

Figure 4-22. Auxiliary grounding electrodes supplement the required EGC of the circuit supplying equipment.

industrial machines or electronic equipment where a manufacturer specifies a local grounding electrode connection in addition to the required EGC for the branch circuit. **See Figure 4-23.**

Items Not Permitted as Grounding Electrodes

Some conductive materials are not compatible with the Earth and are vulnerable to corrosion and deterioration. Section 250.52(B) provides restrictions for using aluminum materials and metal gas piping systems as grounding electrodes.

There are obvious concerns about corrosion that could result in deterioration of the electrode over time. Life expectancy of grounding electrodes is not directly addressed in the *NEC*. However, experience has resulted in the prohibition of metals that are susceptible to corrosion when in contact with the Earth, such as aluminum.

The other item addressed in this section is metal gas piping systems. These systems are not permitted to be used as grounding electrodes. This restriction is

bases in parking lots. Engineers and designers often specify these auxiliary electrodes to help dissipate lightning events into the Earth at the point of a strike. **See Figure 4-22.**

Another common use for auxiliary electrodes is for certain types of

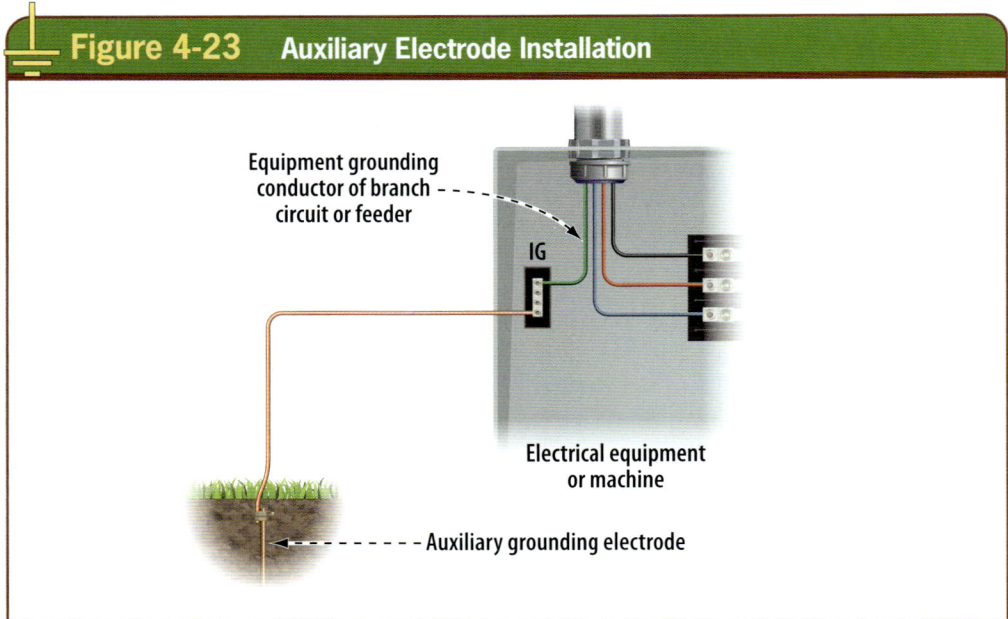

Figure 4-23 **Auxiliary Electrode Installation**

Equipment grounding conductor of branch circuit or feeder

IG

Electrical equipment or machine

Auxiliary grounding electrode

Figure 4-23. Auxiliary electrodes are permitted to supplement the required equipment grounding conductor of the feeder or branch circuit supplying industrial machines that have computers or electronic control systems.

Figure 4-24   Gas Piping Not Permitted as Electrode

Figure 4-24. Gas piping systems are not permitted as grounding electrodes, but they are required to be bonded under certain conditions.

in parallel with similar restrictions in *NFPA 54: National Fuel Gas Code.* (Installers should note that the restriction concerns the use of these systems as grounding electrodes, but *NFPA 70* and *NFPA 54* require these metal piping systems to be bonded under certain conditions.) The restrictions in 250.52(B) are very clear that metal gas piping is not permitted for use as a grounding electrode. Typically, gas service utilities install a dielectric fitting that creates isolation between the customer side of a gas meter and the portion of the metal gas system that is on the supply side of the meter and in contact with the Earth. **See Figure 4-24.**

List item (3) in 250.52(B) indicates that structures and structural rebar for in-ground pools as described in 680.26(B)(1) and (2) are not permitted to be used in the grounding electrode system, even though they are often very conductive and perform grounding electrode functions.

Grounding Electrodes and Utility Services

The *NEC* permits multiple services and systems to be installed for buildings or structures under specific conditions. Small buildings, such as dwelling occupancies, are typically supplied with only one service from a utility. Large buildings often have needs that force the design to include more than one service. Some of the factors that result in more than one service being installed on a building or structure are provided in 230.2(A).

When an AC system is connected to a grounding electrode in or at a building or structure, the *NEC* requires the same electrode be used for grounding conductor enclosures and equipment installed at that building or structure. When more than one service, feeder, or branch circuit supplies a building, they must be connected to the same grounding electrode(s). The reason is to ensure that the same grounding potential is established among all grounded systems and equipment installed in a single building or structure. This *NEC* requirement can be satisfied by bonding two or more grounding electrodes or grounding electrode systems together as specified in 250.50 and 250.58. **See Figure 4-25.**

It is recognized that two or more individual electrodes that are effectively bonded together are considered a single electrode and meet the common grounding electrode requirements in the *NEC*.

An example of this type of installation is if two services of different voltage characteristics supply a single building or structure. If a 480Y/277-volt, 3-phase, 4-wire service and a 120/240-volt, 3-phase, 4-wire delta-connected service supply the same building, they must be grounded using the same grounding electrode system.

Bonding of Lightning Protection Systems to Service Electrode Systems

Section 250.60 prohibits using a lightning protection system as a grounding electrode for electrical power systems. However, the requirement to bond the two systems together must be adhered to, as required in 250.106. This rule specifies that the grounding electrode system (grounding network) of the lightning protection system be bonded to the electrical service grounding electrode system, but it does not specify a size for the bonding conductor. **See Figure 4-26.**

NFPA 780: Standard for the Installation of Lightning Protection Systems provides the requirements for lightning protection systems installed on buildings or structures. This lightning protection standard contains important detailed information on grounding, bonding, and side-flash distance from lightning protection systems. Bonding may be necessary between the lightning protection system and the electrical system based on proximity and whether separation between the systems is through air or building materials. The size of the bonding conductor is typically determined by the requirements in *NFPA 780* and is related to the size of the lightning protection system conductors.

It is important to remember the scope of the *NEC* and its limitations to protect people and property from hazards associated with the use of electricity. Lightning is an uncontrollable force and is not directly covered by the rules in the *NEC*, other than where specific interconnections and bonding are required between the systems covered in *NFPA 70* and *NFPA 780*. For additional information about lightning protection systems and equipment, refer to **Annex E** in this textbook.

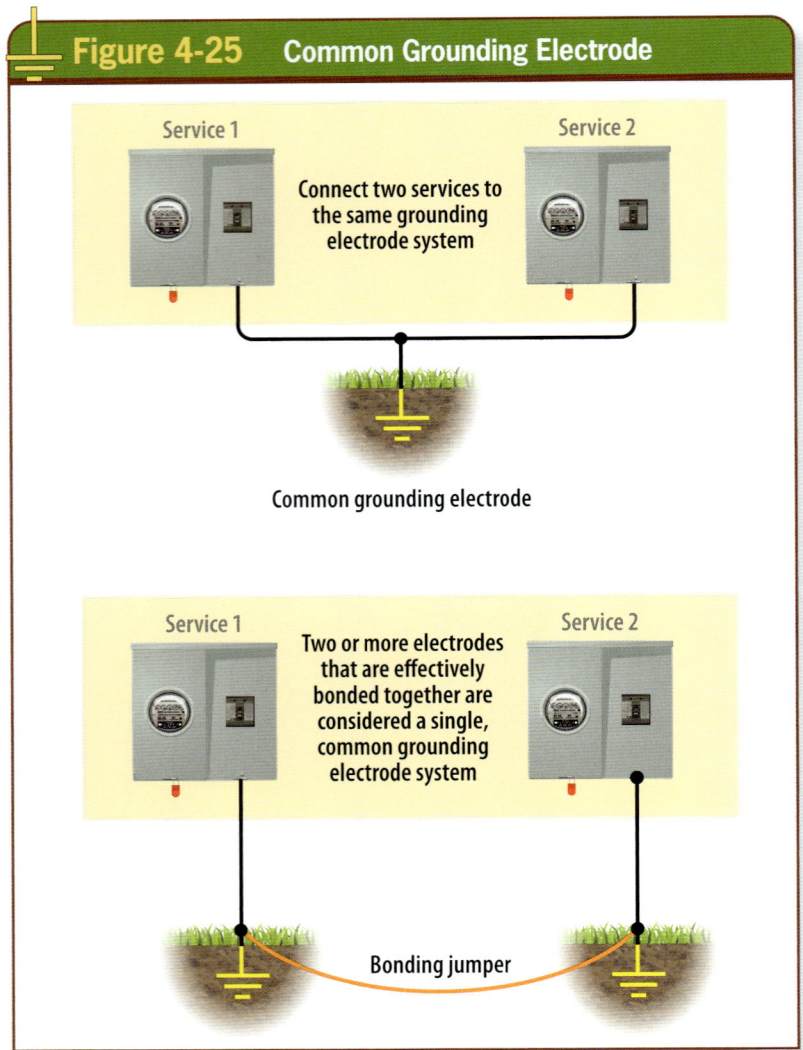

Figure 4-25 **Common Grounding Electrode**

Service 1   Service 2

Connect two services to the same grounding electrode system

Common grounding electrode

Service 1   Service 2

Two or more electrodes that are effectively bonded together are considered a single, common grounding electrode system

Bonding jumper

Figure 4-25. When more than one service is installed on a structure, the same (common) grounding electrode must be used.

Figure 4-26   **Bonding to Service Grounding Electrode**

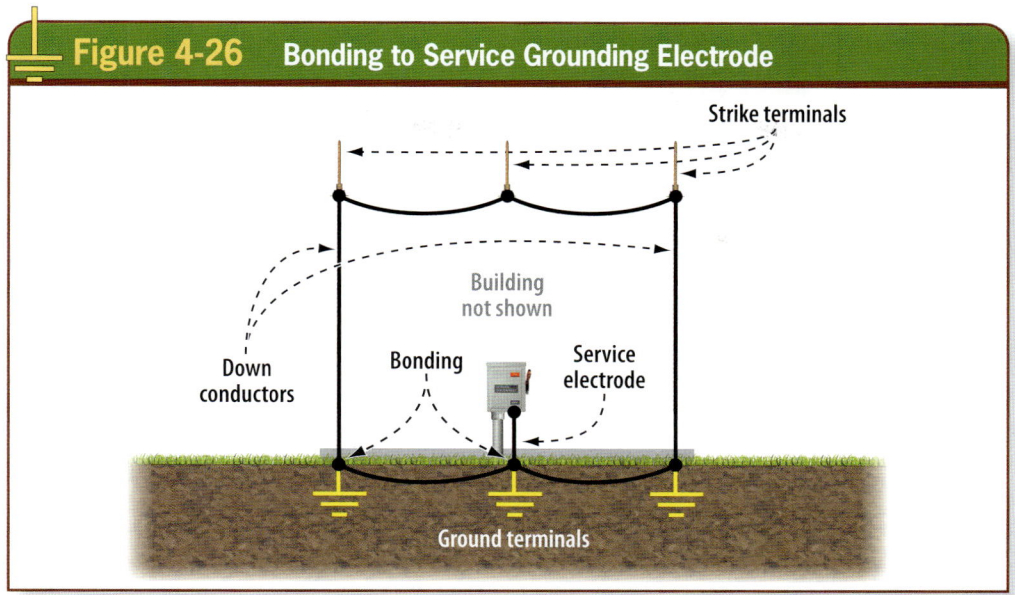

Strike terminals

Building
not shown

Down
conductors

Bonding

Service
electrode

Ground terminals

Figure 4-26. *The electrode system for the electrical supply to a building must be bonded to the ground network of a lightning protection system.*

SUMMARY

Grounding electrodes have an important role in the grounding and bonding scheme. A grounding electrode is a conductive object that establishes and maintains a connection to Earth. Grounding electrode systems are multiple electrodes used together to form a connection to ground through multiple Earth connection points. Grounding electrodes connect systems and equipment to the Earth and function to maintain those conductive parts at or as close to Earth potential as possible.

All grounding electrodes that are present at a building or structure served must be included in the grounding electrode system. Various types of grounding electrodes are recognized for use by the *NEC*. A grounding electrode system must be established at each building or structure served by electricity. The details and descriptions of grounding electrodes are provided in the *NEC*, as are the applicable electrode installation rules for those that are installed by Electrical Workers.

Grounding electrodes have little or no effect in the operation of circuit breakers or fuses. While the Earth is a conductor, it is not suitable for use as an effective ground-fault current path.

Lightning protection systems must be installed in accordance with specific requirements of *NFPA 780* and bonded to the grounding electrode system used for the electrical service serving the building or structure.

REVIEW QUESTIONS

1. The Earth is best defined in the *NEC* by which of the following?
   a. Dirt
   b. Ground
   c. Has only one moon
   d. Third planet from the sun

2. Which is best defined as a conducting object through which a direct connection to Earth is established?
   a. A grounding substitute
   b. A system bonding connection
   c. An equipment grounding connection
   d. Grounding electrode

3. As buildings or other structures are constructed, there are often grounding electrodes that are inherent to the building or structure when it is complete.
   a. True   b. False

4. A footing or foundation of a building or structure is typically a grounding electrode that must be used for the building or structure served by electrical power.
   a. True   b. False

5. If present, which of the following must be used to form a grounding electrode system at a building or structure?
   a. Ground rods
   b. Building or structure underground supports such as structural metal building frame(s) in direct contact with the earth for 10 feet or more, with or without concrete-encasement
   c. Underground metal water pipe
   d. All of the above

6. For new construction, a concrete-encased electrode is generally always required to be used in the grounding electrode system.
   a. True   b. False

7. The purpose of a grounding electrode is to function as the connection between grounded electrical systems and the non–current-carrying conductive parts of equipment and the Earth.
   a. True   b. False

8. The *NEC* requires that more than one grounding electrode be installed to form a system of installed electrodes when none are present.
   a. True   b. False

9. Which of the following is not permitted for use as a grounding electrode?
   a. Galvanized piping systems
   b. Metal underground structures
   c. Metal underground water piping
   d. Underground metal gas piping

10. A metal underground water pipe must have a minimum of __?__ in contact with the Earth to be acceptable as a grounding electrode.
    a. 6'
    b. 8'
    c. 10'
    d. 20'

11. The *NEC* does not specify a burial depth for an underground metal water pipe electrode.
    a. True   b. False

12. Aluminum piping is permitted to be used as a grounding electrode only if other electrodes specified in 250.52(A) are not available for use.
    a. True   b. False

13. The *NEC* generally prohibits the use of an interior metal water piping system that extends more than __?__ from the point of entrance into the building to interconnect grounding electrodes and the grounding electrode conductor.
    a. 2'
    b. 5'
    c. 10'
    d. 20'

14. A metal in-ground support structure for a building or structure is acceptable as a grounding electrode if it is in direct contact with the Earth for 10 feet or more, with or without concrete encasement.
    a. True   b. False

15. Structural metal building frames can serve as grounding electrodes even if the anchor bolts securing the structural steel columns are not connected to a concrete-encased electrode that complies with 250.52(A)(3) located in the support footing or foundation.
    a. True   b. False

REVIEW QUESTIONS

The *NEC* describes the concrete-encased electrode as one or more bare or zinc-galvanized or other electrically-conductive coated rebar of not less than __?__ in diameter, installed in one continuous __?__ length or, if in multiple pieces, connected together by the usual steel tie wires, exothermic welding, welding, or other effective means.

b. $3/8$" / 10'
c. $1/2$" / 20'
d. $5/8$" / 30'
e. $1/2$" / 50'

A concrete-encased electrode can be created using 20 feet or more of bare copper conductor not smaller than __?__ or using the normal 20 feet of $1/2$-inch rebar installed in accordance with 250.52(A)(3)(1).

f. 4 AWG
g. 3 AWG
h. 2 AWG
i. 1 AWG

16. A concrete-encased grounding electrode can be in a horizontal or vertical orientation as long as there is a minimum __?__ of contact with the Earth.

a. 5'
b. 10'
c. 20'
d. 30'

17. Who was responsible for the development of the concrete-encased grounding electrode?

a. Eustace Soares
b. George Simon Ohm
c. Herbert G. Ufer
d. Thomas Edison

18. Ground ring electrode(s) have to encircle the building or structure completely and cannot be smaller than __?__ .

a. 4 AWG copper
b. 3 AWG copper
c. 2 AWG copper
d. 1 AWG copper

19. What is the minimum burial depth required for a ground ring electrode installed around a building?

a. 1'
b. 2'
c. 2.5'
d. 3'

20. When pipe or conduit is installed as a grounding electrode, it must not be smaller than which trade size?

a. $1/2$"
b. $3/4$"
c. 1"
d. 1 $1/2$"

21. A ground rod-type electrode must be driven to a depth of not less than __?__ .

a. 4'
b. 6'
c. 8'
d. 10'

22. A ground rod-type electrode can be buried in a $2 1/2$-foot-deep trench only if rock bottom is encountered, preventing it from being driven either vertically or at an angle not exceeding 45 degrees from vertical.

a. True   b. False

23. As a rule, when installing ground rod electrodes, two always must be installed unless a single ground rod provides not more than 25 ohms resistance to ground.

a. True   b. False

24. If rod, pipe, or plate electrodes are installed, each electrode must be spaced a minimum distance of __?__ from another electrode or another grounding electrode system.

a. 5'
b. 6'
c. 8'
d. 16'

25. The structures and structural rebar used for an in-ground swimming pool shall be permitted to be used as a grounding electrode for a building or structure only if other electrodes specified in 250.52(A) are not present for use.

a. True   b. False

Requirements for Grounded Conductors at Services

Premises wiring systems are typically supplied by an electric utility through conductors and equipment that make up an electrical service. Service equipment is required to be listed and suitable for use as service equipment and is typically comprised of equipment enclosures that contain switches, circuit breakers, fuses, and other accessories. The service equipment is where the service conductors supplying the building or structure are connected. The first point of grounding and bonding for a premises wiring system typically occurs at or within the service equipment.

Objectives

» Understand the roles of the grounded conductor(s) at the service equipment.

» Determine the required grounding electrode conductor connection location(s) at the service equipment and outside the building or structure served.

» Understand the installation and minimum sizing requirements for the grounded conductor at services supplied by grounded systems.

» Understand the physical characteristics and minimum sizing requirements for main bonding jumpers in service equipment.

» Understand the purpose and location of the grounded conductor disconnecting means (neutral disconnecting link) in service equipment enclosures.

» Understand grounding requirements for equipment installed on the supply side of the service disconnect and not supplied by a utility source.

Chapter 5

Table of Contents

Grounded Utility Supply Systems 100

First Line of Defense 101

Grounding Scheme for Services 102

Grounded Conductor Routing and
Connections 104

Main Bonding Jumpers in
Service Equipment 105

Dual-Fed Service Equipment 107

Minimizing Impedance in Service
Grounded Conductors 107

Functions (Purposes) of the Grounded
Service Conductor 109

Grounded Neutral Conductor 110

    Grounded Conductor Sizing
    Requirements 111

    Grounded Conductor Sizing for
    Parallel Installations 112

    Grounded Conductor (Load-Side Use) 113

    Grounded Conductor Identification 115

Requirements for Service Equipment
(Listing) 116

Grounded Conductor (Neutral) Disconnect
Requirement for Services 119

Requirements for Services Supplied by
Ungrounded Systems 120

Marking Equipment for Ungrounded
Systems 121

Grounding of Service Raceways
and Enclosures 122

Supply-Side Grounding at Other
Than a Service 122

Summary 123

Review Questions 124

GROUNDED UTILITY SUPPLY SYSTEMS

The service equipment is installed by Electrical Workers fairly early during construction. The terms *service, service conductors, service equipment,* and *grounded conductor* are all defined in Article 100 of the *NEC*. Referring to these definitions is essential in understanding this electrical equipment and the important grounding connection points associated with electrical services supplying the premises. **See Figure 5-1.**

Service. The conductors and equipment connecting the serving utility to the wiring system of the premises served. (CMP-10)

Service Conductors. The conductors from the service point to the service disconnecting means. (CMP-10)

Service Equipment. The necessary equipment, consisting of a circuit breaker(s) or switch(es) and fuse(s) and their accessories, connected to the serving utility and intended to constitute the main control and disconnect of the serving utility. (CMP-10)

Grounded Conductor. A system or circuit conductor that is intentionally grounded. (CMP-5)

Premises wiring systems are generally supplied by utility systems (sources) that are grounded. The source is usually a transformer. Where the service is supplied from a grounded system, the service conductors routed to the premises wiring equipment and systems must include a conductor that is grounded. The grounded conductor must be electrically connected to the service equipment, as opposed to being connected through induction or magnetic coupling. The term *electrically connected* implies that there is solid connectivity and current-carrying capacity in electrically connected systems (Sections 200.2 and 200.3). **See Figure 5-2.**

Utilities have their own rules and regulations for grounding electrical systems that supply customer-owned premises wiring. The most commonly-used method of grounding for these systems is to solidly ground them, meaning no intentional impedance or resistance is used in the grounding circuit. System grounding methods and requirements for wiring systems on the load side of the service point are not the responsibility of the serving utility. Most utilities will provide service using only grounded systems. In a few areas in the country, utilities may still supply services using an ungrounded system, but this is becoming uncommon. It is always best to verify with the serving utility what their grounding requirements are for services and service equipment. Many utilities publish electrical service requirements that should be used in conjunction with the rules in the

Figure 5-1. Service Equipment

Figure 5-1. Service equipment is typically the first point of grounding after the utility grounding connection outside the building or structure.

NEC when locating, installing, and connecting electrical service equipment.

FIRST LINE OF DEFENSE

A premises wiring system must generally be grounded, so the grounded system must include a grounded service conductor(s) in addition to the ungrounded conductors that are connected to the service.

The grounded system (source) is typically a pad-mounted transformer or one or more transformers mounted on a utility pole. This is the usual location for the first system grounding connection, either at the transformer pad or at the base of the pole supporting a transformer(s) for overhead services. The most common grounding electrode used to connect the system to Earth is a ground rod, but the electrode could be another type. In most areas, the serving utility is responsible for this system grounding connection. This grounding connection is typically outside the footprint of the building or structure served. **See Figure 5-3.**

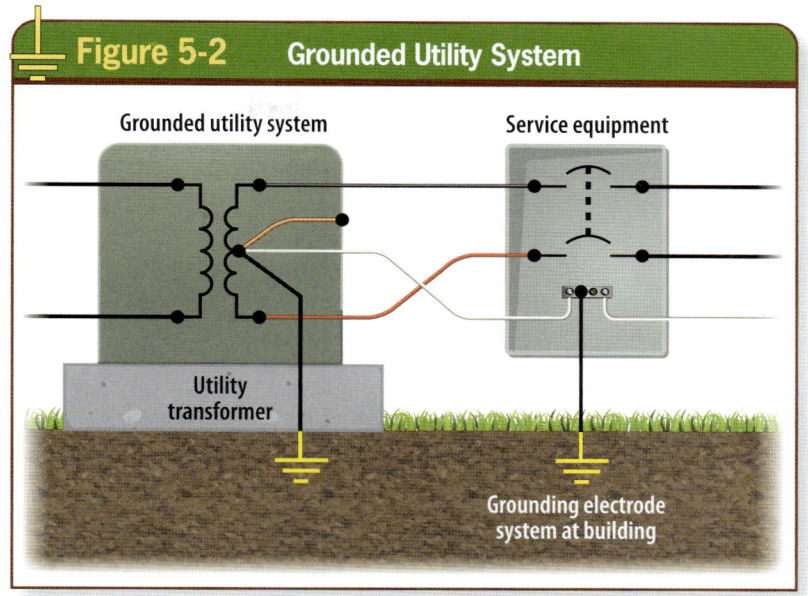

Figure 5-2   Grounded Utility System

Figure 5-2. *Utility service is typically delivered as a grounded system.*

This system grounding connection is a first line of defense or protection for premises wiring systems served by grounded utility sources. This grounding connection provides a level of

Figure 5-3   Pole-Mounted Utility Transformer

Figure 5-3. *A pole-mounted utility transformer can be a single transformer supplying a single-phase, 3-wire, 240/120-volt system, or a bank of transformers supplying a 3-phase, 4-wire, 208Y/120-volt system; both are typically grounded using an electrode (ground rod) installed at the base of the pole.*

Courtesy of IBEW Local 26 Training Center

protection from unintentional contact with higher-voltage lines, line surges, overvoltages from lightning events, and so forth. Although the *NEC* addresses this grounding connection requirement for outdoor transformers supplying services, this grounding is usually the responsibility of the utility company. Since this is an *NEC* requirement, it is a good practice to verify the outside grounding location and method used by the serving utility. The outside electrode is a requirement that must be met by either the utility company or the installer, depending on the specific utility requirements in the area served.

Safety is enhanced by establishing a path to ground (Earth) outside the building or structure in addition to the grounding electrode conductor connection installed for the service, which is usually inside or adjacent to the service equipment enclosure(s). When transformers supplying the service are located outside buildings, 250.24(A)(2) requires at least one additional grounding connection from the grounded system conductor to a grounding electrode. This connection can be made at the transformer or at another point outside the building. **See Figure 5-4.**

Figure 5-4   Pad-Mounted Utility Transformers

Figure 5-4. *A pad-mounted utility transformer is typically grounded using an electrode installed at the pad for the transformer (usually a ground rod).*

GROUNDING SCHEME FOR SERVICES

The grounding requirements on the load side of the service point are usually the responsibility of the electrical contractor and are accomplished when installing the service equipment. Installers must verify that the equipment installed for the service has markings indicating it is either "suitable for use as service equipment" or "suitable for use only as service equipment."

The *NEC* requires a grounding electrode conductor connection to be made at an accessible point anywhere from the load side of the overhead service conductors, the service drop, or the underground service conductors or service lateral, up to the terminal or bus for the grounded service conductor located in the service equipment enclosure or at the service disconnecting means. This long-standing requirement permits the grounding connection to be made at locations such as at the service weather head or in an enclosure used as service equipment. This is not a common location for this connection, but it is not prohibited by the *Code*.

The term *accessible* is defined in Article 100, and as used in this rule, it relates to the feasibility of access after installation. Locations such as at the service head, at meter socket enclosures, in a wireway, in an auxiliary gutter, or within a service disconnecting means enclosure all meet the accessibility requirement. **See Figure 5-5.**

Meter socket enclosures are usually sealed by the serving utility to restrict access to unmetered conductors. This does not make the inside of meter socket enclosures inaccessible according to the *NEC*. They can be accessed by legal means in coordination with the serving utility. Some areas have a local restriction prohibiting the grounding electrode conductor connection in a meter enclosure, but this is not a national requirement or restriction. As far as the *NEC* is concerned, a meter socket enclosure is an accessible

Figure 5-5 **Grounding Electrode Conductor Connection**

Figure 5-5. The grounding electrode conductor connection can be made at any accessible location.

location for making such a grounding electrode conductor connection.

Service equipment typically contains a terminal bar and lug that are intended for the grounding electrode conductor connection and are marked as a terminal for the grounding electrode conductor. The most common location for this connection is within listed service equipment enclosures. This location is the point of convergence for four conductors in the grounding scheme for grounded premises wiring systems. The four conductors that must be connected together within the service equipment are the grounded service conductor, the main bonding jumper, the grounding electrode conductor, and the equipment grounding conductor (EGC), which could be the equipment enclosure. A supply-side bonding jumper might also be present. **See Figure 5-6.**

The conductor used to accomplish the grounding is the grounding electrode conductor. This conductor connects the EGC, the service equipment enclosure, and the grounded conductor to ground (Earth) through a grounding electrode or grounding electrode system.

Where a grounding electrode conductor is connected to the grounded conductor within a service equipment enclosure, it is required to be connected to the grounded (neutral) conductor

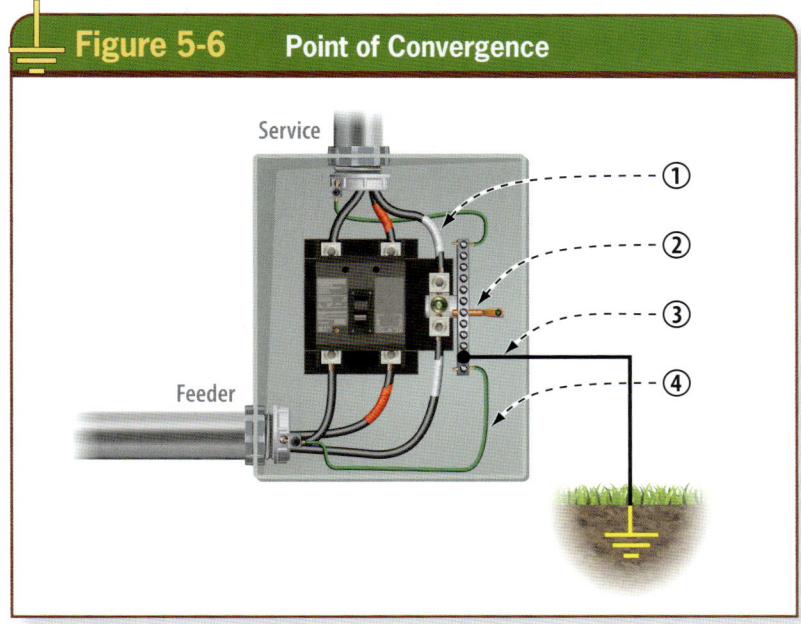

Figure 5-6 **Point of Convergence**

Figure 5-6. Four conductors are connected together in the service equipment. The connections are made at (1) the grounded conductor, (2) the main bonding jumper, (3) the grounding electrode conductor, and (4) the EGC.

terminal bus. If a main bonding jumper (wire or busbar type) is installed between the grounded conductor busbar and the equipment grounding terminal bus in service equipment, the

Figure 5-7   **Grounded Conductor Routing**

Service

The grounded conductor must run to each service disconnecting means enclosure and must be bonded to each service disconnecting means enclosure.

Feeder

Figure 5-7. A grounded conductor is required to be routed to the service disconnecting means enclosure and bonded to it.

Figure 5-8   **Main Bonding Jumper Requirement**

Figure 5-8. A main bonding jumper (MBJ) is required in each separate service disconnecting means.

grounding electrode conductor is permitted to be connected to the equipment grounding terminal bar. This type of arrangement is often provided in larger switchboards or equipment containing larger ground-fault protection of equipment (GFPE) features as required by 230.95.

GROUNDED CONDUCTOR ROUTING AND CONNECTIONS

The *NEC* includes a rule in 250.24(D) that requires the grounded conductor to be routed to the service equipment and be connected to the equipment enclosure. This requirement applies when the service conductors are installed in raceways or cable assemblies. This rule is one of paramount importance in the grounding and bonding scheme for service equipment. This requirement applies to services operating at 1,000 volts or less and those services over 1,000 volts. **See Figure 5-7.**

The means of connecting the grounded conductor to the service equipment enclosure is typically through a conductor identified in the *NEC* as a main bonding jumper. Whether the service is supplied using an underground service lateral or is supplied through a set of service conductors in a riser up to a weatherhead, the grounded conductor must be bonded to the equipment enclosure using a main bonding jumper or by direct connection to service equipment that is identified as "suitable for use only as service equipment."

The main bonding jumper requirement applies to each service disconnecting means on the premises served, whether the service disconnecting means is a single main or is a group of service disconnects as permitted in 230.71. **See Figure 5-8.**

Multiple service disconnects could be installed in a single assembly enclosure. In this case, the equipment must be listed as service equipment and must include a grounded conductor terminal bus and a single main bonding jumper to connect the grounded conductor to the equipment enclosure. This exception typically

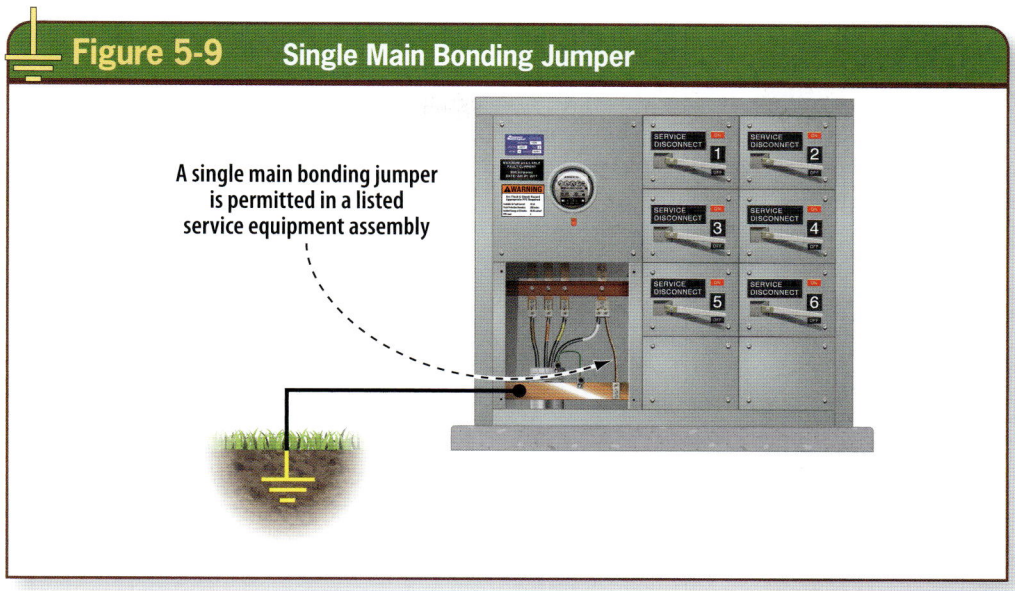

| Figure 5-9 | Single Main Bonding Jumper |

A single main bonding jumper is permitted in a listed service equipment assembly

Figure 5-9. Multi-section service switchboard assemblies are permitted to be equipped with a single main bonding jumper.

applies to multi-section service switchboards or service switchgear that include two to six switches or circuit breakers as the service disconnecting means. In this case, the *NEC* relaxes the requirement for a service grounded conductor at each of the service disconnects; however, it requires the service grounded conductor to be connected to a listed service equipment assembly enclosure, rather than to each of the disconnects contained in the assembly. **See Figure 5-9.**

A similar exception applies to the main bonding jumper when multiple service disconnects are included in a listed service equipment assembly. In this case, a main bonding jumper is not required in each disconnect because a single main bonding jumper is provided in the listed assembly.

MAIN BONDING JUMPERS IN SERVICE EQUIPMENT

The term *main bonding jumper* is defined in Article 100 as the connection between the grounded conductor and the EGC, the supply-side bonding jumper, or both at the service. By definition, this connection is made only at the service. **See Figure 5-10.**

Section 250.28 provides information related to the physical characteristics, connections, and sizing of main bonding jumpers. Main bonding jumpers must be copper, aluminum, copper-clad aluminum, or another corrosion-resistant material, and can be in

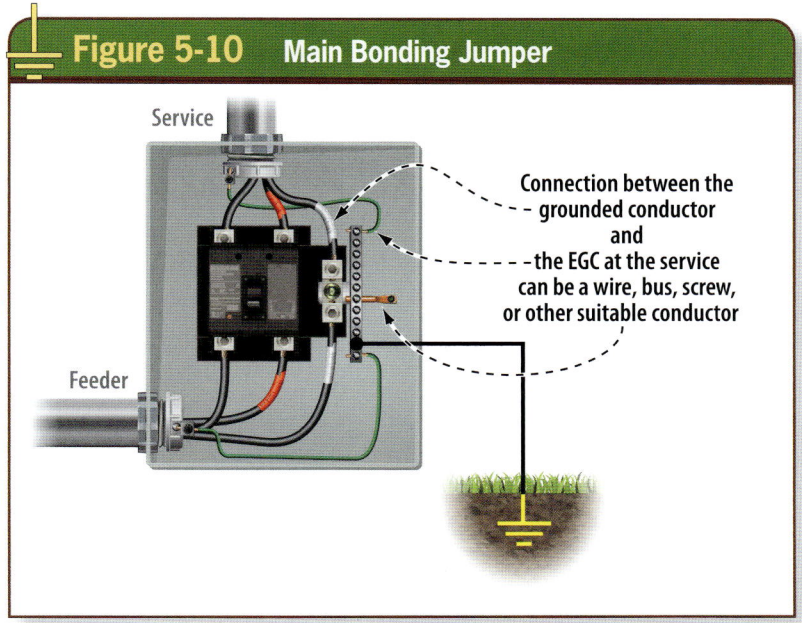

| Figure 5-10 | Main Bonding Jumper |

Service

Feeder

Connection between the grounded conductor and the EGC at the service can be a wire, bus, screw, or other suitable conductor

Figure 5-10. The main bonding jumper connects the grounded conductor to the EGC at the service.

the form of a screw, bus, wire, or other suitable conductor. **See Figure 5-11.**

The main bonding jumper in smaller service equipment is often in the form of a screw that connects the grounded

Figure 5-11   **Main Bonding Jumper – Busbar**

Figure 5-11. The main bonding jumper can be in the form of a busbar in listed equipment.

Courtesy of IBEW Local 26 Training Center

Figure 5-12   **Main Bonding Jumper – Screw**

Figure 5-12. A main bonding jumper in the form of a screw is required to be identified using the color green.

conductor terminal bar to the service equipment enclosure. This is more common in service equipment in sizes 225 amperes and smaller. If the main bonding jumper is a screw, it must have a green finish that is visible after it is installed. **See Figure 5-12.**

The color green helps installers and inspectors readily identify the main bonding jumper among the many other terminal screws that may also be provided on the grounded conductor terminal busbar in the service equipment enclosure. It is common for a screw-type main bonding jumper to be provided in service equipment rated up to 225 amperes. The main bonding jumper in service equipment with larger ratings is usually in the form of a busbar or wire.

When a main bonding jumper is installed, connections must meet the applicable requirements in 250.8. If a main bonding jumper is a wire type, it would be sized using Table 250.102(C)(1); if the service conductor size exceeds 1,100-kcmil copper or 1,750-kcmil aluminum or copper-clad aluminum, it would be sized using 12.5%, as stated in the first note to that table. When a service includes multiple service disconnect enclosures, a main bonding jumper is required in each such enclosure. **See Figure 5-13.**

The size for wire-type main bonding jumpers in each enclosure must be in accordance with 250.28(D)(1), which bases size on the largest ungrounded service conductor serving that individual enclosure.

Use Table 250.102(C)(1) or 12.5% as appropriate for this sizing requirement. If the ungrounded service-entrance phase conductors and the wire-type main bonding jumper are made of different materials—for example, one is copper and one is aluminum—the minimum-size main bonding jumper is based on the assumed use of phase conductors of the same material that have an ampacity equal to that of the installed ungrounded service-entrance phase conductors.

Consider the following examples:
1. Service size: 400 amperes
   Aluminum service-entrance conductor size: 750 kcmil
   Aluminum main bonding jumper: 3/0 AWG; copper main bonding jumper: 1/0 AWG
2. Service size: 800 amperes
   Copper service-entrance conductor size: two 600-kcmil conductors or one 1,200-kcmil conductor
   Aluminum main bonding jumper: 250 kcmil; copper main bonding jumper: 3/0 AWG

Main bonding jumpers supplied with listed service equipment, such as switchboards and panelboards, can be installed without calculating the size. The manufacturer has built the equipment to meet or exceed the requirements in the applicable product safety standards, which includes grounding and bonding provisions. For instance, the UL standard for Panelboards is *UL 67* and the UL standard for service switchboards is *UL 891*.

DUAL-FED SERVICE EQUIPMENT

When service equipment in a single enclosure or a group of separate enclosures is fed from two sources (dual-fed or double-ended), the grounding electrode conductor connection is permitted to be made by a single grounding electrode conductor. Service equipment arranged in this fashion is typically equipped with a tie breaker. This single connection is permitted to be made at the tie point of the grounded conductor terminal bars supplied by separate power sources. **See Figure 5-14.**

MINIMIZING IMPEDANCE IN SERVICE GROUNDED CONDUCTORS

The grounded conductor for a service must be routed with its associated ungrounded conductors, whether installed using a raceway or cable wiring method. This is also a general requirement in the *NEC*, as covered in 300.3(B). The reason

Figure 5-13 **Main Bonding Jumper – Wire**

Figure 5-13. The main bonding jumper can be a wire type.

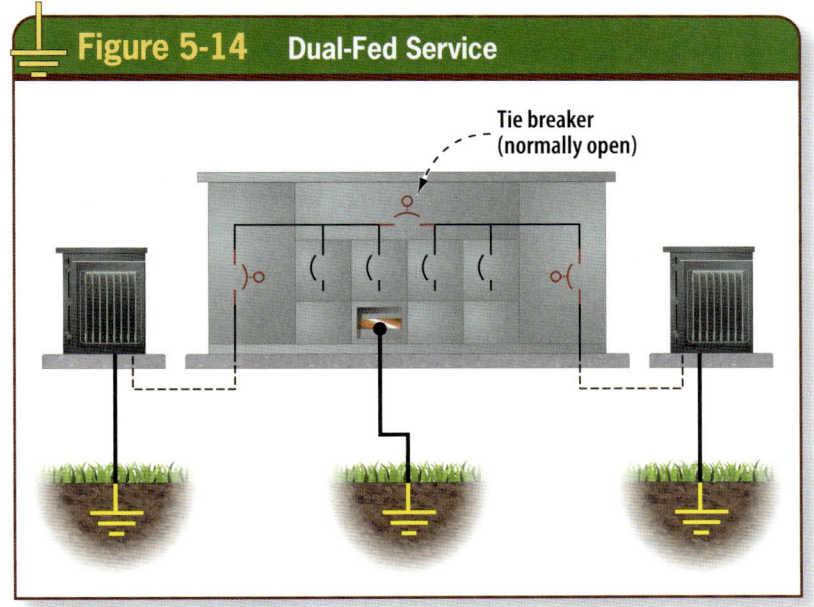

Figure 5-14 **Dual-Fed Service**

Figure 5-14. A single grounding electrode conductor connection is permitted at the tie point for the grounded conductor(s) of dual-fed or double-ended services.

Figure 5-15   **Routing with Ungrounded Service Conductors**

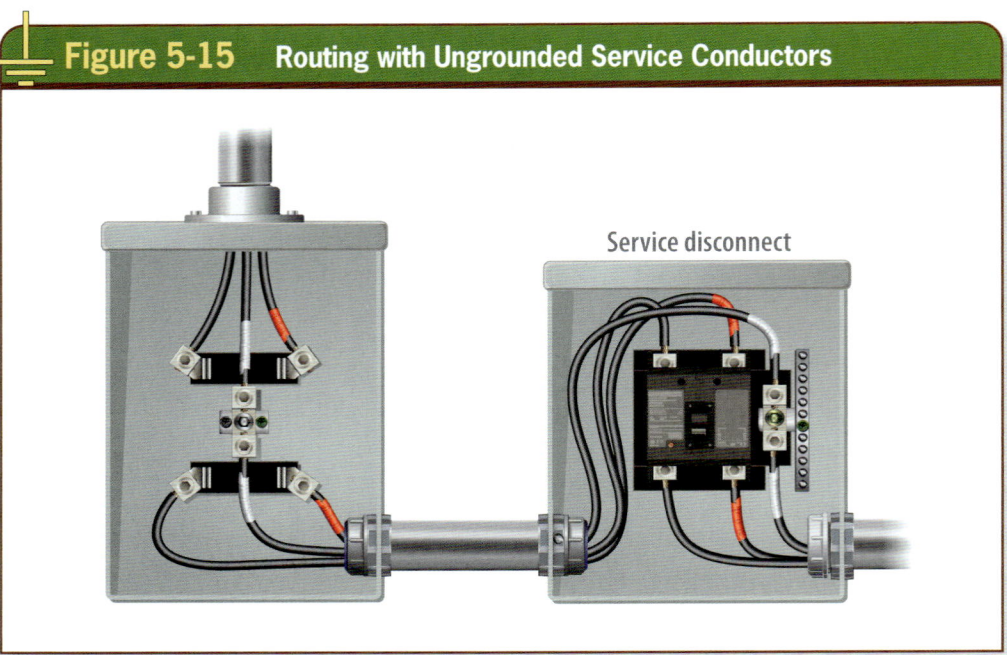

Figure 5-15. Route the grounded service conductor with its associated ungrounded service conductors.

Figure 5-16   **Routing with Service-Entrance Conductors**

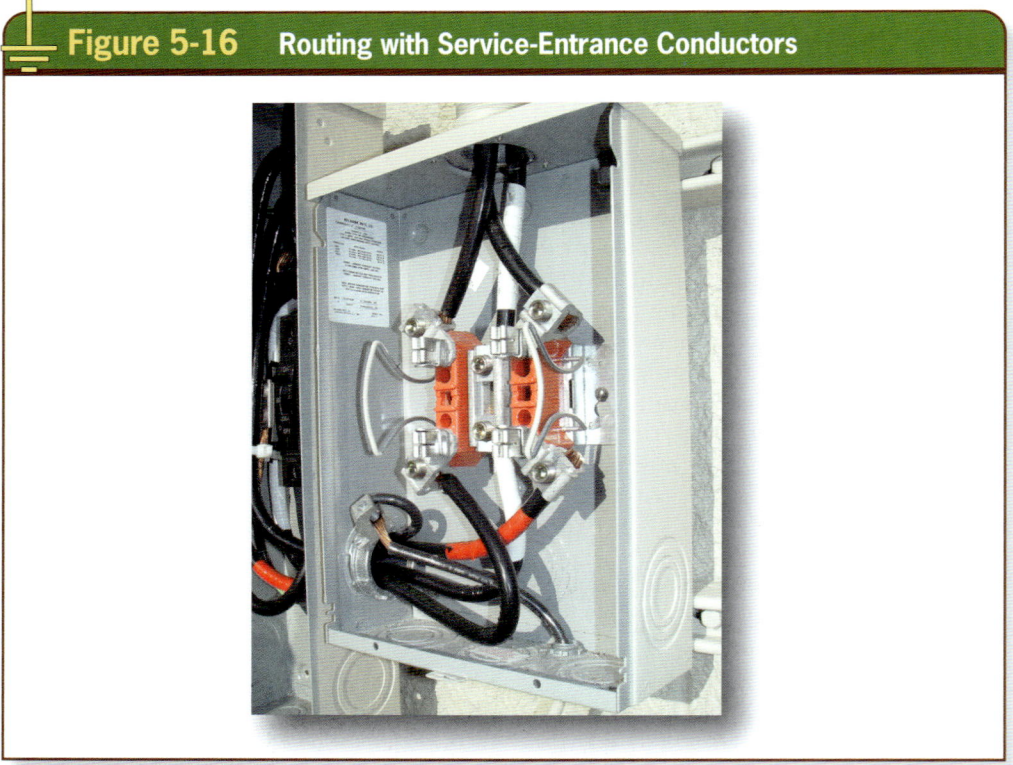

Figure 5-16. The grounded conductor is required to be routed with the service-entrance conductors.

Courtesy of Donald R. Cook, Shelby County, AL

for routing the grounded conductor with its associated ungrounded conductors is to keep the impedance of the circuit as low as possible during normal operation and during abnormal events such as ground faults or short circuits. **See Figure 5-15.**

Separation of the grounded service conductor from its associated ungrounded conductors increases the effects of inductive reactance, thus increasing circuit impedance. This goes against the philosophies of intentionally constructing an effective ground-fault current path as required in 250.4(A)(5). For this reason, the *NEC* requires the grounded service conductor to be routed with the ungrounded phase conductors. **See Figure 5-16.**

FUNCTIONS (PURPOSES) OF THE GROUNDED SERVICE CONDUCTOR

The grounded conductor(s) at the service provides two essential functions for the premises wiring system. The first function is to serve as a current-carrying conductor for the load supplied. The grounded conductor of a service is usually a neutral conductor, but it can also be a phase conductor depending on the type of system supplied. A corner-grounded delta system is an example of a system with a grounded phase conductor and no neutral conductor present. The grounded neutral conductors typically carry the maximum unbalanced neutral current to the system neutral point. During normal operation, the grounded (neutral) conductor is carrying current that returns to the source windings. The same load profile characteristics apply to grounded phase conductors, except that the grounded phase conductor typically carries the same current as the ungrounded phase conductors, as would be the case for a 3-phase motor supplied by a corner-grounded system. Section 250.24(D)(3) requires the grounded conductor of a 3-phase, 3-wire, delta-connected service to have an ampacity not less than that of the ungrounded conductors of this service.

The second essential function of the grounded conductor is to perform as an effective ground-fault current path during a ground-fault event at the service or at any point on the load side of the service equipment. **See Figure 5-17.**

The grounded conductor at the service is used for the intentionally constructed, low-impedance, effective ground-fault current path addressed in 250.4(A)(5). This is an essential element of effective overcurrent protective device operation for any ground fault that occurs on the load side. This is

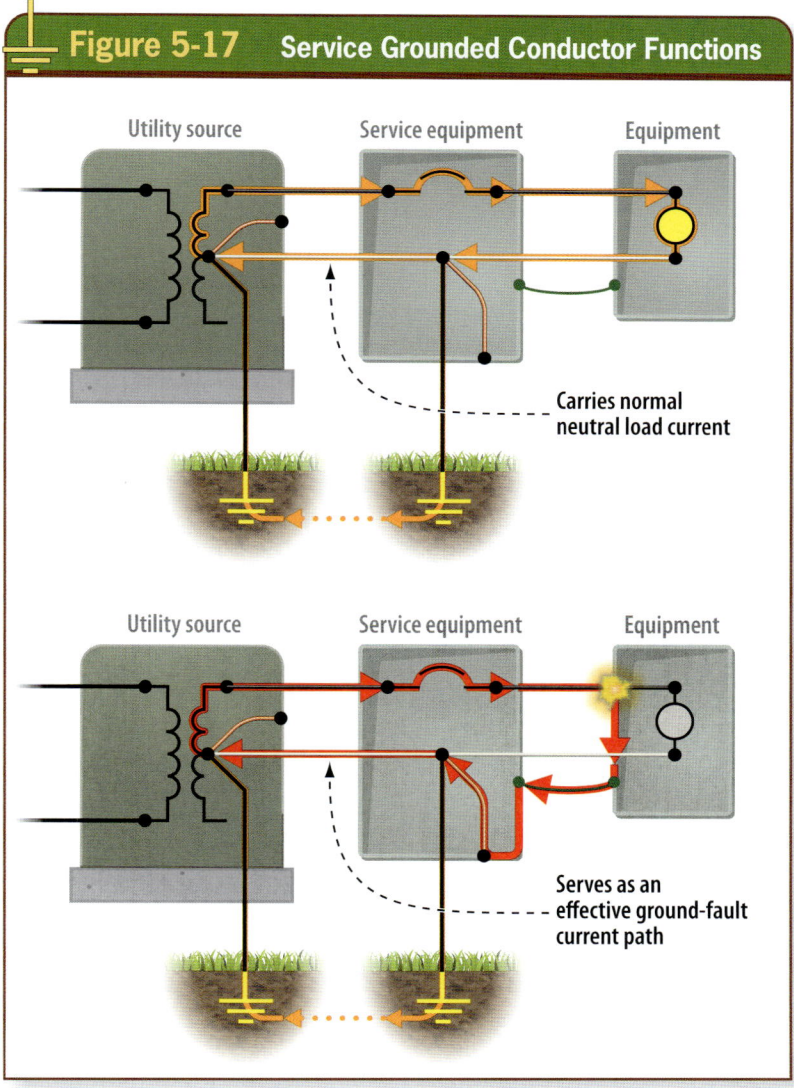

Figure 5-17. *Service grounded conductors at the service equipment enclosure typically perform two essential functions.*

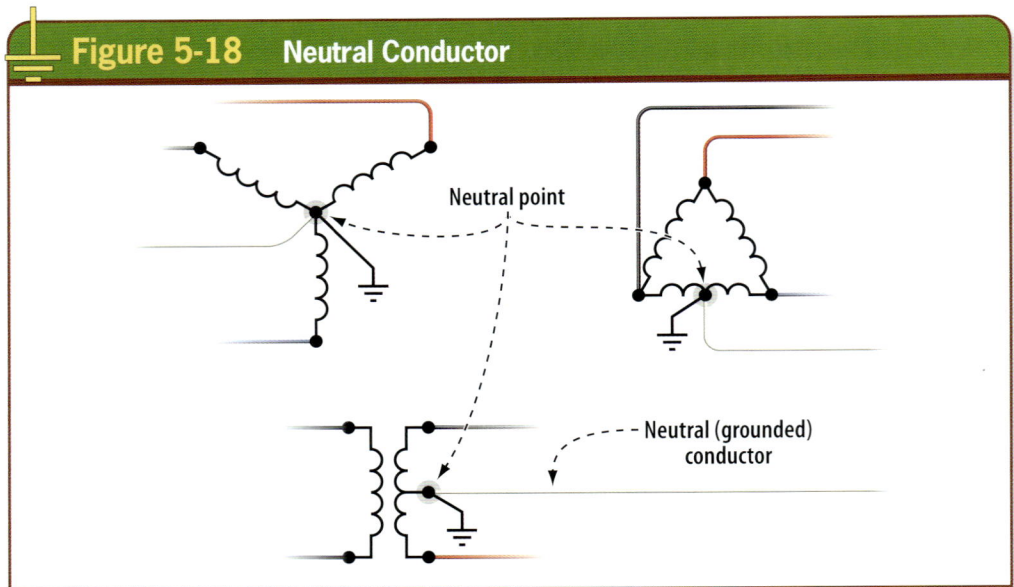

Figure 5-18   Neutral Conductor

Neutral point

Neutral (grounded) conductor

Figure 5-18. The neutral conductor is connected to the neutral point of the system.

why 250.24(D) requires that the grounded conductor be brought to a service disconnecting means enclosure and bonded to the enclosure. This requirement applies for grounded systems whether or not the grounded conductor supplies a load.

GROUNDED NEUTRAL CONDUCTOR

The term *neutral* can relate to either a conductor or a system connection point. The terms *neutral conductor* and *neutral point* are defined in Article 100. These definitions help distinguish between what constitutes system neutrals and the conductors to which they are connected. **See Figure 5-18.**

System neutral conductors are usually grounded, but not all grounded conductors are system neutrals. At the neutral point of the system, the vectorial sum of the nominal voltages from all other phases within the system that utilize the neutral, with respect to the neutral point, is zero potential. **See Figure 5-19.**

Figure 5-19   Connection to the Neutral Point

Figure 5-19. The neutral conductor is connected to the neutral point of a separately derived system (transformer). Grounding and bonding connections must comply with 450.10.

Courtesy of Bill McGovern, City of Plano, TX

Neutral Conductor. The conductor connected to the neutral point of a system that is intended to carry current under normal conditions. (CMP-5)

Neutral Point. The common point on a wye-connection in a polyphase system or midpoint on a single-phase, 3-wire system, or midpoint of a single-phase portion of a 3-phase delta system, or a midpoint of a 3-wire, direct-current system. (CMP-5)

Figure 5-20 Minimum Size Requirements

Size of Largest Service-Entrance Conductor (Copper)	Minimum Size of Grounded Conductor (Copper)	Minimum Size of Grounded Conductor (Aluminum)
1 AWG	6	4
4/0 AWG	2	1/0
500 kcmil	1/0	3/0
750 kcmil	2/0	4/0

Figure 5-20. The minimum sizes required for service grounded conductors are established using Table 250.102(C)(1).

Grounded Conductor Sizing Requirements

The grounded conductor(s) installed in a service could be in a raceway or cable wiring method. The same sizing requirements in 250.24(D) apply to grounded conductors installed using either of these (or any other) wiring methods. The grounded service conductor(s) is generally a current-carrying conductor during normal conditions, but during abnormal conditions, it must be capable of carrying fault current. Therefore, it must meet minimum sizing requirements to ensure adequate capacity to serve both

functions. The neutral conductor for services or feeders must have adequate capacity for the load served, as indicated in 220.61. Table 250.102(C)(1) is also used to determine the minimum size required for service grounded conductors for fault current. This rule applies whether there is a load on the neutral or not. **See Figure 5-20.**

The 12.5% requirement is applicable for larger services that are served by service conductors exceeding 1,100-kcmil copper or 1,750-kcmil aluminum or copper-clad aluminum, as stated in Note 1 to Table 250.102(C)(1). **See Figure 5-21.**

Figure 5-21 The 12.5% Sizing Requirement

Size of Largest Service-Entrance Conductor (Copper)	Calculated Value (Copper)	Minimum Size of Grounded Conductor (Copper)
1200 kcmil	150 kcmil	3/0 copper (167,800 circular mils in Table 8)
1800 kcmil	225 kcmil	250 kcmil
2500 kcmil	312,500 cm	350 kcmil
3000 kcmil	375 kcmil	400 kcmil

Figure 5-21. The 12.5% value must be used for sizing the grounded conductor where the size of the largest ungrounded service conductor exceeds 1,100-kcmil copper or 1,750-kcmil aluminum or copper-clad aluminum.

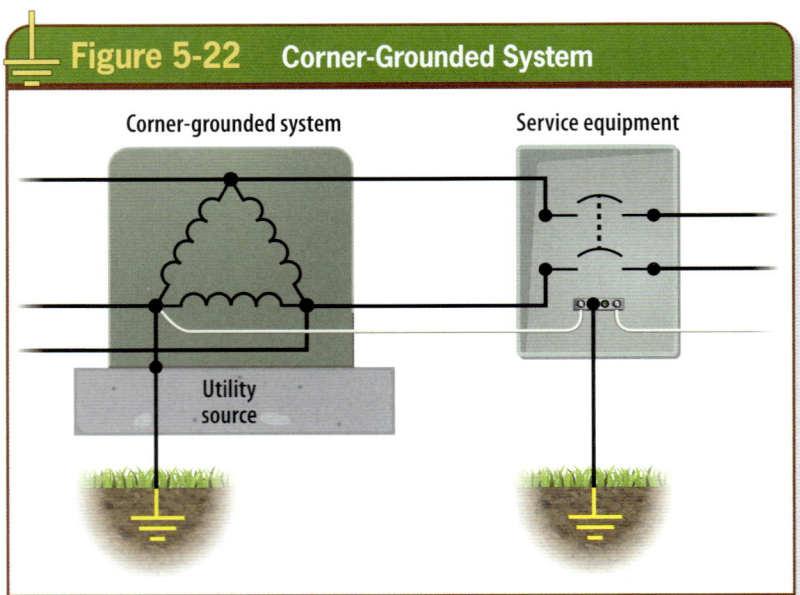

Figure 5-22   Corner-Grounded System

Corner-grounded system

Service equipment

Utility source

Figure 5-22. The grounded conductor of a corner-grounded system must be the same size as the ungrounded system conductors.

If a 3-phase, 3-wire, corner-grounded delta service is supplied, the grounded conductor must be sized the same as the ungrounded phase conductors. **See Figure 5-22.**

The conductor must be sized this way because the current on the grounded conductor will typically be the same as the current on the ungrounded phase conductors during normal operation,

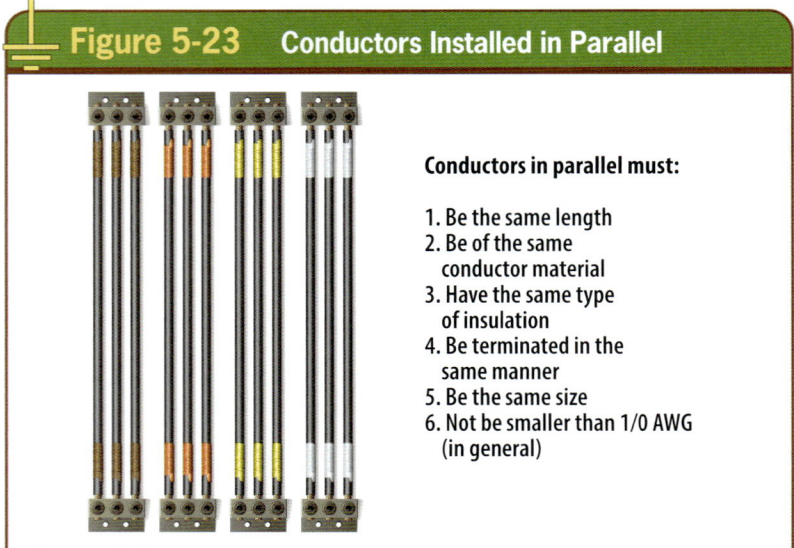

Figure 5-23   Conductors Installed in Parallel

Conductors in parallel must:

1. Be the same length
2. Be of the same conductor material
3. Have the same type of insulation
4. Be terminated in the same manner
5. Be the same size
6. Not be smaller than 1/0 AWG (in general)

Figure 5-23. There are specific rules for parallel conductor installations provided in 310.10(G).

compared with a grounded neutral conductor that may carry only the maximum unbalanced return current from associated phase conductors sharing that neutral conductor. This is usually the case for neutral current associated with a multiwire branch circuit or a feeder that includes a neutral conductor.

When 12.5% is applied, Table 8 in Chapter 9 of the *NEC* must be used. This table provides conductor properties and includes the information for converting sizes such as 2/0 or 4/0 AWG to circular mils (see the third column).

Once 12.5% is applied, the minimum size required is determined by referencing *NEC* Table 8 and rounding to the next higher size than the kcmil value determined by the calculation.

Grounded Conductor Sizing for Parallel Installations

The general requirements for conductors installed in parallel are provided in 310.10(G). **See Figure 5-23.**

Conductors in parallel generally must be the same size, same length, and same conductor material; must have the same insulation type; and must be terminated in the same manner. **See Figure 5-24.**

Parallel conductors are generally not permitted to be installed in sizes smaller than 1/0 AWG.

If a service is supplied by service conductors that are installed in a parallel arrangement, the rules in 250.24(D)(2) apply. The minimum size must be capable of carrying the anticipated normal neutral load current and be capable of carrying the maximum anticipated fault current during a ground-fault event. The minimum-size grounded service conductor for parallel arrangements is based on the total circular mil area of all the ungrounded service conductors in parallel. Once the total circular mil area is determined for the largest ungrounded service-entrance conductor, Table 250.102(C)(1) is then used to determine the minimum-size grounded service conductor required to be installed in each raceway. **See Figure 5-25.**

The 12.5% requirement applies to parallel service conductor arrangements that are larger than 1,100-kcmil copper or 1,750-kcmil aluminum or copper-clad aluminum and is used to determine the minimum-size grounded conductor. These sizing requirements anticipate that all ungrounded conductors and grounded conductors are installed in the same raceway, as would be the case for a wireway or auxiliary gutter installation. If service conductors are installed in parallel using multiple raceways or cables (which is fairly common) the minimum-size grounded conductor in each raceway or cable is based on the size of the ungrounded service conductor in each raceway or cable, but it must not be smaller than 1/0 AWG for parallel runs, in accordance with the requirements in 310.10(G)(1).

Grounded Conductor (Load-Side Use)

The grounded conductor carries current during normal operation and when functioning as an effective ground-fault current path during a ground-fault event. For this reason, the

Figure 5-24   **Parallel Conductor Installations**

Figure 5-24. Conductors are permitted to be installed in parallel arrangements according to 310.10(G).

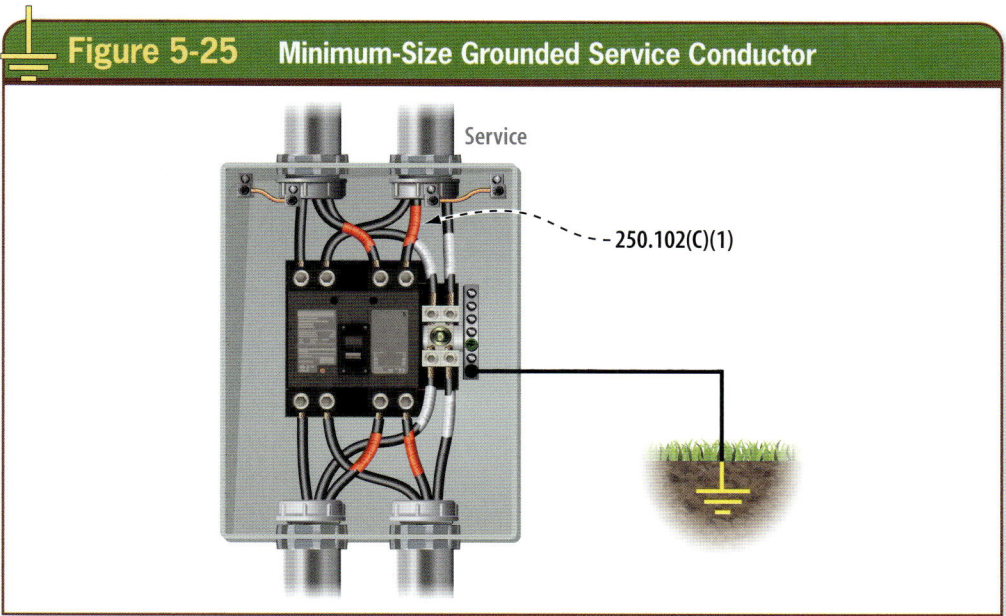

Figure 5-25   **Minimum-Size Grounded Service Conductor**

Service

250.102(C)(1)

Figure 5-25. Use Table 250.102(C)(1) to determine the appropriate size; use 12.5% for larger services. Grounded service conductors cannot be smaller than 1/0 AWG for parallel runs.

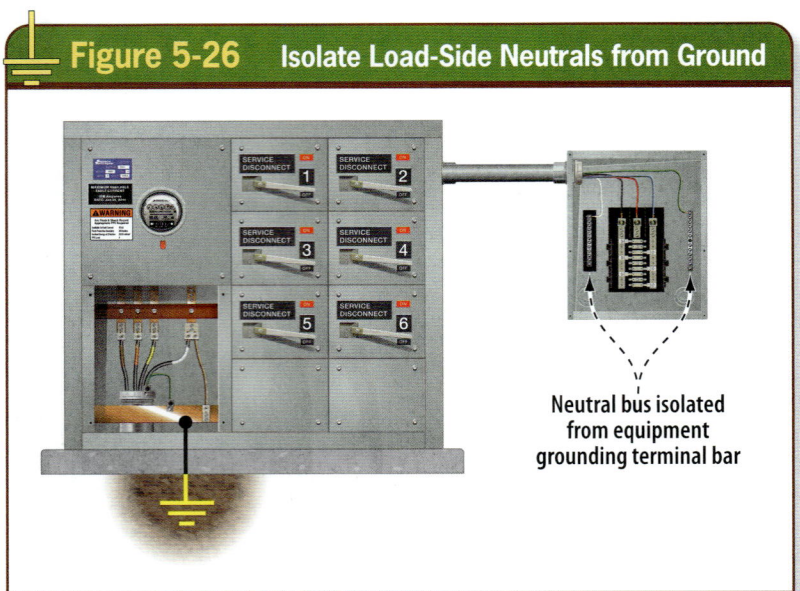

Figure 5-26 **Isolate Load-Side Neutrals from Ground**

Neutral bus isolated from equipment grounding terminal bar

Figure 5-26. Generally, no load-side grounding connections of the grounded conductor are permitted.

NEC generally restricts grounding electrode conductor connections to only those required or permitted on the line side of and up to the service disconnecting means enclosure. This restriction applies on the load side of the grounding point at the service equipment, or the point where the main bonding jumper connection is made in the service equipment. Section 250.24(B) restricts load-side

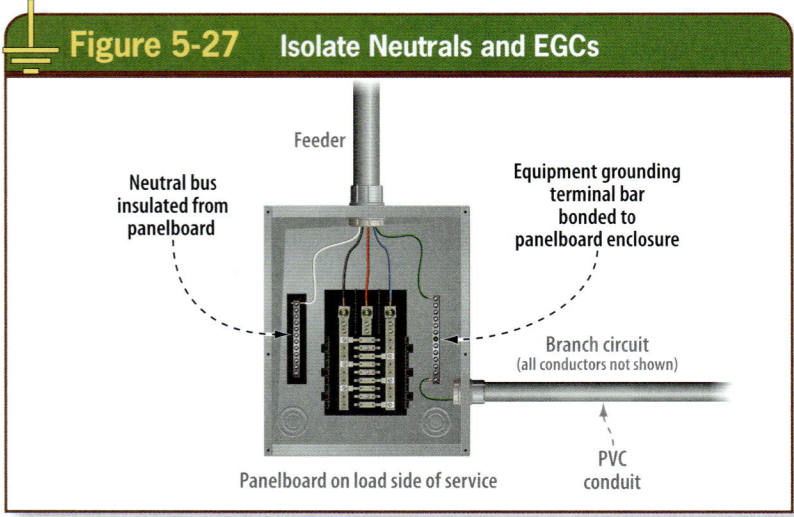

Figure 5-27 **Isolate Neutrals and EGCs**

Feeder

Neutral bus insulated from panelboard

Equipment grounding terminal bar bonded to panelboard enclosure

Branch circuit (all conductors not shown)

Panelboard on load side of service

PVC conduit

Figure 5-27. Grounded conductors (usually neutral conductors) generally must be isolated EGCs in load-side equipment to prevent multiple paths for normal neutral current.

grounding connections to the grounded conductor; essentially, connections on the load side of the service disconnecting means are not permitted. The same restriction is included for separately derived systems, as provided in 250.30(A).

The informational notes following 250.24(B) and 250.30(A) indicate a few other installations in which using the grounded conductor for grounding is permitted. **See Figure 5-26.**

Load-side grounding of the grounded conductor is generally prohibited to minimize the paths that current can divide between while returning to the source windings. Once conductors leave the service equipment enclosure in the form of feeders or branch circuits, installers must not connect the grounded (usually a neutral) conductor to ground or to grounded metal enclosures. Doing so creates multiple paths for current to return to the source. This is often referred to in the field as *parallel paths for neutral current*. The goal is to isolate neutrals from ground and grounded parts everywhere except where the electrical system is initially grounded. **See Figure 5-27.**

If this rule is not complied with, objectionable current will be introduced on conductive parts, raceways, and equipment that are not intended to carry current during normal operation. This condition is also one of the leading causes of many power quality problems experienced today.

There are very few allowances for such load-side grounding connections to the grounded conductor of services and premises wiring systems. These are permitted for separately derived systems because a new system grounding point is established at the source. There are also allowances for use of the grounded conductor to ground equipment, such as existing ranges and dryer installations, as provided in 250.140. Using the grounded conductor for grounding at separate buildings or structures was recognized in the *NEC* prior to the 2008 edition. However, the trend has been to migrate away from using the grounded conductor for equipment grounding purposes on the load side of the service disconnect or the load side of a grounding point for a separately derived system.

Electrical Workers should exercise care in the connections of grounded (neutral) conductors and respect the performance concepts discussed by keeping the neutrals and grounding connections separated in wiring installations on the load side of the service main bonding jumper or system bonding jumper for a separately derived system.

Grounded Conductor Identification

Certain *NEC* requirements apply specifically to grounded conductors, whether they are grounded neutral conductors or grounded phase conductors. One such requirement is identification. Grounded conductors must be identified according to 200.6. **See Figure 5-28.**

This rule generally indicates that grounded conductors in sizes 6 AWG and smaller must be identified using the colors white or gray. **See Figure 5-29.**

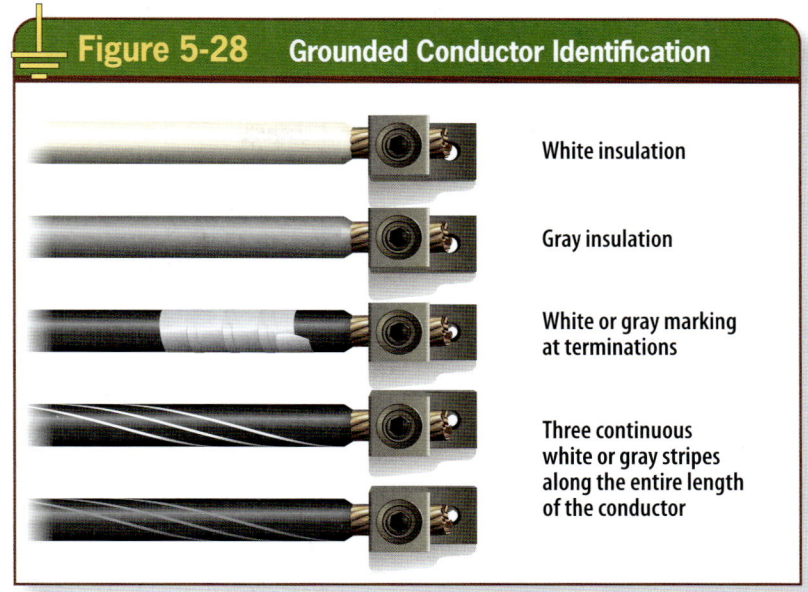

Figure 5-28   Grounded Conductor Identification

White insulation

Gray insulation

White or gray marking at terminations

Three continuous white or gray stripes along the entire length of the conductor

Figure 5-28. Generally, 200.6 and 200.7 require grounded conductors to be identified by one of several methods.

In some service installations, such as those using service entrance (SE) cable assemblies to supply the service equipment, the grounded conductor at a service could be a bare conductor.

Wire manufacturers have made it more convenient for electrical contractors to

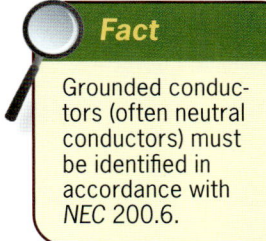

Fact

Grounded conductors (often neutral conductors) must be identified in accordance with *NEC* 200.6.

Figure 5-29   Grounded Conductors – Smaller than 6 AWG

Figure 5-29. Grounded conductors in sizes 6 AWG and smaller are identified using white or gray conductor insulation.

obtain conductors with insulations in a variety of colors. However, it is recognized that for larger conductor sizes, insulation has not been as readily available in white or gray. The *NEC* has, therefore, amended this general requirement by allowing sizes 4 AWG or larger to be identified using any of the following three methods:

1. A continuous white or gray outer finish
2. Three continuous white or gray stripes along the conductor's entire length on other than green insulation
3. A distinctive white or gray marking that encircles the conductor at terminations

In list item (3), the allowance is for applying identification in the form of a distinctive white or gray marking at the terminations. **See Figure 5-30.** Note that the *NEC* does not specify how far from the terminations the markings must be or the size of the marking that must be used. This requirement is usually satisfied when workers apply a sufficient amount of white or gray vinyl marking tape close to where the terminal is located.

It is important that this marking completely encircle the conductor. This provides reasonable assurances that the marking (tape) will remain on the grounded conductor after the installation is complete and will remain visible. Marking tape is a typical method of accomplishing identification of conductors 4 AWG or larger.

REQUIREMENTS FOR SERVICE EQUIPMENT (LISTING)

Service equipment must generally be identified for use as service equipment, especially when installed and used in the service position in the premises wiring system. These requirements are found in 230.66. Equipment that is suitable for use as service equipment has been manufactured and evaluated to meet product safety standards. Some equipment that could be identified as suitable for service use are switchboards, enclosed switches, panelboards, motor control centers, and power outlets, among others. The product standards apply to these types of equipment. **See Figure 5-31.**

Figure 5-30 Grounded Conductors – 4 AWG or larger

Figure 5-30. Identification of grounded conductors 4 AWG or larger is permitted at the termination points.

⏚ **Figure 5-31**	**Product Standards for Service Equipment**
Type of Service Equipment	**Applicable Product Safety Standard**
Switchboards	*UL 891* Deadfront Switchboards
Panelboards	*UL 67* Panelboards
Service power outlets	*UL 231* Power Outlets
Enclosed switches	*UL 98* Enclosed Switches
Motor control centers	*UL 845* Motor Control Centers

Figure 5-31. Some UL product safety standards include information related to products that are suitable for use as service equipment.

The designation of being "suitable for service equipment use" identifies the equipment for use in the service position of premises wiring installations. It does not limit the use of this equipment to only service use, as would be the case for service equipment marked "suitable for use only as service equipment." Equipment that is suitable for service equipment use has been evaluated for short-circuit current capabilities, and the equipment includes provisions for connection of a grounding electrode conductor, a grounded conductor disconnecting means (neutral link or terminal), and a main bonding jumper, to name a few.

Because service equipment includes grounding and bonding capabilities, the requirements in 250.24(D) are satisfied by installing the grounding and bonding connections using the means provided with the equipment. The main bonding jumper in listed service equipment is usually a bus, screw, strap, wire, or other conductive material that has been evaluated by a qualified electrical testing laboratory for size and performance. **See Figure 5-32.**

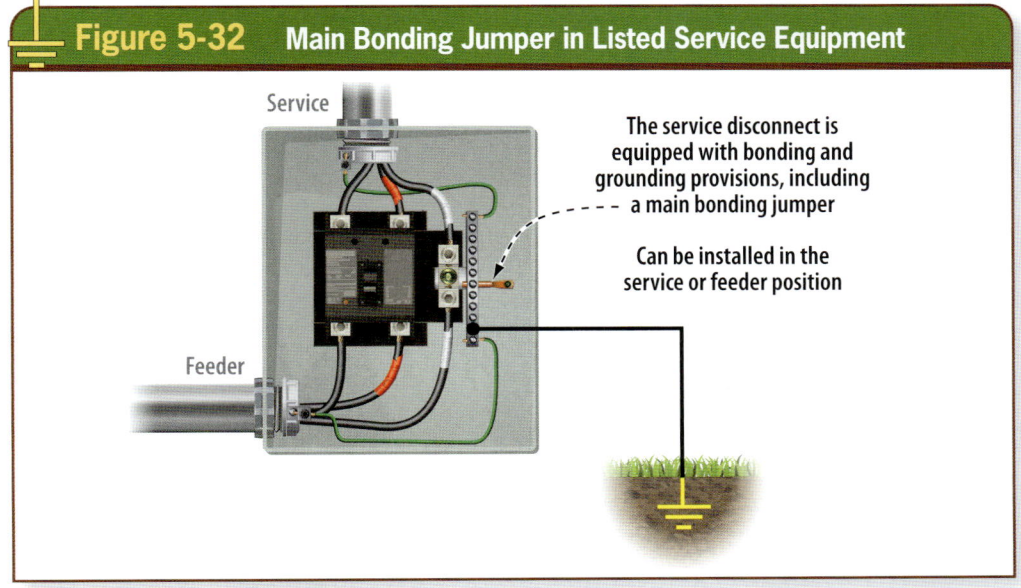

⏚ **Figure 5-32** **Main Bonding Jumper in Listed Service Equipment**

Service

The service disconnect is equipped with bonding and grounding provisions, including a main bonding jumper

Can be installed in the service or feeder position

Feeder

Figure 5-32. Equipment that is suitable for use as service equipment includes a main bonding jumper.

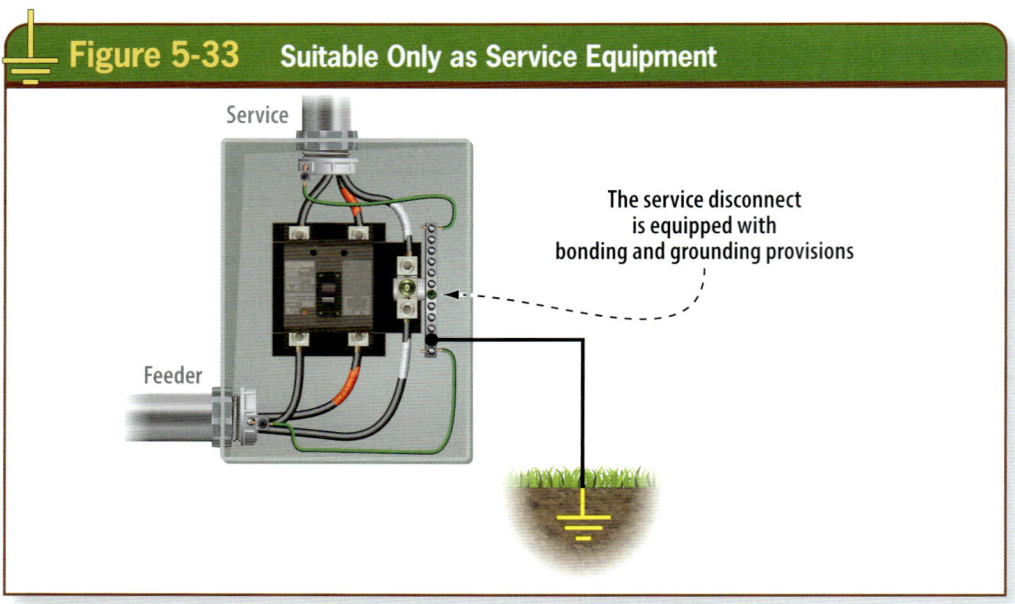

Figure 5-33   Suitable Only as Service Equipment

Service

The service disconnect is equipped with bonding and grounding provisions

Feeder

Figure 5-33. Equipment that is suitable for use as service equipment can be used in the service position or on the load side of the service disconnecting means.

This means the main bonding jumper in listed service equipment can be installed without calculating size. Usually, this has already been done by the manufacturer. **See Figure 5-33.**

If equipment is marked "suitable for use only as service equipment," it typically has a grounded conductor terminal busbar that is bolted or otherwise bonded to the enclosure, and no main bonding jumper is provided. **See Figure 5-34.** This type of equipment is suitable for use only in the service position because the grounded conductor cannot be isolated from the enclosure. **See Figure 5-35.**

If a service is constructed using a wireway or an auxiliary gutter arrangement, then the main bonding jumper of a wire type would be sized using Table 250.102(C)(1); alternatively, 12.5% must be used if the largest ungrounded service conductor size exceeds 1,100-kcmil copper or 1,750-kcmil aluminum or copper-clad aluminum (see the first note of that table).

Figure 5-34   Suitable for Use Only as Service Equipment

BOTTOM FEED ONLY

Figure 5-34. Equipment that is suitable for use only as service equipment does not include a main bonding jumper.

Figure 5-35   Suitable for Use Only in the Service Position

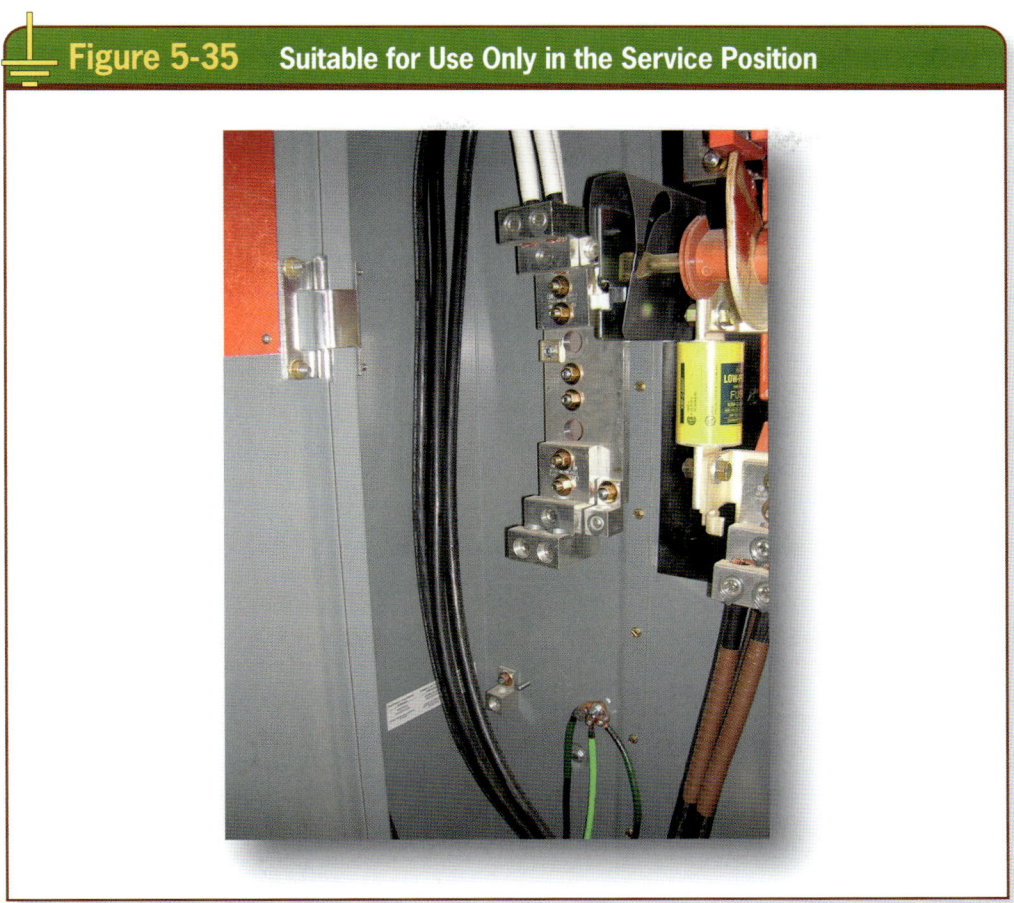

Figure 5-35. Equipment that is suitable for use only as service equipment does not include a main bonding jumper and should generally be used only in the service position.

GROUNDED CONDUCTOR (NEUTRAL) DISCONNECT REQUIREMENT FOR SERVICES

An important requirement for service equipment includes provisions for disconnecting the grounded conductor (usually a neutral) as required by 230.75. Listed service equipment includes such provisions either in the form of a busbar or terminal. **See Figure 5-36.**

In smaller service equipment such as a panelboard, the means to disconnect the grounded conductor is the terminal to which it is connected. In larger equipment such as a switchboard or motor control center, the means to disconnect a grounded conductor is a busbar that is identified as the "neutral disconnect link" in accordance with the

Figure 5-36   Neutral Disconnect Link

Figure 5-36. A neutral disconnect link is typically in the form of a busbar in larger service equipment.

applicable product safety standard. **See Figure 5-37.**

The product standard requires that the enclosure identify the vertical section in which the grounded conductor (neutral) disconnect link and main bonding jumper are located. **See Figure 5-38.**

The neutral disconnect link provides a means for disconnecting and isolating the neutral bus from the equipment

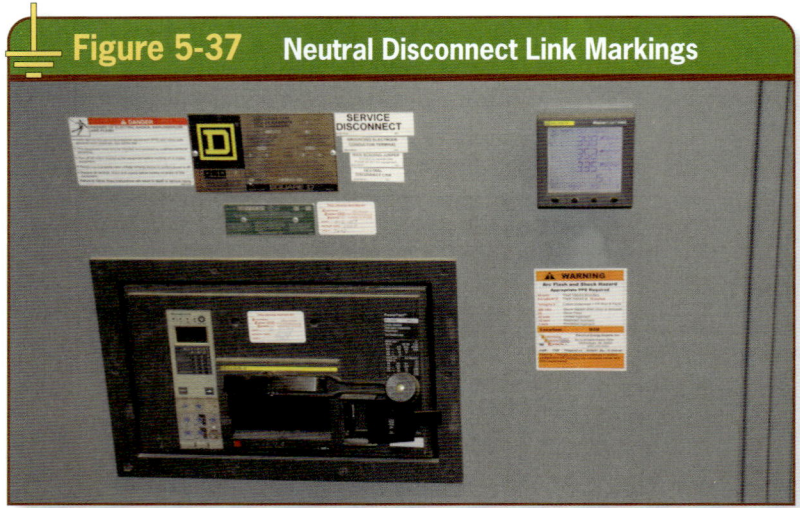

Figure 5-37   Neutral Disconnect Link Markings

Figure 5-37. Vertical sections of switchboards must be identified on the vertical section in which the main bonding jumper and neutral disconnect link are located.

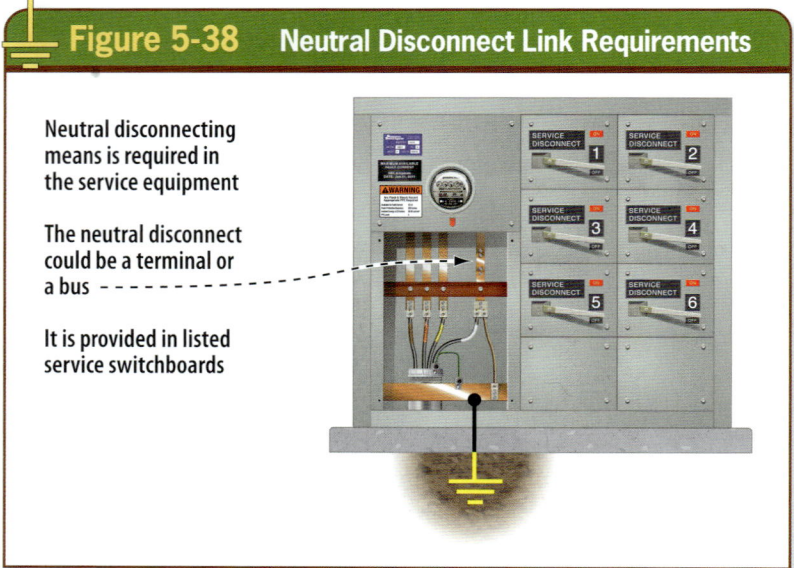

Figure 5-38   Neutral Disconnect Link Requirements

Neutral disconnecting means is required in the service equipment

The neutral disconnect could be a terminal or a bus

It is provided in listed service switchboards

Figure 5-38. Neutral disconnect links are often a removable portion of bus installed in the neutral (grounded conductor) busbar of the service equipment.

grounding terminal bus and the enclosure. It is important that a main bonding jumper be connected on the supply side of the neutral disconnect link so that if the link is removed for any reason, the grounded conductor remains grounded and connected to the service equipment enclosure.

In larger equipment that has ground-fault protection, the neutral disconnect link is an important part of the ground-fault protection of equipment (GFPE) performance testing required in various *NEC* rules that address ground-fault protection for equipment, such as those in Articles 210, 215, 230, 240, and 517.

REQUIREMENTS FOR SERVICES SUPPLIED BY UNGROUNDED SYSTEMS

Some utilities may serve a premises wiring system from an ungrounded source, but this practice is uncommon. Utilities understand the safety benefits of supplying services with systems that are grounded. However, the *NEC* still includes rules for services supplied by ungrounded utility sources in 250.24(F). In these installations, although the system is not grounded, grounding and bonding requirements still apply to metal raceways and enclosures that contain conductors and equipment used with ungrounded systems. If a service is supplied from an ungrounded utility source, the *NEC* requires that the metal enclosure have a grounding electrode conductor connection to a grounding electrode or grounding electrode system as provided in Part III of Article 250. **See Figure 5-39.**

Even though a system could be supplied ungrounded, grounding non–current-carrying metal equipment is still necessary for many of the same reasons grounding is required for enclosures containing conductors supplied from grounded systems. The grounding electrode conductor size is generally determined using Table 250.66, which bases sizing on the size of the largest ungrounded service-entrance conductor. The general performance requirements for un-grounded systems are covered in 250.4(B), which provides all grounding and bonding functions except for system grounding. No system conductor is connected to ground other than through magnetic coupling effects and distributed leakage capacitance.

The grounding electrode conductor installed for a service that is supplied by an ungrounded system must be connected to the metal service enclosure at any accessible point from the load end of the service drop or lateral to the service disconnecting means. The marking requirements in 230.66 are still applicable for these services, except that there will not be a grounded conductor supplied with the utility service conductors. This means that this equipment must be suitable for use as service equipment. The grounding electrode conductor must be installed according to applicable installation rules in 250.64.

Another important requirement for services supplied by ungrounded systems is ground detection. Ground detection equipment must be provided for ungrounded systems as required in 250.21(B).

This equipment can either be part of the service equipment or be field installed. The ground detection sensing devices must be installed as close as practicable to the service or source. This requirement ensures that ground detection monitoring remains active even if feeder or branch circuits are disconnected. Connection of ground detection systems at points downstream from the source increases the possibility of monitor loss resulting from the opening of feeder or branch circuits.

MARKING EQUIPMENT FOR UNGROUNDED SYSTEMS

Section 250.21(C) provides a general requirement that ungrounded systems be marked. **See Figure 5-40.**

Equipment containing ungrounded electrical systems is required to be

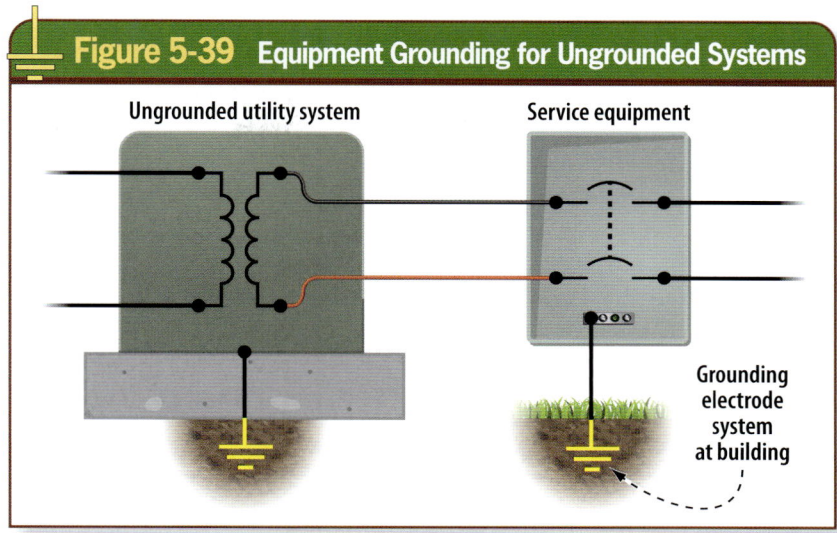

Figure 5-39   Equipment Grounding for Ungrounded Systems

Ungrounded utility system

Service equipment

Grounding electrode system at building

Figure 5-39. Equipment grounding is required for services supplied by an ungrounded system or supply source.

legibly and permanently field marked as follows:

> **Caution: Ungrounded System Operating _____ Volts Between Conductors**

The installer is responsible for adding this marking and inserting the correct operating voltage of the system.

As an example, if the system were an ungrounded 240-volt delta system, the

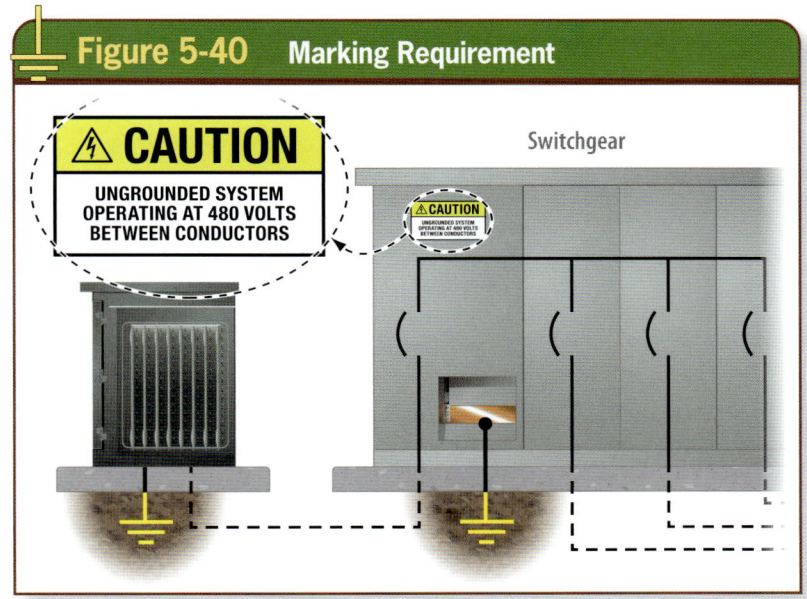

Figure 5-40   Marking Requirement

⚠ **CAUTION**

UNGROUNDED SYSTEM OPERATING AT 480 VOLTS BETWEEN CONDUCTORS

Switchgear

⚠CAUTION
UNGROUNDED SYSTEM
OPERATING AT 480 VOLTS
BETWEEN CONDUCTORS

Figure 5-40. The marking requirement applies to ungrounded system equipment enclosures, including switchboards and panelboards.

marking would read, "Caution: Ungrounded System Operating at 240 Volts Between Conductors." If it were an ungrounded 480-volt delta system, the marking would read, "Caution: Ungrounded System Operating at 480 Volts Between Conductors." This marking must be suitable (durable) for the environment in which it is installed.

This marking requirement is an improvement in safety for workers. Ungrounded electrical systems are less common than grounded systems, especially for premises wiring inside of buildings or structures. This cautionary marking on equipment identifies the unique voltage characteristics of ungrounded systems to raise awareness of the need to apply caution when testing for voltages and troubleshooting such systems.

GROUNDING OF SERVICE RACEWAYS AND ENCLOSURES

Section 250.80 provides requirements for grounding metal raceways and enclosures for service conductors and equipment. They must be connected to the grounded system conductor if the service is supplied by a grounded

> **TechTip!**
>
> Important Information: Ground detectors are required to be installed for ungrounded systems, and the detection-sensing devices must be installed as close as practicable to the service or source location.

electrical supply system; otherwise, they must be connected directly to a grounding electrode if the service is supplied by a system that is ungrounded.

The *NEC* relaxes the requirement (by exception) for grounding metal components such as metal elbows installed in a run of polyvinyl chloride (PVC) conduit or reinforced thermosetting resin conduit (RTRC) if the metal component or elbow is isolated from possible contact by a minimum cover of 18 inches. **See Figure 5-41.**

This is usually accomplished by a concrete encasement or by the minimum required burial depths of most utility company service laterals. When the sheath or armor of a continuous underground metal-sheathed service-entrance cable system is connected to the supply system grounded conductor, the metal sheath or armor is not required to be connected to the grounded conductor at the service equipment enclosure installed at the building or structure. The same requirements apply to underground service raceways that contain a metal-sheathed or armored cable that is connected to the grounded system conductor on the supply end. In this case, the metal sheath or armor is not required to be connected to the grounded conductor at the building or structure service equipment. The sheath or armor is already grounded at one end and is permitted to be insulated from the interior metal piping or raceway systems (see 250.84).

SUPPLY-SIDE GROUNDING AT OTHER THAN A SERVICE

Section 250.25, titled "Grounding Systems Permitted to Be Connected on the Supply Side of the Disconnect" and located in Part II of Article 250, provides the grounding rules for grounded and ungrounded systems connected to the supply side of the service disconnect as permitted in 230.82. This section addresses systems supplied by grounded and ungrounded systems that are not always supplied from utility sources.

Figure 5-41   Exception – Metal Elbows

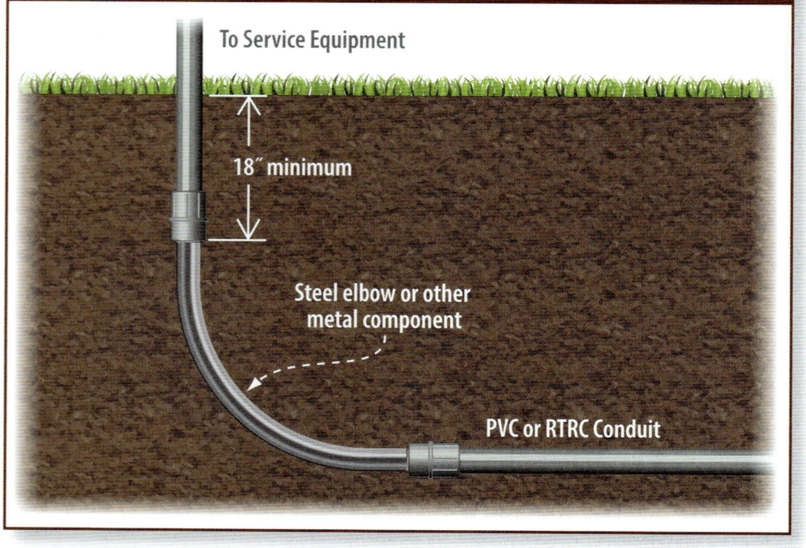

To Service Equipment

18" minimum

Steel elbow or other metal component

PVC or RTRC Conduit

Figure 5-41. Grounding is not required for metal components such as metal elbows buried not less than 18 inches from any part of the metal component and isolated from possible contact by persons.

Not all equipment connected to the utility is considered a service, such as parallel power production equipment, but it should be grounded and bonded according to the same requirements. In some instances, such as solar installations, the power production equipment is connected in parallel with a service, and in some cases is connected directly to the utility solely for power production to the utility with no service equipment in parallel.

The grounding of systems connected on the supply side of the service disconnect in enclosures separate from the service equipment enclosure, as permitted in 230.82, must comply with 250.25(A) or (B). The addition of 250.25 referencing the appropriate parts of Section 250.24 ensures that supply-side equipment is connected to a grounding electrode system, has

an effective ground-fault current path whether it is a grounded service conductor or supply-side bonding jumper, and is bonded using the more robust requirements of 250.92 and 250.102(C).

The existing structure and content of Article 250 already includes the rules necessary to apply to grounding and bonding installations that are from a source to a first system overcurrent protective device and that are not supplied by a utility. Section 250.25 refers to 250.24(A) if the supply system (source) is grounded and refers to 250.24(B) if the supply system (source) is ungrounded. There is no need to repeat requirements for these applications, as all necessary grounding and bonding requirements are already contained within the current content of Article 250.

SUMMARY

When a utility transformer is located outside a building, an additional grounding connection is required at the transformer or elsewhere outside the building. A grounded conductor is required to be brought to each service disconnecting means enclosure and connected to the enclosure. Equipment used as service equipment must be listed, and is generally required to be suitable for use as service equipment. The grounded conductor minimum size is determined based on the load served using 220.61; at a minimum it is sized using Table 250.102(C)(1) or 12.5% per the first note to that table, as required for larger services. The grounded conductor is an essential part of the effective ground-fault current path back to the source for grounded systems, and must meet minimum size requirements whether installed in a raceway, cable, or any other wiring method.

Main bonding jumpers are another essential component in the effective ground-fault current path. Main bonding jumpers provided with listed service equipment can be used without calculating the size. Although services are generally supplied by grounded utility sources, when an ungrounded system is used, there is an equipment marking requirement. The equipment enclosures for ungrounded systems are required to be grounded using a grounding electrode conductor that connects to a grounding electrode in accordance with Part III of Article 250.

Not all electrical systems (sources) are always supplied by a utility. Some systems (sources) are generators, energy storage systems, photovoltaic systems, wind turbine systems, and so forth. The same grounding and bonding requirements that apply on the supply-side of the service disconnecting means also apply on the supply side of a non-service disconnecting means. Section 250.25 addresses these types of applications and provides the necessary clarification of rules that apply to these supply-side applications that are not at the service.

REVIEW QUESTIONS

1. The first point of grounding for a building or structure supplied by a utility service is usually within the service equipment enclosure(s).

   a. True   b. False

2. Premises wiring systems generally are required to be supplied by a system that is __?__ and includes a grounded conductor.

   a. grounded
   b. impedance grounded
   c. reactance grounded
   d. ungrounded

3. In addition to the grounding connection at the service, one additional grounding connection from the grounded system conductor to a grounding electrode must also be made at the transformer or at another point outside the building.

   a. True   b. False

4. The equipment installed on the load side of the service disconnecting means enclosure, such as switchboards and panelboards, have to be identified as either suitable for use as service equipment or suitable for use only as service equipment.

   a. True   b. False

5. When a service is supplied by a grounded system (source), the __?__ is required to be routed from the utility supply system or source to the service disconnecting means enclosure and bonded to the service disconnecting means enclosure.

   a. equipment bonding jumper
   b. equipment grounding conductor
   c. grounded conductor
   d. grounding electrode

6. The grounding electrode conductor connection must be made at an accessible point located anywhere from the load side of the service drop or lateral up to the service equipment enclosure. Which of the following locations is not acceptable for this grounding electrode conductor connection?

   a. Connected at an auxiliary gutter
   b. Connected in meter socket enclosure
   c. Connected in the service disconnecting means enclosure
   d. Directly buried in the Earth

7. If equipment is identified as suitable for use only as service equipment, it typically includes a terminal or bus for the grounded conductor that is fastened directly to the enclosure; there typically is no main bonding jumper.

   a. True   b. False

8. The *NEC* requires the grounded conductor to be routed with the ungrounded service conductors to the service disconnecting means enclosure and bonded to that enclosure. For equipment that is identified as suitable for use as service equipment, this is typically accomplished using a(n) __?__ .

   a. bonding conductor
   b. equipment grounding conductor
   c. grounding conductor
   d. main bonding jumper

9. The connection between the grounded conductor and the equipment grounding conductor or the supply side bonding jumper (or both) at the service best defines which of the following components of the grounding and bonding system?

   a. Equipment bonding jumper
   b. Main bonding jumper
   c. Neutral disconnect link
   d. System bonding jumper

10. The main bonding jumper must be copper, aluminum, copper-clad aluminum, or another corrosion-resistant material and can be in which of the following forms?

    a. A bus
    b. A screw
    c. A wire or other suitable conductor
    d. Any of the above

11. A main bonding jumper in listed service equipment can be used without calculation of size because it is provided by the manufacturer and the equipment is listed and identified as suitable for use as service equipment.

    a. True   b. False

12. When a main bonding jumper is a wire, it must be sized using __?__ ; alternatively, 12.5% is used when the size of the largest ungrounded service-entrance conductor exceeds 1,100-kcmil copper or 1,750-kcmil aluminum or copper-clad aluminum.

    a. Table 8, Chapter 9
    b. Table 250.102(C)(1)
    c. Table 250.122
    d. Table 310.16

13. **If the largest ungrounded service-entrance conductor for a service is 750-kcmil copper, what is the minimum size required for a wire-type main bonding jumper for the service?**
    a. 3/0 AWG copper
    b. 2/0 AWG copper
    c. 1/0 AWG copper
    d. 1 AWG copper

14. **If the total circular mil area of the largest ungrounded service-entrance conductor for a service is 2,000-kcmil copper, what is the minimum size required for a wire-type main bonding jumper for the service?**
    a. 4/0 AWG copper
    b. 250-kcmil copper
    c. 350-kcmil copper
    d. 500-kcmil copper

15. **The minimum size grounded conductor for a service supplied by a grounded system shall not be smaller than the values in which of the following tables?**
    a. Table 250.2
    b. Table 250.66
    c. Table 250.102(C)(1)
    d. Table 250.122

16. **What is the minimum size required for the grounded conductor of a 3-phase, 3-wire, corner-grounded delta system that supplies a grounded service?**
    a. The grounded conductor cannot be smaller than the equipment grounding conductor.
    b. The grounded conductor cannot be smaller than the required grounding electrode conductor.
    c. The grounded conductor can be 12.5% of the largest ungrounded service conductor supplied by the system.
    d. The grounded conductor must have an ampacity not less than the ungrounded conductors supplied by the system.

17. **The ? is the common point on a wye connection in a polyphase system or midpoint on a single-phase, 3-wire system; the midpoint of a single-phase portion of a 3-phase delta system; or a midpoint of a 3-wire, direct-current system.**
    a. grounded conductor
    b. neutral conductor
    c. neutral point
    d. phase conductor

18. **The grounded conductor at the service carries normal line-to-neutral load current and functions as an intentionally constructed, low-impedance, effective ground-fault current path, as required in 250.4(A)(5).**
    a. True   b. False

19. **Parallel service conductors are generally not permitted in sizes smaller than ? .**
    a. 3/0 AWG
    b. 2 AWG
    c. 1/0 AWG
    d. 1 AWG

20. **Grounded service conductors in sizes 4 AWG and larger are required to be identified using which of the following methods?**
    a. A continuous white or gray outer finish
    b. A distinctive white or gray marking at terminations that encircles the conductor
    c. Three continuous white or gray stripes along the conductor's entire length on other than green insulation
    d. Any of the above

21. **An important requirement for service equipment is that provisions for disconnecting the grounded conductor (usually a neutral) are provided within the service equipment enclosure.**
    a. True   b. False

22. **The grounding electrode conductor for ungrounded services and grounded services is generally required to be sized based on the size of the largest ungrounded service-entrance conductor using ? .**
    a. Table 8 in Chapter 9
    b. Table 250.122
    c. Table 250.66
    d. The rating of the overcurrent protective device in the service disconnect

23. **A metal component such as a metal elbow in a run of nonmetallic service raceway is required to be grounded if it is buried not less than 18 inches below grade.**
    a. True   b. False

24. **If the largest ungrounded derived phase conductor for an AC generator is 750-kcmil copper, what is the minimum size required for a wire-type system bonding jumper?**
    a. 3/0 AWG copper
    b. 2/0 AWG copper
    c. 1/0 AWG copper
    d. 1 AWG copper

Grounding Electrode Conductors

Where services, systems, and equipment are installed on the premises and are grounded, a connection to ground (Earth) must be established. The grounding electrode conductor provides the intentionally-installed conductive connection to ground for grounded electrical systems and other equipment that must be grounded as required by the *NEC*. Grounding of equipment is also required for ungrounded systems. There are minimum sizing requirements and specific installation requirements that apply to grounding electrode conductor installation. These requirements relate to how this conductor must perform and how it must be protected from possible damage.

Objectives

» Understand the purpose of the grounding electrode conductor.

» Determine the grounding electrode conductor connection locations at a service.

» Recognize the materials permitted for a grounding electrode conductor.

» Understand requirements for grounding electrode conductor installations.

» Determine minimum sizes required for grounding electrode conductors.

» Understand the requirements for grounding electrode conductor connections.

» Understand the requirements for protecting grounding electrode conductors from physical damage and the effects of magnetic fields.

Chapter 6

Table of Contents

Grounding Electrode Conductor Basics 128

Purpose of Grounding
Electrode Conductors 128

Current in Grounding
Electrode Conductors 130

Grounding Electrode Conductor Material 130

Sizing Grounding Electrode Conductors 131

Using Table 250.66 133

Grounding Electrode Conductors
for DC Systems 134

Grounding Electrode Conductor Installation... 135

Securing and Protecting
from Physical Damage 136

Methods of Installing Grounding
Electrode Conductors 137

Connection Point for Grounding
Electrode Conductors 139

Methods of Connection at Service
Disconnecting Means 140

Single Disconnecting Means 140

Multiple Service Disconnecting
Means in Separate Enclosures 140

Individual Grounding Electrode
Conductors to Multiple
Service Disconnects 142

Effectiveness (Integrity) of the
Grounding Path 144

Grounding Electrode Conductor
Connection Locations 146

Grounding Electrode
Conductor Connections 148

Magnetic Field Concerns 150

The Choke Effect 150

Bonding Ferrous Metal Enclosures for
Grounding Electrode Conductors 151

Summary 153

Review Questions 154

Figure 6-1   GEC Connections to Electrodes

Figure 6-1. Grounding electrode conductors are connected to grounding electrodes. Eight feet of ground rod must be in contact with the earth, and protection of the connection is required.

GROUNDING ELECTRODE CONDUCTOR BASICS

The conductive path referred to in the *NEC* as a grounding electrode conductor has one end typically connected to a grounding electrode or another conductive object that completes a connection to the Earth. Grounding electrode conductors are required for services, separately derived systems, and limited-energy systems such as communications systems and equipment and information technology grounding networks. **See Figure 6-1.**

Grounding electrode conductors provide the connection between the Earth and the object or system conductor that is required to be connected to the ground. These conductors are conductive bodies that extend the ground connections indicated in the definition of *grounded (grounding)*.

Grounding Electrode Conductor (GEC). A conductor used to connect the system grounded conductor or the equipment to a grounding electrode or to a point on the grounding electrode system. (CMP-5)

The definition of *grounding electrode conductor* describes a conductor that performs grounding functions and how it relates to the systems or equipment with which it is used. Grounding electrode conductors are not limited for use with just services or separately derived power systems. The grounding functions required for low-voltage and limited-energy communications systems require a connection to the ground (Earth), which is accomplished by installing a grounding electrode conductor.

The sizing requirements and installation rules for grounding electrode conductors installed and used with communications systems are specified within Chapters 7 and 8 of the *NEC*. This provides relief from having to size these grounding electrode conductors according to Table 250.66, which for these systems would be excessive. The *Code* requirements for these systems are found in *NEC* Article 770 and Chapter 8.

PURPOSE OF GROUNDING ELECTRODE CONDUCTORS

An important step in grounding systems and equipment is the installation of the grounding electrode conductor(s). The key aspects of this installation are securing, sizing, and physical protection. This phase of construction of the grounding system is usually completed after the service equipment or separately derived system is installed and the service or feeder conductors are connected to the equipment. Grounding electrode conductors complete a conductive path to ground (Earth) from systems and equipment that are required to be grounded or are otherwise grounded by choice. This conductor serves to establish and maintain an equal potential between the Earth and the equipment or system conductor being grounded, or as close to the same potential as electrically possible. **See Figure 6-2.**

Grounding electrode conductors can carry varying amounts of current during normal operation (although ideally there would be no current). The

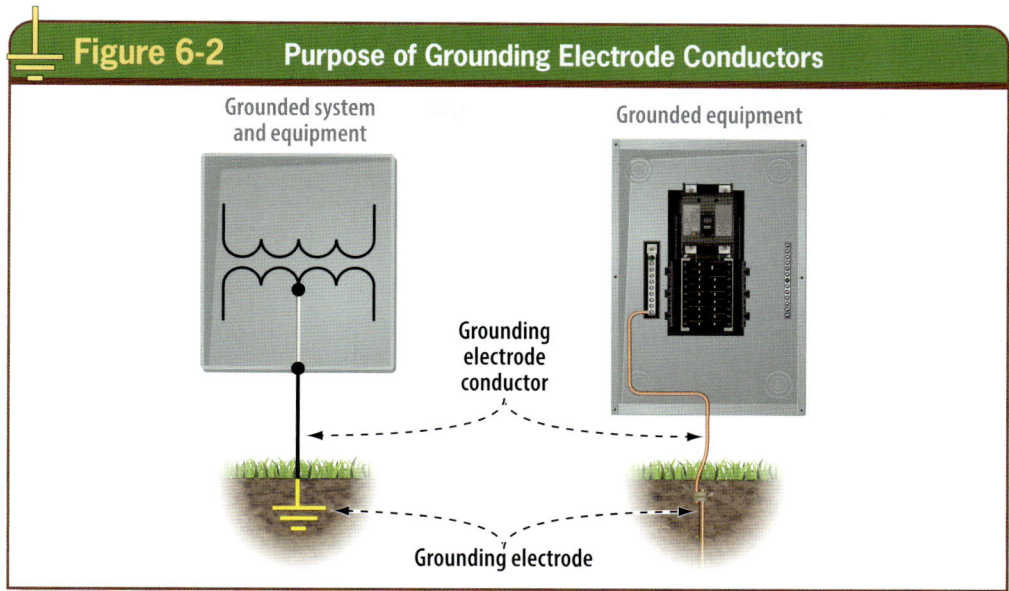

Figure 6-2 **Purpose of Grounding Electrode Conductors**

Grounded system and equipment

Grounded equipment

Grounding electrode conductor

Grounding electrode

Figure 6-2. Grounding electrode conductors connect equipment and systems to the Earth.

amount of current present in a grounding electrode conductor is usually low during normal operation. During a ground-fault event, the amount of current in the grounding electrode conductor can increase for a short period of time. This current will be present only for the time it takes an overcurrent protective device to open the faulted circuit. The amount of current in the grounding electrode conductor during a ground-fault event is typically low because of the high amount of impedance between the fault location and the grounding electrode at the source location. **See Figure 6-3.**

The Earth is a fault current path that offers substantial impedance for current. The purpose of the grounding electrode conductor is not to facilitate operation of the overcurrent protective device. Even though grounding electrode conductors are in the grounding and bonding circuit, the grounding electrode conductors are not intended as effective ground-fault current paths, as clarified in 250.4.

Grounding electrode conductors are connected to electrodes placing Earth in the return path for ground-fault current. The Earth cannot ever be depended upon to serve as an effective ground-fault current path. The grounding electrode conductors are intended to experience lower levels of ground-fault current for the duration of a ground-fault condition due to multiple paths over which the current will divide, including the

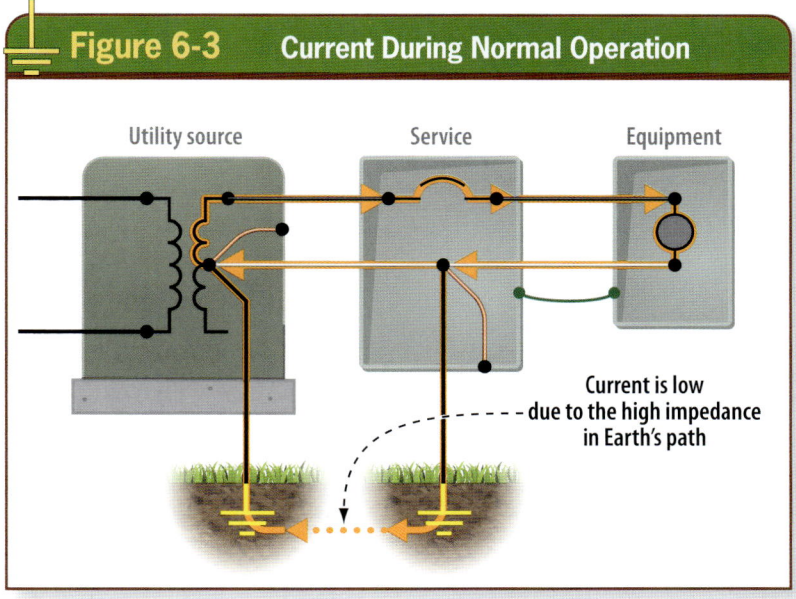

Figure 6-3 **Current During Normal Operation**

Utility source

Service

Equipment

Current is low due to the high impedance in Earth's path

Figure 6-3. There is usually a small amount of current present in grounding electrode conductors during normal system operation.

Earth. This accounts for the differences in sizing requirements for grounding electrode conductors as compared with equipment grounding conductors (EGCs). These two different conductors are used in the grounding and bonding scheme, and they each have different roles.

CURRENT IN GROUNDING ELECTRODE CONDUCTORS

The ground (Earth) is part of an electrical circuit or equipment that is grounded, and varying amounts of current will be present in a circuit as long as the circuit is complete. Current will also be present in all paths available between the source and point of use or ground-fault condition. The amount of current present in each path is directly related to the amount of impedance in each path. The Earth is a conductor, but it is a very large "semi-conductive" body, and the resistance (impedance) values of its conductivity vary greatly, both geographically and seasonally.

These are the main reasons that the amount of current present in grounding electrode conductors is typically low during normal operation and can rise and fall during ground-fault events

on the system. The amount of current a grounding electrode conductor handles is limited because of the high level of impedance in the path between the service or system grounding point and the point of grounding at the source. This is one of the reasons why a grounding electrode conductor connected to a single rod, pipe, or plate electrode never has to be larger than a 6 AWG copper conductor.

The amount of current the grounding electrode conductor must carry during normal operation and under abnormal conditions such as ground faults is kept low because of the high impedance in that path through the Earth. The components in this path are usually the grounding electrode conductor (at the service or system), the grounding electrode, the Earth (between the two grounding connections), the electrode (at the system or source grounding point), and the grounding electrode conductor (at the system or source grounding point). **See Figure 6-4.**

GROUNDING ELECTRODE CONDUCTOR MATERIAL

Grounding electrode conductors are required to be made of copper, copper-clad aluminum, or aluminum. Section 250.62 indicates that grounding electrode conductors must be stranded or solid, and they can be insulated, covered, or bare.

Copper-clad conductors are those manufactured with a thin copper coating over an aluminum conductor core. Copper-clad aluminum was used years ago in limited amounts in an effort to reduce electrical material costs during wartime periods, but it is less common in electrical installations today because copper and aluminum conductors are readily available for this use. The most commonly specified and installed grounding electrode conductors today are insulated or bare copper.

Copper-clad aluminum and aluminum conductors are more vulnerable to the effects of corrosion and deterioration from Earth contact or contact

Figure 6-4 **Current During a Ground-Fault Condition**

Utility source   Service equipment   Equipment

Current is low due to the high impedance in Earth's path

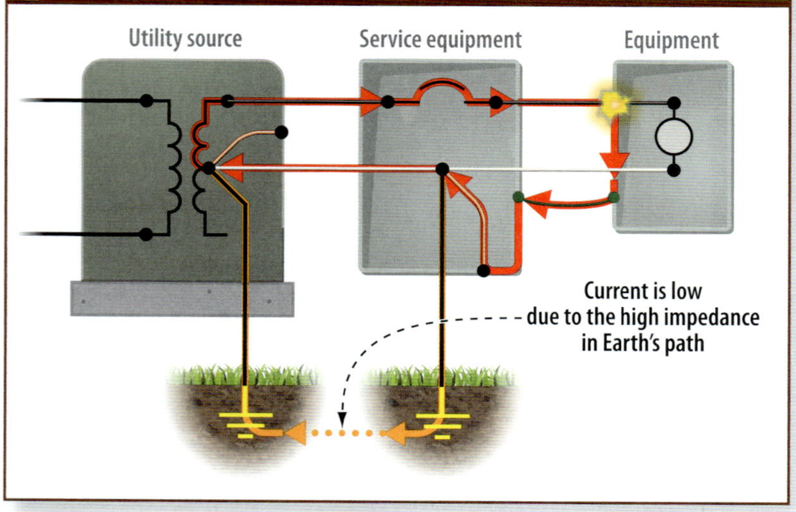

Figure 6-4. *Current is usually relatively low in grounding electrode conductors during a ground-fault condition.*

with masonry. The *NEC* includes installation and sizing requirements for aluminum and copper-clad aluminum grounding electrode conductors. One such restriction is the prohibition of terminating grounding electrode conductors within 18 inches of the Earth outside. Section 250.64(A)(2) relaxes the 18-inch restriction where an aluminum or copper-clad aluminum grounding electrode conductor is terminated within an enclosure that is suitable for the environment, such as outdoor-rated switchgear or switchboards with open bottoms. Aluminum or copper-clad aluminum conductors external to buildings or equipment shall not be terminated within 18 inches of the Earth. **See Figure 6-5.** Bare or covered grounding electrode conductors without an extruded polymeric covering are not permitted to be in direct contact with the Earth or concrete and cannot be subjected to other corrosive conditions.

Insulated copper-clad or aluminum grounding electrode conductors offer protection from the previously-mentioned corrosive conditions. A marking, such as a "W," should be used in the suffix of the insulation type to indicate suitability for wet locations.

SIZING GROUNDING ELECTRODE CONDUCTORS

Sizing grounding electrode conductors is covered in 250.66 of the *NEC*. The size of a grounding electrode conductor for a service or separately derived system is generally based on the size of the largest ungrounded supply conductor. The grounding electrode conductor is never sized based on an overcurrent protective device of a service or system. **See Figure 6-6.**

Section 250.66 requires that grounding electrode conductors installed for services or separately derived systems and those installed at separate buildings or structures be sized according to the values in Table 250.66. This sizing rule applies to both grounded and ungrounded alternating current (AC)

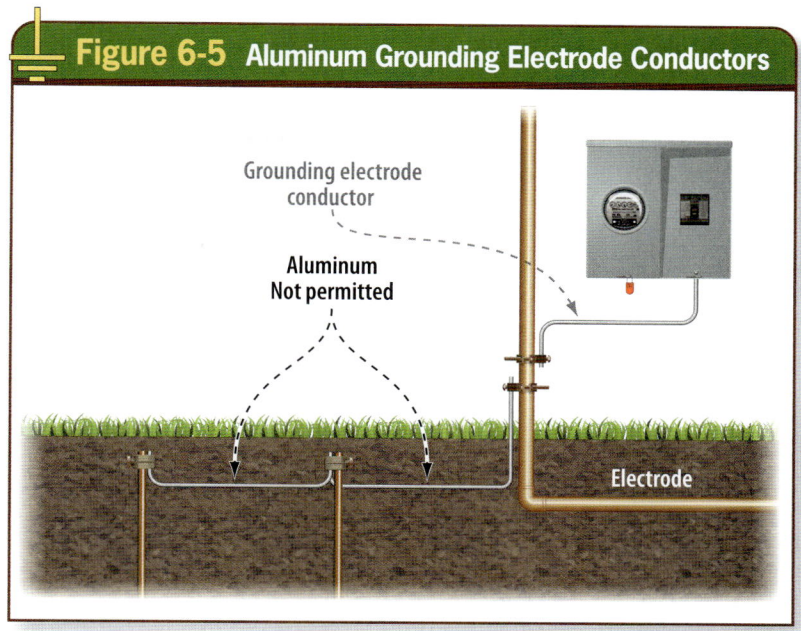

Figure 6-5. Aluminum grounding electrode conductor connections to electrodes are not permitted within 18 inches of the Earth outside.

Figure 6-6. Grounding electrode conductors at services are generally sized according to Table 250.66 based on the circular mil area of the largest ungrounded service-entrance conductor.

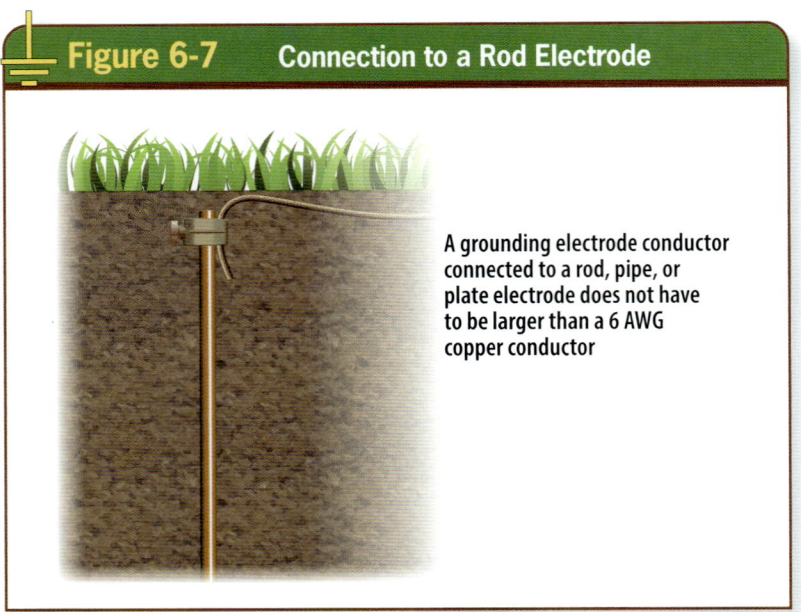

Figure 6-7   **Connection to a Rod Electrode**

A grounding electrode conductor connected to a rod, pipe, or plate electrode does not have to be larger than a 6 AWG copper conductor

Figure 6-7. A connection of a grounding electrode conductor to a single ground rod is addressed in 250.66(A).

Sections 250.66(A), (B), and (C) permit a reduction in size of the grounding electrode conductor where it is connected to any of the electrode types specifically identified in those subdivisions. For example, 250.66(A) indicates that if a grounding electrode conductor is connected to a rod, pipe, or plate electrode, that portion of the grounding electrode conductor that is connected to the rod-type electrode is not required to be larger than a 6 AWG copper or 4 AWG aluminum conductor. If multiple rod, pipe, or plate electrodes are used, the largest grounding electrode conductor required is 6 AWG copper or 4 AWG aluminum. The sizing provisions of this rule relate to the type of electrode installed and where each electrode of the grounding system is located in relation to the bonding jumper connections and the grounding electrode conductor. **See Figure 6-7.**

Section 250.66(B) indicates that when the grounding electrode conductor is connected only to a concrete-encased electrode or multiple separate concrete-encased electrodes, the portion of the grounding electrode conductor that is connected only to the concrete-encased

systems. Bonding jumpers used to connect grounding electrodes together and form a grounding electrode system must be sized in accordance with 250.66 and must meet the installation requirements in 250.64(A), (B), and (E).

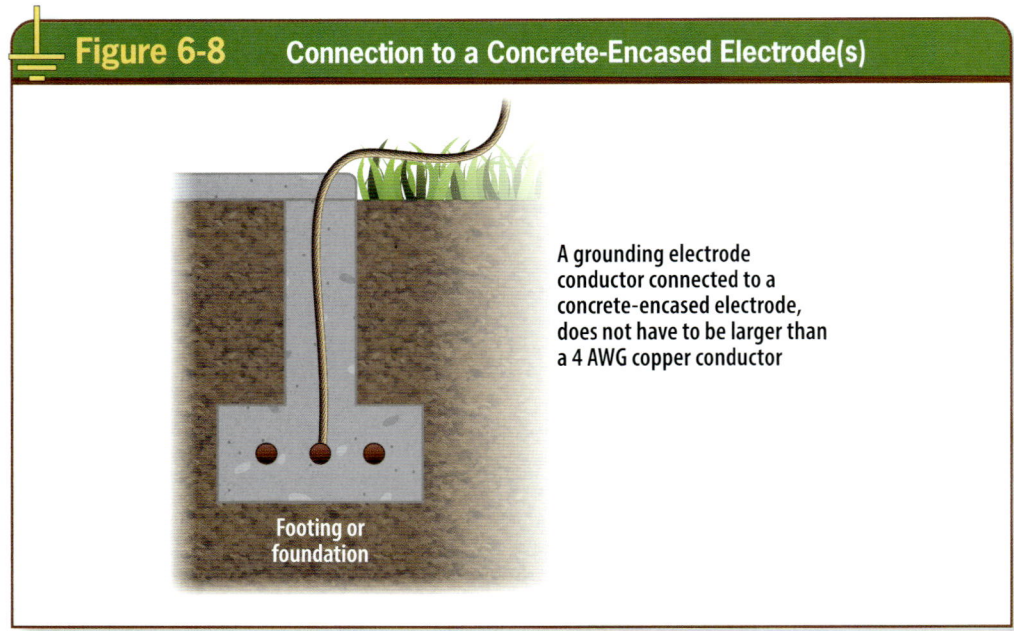

Figure 6-8   **Connection to a Concrete-Encased Electrode(s)**

A grounding electrode conductor connected to a concrete-encased electrode, does not have to be larger than a 4 AWG copper conductor

Footing or foundation

Figure 6-8. A connection to a concrete-encased electrode is addressed in 250.66(B).

electrode(s) is not required to be larger than a 4 AWG copper. If multiple concrete-encased electrodes are used, only one connection is required, and the largest grounding electrode conductor required is 4 AWG copper. **See Figure 6-8.**

Section 250.66(C) indicates that when the grounding electrode conductor is connected to a ground ring, the portion of the grounding electrode conductor that is connected only to the ring electrode is not required to be larger than the ground ring. The minimum-size conductor permitted for use as a ground ring electrode is 2 AWG copper, as provided in 250.52(A)(4). **See Figure 6-9.**

USING TABLE 250.66

Table 250.66 is titled "Grounding Electrode Conductor for Alternating-Current Systems." (Sizing grounding electrode conductors for direct current (DC) systems is covered in 250.166.) Table 250.66 is easy to use and is based on conductor sizes rather than the sizes of overcurrent protective devices. To use Table 250.66, the size (in AWG or kcmil) of the largest ungrounded supply conductor or equivalent area for parallel conductors must first be determined. Then the range of conductor sizes within which the supply conductor falls should be located in Column 1 or 2, depending on whether the conductors are copper or aluminum. Last, Column 3 or 4 (depending on whether the conductors are copper or aluminum) should be examined to determine the grounding electrode conductor size. **See Figure 6-10.**

As an example, if a 400-ampere, 3-phase, 4-wire, 480Y/277-volt service was supplied by four 750-kcmil aluminum conductors (XHHW), the minimum-size copper grounding electrode conductor must be 1/0 AWG copper or 3/0 AWG aluminum. If an 800-ampere, 3-phase, 4-wire, 480Y/277-volt service was supplied by two sets of four 750-kcmil aluminum conductors (XHHW) in parallel, the copper grounding electrode conductor must not be smaller than 2/0 AWG copper

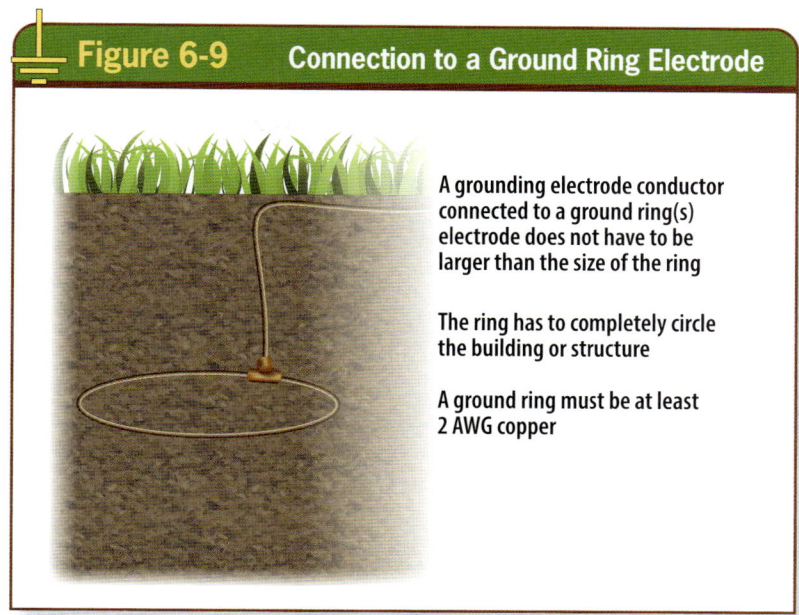

Figure 6-9 **Connection to a Ground Ring Electrode**

A grounding electrode conductor connected to a ground ring(s) electrode does not have to be larger than the size of the ring

The ring has to completely circle the building or structure

A ground ring must be at least 2 AWG copper

Figure 6-9. *A connection to a ground ring electrode is addressed in 250.66(C).*

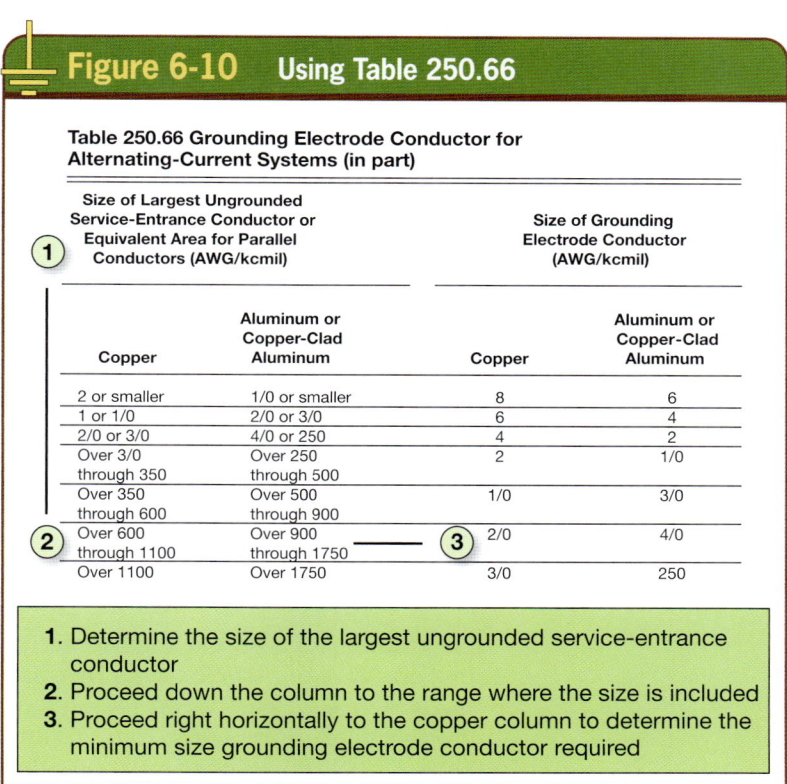

Figure 6-10 **Using Table 250.66**

Table 250.66 Grounding Electrode Conductor for Alternating-Current Systems (in part)

Size of Largest Ungrounded Service-Entrance Conductor or Equivalent Area for Parallel Conductors (AWG/kcmil)		Size of Grounding Electrode Conductor (AWG/kcmil)	
Copper	Aluminum or Copper-Clad Aluminum	Copper	Aluminum or Copper-Clad Aluminum
2 or smaller	1/0 or smaller	8	6
1 or 1/0	2/0 or 3/0	6	4
2/0 or 3/0	4/0 or 250	4	2
Over 3/0 through 350	Over 250 through 500	2	1/0
Over 350 through 600	Over 500 through 900	1/0	3/0
Over 600 through 1100	Over 900 through 1750	2/0	4/0
Over 1100	Over 1750	3/0	250

1. Determine the size of the largest ungrounded service-entrance conductor
2. Proceed down the column to the range where the size is included
3. Proceed right horizontally to the copper column to determine the minimum size grounding electrode conductor required

Figure 6-10. *Table 250.66 of the NEC is used to determine the appropriate size for grounding electrode conductors. [Reproduction of Table 250.66 (in part)]*

or 4/0 AWG aluminum. When parallel conductors are installed, add the circular mil area of all conductors installed in parallel to make up one phase. In this case, two 750-kcmil service conductors are the equivalent of one 1,500-kcmil aluminum conductor, and therefore require a grounding electrode conductor not smaller than 2/0 AWG copper or 4/0 AWG aluminum.

Smaller grounding electrode conductors are permitted when they are only connected to rod, pipe, plate, concrete-encased, and ground ring electrodes as detailed in 250.66(A), (B), and (C). The grounding electrode conductor does not need to be sized larger than the values provided in Table 250.66, unless the engineering design specifies larger sizes.

Sometimes grounding electrode conductors are installed in lengths that are excessive (for example, 200 to 300 feet). In these cases, the size can be evaluated and possibly increased to compensate for the additional length. The *NEC* does not address this issue, but in the 1950s, Eustace Soares did extensive work on the subject of sizing grounding electrode conductors to compensate for the effects of voltage drop. The withstand capabilities of grounding electrode conductors when installed in excessive lengths are obviously a concern, and the length of grounding electrode conductors and how length relates to performance are design elements that should be addressed in the planning and design phases of projects.

GROUNDING ELECTRODE CONDUCTORS FOR DC SYSTEMS

Grounding for DC systems is covered in Part VIII of Article 250. This part includes rules specific to DC system grounding requirements, and it must be applied in addition to the other parts of Article 250. A DC system is required to be grounded as follows:

1. Two-wire DC systems supplying premises wiring operating at greater than 60 volts but not exceeding 300 volts are required to be grounded [note the three exceptions to 250.162(A)].
2. All 3-wire DC systems are required to be grounded by connecting the neutral conductor of the system to ground.

The connection point of the grounding electrode conductor for a DC system is covered in 250.164. If the DC source is not located on the premises, this connection must be made at one or more supply stations. The grounding electrode conductor connection is not permitted at individual services or at any point on the premises wiring if the source is not located on the premises. If the DC system (source) is located on the premises, the grounding electrode conductor connection must be made either:

1. At the source
2. At the first system disconnecting means or overcurrent protective device
3. By other means that provide equivalent protection and utilize listed and identified equipment

The minimum-size grounding electrode conductor for a DC system is generally required to be not less than the sizes indicated in 250.166(A) and (B), but it is not required to be larger than 3/0 AWG copper or 250-kcmil aluminum.

When the DC system consists of a 3-wire balancer set or a balancer winding with overcurrent protection, as provided in 445.12(D), the grounding electrode conductor shall be sized no smaller than the neutral conductor and can never be smaller than 8 AWG copper or 6 AWG aluminum. When the DC system is other than as described in 250.166(A), the grounding electrode conductor cannot be smaller than the largest conductor supplied by the system and can never be smaller than 8 AWG copper or 6 AWG aluminum. **See Figure 6-11.**

When the grounding electrode conductor for a DC system is only connected to a rod, pipe, or plate electrode, the grounding electrode conductor does not have to be sized larger than 6 AWG copper or 4 AWG aluminum. If the grounding electrode conductor for a DC system is only connected to a concrete-encased electrode, it never has to be larger than 4 AWG copper. If the grounding electrode conductor for a DC system is only connected to a ground ring, it never has to be sized larger than the size of the ground ring electrode. These provisions for connections of grounding electrode conductors installed for DC systems are identical to the allowances in 250.66(A), (B), and (C) for AC systems. The installation requirements in 250.64 apply to grounding electrode conductors for DC systems.

The requirement in 250.160 indicates that the other parts of the article have general application in addition to the rules in Part VIII, unless a rule is specifically intended to apply to AC systems.

GROUNDING ELECTRODE CONDUCTOR INSTALLATION

Section 250.50 requires all grounding electrodes present at a building or structure to be connected together to form a grounding electrode system. There are only two exceptions to this rule.

The first exception, which is not written in the *NEC*, is relevant when a building or structure has no power supplying it. There is no requirement for a grounding electrode system for that building or structure.

The second exception follows 250.32 and relaxes the grounding electrode requirement for structures supplied by a single branch circuit that includes an EGC. An example of such an installation is a lighting pole in a parking lot. Typically, an auxiliary grounding electrode is installed for these poles anyway, but this is not an *NEC* requirement. It is a design consideration that is often included in plans and specifications. Another example is a detached garage or storage building supplied by only one branch circuit. An individual branch

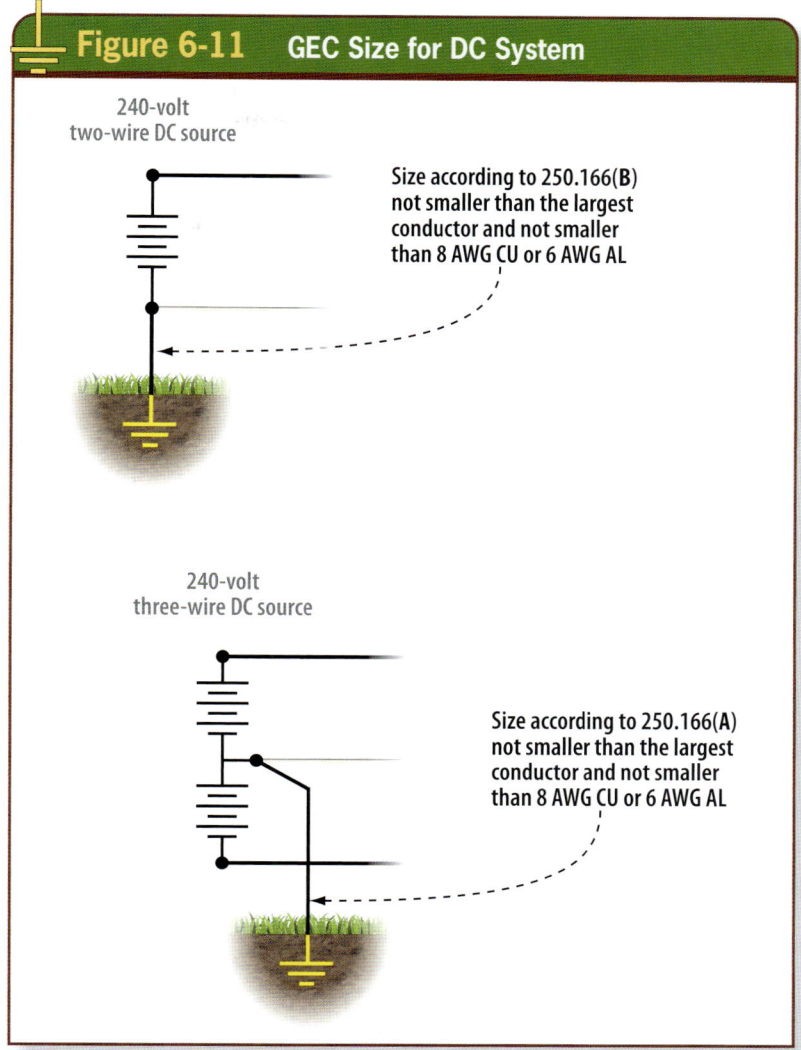

Figure 6-11. *The grounding electrode conductor for a DC system is sized using 250.166.*

circuit or a multiwire branch circuit can qualify for this exception.

Workers installing grounding electrode conductors must understand several important performance characteristics and requirements specified in Part III of Article 250. The grounding electrode conductor is the conductive path to ground for equipment and systems. Specific rules apply to installations of the grounding electrode conductor. No identification requirement for grounding electrode conductors exists. They can be insulated, covered, or bare, but they are not required to be identified using the color green or green with yellow stripes, as is required for an EGC. There is no prohibition against using the color green for identifying the grounding electrode conductor, but it is not a requirement. The colors white or gray must not be used for grounding electrode conductors.

Verify any specific identification requirements that may be included in the project specifications. The installation requirements for grounding electrode conductors in the *NEC* are concerned with sizing, connections, protection of the conductor from physical damage, and protection from external influences such as the effects of magnetic fields.

Securing and Protecting from Physical Damage

When grounding electrode conductors are installed exposed, they are required to be securely fastened to the surface on which they are placed. This means that they should be secured in a similar fashion to flexible wiring methods, such as any of the nonmetallic or metallic cable assemblies.

If a grounding electrode conductor is installed in a conduit or other armor, the conduit or armor also must be secured to the surface. Grounding

Figure 6-12   **Protecting from Physical Damage**

Listed grounding electrode conductor connection

Grounding Electrode

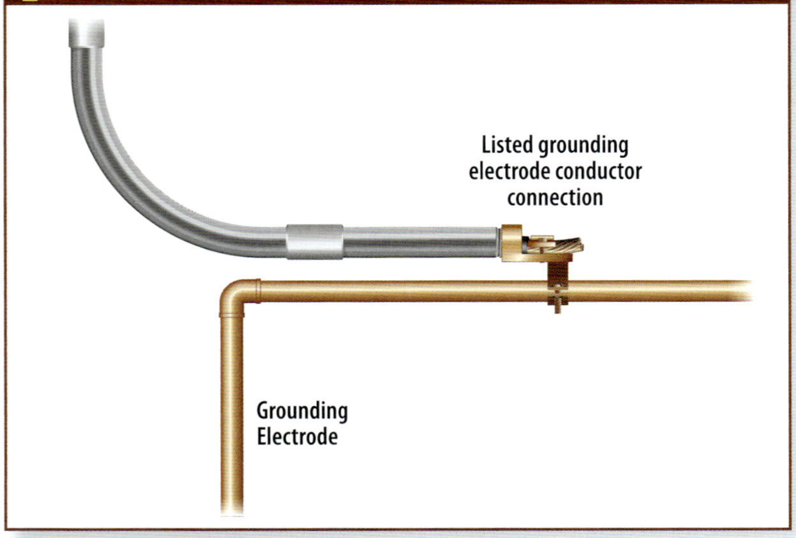

Figure 6-12. Grounding electrode conductors must be protected from physical damage.

electrode conductors in sizes 6 AWG or larger must be protected where installed in locations that expose them to physical damage. Various protection methods can also be used, such as installing in the covered partitions of a building or burial in the Earth. **See Figure 6-12.**

The *NEC* also recognizes that if grounding electrode conductors of 6 AWG or larger are installed such that they are not exposed to potential physical damage, they can be run without being placed in a raceway or armor if they are securely fastened. The authority having jurisdiction is typically responsible for judging whether a grounding electrode conductor is exposed to potential damage. Grounding electrode conductors can also be installed on or run through wood or metal framing members. Physical protection in this case is usually inherently provided by this location within the construction framing. Other installation locations that provide the necessary physical protection can also be selected for grounding electrode conductors without the use of conduit or armor. Grounding electrode conductors and grounding electrode bonding jumpers are not required to comply with the burial depths in 300.5.

Grounding electrode conductors are generally required to be installed in a continuous length without a splice or joint unless splices are made using irreversible compression connectors or the exothermic welding process. **See Figure 6-13.**

This is the general rule in 250.64(C). The exothermic welding process is an acceptable method of splicing grounding electrode conductors together or making a connection to a grounding electrode.

The irreversible compression connectors must be listed as grounding and bonding equipment. This means they have been evaluated for the anticipated performance criteria in accordance with *UL 467, Grounding and Bonding Equipment.* Busbars are also permitted to be connected together to form grounding electrode conductors.

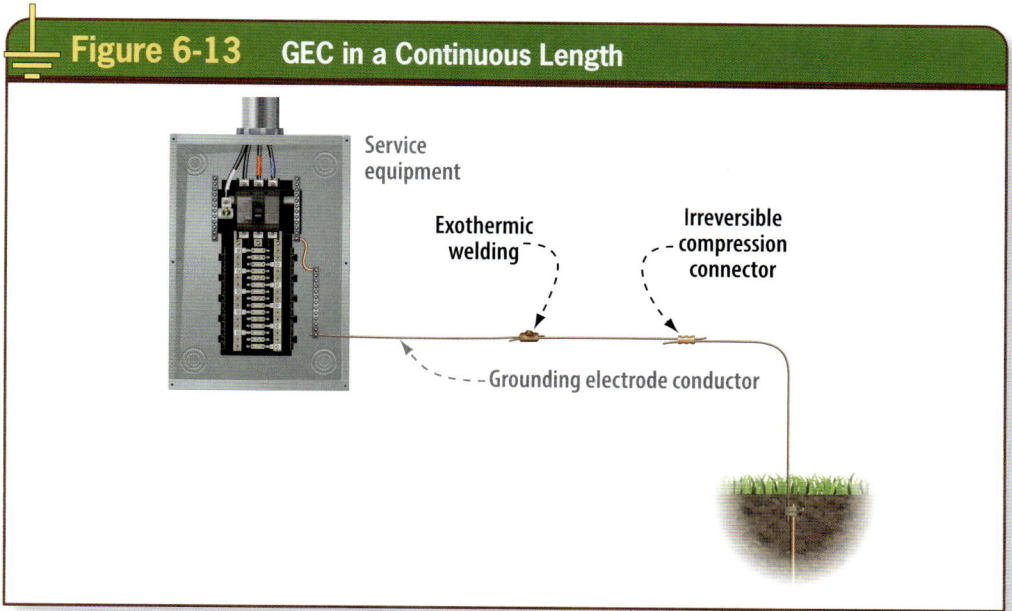

Figure 6-13   **GEC in a Continuous Length**

Figure 6-13. Wire-type grounding electrode conductors are generally required to be installed without splices or joints other than as permitted in 250.64(C)(1).

Methods of Installing Grounding Electrode Conductors

Individual grounding electrode conductors can be installed from each electrode back to the service, or a grounding electrode conductor can be run to one or more electrodes and bonding jumpers can interconnect the rest of the electrodes to form the grounding electrode system. Section 250.64(F) describes installation methods that can be used when constructing the electrode system. **See Figure 6-14.**

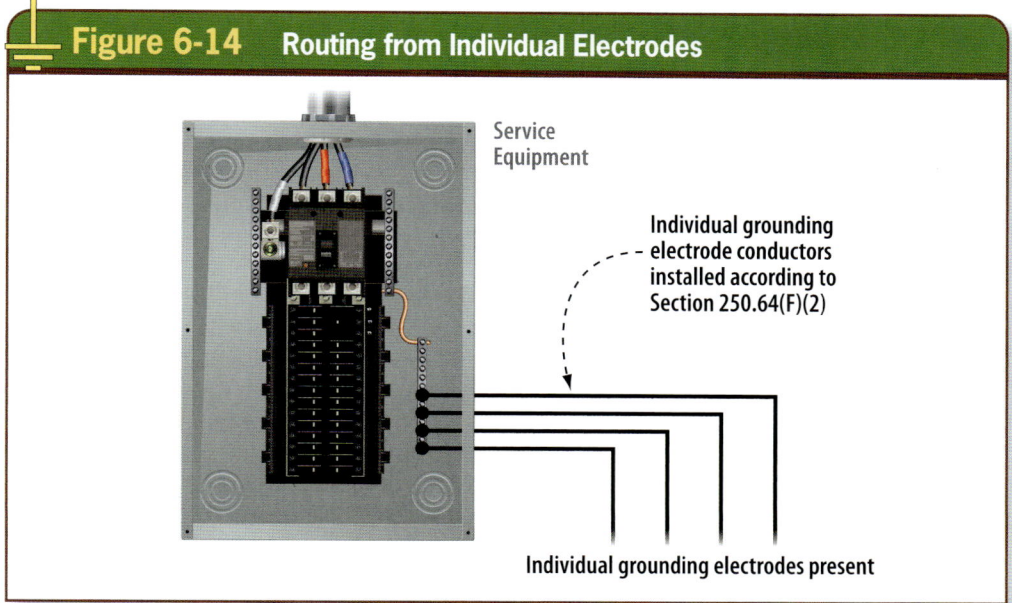

Figure 6-14   **Routing from Individual Electrodes**

Figure 6-14. Grounding electrode conductors may be routed from individual electrodes to the service equipment.

This section also provides a sizing requirement that relates to where the grounding electrode conductor or

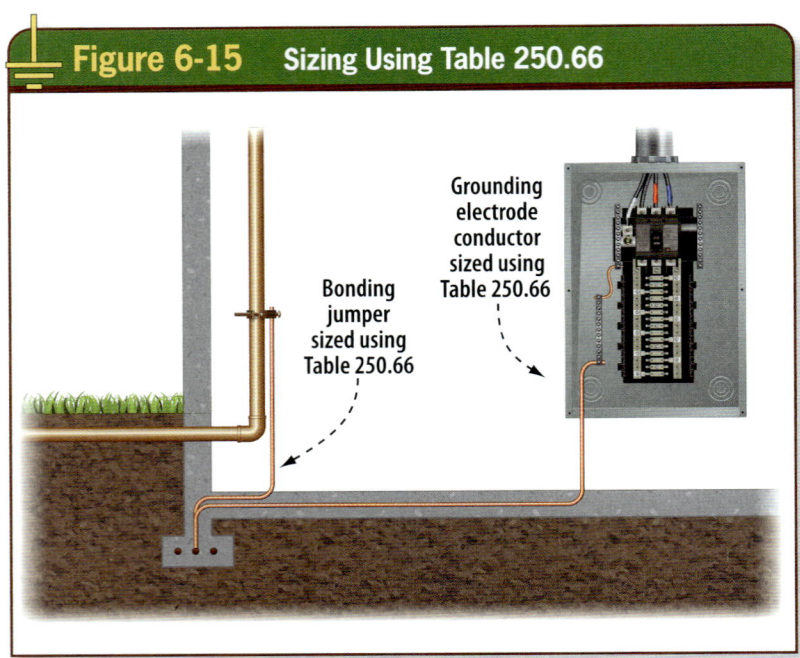

Figure 6-15   Sizing Using Table 250.66

Grounding electrode conductor sized using Table 250.66

Bonding jumper sized using Table 250.66

Figure 6-15. Table 250.66 is used for sizing the grounding electrode conductor and the bonding jumper(s) of a grounding electrode system, unless using the provisions of 250.66(A), (B), or (C).

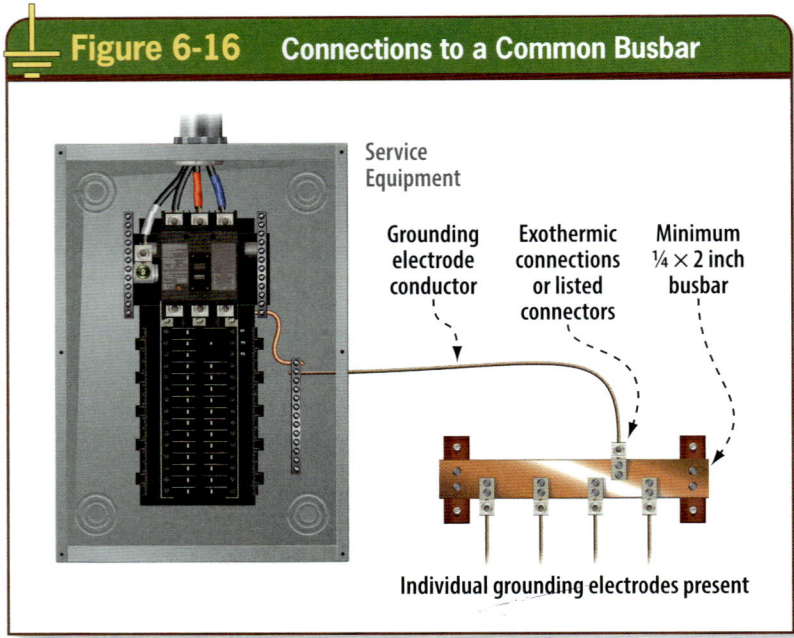

Figure 6-16   Connections to a Common Busbar

Service Equipment

Grounding electrode conductor

Exothermic connections or listed connectors

Minimum ¼ × 2 inch busbar

Individual grounding electrodes present

Figure 6-16. Grounding electrode conductors and bonding jumpers are permitted to be connected to a common busbar.

bonding jumper is connected in the grounding electrode system. If multiple grounding electrodes form a system of electrodes, the size required for the grounding electrode conductor must be for the largest grounding electrode conductor connected in the system. For example, if a grounding electrode conductor is installed from a service to a concrete-encased electrode and a bonding jumper is installed from the concrete-encased electrode to a water pipe electrode, then both would have to be sized using Table 250.66 based on the sizing requirement for connection to a water pipe electrode. **See Figure 6-15.**

Grounding electrode conductors are permitted to be run to any electrode in the grounding electrode system if the other grounding electrodes are connected using bonding jumpers as indicated in 250.64(F). These bonding jumpers must be installed in accordance with 250.53(C), which indicates that sizing must be per 250.66 and the connections must be made according to the requirements in 250.70.

Sometimes it is convenient and practical to install multiple grounding electrode conductors to each electrode of the system. Section 250.64(F)(2) recognizes this practice, and the same sizing requirements apply to each individual grounding electrode conductor.

Another method of interconnecting grounding electrodes to form the grounding electrode system is to connect the grounding electrode conductor(s) and bonding jumper(s) from each electrode to a copper or aluminum busbar. **See Figure 6-16.** The busbar must be at least ¹/₄ inch thick by two inches wide in the cross-sectional area. These connections must be listed, and the busbar must be securely fastened to the structure in an accessible location. The thickness and width of the busbar is specified in the *NEC*, but the length is not. The *Code* only requires that the busbar be long enough to accommodate all grounding electrode conductors and bonding jumpers that must be connected to it. **See Figure 6-17.**

When installing aluminum busbars, installers must follow the restrictive criteria in 250.64(A), which prohibits terminating aluminum conductors within 18 inches of the Earth. In this case, the busbar is aluminum and must meet the same requirement due to the same concerns about corrosive influences.

Connection Point for Grounding Electrode Conductors

A single grounding electrode conductor is permitted to be run to the grounding electrode system. A common method of installing grounding electrode conductors is to run them directly from the service equipment enclosure to the grounding electrode system without a splice or joint. **See Figure 6-18.**

Section 250.24(A)(1) indicates that the grounding electrode conductor connection to the system may be made at any point from the load end of the overhead service conductors or service drop or from the underground service conductors or lateral up to the service equipment. Typically, a specific termination means is provided within the

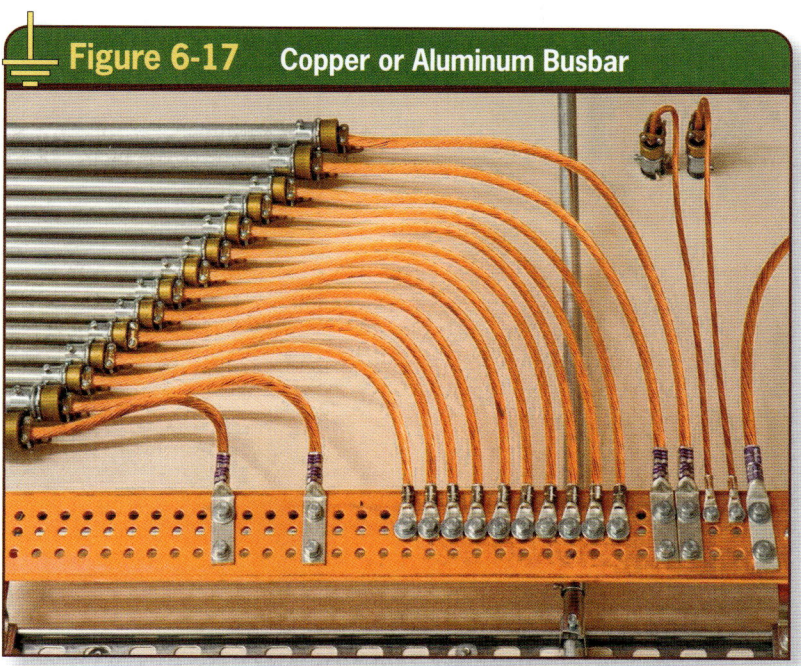

Figure 6-17 Copper or Aluminum Busbar

Figure 6-17. Grounding electrode conductors and bonding jumpers for grounding electrode systems are permitted to be connected by attachment to copper or aluminum busbars.

listed service equipment, such as within a single unit of service equipment or in assemblies such as switchboards,

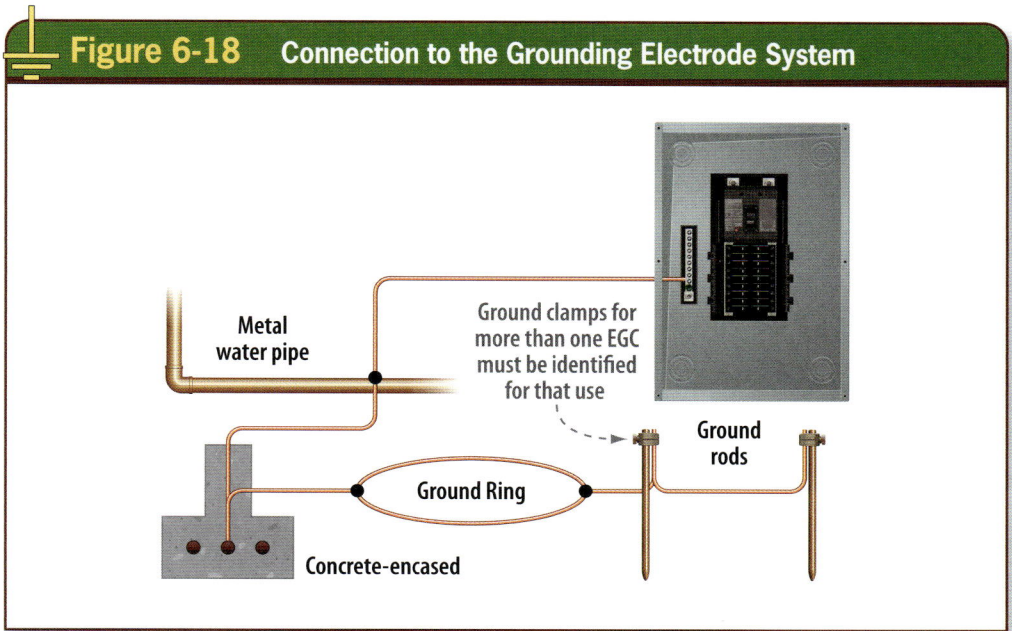

Figure 6-18 Connection to the Grounding Electrode System

Metal water pipe

Ground clamps for more than one EGC must be identified for that use

Ground rods

Ground Ring

Concrete-encased

Figure 6-18. A single grounding electrode conductor is permitted to be run to a grounding electrode system. It is usually connected at the service equipment enclosure.

panelboards, motor control centers, and other equipment that is suitable for use as service equipment. **See Figure 6-19.**

Figure 6-19   Grounding Electrode Conductor Terminal

Figure 6-19. A grounding electrode conductor terminal is often provided on the terminal bar for the grounded conductor within the listed service equipment.

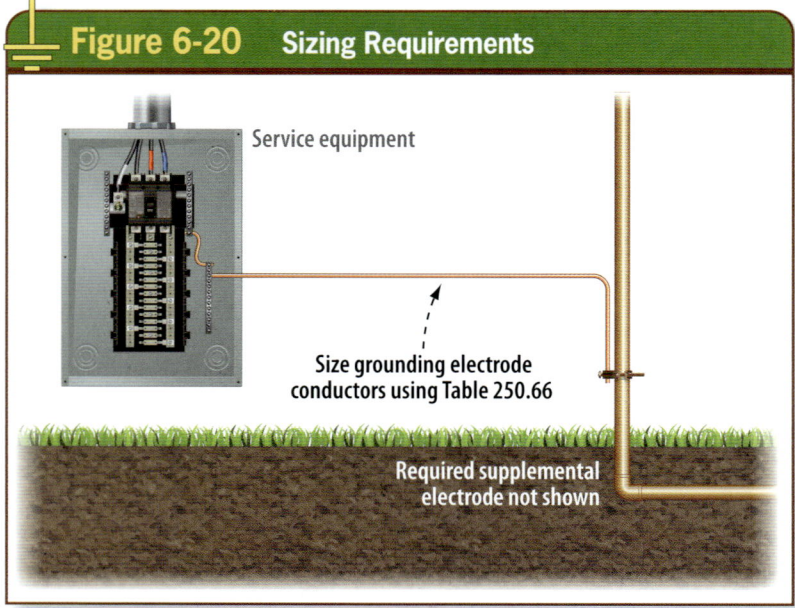

Figure 6-20   Sizing Requirements

Service equipment

Size grounding electrode conductors using Table 250.66

Required supplemental electrode not shown

Figure 6-20. The general sizing rule for grounding electrode conductors is provided in 250.66.

Methods of Connection at Service Disconnecting Means

Services must be provided with a service disconnecting means to disconnect all ungrounded circuit conductors at the building or structure served. The service disconnecting means can be either a single (main) service disconnect in a panelboard, a motor control center (MCC) or switchboard, or a separate individual disconnecting means enclosure (fused switch or enclosed circuit breaker). The service disconnecting means can also consist of up to six disconnects (each suitable for use as service equipment) and can be installed in separate enclosures and grouped as required in 230.71(A).

Single Disconnecting Means

For installations of service equipment that include only a single service disconnecting means, either as a separate enclosure or as part of an assembly that is suitable for use as service equipment, a single grounding electrode conductor can be installed from the grounded conductor terminal bus in the equipment to the grounding electrode system.

The minimum size of the grounding electrode conductor is determined by using Table 250.66; sizing is based on the size of the largest service-entrance conductor. For example, if the size of the largest ungrounded service-entrance conductor for a 400-ampere service is 500 kcmil copper, then the minimum-size grounding electrode conductor is 1/0 AWG copper or 3/0 AWG aluminum or copper-clad aluminum. **See Figure 6-20.**

Remember that Section 250.66 permits smaller sizes to be used for certain types of electrodes. **See Figure 6-21.**

Multiple Service Disconnecting Means in Separate Enclosures

A popular method of installing grounding electrode conductors at services is by using the common grounding electrode conductor tap concept. This concept is typically used where multiple service disconnecting means are installed in separate enclosures as

Figure 6-21 Sizing Using Section 250.66

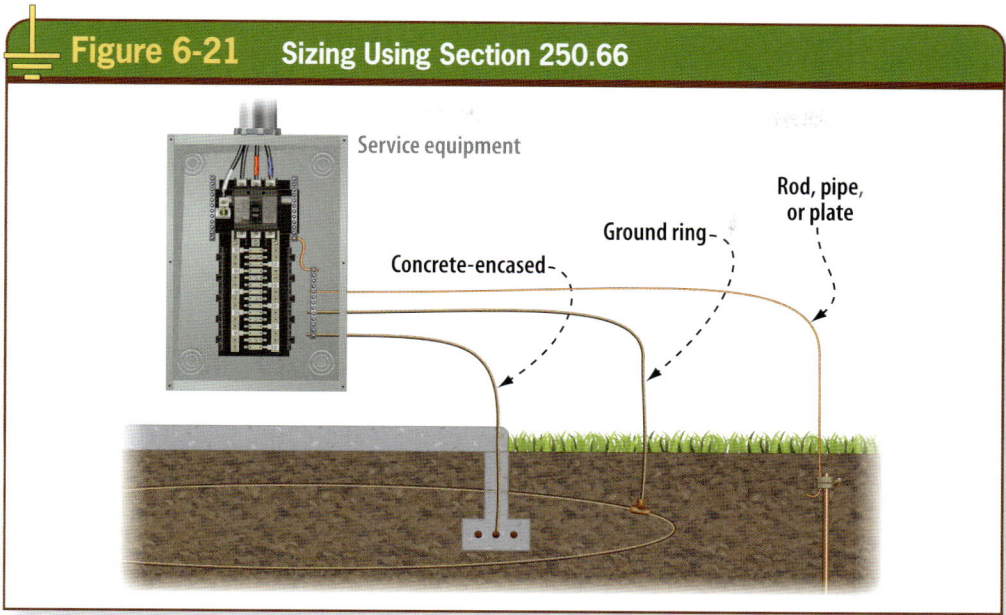

Figure 6-21. Grounding electrode conductor types addressed in Section 250.66 offer smaller sizing options as an alternative to Table 250.66.

permitted by 230.71. In these types of installations, a single large common grounding electrode conductor is installed, and multiple grounding electrode conductor taps are connected to the common grounding electrode conductor. **See Figure 6-22.**

The term *common grounding electrode conductor* refers to the conductor that is not permitted to have any splices or joints, as described in the general rule in 250.64(D)(1). The term *grounding electrode conductor tap* is not a defined term in the *NEC* and is therefore not

Figure 6-22 Grounding Electrode Conductor Taps

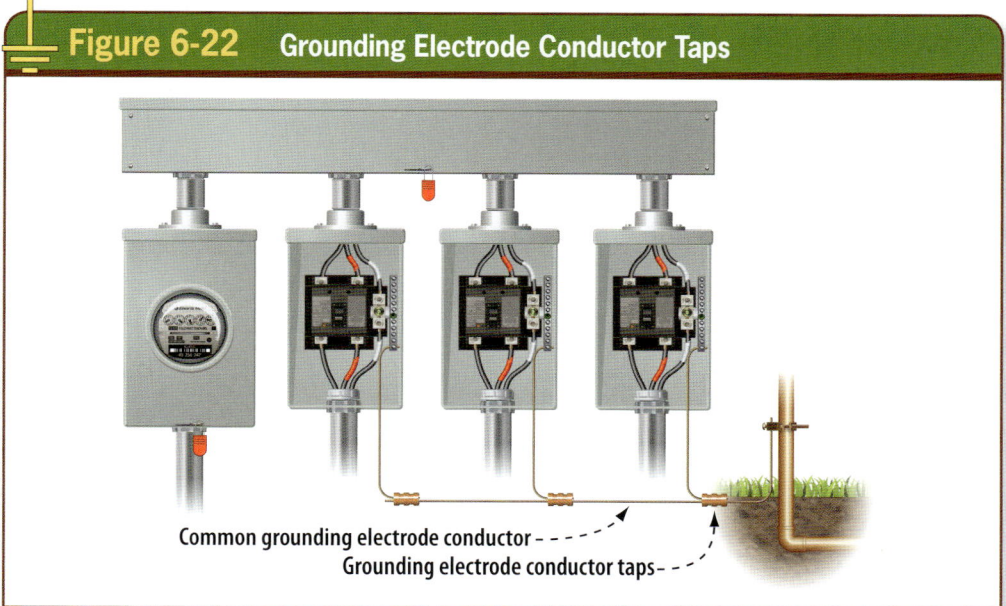

Figure 6-22. Grounding electrode conductor taps are permitted to be installed to separate service disconnects.

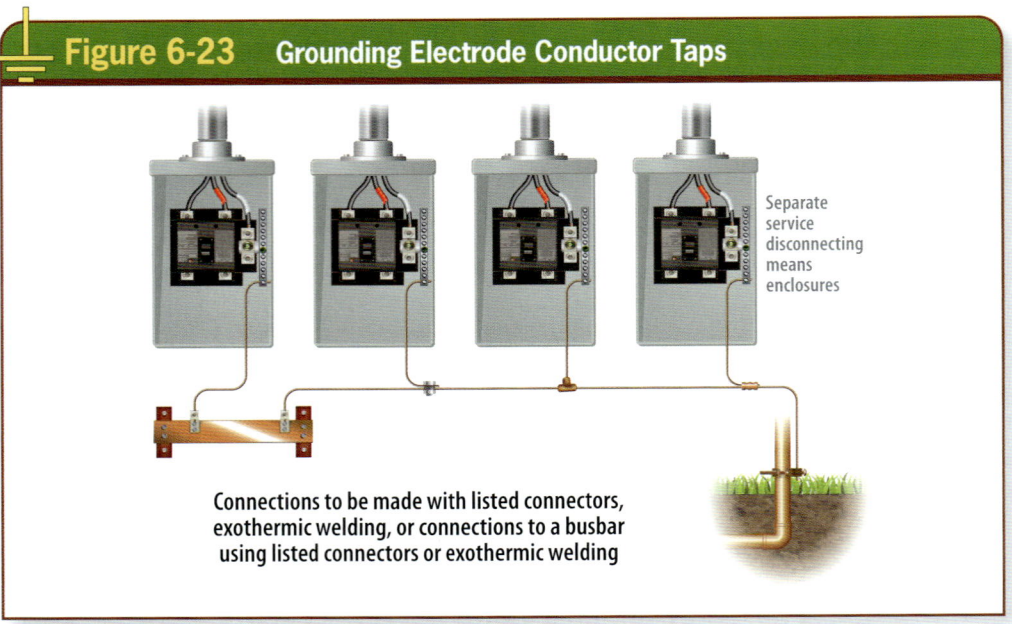

Figure 6-23   **Grounding Electrode Conductor Taps**

Separate service disconnecting means enclosures

Connections to be made with listed connectors, exothermic welding, or connections to a busbar using listed connectors or exothermic welding

Figure 6-23. Grounding electrode conductor taps must be connected to a common grounding electrode conductor in a manner such that the common grounding electrode conductor remains without a splice.

subject to this same restriction. The connections of the grounding electrode conductor taps to a common grounding electrode conductor must be made using exothermic welding or by listed grounding and bonding connections. A 1/4-inch thick by two-inch wide aluminum or copper busbar may also be used with exothermic or listed connections. **See Figure 6-23.**

The minimum size of the common grounding electrode conductor is based on the sum of the circular mil area of the largest ungrounded service-entrance conductor(s).

If the service-entrance conductors connect directly to a service drop or service lateral, the common grounding electrode conductor shall be sized in accordance with Table 250.66 Note 1. The grounding electrode conductor tap then is extended to the inside of each service disconnecting means enclosure. Grounding electrode tap conductors must be sized using Table 250.66; sizing is based on the size of the largest phase conductor serving each service disconnecting means enclosure. **See Figure 6-24.**

The common grounding electrode conductor from which the taps are made is sized using Table 250.66; sizing is based on the sum of the cross-sectional areas of the largest ungrounded service-entrance conductors or equivalent cross-sectional area for parallel service-entrance conductors that supply the multiple separate service disconnecting means. The common grounding electrode conductor tap concept is popular because it eliminates the difficulties in looping a single larger grounding electrode conductor from one enclosure to another. **See Figure 6-25.**

Individual Grounding Electrode Conductors to Multiple Service Disconnects

Another method of installing grounding electrode conductors for individual separate service disconnecting means enclosures is to run a grounding electrode conductor from the grounding electrode to each separate enclosure. For example, a service could be installed using up to six separate service disconnecting means enclosures on the

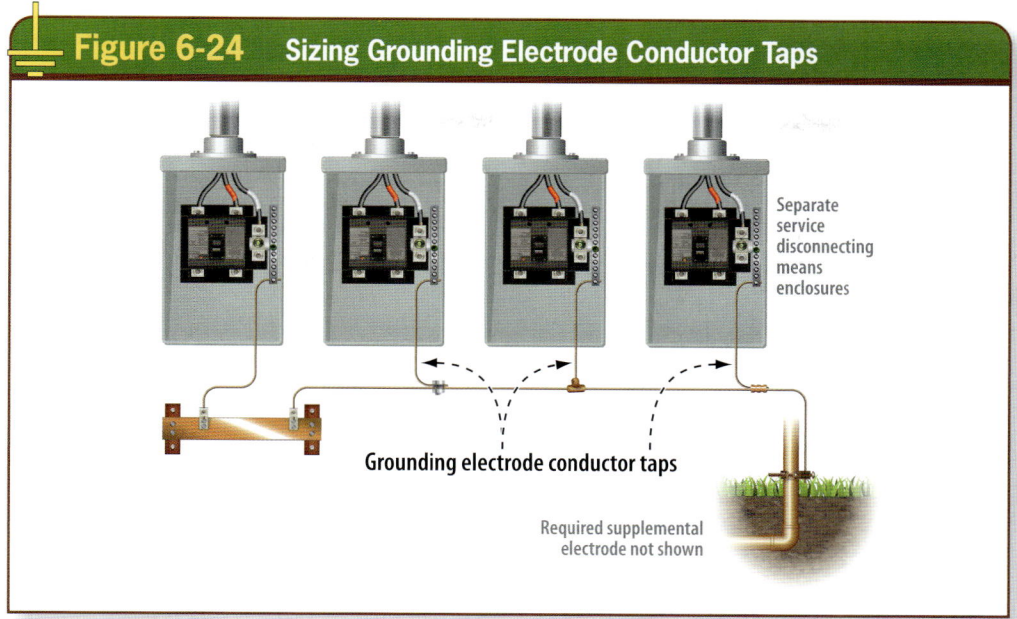

Figure 6-24. *Use Table 250.66 to size grounding electrode conductor taps. The size to be used is based on the largest service conductor in each separate service disconnect.*

same building with a separate riser from each disconnect. The size of the grounding electrode conductor run to each individual service disconnect enclosure is determined using 250.66 and is based on the largest ungrounded service conductor in each separate enclosure. In these cases, the same grounding electrode system is inherently used for all service disconnects

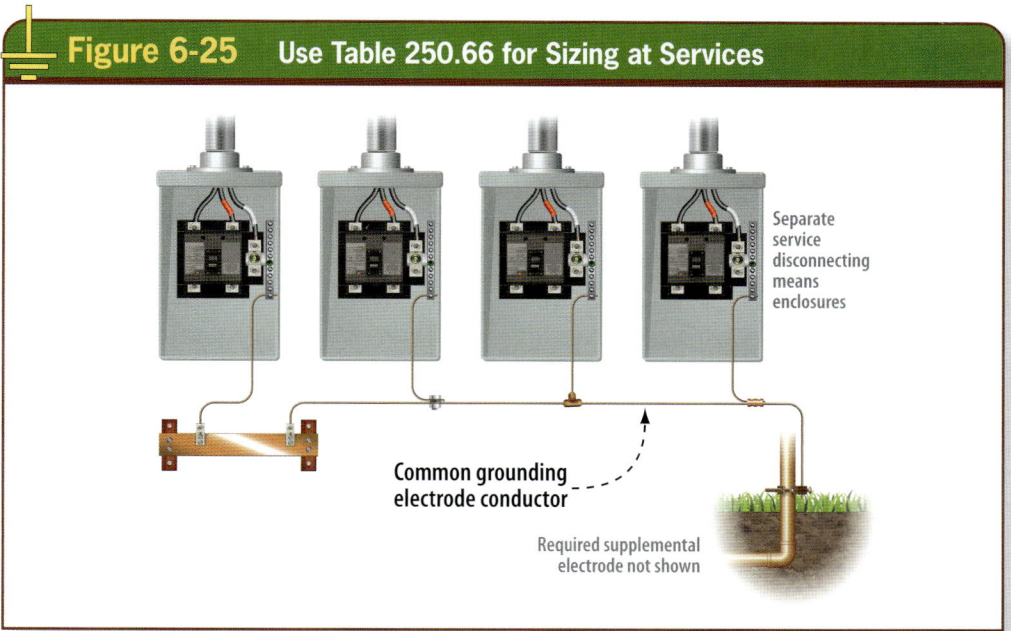

Figure 6-25. *Use Table 250.66 to size the common grounding electrode conductor at services. The size to be used is based on the sum of the circular mil area of the largest service conductor.*

Figure 6-26   Separate Service Disconnects

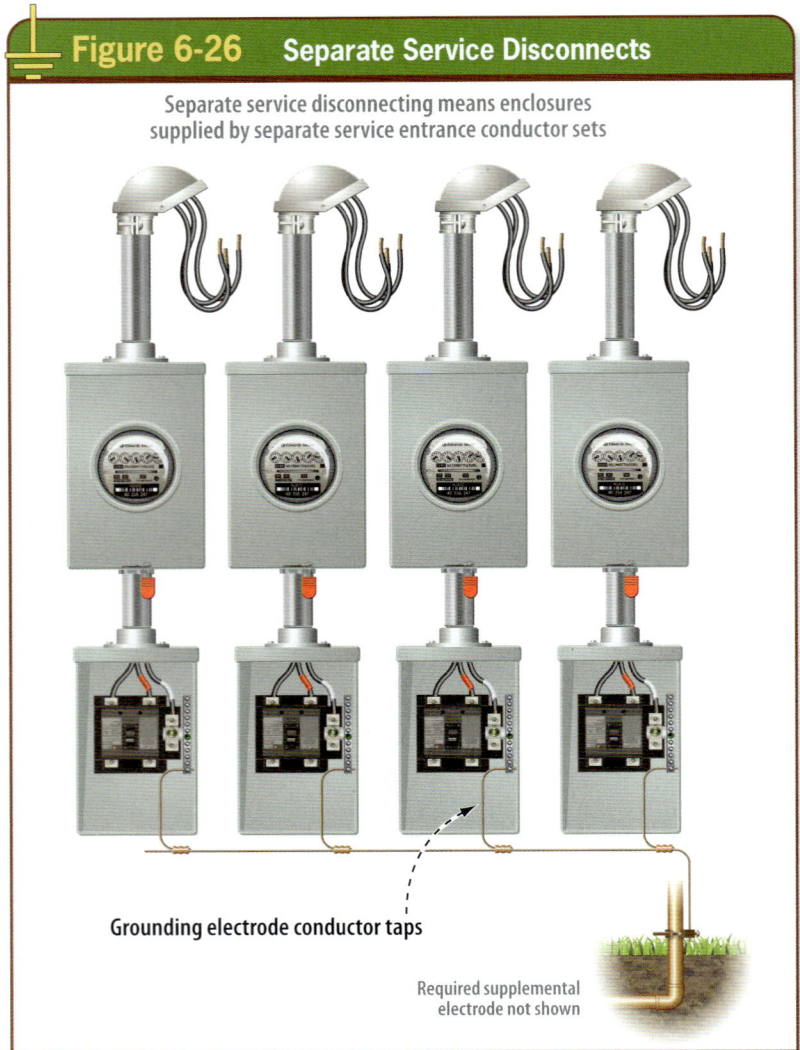

Separate service disconnecting means enclosures
supplied by separate service entrance conductor sets

Grounding electrode conductor taps

Required supplemental
electrode not shown

Figure 6-26. *Individual grounding electrode conductors run to separate service disconnects are sized according to 250.66. The size to be used is based on the largest service conductor in each separate service disconnect.*

on the same building or structure. **See Figure 6-26.**

EFFECTIVENESS (INTEGRITY) OF THE GROUNDING PATH

Section 250.68 generally requires that the mechanical connections to a grounding electrode be accessible. **See Figure 6-27.**

There are some exceptions to Section 250.68 that address conditions or installations that relax this grounding electrode conductor connection accessibility requirement:

1. Grounding electrode connections that are buried, such as those used for rods, pipes, or plate electrodes
2. Concrete-encased electrodes where the connection is encased in concrete
3. If exothermic welding is used as the connection means or if the connections are made using suitable irreversible compression connectors to fireproofed structural metal—**See Figure 6-28.**

The connections of the grounding electrode conductor(s) or bonding jumper(s) to a grounding electrode must ensure an effective path to ground (Earth). If it is necessary to ensure the grounding path for a metal piping system electrode, bonding must be provided around any insulated joints and

Figure 6-27   Accessible Connections

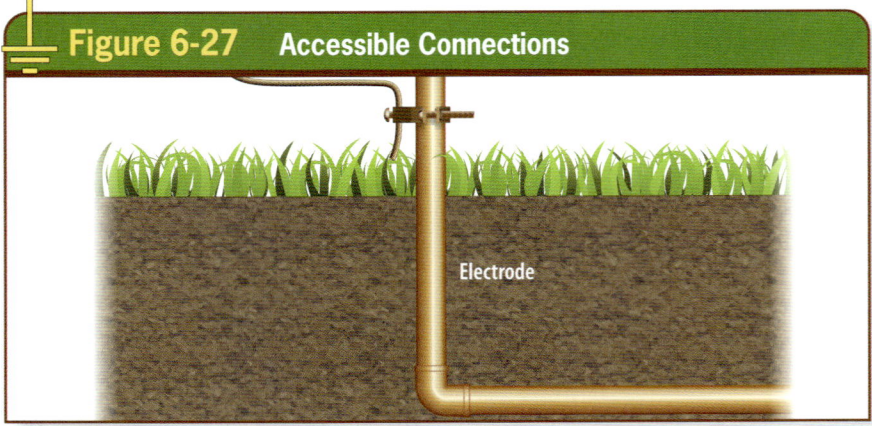

Electrode

Figure 6-27. *Grounding electrode conductor connections are generally required to be accessible.*

Figure 6-28 **Accessibility Exception**

Figure 6-28. Grounding electrode conductor connections are permitted to be buried in fireproofing material on building steel electrodes.

around equipment that is likely to be disconnected or removed for repairs or replacement.Bonding jumpers must be of sufficient length to permit removal of such equipment while retaining the integrity of the grounding path. An example of an installation where a bonding jumper is required to ensure the grounding path is at a water meter or other equipment such as a water softener or filter that is installed in series with a metal water piping system used as a grounding electrode. **See Figure 6-29.**

Figure 6-29 **Integrity of the Grounding Connection**

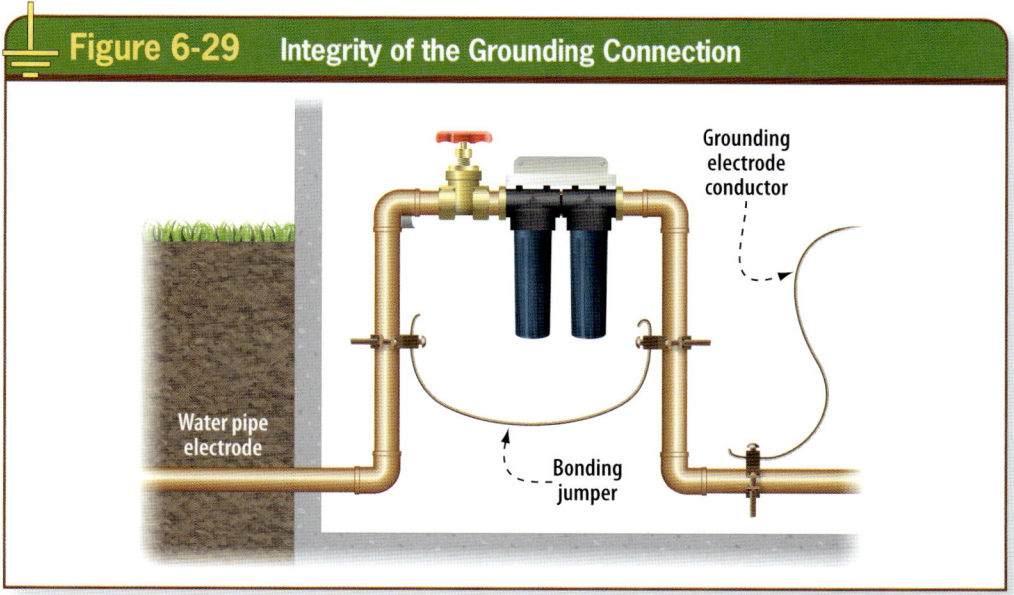

Grounding electrode conductor

Water pipe electrode

Bonding jumper

Figure 6-29. The integrity of the grounding connection must be ensured by installing a bonding jumper around any insulating joints or equipment likely to be removed for replacement or repairs.

GROUNDING ELECTRODE CONDUCTOR CONNECTION LOCATIONS

The connections of grounding electrode conductors can be made at a variety of locations throughout a building or structure. If a metal water pipe in direct contact with the earth is used as a grounding electrode, the grounding electrode conductors or bonding jumpers from a grounding electrode system are permitted to be connected to the interior metal water piping at any point on the exposed piping within five feet from the point of entrance into the building or structure. The reason for the restriction is to limit the length of water piping in a building that functions as a grounding electrode conductor. **See Figure 6-30.**

Section 250.68(C)(1) also prohibits the use of that portion of the interior metal water piping system that extends more than five feet beyond the point of entrance into the building to be used as a conductor to interconnect grounding electrodes. This is because nonmetallic piping or fittings used for replacements would cause an interruption in the electrical continuity of the metal water piping.

There is an exception for commercial, industrial, and institutional facilities. This exception allows the grounding electrode conductor connection or connections of bonding jumpers of the grounding electrode system to be made at locations farther than five feet from the point where the water piping enters the building or structure if the following specific restrictive conditions are met:

1. Conditions of maintenance and supervision ensure qualified persons service the installation.
2. The entire length of the piping is exposed other than where it passes perpendicularly through walls or floors.

Buildings or structures often include conductive paths that connect to grounding electrodes. These conductive paths perform the function of a grounding electrode conductor, but they are not always wire-type conductors. A couple of examples are interconnected metal building frames and metal water piping systems that do not qualify as grounding electrodes by definition or by the details and descriptions provided in 250.52(A). Previous editions of the *NEC* referred to these conductive objects as grounding electrodes, but they do not meet the definition of the term *grounding electrode,* and they may function as grounding electrode conductors in this case. Section 250.68(C) addresses locations of connections permitted for grounding electrode conductors and bonding jumpers of a grounding electrode system. **See Figure 6-31.**

This section includes provisions recognizing these conductive paths that are ultimately connected to grounding electrodes but are not grounding electrodes themselves. Metal building frames that are electrically connected to a grounding electrode or a system of grounding electrodes are permitted to function as conductive paths to ground as provided in either (1) or (2) below:

1. The structural metal member is in direct contact with the Earth for a minimum of 10 feet, with or without encasement.
2. Hold-down bolts (anchor bolts) that secure the structural steel are connected to a concrete-encased electrode that complies with 250.52(A)(3). The anchor bolts must be connected to the rebar using welding or exothermic welding, by the usual tie wires, or by other approved means.

Either the reinforcing rod or bars or the copper conductor of a concrete-encased electrode shall be permitted to extend from the foundation or footing up to an accessible location above the concrete for connection of the grounding electrode conductor. This practice is fairly common because the concrete-encased electrode is often installed before the electrical contractor is on site. Section 250.68(C)(3) provides important criteria that must be met when

Figure 6-30 **Water Pipe Electrode Connection Requirements**

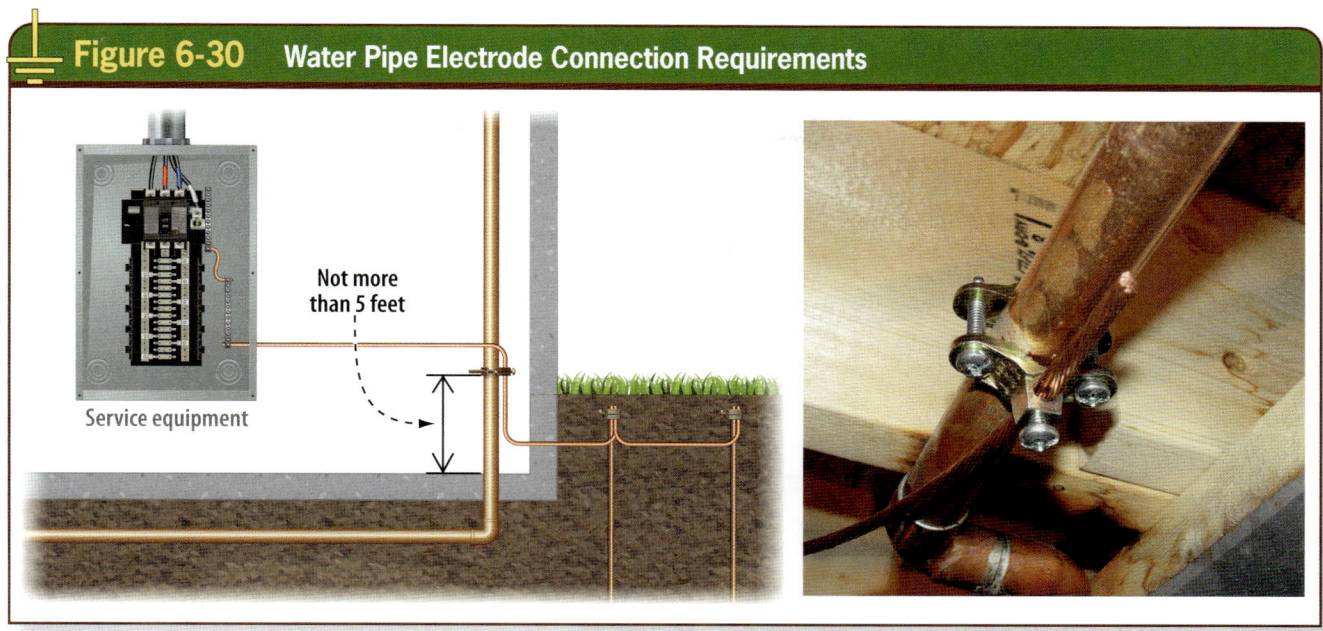

Figure 6-30. The grounding electrode conductor connection to a water pipe electrode must be made within five feet of the water pipe point of entry at the building or structure served.

installing a rebar extension from the foundation or footing to an accessible location for the grounding electrode conductor connection. First, the rebar extension must be connected to the rebar in the foundation or footing. Second, the rebar extension shall not be exposed to Earth contact without corrosion protection. Third, the rebar extension shall not be used to interconnect electrodes of the grounding electrode system.

Figure 6-31 **Conductive Paths to Grounding Electrodes**

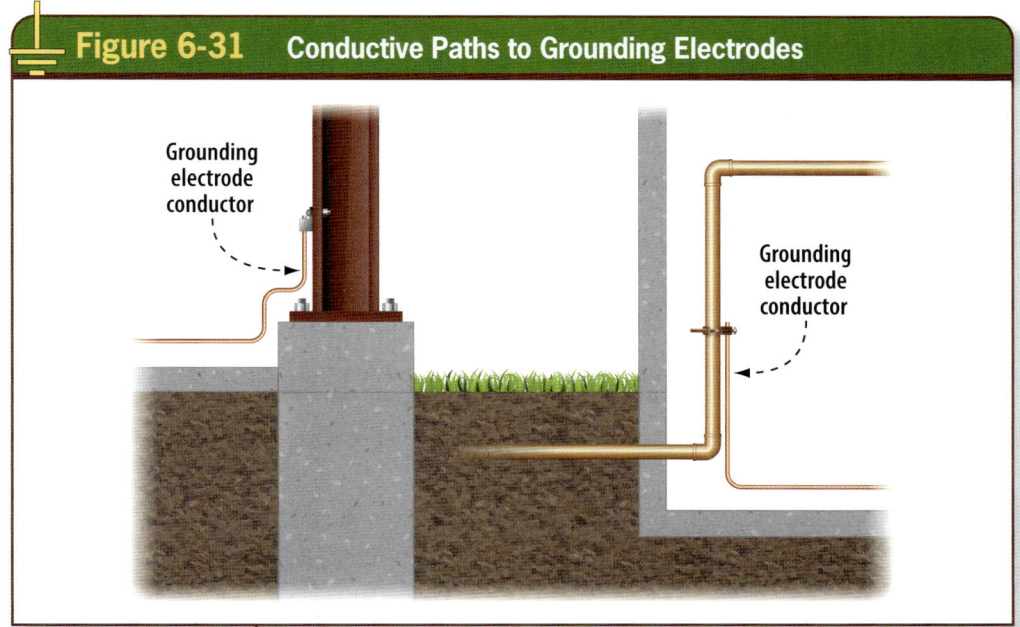

Figure 6-31. Building steel and metal water piping systems often perform as conductive paths to the grounding electrode system but are not grounding electrodes themselves.

GROUNDING ELECTRODE CONDUCTOR CONNECTIONS

Section 250.70 provides information about acceptable methods of grounding electrode conductor connections. **See Figure 6-32.**

Connections of grounding electrode conductors or bonding jumpers of the grounding electrode system can be made using one of the following methods:

1. Listed lugs
2. Exothermic welding processes
3. Listed pressure connectors
4. Listed clamps
5. Other listed means

The integrity of the connection to a grounding electrode system is important for safety and proper operation of the electrical system. Solder is obviously not permitted to be relied on for

---

Figure 6-32   Acceptable Grounding Electrode Conductor Connections

Listed lugs used for grounding electrode conductor connections.

Exothermic welding process used for grounding electrode conductor connections. (10-foot rod shown)

Listed pressure connectors used for grounding electrode conductor connections.

Listed clamps used for grounding electrode conductor connections.

Figure 6-32. *Grounding electrode conductor connections can be made using a variety of methods and materials that are listed and/or identified for this purpose.*

Courtesy of ABB

grounding electrode conductor connections because the integrity of soldered connections can be negatively affected by heat caused by the current during a ground-fault event. Listed grounding clamps must be compatible with the grounding electrode conductor material and must be suitable for the type of piping or material to which they are connected. This is important because dissimilar metals can cause a deterioration effect that can compromise the connection.

For grounding electrode conductors that are connected to rod, pipe, or plate  electrodes, the connection device must be listed for direct burial. Clamps that are listed for direct burial are also suitable for encasement in concrete as indicated in category KDER of UL Product iQ. **See Figure 6-33.**

For additional information, visit qr.njatcdb.org Item #5333

Grounding or bonding clamps must be listed and are typically suitable for connecting only one conductor. **See Figure 6-34.**

More than one grounding electrode conductor is permitted to be used with a single clamp where the clamp is identified for multiple conductors. Listed grounding clamps for grounding electrode conductors used with communications systems or other communications equipment are available in a strap-type configuration for indoor use only. These clamps must be listed and must have a rigid metal base and strap that encircles the piping system that is not likely to stretch after installation.

These types of grounding electrode conductor connections are not permitted for use on power system installations. Category KDSH in UL Product iQ provides additional information about communications system grounding electrode conductor connection devices. Grounding clamps are required to be protected from physical damage unless the fittings (clamps) are approved for use without additional protection. **See Figure 6-35.**

Figure 6-33. A ground clamp listed for direct burial is also suitable for concrete encasement.

Courtesy of ABB

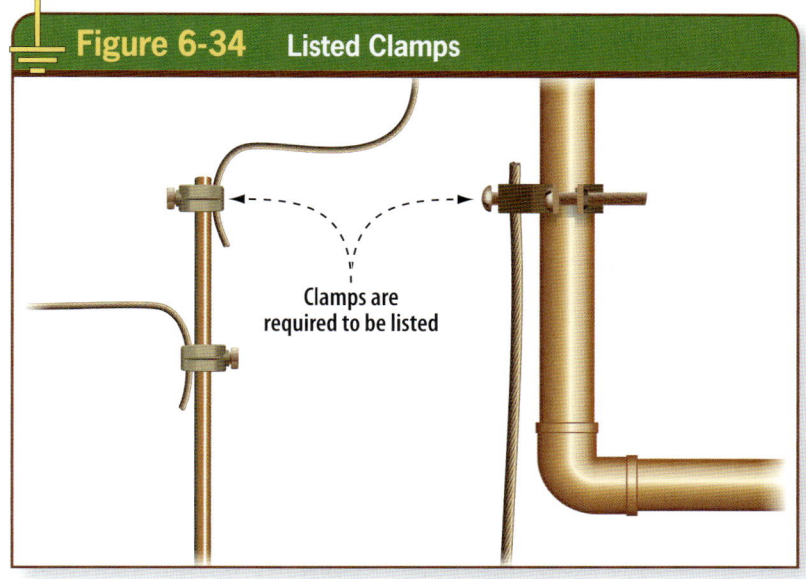

Clamps are required to be listed

Figure 6-34. Grounding clamps are generally limited to connection of one grounding electrode conductor unless identified for more than one.

Figure 6-35. Ground clamps are specifically listed for use with communications system grounding electrode conductors and bonding conductors.

Courtesy of ABB

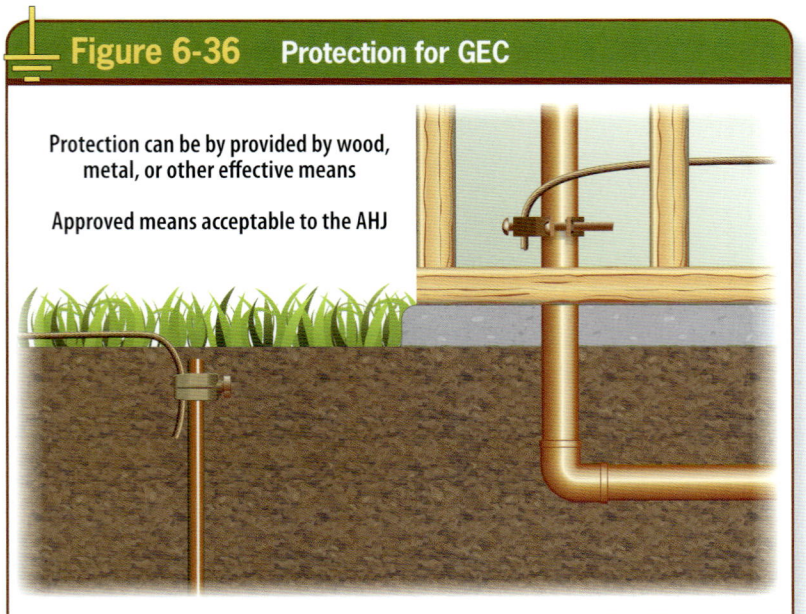

Figure 6-36 **Protection for GEC**

Protection can be by provided by wood, metal, or other effective means

Approved means acceptable to the AHJ

Figure 6-36. Protection is required for grounding electrode conductor connections unless the clamp or connection is approved for use without protection from damage.

Otherwise, the grounding clamp must be protected using metal, wood, or other equivalent materials as described in 250.10. **See Figure 6-36.**

MAGNETIC FIELD CONCERNS

Grounding electrode conductors must be protected if subject to physical damage. In addition to concerns about physical damage from external causes, grounding electrode conductors can be affected by magnetic fields. This is not a concern for grounding electrode conductors that are not installed in ferrous (magnetic) metal raceways or enclosures as a means for protection against physical damage, such as installation within Schedule 80 PVC conduit. However, if a grounding electrode conductor is installed in a ferrous metal raceway or enclosure, the raceway or enclosure must be electrically continuous from the point of attachment to the cabinet or equipment to the grounding electrode and must be securely fastened to the ground clamp or fitting.

Ferrous metal raceways or enclosures are those with iron or steel content

such as rigid metal conduit (RMC), intermediate metal conduit (IMC), or electrical metallic tubing (EMT). These conduits and tubing have a magnetic property that reacts to rising and falling magnetic fields present in AC systems. During a ground-fault event, the current in a grounding electrode conductor can be relatively high for the duration of the event.

The strength of the magnetic field will also increase in direct proportion to the amount of current in the conductor. In many cases, the magnetic lines of force in the conductor are induced into the conduit enclosing the grounding electrode conductor; they can even surpass the saturation point of the steel raceway. At the point where the grounding electrode conductor exits the conduit, the magnetic lines of force generated by the fault current in the conductor will try to be induced on the end of the conduit, creating a saturation point that exceeds the capacity of the conduit. The steel conduit in this instance is acting like a steel core of a coil to concentrate the magnetic lines of force. This creates what the industry refers to as a *choke effect*. Specific bonding requirements are necessary for ferrous metal raceways that contain grounding electrode conductors.

The Choke Effect

Ferrous metal raceways and enclosures are bonded to the contained grounding electrode conductor to reduce the effects of magnetic fields that are present while the system is energized and in use. The grounding electrode conductor for an AC system or service is an alternating current-carrying conductor. There are varying amounts of current in a grounding electrode conductor during normal operation. This current can also rise and fall significantly depending on events such as ground faults, short circuits, or line surges. As the current rises and falls, the magnetic field of the contained conductor gets larger and smaller accordingly. This means the stresses on the contained grounding electrode conductor increase and decrease as the current rises and falls. Because the ferrous metal raceway or enclosure surrounds this conductor, there is an inductive reactance between the ferrous metal raceway or enclosure and the contained grounding electrode conductor. This inductive reactance is one component of impedance, and actually impedes current in the grounding electrode conductor. The magnetic field and the capacitance result in a coupling effect between the current in the conductor and the surrounding ferrous metal raceway or enclosure.

In actuality, much of the current would be present in the ferrous raceway or enclosure rather than the contained grounding electrode conductor. This condition is often referred to as the *choke effect* because it restricts the grounding electrode conductor from performing its function.

Bonding Ferrous Metal Enclosures for Grounding Electrode Conductors

Ferrous metal raceways or enclosures for grounding electrode conductors that are not physically continuous from cabinets or equipment to the grounding electrode must be made electrically continuous by bonding each end of the raceway to the contained grounding electrode conductor. This action puts the contained grounding electrode conductor in parallel with the enclosing ferrous metal raceway or enclosure so that the two work together when the current in the grounding electrode conductor rises and falls in response to various events occurring on the system. **See Figure 6-37.**

The methods required for bonding each end of the raceway or enclosure are provided in 250.92(B)(2) through (B)(4). **See Figure 6-38.**

These methods apply to all intervening ferrous raceways, boxes, and enclosures containing the grounding electrode conductor. If a bonding jumper is used to accomplish this bonding to intervening metal raceways and enclosures, the size of the bonding jumper must not be smaller than the required

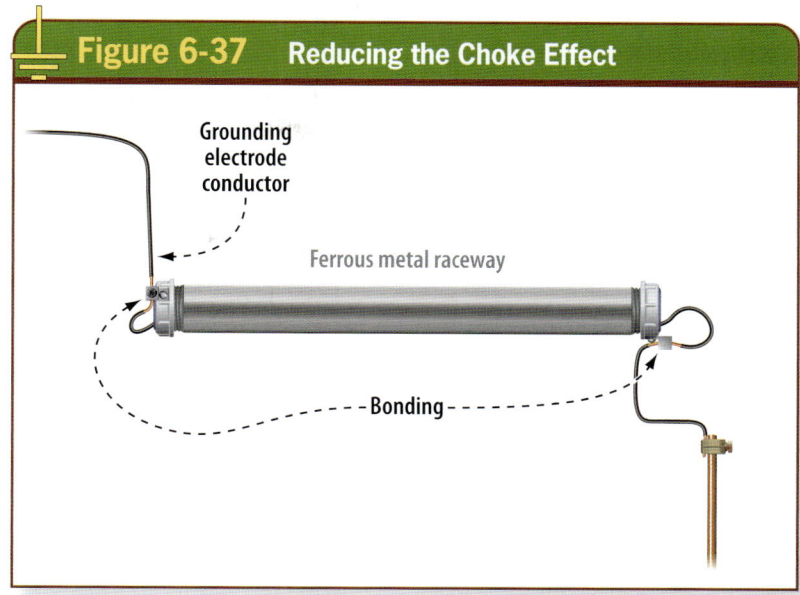

Figure 6-37 **Reducing the Choke Effect**

Figure 6-37. *If ferrous metal raceways are not physically continuous from the cabinet or equipment to the grounding electrode, bond both ends of the ferrous metal raceways that enclose grounding electrode conductors.*

enclosed grounding electrode conductor. If a raceway is used as protection for the grounding electrode conductor, the installation rules in Chapter 3 of the *NEC* are applicable for such raceways.

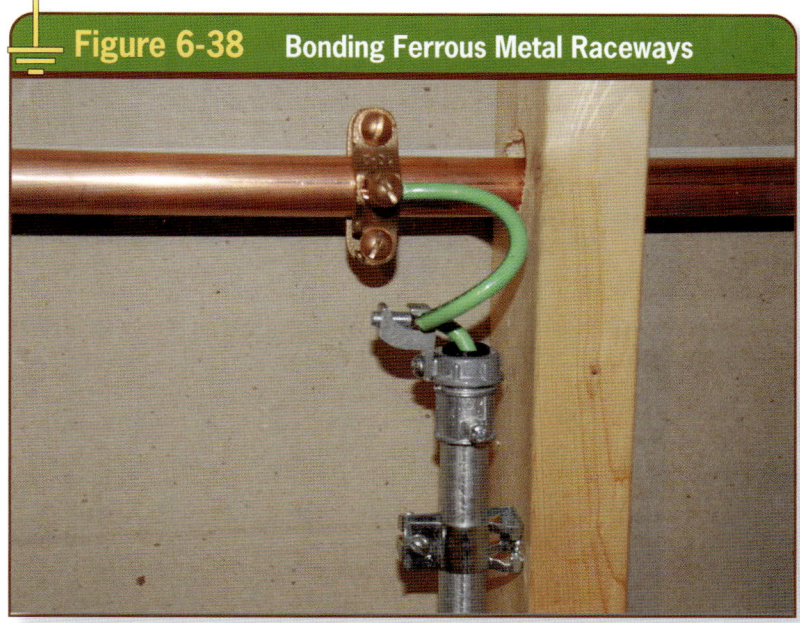

Figure 6-38 **Bonding Ferrous Metal Raceways**

Figure 6-38. *Bond both ends of ferrous metal raceways that enclose grounding electrode conductors.*

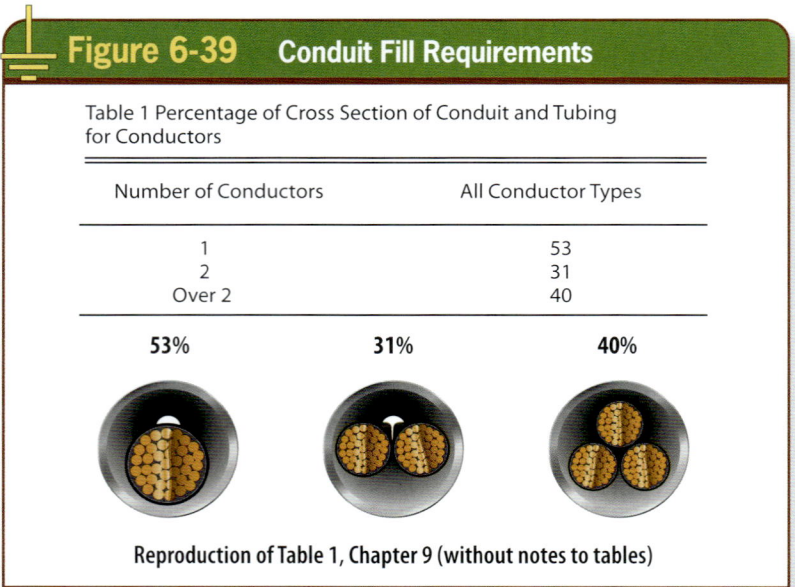

Figure 6-39   Conduit Fill Requirements

Table 1 Percentage of Cross Section of Conduit and Tubing for Conductors

Number of Conductors	All Conductor Types
1	53
2	31
Over 2	40

53%   31%   40%

Reproduction of Table 1, Chapter 9 (without notes to tables)

Figure 6-39. *Table 1 in Chapter 9 limits a single conductor to fill not more than 53% of the raceway. This reproduction does not include the notes to the table.*

Nonferrous conduits, such as brass and aluminum types, do not have magnetic content and do not have to be electrically continuous. Another important installation requirement to consider is the conduit and tubing fill percentage restrictions in Table 1 of Chapter 9 of the *NEC*. If a single conductor (grounding electrode conductor in this case) is installed in conduit or tubing, the percentage of fill must not exceed 53%. **See Figure 6-39.**

For smaller grounding electrode conductors installed in conduit or tubing, it is relatively easy to use a bonding bushing on the ends of the raceway and run the grounding electrode conductor back through it to complete the bonding connection. However, with larger grounding electrode conductors, it is more difficult to meet these bonding requirements. Because the bonding must be accomplished using the same size jumper as the grounding electrode conductor, two methods can be used for larger conductors and conduit combinations. One method is to use bonding bushings at each end of the raceway and use an irreversible compression connector to crimp on a short bonding jumper that connects to the bushing lug. **See Figure 6-40.**

Another method is to use bonding fittings manufactured specifically for this purpose. In this case, the hub-type fitting is installed on the end(s) of the conduit or tubing, and it includes a conductor clamp for completing an effective bonding connection between the hub and the grounding electrode conductor. Remember that per 250.64(E)(1), the bonding at each end must be accomplished using a conductor that is the same size as the grounding electrode conductor contained in the raceway. **See Figure 6-41.**

The requirement to bond the contained grounding electrode conductor at both ends reduces the choke effect by putting the contained conductor in parallel with the ferrous metal raceway or enclosure surrounding it. In this case, the ferrous metal raceway or enclosure and the contained grounding electrode conductor work together. Test data have demonstrated that for practical purposes, the impedance of an AC grounding electrode conductor enclosed in a ferrous metal conduit when the conduit is bonded at both

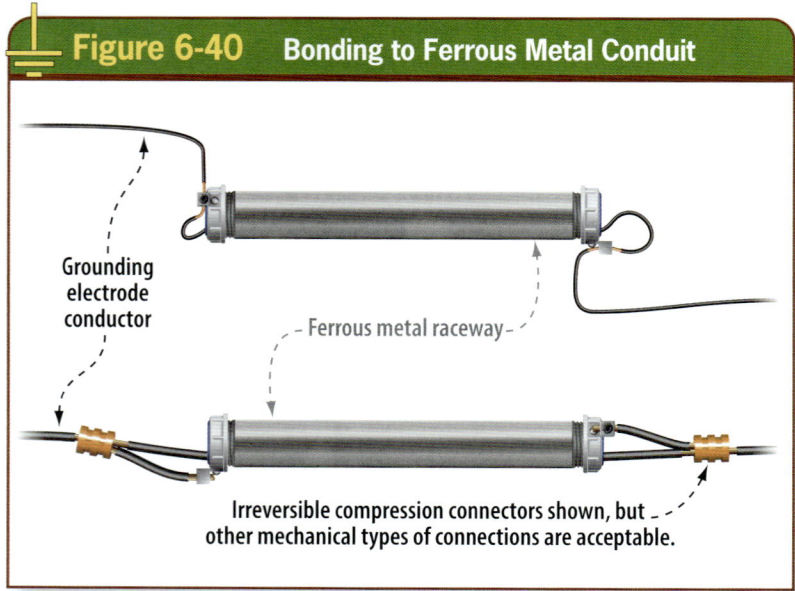

Figure 6-40   Bonding to Ferrous Metal Conduit

Grounding electrode conductor

- Ferrous metal raceway -

Irreversible compression connectors shown, but other mechanical types of connections are acceptable.

Figure 6-40. *Grounding electrode conductors are required to be bonded to enclosing ferrous metal conduit.*

ends is approximately equal to the impedance of the conduit itself.

Another simple solution to avoid the choke effect in grounding electrode conductors is to use Schedule 80 polyvinyl chloride (PVC) conduit as a means of providing physical protection. Be sure that the type of construction for the project does not restrict the use of non-metallic (plastic) wiring methods.

Because a grounding electrode conductor carries varying amounts of current during normal operation, it is generally not permitted to function simultaneously as an EGC, in accordance with 250.118(B). An exception recognizes that a wire-type equipment grounding conductor could simultaneously serve as a grounding electrode conductor if all the requirements in Parts II, III, and VI of Article 250 have been met. This method of installation must also comply with 250.6(A), which requires that grounding must be arranged in a manner that will prevent objectionable current over normally non–current-carrying conductors and metal parts of equipment.

Figure 6-41 **Conduit Hub and Clamp**

Figure 6-41. *A conduit hub fitting can be used for bonding the grounding electrode conductor to the ferrous metal raceway that contains it.*

Courtesy of ABB

SUMMARY

The grounding electrode conductor is an important component of the grounding and bonding system. This conductor performs the grounding function and maintains the connection to the Earth for systems and equipment. The grounding electrode conductor can carry varying degrees of current during normal operation and does not perform as an effective ground-fault current path. Grounding electrode conductor sizing is generally determined by using Table 250.66, unless the grounding electrode conductor is only connected to the type(s) of grounding electrodes covered in 250.66(A), (B), or (C). Grounding electrode conductor connections must be generally accessible unless they are buried, encased in concrete, or placed under fireproofing materials. Grounding electrode conductors are each generally required to be installed in a continuous length without a splice and in a manner that protects them from physical damage; otherwise, physical protection must be provided. Grounding electrode conductors can be copper, aluminum, or copper-clad aluminum. Aluminum and copper-clad aluminum grounding electrode conductors must not be terminated within 18 inches of the Earth unless such connections are made within enclosures suitable for the environment, such as open-bottom switchgear or switchboards where the busbar is within 18 inches of the Earth as indicated in 250.64(A)(2).

REVIEW QUESTIONS

1. **A grounding electrode conductor can be made of which of these materials?**
   a. Aluminum
   b. Copper
   c. Copper-clad aluminum
   d. Any of the above

2. **A conductor used to connect the system grounded conductor or the equipment to a grounding electrode or to a point on the grounding electrode system best defines which of the following?**
   a. Bonding jumper
   b. Equipment grounding conductor
   c. Grounded conductor
   d. Grounding electrode conductor

3. ***NEC* Table 250.66, used for sizing grounding electrode conductors, uses the largest __?__ to determine the minimum size grounding electrode conductor required for a service.**
   a. disconnect
   b. overcurrent protective device
   c. service equipment rating
   d. ungrounded service conductor size or equivalent area for parallel conductors

4. **Grounding electrode conductors provide the connection between the Earth and the object or system conductor required to be connected to ground.**
   a. True   b. False

5. **A grounding electrode conductor is an effective ground-fault current path that will facilitate operation of overcurrent protective devices during ground-fault conditions.**
   a. True   b. False

6. **A grounding electrode conductor has little or no effect on facilitating overcurrent protective device operation.**
   a. True   b. False

7. **A grounding electrode conductor is never sized based on an overcurrent protective device of a service or separately derived system.**
   a. True   b. False

8. **If the grounding electrode conductor is connected to a concrete-encased electrode, the portion of the grounding electrode conductor that is connected only to the concrete-encased electrode is not required to be larger than a __?__ copper conductor.**
   a. 4 AWG
   b. 3 AWG
   c. 2 AWG
   d. 1 AWG

9. **If the grounding electrode conductor is connected to a ground ring that is a 1 AWG copper conductor, the portion of the grounding electrode conductor that is connected only to the ground ring electrode is not required to be larger than __?__ copper.**
   a. 4 AWG
   b. 3 AWG
   c. 2 AWG
   d. 1 AWG

10. **If the DC system (source) is located on the premises, the grounding electrode conductor connection to the DC system must be made at which of the following locations?**
    a. At the first system disconnecting means or overcurrent protective device
    b. At the source
    c. Either a. or b.
    d. Neither a. nor b.

11. **A grounding electrode conductor for a DC system is generally required to be sized using Table 250.66.**
    a. True   b. False

12. **The requirements for installing grounding electrode conductors for AC systems are provided in __?__ of Article 250.**
    a. Part II
    b. Part III
    c. Part IV
    d. Part V

13. **A grounding electrode conductor smaller than 6 AWG copper generally must be protected by installation in a cable armor or raceway.**
    a. True   b. False

14. **What is the minimum size grounding electrode conductor for a 400-ampere service (600-kcmil copper service conductors) if the grounding electrode is an in-ground structural steel building frame as provided in 250.52(A)?**
    a. 3 AWG copper
    b. 2 AWG copper
    c. 1/0 AWG copper
    d. 1 AWG copper

15. **A grounding electrode conductor is required to be identified by the color green.**
    a. True   b. False

16. **Grounding electrode conductors in sizes 6 AWG and larger that are installed such that they are not exposed to physical damage can be run without being placed in a raceway or armor as long as they are securely fastened.**
    a. True   b. False

17. **Grounding electrode conductors generally must be installed in a continuous length without a splice or joint unless splices are made using __?__.**
    a. bars using listed connectors
    b. exothermic welding
    c. irreversible compression connectors listed as grounding and bonding equipment
    d. any of the above

18. **A grounding electrode conductor is required to be connected to a grounding electrode using which of the following methods?**
    a. Exothermic welding process
    b. Listed lugs or listed clamps
    c. Listed pressure connectors
    d. Any of the above

19. **If a single ground rod is installed as the grounding electrode for a service, what is the maximum size grounding electrode conductor required for the installation?**
    a. 8 AWG copper
    b. 6 AWG copper
    c. 4 AWG copper
    d. 2 AWG copper

20. **The mechanical connections to a grounding electrode are required to be accessible, unless __?__.**
    a. exothermic welding is used as the connection means or the connections are made by irreversible compression connectors to fireproofed structural metal
    b. grounding electrodes are buried, as with rods, pipes, or plate electrodes
    c. the electrodes at the connection are encased in concrete
    d. any of the above

21. **If a grounding electrode conductor or bonding jumper is connected to a busbar, the busbar must be a minimum of ¼ inch thick by two inches wide and not less than 12 inches in length.**
    a. True   b. False

22. **If a grounding electrode conductor is installed in a ferrous metal raceway or enclosure, the raceway or enclosure must be electrically continuous from the point of attachment to the cabinet or equipment to the grounding electrode and must be securely fastened to the ground clamp or fitting.**
    a. True   b. False

23. **If a grounding electrode conductor is protected from physical damage with PVC conduit, the conduit is required to be Schedule 80.**
    a. True   b. False

Bonding Requirements

Bonding is the process of connecting conductive parts or equipment together. Section 250.4 contains important performance criteria clearly describing what electrical bonding is intended to accomplish. Bonding methods and sizing of equipment bonding jumpers are covered in Part V of Article 250, along with the requirements for bonding piping systems, structural metal building frames, and other conductive parts within or attached to buildings or structures. Electrical bonding requires effective electrical connections to ensure that optimal performance is achieved at any point in the electrical system.

Objectives

» Understand the *NEC* definitions related to electrical bonding requirements.

» Understand the requirements for bonding on the supply side of the service disconnecting means.

» Determine where equipment bonding jumpers are required.

» Understand the difference between supply-side and load-side bonding requirements and establish appropriate sizes for supply-side bonding jumpers and load-side bonding jumpers.

» Understand the importance of bonding in the effective ground-fault current path.

» Determine the requirements for bonding metal piping systems and structural metal building framing.

Chapter 7

Table of Contents

Definitions of Bonding Terms 158

Bonding Performance Criteria 158

Maintaining Continuity 160

Bonding Connections
(Wire-Type Conductors) 161

Cleaning Coated Surfaces 163

Bonding Jumper and Bonding
Conductor Length 163

Equipment Bonding Jumpers
(Function and Purpose) 164

Sizing Requirements for the Supply Side
and Load Side Applications 164

Service Bonding Rules (Supply Side) 166

Reducing Washers 168

Boxes With Concentric or
Eccentric Knockouts 168

Sizing Supply-Side Bonding Jumpers
(Wire Types) 169

Load Side Bonding Rules 170

   General Equipment Bonding Rules 170

   Installation and Size of Equipment
   Bonding Jumpers 171

Bonding Around Expansion Fittings and
Loose-Joined Metal Raceways 173

Bonding Other Metal Parts 174

   Bonding Metal Piping Systems 174

      Bonding and Accessibility 174

      Metal Water Piping in Multiple
      Occupancy Buildings 175

      Multiple Buildings or Structures
      Supplied by a Feeder
      or Branch Circuit 176

      Bonding Other Metal Piping Systems .. 176

   Bonding Structural Metal
   Building Frames 178

   Bonding Lightning Protection Systems 179

Summary 179

Review Questions 180

DEFINITIONS OF BONDING TERMS

To understand and properly apply electrical bonding requirements in the *NEC*, a thorough understanding of defined bonding terms must be developed. When a term is used in a particular *NEC* rule, the definition helps clarify the intent and meaning of that rule. The following are defined bonding terms discussed in this chapter:

Bonded (Bonding). Connected to establish electrical continuity and conductivity. (CMP-5)

Bonding Conductor (Bonding Jumper). A reliable conductor to ensure the required electrical conductivity between metal parts required to be electrically connected. (CMP-5)

Bonding Jumper, Equipment (Equipment Bonding Jumper). The connection between two or more portions of the equipment grounding conductor. (CMP-5)

Bonding Jumper, Main (Main Bonding Jumper). The connection between the grounded circuit conductor and the equipment grounding conductor, or the supply-side bonding jumper, or both, at the service. (CMP-5)

Bonding Jumper, Supply-Side (Supply-Side Bonding Jumper). A conductor installed on the supply side of a service or within a service equipment enclosure(s), or for a separately derived system, that ensures the required electrical conductivity between metal parts required to be electrically connected. (CMP-5)

Bonding Jumper, System (System Bonding Jumper). The connection between the grounded circuit conductor and the supply-side bonding jumper, or the equipment grounding conductor, or both, at a separately derived system. (CMP-5)

Service Point. The point of connection between the facilities of the serving utility and the premises wiring. (CMP-10)

BONDING PERFORMANCE CRITERIA

In electrical systems, grounding and bonding functions are often accomplished at the same time and are usually accomplished through a single action, often through a single conductive path or conductor. In other words, if a system and equipment are grounded and other conductive parts, such as conduit, are connected to such grounded equipment, the function of bonding has taken place and grounding is accomplished simultaneously. The result is grounded and bonded equipment. **See Figure 7-1.**

Section 250.4 provides general performance requirements that explain what electrical grounding and bonding are intended to accomplish and how these functions are expected to perform during normal and abnormal conditions. Section 250.4 consists of two subdivisions: (A) and (B). The difference between these two subdivisions is that (A) deals with grounding and bonding performance criteria for grounded systems and associated equipment, whereas (B) provides the performance criteria for systems that are ungrounded. General performance requirements for electrical bonding are provided in both subdivisions (A) and (B) of 250.4, specifically (A)(3), (A)(4), (B)(2), and (B)(3). The goals of bonding are the same for grounded systems and ungrounded systems. **See Figure 7-2.**

The *NEC* also provides bonding requirements for conductive materials foreign to electrical equipment. Examples of such conductive materials within buildings or structures are structural metal building frames and metal piping systems. The bonding required by 250.4(A)(4) is intended to establish a direct path back to the source from materials that are "likely to become energized." The term *Energized, Likely to Become* is now defined in the 2023 *NEC* to assist in the requirements of the *NEC*. This phrase appears multiple times and requires careful judgment when it is applied to installations and systems.

Energized, Likely to Become. (Likely to Become Energized) Conductive material that could become energized because of the failure of electrical insulation or electrical spacing. (CMP-5)

Bonding takes place at each and every connection of conductive materials or equipment. For bonding to be effective and perform properly during ground-fault conditions, each fitting, bushing, connector, coupling, and so forth must be made up tight to keep impedance low if a fault should occur and cause current through the conduit, enclosures, and fittings. The effectiveness of all bonding connections is directly related to a skillfull installation. Good skill and quality are essential in all aspects of electrical installations. Loose fittings and poorly installed wiring methods that are not supported properly can impair bonding connections in metal wiring methods installed for services, feeders, and branch circuits.

Fact

The National Electrical Contractors Association (NECA) publishes a family of National Electrical Installation Standards (NEIS) that help define what is meant by skillful installation in electrical construction. The *NEC* requires electrical wiring and equipment to be installed in a professional and skillfull manner, but it leaves this requirement open-ended and subject to interpretation. The NEIS establishes a benchmark of quality in multiple aspects of the electrical contracting business, covering subjects such as installing generators, switchboards, panelboards, conduit, fire alarm systems, transformers, security systems, and many more. More information and the full library of NEIS is available at www.neca-neis.org.

For additional information, visit qr.njatcdb.org Item #2281

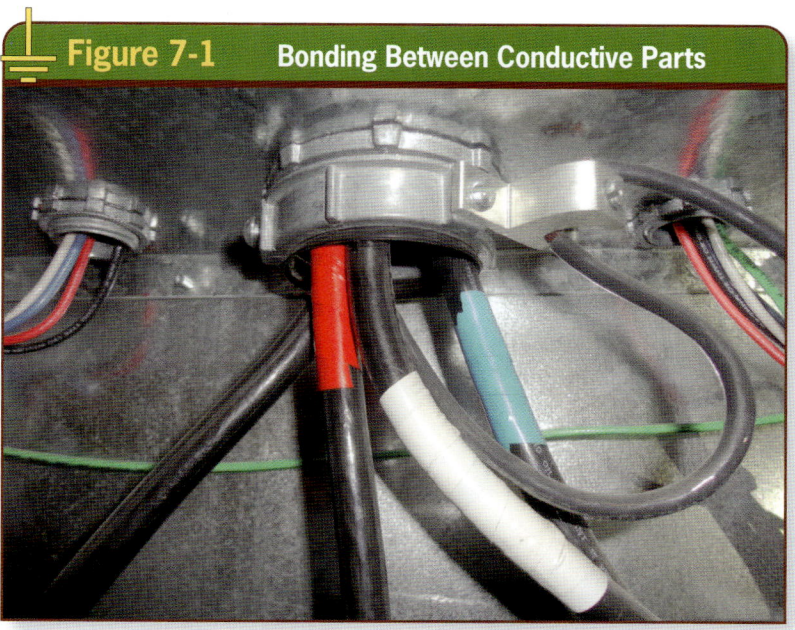

Figure 7-1. Bonding establishes continuity and conductivity between conductive parts of equipment required to be electrically bonded together.

It is important to understand the general intent of bonding so that the more prescriptive requirements in Article 250 are easily understood and applied to installations and systems. Specific bonding requirements are provided in Part V of *NEC* Article 250.

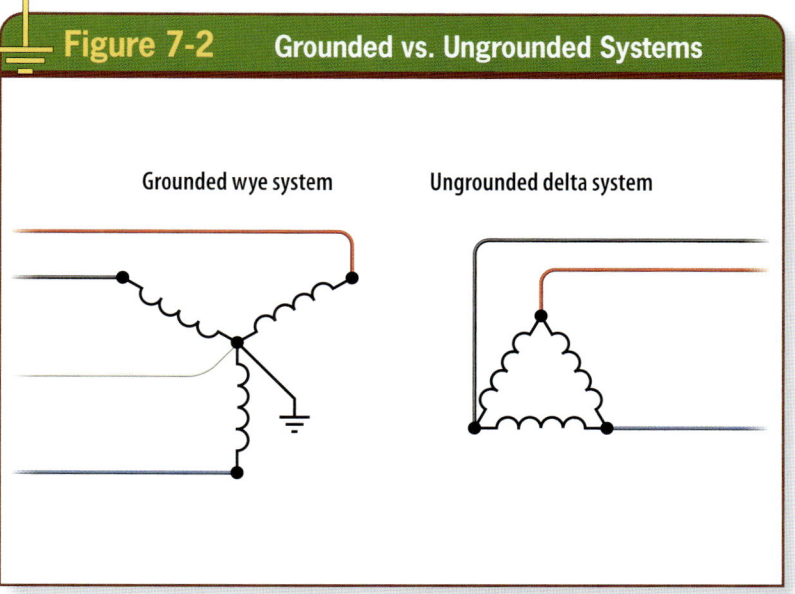

Figure 7-2. Section 250.4(A) covers grounded systems. Section 250.4(B) covers ungrounded systems.

Figure 7-3. Fittings (couplings) are used with EMT to connect (bond) 10-foot sections.

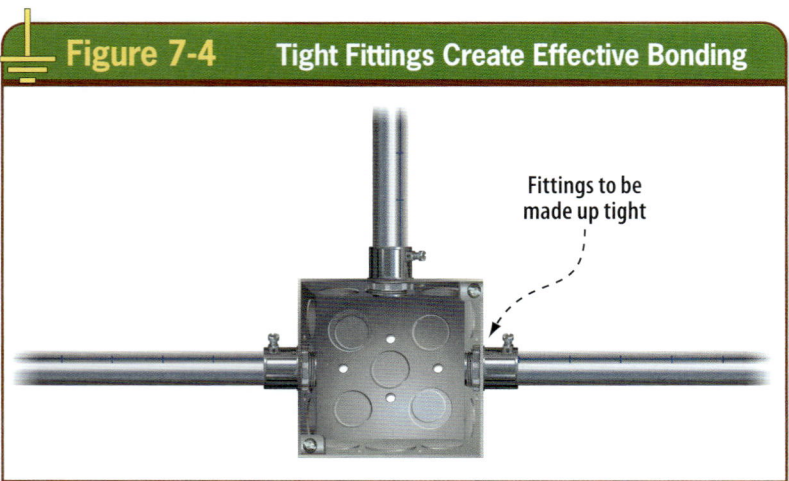

Fittings to be made up tight

Figure 7-4. Listed fittings provide bonding continuity between conductive parts of equipment and raceways.

Figure 7-5. Fittings for electrical conduit and tubing systems must be made up tight and installed in a professional and skillful fashion.

The function of bonding occurs throughout the entire electrical installation, as is clearly illustrated in Figure 250.1 of Article 250. Notice how bonding is related to all the other parts of the article. When there is an electrical connection, bonding happens.

A good example of bonding as an ongoing function is the installation of rigid metal conduit (RMC) or electrical metallic tubing (EMT) from one point to another. As the conduit is connected to an enclosure (panelboard, switchboard, box, and so forth), bonding occurs. The run of conduit is then assembled using couplings (fittings) to connect the 10-foot sections. **See Figure 7-3.**

This is a bonding function because continuity and conductivity are established between the sections or lengths of conduit or tubing connected by listed couplings and connectors. Fittings used with electrical wiring methods, specifically metal raceways, are required to be listed and to have been evaluated for grounding and bonding purposes. Testing laboratories have evaluated the effectiveness of various metal wiring fittings for conduit and tubing, and have determined that the fittings typically provided satisfactory performance when good metal-to-metal contact was accomplished. **See Figure 7-4.**

Listed metal raceway fittings will only perform effectively when skillfully installed by a professional with proper securement and support. The supports must meet the minimum requirements based on the applicable wiring method article in the *NEC*; furthermore, installed fittings must be made up tight using suitable tools, as required in 250.120(A). **See Figure 7-5.**

MAINTAINING CONTINUITY

The process of mechanically and electrically connecting metal parts or enclosures to conductors or components provides bonding. This establishes the required conductivity between them. In general, bonding can be accomplished by metal-to-metal contact using suitable listed fittings or by using bonding bushings and jumpers. **See Figure 7-6.**

The primary objective of electrical bonding is to make two conductive objects become one electrically. Bonding conductive parts of an electrical system together puts the parts at the same or nearly the same potential. Section 250.90 provides a general requirement to bond where necessary to ensure electrical continuity and the capacity to conduct safely any imposed fault current. Fault current (short-circuit current) is typically at higher levels than the normal operating current of a system. **See Figure 7-7.**

The amount of fault current that bonding connections must endure is related to the amount of available fault current delivered by the system or electric utility. Bonding connections must endure the higher levels of fault current for a duration long enough to cause overcurrent protective devices to operate. This requires all bonding connections in the ground-fault current path to be mechanically and electrically effective during ground-fault events. Maintaining effective bonding continuity is required from the service equipment all the way to the farthest outlet supplied by the system. Any loose or improper bonding connections in this path can compromise the performance of the electrical safety system (the grounding and bonding circuits).

Understanding the "weakest link in the chain" concept is important. The entire electrical safety circuit is only as good as the weakest link. All bonding connections, such as locknuts and fitting setscrews, must be made using suitable fittings and connections and adequate sizes of jumpers or bonding conductors to result in effective operation during both normal and ground-fault conditions. Section 250.96 requires that bonding be provided around connections of metal raceways, cable trays, cable armor, cable sheaths, enclosures, frames, fittings, and other

Figure 7-6. *A bonding jumper is used to connect a cable tray to EMT at a transition point.*

normally non–current-carrying conductive parts used as equipment grounding conductors (EGCs) or that are otherwise in the effective ground-fault current path.

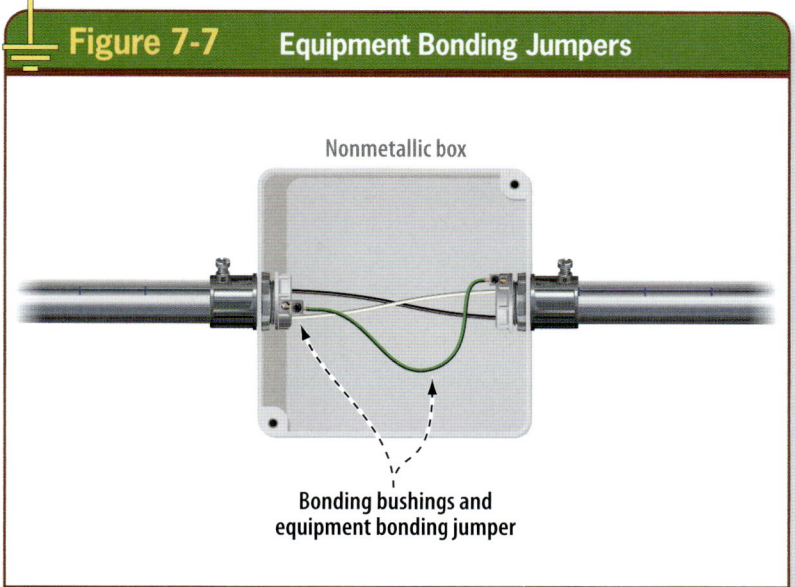

Figure 7-7. *Bonding establishes continuity and conductivity to bring conductive parts to the same potential.*

Figure 7-8   Various Methods for Accomplishing Bonding

Listed wire pressure connectors are suitable for grounding and bonding connections.

Available separately or as accessories for switchboards, panelboards, and other equipment
Equipment grounding terminal bars are often installed in panelboards and switchboards.

Pressure connectors and irreversible compression connectors are available and listed for use in electrical connections, and they can also be listed as grounding and bonding equipment.

Exothermic welding processes are suitable for grounding and bonding connections. They are not required to be listed, but the manufacturer's instructions must be followed.

Metal building frame

Machine screw fasteners can be used to connect (such as grounding screws) as long as no fewer than two full threads are engaged. Machine screws that are secured with a nut are also suitable for grounding and bonding connections.

Figure 7-8. Bonding connections can be accomplished using any of the various methods addressed in the NEC.

BONDING CONNECTIONS (WIRE-TYPE CONDUCTORS)

For EGCs, bonding conductors, and bonding jumpers to perform effectively, tight connections must be made. Section 250.8 provides the acceptable methods of making grounding and bonding connections and applies generally to connections for wire-type conductors. Several methods and various devices provide many choices for grounding and bonding connections. The type of connection made using one or more of the means provided in 250.8 is usually determined based on the equipment type and location in the system. **See Figure 7-8.**

The following connection means are provided in 250.8:
1. Listed pressure connectors
2. Terminal bars
3. Pressure connectors listed as grounding and bonding equipment
4. Exothermic welding processes
5. Machine screw-type fasteners engaging not less than two threads or secured with a nut
6. Thread-forming machine screws engaging not less than two threads in the enclosure
7. Connections that are part of a listed assembly
8. Other listed means

CLEANING COATED SURFACES

An important aspect of any electrical connection is the contact made between the conductive parts. For effective electrical connections, the joint or termination must offer little to no opposition in the electrical circuit.

Section 250.12 of the *NEC* provides specific language that addresses cleaning of surfaces. Coated electrical products such as painted enclosures and nonconductively coated raceways can introduce additional impedance in the grounding and bonding system. For this reason, the *NEC* requires that coated or painted surfaces be cleaned to remove coatings such as paint, lacquer, and enamel from threads and contact surfaces. This helps ensure electrical continuity.

Figure 7-9. *Coatings must be removed for effective bonding connections.*

To accomplish this, the coatings on some enclosures must be removed to establish metal-to-metal contact between parts required to be connected (bonded). Section 250.12 indicates that removal of paint or coatings is not necessary if the fittings used for the installation are designed to make such removal unnecessary. It is a good approach to verify this capability with the fitting manufacturer. **See Figure 7-9.**

Obviously, if the enclosures are uncoated, allowing metal-to-metal contact to be established, the bonding required by the *NEC* is accomplished. Many panelboards and other electrical enclosures are manufactured without paint or nonconductive coatings that could impair proper continuity and conductivity at connection points in the system. With this type of equipment, additional steps to achieve contact between metal parts are not necessary. **See Figure 7-10.**

Figure 7-10. *Bonding is established by tightened locknuts, which create metal-to-metal contact because the enclosure is not coated or painted.*

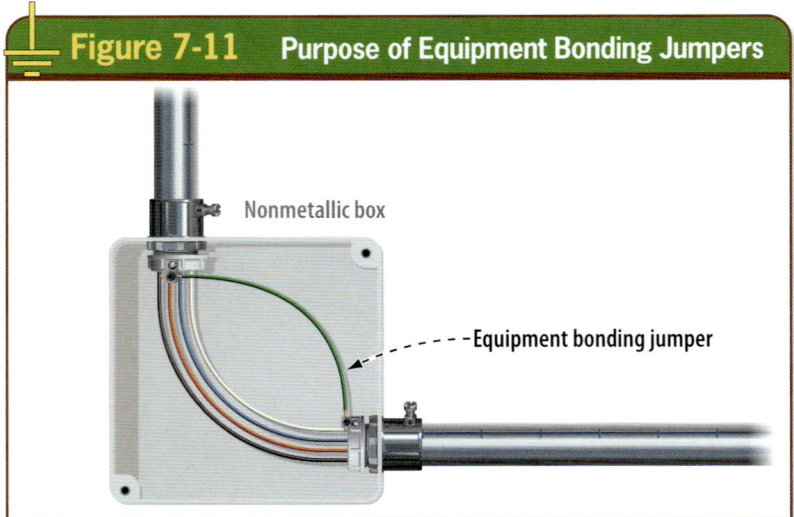

Figure 7-11   **Purpose of Equipment Bonding Jumpers**

Nonmetallic box

Equipment bonding jumper

Figure 7-11. Bonding is accomplished by using an equipment bonding jumper connected by bonding-type bushings. Equipment bonding jumpers are required in 314.3 Exception No. 1 to be installed between sections of EMT in nonmetallic boxes.

BONDING JUMPER AND BONDING CONDUCTOR LENGTH

The defined terms *bonding conductor* or *bonding jumper* address two concepts that accomplish the same purpose: to ensure electrical conductivity and continuity between metal parts.

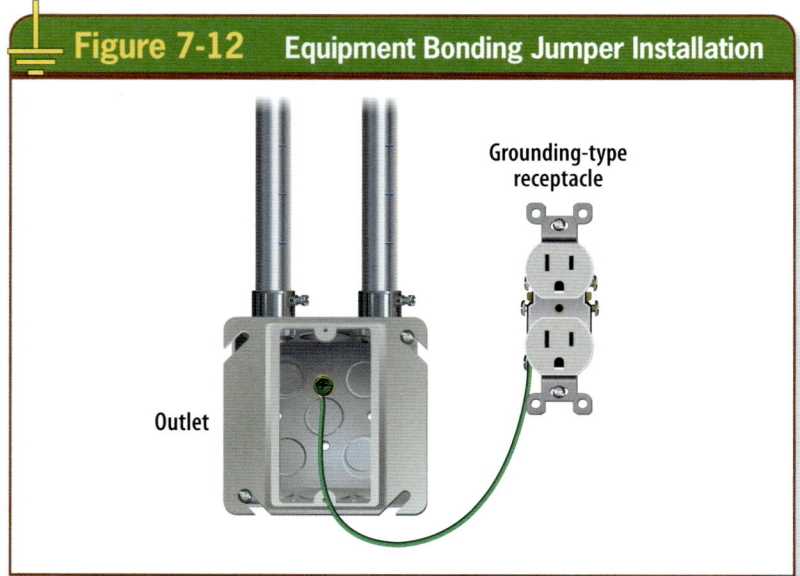

Figure 7-12   **Equipment Bonding Jumper Installation**

Grounding-type receptacle

Outlet

Figure 7-12. A short equipment bonding jumper is installed from a grounded box to a grounding-type receptacle.

A *bonding jumper* is generally understood to be a relatively short bonding means, while a *bonding conductor* has generally been understood to be a conductor with a length exceeding that of a jumper.

For example, if a bonding jumper is installed from a grounded metal box to a grounding-type receptacle, as covered in 250.146, it is relatively short in length—about six to ten inches. On the other hand, if a bonding jumper is installed to connect two grounding electrodes to form a grounding electrode system, as required in 250.50 and 250.53(C), it will be much longer, depending on the particular installation and the relative location of, and distance between, the grounding electrodes in the system.

EQUIPMENT BONDING JUMPERS (FUNCTION AND PURPOSE)

Equipment bonding jumpers are described in the definition as the connection between two or more portions of the EGC. Bonding jumpers are connections between two conductive objects that establish continuity and conductivity between them. An example of an equipment bonding jumper installation is one that is used to connect two sections of EMT attached to a nonmetallic enclosure. An equipment bonding jumper is installed between the two metal conduits using bonding bushings. **See Figure 7-11.**

Another example of an equipment bonding jumper installation is the connection between a grounded metal outlet box and a grounding-type receptacle. **See Figure 7-12.**

Equipment bonding jumpers are often used for metal conduits that emerge from the slab under an open-bottom switchboard or motor control center. An equipment bonding jumper(s) is installed from the bonding bushing on the conduit to the equipment grounding terminal bar within the equipment enclosure. **See Figure 7-13.**

Figure 7-13 Bonding Bushings

Figure 7-13. Bonding bushings are installed on conduits entering a switchboard enclosure, and bonding jumpers connect to the equipment grounding terminal bar.

SIZING REQUIREMENTS FOR THE SUPPLY SIDE AND LOAD SIDE APPLICATIONS

Determining sizing for bonding jumpers and bonding conductors is accomplished by following the prescriptive language in a particular *NEC* rule that provides bonding jumper or conductor sizing. The *NEC* includes two general sizing requirements for these conductors and jumpers.

The applicable sizing process for each depends on whether the bonding jumper or conductor is on the supply side of a service or source overcurrent protective device or if it is on the load side of an overcurrent protective device, as covered in 250.102. The definition of the term *service point* clarifies which portion of the service installation is covered by the *NEC* and which portion is the responsibility of the electric utility.

An example of supply-side bonding is often found at the service equipment enclosure for metal conduit containing service-entrance conductors supplying the service disconnecting means. Supply-side bonding is also required on the supply side of a source or system, such as a separately derived system, that is not a service. The conductors from the service point to the service disconnect are usually not protected at their ampacity. In this case, line- or supply-side bonding rules apply.

Sizing bonding jumpers on the supply side of the service overcurrent protective devices is accomplished using Table 250.102(C)(1). Sizing is based on the size of the largest ungrounded service-entrance conductor supplying the service disconnecting means. **See Figure 7-14.**

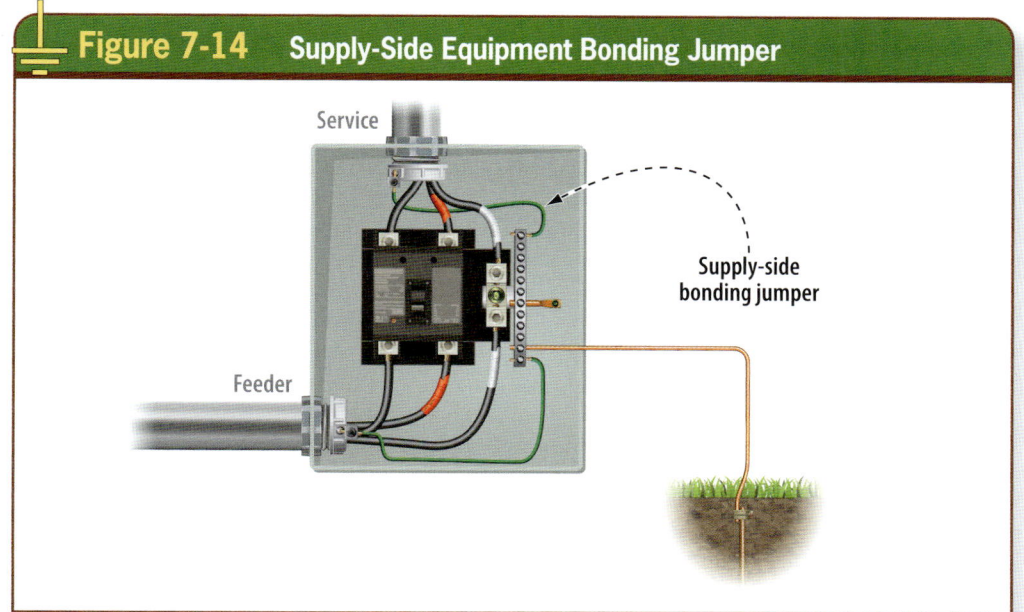

Figure 7-14   Supply-Side Equipment Bonding Jumper

Service

Supply-side bonding jumper

Feeder

Figure 7-14. Supply-side bonding jumpers are sized in accordance with 250.102(C), specifically Table 250.102(C)(1).

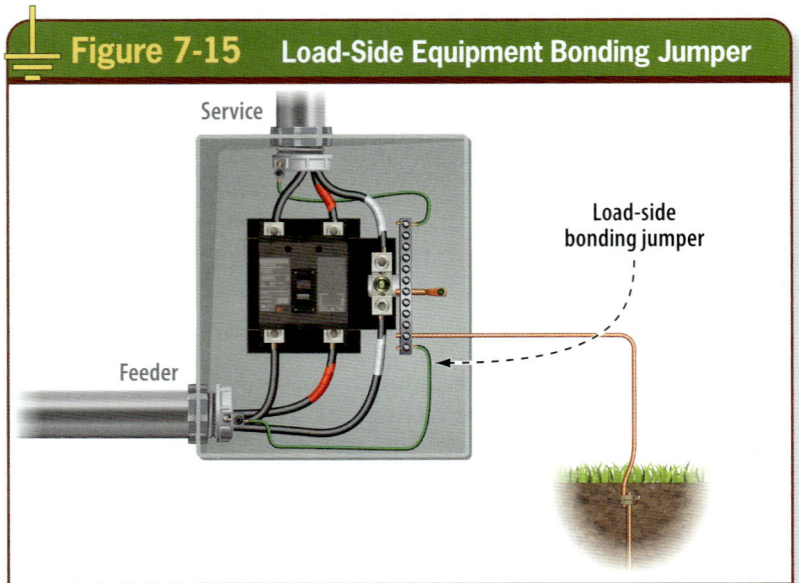

Figure 7-15   Load-Side Equipment Bonding Jumper

Service

Load-side
bonding jumper

Feeder

Figure 7-15. Load-side equipment bonding jumpers are sized per 250.102(D). Refer to Table 250.122.

Remember that the Notes to tables are mandatory. For conductors larger than 1,100 kcmil copper or 1,750 kcmil aluminum, use 12.5% as stated in Note 1 to that table.

Conversely, equipment bonding jumpers for a feeder supplying a panelboard from the service equipment enclosure(s) are on the load side of an overcurrent protective device. In this case, the equipment bonding jumper is sized using Table 250.122 based on the rating of the overcurrent protective device for the feeder circuit. Notice that *supply-side bonding jumper* is the term typically used on the line side; the term *equipment bonding jumper* is used on the load side of overcurrent protective devices. **See Figure 7-15.**

SERVICE BONDING RULES (SUPPLY SIDE)

Section 250.92 provides rules related to bonding methods for enclosures, raceways, and other normally non–current-carrying metal parts at the service and on the supply side of the service disconnecting means and overcurrent protective device. These conductive enclosures contain conductors supplied directly from a utility source that are usually not protected at their ampacity. During ground-fault conditions, these metal enclosures and raceways carry high levels of fault current for the duration of time it takes for the overcurrent protective device on the primary side of the utility transformer to open the circuit. For this reason, bonding requirements for metal parts on the supply side of the service disconnect are more restrictive and result in more robust or strengthened bonding installations.

Section 250.92 is divided into two parts. Subdivision (A) addresses the bonding requirement generally and the service parts (equipment) required to be bonded; subdivision (B) describes methods by which the bonding requirements in (A) can be accomplished.

The following metal parts of equipment containing service conductors are required to be bonded together:
1. Raceways
2. Cable trays
3. Cable bus frames
4. Auxiliary gutters
5. Service cable armor or sheath

In addition, all enclosures containing service conductors, such as meter enclosures, panelboards, switchboards, boxes, and so forth, are considered interposed in (set within) the service raceway installation and must be bonded together. **See Figure 7-16.**

Bonding the metal parts at services can be accomplished using any of the following methods:
1. Use of the grounded conductor (often a neutral conductor) for bonding enclosures together
2. Connections using threaded couplings or listed threaded hubs on enclosures if made up wrench-tight
3. Threadless couplings or connectors for metal raceways and metal-clad cables (where made up tight)
4. Other listed devices, such as bonding-type locknuts or bonding bushings

It is important to bond around any compromised entries to enclosures, such as reducing washers and oversized, concentric, or eccentric knockouts. Often, the process of removing concentric or eccentric knockouts

Figure 7-16   Bonding Service Raceways or Enclosures

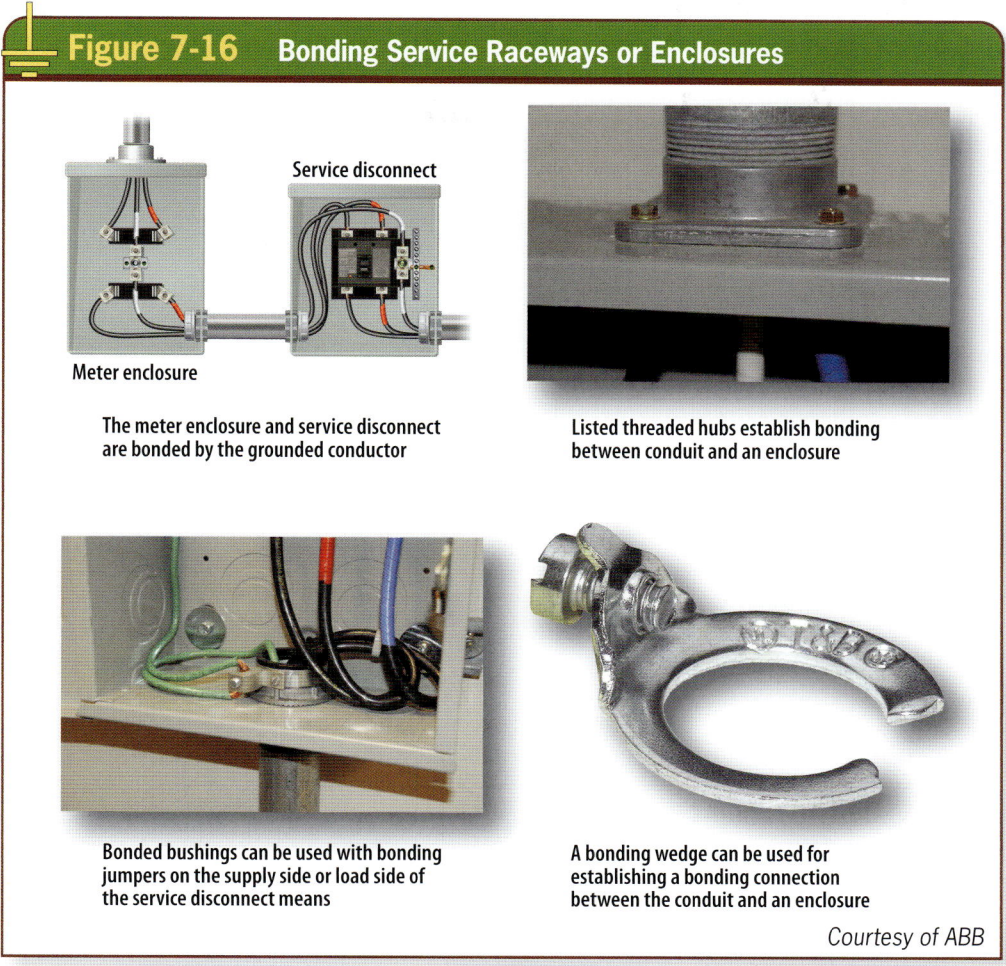

The meter enclosure and service disconnect are bonded by the grounded conductor

Listed threaded hubs establish bonding between conduit and an enclosure

Bonded bushings can be used with bonding jumpers on the supply side or load side of the service disconnect means

A bonding wedge can be used for establishing a bonding connection between the conduit and an enclosure

Courtesy of ABB

Figure 7-16. *The bonding required for metal raceways or enclosures containing service conductors can be accomplished by any of the methods included in 250.92(B) of the* NEC.

results in the loosening of the other concentric or eccentric rings in the punched enclosure, which can compromise the integrity of the bonding connection to the enclosure.

For services, standard locknuts or bushings are not permitted as the only means to accomplish the bonding required for service raceways and enclosures. Standard locknuts, on both sides of the enclosure, are acceptable for making the mechanical connection of the service raceway to the enclosure, but they are not suitable for the heavier bonding prescribed in 250.92. In this case, additional bonding means such as bonding bushings and jumpers, bonding wedges, bonding locknuts, or other methods must be used in addition to the standard locknuts.

Bonding locknuts are effective because, in addition to the teeth of the locknut, the setscrew provides an effective and strengthened bonding connection between the locknut and the enclosure.

Use of bonding wedges is another effective method to establish a bonding connection between a fitting or raceway and an enclosure. Bonding wedges provide two ways to ensure effective bonding between a raceway or fitting and the enclosure to which it is fastened. The first method of installing a bonding wedge is effective for bonding without the use of a bonding jumper, but only where concentric or eccentric knockouts are not encountered. The second method of installing a bonding wedge is used when concentric knockouts are encountered. In this scenario, a bonding jumper must be installed from one terminal screw on the wedge to the enclosure.

Strengthened bonding methods are also required for installations in health care facilities and hazardous locations. Section 250.100 provides the requirements for equipment and enclosures installed in hazardous locations. This section requires the methods of bonding prescribed in 250.92(B)(2) through (B)(4), regardless of the voltage of the circuit.

Where concentric, eccentric, or oversized knockouts are encountered in the installation, the bonding requirements in 250.97 apply if the circuit voltage exceeds 250 volts phase-to-ground. In this case, the electrical continuity of metal raceways and cables with metal sheaths that contain any conductor other than service conductors must be ensured by one or more of the methods specified for services in 250.92(B)(2) through (B)(4).

The exception to 250.97 relaxes this bonding requirement for installations where oversized, concentric, or eccentric knockouts are not encountered or where a box or enclosure with concentric or eccentric knockouts is listed. Listed outlet boxes have been evaluated for the bonding required by this section.

If additional bonding is required, one of the following methods can be used to establish an effective bonding connection:

1. Threadless couplings and connectors for cables with metal sheaths
2. Two locknuts on RMC or intermediate metal conduit—one inside and one outside the box or cabinet
3. Fittings with shoulders that seat firmly against the box or cabinet, such as EMT connectors, flexible metal conduit connectors, and cable connectors, with one locknut on the inside of the box or cabinet
4. Listed fittings

Figure 7-17. *Listed reducing washers are suitable for bonding when they meet listing requirements.*

REDUCING WASHERS

When oversized knockouts are encountered, the common solution is to use reducing washers to connect smaller conduits, tubing, and cables into larger knockouts. A common question is whether reducing washers are suitable for bonding raceways or cables to boxes or other enclosures. **See Figure 7-17.**

The QCRV category in UL Product iQ indicates that listed metal reducing washers are considered suitable for grounding in circuits over and under 250 volts when installed according to the *NEC.* An important qualifier is that this UL information also goes on to indicate that reducing washers are intended for use with conductors other than service conductors, meaning they do not meet the bonding requirements for services in 250.92.

Reducing washers can only be used on enclosures with a wall thickness of 0.053 inches or greater. Reducing washers can be installed in boxes or other enclosures where the concentric or eccentric knockouts have been completely removed, and they are suitable for use on concentric or eccentric knockout rings that are not all removed. Be sure to remove any nonconductive coatings if reducing washers are installed in equipment that is coated with paint or another type of coating; otherwise, additional bonding is necessary to

For additional information, visit qr.njatcdb.org Item #5333

ensure a thorough bonding connection. For the most effective bonding connections, it is highly recommended that equipment bonding jumpers around reducing washers be used when installed in coated metal enclosures, unless the coating is removed during installation.

BOXES WITH CONCENTRIC OR ECCENTRIC KNOCKOUTS

Bonding around punched concentric or eccentric knockouts is not required in all cases if the box or enclosure containing the pre-punched concentric or eccentric knockouts has been tested and is listed as suitable for bonding. The information from the QCIT category in the UL Product iQ indicates that concentric and eccentric knockouts of all metallic outlet boxes evaluated in accordance with *UL 514A, Metallic Outlet Boxes* are suitable for bonding in circuits of above or below 250 volts to ground without the use of additional bonding equipment. Be sure that metal-to-metal contact is made according to 250.12. **See Figure 7-18.**

For additional information, visit qr.njatcdb.org Item #5333

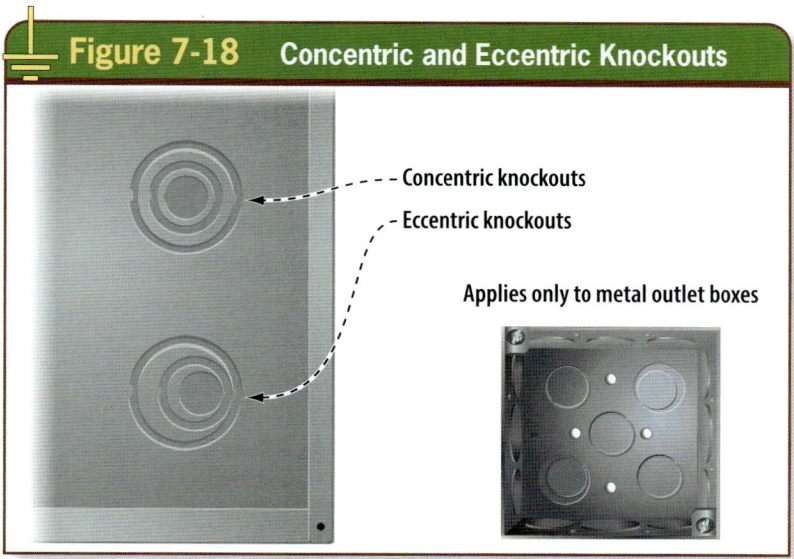

Figure 7-18   Concentric and Eccentric Knockouts

- Concentric knockouts
- Eccentric knockouts

Applies only to metal outlet boxes

Figure 7-18. Boxes with concentric or eccentric knockouts are listed for bonding in circuits over 250 volts to ground and in circuits 250 volts and less to ground.

SIZING SUPPLY-SIDE BONDING JUMPERS (WIRE TYPES)

When wire-type bonding jumpers are installed on the supply side of the service disconnecting means, they must be sized according to the requirements in 250.102(C). Wire-type supply-side bonding jumpers are typically installed from bonding bushings or fittings to enclosures and are usually short in length. For a single raceway or cable, the minimum size required for supply-side bonding jumpers must not be smaller than the size provided in Table 250.102(C)(1), which bases sizing on the size of the largest ungrounded service-entrance conductor. If the largest ungrounded service-entrance conductor exceeds 1,100-kcmil copper or 1,750-kcmil aluminum or copper-clad

aluminum, the minimum size required must not be less than 12.5% of the total circular mil area of the largest ungrounded service-entrance phase conductor as stated in Note 1 of Table 250.102(C)(1). **See Figure 7-19.**

If the ungrounded supply conductors are paralleled in two or more raceways

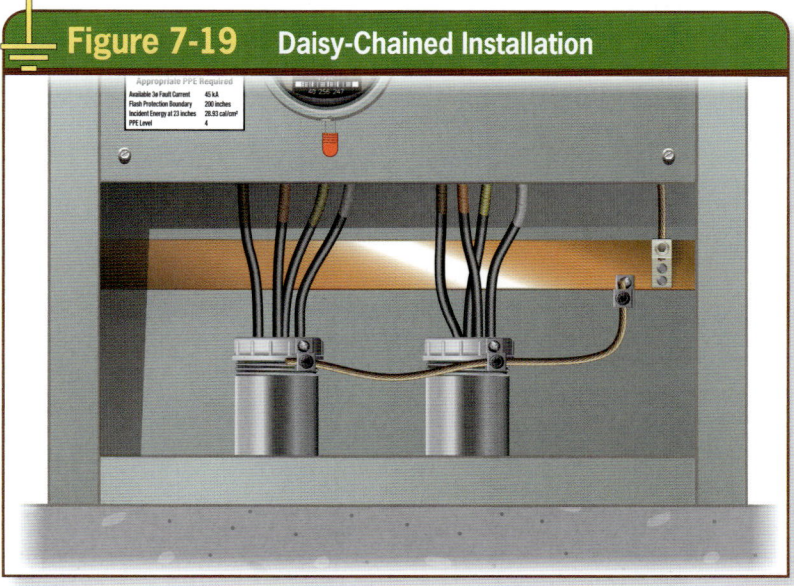

Figure 7-19   Daisy-Chained Installation

Figure 7-19. A single supply-side bonding jumper can be installed in a "daisy-chain" arrangement. Size supply-side bonding jumpers using Table 250.102(C)(1). Use the total circular mil area of the largest ungrounded service conductor in both raceways.

or cables, the supply-side bonding jumper(s), if routed with the raceways or cables, shall be installed in parallel. The size of the supply-side bonding jumpers for each of the individual

Figure 7-20   Individual EBJ Installation

Figure 7-20. Separate equipment bonding jumpers can be installed individually from each raceway to the enclosure. Size supply-side bonding jumpers using Table 250.102(C)(1). Use the total circular mil area of the largest conductor in each of the raceways.

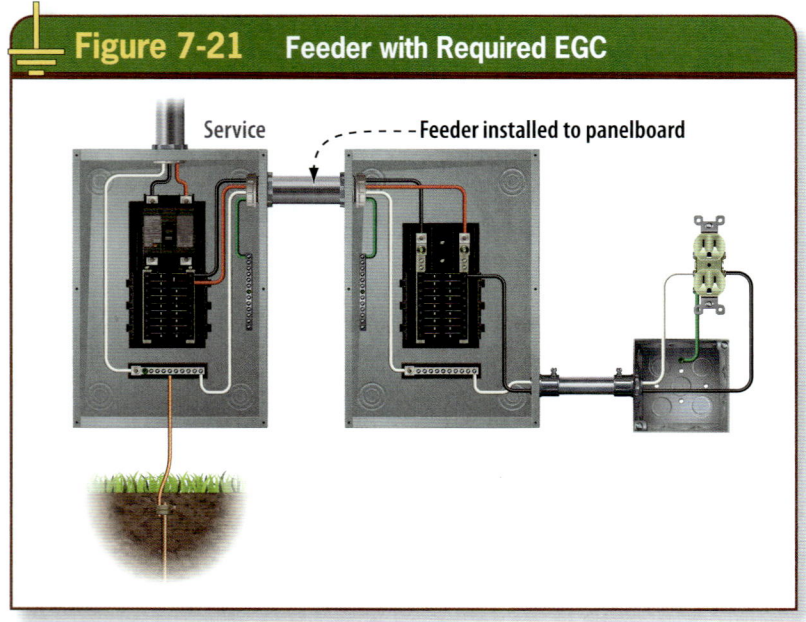

Figure 7-21   Feeder with Required EGC

Service — — — - **Feeder installed to panelboard**

Figure 7-21. The feeder EGC, either a wire type or raceway, performs bonding and grounding functions.

raceways in the parallel set is required to be based on the size of the largest ungrounded service-entrance conductor in each raceway or cable. **See Figure 7-20.**

If the ungrounded supply conductors and the supply-side bonding jumpers are of different materials (copper or aluminum), the minimum size of the bonding jumpers is determined by the assumed use of ungrounded conductors of the same material as the supply-side bonding jumper and with an ampacity equivalent to that of the installed ungrounded supply conductors. The ampacity equivalents can be determined from the values in Table 310.16.

LOAD SIDE BONDING RULES

Load-side bonding requirements in the *NEC* apply to bonding connections downstream (on the load side) of a circuit overcurrent protective device.

General Equipment Bonding Rules

Electrical bonding is also required on the load side of the service disconnect. Bonding functions occur from the point of delivery at the service equipment all the way to the final outlets in the branch circuits. Once the service bonding is complete, bonding is required for conductive parts of equipment associated with feeders and branch circuits that are connected to the service equipment enclosure, as well as all other equipment downstream of the service equipment enclosure. The bonding requirements are less restrictive on the load side of the service overcurrent protective device, but they are just as important.

Bonding requirements apply to metal raceways and enclosures for feeders and branch circuits. Each feeder and branch circuit is required to be provided with an equipment grounding conductor (EGC). This conductor can be a wire type or can be in the form of a metal raceway or other metal enclosure that is part of the effective ground-fault current path. An example is a feeder conduit installed from the service enclosure to a panelboard elsewhere in the building. The feeder includes an

Figure 7-22 Bonding with EGC

Figure 7-22. Equipment bonding jumpers connect two or more portions of the EGC.

EGC that performs bonding functions by being connected at both ends of the feeder. **See Figure 7-21.**

The same is true for the EGC of branch circuits. Bonding functions are accomplished by the EGC installed from enclosure to enclosure in the entire length of the branch circuit. The definition of the term *equipment grounding conductor* includes bonding functions, as indicated in Informational Note No. 1 following the definition of the term *equipment ground-*

ing conductor (EGC).

Installation and Size of Equipment Bonding Jumpers

Equipment bonding conductors or jumpers are used to connect two or more portions of the EGC as indicated in the definition of *equipment bonding jumper.* The general requirements for sizing equipment bonding jumpers are provided in 250.102. **See Figure 7-22.**

Equipment bonding jumpers must be made of copper, aluminum, copper-clad aluminum, or another corrosion-resistant material. Bonding jumpers can be in the form of a bus, wire, or other suitable conductor. Bonding jumper connections must meet the applicable provisions in 250.8. Connections for the grounding electrode system, covered in 250.53(C), must be made in accordance with 250.70.

Bonding jumpers are permitted to be installed inside or outside the raceway or enclosure. If installed on the outside of the raceway, the bonding jumper or conductor is generally limited to a maximum of six feet in length. An exception permits longer lengths for outside pole installations. **See Figure 7-23.**

An example of installing an equipment bonding jumper on the outside of

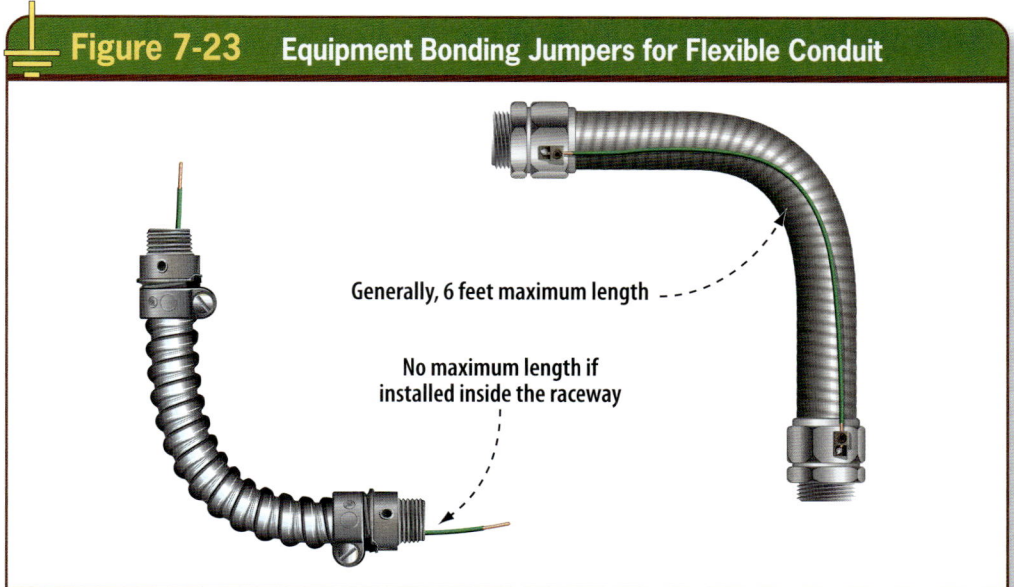

Figure 7-23 Equipment Bonding Jumpers for Flexible Conduit

Generally, 6 feet maximum length

No maximum length if installed inside the raceway

Figure 7-23. Equipment bonding jumpers are permitted to be installed on the outside or on the inside of raceways.

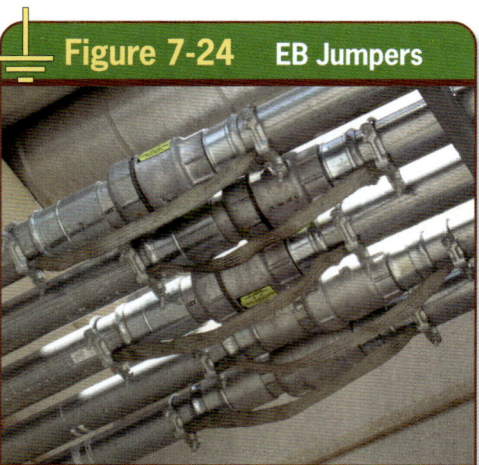

Figure 7-24. *Equipment bonding jumpers are often installed around expansion-deflection conduit fittings to maintain the required continuity.*

a raceway can be found in conduit runs where expansion, expansion- deflection, or deflection fittings are necessary for the installation. If equipment bonding jumpers or conductors are installed outside raceways or enclosures, they can be vulnerable to damage. Where they are installed in locations subject to physical damage, 250.102(E)(3) requires that they be protected in accordance with the provisions in 250.64(A) and (B). **See Figure 7-24.**

Section 250.148 addresses required connections to boxes or enclosures. Sizing equipment bonding jumpers on the load side of an overcurrent protective device is covered in 250.102(D). The size of wire-type equipment bonding jumpers or conductors is determined from Table 250.122, which bases sizing on the rating of the circuit breaker or fuse protecting the circuit in which the equipment bonding jumper is installed. An example of a requirement for sizing equipment bonding jumpers on the load side of the service disconnecting means is provided in 250.146. Section 250.146 requires equipment bonding jumpers from grounded metal boxes to grounding-type receptacles to be sized based on the rating of the circuit breaker or fuse protecting the circuit. **See Figure 7-25.**

Fact

The size of wire-type equipment bonding jumpers or conductors is determined from Table 250.122, which bases sizing on the rating of the circuit breaker or fuse protecting the circuit in which the equipment bonding jumper is installed.

Figure 7-25   Sizing Receptacle Equipment Bonding Jumpers

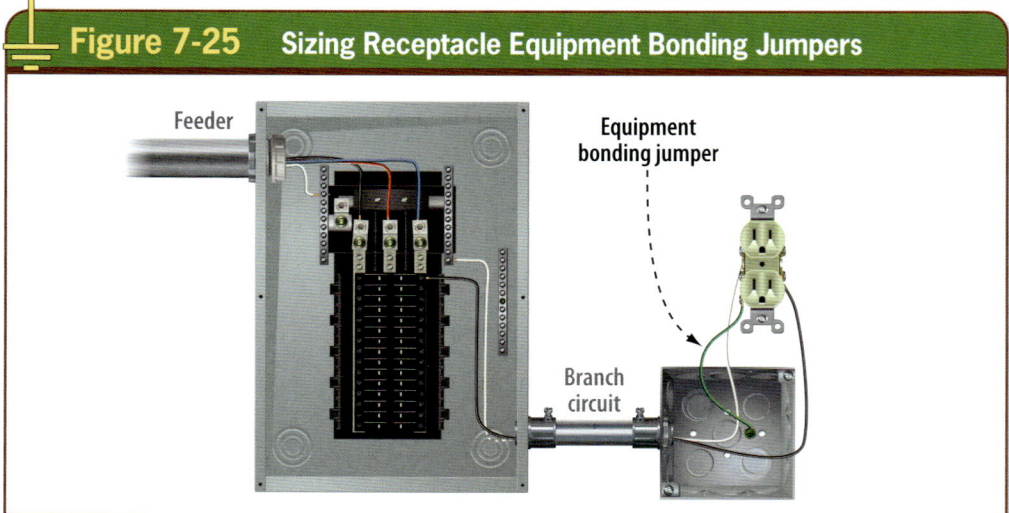

Figure 7-25. *To size an equipment bonding jumper installed from a grounded metal box to a grounding receptacle, determine the size of the circuit breaker or fuse and then refer to Table 250.122.*

Another example of sizing equipment bonding jumpers based on the rating of an overcurrent protective device is when conduits emerge from a slab into an open-bottom enclosure, such as a switchboard. The equipment bonding jumpers provide the electrical continuity and conductivity between the exposed portion of the metal raceway and the equipment grounding terminal bus in the switchboard assembly. The equipment bonding jumper can be installed in a daisy-chain fashion or as individual jumpers from each conduit bushing to the terminal bar. **See Figure 7-26.**

BONDING AROUND EXPANSION FITTINGS AND LOOSE-JOINED METAL RACEWAYS

If expansion, expansion-deflection, or deflection fittings are installed in the run of a metal raceway, the electrical continuity and conductivity must be ensured, either by installing an equipment bonding jumper or through other means. Listed fittings are available for raceways that cross expansion joints in building construction projects. These

listed fittings must be installed according to the manufacturer's installation instructions. Failure to install these fittings without the necessary amount of travel left on either side could limit movement and result in damage to the conduit system or an ineffective bond around the fitting. These fittings often include the required wire-type equipment bonding jumpers and connections to be used across the fitting after they are inserted in the raceway. **See Figure 7-27.**

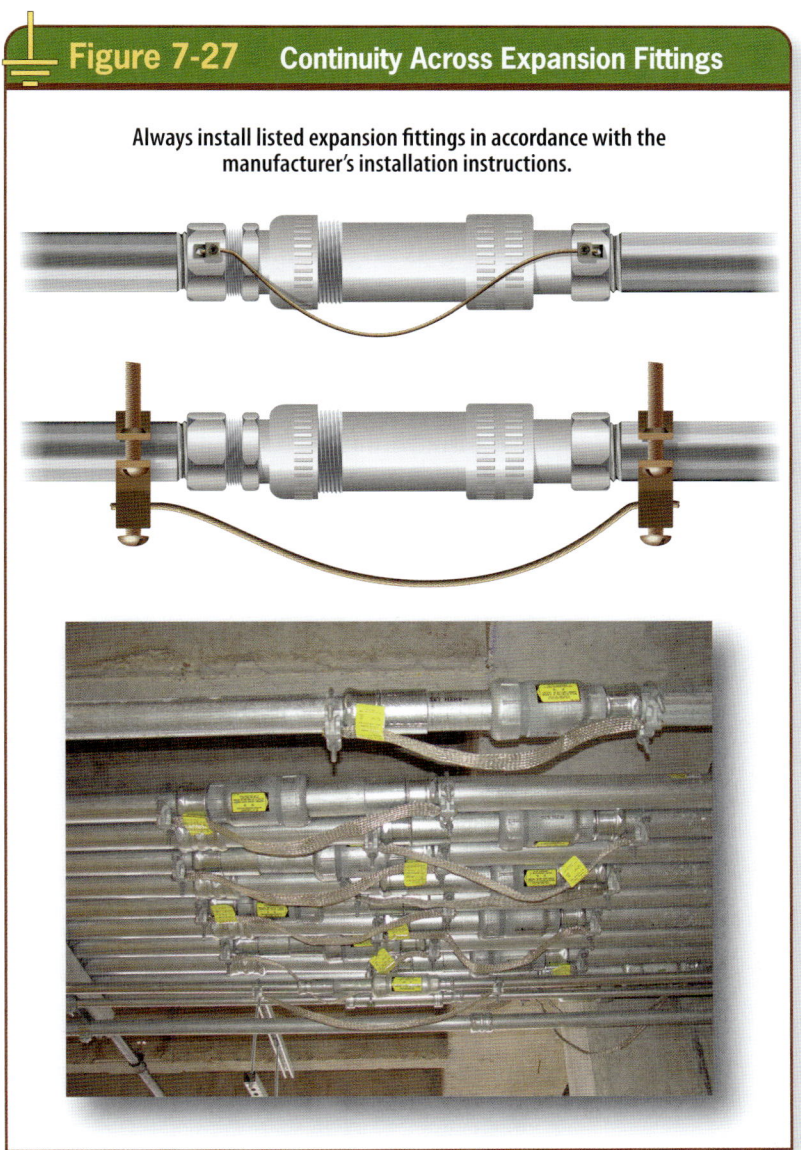

Figure 7-27 Continuity Across Expansion Fittings

Always install listed expansion fittings in accordance with the manufacturer's installation instructions.

Figure 7-27. Bonding jumpers are required to be used with fittings to maintain continuity across the fitting.

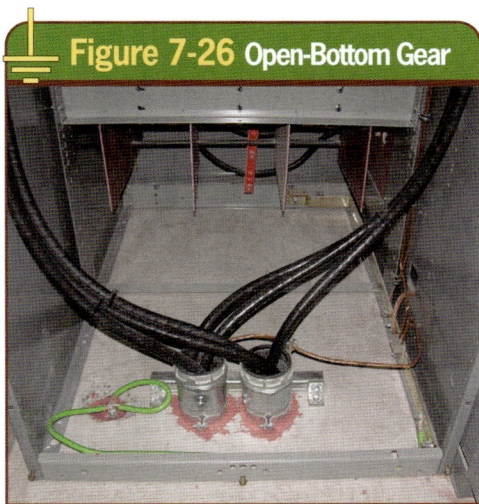

Figure 7-26 Open-Bottom Gear

Figure 7-26. Equipment bonding jumpers are installed from conduits in the open bottom of a switchboard to the equipment grounding terminal bar of the equipment.

Courtesy of Jim Dollard, IBEW Local 98

Figure 7-28   Listed Expansion Fittings

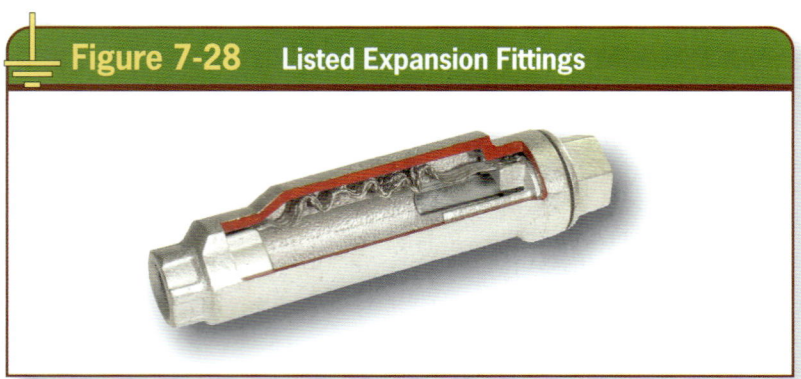

Figure 7-28. Fittings that include a bonding jumper that is internal to the assembly are available.

Courtesy of ABB

Fact

Water piping installed inside a building, on a roof, or on an exterior wall is subject to the bonding requirements in 250.104(A).

Listed fittings that provide the equipment bonding jumper as an internal, integral part of the overall assembly are also available. Installing an equipment bonding jumper on the outside is not necessary if these types of fittings are installed. **See Figure 7-28.**

BONDING OTHER METAL PARTS

There are some conductive parts within buildings that are not a part of the electrical installation that are required to be bonded.

Figure 7-29   Water Piping System Bonding

Figure 7-29. Metal water piping systems are required to be bonded in accordance with 250.104(A).

Bonding Metal Piping Systems

The *NEC* contains requirements for bonding metal piping systems. Metal water piping systems have one set of bonding rules, and other metal piping systems have a different set. The bonding requirements for metal piping systems are provided in 250.104. The purpose of bonding metal piping systems is to place the metal piping at the same potential as the grounded metal components of the electrical system and to provide a path for any fault current likely to be imposed on them. **See Figure 7-29.**

A metal piping system should not be left isolated in a building or structure because it is possible that the metal piping could become energized and present a shock or electrocution hazard. In addition, the risk of fire or property damage is increased.

Section 250.104(A) provides the bonding rules for water piping systems only. There are two key points to understand in the wording of this rule. First, the system must be metal. An installation of water piping that uses nonmetal piping and short metal sections at the points of connection is not a metal piping system; it is a nonmetal system with short metal nipples or short sections of metal piping installed. Second, this rule applies to all water piping systems in or attached to the building or structure. Water piping installed inside a building, on a roof, or on an exterior wall is subject to the bonding requirements in 250.104(A). This includes potable water piping systems, sprinkler system piping, and chilled water system piping. Many make the mistake of applying this bonding rule to the only domestic potable hot- and cold-water systems in the building.

Bonding and Accessibility

The *NEC* requires metal water piping systems to be bonded to the service equipment enclosure, to the grounded conductor at the service, to the grounding electrode conductor where it is of sufficient size, or to one or more of the grounding electrodes of the building grounding electrode system. This bonding jumper must not be smaller than the values listed in Table 250.102(C)(1), except that it is not required to be larger than 3/0 AWG copper or 250-kcmil aluminum or copper-clad aluminum. For example, if the size of the largest ungrounded service-entrance phase conductor is 4/0 AWG copper, then a bonding jumper that is at least 2 AWG copper or 1/0 AWG aluminum is required. **See Figure 7-30.**

As a sizing example, if a 500-ampere service is supplied with 750-kcmil copper as the largest ungrounded service-entrance conductor, the minimum-size bonding jumper for the water piping system is 2/0 AWG copper or 4/0 AWG aluminum.

The points of attachment of the bonding jumper for water piping systems are required to be accessible. The phrase *accessible (as applied to wiring methods)* is defined in Article 100 as "capable of being removed or exposed without damaging the building structure or finish or not permanently closed in or blocked by the structure, other electrical equipment, other building systems, or finish of the building."

Metal Water Piping in Multiple Occupancy Buildings

Section 250.104(A)(2) offers a water piping system bonding alternative for multiple-occupancy buildings. In buildings of multiple occupancy where the metal water piping system or systems are installed in or attached to a building or structure, the metal piping system in each occupancy must be bonded.

If the metal piping system for each individual occupancy is metallically isolated from all other occupancies by use of nonmetal water piping, the metal water piping system or systems for each occupancy are permitted to be bonded to the equipment grounding terminal bus of the panelboard or switchboard enclosure that supplies the individual occupancy. This alternative is in lieu of installing a larger water piping bonding conductor to the service equipment enclosure as required by 250.104(A).

The bonding jumper installed for the water piping system in each individual occupancy must be sized in accordance with Table 250.122, which bases sizing on the rating of the overcurrent protective device for the

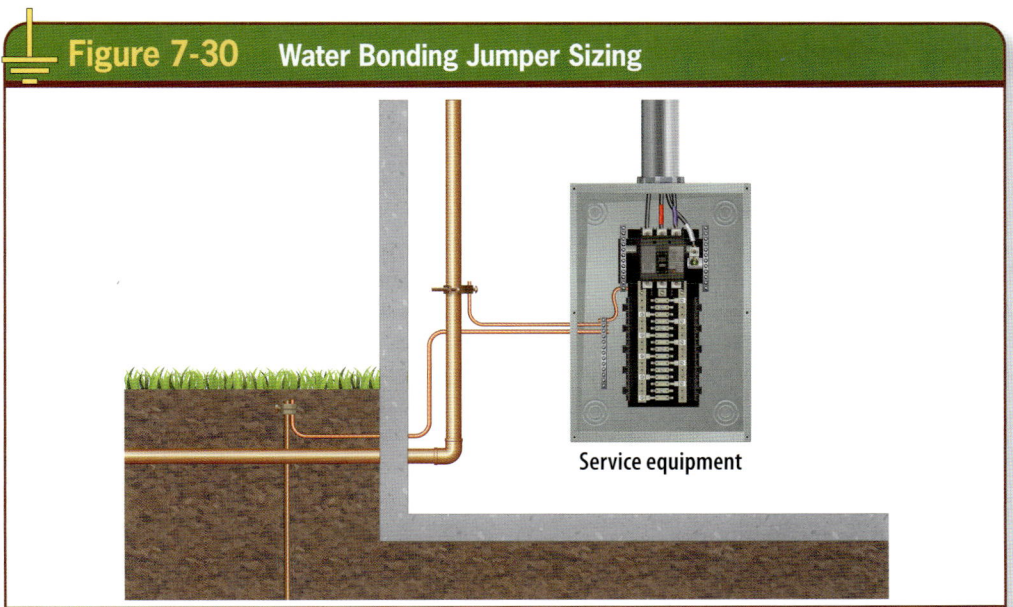

Figure 7-30   Water Bonding Jumper Sizing

Service equipment

Figure 7-30. Size water bonding conductors using Table 250.102(C)(1) based on the size of the largest ungrounded service-entrance conductor.

circuit supplying the occupancy. **See Figure 7-31.**

An example of this type of bonding is often found in apartment complexes.

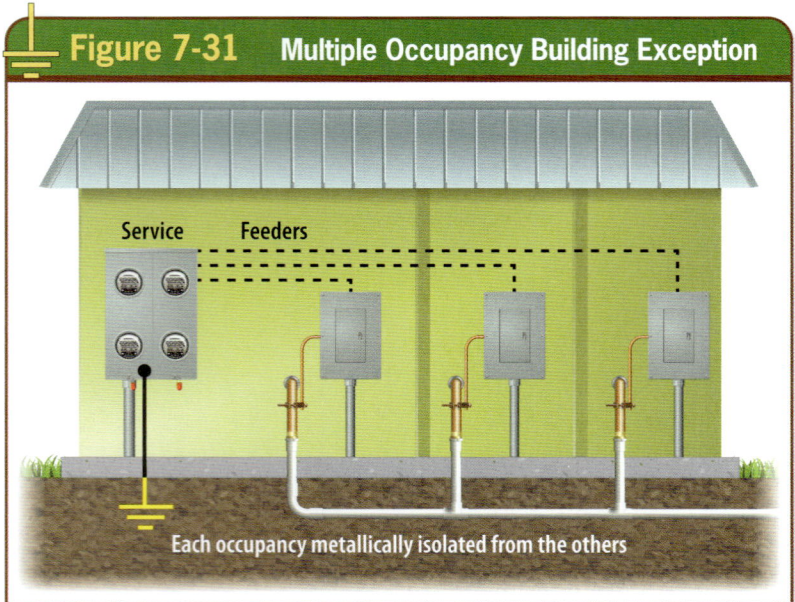

Figure 7-31. Water pipe bonding in multiple-occupancy buildings is accomplished within each occupancy. Size bonding conductors based on the rating of the feeder overcurrent protective device for each individual unit.

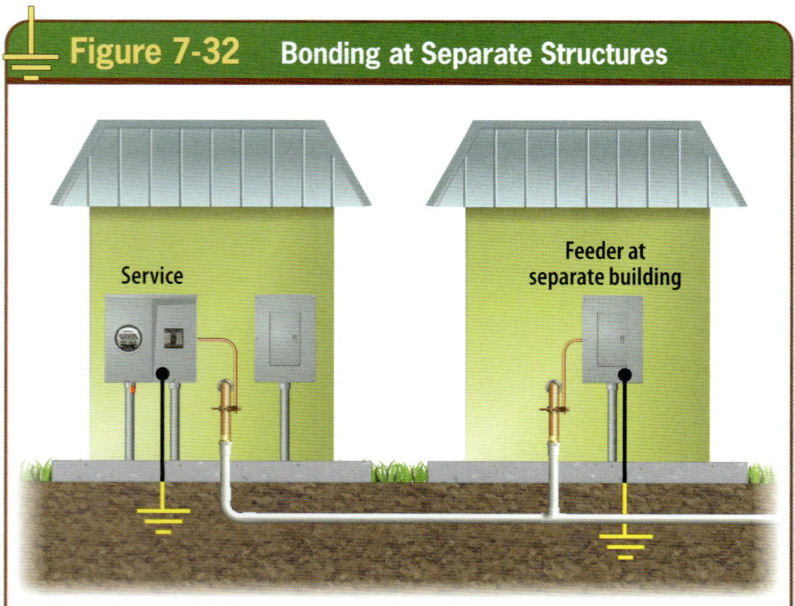

Figure 7-32. Size water pipe bonding jumper(s) at separate buildings or structures using Table 250.102(C)(1), noting that the bonding jumper is not required to be larger than 3/0 AWG copper or 250-kcmil aluminum or copper-clad aluminum.

The key to being able to use this sizing option and connecting to the panelboard in the apartment is that the metal water piping system within each individual unit must be isolated from all other apartments or other occupancies.

Multiple Buildings or Structures Supplied by a Feeder or Branch Circuit

A metal water piping system or systems installed in or attached to a building or structure are required to be bonded either to the building or structure disconnecting means enclosure where located at the building or structure, to the EGC run with the supply conductors, or to the one or more grounding electrodes used. The bonding jumpers shall be sized in accordance with Table 250.122, using the size of the largest overcurrent protective device for the feeder or branch circuit conductors that supply the building. The bonding jumper is not required to be larger than the largest ungrounded feeder or branch circuit conductor supplying the building or structure. **See Figure 7-32.**

Bonding Other Metal Piping Systems

There are other metal piping systems in buildings or structures that must be bonded to the electrical service supplying the building or structure under certain conditions. The requirements for bonding metal piping systems other than metal water piping systems are provided in 250.104(B). Some of the other metal piping systems include compressed air piping, piping systems for lubricants and fuels, and pneumatic systems for controls. One of the key drivers of this requirement is that the other metal piping systems are installed in or attached to a building or structure. The *NEC* requires bonding only if these other metal piping systems (either in or attached to the building or structure) are *likely to become energized*, which is now defined in the *NEC*. **See Figure 7-33.**

Figure 7-33. *Bonding for other metal piping systems must be in accordance with 250.104(B).*

Figure 7-34. *A gas water heater supplied by a branch circuit and EGC performs the bonding required in 250.104(B).*

The bonding jumper must be connected to any of the following:

1. Equipment grounding conductor for the circuit that is likely to energize the piping system
2. Service equipment enclosure
3. Grounded conductor at the service
4. Grounding electrode conductor, if of sufficient size
5. One or more grounding electrodes used

The bonding conductor or jumper must be sized in accordance with Table 250.122, which is based on the rating of the overcurrent protective device for the circuit likely to energize the piping system.

Consider a gas water heater and a gas furnace. If the branch circuit supplying the furnace is protected by a 30-ampere overcurrent protective device, then the 10 AWG EGC of the 30-ampere branch circuit can serve as the bonding means. In this case, the branch circuit supplying the appliance is viewed as the circuit likely to energize the piping system.

It is difficult to say with absolute certainty that there is no possibility of isolated piping becoming energized. Many engineering designs require a bonding jumper that is sized according to Table 250.102(C)(1), which is based on the size of the largest ungrounded service-entrance conductor. This, of course,

exceeds the minimum requirements in the *NEC.* **See Figure 7-34.**

Another good example of metal piping systems that are required to be bonded is metal piping used for compressed air systems. At a minimum, the EGC supplying this equipment can serve as the required bonding means because this is the circuit that would be likely to energize the piping system. **See Figure 7-35.**

The points of attachment of the bonding jumper or conductor for the piping systems are required to be accessible,

Figure 7-35. *Metal piping systems for compressed air systems are required to be bonded in accordance with 250.104(B).*

Courtesy of Cogburn Bros. Inc.

Figure 7-36   Building Frames

Figure 7-36. *Structural metal building framing, if likely to become energized, must be bonded.*

Fact

Before drilling or modifying structural metal building frames, it is necessary to get approval from the building official or person responsible for the design. Some structural elements are part of engineered systems and are not allowed to be modified.

just as is required for water piping system bonding jumper connections.

Some gas piping systems have been known to be vulnerable to damage from lightning. As a result, some manufacturers of corrugated stainless-steel tubing (CSST) gas piping have provided instructions for more restrictive bonding requirements than those contained in the *NEC*. *NFPA 54: National Fuel Gas Code* requires that manufacturer's installation

instructions be followed, and these bonding requirements for CSST piping systems may exceed those in the *NEC*. Prudent design dictates that a coordinated effort be made to address the bonding of CSST piping in a manner that satisfies the *NEC*, the *National Fuel Gas Code*, and the manufacturer's installation instructions. This may require coordination between the plumbing/mechanical contractor, the electrical contractor, and the inspection authority.

NFPA 780: Standard for the Installation of Lightning Protection Systems also provides bonding information and an Annex that describes bonding methods to use for CSST piping systems.

An informational note following 250.104(B) references additional bonding requirements of the *National Fuel Gas Code*. Another informational note following 250.104(B) advises about additional safety being achieved by bonding all piping and metal air ducts within the premises. This is an advisement only and not a requirement of the *NEC*.

Bonding Structural Metal Building Frames

Any exposed structural metal that is not acceptable as an electrode, that is interconnected to form a building frame, and that is not already intentionally grounded or bonded is required to be bonded to the electrical supply service of the building or structure as indicated in 250.104(C). **See Figure 7-36.**

This requirement also applies only if the structural metal building framing is likely to become energized. The bonding jumper must be connected to one of the following locations:

1. The service equipment enclosure
2. The grounded conductor at the service
3. The grounding electrode conductor, where large enough
4. The disconnecting means enclosure supplied by a feeder or branch circuit
5. Any of the grounding electrodes in the grounding electrode system

The bonding jumper is required to be sized in accordance with Table

Figure 7-37   Bonding Structural Metal Frames

- - - Interconnected - - -

Service

Building steel

Figure 7-37. *Bonding jumpers for interconnected structural metal building frames must be sized according to Table 250.102(C)(1) except that they are not required to be larger than 3/0 AWG copper or 250 kcmil aluminum or copper-clad aluminum.*

250.102(C)(1), based on the size of the largest ungrounded service-entrance conductor supplying the service equipment. The maximum size required for the structural metal frame bonding jumper is established using Table 250.102(C)(1), except that it is not required to be larger than 3/0 AWG copper or 250-kcmil aluminum or copper-clad aluminum. **See Figure 7-37.**

Just like the points of attachment for the water piping system, the points for attachment to the building metal frame have to be accessible unless the connection is covered with fireproofing material and the installation meets the applicable provisions in 250.68(A) Exception No. 2. Structural metal building frame sections that are isolated by expansion joints must be bonded together using a bonding jumper that meets the sizing requirements of 250.102(C). Remember that the connection of the bonding jumper to the building steel must be in accordance with 250.8 and the metal surface must be cleaned to bare metal in accordance with 250.12. **See Figure 7-38.**

Bonding Lightning Protection Systems

Requirements for lightning protection systems are provided in *NFPA 780:* *Standard for the Installation of Lightning Protection Systems.* Section 250.106 of the *NEC* requires that the ground terminals of a lightning protection system be bonded to the building or structure grounding electrode system. This ensures that any rise or fall of potential on conductive objects in or on the building or structure will occur at the same potential and reduce flashover possibilities.

Figure 7-38 Remove Coating

Figure 7-38. Coating or paint on building frames must be removed for effective metal-to-metal contact.

SUMMARY

Bonding is the process of connecting conductive parts or equipment together. In electrical systems, grounding and bonding functions go hand-in-hand and usually occur simultaneously through a single bonding action. In other words, if a system and equipment are grounded and other conductive parts such as conduit are connected to such grounded equipment, the function of bonding has taken place and grounding is accomplished simultaneously. The result is grounded and bonded equipment.

Bonding methods and sizing of equipment bonding jumpers must be understood, along with requirements for bonding metal piping systems, structural metal building frames, and other conductive parts within or attached to buildings or structures. Electrical bonding results in conductive parts being connected together and establishes conductivity and continuity between them in a manner that functions electrically to provide safety for the overall electrical system.

REVIEW QUESTIONS

1. Bonding is the process of connecting conductive objects together.
   a. True   b. False

2. Bonding establishes continuity and conductivity between conductive parts that are connected together.
   a. True   b. False

3. The general requirements for bonding are provided in __?__ of Article 250.
   a. Part II
   b. Part III
   c. Part IV
   d. Part V

4. By definition, a(n) __?__ is a reliable conductor used to ensure the required electrical conductivity between metal parts required to be electrically connected.
   a. bonding conductor or jumper
   b. equipment bonding jumper
   c. equipment grounding conductor
   d. main bonding jumper

5. A __?__ is a conductor installed on the supply side of a service or separately derived system to ensure the required electrical conductivity between metal parts required to be electrically connected.
   a. bonding busbar
   b. load-side bonding jumper
   c. supply-side bonding jumper
   d. system bonding jumper

6. The *NEC* defines *equipment bonding jumpers* as the connection between two or more portions of the equipment grounding conductor.
   a. True   b. False

7. Section __?__ provides rules related to bonding methods for enclosures, raceways, and other normally non–current-carrying metal parts at the service and on the supply side of the service disconnecting means and overcurrent protective device.
   a. 250.120
   b. 250.4
   c. 250.66
   d. 250.92

8. Which of the following products are not suitable for bonding on the supply side of the service disconnecting means?
   a. Bonding bushings with jumpers
   b. Standard locknuts
   c. Listed threaded hubs
   d. Threadless connectors made up tight

9. Which of the following metal parts of equipment containing service conductors are required to be bonded together?
   a. Auxiliary gutters
   b. Cable tray
   c. Raceways
   d. All of the above

10. All enclosures containing service conductors, such as meter enclosures, boxes, and so forth that are interposed in the service raceway installation are required to be bonded in accordance with 250.92(A).
    a. True   b. False

11. When concentric, eccentric, or oversized knockouts are encountered in the installation, the bonding requirements in 250.97 apply where the circuit voltage exceeds __?__ phase-to-ground.
    a. 100 V
    b. 120 V
    c. 220 V
    d. 250 V

12. Reducing washers are suitable for bonding on the supply side of the service when they meet all listing requirements and no paint or coatings are provided on the service equipment enclosure.
    a. True   b. False

13. What is the minimum-size supply-side bonding jumper required if installed on the line side of a 300-ampere service disconnecting means if the service conductors are sized at 400-kcmil copper?
    a. Parallel 2 AWG copper conductors
    b. 3 AWG copper
    c. 2 AWG copper
    d. 1/0 AWG copper

14. An equipment bonding jumper is the connection between two or more portions of an equipment grounding conductor.
    a. True   b. False

15. Bonding jumpers can be in the form of a(n) ___?___.
    a. bus
    b. other suitable conductor
    c. wire
    d. all of the above

16. A metal water piping system installed in a building or structure is required to be bonded to the service equipment enclosure, to the grounded conductor at the service, to the grounding electrode conductor if of sufficient size, or to one or more grounding electrodes installed for the service and sized in accordance with ___?___.
    a. Section 250.122
    b. Table 9 in Chapter 8
    c. Table 250.102(C)(1), except that it is not required to be larger than 3/0 AWG copper or 250 kcmil aluminum or copper-clad aluminum
    d. Table 250.122

17. When expansion, expansion-deflection, or deflection fittings are installed in a run of metal raceway, the electrical continuity and conductivity must be ensured by installing an equipment bonding jumper or by other means. This means is permitted to be either inside or outside the raceway.
    a. True   b. False

18. The equipment grounding conductor of a circuit likely to energize a copper compressed air piping system can be used as the bonding means.
    a. True   b. False

19. What is the minimum-size equipment bonding jumper required from a grounded metal outlet box to a 40-ampere receptacle?
    a. 12 AWG copper
    b. 10 AWG copper
    c. 8 AWG copper
    d. 6 AWG copper

20. If two lengths of conduit for a 200-ampere feeder have to be bonded together, what is the minimum-size aluminum equipment bonding jumper required?
    a. 10 AWG
    b. 8 AWG
    c. 6 AWG
    d. 4 AWG

21. Points of attachment of bonding jumpers for metal water piping systems and structural metal building frames are required to be ___?___.
    a. accessible
    b. guarded
    c. inaccessible
    d. not accessible to the public

22. The grounded conductor is not permitted to be used for bonding between a meter enclosure and a service disconnecting means enclosure.
    a. True   b. False

23. For circuits over 250 volts to ground, bonding is accomplished when concentric, eccentric, or oversized knockouts are encountered by which of the following means?
    a. Fittings with shoulders that seat firmly against the box or cabinet, such as electrical metallic tubing connectors, flexible metal conduit connectors, and cable connectors, with one locknut on the inside of boxes and cabinets
    b. Threadless couplings and connectors for cables with metal sheaths
    c. Two locknuts on rigid metal conduit or intermediate metal conduit—one inside and one outside of boxes and cabinets
    d. Any of the above

24. Structural metal building framing that is bonded to the service must be bonded using a bonding jumper not smaller than the sizes required in ___?___.
    a. Table 9 in Chapter 8
    b. Table 250.66
    c. Table 250.102(C)(1), except that it is not required to be larger than 3/0 AWG copper or 250 kcmil aluminum or copper-clad aluminum
    d. Table 250.122

25. A bonding jumper for a metal water piping system must be sized according to ___?___.
    a. Table 9 in Chapter 8
    b. Table 250.66
    c. Table 250.102(C)(1), except that it is not required to be larger than 3/0 AWG copper or 250 kcmil aluminum or copper-clad aluminum
    d. Table 250.122

Equipment Grounding Conductors

The equipment grounding conductor (EGC) is another very important component in the grounding and bonding system. EGCs are typically installed with the feeders and branch circuits of electrical systems. They perform grounding and bonding and serve as effective ground-fault current paths. EGCs are electrically conductive paths that extend the ground (Earth) connection to equipment that is required to be grounded. The performance of EGCs is directly related to the integrity of this conductive path, which is ensured through professional and skilled mechanical execution of the work.

Objectives

» Understand the definition of the term *equipment grounding conductor (EGC)*.

» Identify the types of equipment grounding conductors recognized by the *NEC*.

» Understand the purpose and performance of the equipment grounding conductor in the electrical system.

» Determine the requirements for identification and connections of equipment grounding conductors.

» Understand installation requirements and determine minimum sizes for wire-type equipment grounding conductors.

Chapter 8

Table of Contents

Equipment Grounding Conductor Defined 184

Equipment Grounding Conductor Basics...... 184

  Performance of Equipment
Grounding Conductors 184

  Types of Equipment
Grounding Conductors 186

  Metallic Cable Assemblies 189

    Metal-Clad Cable 189

    Armored-Clad Cable.........................191

Equipment Grounding
Conductor Installations 192

  Protection from Physical Damage.......... 193

  Installation with Circuit Conductors........ 193

Equipment Grounding
Conductor Connections 194

Equipment Grounding
Conductor Identification 195

  Multiconductor Cables......................... 196

  Sizing Criteria 196

Equipment Grounding Conductor Sizing....... 197

  Increases in Size................................. 198

  Multiple Circuits in a Single Raceway
or Cable Tray..................................... 199

  Equipment Grounding Conductors
for Parallel Runs................................. 199

  Cable Assemblies in Parallel 201

  Sizing Examples 201

  Equipment Grounding Conductors
for Motor Circuits................................ 202

  Equipment Grounding Conductors
with Feeder Taps................................. 202

Current in Equipment Grounding
Conductors.. 202

Summary... 203

Review Questions 204

EQUIPMENT GROUNDING CONDUCTOR DEFINED

The general requirements for EGCs are provided in Part VI of *NEC* Article 250. These rules cover EGC types, identification, installation, and sizing. An EGC is required to be installed with feeders and branch circuits and can be in the form of a wire-type conductor. **See Figure 8-1.**

Figure 8-1 Wire-Type EGC

Figure 8-1. *Wire-type EGCs are connected to the equipment grounding terminal bar in an enclosure.*

Grounding Conductor, Equipment (EGC). (Equipment Grounding Conductor) A conductive path(s) that provides an effective ground-fault current path and connects normally non–current-carrying metal parts of equipment together and to the system grounded conductor or to the grounding electrode conductor, or both. (CMP-5)

Informational Note No. 1: It is recognized that the equipment grounding conductor also performs bonding.

Informational Note No. 2: See 250.118 for a list of acceptable equipment grounding conductors.

Wiring methods such as conduit, tubing, and cable armor can also be acceptable EGCs. Whether an EGC is a wire type, conduit or tubing, or cable armor, it must be able to perform as an effective ground-fault current path. **See Figure 8-2.**

EGCs perform multiple functions, one of which is to ground equipment. To understand the role of the EGC in the safety system, it is important to review related *NEC* definitions.

Grounded (Grounding). Connected (connecting) to ground or to a conductive body that extends the ground connection. (CMP-5)

Ground-Fault Current Path, Effective. (Effective Ground-Fault Current Path) An intentionally constructed, low-impedance electrically conductive path designed and intended to carry current during ground-fault events from the point of a ground fault on a wiring system to the electrical supply source and that facilitates the operation of the overcurrent protective device or ground-fault detectors. (CMP-5)

EQUIPMENT GROUNDING CONDUCTOR BASICS

The term *equipment grounding conductor* inherently implies that grounding is a function accomplished by this component of the grounding and bonding system.

Performance of Equipment Grounding Conductors

The EGC performs three important functions in the electrical safety system. The first of those functions is grounding. EGCs are intended to act as a conductive path that connects equipment to ground (the Earth). The EGC extends the ground connection to various points in the electrical system because it is generally installed with feeders or branch circuits. The role of grounding is to place a conductive object (equipment) at or as close to Earth (ground) potential

Figure 8-2 EMT as EGC

Figure 8-2. Electrical metallic tubing is recognized as an EGC in accordance with 250.118(4).

Courtesy of IBEW Local 26 Training Center

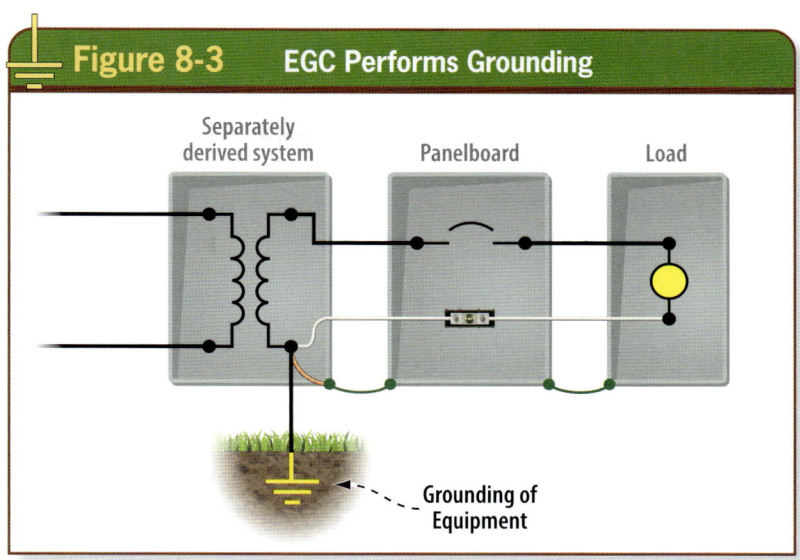

Figure 8-3 EGC Performs Grounding

Figure 8-3. EGCs perform grounding functions.

as possible. The EGC limits voltages above ground potential on conductors and equipment enclosures during normal operation and during abnormal conditions, such as a ground-fault or short-circuit event. **See Figure 8-3.**

Another important function performed by EGCs is bonding. In the definition of EGC, the words *connect* and *together* are used. The act of connecting together is a bonding function. For example, when an EGC is installed from one metal outlet box to another and is connected to the metal box at both ends, the boxes not only become grounded, but they are also bonded electrically together. **See Figure 8-4.**

The third important role of the EGC is to serve as part of an effective ground-fault current path during abnormal events such as ground faults. The conductor must be capable of carrying the fault current back to the source for the time it takes the overcurrent protective device to open and clear the event from the circuit. Section 250.4(A)(5) clearly provides the performance requirements and criteria for an effective ground-fault current path. Ideally, there will never be a ground fault on the circuit, but insulation breakdown or failure can occur and result in a ground fault. Insulation

can be in the form of a dielectric material or air space, such as the space between busbars in a switchboard or panelboard.

Ground faults are typically unintentional. In addition to wire insulation failure, a ground fault can result from human error or accidents, such as the dropping of a conductive tool in an electrical enclosure during work on energized equipment. During those events, the EGC must be capable of

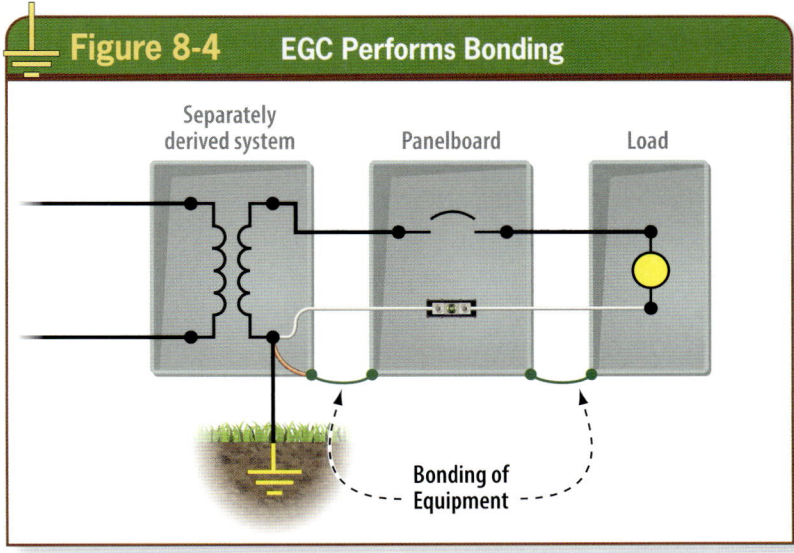

Figure 8-4 EGC Performs Bonding

Figure 8-4. EGCs perform bonding functions.

Figure 8-5 EGC as Effective Ground-Fault Current Path

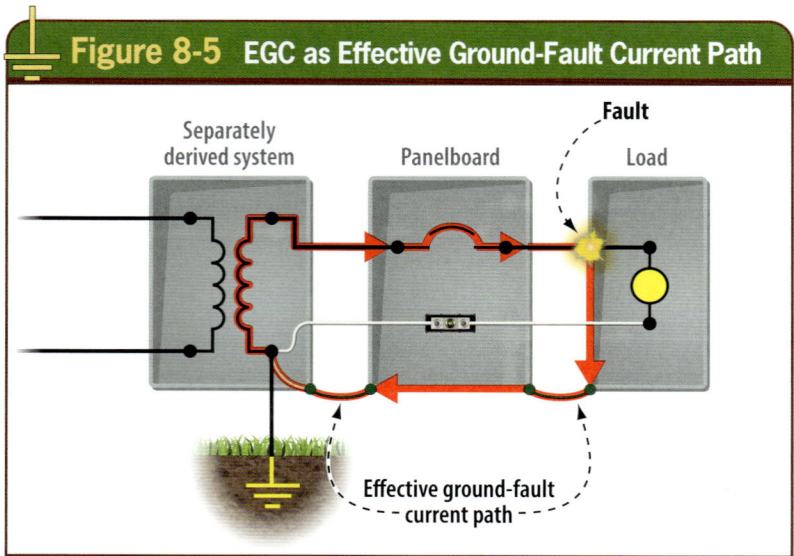

Figure 8-5. EGCs are part of an effective path for ground-fault current and facilitate overcurrent protective device operation during ground-fault conditions.

withstanding the higher level of current to perform its all-important safety function. **See Figure 8-5.**

The functions of EGCs are quite simple. They provide grounding for equipment, perform bonding functions, and facilitate overcurrent protective device operation, making them an important component of the electrical safety system.

Section 250.118 provides the conductive materials and wiring methods that can be used as EGCs with feeders and branch circuits. List item (1) in 250.118 states EGCs shall be made of copper, aluminum, or copper-clad aluminum material. These conductor materials can be in the form of a wire (stranded or solid) or busbar of any shape, and they can be insulated, covered, or bare.

Types of Equipment Grounding Conductors

Section 250.118 provides a list of recognized EGCs that must be run with the circuit conductors. EGCs can be in various forms. They are required to be any one or combination of several types. **See Figure 8-6.**

Electrical conduit and tubing are acceptable as EGCs according to 250.118. These wiring methods are generally suitable as EGCs without a wire-type conductor being installed. **See Annex B.**

When conduit or tubing is installed, all fittings (for example, connectors, couplings, and locknuts) must be made up tight in a professional and skilled manner, and the wiring system must be secured and supported properly in accordance with the applicable rules in Chapter 3 of the *NEC*. This requirement applies whether a wire-type EGC is installed in the raceway or not.

Proper supports for wiring methods are often directly related to maintaining the integrity of the effective ground-fault current path. For example, rigid metal conduit must generally be supported at intervals not exceeding 10 feet, and also must generally be secured within three feet of outlets, junction boxes, conduit bodies, etc.

Section XXX.30 in each respective wiring method article provides securing and supporting requirements. For example, 358.30 requires that electrical metallic tubing (EMT) be installed as a complete system and securely fastened in place and supported in accordance with 358.30(A) and (B). The general requirement in this section calls for secure fastening of the tubing at intervals not exceeding 10 feet and within three feet of each outlet box, junction box, device box, cabinet,

Figure 8-6   *Code* Recognized EGCs

- A copper, aluminum, or copper-clad aluminum conductor. This conductor shall be solid or stranded; insulated, covered, or bare; and in the form of a wire or busbar of any shape
- Rigid metal conduit (RMC)
- Intermediate metal conduit (IMC)
- Electrical metallic tubing (EMT)
- Armor of Type AC cable, as provided in Section 320.108
- The copper sheath of mineral-insulated, metal-sheathed cable, Type MI
- Cable trays, as permitted in Sections 392.10 and 392.60
- Cable bus framework, as permitted in Section 370.60(1)
- Other listed electrically continuous metal raceways and listed auxiliary gutters
- Surface metal raceways listed for grounding

Figure 8-6. NEC Section 250.118 lists the types of acceptable EGCs.

Figure 8-7. *Secure and supported metal raceway installations help provide an effective ground-fault current path.*

Figure 8-8. *Flexible metal conduit is permitted as an EGC in accordance with the conditions in 250.118(5)(a) through (f).*

conduit body, or other tubing termination. **See Figure 8-7.**

See Annex B for more information about conduit, tubing, and wire-type equipment grounding conductor performance and the GEMI analysis reports developed by the Georgia Institute of Technology.

Listed flexible metal conduit can serve as an EGC, but only in limited applications. It is important to understand that the interlocking metal-tape construction of flexible metal conduit provides a high-impedance path for ground-fault current. Thus, it is limited in length and by the overcurrent protective device protecting the contained ungrounded conductors (if serving as an EGC). **See Figure 8-8.**

Section 250.118(5) indicates that listed flexible metal conduit is only suitable as an EGC if it meets all of the following conditions:

1. The conduit is terminated in listed fittings.
2. The circuit contained in the conduit is protected by an overcurrent protective device rated 20 amperes or less.
3. The size does not exceed 1 1/4 inch trade size.
4. The combined length of flexible metal conduit, flexible metallic tubing, and liquidtight flexible metal conduit in the same ground-fault-current path does not exceed six feet.
5. If used to connect equipment in

which flexibility is necessary to minimize the transmission of vibration from equipment or to provide flexibility for equipment that requires movement after installation, an EGC must be installed.

6. If the flexible metal conduit is constructed of stainless steel, a wire-type EGC or bonding jumper is required to be installed. **See Figure 8-9.**

An example of flexible metal conduit installed where flexibility is necessary after installation is a flexible conduit "whip" that supplies chain-hung

For additional information, visit qr.njatcdb.org Item #2610

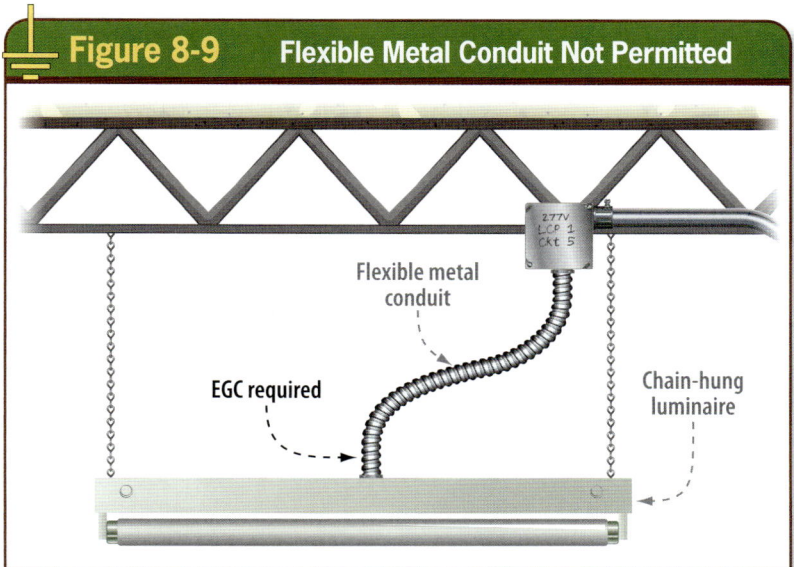

Figure 8-9. *Listed flexible metal conduit is not permitted as an EGC if flexibility or movement is necessary after installation.*

Figure 8-10. Listed liquidtight flexible metal conduit is permitted as an EGC in accordance with the conditions in 250.118(6)(a) through (f).

fluorescent luminaires. In this case, there is anticipated movement in the flexible metal conduit, so an EGC is required. Examples of equipment that introduces vibration are motors or transformers, which are often connected using flexible metal conduit. EGCs are required in these types of installations.

If flexibility is not necessary after installation and the installation meets the provisions in 250.118(5)(a) through (d), a wire-type EGC is not required. An example of this type of installation could be a commercial garage door opener wired using flexible metal conduit.

Some localities and some specifications require an EGC in all installations of flexible metal conduit. This is more restrictive than the *NEC* minimum requirements, so it is important to verify any local electrical code requirements with the authority having jurisdiction.

Listed liquidtight flexible metal conduit can also serve as an EGC, but only in limited applications. Understanding that the interlocking metal-tape construction of liquidtight flexible metal conduit also provides a high-impedance path for ground-fault current is important. Thus, it has restrictions that must be met to qualify as an EGC. **See Figure 8-10.**

Section 250.118(6) indicates that listed liquidtight flexible metal conduit is suitable for use as an EGC under the following conditions:

1. The conduit is terminated in listed fittings.

2. For trade sizes ⅜ through ½, the circuit conductors contained in the conduit are protected by overcurrent protective devices rated at 20 amperes or less.

3. For trade sizes ¾ through 1 ¼, the circuit conductors contained in the conduit are protected by overcurrent protective devices rated not more than 60 amperes and there is no flexible metal conduit, flexible metallic tubing, or liquidtight flexible metal conduit in trade sizes ⅜ through ½ in the effective ground-fault current path.

4. The combined length of flexible metal conduit, flexible metallic tubing, and liquidtight flexible metal conduit in the same effective ground-fault current path does not exceed six feet.

5. If used to connect equipment in which flexibility is necessary to minimize the transmission of vibration from connected equipment or to provide flexibility for equipment that requires movement after installation, a wire-type EGC must be installed.

6. If the liquidtight flexible metal conduit contains a stainless steel core, a wire-type EGC or bonding jumper is required to be installed. **See Figure 8-11.**

An example of liquidtight flexible metal conduit being used to minimize the effects of vibration is when it is used to supply air-conditioning and refrigeration equipment on a rooftop or a large dry-type transformer. An example of a connection to equipment that might also require more movement after installation is a large commercial refrigerator that is "direct- or hard-wired" with liquidtight flexible metal conduit and is on wheels so it can be rolled away from a wall for cleaning or service operations. **See Figure 8-12.**

Flexible metallic tubing is not commonly used anymore, but it is still recognized in the *NEC* as an EGC when the tubing is terminated in listed fittings and the installation meets the following limitations:

Figure 8-11 Liquidtight Flexible Metal Conduit Restrictions

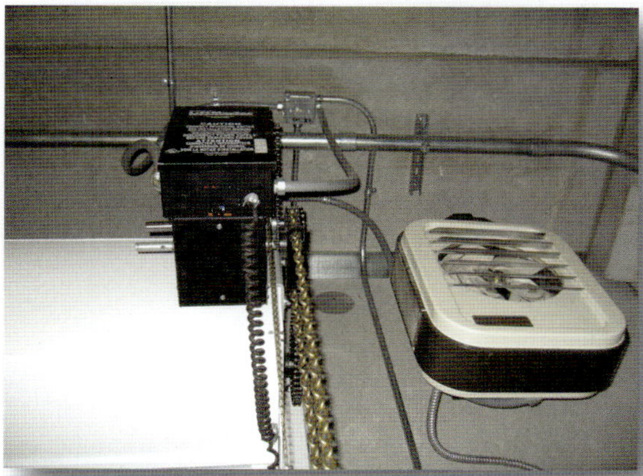

Listed liquid-tight flexible metal conduit can be used as an EGC if all the conditions of Section 250.118(6) have been met.

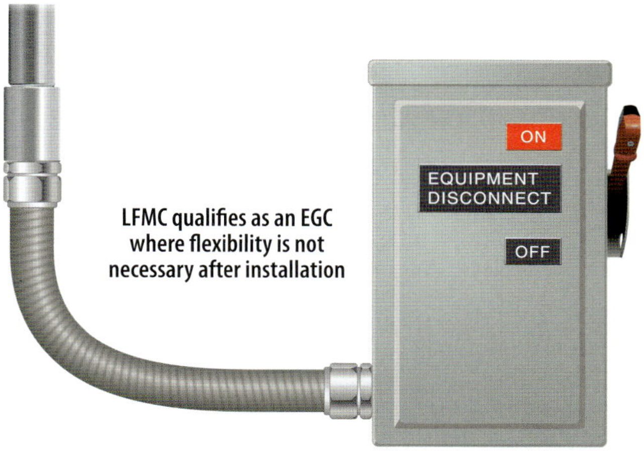

LFMC qualifies as an EGC where flexibility is not necessary after installation

ON

EQUIPMENT DISCONNECT

OFF

Figure 8-11. Listed liquidtight flexible metal conduit is suitable as an EGC when installed and used within the limitations of 250.118(6).

1. The circuit conductors contained in the tubing are protected by overcurrent protective devices rated at 20 amperes or less.
2. The combined length of flexible metal conduit, flexible metallic tubing, and liquidtight flexible metal conduit in the same ground-fault current path does not exceed six feet.

Metallic Cable Assemblies

Metallic cable assemblies can be suitable as equipment grounding conductors if they meet certain and specific construction criteria.

Figure 8-12 Vibration

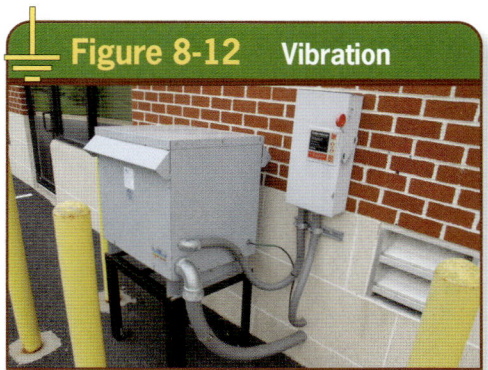

Figure 8-12. A dry-type transformer is wired using liquidtight flexible metal conduit to reduce noise transmission and vibration.

Figure 8-13 Metal-Clad Cable

Figure 8-13. *The armor of conventional metal-clad cable Type MC (interlocking metal-tape construction) is not an effective ground-fault current path.*

Metal-Clad Cable

Metal-clad (Type MC) cable is available with three types of armor: the interlocking metal-tape armor (the most common type), the corrugated tube-type, and the smooth tube-type. The physical characteristics of interlocking metal tape-style MC cable limit its performance characteristics. **See Figure 8-13.**

This type of MC cable assembly must contain an EGC. The armor of the smooth tube-type or corrugated tube-type can serve as an EGC when used with fittings listed for that particular cable. **See Figure 8-14.**

Metal-clad (Type MC) cable is also suitable for use as an EGC in accordance with any of the following:

1. It contains an insulated or uninsulated EGC in compliance with 250.118(1).
2. The combined metallic sheath and uninsulated equipment grounding/bonding conductor of interlocked metal tape-type MC cable is listed and identified as an EGC.
3. The metallic sheath or the combined metallic sheath and EGCs of the smooth or corrugated tube-type MC cable is listed and identified as an EGC.

There is a type of MC cable available with interlocking metal tape-type construction that has an armor listed as an EGC. It is similar in construction to Type AC cable in that it includes a separate bare conductor (typically 10 AWG aluminum) within the assembly installed on the outside of the plastic-wrapped conductors of the cable assembly and in intimate contact with the cable armor. **See Figure 8-15.**

This MC cable assembly is manufactured with and without contained

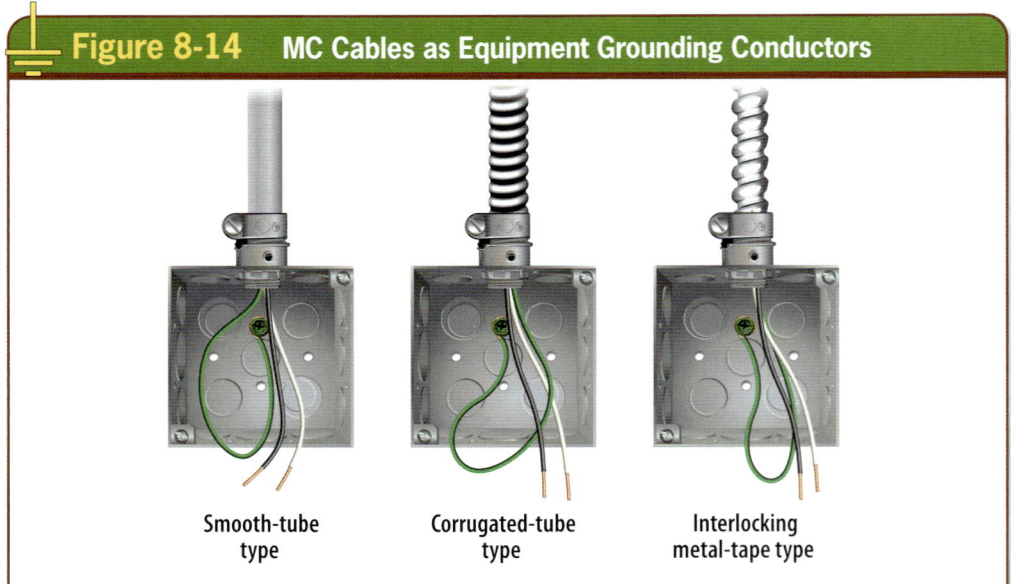

Figure 8-14 MC Cables as Equipment Grounding Conductors

Smooth-tube type

Corrugated-tube type

Interlocking metal-tape type

Figure 8-14. *MC cable must be installed with listed fittings.*

Figure 8-15   **Armor of AC Cable Suitable as EGC**

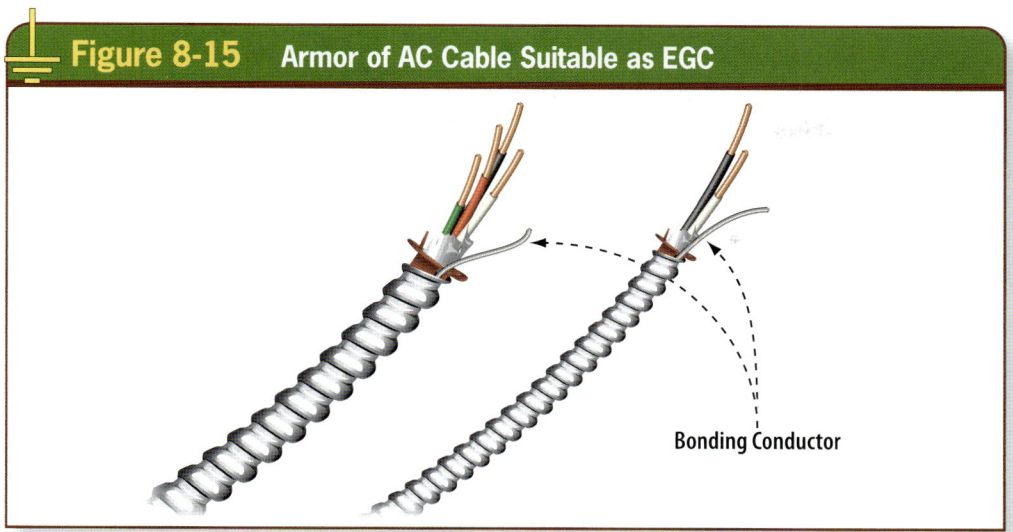

Bonding Conductor

Figure 8-15. MC cable assemblies may have an armor that is suitable as an EGC. The bonding conductor is in contact with armor between cable terminations.

insulated copper EGCs, but the armor plus the bare bonding conductor serves as an EGC when installed using listed fittings. To ensure compliance with 110.3(B) of the *NEC*, follow the manufacturer's installation instructions when installing this and any other product. If this type of cable is installed in patient care locations of health care facilities, it must include an insulated copper EGC internal to the cable assembly to satisfy the requirements for two EGC paths as required in 517.13.

Armored-Clad Cable
Section 250.118(8) recognizes the armor sheath of armored-clad (Type AC) cables as an EGC. This cable armor qualifies as an EGC because of the bare internal bonding strip that remains in intimate contact with the armor from fitting to fitting. **See Figure 8-16.**

The combination of an internal bonding strip in the assembly, together with an interlocking metal tape-type armor, is acceptable and recognized as an effective ground-fault current path. Some armored-clad cable assemblies are

Figure 8-16   **Standard Armored-Clad (AC) Cable**

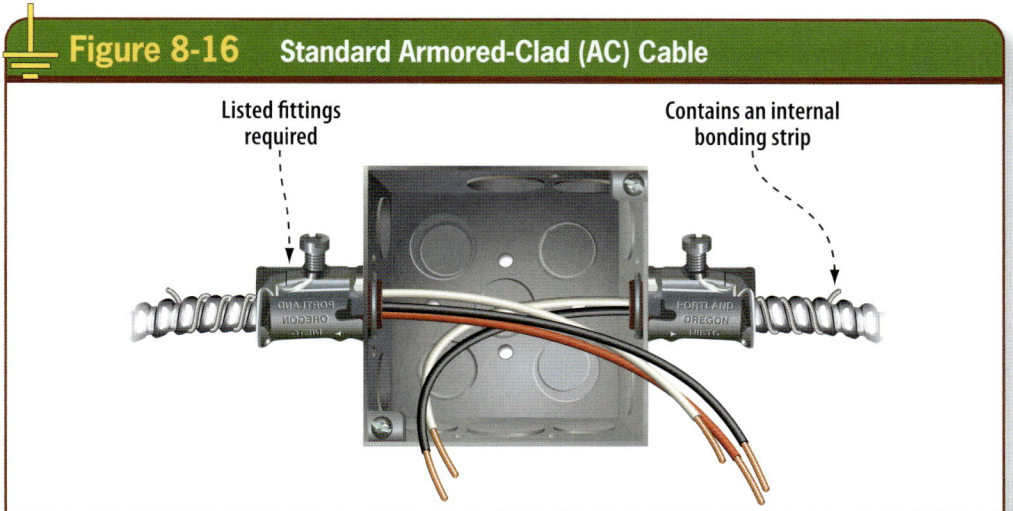

Listed fittings required

Contains an internal bonding strip

Figure 8-16. The armor of armored-clad cable is suitable as an EGC because of its contained internal bonding strip in contact with the armor.

Figure 8-17 AC Cable with Listed Fittings

Figure 8-17. Listed armored-clad cable (Type AC) is suitable as an EGC when used with listed fittings.

EQUIPMENT GROUNDING CONDUCTOR INSTALLATIONS

Section 110.12 addresses mechanical execution of the work and requires electrical conductors and equipment to be installed in a professional and skillful manner. This general requirement also applies to EGCs installed with feeders and branch circuits. Where the wiring method, such as rigid metal conduit or electrical metallic tubing, is installed and serves as the required EGC, the installation must conform to the provisions in 110.12. **See Figure 8-18.**

This includes, but is not limited to, sufficient support and methods of securing the wiring method. If the conduit or tubing is secured properly, it should not be subject to movement that could compromise its performance as an effective ground-fault current path due to loosening of fittings.

manufactured with an insulated copper EGC and therefore meet the criteria for isolated grounding circuits and redundant EGCs specified in 517.13(A) and (B) for use in branch circuits serving patient care locations. These cable assemblies are all required to be installed using listed fittings. **See Figure 8-17.**

Section 250.120 provides installation requirements for EGCs of all types. Essentially, whatever EGC type is installed, the applicable provisions in the *NEC* must be met. This means that if a cable tray is the EGC, it must meet the provisions in Articles 250, 300, and 392 relative to cable trays performing as EGCs. If a raceway such as electrical metallic tubing (EMT) is installed, it must meet the applicable installation requirements in Articles 250, 300, and 358. The fittings and terminations used with the wiring method chosen must be suitable for use with the type of wiring method installed. For example, fittings (such as connectors, couplings, and locknuts) used with EMT must be listed for use with EMT, as required by 358.6. **See Figure 8-19.**

All connections and joints must be made tight using suitable tools. This goes back to the basic mechanical execution of the work requirement in 110.12. It is important to tighten fittings because of the functions they are expected to perform in both normal operation and during abnormal conditions such as ground faults.

Loose electrical fittings such as set-screw couplings, connectors, and lock-nuts introduce impedance into the

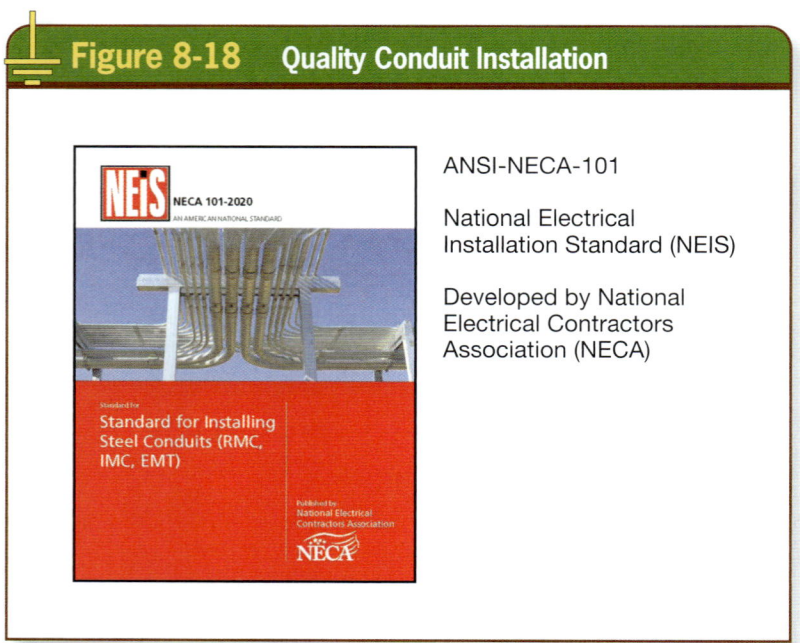

Figure 8-18 Quality Conduit Installation

ANSI-NECA-101

National Electrical Installation Standard (NEIS)

Developed by National Electrical Contractors Association (NECA)

NEiS
NECA 101-2020
AN AMERICAN NATIONAL STANDARD

Standard for
Installing Steel Conduits (RMC, IMC, EMT)

Published by
National Electrical
Contractors Association
NECA

Figure 8-18. NECA-101: Standard for Installing Steel Conduit (Rigid, IMC, EMT) provides specific details about installing metal conduit and tubing installations in a professional and skillful manner.

ground-fault current path and could affect the quick operation of overcurrent protective devices. Chapter 3 of the *NEC* also includes requirements for securing and supporting conduit and other raceways that are included in 250.118 as EGCs. The integrity of the effective ground-fault current path established by the wiring method itself depends on effective, *NEC*-compliant support and securing of the raceway system. A quality intallation is important for many reasons. **See Figure 8-20.**

Protection from Physical Damage

An important installation requirement applies to aluminum or copper-clad aluminum conductors. These conductors are more vulnerable to the effects of corrosion and deterioration than copper conductors. For this reason, bare EGCs are not permitted to come into contact with masonry or the Earth and are not permitted where subject to corrosive conditions. These types of EGCs must not be terminated within 18 inches of the Earth, if external to buildings or enclosures, unless listed wire connector systems are used. Section 250.120(C) indicates that EGCs smaller than 6 AWG are required to be protected from physical damage by a raceway or cable armor. This damage protection rule is relaxed for EGCs installed in hollow spaces of walls or partitions that protect them from physical damage. An example of this is in the rare case when EGCs are installed separately from the circuit conductors as permitted in 250.130(C).

Installation with Circuit Conductors

Sections 300.3(B) and 250.134(2) provide important information about installing EGCs. One of the most important considerations is to keep the EGC as close to its associated ungrounded circuit conductors as possible. This keeps the impedance values at a minimum during normal operation and during ground-fault conditions. Section 300.3(B) requires all conductors

of the circuit, including any grounded conductor or EGC, to be run in the same raceway, cable, or trench. When the EGC of the circuit is rigid metal conduit, intermediate metal conduit, electrical metallic tubing, or any of the other raceway types mentioned in 250.118, the EGC is automatically

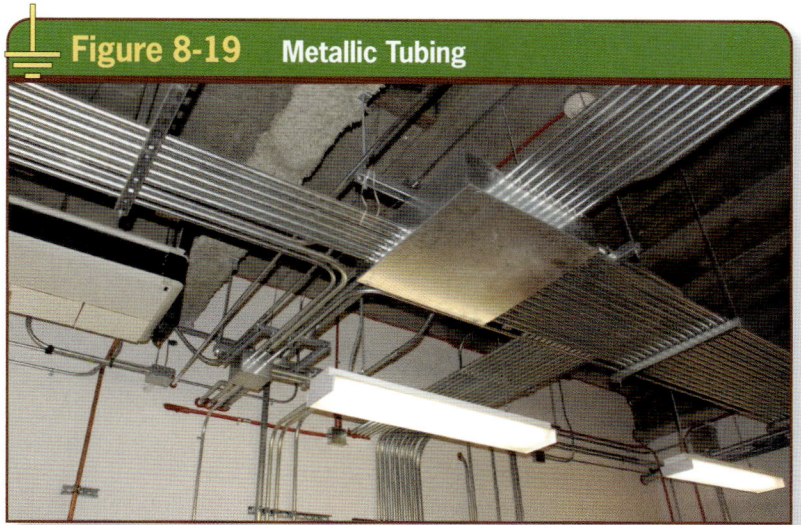

Figure 8-19. *Electrical metallic tubing is suitable as an EGC when used with listed fittings and properly secured and supported according to the methods required in 358.30.*

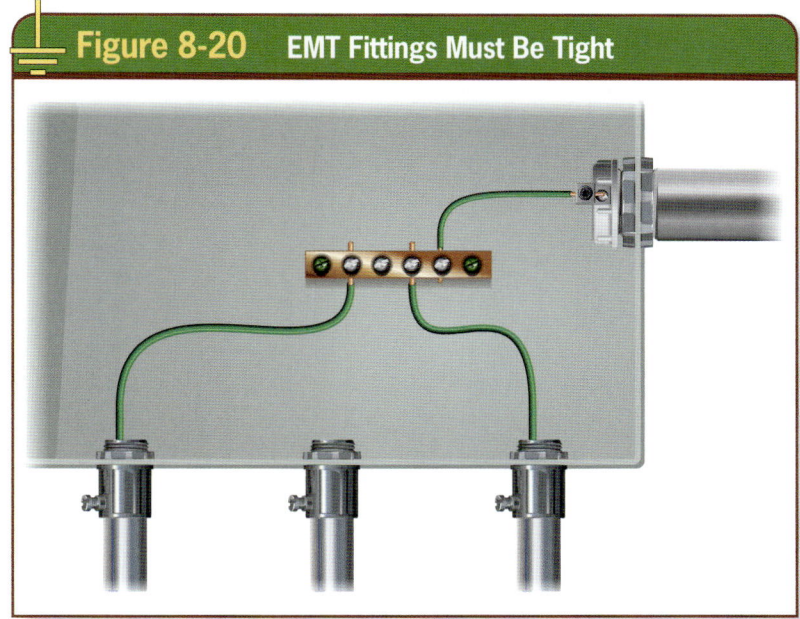

Figure 8-20. *Tighten fittings in conduit and tubing system installations to ensure an effective ground-fault current path.*

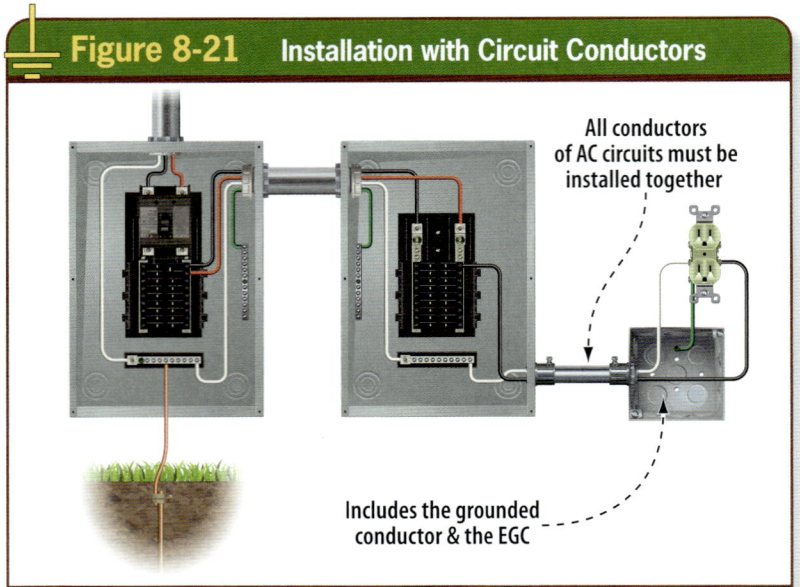

Figure 8-21 **Installation with Circuit Conductors**

All conductors of AC circuits must be installed together

Includes the grounded conductor & the EGC

Figure 8-21. Install EGCs together with the associated ungrounded conductors of the AC circuit.

run with the circuit conductors and is integral to the wiring method. **See Figure 8-21.**

Section 250.134(2) has a similar requirement that addresses equipment connected to an EGC that is run with the circuit conductors and contained in the same raceway or cable, or otherwise installed with the circuit conductors. If EGCs are separated from the associated

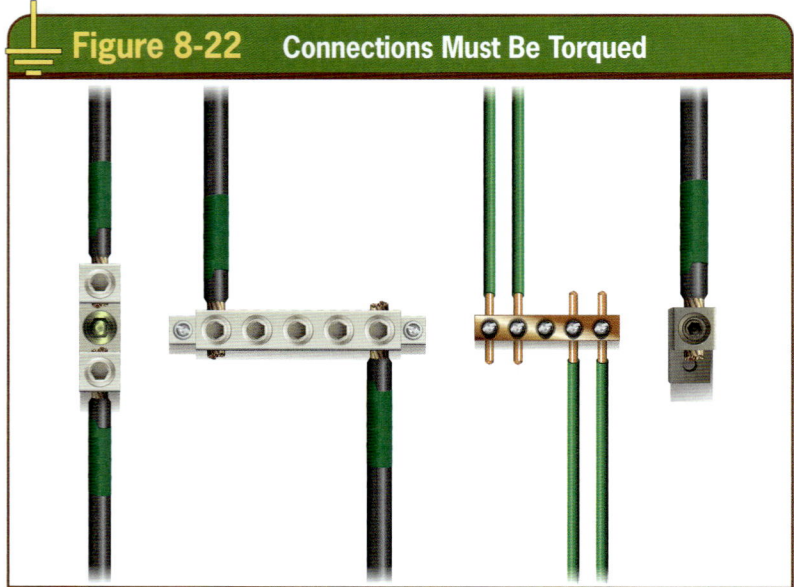

Figure 8-22 **Connections Must Be Torqued**

Figure 8-22. Terminals for grounding and bonding connections must be torqued to the manufacturer's requirements.

circuit conductors, the inductive reactance of the circuit is also increased; thus, circuit impedance is added. This can become problematic during ground-fault events, when the objective is to keep the impedance as low as possible to allow the high level of fault current to quickly operate the circuit breaker or fuse protecting the circuit.

EQUIPMENT GROUNDING CONDUCTOR CONNECTIONS

Connections of grounding and bonding conductors are covered in 250.8. The list of connection means is provided to clarify how EGCs must be terminated. Connection methods not mentioned in 250.8 are not recognized by the *NEC*.

The connections must be tight, and if the EGC is of the wire type, installers must torque connections at terminal lugs to the manufacturer's requirements. **See Figure 8-22.**

Terminal lugs and equipment generally provide torque values for installers to attain the correct tightness of wire-type conductors. It is important not to under-tighten or over-tighten electrical connections, so manufacturers provide specific torque values. This requirement applies not only to the ungrounded circuit conductors, but also to the grounding and bonding conductor connections. Obviously, loose or otherwise impaired connections or terminal lugs in the EGC path will introduce impedance into that path. **See Figure 8-23.**

It is vitally important to tighten all EGC connections to establish an effective path for ground-fault current. When fittings such as couplings, connectors, and locknuts are used to install wiring methods such as conduit, the fittings should be made tight for the same reason—integrity of the ground-fault current path.

TechTip!

The connections must be tight, and if the EGC is of the wire type, installers must torque connections at terminal lugs to the manufacturer's requirements.

Figure 8-23   EGC Termination

Figure 8-23. EGCs terminate on a grounding terminal bar in a switchboard enclosure.

Courtesy of Jim Dollard, IBEW Local 98

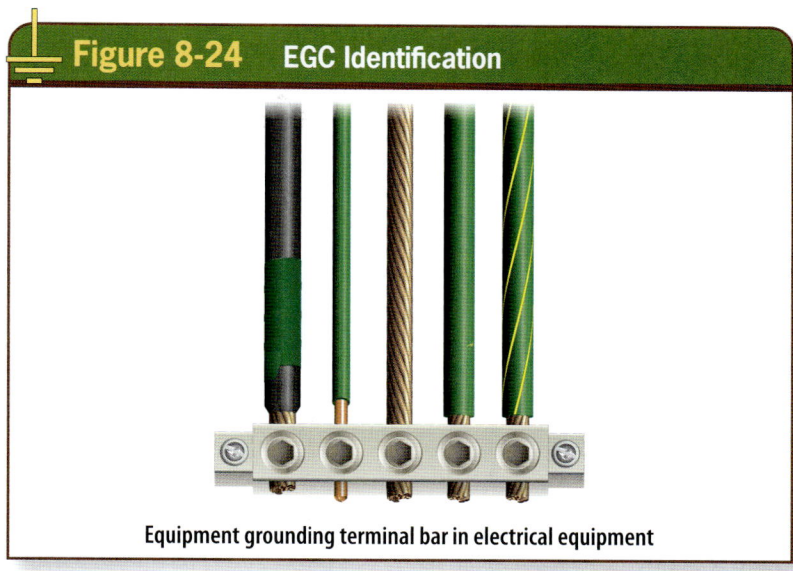

Figure 8-24   EGC Identification

Equipment grounding terminal bar in electrical equipment

Figure 8-24. EGCs can be insulated, covered, or bare and must be identified by any of the methods in 250.119.

Figure 8-25   Use of Tape

Figure 8-25. Green marking tape can be used to identify EGCs in sizes 4 AWG and larger.

Courtesy of IBEW Local 26 Training Center

EQUIPMENT GROUNDING CONDUCTOR IDENTIFICATION

The *NEC* provides specific identification requirements for EGCs of the wire type. EGCs can be insulated, covered, or bare. Section 250.119 recognizes a bare conductor as an EGC. If the conductor is individually insulated or covered, the insulation or covering is required to have a continuous outer finish that is either green or green with one or more yellow stripes. **See Figure 8-24.**

The color green or green with one or more yellow stripes cannot be used for grounded (usually neutral) conductors or ungrounded (phase or hot) conductors. EGCs installed in raceway systems are often green for the entire length or green with one or more yellow stripes.

EGCs in sizes 4 AWG and larger must meet the applicable requirements in 250.119(B)(1) and (B)(2). As indicated in (B)(1), an insulated or covered conductor in sizes 4 AWG and larger is permitted to be identified at each end and at every point where the conductor is accessible. By exception, wire-type EGCs in sizes 4 AWG and larger are not required to be marked in conduit bodies that contain no splices or unused hubs. **See Figure 8-25.**

The identification means chosen must encircle the conductor and must be accomplished by one of the following:

1. Stripping the insulation or covering from the entire exposed length
2. Coloring the insulation or covering green at the terminations
3. Marking the insulation with green tape or green adhesive labels at terminations

Multiconductor Cables

Section 250.119(B) indicates that one or more insulated conductors in a multiconductor cable at the time of installation are permitted to be reidentified as EGCs at each end and at every point where the conductors are accessible by any one of the following methods:

1. Stripping the insulation from the entire exposed length
2. Coloring the exposed insulation green
3. Marking the exposed insulation with green tape or green adhesive labels

Sizing Criteria

Sizing requirements for wire-type EGCs are found in 250.122 and Table 250.122 of the *NEC*. The minimum sizes are provided in Table 250.122 and are related to the short-time withstand capabilities of the conductor. An important note follows Table 250.122 and provides an appropriate reference to 250.4. This is an indication to *NEC* users that the minimum size required for wire-type EGCs could be larger than the sizes provided in the table. **See Figure 8-26.**

The conductor sizes in Table 250.122 of the *NEC* offer an approximate relation to the size of the overcurrent protective device given in the table. The I^2t values (short-time rating or withstand rating) of the EGC sizes are generally between 13 and 28 times their nominal continuous rating based on one ampere for every 42.25 circular mils of conductor. This value was used to develop the five-second withstand rating for insulated conductors. Using this formula, a five-second rating can be established for insulated conductors covered in the *NEC*. This is the proven conductor withstand formula developed through extensive testing and data collection by the Insulated Cable Engineers Association (ICEA). The value of 42.25 circular mils is applicable to insulated conductors. The value is 29.1 circular mils for uninsulated conductors based on the findings of the ICEA.

For example: To determine the five-second withstand rating of a 4 AWG insulated conductor, use the values in the third column of *NEC* Table 8, Chapter 9, to determine the circular mil area of a 4 AWG conductor. **See Annex D.** The value is 41,740 circular mils. Take the value 41,740 circular mils ÷ 42.25 circular mils and the result is 987.9 amperes, which can be carried safely for five seconds by a 4 AWG conductor.

Note that five seconds is a long time for any overcurrent protective device to open. Overcurrent protection typically operates in a few cycles or fractions of a cycle. The faster the overcurrent protective device operates, the more fault current the insulated conductor can carry safely without degradation or annealing of the conductor. This point emphasizes

Figure 8-26 **Minimum Size for Wire-Type EGCs**

Table 250.122 Minimum Size Equipment Grounding Conductors for Grounding Raceway and Equipment (in part)

Rating or Setting of Automatic Overcurrent Device in Circuit Ahead of Equipment, Conduit, etc., Not Exceeding (Amperes)	Size (AWG or kcmil)	
	Copper	Aluminum or Copper-Clad Aluminum*
15	14	12
20	12	10
60	10	8
100	8	6
200	6	4
300	4	2
400	3	1

Note: Where necessary to comply with 250.4(A)(5) or (B)(4), the equipment grounding conductor shall be sized larger than given in this table.

* See installation restrictions in 250.120.

Figure 8-26. The minimum sizes for wire-type EGCs are provided in Table 250.122 (reproduced in part).

that there are factors such as voltage drop and higher amounts of fault current in systems that could result in the EGC sizes in Table 250.122 being insufficient. Good engineering designs require a careful study and selection of overcurrent protection in conjunction with sufficiently-sized EGCs in order to ensure fast and effective operation of overcurrent protective devices in short-circuit and ground-fault conditions.

EQUIPMENT GROUNDING CONDUCTOR SIZING

In addition to the engineering basics of the effective ground-fault current path, the sizing rules in the *NEC* for EGCs are also important. The driving text of 250.122 is that the minimum size required for wire-type EGCs is not to be less than the values in Table 250.122. This rule goes on to state that in no case are the EGCs required to be larger than the circuit conductors supplying the equipment. **See Figure 8-27.**

It is important to mention parallel arrangements and their relationship to the circuit. The EGC is never required to be larger than the circuit conductors. Where circuit conductors are installed in parallel to create the equivalent of a larger circuit conductor, the size of the circuit conductor is the total area of all individual conductors in parallel that create the larger circuit conductor using multiple parallel paths. It is important to understand that the *NEC* does not permit conductors to be installed in parallel to create an EGC. Each EGC in a parallel installation must be sized based on Table 250.122. **See Figure 8-28.**

Section 250.122 also indicates that raceways, cable sheaths, and cable trays serving as EGCs must meet the performance requirements in 250.4(A)(5) or (B)(4), as applicable. For multi-conductor cable, the EGC within the cable assembly can be sectioned where the combined circular mil area of the sectioned EGC meets the size requirement in Table 250.122. Using Table 250.122 requires knowing the rating of the

Fact

In no case are the EGCs required to be larger than the circuit conductors supplying the equipment.

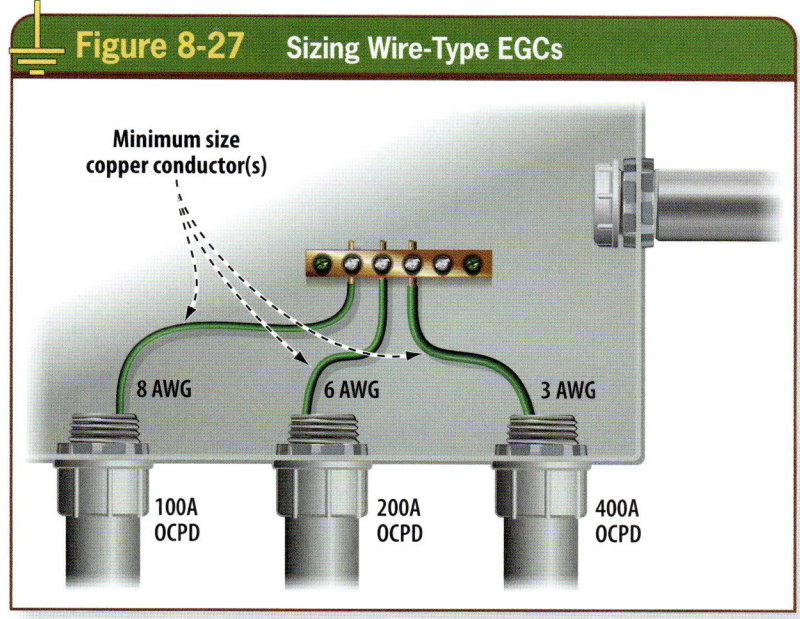

Figure 8-27. *Wire-type EGCs must be sized according to Table 250.122.*

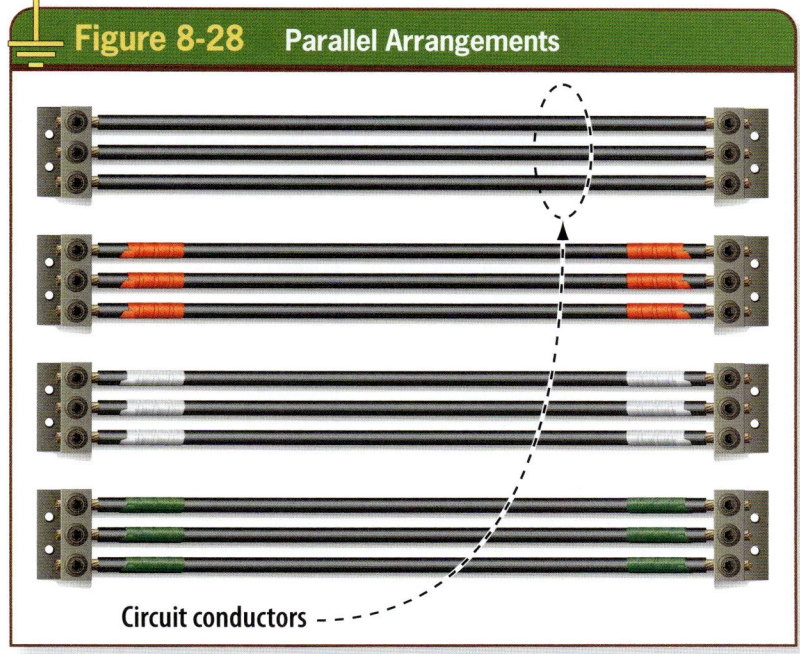

Figure 8-28. *The circuit conductors in parallel arrangements include all the ungrounded conductors of each phase and/or a grounded conductor to create the entire circuit.*

overcurrent protective device protecting the branch circuit or feeder. Once this value is known, the rating (or a rating that does not exceed the value in the left column of the table) should be found. The reader can then move across the table horizontally from left to right to determine the minimum size EGC expressed in AWG or circular mils. Make sure to use the appropriate column for aluminum as compared to copper EGCs.

For example: Consider several feeders and branch circuits installed in the slab of a commercial building. All are installed in PVC conduit. Each circuit requires that an EGC be connected to the equipment. This information can be used to determine the minimum size copper and aluminum EGCs (wire type) for feeders or branch circuits protected at 20, 45, 60, 90, 110, 225, 350, and 450 amperes. Using Table 250.122, the minimum size EGC required for a 4,000-ampere feeder can be found to be not less than 500-kcmil copper or 750-kcmil aluminum or copper-clad aluminum. **See Figure 8-29.**

Increases in Size

EGCs are required to be increased in size to address voltage drop. A proportionate increase of the associated ungrounded circuit conductors is required.

If the ungrounded conductors of a circuit are increased in size for any reason other than as required in 310.15(B) or (C), the wire-type EGCs must also be increased proportionately according to the circular mil area of the ungrounded conductors. This can be verified by following these steps, using a 400-ampere feeder as an example.

The 400-ampere feeder (420 ampacity) is generally installed using 600-kcmil copper circuit conductors and a 3 AWG copper EGC. For voltage drop reasons or due to circuit capacity needs, the 600-kcmil conductor is increased in size from 600-kcmil to two paralleled 400-kcmil copper conductors for each ungrounded phase conductor and the neutral conductor. The circular mil values are added together (400 plus 400) to result in 800-kcmil copper now required for the circuit conductors in this installation. The adjusted size (800 kcmil) is then divided by the originally required size (600 kcmil) to determine the proportionate value of circular mil area adjustment.

$$800 \div 600 = 1.3 \text{ (multiplier)}$$

The minimum size wire-type EGC for a 400-ampere feeder is normally a 3 AWG copper, according to Table 250.122. The circular mil value of a 3 AWG conductor is 52,620 circular mils, as provided in *NEC* Table 8, Chapter 9. Take the value 52,620 circular mils and multiply it by 1.3 to come up with 68,406 circular mils.

$$52,620 \times 1.3 = 68,406 \text{ circular mils}$$

Take this value back to *NEC* Table 8, Chapter 9 and round up to the next higher value (in the third column) to determine the minimum size EGC as adjusted proportionately to the increase in size for the ungrounded phase conductors of the circuit. The next higher circular mil value in *NEC* Table 8 is 83,690. Therefore, the new minimum

Size of Overcurrent Device	Equipment Grounding Conductor (Copper)	Equipment Grounding Conductor (Aluminum)
20	12	10
45	10	8
60	10	8
90	8	6
110	6	4
225	4	2
350	3	1
450	2	1/0

Figure 8-29. Table 250.122

Figure 8-29. *Table 250.122 can be used to determine the minimum size EGC based on the size of the overcurrent protective device installed in the circuit.*

size required for this EGC is a 1 AWG copper, based on the adjustment.

It is always best to perform this simple calculation to verify that the adjusted size of the EGC meets or exceeds the minimum requirements. Simply increasing the size of the EGC to the next higher size is not adequate in all cases. An exception to 250.122(B) indicates that qualified persons shall be permitted to size equipment grounding conductors to provide an effective ground-fault current path.

Multiple Circuits in a Single Raceway or Cable Tray

It is common for installers to economize in construction installation methods. One way to economize is to combine electrical circuits into a single raceway rather than running individual conduits or cables. This combination offers the advantage of requiring only a single EGC in the run. **See Figure 8-30.**

If multiple circuits are installed in a single raceway, cable, trench, or cable tray, a single EGC is permitted. The sizing requirement is based on the rating of the largest overcurrent protective device ahead of any circuit in the raceway, cable, trench, or cable tray. The required ampacity adjustment factors can be a disadvantage, however, because of the number of current-carrying conductors in the same raceway. If single EGCs are installed in a cable tray, they must meet the requirements in 250.122 and 392.10(B)(1)(c) and must be 4 AWG or larger.

Equipment Grounding Conductors for Parallel Runs

Many commercial and industrial electrical designs use parallel arrangements for large feeders or branch circuits to equipment. Installing parallel conductors becomes a necessity when supplying large switchboards and other large electrical equipment simply because large single conductors are not practical, economical, or even available in many cases. Installing feeders or circuits in parallel requires compliance with 310.10(G), which means the

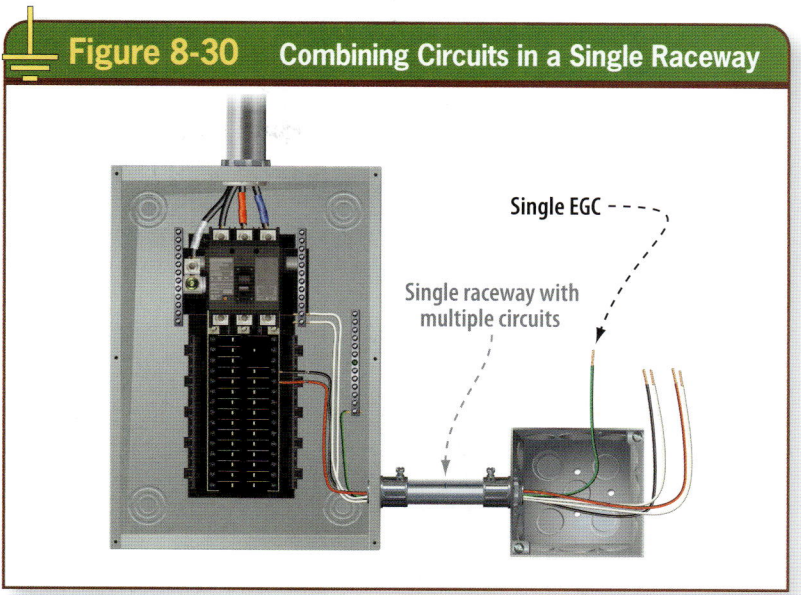

Figure 8-30 Combining Circuits in a Single Raceway

Single EGC

Single raceway with multiple circuits

Figure 8-30. Only one EGC is needed when multiple circuits are combined in the same raceway.

conductors must be the same length, material, and size; have the same insulation; and be terminated in the same manner. **See Figure 8-31.**

Installing conductors in parallel for feeders means that multiple conductors

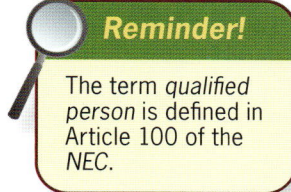

Reminder!

The term *qualified person* is defined in Article 100 of the *NEC*.

Figure 8-31 EGCs Installed in Parallel Runs

Figure 8-31. EGCs in parallel must comply with all of the requirements in 310.10(G) and must be installed in each raceway of the parallel run.

are electrically connected at both ends of the circuit. If the entire parallel arrangement of conductors is installed in a single raceway, cable, or cable tray, a single EGC is permitted for the entire parallel feeder or branch circuit arrangement. The metal raceway or cable tray could also be suitable as an EGC,

Figure 8-32   Single EGC

Figure 8-32. A single EGC is permitted with multiple circuits installed in the same cable tray.

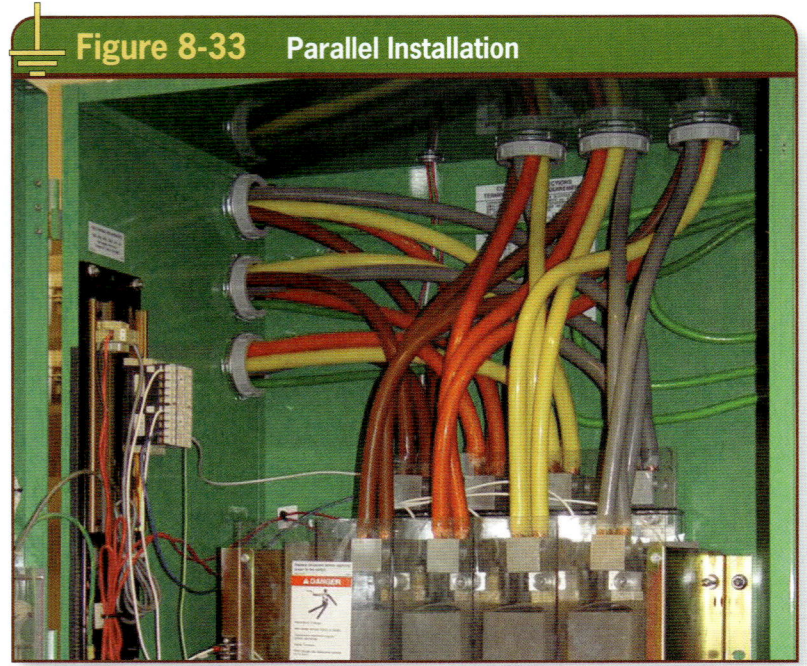

Figure 8-33   Parallel Installation

Figure 8-33. Circuits installed in parallel require that EGCs also be run in parallel.

in accordance with 250.118. In this case, a wire-type EGC is not necessary to be installed to meet the minimum *NEC* requirements because the raceway serves as the required EGC. If installed, it must meet the minimum sizing rules because the *NEC* offers no exception. **See Figure 8-32.**

There are specific rules for wire-type EGCs installed with parallel conductor installations. Section 250.122(F) states that when conductors are run in parallel in separate raceways or cables, any wire-type EGCs are also required to be run in parallel in their respective raceway or cable. **See Figure 8-33.**

If wire-type EGCs are installed with the paralleled, ungrounded feeder conductors, the EGCs must also be installed in parallel, but they can be smaller than 1/0 AWG. This is because the multiple EGCs are not being installed to share ground-fault current in the case of a ground-fault event, but to provide a low-impedance path regardless of where the fault occurs. Each one is already sufficiently sized according to 250.122, as required. When parallel arrangements of conductors are installed in separate raceways, such as in multiple conduits, and wire-type EGCs are required, an EGC must be installed in each of the separate raceways. **See Figure 8-34.**

In this type of installation, each EGC installed in parallel is required to be sized using Table 250.122 based on the rating of the circuit breaker or fuse protecting the entire parallel set. **See Figure 8-35.**

It is important to remember that 300.3(B) generally requires all conductors of the circuit, including the EGCs, to be installed in the same raceway, cable, or trench.

One reason for the requirement to include an EGC in each of the raceways is that the EGC prevents overloading, and damage to a smaller, inadequately-sized EGC is possible should a ground fault occur in one of the parallel branches. Another reason is that it keeps impedance levels low during normal operation and during ground-fault events.

Installing only an EGC in one raceway and no EGC in the raceways of the

Figure 8-34 Table 250.122 Sizing

Figure 8-34. Each EGC for circuits in parallel must be sized using Table 250.122.

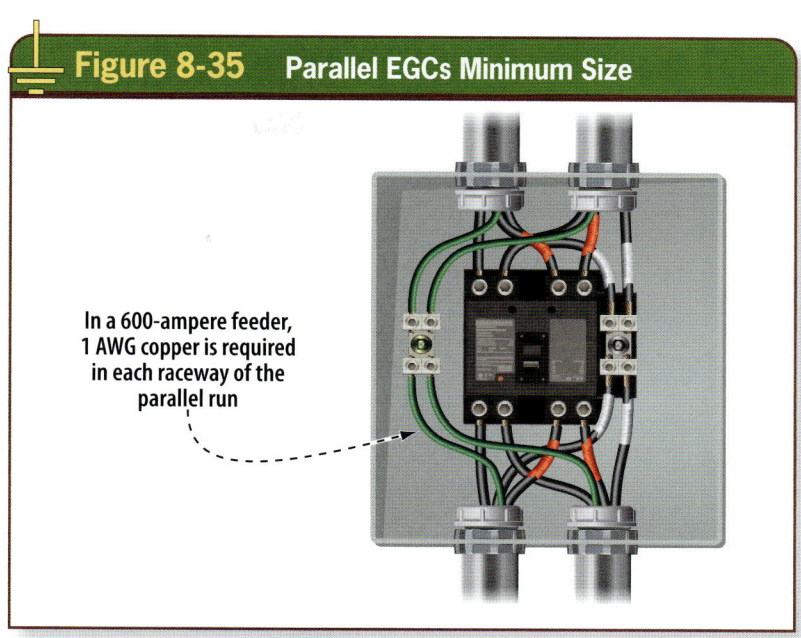

Figure 8-35 Parallel EGCs Minimum Size

In a 600-ampere feeder, 1 AWG copper is required in each raceway of the parallel run

Figure 8-35. When sizing EGCs for circuits in parallel arrangements, the minimum size must be not smaller than the sizes in Table 250.122 based on the rating of the fuse or circuit breaker.

other parallel feeders separates the EGC from its associated ungrounded conductors of the same circuit. This is a violation of 250.122(F), 300.3(B), and 310.10(G). Keeping the impedance in the effective ground-fault current path as low as possible ensures fast, effective operation of circuit breakers or fuses in the case of a ground fault.

Cable Assemblies in Parallel

Cable assemblies, such as those using metal-clad (Type MC) cable, are manufactured in standard conductor size configurations unless they are a special-order item. Some cable manufacturers can produce cable assemblies with larger EGCs when parallel arrangements are planned for a project. The EGC in a

standard cable is typically sized adequately for single circuit use, but it is probably not adequate for all parallel circuit installations. If cable assemblies are installed in parallel circuits, it is necessary to verify that the EGC in each of the individual cables of the parallel set is sized as required by Table 250.122, based on the size of the circuit breaker or fuse protecting the entire parallel circuit.

Installing cable assemblies in parallel arrangements may necessitate a special order that includes sizing the EGC in each cable large enough to allow the

Sizing Examples

Questions about sizing EGCs for parallel installations are common. Installing EGCs in parallel arrangements is not overly complicated. The following examples address a couple of sizing exercises:

Example 1: A 4,000-ampere feeder is installed in 10 PVC conduits in a parallel arrangement, each containing four 750-kcmil copper conductors. What is the minimum size copper EGC required in each conduit?

Answer: A 500-kcmil copper EGC is required in each raceway based on the 4,000-ampere overcurrent protective device, in accordance with Table 250.122.

Example 2: If an 800-ampere feeder is installed in two raceways in a parallel arrangement, each containing four 700-kcmil copper conductors, what is the minimum size wire-type EGC for this circuit?

Answer: A 1/0 AWG copper EGC is required in each raceway.

cable to be connected in a parallel installation. Section 250.122(F)(2)(b) allows a single EGC to be installed in a cable tray containing multiple multiconductor cables. The single EGC is sized using 250.122 and must be connected to and in combination with the other EGCs contained in the individual cables.

Section 250.122(A) indicates that EGCs never have to be larger than the ungrounded conductors of the circuit. In a parallel feeder, the ungrounded conductors of the circuit include all of the conductors in parallel that are added together to make a single conductor. For example, if four 600-kcmil conductors are installed in parallel to create a 1,600-ampere feeder circuit, the total circular mil value for the circuit conductors is 2,400-kcmil. The minimum size EGC required in each raceway is 4/0 AWG copper, based on the 1,600-ampere overcurrent protective device.

Equipment Grounding Conductors for Motor Circuits

Sizing requirements for EGCs in motor circuits are provided in 250.122(D)(1)

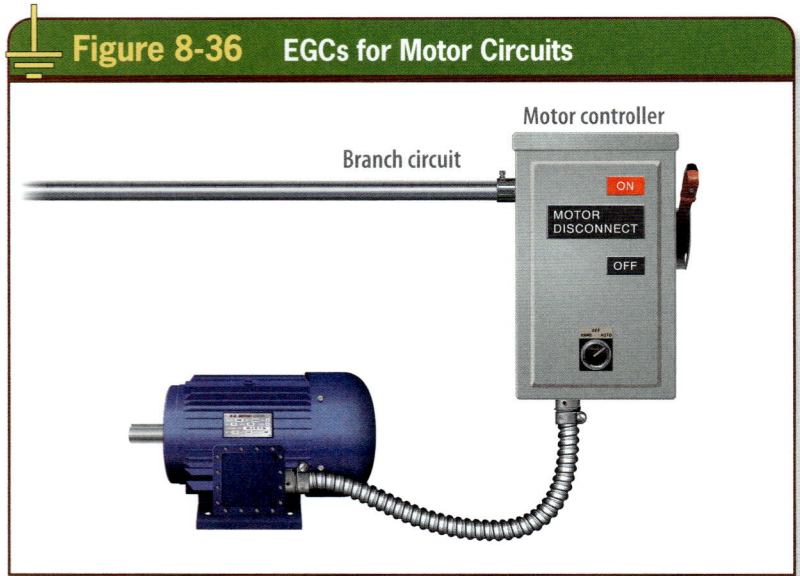

Figure 8-36   EGCs for Motor Circuits

Motor controller

Branch circuit

ON

MOTOR DISCONNECT

OFF

Figure 8-36. Size EGCs for motor circuits using Table 250.122 based on the rating of the branch-circuit short-circuit and ground-fault protective device.

and (D)(2). The basic rule here is that the EGC (wire-type) be sized not smaller than determined by 250.122(A) based on the rating of the branch-circuit short-circuit and ground-fault protective device of the motor circuit, as determined based on Section 430.52. **See Figure 8-36.**

It is important to remember that the branch-circuit short-circuit and ground-fault protective device is usually sized larger to carry the starting current of the motor, which affects the size of a wire-type EGC. When an instantaneous-trip circuit breaker or motor short-circuit protector is selected as the overcurrent protective device for a motor circuit, a wire-type EGC is required to be sized no smaller than provided in 250.122(A), using the maximum rating of a dual-element time-delay fuse selected for branch-circuit short-circuit and ground-fault protection in accordance with 430.52(C)(1) Exception No. 1.

Equipment Grounding Conductors with Feeder Taps

When feeder taps are installed in accordance with the provisions in 240.21, the size of the EGC with the tap conductors must not be smaller than the size required based on the rating of the overcurrent protection for the feeder to which the tap is connected. For example, a 400-ampere feeder tapped by two 200-ampere feeders would require a 3 AWG copper EGC in each of the feeder tap raceways. **See Figure 8-37.**

CURRENT IN EQUIPMENT GROUNDING CONDUCTORS

EGCs should only carry current during abnormal conditions such as a ground-fault event, when fault current is present in the EGC for the duration of the time it takes the circuit breaker or fuse to clear the fault. In normal operation, no current should be present in the EGC. If current is present, then there probably is a noncompliant neutral-to-ground connection on the load side of the service disconnect or the load side of the first system overcurrent protective device of a separately derived system. Sections 250.24(A)(5) and 250.30(A) state these restrictions.

The *NEC* does not generally permit the EGC to perform simultaneously as a grounding electrode conductor and EGC, even though this has been practiced in the past, particularly with separately derived system installations. Section 250.118(B)(1) prohibits using a grounding electrode conductor as an equipment grounding conductor. An exception to this restriction recognizes that only a wire-type equipment grounding conductor could simultaneously serve as a grounding electrode conductor if all the requirements in Parts II, III, and VI of Article 250 have been met. This method of installation must also comply with 250.6(A), which requires that grounding must be arranged in a manner that will prevent objectionable current through normally non–current-carrying conductors and metal parts of equipment. Section 250.118(B)(2) provides another restriction from using metal building framing (structural steel) as an equipment grounding conductor.

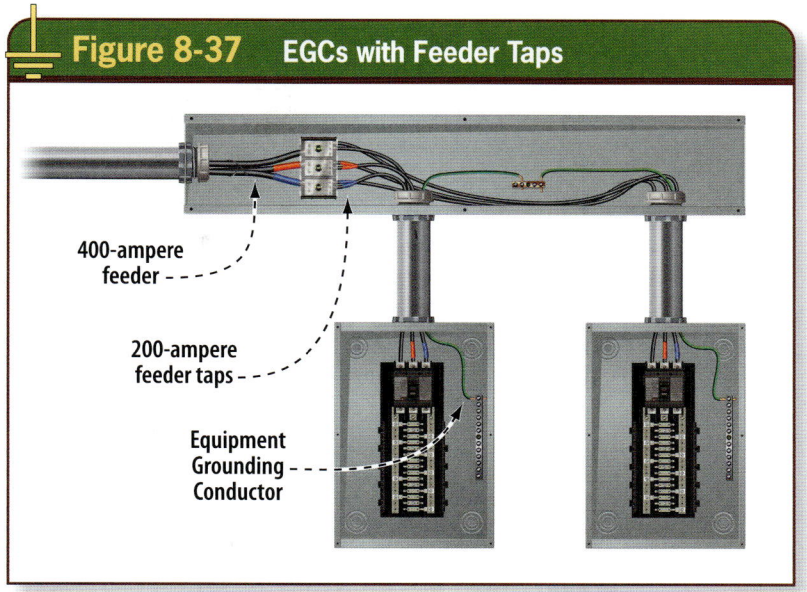

Figure 8-37   **EGCs with Feeder Taps**

400-ampere feeder

200-ampere feeder taps

Equipment Grounding Conductor

Figure 8-37. EGCs installed with feeder taps are sized using Table 250.122 based on the rating of the overcurrent protection on the supply side of the tap conductors.

Fact

EGCs should only carry current during abnormal conditions, such as a ground-fault event, when heavy fault current is present in the EGC for the duration of time it takes the overcurrent protective device to clear the fault.

SUMMARY

EGCs perform three essential functions in the grounding and bonding safety system: they perform a grounding function for equipment required to be grounded, they perform a bonding function because they connect equipment together, and they facilitate overcurrent protective device operation while performing as part of an effective ground-fault current path during ground-fault events.

EGCs are available in a variety of *NEC*-recognized types, including raceways, cable assemblies, cable tray, wireways, flexible metal conduit, and liquidtight flexible metal conduit. The minimum size required for wire-type EGCs is determined using Table 250.122 on the basis of the overcurrent protective device protecting the circuit.

REVIEW QUESTIONS

1. The conductive path installed to connect normally non–current-carrying metal parts of equipment together and to the system grounded conductor, the grounding electrode conductor, or both best defines which of the following components of the electrical grounding system?
   a. Equipment bonding jumper
   b. Equipment grounding conductor
   c. Grounding electrode conductor
   d. System bonding jumper

2. *NEC* 250.118 includes a list of acceptable equipment grounding conductors.
   a. True   b. False

3. Equipment grounding conductors can be copper or copper-clad, but not aluminum.
   a. True   b. False

4. Equipment grounding conductors perform bonding functions, among other functions.
   a. True   b. False

5. Which of the following is not acceptable by the *NEC* as a suitable equipment grounding conductor?
   a. Electrical metallic tubing
   b. Flexible metal conduit in lengths not longer than 8'
   c. Rigid metal conduit
   d. The armor of Type AC cable

6. Flexible metal conduit longer than 10 feet is acceptable as an equipment grounding conductor as long as the overcurrent protective device protecting the circuit does not exceed 20 amperes.
   a. True   b. False

7. When flexible metal conduit is installed for flexibility, a wire-type equipment grounding conductor is required.
   a. True   b. False

8. Which of the following characteristics is not related to an effective ground-fault current path?
   a. Electrically conductive
   b. High impedance
   c. Intentionally constructed
   d. Sufficient capacity for any fault current likely to be imposed on it

9. Which of the following conditions is not required for flexible metal conduit to qualify as an equipment grounding conductor?
   a. Length is limited to 6' in the ground-fault return path
   b. Listed fittings must be used
   c. The circuit contained is protected at no greater than 20 A
   d. The flexible metal conduit is installed for flexibility

10. When rigid metal conduit is installed and used as an equipment grounding conductor, all fittings and joints are required to be __?__.
    a. approved
    b. made up tight using suitable tools
    c. sealed
    d. tested

11. The steel alloy sheath of mineral-insulated cable (Type MI) is suitable as an equipment grounding conductor.
    a. True   b. False

12. When listed liquidtight flexible metal conduit is installed for equipment, and flexibility of the conduit is necessary after installation, an equipment grounding conductor (wire-type) must be installed in the conduit.
    a. True   b. False

13. Flexible metal conduit is never allowed as an equipment grounding conductor when flexibility is not necessary after installation.
    a. True   b. False

14. Which of the following does not qualify as an equipment grounding conductor?
    a. A copper conductor sized at 12 AWG
    b. Electrical metallic tubing
    c. Flexible metallic tubing 3' in length in which the circuit contained is protected by a 30A overcurrent protective device
    d. Intermediate metal conduit

15. What table in the *NEC* is used to size equipment grounding conductors of the wire type?
    a. Table 1, Chapter 9
    b. Table 250.122
    c. Table 250.66
    d. Table 310.15(B)(16)

16. Table 250.122 establishes minimum sizes for wire-type equipment grounding conductors based on the size of the __?__ of the circuit.
    a. largest ungrounded service conductor
    b. overcurrent protective device
    c. overload protective device
    d. smallest ungrounded conductor

17. An equipment grounding conductor is never permitted to perform as both the equipment grounding conductor and grounding electrode conductor simultaneously for a separately derived system.
    a. True   b. False

18. What is the minimum size wire-type equipment grounding conductor for a 4,000-ampere feeder circuit?
    a. 250-kcmil aluminum
    b. 3/0 AWG copper
    c. 400-kcmil copper
    d. 500-kcmil copper

19. What is the minimum size equipment grounding conductor for a motor circuit if the motor branch-circuit short-circuit and ground-fault protective device is sized at 1,200 amperes?
    a. 3/0 AWG copper
    b. 2/0 AWG copper
    c. 1/0 AWG copper
    d. 250-kcmil copper

20. If the ungrounded circuit conductors for a 400-ampere circuit are increased in size from 600-kcmil copper to 1,000-kcmil copper due to length and excessive voltage drop, what is the minimum size required for a wire-type equipment grounding conductor?
    a. 3 AWG copper
    b. 2 AWG copper
    c. 1/0 AWG copper
    d. 1 AWG copper

21. Green insulation along the entire length of a conductor is one method of identification for equipment grounding conductors of branch circuits.
    a. True   b. False

22. Table 250.122 provides the minimum sizes for wire-type equipment grounding conductors, which may need to be larger to meet the performance requirements in 250.4.
    a. True   b. False

23. When a 400-ampere tap is made to a 1,000-ampere feeder, what is the minimum size equipment grounding conductor (wire-type) required to be run with the tap conductors to the equipment?
    a. 4/0 AWG aluminum
    b. 3 AWG copper
    c. 2/0 AWG copper
    d. 1 AWG copper

24. If six 400-kcmil copper conductors are run in parallel (per phase) as part of a 2,000-ampere feeder, what is the minimum size wire-type equipment grounding conductor required in each PVC raceway?
    a. 4/0 AWG copper
    b. 3/0 AWG copper
    c. 250-kcmil aluminum
    d. 250-kcmil copper

25. Flexible metal conduit in sizes greater than 1 1/4 inch is a suitable equipment grounding conductor if limited to not more than six feet in length.
    a. True   b. False

Grounding Electrical Equipment

The *NEC* includes several rules for grounding electrical equipment. Grounding equipment is accomplished by a direct connection to ground (Earth), by connection to a conducting body that extends the grounding connection, or both. Specific *NEC* rules apply to fixed equipment that must be grounded with very few exceptions. Auxiliary grounding electrodes are addressed in the *NEC* as optional. If installed, they must meet specific requirements, but they are not permitted as the only grounding means for equipment. Grounded conductors are permitted for grounding equipment such as on the line side of the service disconnect or at appliances, but only in existing installations.

Objectives

» Understand the reasons for grounding electrical equipment.

» Identify the methods of equipment grounding for feeders and branch circuits and the specific conditions that provide exemptions from grounding equipment.

» Understand the methods for installing equipment grounding conductors for devices such as receptacles and switches.

» Understand the requirements to isolate grounded conductors (neutrals) from ground and grounded metal parts.

» Determine the requirements for auxiliary grounding electrodes installed for equipment.

Chapter 9

Table of Contents

Purpose of Grounding Equipment................ 208

General Grounding Rules for Equipment 208

Conductor Enclosure and Raceway Grounding Requirements..................... 212

Methods of Grounding Equipment (Part VII of Article 250) 213

    Feeders and Branch Circuits 213

    Panelboard Equipment Grounding Terminal Bars 214

    Isolate the Grounded Conductor from Ground........................... 215

Connections of Equipment Grounding Conductors.................................216

Receptacle Grounding Connections.............216

    Equipment Grounding Conductor Continuity 219

    Grounding Snap Switches..................... 220

    Identification of Wiring Device Terminals ...221

    Continuity between Attachment Plugs and Receptacles 222

    Equipment Grounding Conductor Connections 222

Receptacle Replacements 223

    Replacing Grounding-Type Receptacles ... 223

    Replacing Non–Grounding-Type Receptacles 223

Grounding Appliances Using the Grounded Conductor..................... 224

Auxiliary Grounding Electrode Requirements............................ 225

Grounding Nonelectrical Equipment 227

Equipment Grounded by Secure Metal Supports 227

Use of the Grounded Conductor for Grounding............................. 228

Summary..................................... 229

Review Questions 230

PURPOSE OF GROUNDING EQUIPMENT

Grounding is necessary to establish an Earth reference (connection) for connected systems and equipment. Grounding equipment places it as close to Earth potential as possible, thereby minimizing shock hazard possibilities. Grounding also limits the voltage to ground during line surge events, lightning events, and unintentional contact with higher-voltage lines. **See Figure 9-1.**

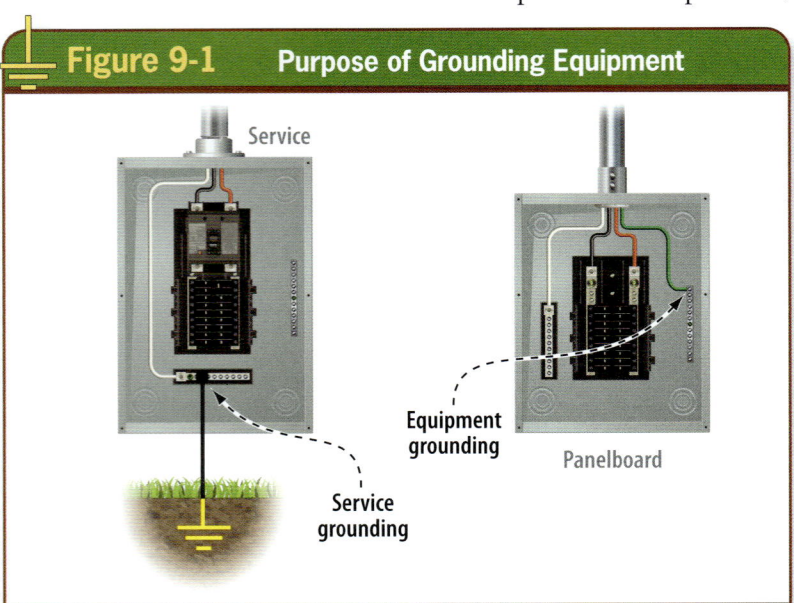

Figure 9-1   Purpose of Grounding Equipment

Figure 9-1. *Grounding equipment places its potential as close to Earth as possible, reducing shock hazards.*

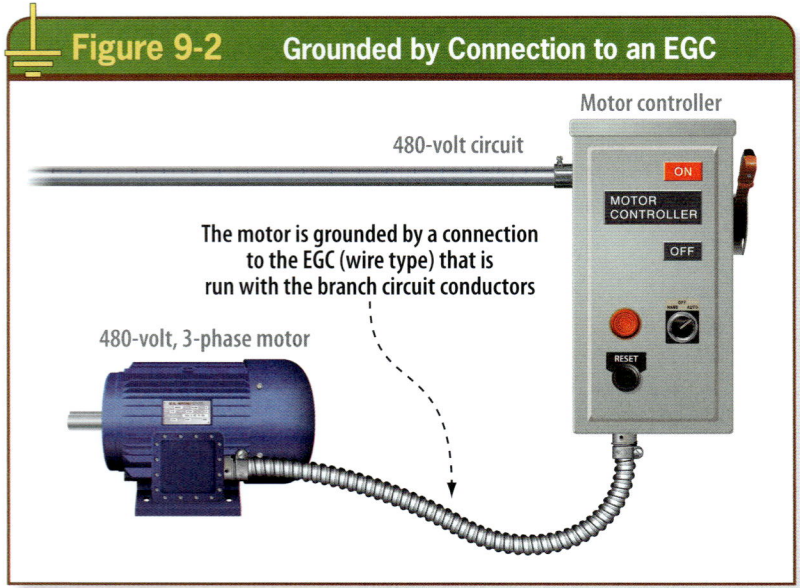

Figure 9-2   Grounded by Connection to an EGC

Figure 9-2. *Equipment operating at more than 150 volts to ground is generally required to be grounded by a connection to an EGC of the supply circuit.*

Equipment grounding as required by the *NEC* is not intended to provide total protection from lightning strikes. The goal of the *NEC* is to protect people and property from the hazards arising from the use of electricity. Lightning is not electricity that is used; it is an unpredictable force. Grounding requirements in the *NEC* provide ancillary benefits of providing a path to ground for lightning events, but that is not the primary purpose. Lightning protection systems installed according to *NFPA 780: Standard for the Installation of Lightning Protection Systems* provide an increased level of protection for buildings and structures where they are installed.

The grounding process involves connecting a system of conductive parts to ground or to a conductive body that extends the ground connection. Grounding electrical equipment is generally accomplished by connection to an equipment grounding conductor (EGC) of a feeder or branch circuit. The Earth is never permitted as an effective path for ground-fault current.

GENERAL GROUNDING RULES FOR EQUIPMENT

The general requirements for grounding equipment that is fastened in place or connected by permanent wiring methods (fixed) are provided in Part VI of Article 250. There are instances in which substitutes for grounding such as isolation, insulation, or guarding can be applied.

These alternatives to grounding are addressed in 250.1(6), and some are addressed more specifically in 250.110. If the equipment operates at more than 150 volts to ground, it is generally required to be grounded by connection to an EGC. **See Figure 9-2.**

Fixed equipment with exposed non–current-carrying metal parts that are likely to become energized is required to be grounded (connected to an EGC) under several conditions. For example, if the equipment is within eight feet vertically or five feet horizontally of the ground or grounded metal objects that are subject to contact by people, it must be grounded. There is a condition that relaxes the grounding requirement due to isolation or elevation. If the equipment is 10 feet above the ground and not subject to contact by people, grounding is optional as long as the location is not wet or damp.

Where the equipment is in a wet or damp location and is not isolated, the equipment also must be grounded if it is in contact with other metal material. The idea is to ground all metal surfaces and not leave any isolated or "floating" at a potential above ground. Electrical equipment in a hazardous (classified) location is required to be grounded as covered in Articles 500 through 517.

Equipment that is supplied by metal raceways, metal-sheathed or metal-clad cables, or another wiring method that provides an EGC must be grounded by connection to the EGC. **See Figure 9-3.**

Three exceptions follow the grounding requirements provided in 250.110. The first exception (Exception No. 1) relaxes the grounding requirement for frames of electrically-heated appliances that are permanently and effectively insulated from ground. Another exception (Exception No. 2) from the grounding requirement applies to distribution apparatuses such as transformers and capacitor enclosure cases mounted on wood poles and elevated to a height that exceeds eight feet above ground. **See Figure 9-4.** Another exception (Exception No. 3) applies to installations in which insulation is used as a substitute for grounding and offers equal and effective safety by using double-insulated equipment.

Section 250.112 provides a list of specific equipment that is connected by permanent wiring and is required to be grounded by connection to an EGC. Equipment covered by the equipment

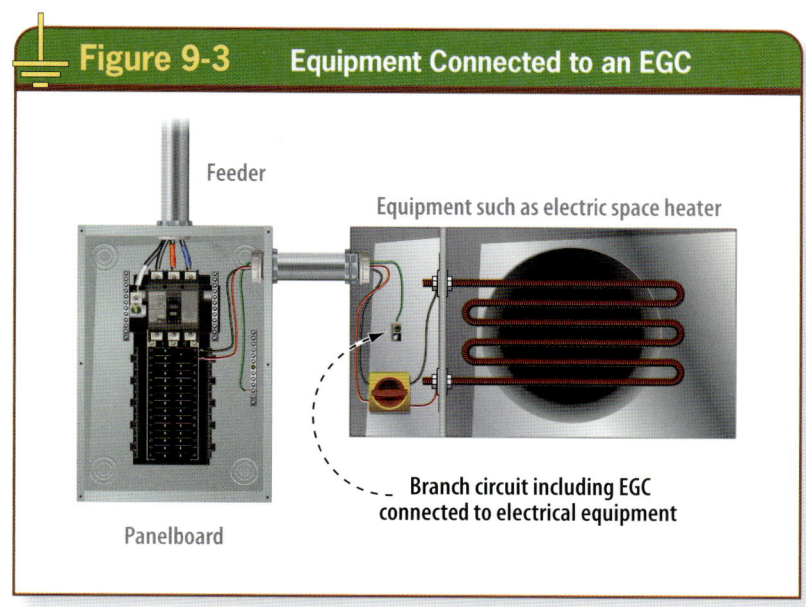

Figure 9-3. Equipment is grounded by being connected to an EGC run with the circuit conductors.

Figure 9-4. Grounding is not required for pole-mounted transformers or capacitors.

grounding rules in 250.112 consists of the following:

- Switchgear and switchboard frames and structures
- Enclosures for motor controllers
- Electric signs
- Luminaires (lighting fixtures)
- Pipe organs
- Elevators and cranes
- Motion picture projection equipment
- Skid-mounted equipment

- Motor frames
- Garages, theaters, and motion picture studios
- Remote-control, signaling, and fire alarm circuits
- Motor-operated water pumps and metal well casings

The exposed, normally non–current-carrying parts of equipment described in 250.112(A) through (M) are required to be grounded by connection to an EGC, regardless of the operating voltage applied. Switchgear or switchboard frames and structures supporting switching equipment, except frames of 2-wire DC switchgear or switchboards that are effectively insulated from ground, must be grounded. Section 250.112(A) clarifies that DC switchgear or switchboards insulated from ground are not required to be grounded. Generator and motor frames in an electrically-operated pipe organ, unless effectively insulated from ground, are required to be grounded, as is the motor for the pipe organ. **See Figure 9-5.**

Motor frames are required to be grounded by connection to an EGC as specified in Part XIII of Article 250, which addresses both stationary and portable motors. **See Figure 9-6.**

Figure 9-5   Pipe Organs

Figure 9-5. Motor frames in electrically-operated pipe organs are required to be grounded.

Figure 9-6   Motor Frames

Figure 9-6. Motor frames are required to be grounded.

Figure 9-7   Motor Controllers

Figure 9-7. Motor controllers are generally required to be grounded.

Enclosures for motor controllers must be grounded unless attached to ungrounded portable equipment. These could be part of a motor control center assembly. **See Figure 9-7.**

Electrical equipment for elevators and cranes is required to be grounded by connection to the EGC of the supply circuit. **See Figure 9-8.**

Electrical equipment in commercial garages, theaters, and motion picture studios, except pendant lampholders, all must be grounded, except when they are supplied by circuits not more than 150 volts to ground. Electric signs, outline lighting, and associated equipment are all required to be grounded by connection to an EGC as provided in 600.7. **See Figure 9-9.**

All motion picture projection equipment must be grounded. Equipment supplied by Class 1 circuits shall be grounded unless the equipment operates at less than 50 volts. Equipment supplied by Class 1 power-limited circuits, by Class 2 and Class 3 remote-control and signaling circuits, and by fire alarm circuits must be grounded

Figure 9-8 **Grounding Elevators and Cranes**

Figure 9-8. Electrical equipment for cranes and elevators is required to be grounded.

only when system grounding is required by Part II or Part VIII of Article 250. In other words, if the supply system is grounded, the equipment it supplies must also be grounded. Luminaires (lighting fixtures) are required to be grounded as provided in Part V of Article 410. **See Figure 9-10.**

Figure 9-9 **Electric Signs**

Figure 9-9. Electric signs and equipment associated with outline lighting are required to be grounded with the exception of equipment supplied by Class 2 power supplies, such as LED power supplies.

Figure 9-10 **Luminaires**

Figure 9-10. Luminaires are required to be grounded in accordance with Part V of Article 410.

Figure 9-11   Water Pumps

Figure 9-11. A submersible well pump casing must be bonded to the EGC of the branch circuit.

Permanently-mounted electrical equipment and skids are also required to be connected to the branch circuit EGC sized as required by Table 250.122. Motor-operated water pumps, including the submersible type, are required to be grounded by connection to an EGC.

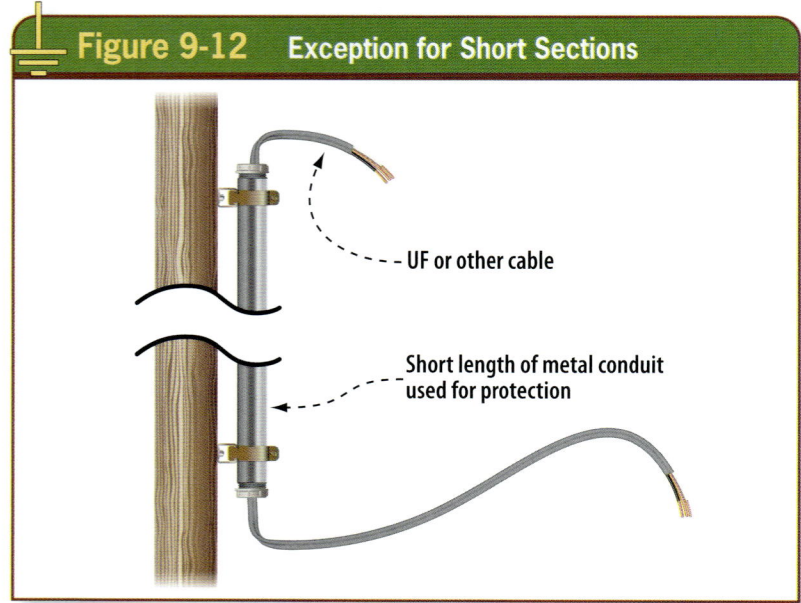

Figure 9-12   Exception for Short Sections

- - UF or other cable

Short length of metal conduit used for protection

Figure 9-12. Short sections of metal raceways are not required to be grounded.

When a submersible pump is used in a metal well casing, the well casing must be connected to the pump circuit EGC. **See Figure 9-11.**

CONDUCTOR ENCLOSURE AND RACEWAY GROUNDING REQUIREMENTS

Section 250.86 provides grounding requirements for metal raceways and enclosures other than those for service conductors. The general requirement in 250.86 is that these metal enclosures be grounded by connecting them to an EGC.

There are three exceptions that relax this requirement. The first exception is for metal enclosures and raceways used for conductors that extend existing installations of old knob-and-tube systems and nonmetallic sheathed cable system installations. The following conditions must be met to qualify for this exemption from grounding requirements:

1. The circuits do not contain or include an EGC. If they do include an EGC, grounding these raceways and enclosures is required.
2. The runs are in lengths less than 25 feet. For longer runs, the grounding requirement applies.
3. The installation is isolated and free from possible contact with ground, grounded objects, metal lath, or other conductive material.
4. The installation is guarded from contact by people.

If these conditions are met, then grounding is not required for raceways or enclosures that are installed to extend existing knob-and-tube installations or existing nonmetallic sheathed cable installations.

Another exception is for short sections of metal raceway or metal enclosures used for support or to provide protection from physical damage for cable assemblies. An example of a short section of metal not required to be grounded would be a length of conduit used to provide physical protection for UF cable emerging from the ground at a pole. **See Figure 9-12.** The first eight feet of this UF cable must be protected from physical damage, as required by 300.5(D)(1).

Another example is when a metal conduit sleeve is installed for a run of nonmetallic sheathed cable that penetrates a masonry wall. The metal sleeve is not required to be grounded. Note that the *NEC* does not define the term *short sections* in terms of a maximum distance. This is typically left up to the judgment of the authority having jurisdiction (AHJ).

Metal components, such as metal elbows, that are isolated from public contact by a minimum of not less than 18 inches when buried or encased in not less than two inches of concrete are not required to be grounded. **See Figure 9-13.** See Exception No. 3 following 250.86 for the actual *NEC* provisions.

Section 250.132 requires isolated sections of metal raceway or cable armor to be connected to an EGC if they are required to be grounded in accordance with the provisions in 250.134.

METHODS OF GROUNDING EQUIPMENT (PART VII OF ARTICLE 250)

As the feeder and branch circuit wiring is installed on the job site, EGCs should be installed as required. Section 215.6 requires EGCs to be provided with feeder conductors if they supply branch circuits that require EGCs. Because most equipment is required to be grounded by 250.110 and 250.112, nearly all feeders and branch circuits will have an EGC installed with the circuit conductors.

The requirements of 250.134 indicate that if equipment such as raceways and other enclosures are required to be grounded, then a method in 250.134(1) or (2) must be used. This rule states that equipment grounding can be accomplished by connection to any EGC type specified by 250.118, which could be a wire, raceway, or other type mentioned in the list. In general, the EGC must be installed with its associated

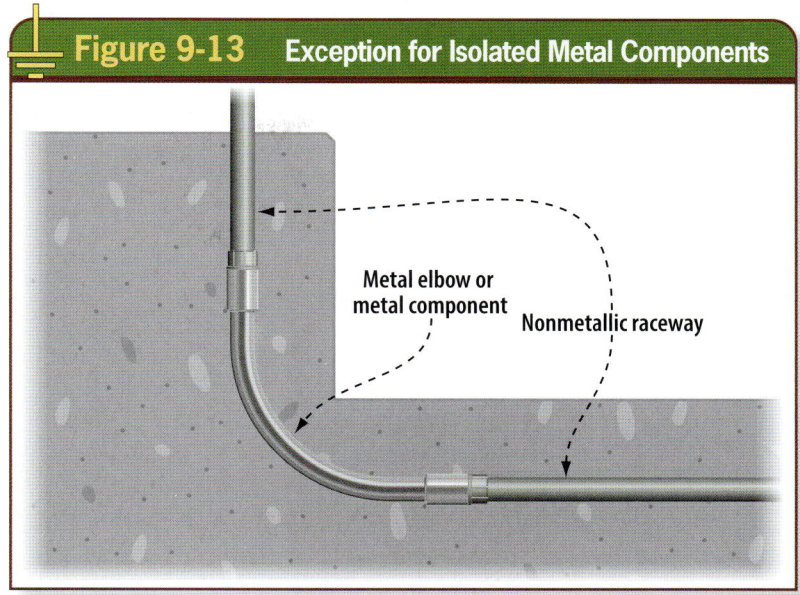

Figure 9-13 Exception for Isolated Metal Components

Figure 9-13. Metal components such as metal elbows encased in not less than two inches of concrete and isolated from contact by people are not required to be grounded.

circuit conductors. It could be a separate conductor contained in the wiring method of the circuit, or it could be the wiring method itself, such as a metal raceway or cable armor assembly (such as Type AC cable).

Two cases in which EGCs are permitted to be installed separately from the associated circuit conductors are addressed in the exceptions to 250.134(2). Exception No. 1 relaxes this requirement for separate EGCs installed to address grounding conductors for existing branch circuit extensions of circuits without an EGC as provided in 250.130(C). Exception No. 2 indicates that EGCs of DC circuits are permitted to be run separately from the associated circuit conductors.

Feeders and Branch Circuits

As the power circuits are distributed on the project, feeders typically are run from the service equipment to the panelboards or other distribution equipment installed at various locations throughout the facility. From the panelboards or other distribution equipment, such as motor control centers or other control panels, branch

Figure 9-14   EGCs With Feeder Conductors

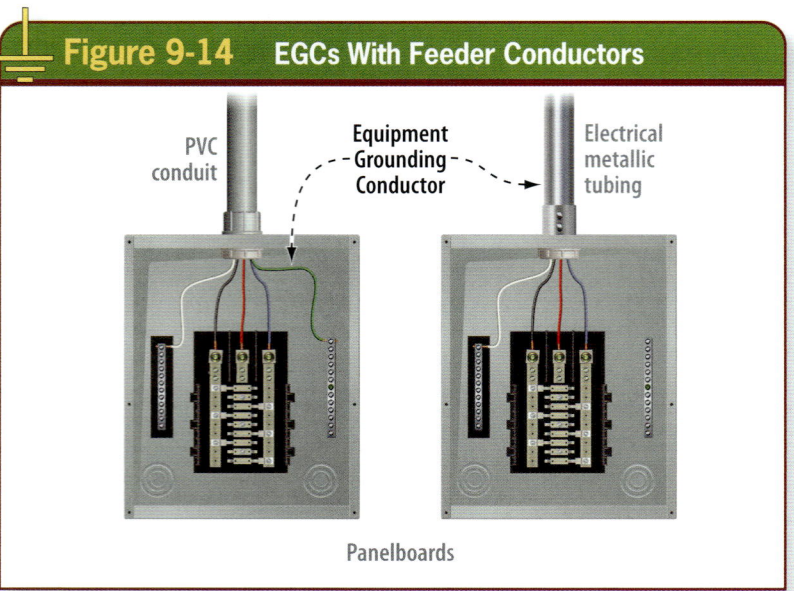

Figure 9-14. Section 215.6 generally requires an EGC to be installed with the feeder conductors supplying switchboards, panelboards, or other distribution equipment.

circuits are installed to the final outlets or are connected directly to utilization equipment. As the feeder wiring is connected to the panelboard enclosure, an EGC is connected as well, which complies with the general provisions in 215.6. Sometimes the EGC is a wire type, but often it is the conduit, tubing, or qualifying cable armor containing the feeder conductors. **See Figure 9-14.**

If the feeder and branch circuits connected at a panelboard enclosure are all installed using rigid metal conduit (RMC) or intermediate metal conduit (IMC), by the minimum *NEC* requirements, separate (wire-type) EGCs are not required. In this case, the connection between the feeder EGC and the EGCs of the branch circuits is accomplished through the metal panelboard enclosure. **See Figure 9-15.**

It is important that suitable locknuts or other fittings be installed properly to maintain the integrity of the EGC connections and path from the feeder to the branch circuit.

Panelboard Equipment Grounding Terminal Bars

Section 408.40 requires an equipment grounding terminal bar in the panelboard if nonmetallic raceways or cables are used, or where separate EGCs of the wire type are provided. The equipment grounding terminal bar is typically an accessory feature provided by the panelboard manufacturer. **See Figure 9-16.**

The installation and use of equipment grounding terminal bars for EGCs is driven by panelboards used with nonmetallic raceways or cables, or when separate wire-type EGCs are

Figure 9-15   Rigid Metal Conduit EGC

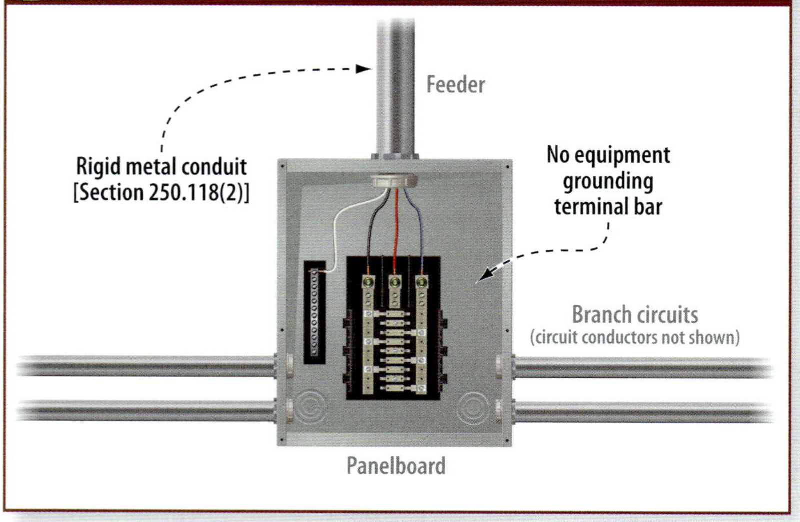

Figure 9-15. Rigid metal conduit qualifies as an EGC for the feeder and the branch circuits.

Figure 9-16 EG Terminal Bars

Figure 9-16. An equipment grounding terminal bar is often an accessory feature provided by the manufacturer.

provided. This terminal bar may be installed on the panelboard or its enclosure. A terminal bar assembly kit must include instructions for installation and panelboard or enclosure markings. **See Figure 9-17.**

Note that if no wire-type EGCs are present with the branch circuits or feeders, the equipment grounding terminal bar is not required or necessary. However, this is rarely the case.

Isolate the Grounded Conductor from Ground

Another important requirement for panelboard wiring on the load side of the service equipment is that the grounded (usually the neutral) conductor of the panelboard must not be connected to ground (the enclosure). This restriction also applies to panelboards on the load side of the grounding point for a separately derived system. Sections

Figure 9-17   Panelboard Equipment Grounding Terminal Bars

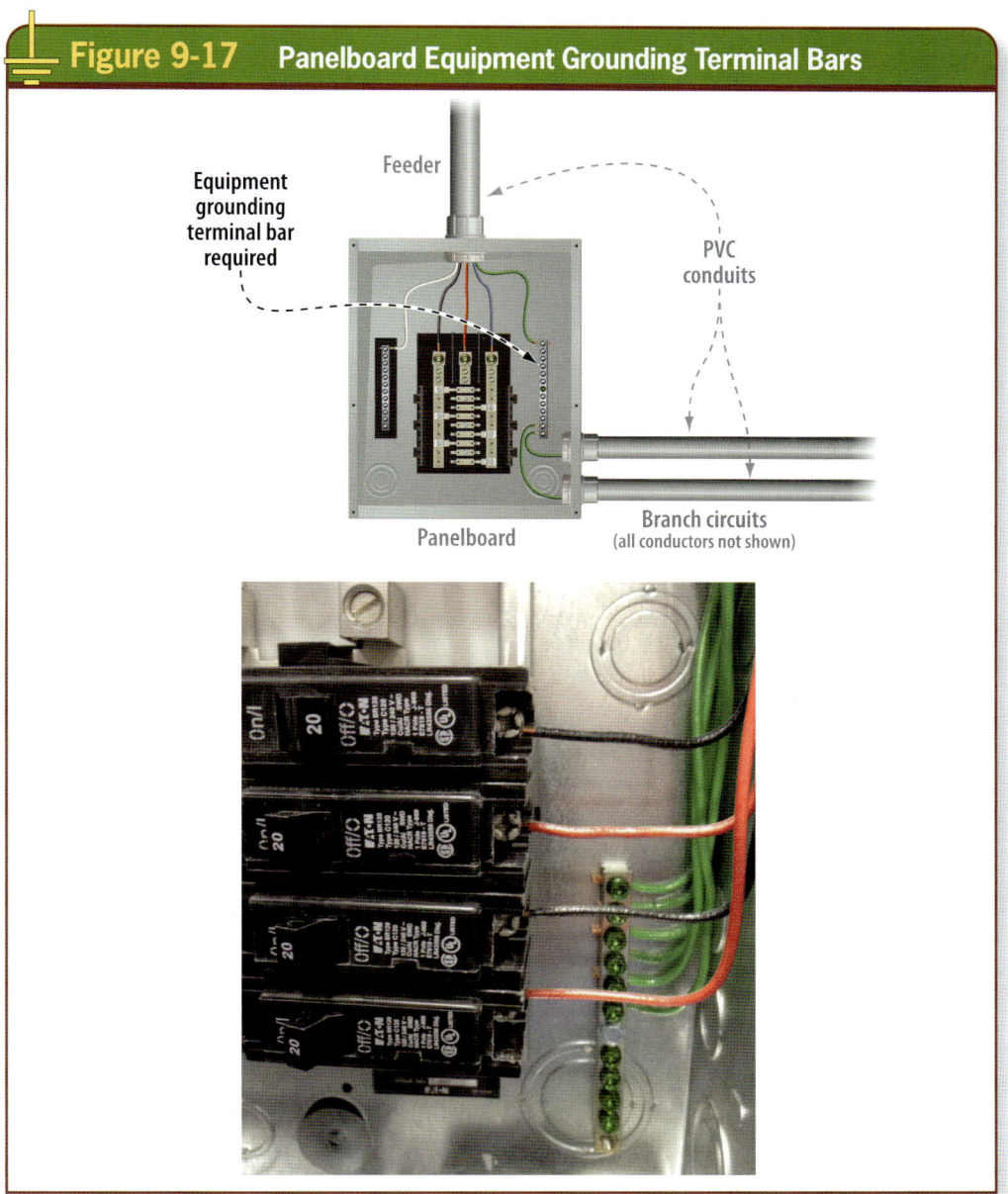

Figure 9-17. Equipment grounding terminal bars must be installed in panelboard enclosures where there are wire-type EGCs contained in the circuits entering the enclosure.

250.24(B), 250.142(B), and 250.30(A) provide clear direction on the requirement to isolate and separate the neutrals and EGCs. **See Figure 9-18.**

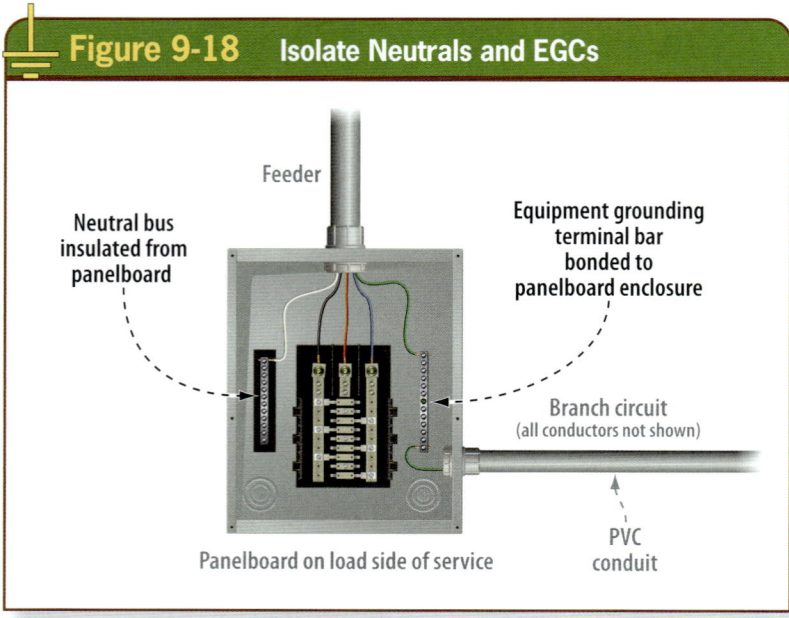

Figure 9-18   Isolate Neutrals and EGCs

Feeder

Neutral bus insulated from panelboard

Equipment grounding terminal bar bonded to panelboard enclosure

Branch circuit (all conductors not shown)

PVC conduit

Panelboard on load side of service

Figure 9-18. Grounded conductors (usually neutral conductors) generally must be isolated from EGCs in load-side equipment to prevent multiple paths for normal neutral current.

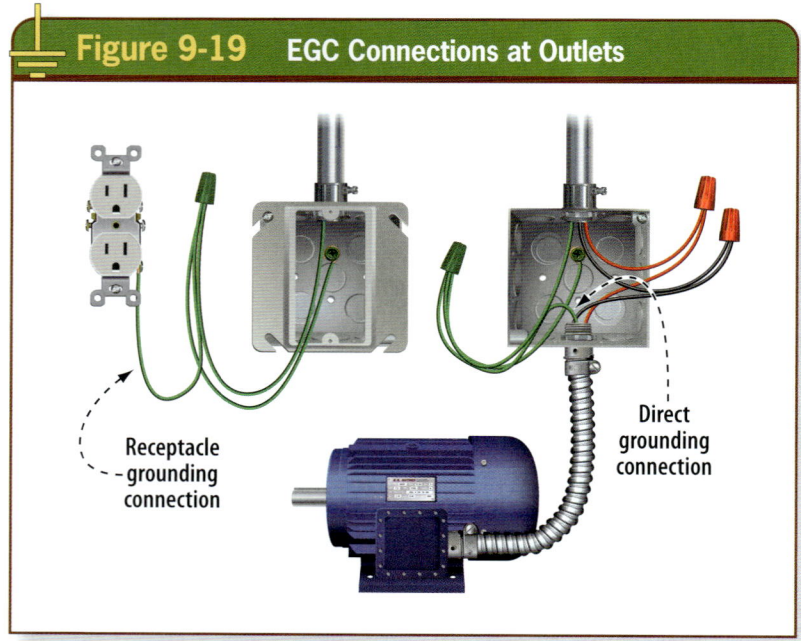

Figure 9-19   EGC Connections at Outlets

Direct grounding connection

Receptacle grounding connection

Figure 9-19. EGC connections can be made through a direct wiring connection or through a receptacle connection.

The exceptions to this general restriction are for meter enclosures and the frames of existing ranges or dryers in which the grounded conductor can be used for grounding the equipment.

There are also exceptions to this requirement for DC systems and electrode-type boilers that operate at more than 1,000 volts. The requirements for grounding this type of equipment are provided in 495.72(E)(1) and 495.74.

CONNECTIONS OF EQUIPMENT GROUNDING CONDUCTORS

Once the panelboards, motor control centers, and other equipment and feeders are installed, branch circuit wiring is then connected and distributed to outlets and utilization equipment within the building, structure, or elsewhere on the same premises. The outlet is a point on the wiring system at which current is taken to supply utilization equipment. The wiring from the final overcurrent protective device to an outlet(s) is a branch circuit, by definition. Receptacles are typically installed at outlets, but utilization equipment can also be directly wired to the outlets without a receptacle. Sections 250.146 and 250.148 provide the base requirements for attachment of EGCs at outlets and other boxes. **See Figure 9-19.**

RECEPTACLE GROUNDING CONNECTIONS

Section 406.4(C) requires that branch circuit wiring methods provide an EGC to which the EGC terminal of receptacles is to be connected. Section 250.146 requires an equipment bonding jumper to be installed from a grounded metal box to the grounding terminal on a grounding-type receptacle. The equipment bonding jumper must be sized from Table 250.122 based on the rating of the circuit breaker or fuse protecting the branch circuit. **See Figure 9-20.**

Figure 9-20 **Receptacle Grounding Connections**

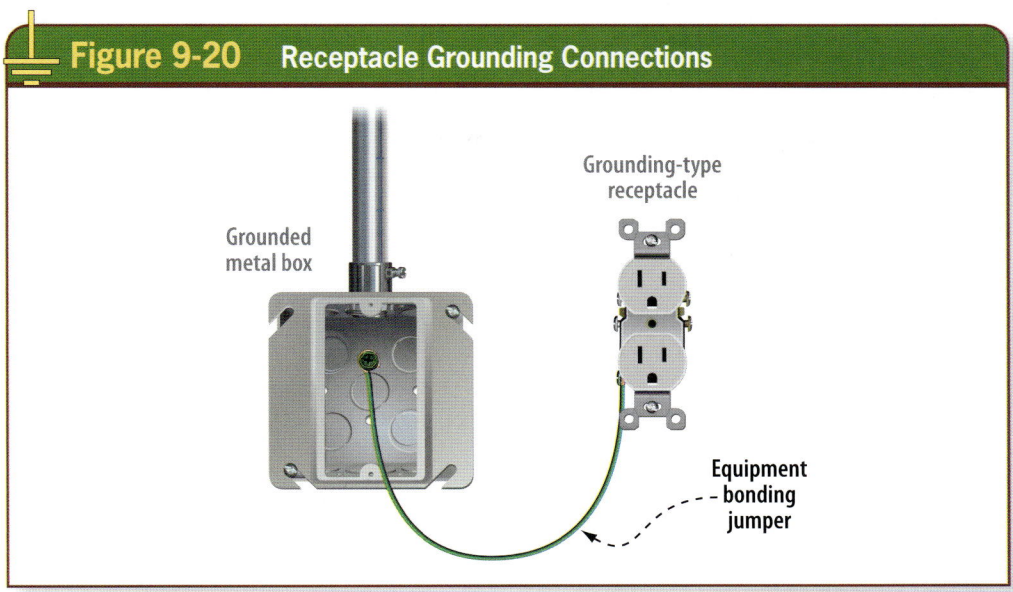

Figure 9-20. *An equipment bonding jumper is generally required from a grounded metal box to a grounding-type receptacle.*

Section 250.146(A) through (D) offers alternatives to this general requirement.

Section 250.146(A) deals with surface metal boxes only. If the grounded metal box is mounted on the surface and there is direct metal-to-metal contact between the mounting strap of the device and the grounded metal box, the equipment bonding jumper is not required. One condition for this allowance is that at least one of the insulating washers (6-32 device screw retainers) be removed to ensure metal-to-metal contact is established. **See Figure 9-21.**

Figure 9-21 **Receptacle Metal-to-Metal Contact**

Figure 9-21. *An equipment bonding jumper is not required for receptacles mounted in surface boxes if metal-to-metal contact is established between the box and receptacle mounting strap.*

Figure 9-22. *Surface covers with a flat, nonraised portion making metal-to-metal contact on surface boxes are excused from the equipment bonding jumper requirement.*

Figure 9-23. *Equipment bonding jumpers are not required when self-grounding receptacles are installed. (Drywall not shown.)*

The equipment bonding jumper is also not required for surface box and cover combinations in which the receptacle cover has provisions to securely fasten the receptacle with rivets, thread locking screws, or a screw and nut combination and the cover has a flat, nonraised portion that seats firmly against the grounded metal box. These provisions apply only to surface-mounted boxes. **See Figure 9-22.**

Another alternative to installing an equipment bonding jumper from the receptacle to the grounded metal box is when self-grounding receptacles are installed. These types of receptacles have a spring tension device that maintains an effective bonding connection between the 6-32 device mounting screw and the grounded metal box. **See Figure 9-23.**

Listed floor boxes that provide satisfactory grounding continuity between the grounding-type receptacle and the grounded metal portion of the assembly do not require an equipment bonding jumper, as provided in 250.146(C). **See Figure 9-24.**

Some designs specify an isolated ground receptacle (sometimes called a quiet ground) for the reduction of electromagnetic interference on the grounding circuit. Section 250.146(D) requires an insulated EGC be connected to a receptacle that is designed specifically to isolate the grounding terminal

Figure 9-24. *Equipment bonding jumpers are not required when listed floor boxes are installed.*

Courtesy of ABB

Figure 9-25. *The mounting strap for isolated grounding-type receptacles is isolated from the grounding connection to the receptacle. (Plaster ring and drywall not shown.)*

and mounting strap from the grounded metal box. These receptacles are generally referred to in the *NEC* as *isolated ground receptacles*. **See Figure 9-25.**

The requirements in 406.3(E), 408.40 Exception, and 250.146(D) must all be followed when installing isolated ground receptacles. In these types of installations, the insulated EGC is run with the circuit conductors but does not connect to the grounded metal box where the receptacle is installed. Another EGC or the metal raceway (often referred to as the dirty ground) is connected to the box to accomplish the grounding required for the metal box and plate, if metal.

The isolated (insulated) EGC used with isolated grounding (IG) receptacles is permitted to pass through boxes, wireways, panelboards, and other enclosures in order to terminate at the grounding point for the applicable service or separately derived system. Installing an IG circuit and receptacle does not relieve the requirement for connecting metal boxes and/or metallic portions of the raceway system supplying the IG receptacle to the required equipment grounding conductor. **See Figure 9-26.**

Equipment Grounding Conductor Continuity

If EGCs are spliced within a box or connected to devices that are secured to the box, all wire-type EGCs associated with any of those circuit conductors are required to be connected to the box or within the box with suitable devices such as grounding screws, listed grounding clips, or other equipment listed for accomplishing this grounding connection. **See Figure 9-27.**

If EGCs are spliced in a box, the connections or splices must meet the requirements in 110.14(B), though insulation of the splice is not required. This means that standard wire nuts can be used to splice EGCs together in outlet boxes. Wire nuts that are specifically listed as grounding and bonding equipment (usually green in color) can also be used for this purpose, but this is not an *NEC* requirement. See the acceptable methods

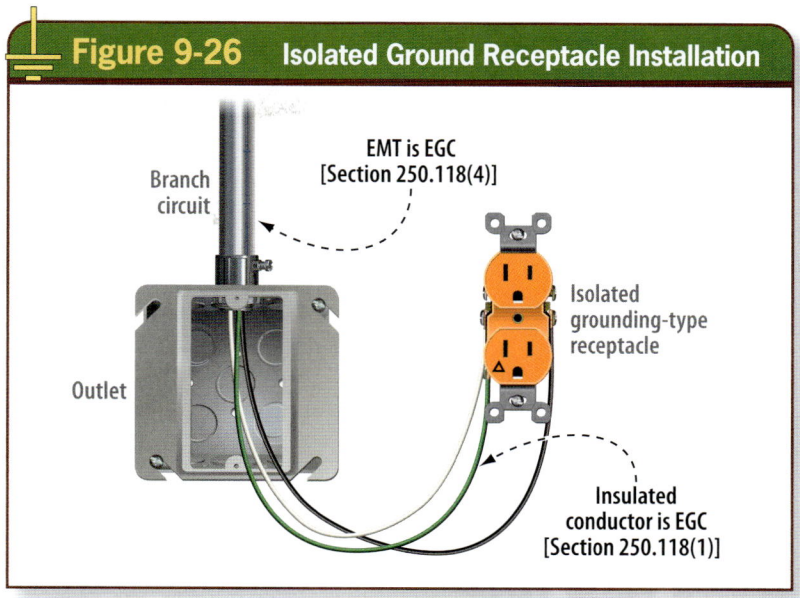

Figure 9-26 | **Isolated Ground Receptacle Installation**

Figure 9-26. Installation of isolated grounding-type receptacles requires two EGCs.

Figure 9-27 | **EGC Continuity**

Figure 9-27. An equipment bonding jumper can be fastened to a grounded metal box using a grounding screw (top) or a grounding clip (bottom). (Plaster rings and drywall not shown.)

of connecting grounding and bonding conductors in 250.8.

Often, EGCs are spliced in boxes. In these cases, an equipment bonding jumper is connected to the metal box and another equipment bonding jumper is connected to the grounding-type receptacle. This method of connecting EGCs and equipment bonding jumpers at outlet boxes results in compliance with 250.148(C). **See Figure 9-28.**

The arrangement and connections of EGCs in outlet boxes must be made in such a way that the removal of a device such as a receptacle, switch, or luminaire will not interfere with the grounding connection or continuity as required by 250.148(B). A common method to accomplish this is to splice together all EGCs associated with any of the circuit conductors and to install an equipment bonding jumper to the metal box using a grounding screw or grounding clip. Another method for connecting EGCs in larger junction and pull boxes is to install terminal lugs or an equipment grounding terminal bar for this purpose. Be sure to remove any coating, such as paint, to ensure that good metal-to-metal contact is established as required in 250.12. **See Figure 9-29.**

If EGCs are installed in nonmetallic boxes, provisions must be made so the EGC can be connected to any device or equipment in the box that requires grounding. If multiple circuits with separate equipment grounding conductors are in a nonmetallic box, all the EGCs in the box do not have to be connected together. Only the EGCs for each circuit (associated with any of those circuit conductors) need to be connected together. This type of wiring at outlets is common in residential installations where nonmetallic-sheathed cable and nonmetallic boxes are installed for the branch circuits.

Grounding Snap Switches

Section 404.9(B) provides specific grounding requirements for snap switches. The grounding rules for switches are similar to those for receptacles because both of these types of equipment (devices) have a mounting strap or yoke. For snap switches, an EGC must be connected to the switch mounting strap so that the grounding is extended to a metal faceplate, even if a nonmetallic faceplate is initially installed. **See Figure 9-30.**

The mounting yoke of a snap switch can serve as an effective ground-fault current path under the following conditions:

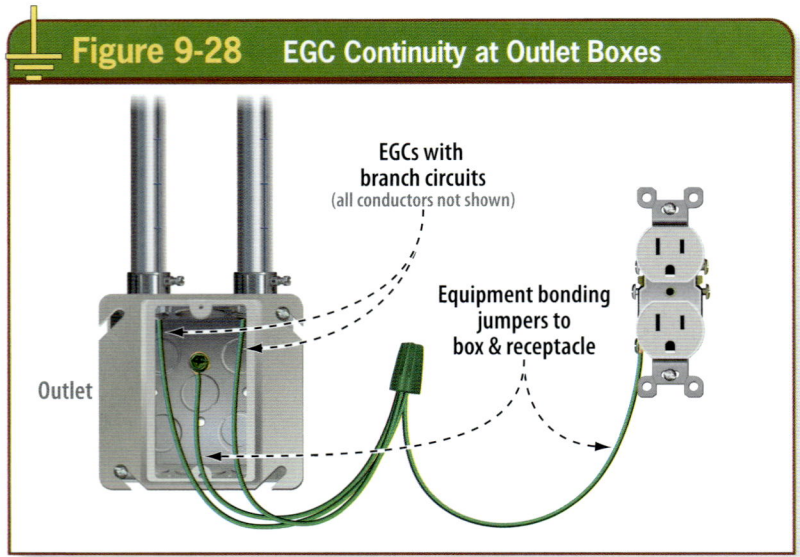

Figure 9-28   EGC Continuity at Outlet Boxes

Figure 9-28. Continuity of EGCs is required at outlet boxes.

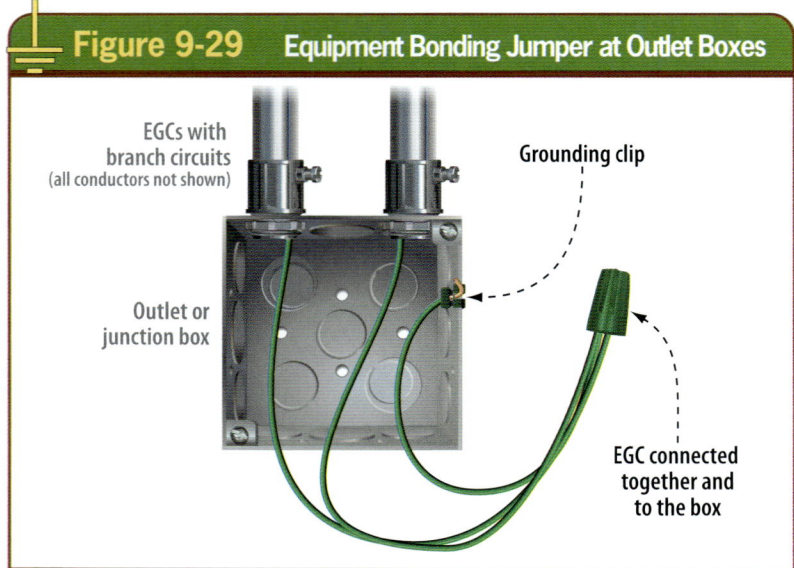

Figure 9-29   Equipment Bonding Jumper at Outlet Boxes

Figure 9-29. Continuity of EGCs is required at outlet boxes, and an equipment bonding jumper is connected to the metal box.

1. The switch is mounted with metal screws to a metal box or metal cover that is connected to an EGC or to a nonmetallic box with integral means for connecting to an EGC.
2. An EGC or equipment bonding jumper is connected to an equipment grounding termination of the snap switch.

There are three exceptions that relax the snap switch grounding requirements:

1. If the wiring method does not include an EGC or if no means in the enclosure exists to connect to one, then an EGC is not required for replacement snap switches. Under this condition, the faceplate and its mounting screws, if within eight feet vertically or five feet horizontally from ground or exposed grounded parts, must be nonmetallic unless either the snap switch strap is nonmetallic or the circuit has GFCI protection.
2. Entirely nonmetallic (accessible parts) listed kits or assemblies with "non-standard" faceplate mounting means do not require an EGC.
3. A snap switch with integral nonmetallic enclosure for nonmetallic-sheathed cable that complies with 300.15(E) does not require an EGC.

Identification of Wiring Device Terminals

Wiring devices such as receptacles and switches are generally required to provide a terminal for connecting an EGC to the device. This grounding terminal must be identified by one of the following methods:

1. A green screw with a hexagon head that is not easily removed from the device.
2. A green hexagonal nut that is not easily removed.
3. A green pressure wire connector.
4. If the connection point for the EGC to the device is not visible, then the EGC entrance hole must be marked with the words *green* or *ground*, the letters *G* or *GR*, or a grounding symbol, or must be

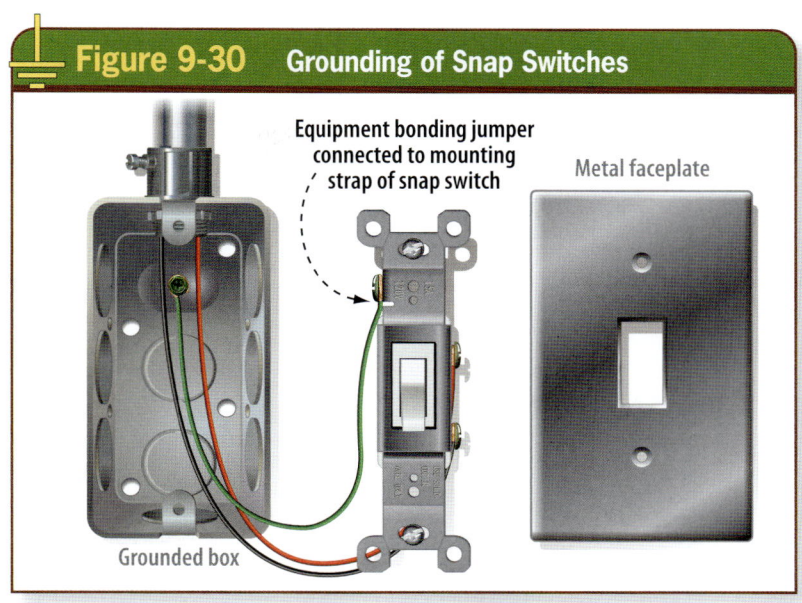

Figure 9-30. *Grounding of snap switches must be accomplished using one of the methods provided in 404.9(B).*

otherwise identified by a distinctive green color. The installer does not have to provide this identification on devices because it is provided by the manufacturer. Informational Note Figure 406.10(B)(4) provides an example of a grounding symbol that is often used to identify a grounding terminal connection point. **See Figure 9-31.**

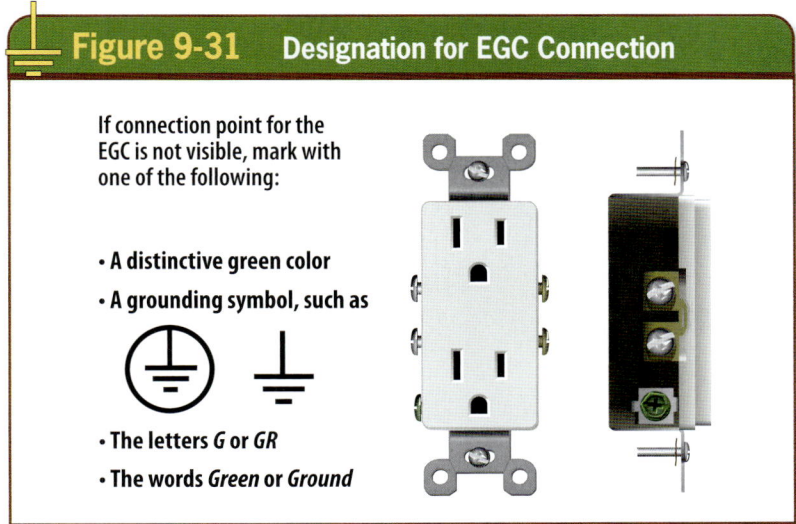

Figure 9-31. *The grounding symbol is one way to identify the terminal for attaching EGCs to receptacles.*

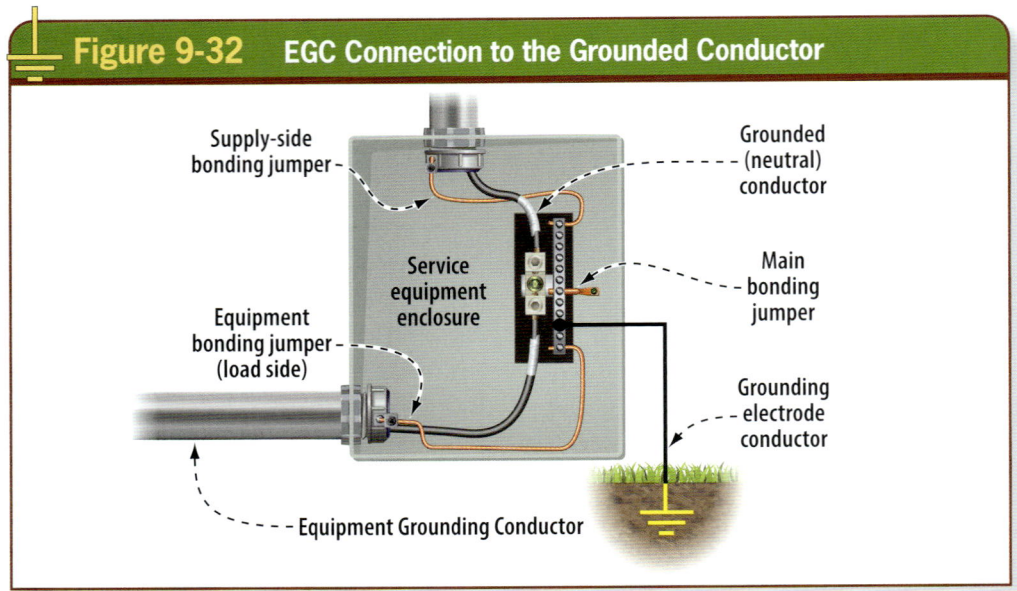

Figure 9-32 EGC Connection to the Grounded Conductor

Figure 9-32. *EGC connections are required to be made to the grounded conductor at service equipment supplied by a grounded system.*

Continuity between Attachment Plugs and Receptacles

EGC continuity is established through receptacles and attachment plugs. These are referred to in the *NEC* as *separable connections.* Section 250.124 requires separable connections to provide for a "first-make, last-break" EGC connection. The first-make, last-break connection means is not a requirement for equipment connections with an interlock that does not allow disconnecting or connecting in the energized state. Standard listed receptacles and attachment plugs are manufactured to provide first-make, last-break grounding conductor connections of the devices. Automatic switches or cutouts are not permitted in the EGC of the premises' wiring unless they disconnect all power sources when operated.

Figure 9-33 EGC Connection to the Enclosure

Figure 9-33. *EGC connections are required to be made to the enclosure at service equipment supplied by an ungrounded system.*

Equipment Grounding Conductor Connections

EGC connections at services must be made in accordance with 250.130(A) or (B). EGC connections for separately derived systems are made in accordance with 250.30. Section 250.130(A) addresses premises wiring that is supplied by grounded systems. This rule requires that the EGCs be connected to the grounded conductor and the grounding electrode conductor at the service. **See Figure 9-32.**

When the wiring system is supplied by a service that is ungrounded, the connection of the EGCs must be made to the enclosure and grounding electrode conductor at the service. **See Figure 9-33.**

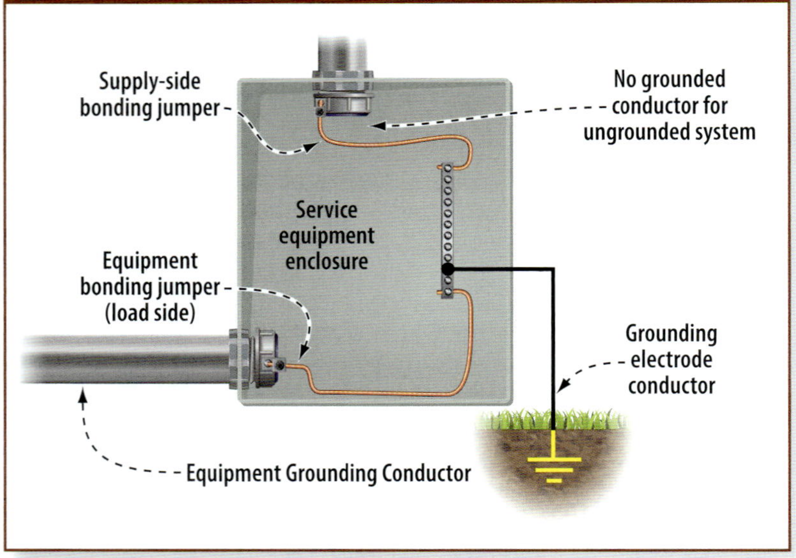

RECEPTACLE REPLACEMENTS

Grounding requirements are provided in the *NEC* for situations requiring receptacle replacements. It is not uncommon to have to replace receptacles over time because of damage or wear. Many older installations of branch circuit wiring were installed without an EGC, while others were installed with an EGC included. There are some options recognized in the *NEC* for receptacle replacements, either by equipment grounding or other methods of ensuring safety.

Replacing Grounding-Type Receptacles

Section 406.4(D)(1) indicates that grounding-type receptacles are required to be replaced only with grounding-type receptacles. Non–grounding-type receptacles are not permitted as replacements when the outlet has an EGC.

Replacing Non–Grounding-Type Receptacles

Section 406.4(D)(2) addresses replacements of non–grounding-type receptacles and provides three alternatives. First, a non–grounding-type receptacle can be replaced with another non–grounding-type receptacle. Non–grounding-type receptacles are still available for this use. **See Figure 9-34.**

The second alternative allows a ground-fault circuit-interrupter (GFCI) receptacle device to replace a non–grounding-type receptacle. If this alternative is chosen, the GFCI device or cover plate must be marked "No Equipment Ground." Using this alternative, an EGC is not permitted to be installed from the GFCI receptacle device to any outlet downstream supplied by the GFCI replacement. **See Figure 9-35.**

The third alternative allows grounding-type receptacles to be installed as replacements for non–grounding-types on the load side of the GFCI replacement in the circuit, provided each grounding-type receptacle on the load side of the GFCI device is marked "No

Equipment Ground." An EGC is not permitted to be connected between the GFCI replacement and the grounding-type receptacles installed on the load

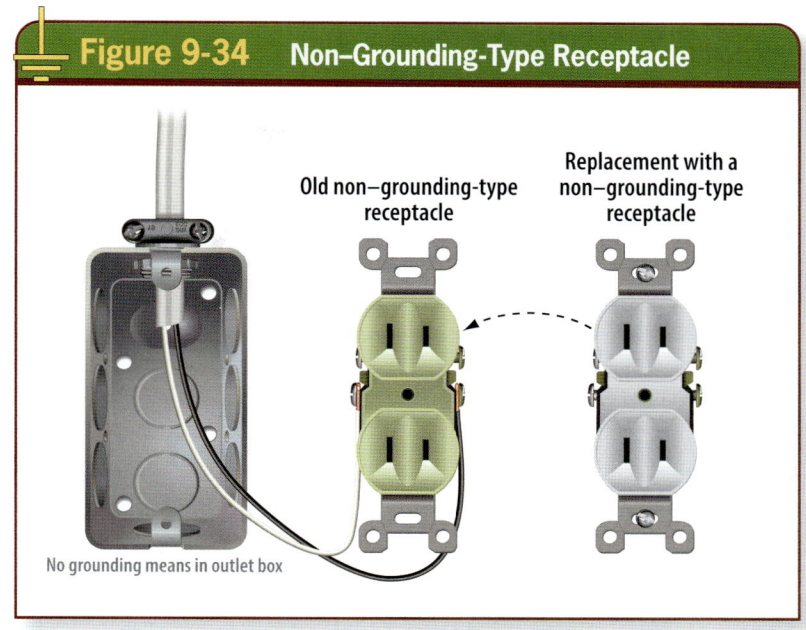

Figure 9-34 Non–Grounding-Type Receptacle

Old non–grounding-type receptacle

Replacement with a non–grounding-type receptacle

No grounding means in outlet box

Figure 9-34. A non–grounding-type receptacle can replace another non–grounding-type receptacle when no equipment grounding means exist at the outlet.

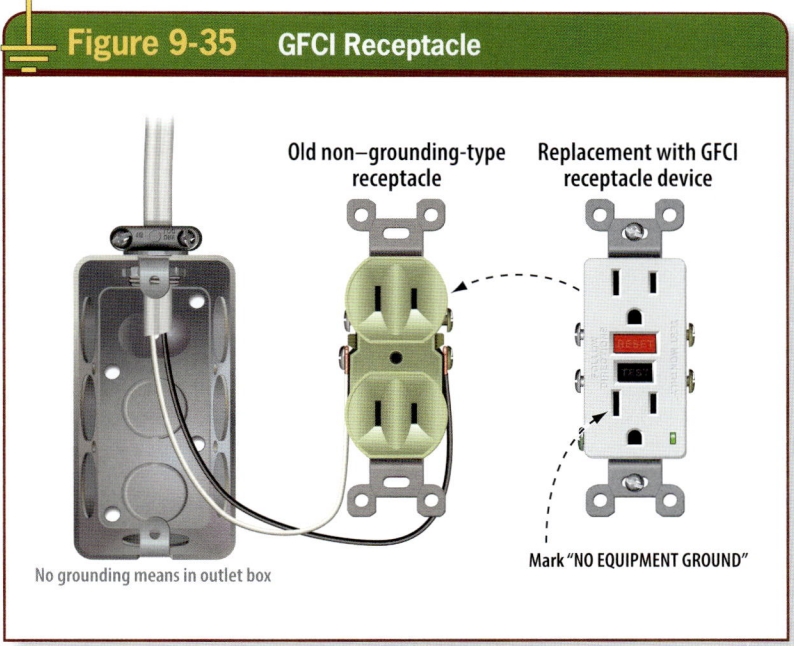

Figure 9-35 GFCI Receptacle

Old non–grounding-type receptacle

Replacement with GFCI receptacle device

No grounding means in outlet box

Mark "NO EQUIPMENT GROUND"

Figure 9-35. A GFCI receptacle is permitted to replace a non–grounding-type receptacle.

Figure 9-36 Grounding-Type Receptacles

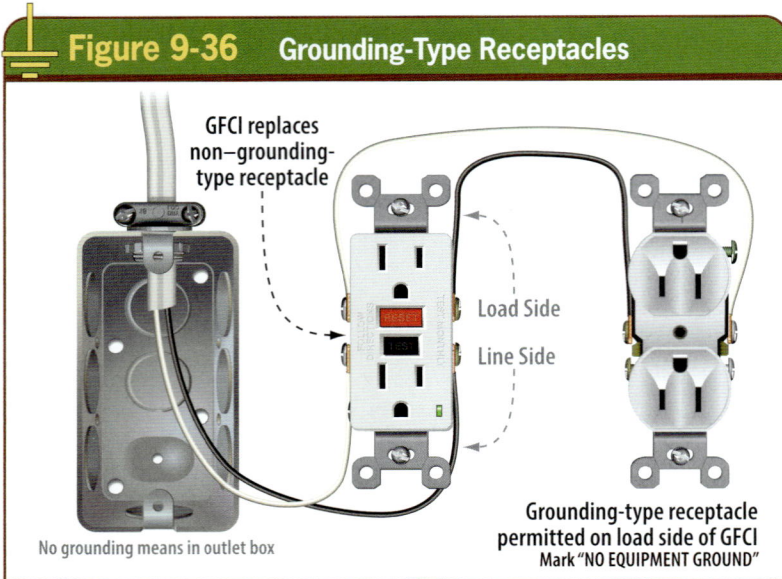

GFCI replaces non–grounding-type receptacle

Load Side

Line Side

No grounding means in outlet box

Grounding-type receptacle permitted on load side of GFCI
Mark "NO EQUIPMENT GROUND"

Figure 9-36. Grounding-type receptacles are permitted as replacements on the load side of a GFCI receptacle device if no EGC is installed at the downstream outlets and the receptacle is marked as such.

side of the GFCI downstream. **See Figure 9-36.**

Section 250.130(C) includes some criteria that can be applied for replacements using grounding-type receptacles if an EGC does not exist at the outlet. An EGC is permitted to be installed separately from the

Figure 9-37 Separate EGC Installation

Separate Equipment Grounding Conductor

No grounding means in outlet box

Figure 9-37. Installation of a separate EGC is permitted under restrictive conditions.

existing branch circuit wiring already in place within the building walls, ceilings, or other inaccessible locations. **See Figure 9-37.**

If a separate, wire-type EGC is used for any reason, such as the need for an isolated grounding receptacle, the EGC of a grounding-type receptacle or a branch circuit extension shall be permitted to be connected to any of the following:

- Any accessible point on the grounding electrode system as described in 250.50
- Any accessible point on the grounding electrode conductor
- The equipment grounding terminal bar within the enclosure where the branch circuit for the receptacle or branch circuit originates
- An equipment grounding conductor that is part of another branch circuit that originates from the enclosure where the branch circuit for the receptacle or branch circuit originates
- For grounded systems, the grounded service conductor within the service equipment enclosure
- For ungrounded systems, the grounding terminal bar within the service equipment enclosure

GROUNDING APPLIANCES USING THE GROUNDED CONDUCTOR

The frames of wall-mounted ovens, counter-mounted cooking units, ranges, dryers, and associated outlet or junction boxes are required to be connected to an EGC and must include an insulated grounded circuit conductor only if it is needed for the load, according to 250.140 and 250.142. **See Figure 9-38.**

The result is a requirement for a 4-wire device being installed at the outlets for this type of equipment, specifically at ranges and clothes dryers. The *NEC* relaxes this requirement in 250.140(B) and allows the grounded conductor to be used for grounding in existing 3-wire circuits supplying this type of equipment, but only if conditions 1, 2, and 3 are met and either condition 4 or 5 is met:

1. The supply circuit is 120/240-volt, single-phase, 3-wire; or 208Y/120-volt derived from a 3-phase, 4-wire, wye-connected system.
2. The grounded conductor is not smaller than 10 AWG copper or 8 AWG aluminum.
3. Grounding contacts of receptacles furnished as part of the equipment are bonded to the equipment.
4. The grounded conductor is insulated, or the grounded conductor is uninsulated and part of a Type SE (service-entrance) cable and the branch circuit originates at the service equipment.
5. The grounded conductor is part of SE cable that originates in equipment other than a service. The grounded conductor is required to be insulated or field-covered within the supply equipment with listed insulating material to prevent accidental contact with non–current-carrying metal parts.

The grounded conductor (neutral) of newly-installed branch circuits supplying ranges and clothes dryers is no longer permitted to be used for grounding the non–current-carrying metal parts of the appliances. Branch circuits installed for new appliance installations are required to provide an EGC sized in accordance with Table 250.122 for grounding the non–current-carrying metal parts.

AUXILIARY GROUNDING ELECTRODE REQUIREMENTS

Auxiliary grounding electrodes are often specified as a design requirement, but they are not required by the *NEC*. However, if an auxiliary electrode is installed, important requirements apply. Section 250.54 indicates that it is permissible to connect one or more grounding electrodes to the EGCs specified in 250.118.

An example of an auxiliary grounding electrode installation is typically found at light-pole bases in parking

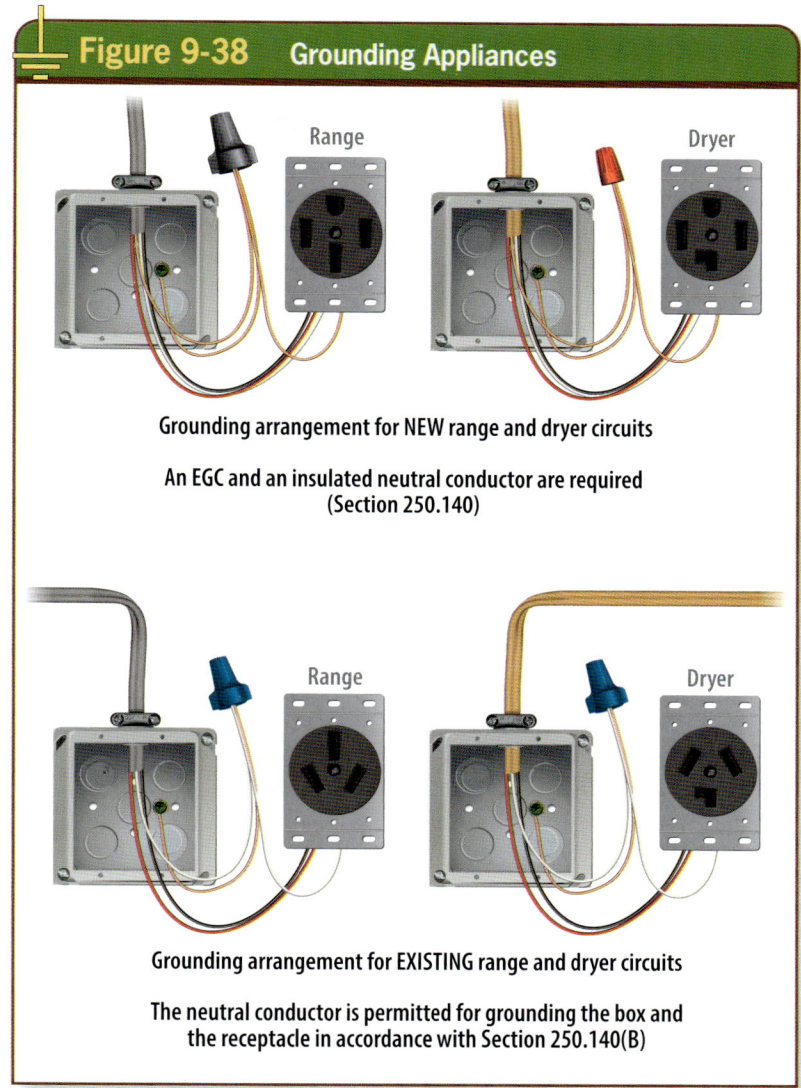

Grounding arrangement for NEW range and dryer circuits

An EGC and an insulated neutral conductor are required
(Section 250.140)

Grounding arrangement for EXISTING range and dryer circuits

The neutral conductor is permitted for grounding the box and
the receptacle in accordance with Section 250.140(B)

Figure 9-38. Four-wire branch circuits are required for wall-mounted ovens, counter-mounted cooking units, ranges, dryers, and associated outlet or junction boxes. These circuits include two ungrounded conductors, an insulated grounded neutral conductor, and an EGC (covered, insulated, or bare).

lots. The EGC is installed with the branch circuit supplying the pole luminaires as a requirement, so the equipment is grounded and an effective path for ground-fault current is provided as required. An auxiliary grounding electrode is often installed and connected to the metal pole in addition to the required EGC. This allows a lightning strike to be dissipated into the Earth locally at the light pole base. The local Earth connection of this equipment also establishes an

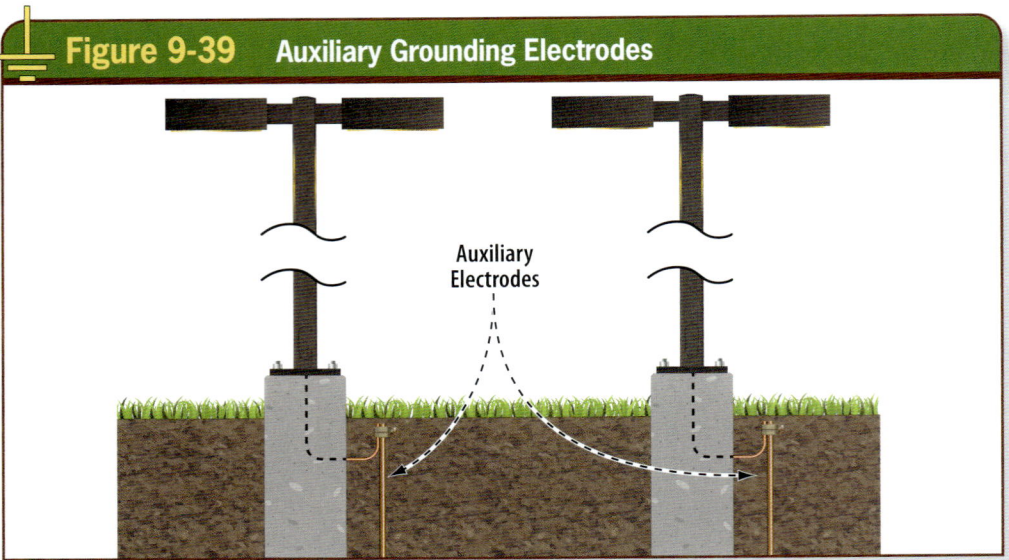

Figure 9-39 **Auxiliary Grounding Electrodes**

Auxiliary Electrodes

Figure 9-39. Auxiliary electrodes supplement the required EGC of the circuit supplying equipment.

equipotential between the conductive pole and the Earth, or as close as possible to equipotential.

Auxiliary grounding electrodes do not have to be bonded to the grounding electrode system for the building or structure as indicated in 250.50 or 250.53(C). **See Figure 9-39.**

The auxiliary electrode is bonded to the grounding electrode system of the supply system by being connected to the EGC of the branch circuit. It is important that installations of auxiliary grounding electrodes for equipment meet the provisions in 250.54 and satisfy the provisions in either 250.4(A)(5) or 250.4(B)(4). The Earth is never permitted as an effective path for ground-fault current. **See Figure 9-40.**

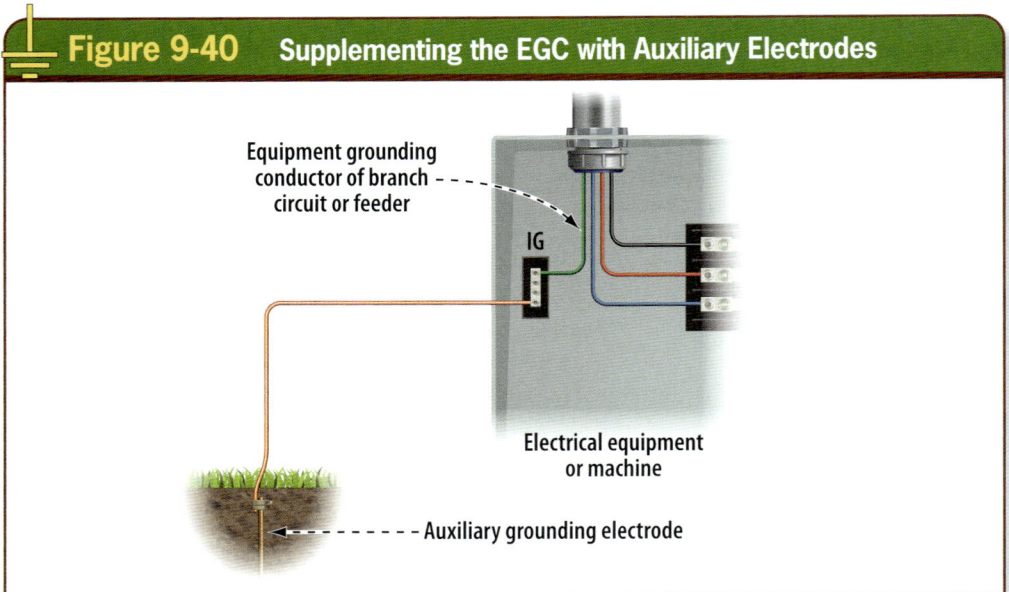

Figure 9-40 **Supplementing the EGC with Auxiliary Electrodes**

Equipment grounding conductor of branch circuit or feeder

IG

Electrical equipment or machine

Auxiliary grounding electrode

Figure 9-40. Auxiliary grounding electrodes are permitted to supplement the required equipment grounding conductor of the circuit supplying equipment.

GROUNDING NONELECTRICAL EQUIPMENT

The metal parts of some nonelectrical equipment are required to be connected to an EGC, in accordance with 250.116. For example, the frames and tracks of electrically-operated cranes and hoists must be grounded by connection to an EGC of the circuit supplying them. The frames of non–electrically-driven elevator cars must be grounded, and hand-operated metal shifting ropes or cables of electric elevators must be grounded.

Grounding of nonelectrical equipment provides a higher level of safety because it reduces the possibility of these metal parts remaining energized and resulting in a shock hazard, which could occur if the parts are not grounded.

EQUIPMENT GROUNDED BY SECURE METAL SUPPORTS

There are installations in which a support frame can provide the equipment grounding required by the *NEC* as long as the metal rack or support structure is connected to an equipment grounding conductor of the circuit supplying the equipment mounted to the rack. For example, a common frame that supports multiple motors can be grounded by a single EGC that grounds all the motors mounted to the common metal frame or rack. The EGC, if of the wire type, must be sized based on the requirements in Table 250.122. Section 250.136 recognizes these types of installations as meeting the minimum requirements for grounding equipment. **See Figure 9-41.**

| Figure 9-41 | Grounding to a Common Metal Frame |

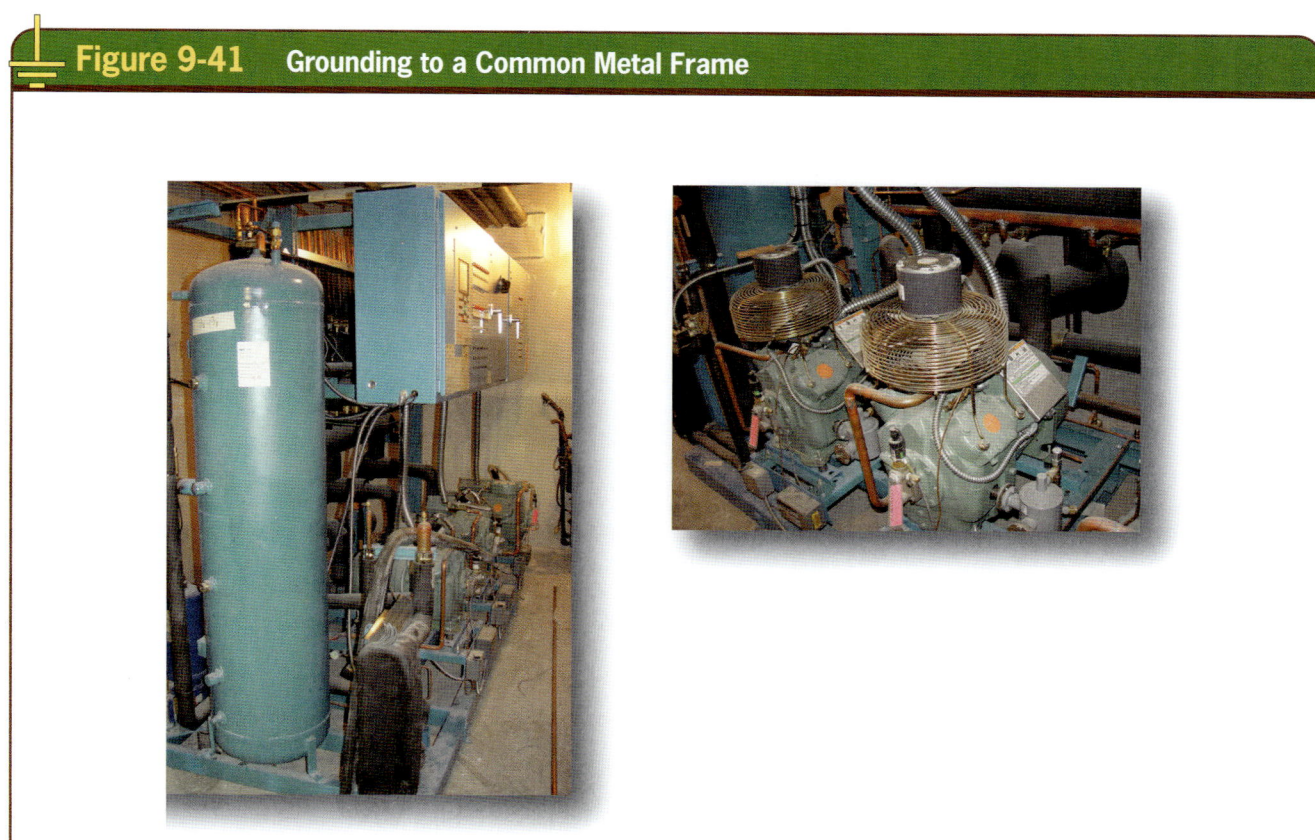

Figure 9-41. Equipment grounding can be accomplished by connection to a common metal frame that is grounded.

Although a structural metal building frame may be used to interconnect or be bonded to one or more electrodes, it is important to understand that the structural metal framing of a building or structure is not permitted to be used as the required EGC for AC circuits or equipment.

USE OF THE GROUNDED CONDUCTOR FOR GROUNDING

Up to this point, with the exception of the provisions for grounding services, the objectives on the load side of the service disconnecting means (grounding point) are to isolate the grounded (neutral) conductor from the EGC and grounded enclosures. Section 250.142 includes a few conditions in which the grounded conductor is permitted to be used for grounding equipment, but these are restrictive conditions. On the supply side or within the enclosure of

the service, a grounded (usually the neutral) conductor can be used for grounding non–current-carrying metal parts of equipment, raceways, and other enclosures. Under the conditions stated, a grounded conductor can also be used for grounding equipment for separate buildings, as provided in 250.32(B) Exception No. 1 and 2, and for separately derived systems, as permitted by 250.30(A)(1) Exception No. 2.

The restrictions of using grounded conductors for grounding are supported by the rules in 250.24(B) and 250.142(B), which both prohibit the grounded (often a neutral) conductor from being used for grounding on the load side of the grounding point at a service or separately derived system. This is related to the objective of keeping normal load current on the path it is intended to be on and limiting normal current from being imposed on the EGC or other conductive paths within the building or structure.

The grounded conductor (usually a neutral) is permitted for grounding on the supply-side of the service disconnect in accordance with 250.142(A).

SUMMARY

Equipment grounding can be accomplished by multiple methods addressed in the *NEC*. Parts VI and VII of Article 250 include the general requirements for equipment grounding. It is important to understand the requirements for specific fixed equipment that must be grounded and how to make suitable grounding connections at the equipment. Equipment grounding connections at standard receptacles and switches and at special receptacles, such as range and dryer receptacles, are also important aspects of proper grounding. Substitutes for grounding such as isolation, insulation, or guarding are addressed in 250.1(6) and more specifically in 250.110. Grounding of equipment can be accomplished using an auxiliary electrode in addition to the required EGC. Separation and isolation is required between the grounded (usually the neutral) conductor of a circuit and the EGC and grounded metal parts on the load side of the grounding point, either at the service equipment or the source of a separately derived system.

REVIEW QUESTIONS

1. Equipment grounding places equipment potential at or as close to Earth as possible, thereby minimizing shock hazard possibilities and limiting the voltage to ground during line surge events, lightning events, and unintentional contact with higher-voltage lines.

   a. True   b. False

2. Lightning protection systems installed according to *NFPA 780: Standard for the Installation of Lightning Protection Systems* provide an increased level of lightning protection for buildings that have them installed.

   a. True   b. False

3. Generally, if equipment operates at more than __?__ to ground, it is required to be grounded by connection to an equipment grounding conductor.

   a. 30 V
   b. 50 V
   c. 120 V
   d. 150 V

4. Equipment is required to be grounded if it is located __?__ vertically or __?__ horizontally from the ground or grounded metal objects that are subject to contact by people.

   a. 5' / 8'
   b. 8' / 5'
   c. 10' / 6'
   d. 15' / 8'

5. If equipment were installed 10 feet above the ground and were not subject to contact by people, grounding would always be required if the location was not wet or damp.

   a. True   b. False

6. The *NEC* relaxes the grounding requirement for short sections of metal enclosures that are used for protecting cable assemblies from physical damage, as long as the short sections do not exceed __?__.

   a. 2'
   b. 4'
   c. 6'
   d. None of the above; no distance is provided in the *NEC* for the term *short section*

7. When a submersible pump is used in a metal well casing, the well casing must be connected to the __?__ of the pump circuit.

   a. equipment grounding conductor
   b. grounded conductor
   c. grounding electrode conductor
   d. system bonding jumper

8. An exception to the grounding requirements in 250.86 is made for short sections of raceway or metal enclosures used for support or to provide some protection from possible physical damage.

   a. True   b. False

9. Metal components such as metal elbows are not required to be grounded if they are isolated from public contact by a minimum cover of not less than 18 inches where buried or encased in not less than __?__ of concrete.

   a. 2"
   b. 6"
   c. 12"
   d. 24"

10. Equipment grounding conductors are generally required to be installed with feeders of electrical systems.

    a. True   b. False

11. The equipment grounding conductor with feeders and branch circuit conductors can be any of the types provided in which of the following *NEC* rules?

    a. Section 250.110
    b. Section 250.112
    c. Section 250.118
    d. Section 250.122

12. If a panelboard is wired using rigid metal conduit for the feeder and branch circuits, and no equipment grounding conductors of the wire type are installed, an equipment grounding terminal bar is required to be installed in the panelboard enclosure.

    a. True   b. False

13. The grounded (usually the neutral) conductor of a panelboard must not be connected to ground (the enclosure) on the load side of the service disconnecting means or on the load side of the grounding point for a separately derived system.

    a. True   b. False

14. An equipment bonding jumper connected from a grounded metal outlet box to a grounding-type receptacle is required to be sized based on __?__, using the rating of the overcurrent protective device ahead of the branch circuit.

    a. The minimum size of the ungrounded conductors
    b. Table 1, Chapter 9
    c. Table 250.122
    d. Table 250.66

15. An equipment bonding jumper is not required to be installed from a grounding-type receptacle to a grounded metal box mounted on the surface if which of the following conditions are met?

    a. The box and cover combination has provisions to securely fasten the receptacle to the cover with rivets, thread locking screws, or a screw and nut combination.

    b. The cover has a flat, non-raised portion that sits firmly against the grounded metal box.

    c. The metal box is grounded by connection to an equipment grounding conductor.

    d. All of the above

16. An equipment bonding jumper is required to be installed from a self-grounding-type receptacle to a flush-mounted grounded metal outlet box.

    a. True   b. False

17. Isolated grounding receptacles are often installed for branch circuits in an effort to reduce the __?__ on the grounding circuit.

    a. electromagnetic interference

    b. excessive voltage

    c. static

    d. stray current

18. An isolated ground-type receptacle is required to have an equipment bonding jumper installed between the grounding terminal of the device and the grounded metal outlet box in which it is installed.

    a. True   b. False

19. When equipment grounding conductors are spliced within a box or connected to equipment such as devices that are secured to the box, any equipment grounding conductors associated with any of those circuit conductors are required to be connected to the box or within the box with which of the following means?

    a. A device such as a grounding screw

    b. A listed grounding clip

    c. Other equipment listed for accomplishing this grounding connection

    d. Any of the above

20. The arrangement and connections of __?__ in outlet boxes must be made in such a fashion that the removal of a device such as a receptacle, switch, or luminaire does not interfere with the grounding connection or continuity.

    a. equipment grounding conductors

    b. grounded conductors

    c. grounding electrode conductors

    d. ungrounded conductors

21. The mounting yoke of a snap switch can serve as an effective ground-fault current path under which of the following conditions?

    a. An equipment grounding conductor or equipment bonding jumper is connected to an equipment grounding termination of the snap switch.

    b. The switch is mounted with metal screws to a metal box or metal cover that is connected to an equipment grounding conductor or to a nonmetallic box with integral means for connecting to an equipment grounding conductor.

    c. Either a. or b.

    d. Neither a. nor b.

22. The branch circuits installed for wall-mounted ovens, counter-mounted cooking units, ranges, dryers, and associated outlet or junction boxes are required to be connected to an equipment grounding conductor and must include an insulated grounded circuit conductor if it is needed based on the load served.

    a. True   b. False

23. Wiring devices such as receptacles and switches are generally required to provide a terminal for connecting an equipment grounding conductor to the device. This grounding terminal can be identified by any one of the following methods except __?__.

    a. a green screw that is not easily removed from the device

    b. a green pressure wire connector

    c. a green screw that has a round head

    d. a green screw with a hexagon head that is not easily removed from the device

24. Standard receptacles and attachment plugs are manufactured to provide first-make, last-break of the grounding conductor connections of the devices.

    a. True   b. False

25. Non–grounding-type receptacles are permitted as replacements when the outlet does not provide an equipment grounding conductor.

    a. True   b. False

Isolated/Insulated Grounding Circuits and Receptacles

Clean power is the main objective when designing and installing electrical systems for electronic equipment. The term *clean power* is not defined in the *NEC*; neither are the terms *isolated ground* or *quiet ground*. There are wiring techniques that can be used to achieve optimal performance in equipment grounding circuits for electronic equipment while maintaining compliance with the *NEC* safety regulations. Wiring isolated grounding circuits and receptacles for information technology (IT) equipment must never compromise safety in the grounding and bonding system.

Objectives

» Identify provisions in the *NEC* that address objectionable current in the grounding and bonding system.

» Understand sources of electromagnetic interference (EMI) that can affect normal operation of electronic equipment.

» Understand various alternatives for reducing objectionable current and electromagnetic interference in the grounding circuits and review the use of surge protective devices (SPDs).

» Understand installation requirements for isolated equipment grounding conductors and isolated ground receptacles.

» Determine specific grounding requirements that apply to information technology equipment and rooms and understand the purpose of signal reference structures (grids).

Chapter 10

Table of Contents

Electromagnetic Interference (EMI)
in Grounding Circuits 234

    Practical Solutions 235

    High-Frequency Effects (Resonance) 235

Purpose of Isolated Grounding Circuits
and Receptacles....................................... 236

Objectionable Currents in Grounding Paths ... 237

Power Quality System Grounding Analysis ... 238

Isolated Grounding Circuits 239

    Isolated Ground Receptacle
    Wiring Rules 239

    Panelboards and Isolated
    Grounding Circuits 241

    Isolated Equipment Grounding
    Circuits for Equipment......................... 242

    Isolated Grounding Circuits in Health
    Care Facilities...................................... 243

Use of Auxiliary Grounding Electrodes......... 244

Grounding and Bonding in Information
Technology Centers.................................... 244

Signal Reference Structures (Grids)............. 246

Surge Protection 248

Summary.. 249

Review Questions 250

ELECTROMAGNETIC INTERFERENCE (EMI) IN GROUNDING CIRCUITS

Some electronic equipment can react negatively to electromagnetic interference (EMI) in the grounding circuit. This interference is often referred to in the industry as "electrical noise." The

Figure 10-1   **IG Receptacles for IT Rooms**

Figure 10-1. IT rooms typically use isolated grounding (IG) receptacles and circuits supplied from power distribution units (PDUs).

Photo from i-Stock

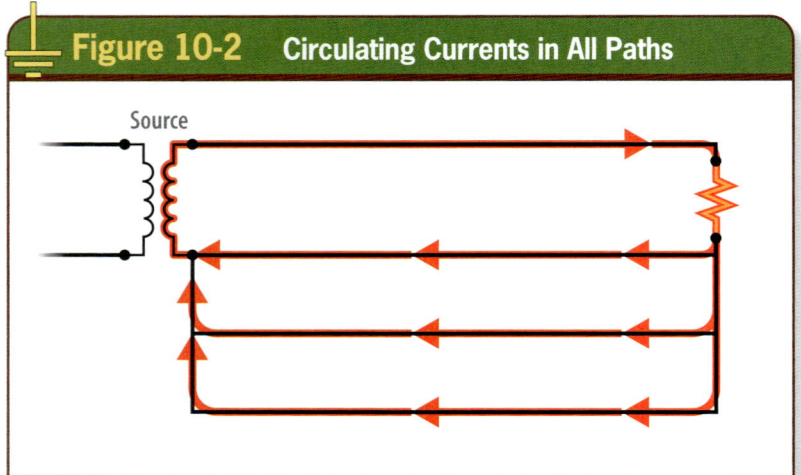

Figure 10-2   **Circulating Currents in All Paths**

Source

Figure 10-2. Circulating currents divide over all available paths to return to the source.

term *sensitive electronics* refers to equipment that is vulnerable to electromagnetic interference or circulating currents (often objectionable current), typically at low current levels.

Other standards that address power quality issues sometimes use the term *susceptible* or *sensitive electronic equipment*. Article 647 of the *NEC* addresses wiring methods and systems for sensitive electronic equipment. A common objective in wiring systems for computer equipment is minimizing "ground loops," or circulating currents, and their effects on electronic equipment. The minimum requirements of the *NEC* must be met for safe electrical installations.

Some alternative equipment grounding techniques can be applied in system designs to address concerns related to normal operation of electronic equipment. Installations of circuits for electronic equipment, such as computer circuits in IT centers, typically include isolated/insulated grounding circuits derived from power distribution units (PDUs). **See Figure 10-1.**

Objectionable current and *ground loops* are not defined in the *NEC*, but in the IT world, these terms often refer to currents circulating through multiple grounding paths. The current in these paths is thought to be moving in circular fashion or over multiple paths while returning to the source. This circular movement of current is commonly called a *ground loop*, a term that is often used in the field but is not mentioned or defined in the *NEC*. **See Figure 10-2.**

Ground loops, or low-level circulating current, are common when computer equipment is connected to circuits supplied from different power sources and those power sources are then interconnected with shielded communication cables. Multiple equipment grounding points or paths can also cause circulating current in the frames of electronic equipment. Because electronic circuits in these types of equipment operate at low voltage levels (typically less than five volts) they are more easily affected by EMI from power circuits that are installed in close proximity. Differences in potential can also cause data errors or losses in electronic equipment.

As potential differences attempt to equalize over multiple grounding paths, circulating currents are present.

Practical Solutions

One solution is to intentionally reduce or minimize potential differences by isolating the grounding circuit supplying the equipment and connecting the frames of all equipment to a common signal reference grid structure. **See Figure 10-3.**

Another method for solving this problem is to supply all equipment from the same power supply or power distribution unit, as is often the case in IT room designs. The equipment grounding means of the branch circuit wiring will help keep the ground potential the same. **See Figure 10-4.**

This solution may be impractical when the computer is far from peripheral equipment and is supplied by a different system. Some other solutions to the problem of circulating currents are:

1. Single-point grounding and use of a single power supply system
2. Modems, which are normally used as interfaces with telephone circuits
3. Fiber optic transmission over completely non-conducting paths or optical isolators
4. Interface devices (surge arresters and surge protection devices)

High-Frequency Effects (Resonance)

Another challenge in IT room installations or systems that include connected electronic equipment is the high-frequency effects in grounding circuits caused by the electronic (signaling) circuitry in IT equipment. Avoiding resonance at high frequency is important and more challenging for IT equipment because of the higher frequencies in today's digital signaling circuits. Processor speeds have increased significantly. Resonance can occur when the length of a conductor and the frequency of alternating current are in tune.

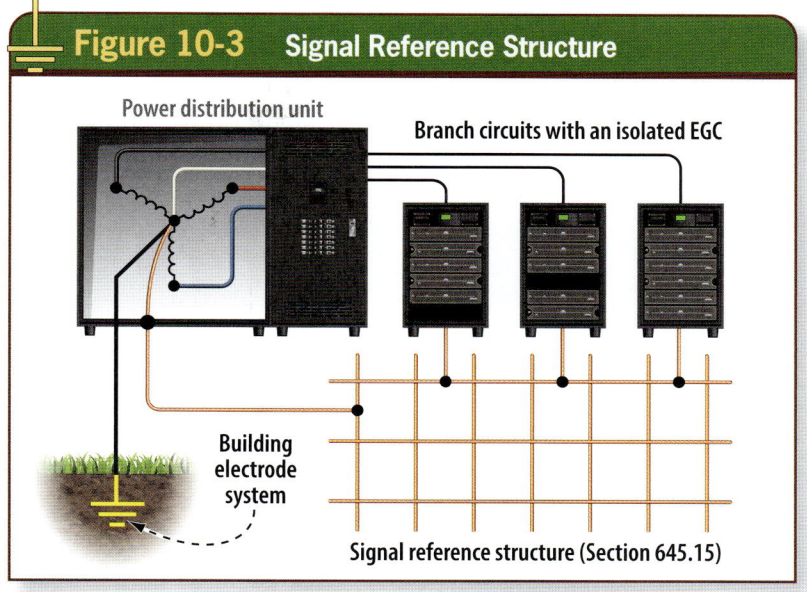

Figure 10-3   Signal Reference Structure

Power distribution unit

Branch circuits with an isolated EGC

Building electrode system

Signal reference structure (Section 645.15)

Figure 10-3. IT room designs typically include an isolated grounding circuit and a signal reference structure connected to the IT equipment.

At frequencies above and below resonance, the partial resonance increases circuit impedance and thus is less effective as a constant ground potential,

Figure 10-4   Power Distribution Units (PDUs)

Figure 10-4. Power distribution units are often installed for supplying branch circuits and equipment in information technology rooms.

Photo from iStock

Figure 10-5 IG Receptacle

Figure 10-5. *Isolated grounding receptacles installed in a branch circuit include an isolated, insulated EGC installed in accordance with 250.146(D).*

Courtesy of Pass and Seymour/Legrand

Figure 10-6   **Minimizing Electromagnetic Interference**

Service

Figure 10-6. *The isolated EGC is installed from the receptacle to the grounding point at the service. No connection is made to the intervening panelboard or any other grounded enclosures such as boxes, wireways, or conduit bodies.*

which is needed for reference and steady-state operation. Good engineering practices and designs typically address this problem by using common signal reference structures (grids) and short bonding straps (usually only a few feet in length) to equalize potential and minimize high-frequency effects in the grounding paths.

PURPOSE OF ISOLATED GROUNDING CIRCUITS AND RECEPTACLES

When sensitive electronic equipment is installed and used in IT rooms, for example, there are electrical circuit designs that can exceed the minimum requirements of the *NEC.* Isolated grounding circuits and receptacles are installed in an effort to reduce EMI that can interfere with data systems and equipment. **See Figure 10-5.**

This type of circuit design can reduce or minimize EMI on the equipment grounding circuits by insulating the conductive paths and reducing the grounding circuit to a single insulated path that extends back to the source grounding point, usually at a service or separately derived system. **See Figure 10-6.**

Electromagnetic interference can negatively affect some electronic equipment and lead to data errors and sometimes data loss. In the IT world, three characteristics are sought regarding electrical power supplying these systems: reliability, power quality, and minimization of EMI in the equipment grounding circuits or other grounding paths. When solutions to these interference problems are being sought, the minimum requirements of the *NEC* must never be compromised. Safety must always come first.

An isolated EGC should remain insulated from connections to ground or grounded parts until it connects to the point of grounding at the applicable service or separately derived system. Section 250.146(D) contains permissive text that allows the insulated EGC to pass through one or more panelboards or other enclosures without a grounding connection so as to terminate at the grounding point at the service or separately derived system. These provisions of the *NEC* do not require the isolated grounding conductor to return all the way to the source grounding point. Similar provisions are included in 408.40 Exception.

From a design perspective, the optimal isolated EGC is one that is kept insulated from all other grounded conductive surfaces. This will minimize the possibility of creating a ground loop and circulating current disturbance for connected electronic equipment. It is obviously better from a design standpoint to keep this EGC isolated (insulated) all the way back to the source or service. As an example, in a remodel project where isolated grounding circuits and receptacles are specified, the point of connection of the EGC might be at an existing panelboard or switchboard located downstream of the point of grounding at the service or separately derived system. The requirements in 250.146(D) are permissive, meaning there is a choice of where the isolated/insulated EGC connects to ground. The important factor is that isolated

grounding circuits provide an effective path for ground-fault current in each completed installation.

The EGC must be part of an effective ground-fault current path, even if it is an isolated/insulated EGC. When isolated/insulated EGCs are installed, two separate EGC paths are necessary for the branch circuit. One will serve as the required EGC of the wiring method and enclosures; the other will be the additional isolated/insulated EGC, which terminates directly on the isolated grounding receptacle without a connection to the grounded metal outlet box. Section 250.118 provides a list of wiring methods that qualify as EGCs. **See Figure 10-7.**

OBJECTIONABLE CURRENTS IN GROUNDING PATHS

The *NEC* addresses objectionable current in 250.6 and provides some alternatives to reduce objectionable current. Objectionable currents through EGCs and other grounding paths can contribute to the overall electromagnetic interference in grounding circuit conductors and other grounding paths.

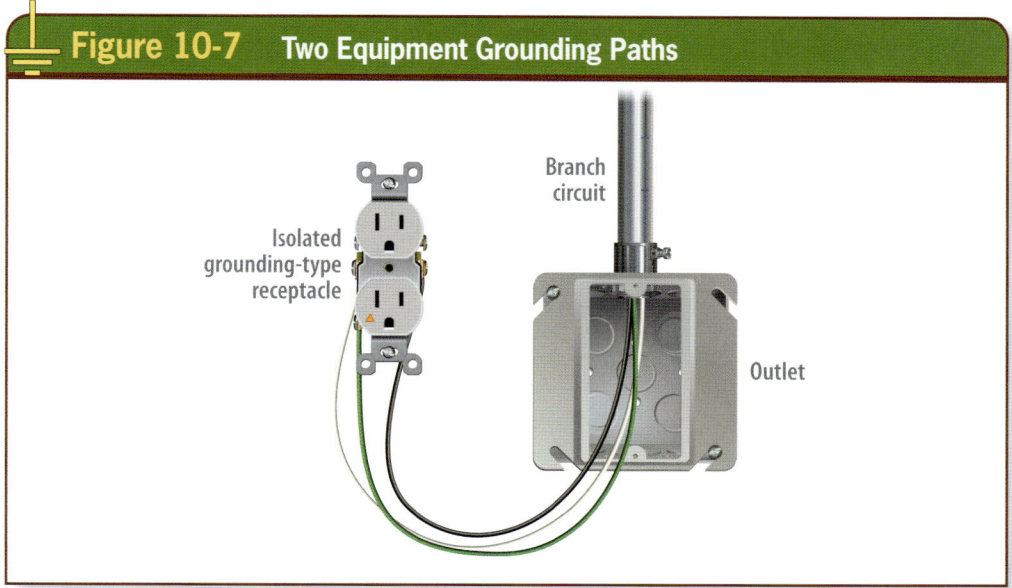

Figure 10-7   Two Equipment Grounding Paths

Isolated grounding-type receptacle

Branch circuit

Outlet

Figure 10-7. Two equipment grounding paths are created for isolated grounding circuits: one is the required EGC for the branch circuit, while the other is the additional isolated/insulated EGC installed according to 250.146(D).

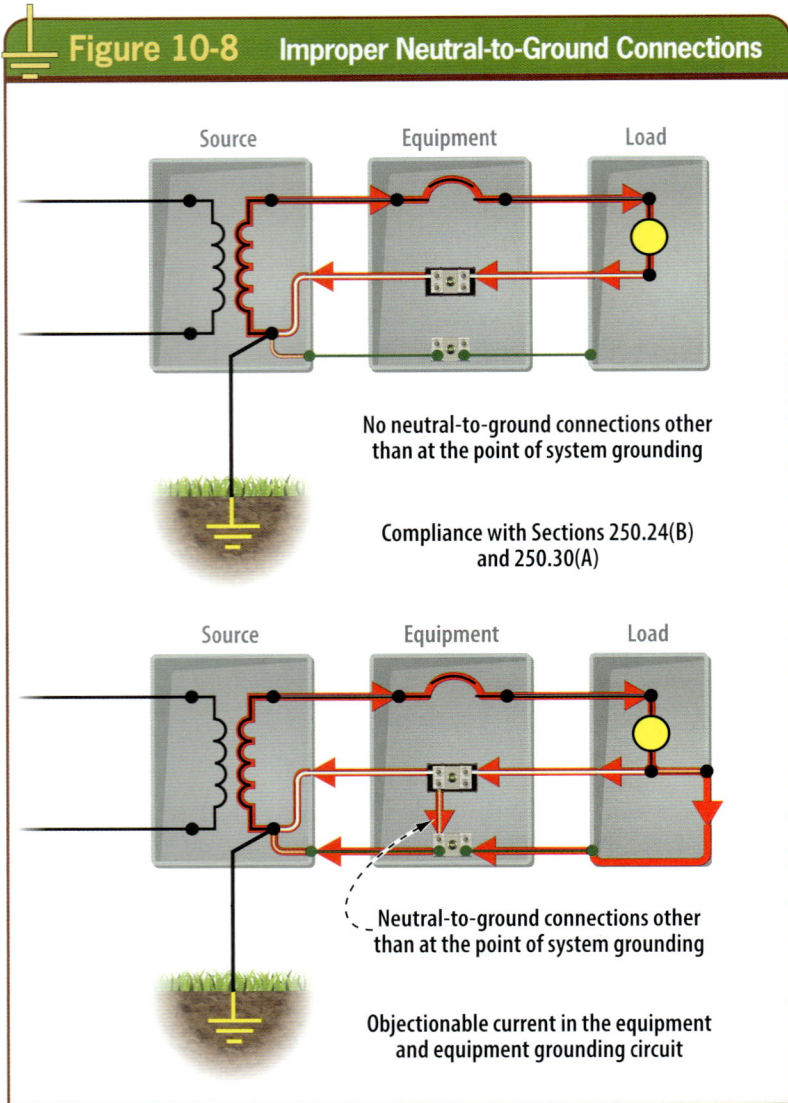

Figure 10-8   Improper Neutral-to-Ground Connections

Source   Equipment   Load

No neutral-to-ground connections other than at the point of system grounding

Compliance with Sections 250.24(B) and 250.30(A)

Source   Equipment   Load

Neutral-to-ground connections other than at the point of system grounding

Objectionable current in the equipment and equipment grounding circuit

Figure 10-8. Improper neutral-to-ground connections create objectionable current in the grounding circuits and violate 250.24(B) and 250.30(A).

Fact

Current that introduces electromagnetic interference or data errors in electronic equipment is not considered objectionable current, as addressed in 250.6.

These objectionable currents in grounding circuits are often a result of improper neutral-to-ground connections on the load side of the service disconnecting means or the load side of the point of grounding for a separately derived system. Note that temporary current, such as that present during ground-fault events, is not considered objectionable current when applying the provisions in 250.6. If objectionable currents are encountered, remedial alternatives for eliminating or minimizing objectionable currents are addressed in 250.6(B), but it is essential that the effective ground-fault current paths remain continuous and functional. **See Figure 10-8.**

If multiple grounding connections result in objectionable current, one or more of the following alterations are permitted as long as the effective ground-fault current path is not interrupted:

1. Disconnection of one or more, but not all, grounding connections
2. Change of the location of the grounding connections
3. Interruption of the continuity of the conductor or conductive path causing objectionable current
4. Other remedial action approved by the authority having jurisdiction (AHJ)

The alternatives provided in 250.6 do not allow for the removal of safety equipment grounding in compliance with *NEC* minimums.

POWER QUALITY SYSTEM GROUNDING ANALYSIS

Good power quality and effective *NEC*-compliant grounding and bonding are essential for proper electronic equipment operation. When a building power quality analysis is performed, it should always include a thorough analysis of the building grounding and bonding system. There should be no neutral-to-ground connections on the load side of the service grounding point or on the load side of the grounding point for a separately derived system, other than those few permitted exceptions in the *NEC* [250.24(B) and 250.30(A)]. Neutral-to-ground connections downstream of a main bonding jumper or system bonding jumper in a separately derived system can cause current in the EGC circuit(s) and in other conductive paths connected to the source.

The other important part of this analysis is to determine that the grounding electrode system meets *NEC* requirements. Before power quality issues can be effectively handled by using any filtering equipment, surge arresters, surge protection devices, or other remedies, the grounding and bonding system for the structure must be *NEC*-compliant.

ISOLATED GROUNDING CIRCUITS

When isolated/insulated EGCs are installed with the branch circuit, there are two EGC paths. The first path is the required EGC for safety; the second path is the desired isolated/insulated EGC for performance. The first path can be metal conduit, tubing, cable armor, and so forth.

The second path must always be an insulated conductor of the wire type. Sometimes cable assemblies such as metal-clad (MC) cable and armored-clad (AC) cable include two insulated conductors of the wire type for use with these circuits. One will generally be identified as green, and the other as green with one or more yellow stripes in accordance with 250.119. **See Figure 10-9.**

Isolated Ground Receptacle Wiring Rules

The *NEC* addresses isolated grounding circuits and receptacles that are installed for the reduction of EMI. The objective of an isolated grounding circuit is to reduce circulating currents through the grounding circuit by insulating it from other grounded conductive paths between the source and the outlet connection point. Isolated ground receptacles are manufactured with a grounding terminal that is deliberately isolated from the mounting strap of the device. Isolated ground-type receptacles must be marked with an orange triangle on the face of the receptacle. **See Figure 10-10.**

There are specific requirements for installing isolated ground receptacles and branch circuits. Section 250.146(D) includes the permissive text that allows an insulated EGC to pass through panelboards and other

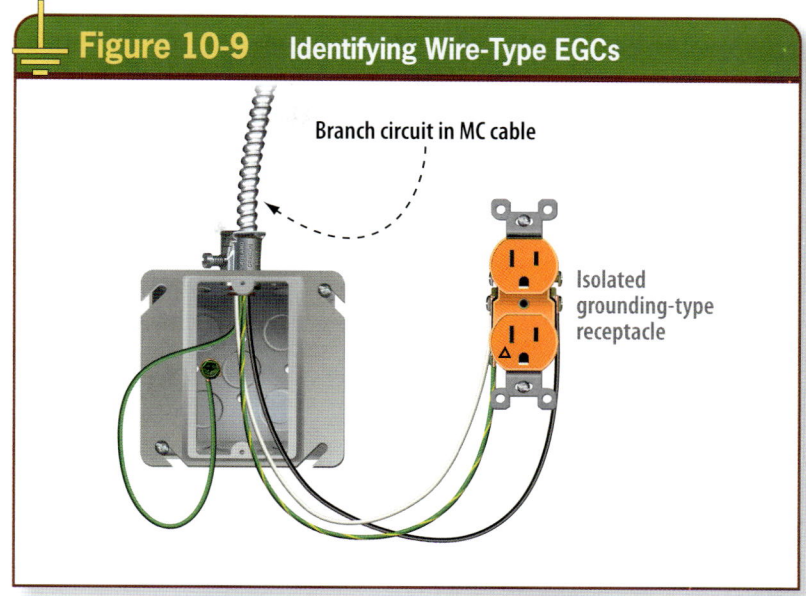

Figure 10-9. Identification of wire-type EGCs can be accomplished using two methods.

enclosures without connecting to them, as long as they are terminated at the point of grounding of the circuit. The point of grounding is either at the service equipment or at a source of a separately derived system, which could be a transformer or PDU in an IT room. This is usually the point where the main bonding jumper or system bonding jumper is installed.

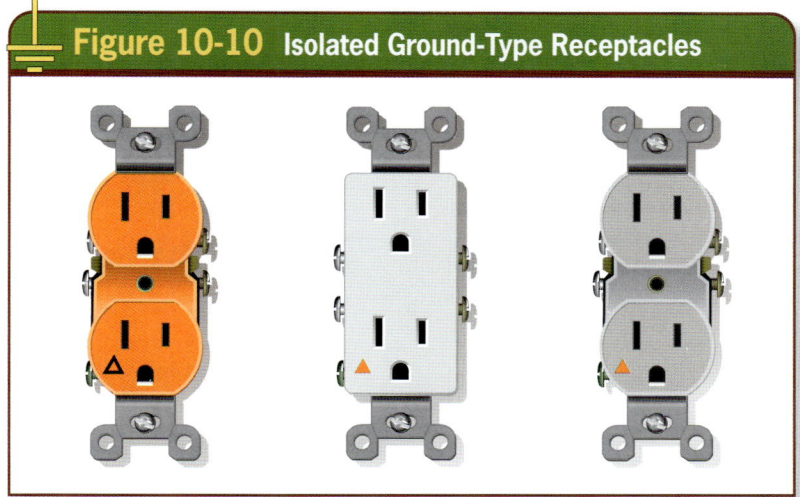

Figure 10-10. An orange triangle must appear on the face of isolated grounding-type receptacles. The ground terminal is isolated from the mounting yoke of the receptacle.

The isolated/insulated EGC (often referred to as a "clean ground") is installed in addition to the normally required EGC ("dirty ground") for the circuit. In the completed wiring installation, two separate EGC paths are established from the outlet to the source grounding point. **See Figure 10-11.**

The required EGC can be achieved by installing a conduit or other raceway system that qualifies as an EGC in accordance with 250.118. Examples are rigid metal conduit (RMC), intermediate metal conduit (IMC), electrical metallic tubing (EMT), and armored-clad cable (Type AC). **See Figure 10-12.**

There are listed armored-clad (Type AC) and metal-clad (Type MC) cable assemblies which provide an outer armor that qualifies as an EGC when used with listed fittings. These cable assemblies may also include a separate insulated EGC (wire type) within the assembly. These cables with two EGC

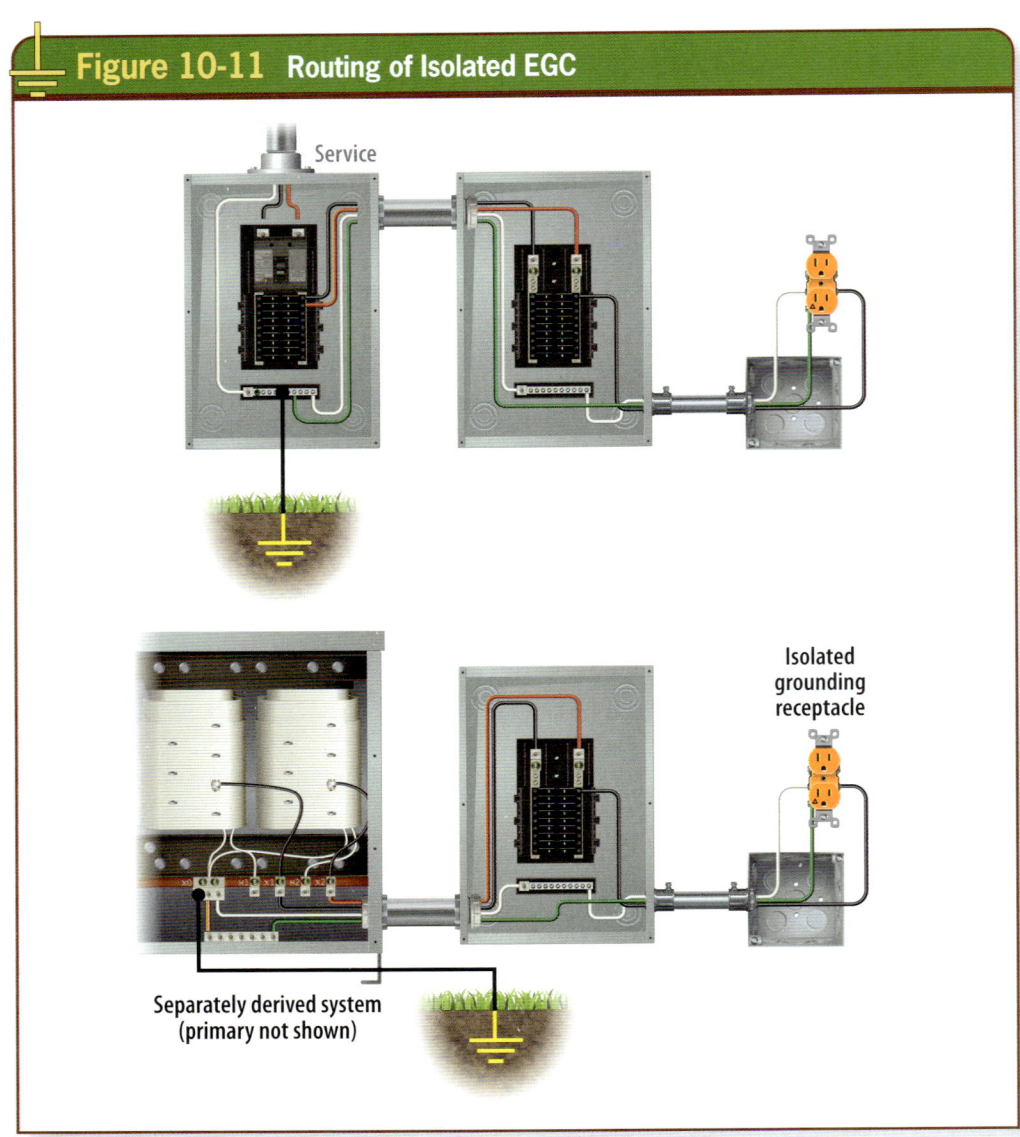

Figure 10-11   Routing of Isolated EGC

Service

Isolated grounding receptacle

Separately derived system (primary not shown)

Figure 10-11. *The isolated/insulated EGC conductor is installed from the receptacle to the grounding point at the service or separately derived system. No connection is made to the intervening panelboard or any other grounded enclosure. (Plaster rings not shown at outlet boxes.)*

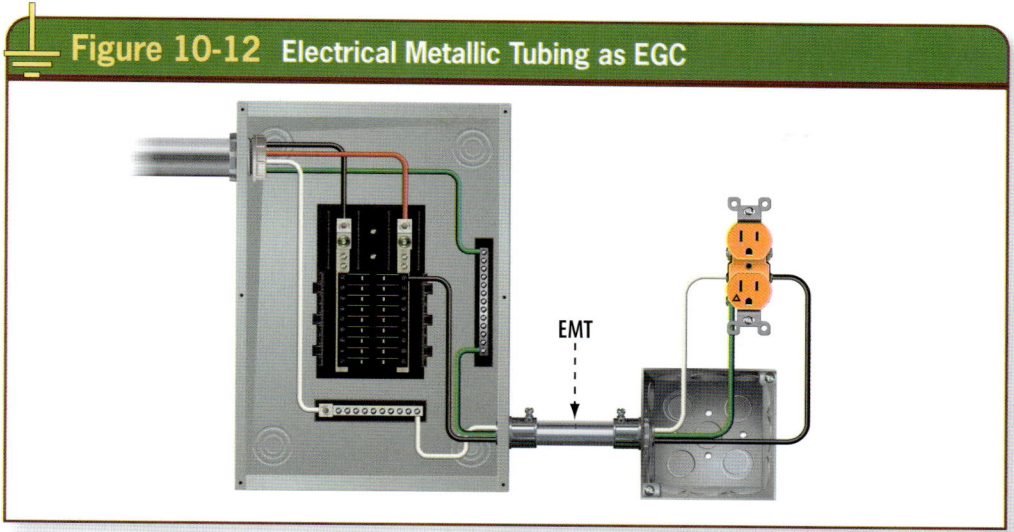

Figure 10-12   **Electrical Metallic Tubing as EGC**

EMT

Figure 10-12. EMT qualifies as an EGC for branch circuits, as provided in 250.118(4). The insulated wire-type EGC connects directly to the isolated grounding-type receptacle. (Plaster ring not shown at outlet box.)

paths are suitable for use in isolated grounding circuits. The metallic path inherent to these wiring methods establishes the required EGC for the circuit. The isolated/insulated EGC connected to the isolated grounding-type receptacle establishes a "clean ground," which is the second EGC path. Both paths terminate together at the grounding point of either the service

equipment or the source of a separately derived system. At the receptacle, however, the two EGCs are insulated from each other. **See Figure 10-13.**

Panelboards and Isolated Grounding Circuits

In some instances, multiple isolated grounding circuits are supplied from a single panelboard. Some designs exceed

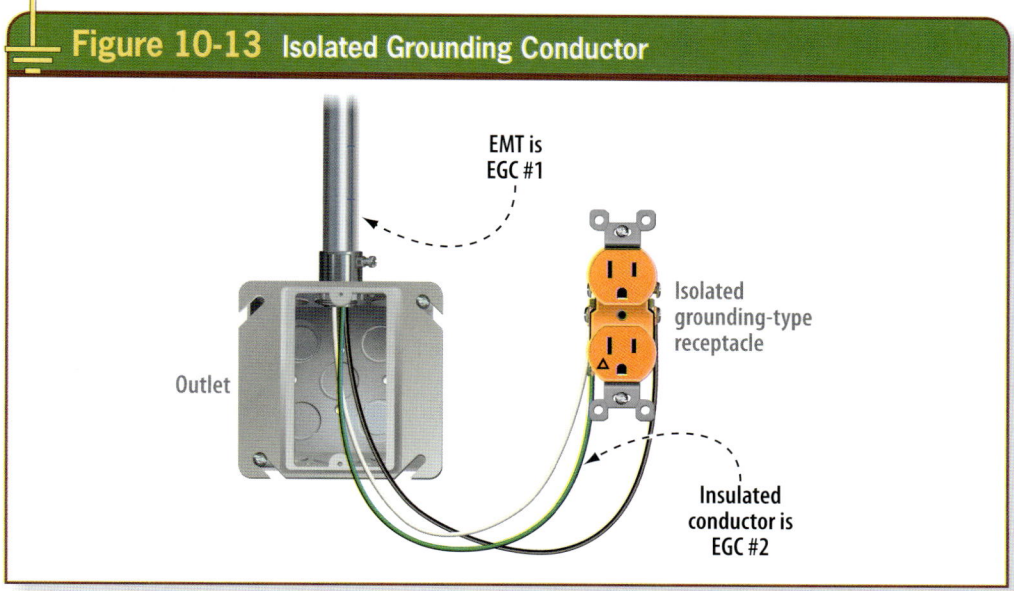

Figure 10-13   **Isolated Grounding Conductor**

EMT is
EGC #1

Outlet

Isolated
grounding-type
receptacle

Insulated
conductor is
EGC #2

Figure 10-13. The isolated grounding conductor is separate from the required EGC at the isolated grounding-type (IG) receptacle.

the minimum *NEC* requirements and specify that an isolated grounding terminal bar be installed in the panelboard for connecting all isolated/insulated EGCs from such branch circuits. In such designs, two EGCs are typically run with the feeder to the panelboard.

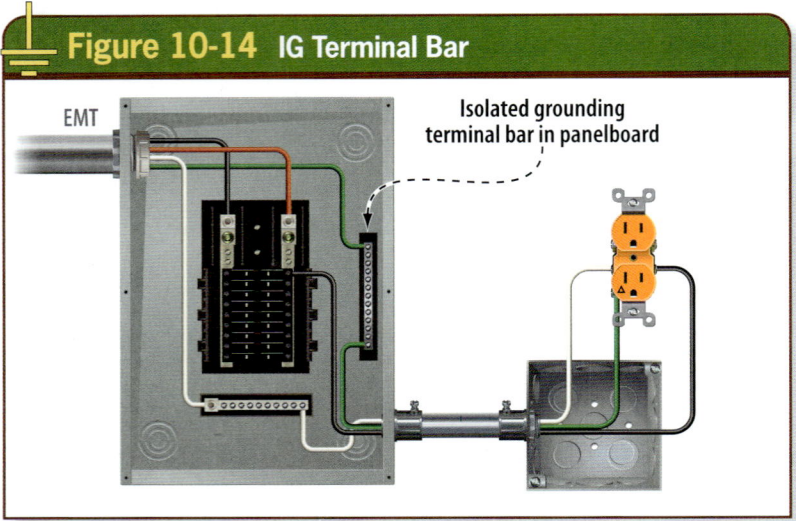

Figure 10-14. *In this panelboard, the EMT and metallic enclosures are EGCs. The insulated EGC connects to the isolated ground receptacle. (Plaster ring not shown at outlet box.)*

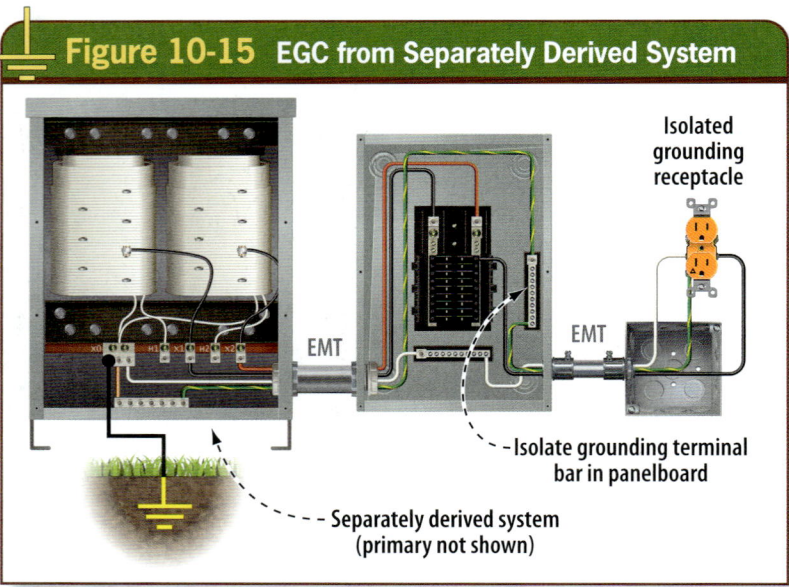

Figure 10-15. *Some designs go beyond* NEC *requirements and specify that an isolated EGC be installed with the feeder supplying a panelboard from a separately derived system (transformer). (Plaster ring not shown at outlet box.)*

One EGC serves as the normal EGC and connects directly to the enclosure; the other (isolated/insulated) EGC connects to an equipment grounding terminal bar that is mounted in but isolated from the panelboard enclosure. Isolated grounding terminal bars built specifically for this purpose are available from various panelboard manufacturers. **See Figure 10-14.**

The size of both EGCs installed with the feeder is based on the overcurrent protective device ahead of the feeder supplying the panelboard; refer to Table 250.122.

Sometimes a separately derived system supplies a panelboard. The same installation techniques and methods are applied to this type of installation, except that the isolated/insulated EGC terminates at the grounding point for the separately derived system, usually within the transformer enclosure. The *NEC* does not specifically address this type of design, but it is logical to conclude it can be installed and sized in the same manner as the normal feeder EGC for the circuit. **See Figure 10-15.**

Isolated Equipment Grounding Circuits for Equipment

The term *clean ground* often refers to the isolated EGC sometimes specified for IT equipment or receptacle outlets by the equipment manufacturer or owner. Section 250.96(B) indicates that when a reduction of electrical noise on grounding circuits is desired, an equipment enclosure (typically an IT equipment enclosure) is permitted to be supplied by a branch circuit containing an insulated EGC that is isolated from metal raceways using listed nonmetallic raceway fittings (such as polyvinyl chloride, or PVC, fittings). Isolation is accomplished by using a nonmetallic fitting between the conductive frame of the equipment and the grounded metal raceway, thus reducing the vulnerability to any EMI or circulating currents that may be present on the metal raceway. **See Figure 10-16.**

The *NEC* does not specifically address who determines what circumstances justify the use of isolated equipment grounding. This is a design issue and not an *NEC* requirement. The owner, design engineer, or equipment manufacturer usually specifies the circuit to use for isolated equipment grounds.

This *NEC* provision does not mandate the use of a metal raceway to supply the IT equipment. If PVC conduit is used, the insulating connector between the metal raceway and the IT equipment is not required. However, an EGC must be installed inside the PVC conduit to serve as the required equipment grounding means for the IT equipment.

Isolated Grounding Circuits in Health Care Facilities

Isolated/insulated equipment grounding circuits must provide an effective ground-fault current path in addition to providing a clean grounding connection for the equipment. The isolated/insulated EGC essentially serves both purposes. The required EGCs must always be in place and effective in addition to any desired isolated/insulated EGC.

When designs call for more specialized equipment grounding means for reducing unwanted electromagnetic interference on the grounding circuit, an additional insulated EGC path is usually installed.

In health care facilities, isolated ground receptacles and circuits are not permitted within a patient care vicinity, as indicated in 517.16(A). Section 517.16(B) provides the requirements for isolated grounding receptacles and circuits that are installed in health care facilities in areas outside of the patient care vicinity. Essentially, the isolated grounding circuits installed in patient care spaces must be installed in accordance with 250.146(D). **See Figure 10-17.**

This means that two equipment grounding conductor paths are required. Path one satisfies 517.13(A) and path two satisfies 517.13(B). This requirement clarifies that the insulated equipment

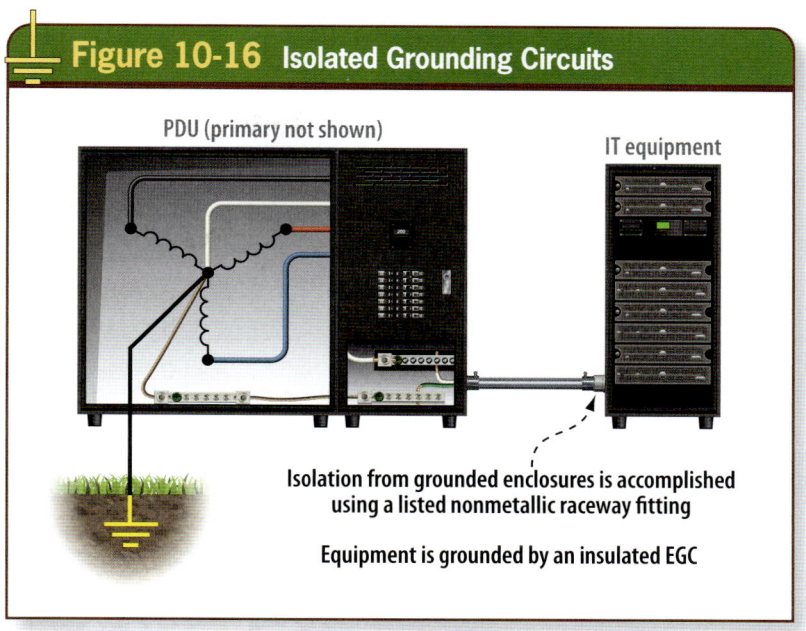

Figure 10-16 Isolated Grounding Circuits

PDU (primary not shown)

IT equipment

Isolation from grounded enclosures is accomplished using a listed nonmetallic raceway fitting

Equipment is grounded by an insulated EGC

Figure 10-16. Some installations of isolated/insulated grounding circuits are provided for equipment that is wired directly to circuits without using an IG receptacle.

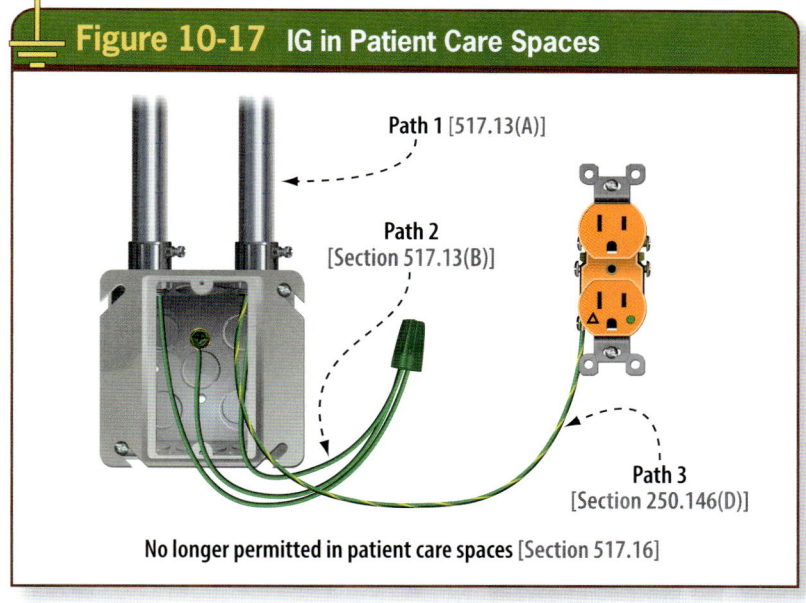

Figure 10-17 IG in Patient Care Spaces

Path 1 [517.13(A)]

Path 2 [Section 517.13(B)]

Path 3 [Section 250.146(D)]

No longer permitted in patient care spaces [Section 517.16]

Figure 10-17. An isolated/insulated grounding circuit in patient care spaces shall be in accordance with 517.16(B).

grounding conductor path installed in accordance with 250.146(D) is in addition to the two grounding paths required in Section 517.13. This revision also brings the *NEC* into alignment with the wiring requirements covered by *NFPA*

99: Health Care Facilities Code. NFPA 99 also requires periodic testing of grounding systems, which includes installations of isolated ground receptacles and circuits in health care facilities.

USE OF AUXILIARY GROUNDING ELECTRODES

When isolated/insulated circuits are installed, there is often a desire to install a separate supplemental connection to the Earth at the equipment location. This grounding electrode is known as an *auxiliary grounding electrode*. Installation of auxiliary grounding electrodes does not relieve the requirement of connecting an EGC; rather, it is installed in addition to the required EGC for the branch circuit. **See Figure 10-18.**

When an auxiliary grounding electrode is installed, both the EGC and the grounding electrode conductor to the electrode must be connected to the equipment. It is a violation of the *NEC* to eliminate the required EGC because it creates a safety hazard. Remember that Sections 250.4(A)(5) and 250.54 both clearly indicate that the Earth is not permitted to be used as an effective ground-fault current path.

Sometimes enhanced grounding electrode systems are installed for IT centers. These grounding electrodes cannot be isolated from the building grounding electrode system; they are installed in addition to and must be bonded to the building grounding electrode system. Otherwise, unwanted and unsafe differences of potential can result.

GROUNDING AND BONDING IN INFORMATION TECHNOLOGY CENTERS

In large facilities such as data centers, it is common for all information technology equipment to be located in a single room or data center within the facility. Sometimes, depending on the extent of the IT needs for the particular facility, an entire building may be constructed to meet these criteria. This space is constructed to meet specific criteria to qualify as an IT room. Section 645.4 of the *NEC* provides specific requirements that must be met before the rules in Article 645 can be applied to an IT room. In other words, to be able to use the provisions in Article 645, all requirements in 645.4 must be complied with.

All of the following must be provided for an IT room to take advantage of the provisions in Article 645:
1. A disconnecting means according to 645.10 must be provided.
2. A separate HVAC system must be provided (dedicated to the room and equipment).
3. Listed communications equipment and IT equipment must be installed.
4. The room must be accessible to and occupied only by operators and maintainers of the equipment/system.
5. The room must be separated from other occupancies in the building by fire-resistant-rated construction.
6. Only equipment and wiring associated with the operation of the IT room can be located in the room.

Note: See 645.4 for complete information about each of these six items.

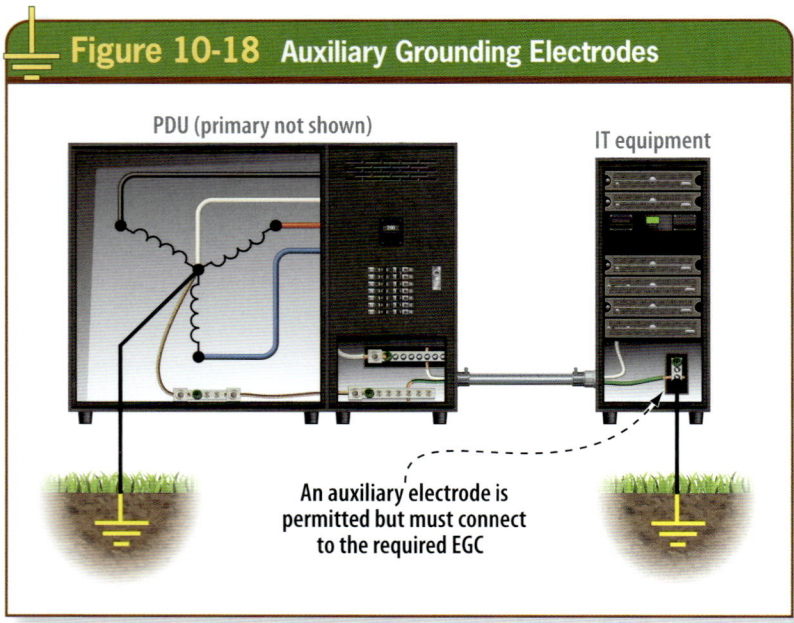

Figure 10-18 Auxiliary Grounding Electrodes

PDU (primary not shown)

IT equipment

An auxiliary electrode is permitted but must connect to the required EGC

Figure 10-18. Auxiliary grounding electrodes must supplement the required EGC of the circuit supplying equipment.

Section 645.4 and *NFPA 75: Standard for the Protection of Information Technology Equipment* both provide more detailed information on the construction requirements for an IT room. If the room is not constructed to meet the criteria, then all the rules in Chapters 1 through 4 of the *NEC* apply to the installation. The driving text in 645.4 indicates that Article 645 shall be permitted to provide alternate wiring methods to the provisions of Chapter 3, Parts I and II of Article 725 for signaling wiring, and Parts I and V of Article 770 for optical fiber cabling, if meeting the conditions in items (1) through (6).

Section 645.15 includes specific equipment grounding and bonding requirements for equipment in an IT system. The primary requirement is that all non–current-carrying metal parts of such equipment be connected to the EGC of the supply branch circuit or feeder in accordance with Parts I, V, VI, VII, and VIII of Article 250. There is an exemption from this grounding requirement, but only where the IT equipment is double-insulated.

To minimize possible differences in potential in the grounding systems for power circuits supplying IT equipment, it is common for these centers to be equipped with a single or multiple PDUs. The term *power distribution unit* is not defined in the *NEC*, but it is described in 645.17. The PDUs that are used for IT equipment are permitted to have multiple panelboards within a single cabinet, provided that the PDU is utilization equipment listed for IT applications. **See Figure 10-19.**

PDUs are typically built with transformers and panelboards in a single enclosure or assembly. This unit is supplied by a feeder, and branch circuits are routed to the IT equipment from the PDU. PDUs usually provide a convenient shunt trip feature that affords easy compliance with the disconnecting means rule in 645.10.

Section 645.14 indicates that the power systems derived in listed IT equipment (PDUs) supplying IT

Figure 10-19. *PDUs often include a single transformer and multiple panelboards in a single enclosure.*

systems through specially constructed receptacles and cable assemblies are not considered as separately derived for the purposes of applying the grounding requirements for separately derived systems. This means that installing, wiring, and grounding these circuits and systems all must be done in accordance with specific instructions provided with the IT equipment.

Some of these PDUs require a connection to the building grounding electrode system while others do not. The listed equipment provides the requirements that installers must follow for grounding of such circuits and systems. The isolated/insulated grounding circuits and receptacles are usually supplied as premanufactured "whips" by the supplier of the PDU. The grounding and bonding connections for all such circuits are made within the PDU because that is the power source. Thus, EMI in the grounding circuits supplying the IT equipment is kept to a minimum, because these circuits are relatively short and do not extend throughout the building or structure.

Figure 10-20. *Signal reference structures are typically installed in an IT room before installing the raised floor (platform).*

Courtesy of Rick Maddox, Clark County, NV

Figure 10-21. *Signal reference structures often incorporate the support frame for the raised floor in the IT room.*

Courtesy of Harger Lightning and Grounding

SIGNAL REFERENCE STRUCTURES (GRIDS)

The term *signal reference grid* refers to a common conductive structure such as a computer floor, a series of interconnected copper sheet strips, or a copper mesh grid installed under the raised floor. This grid provides an effective equipotential bonding structure to which all equipment can be connected. **See Figure 10-20.**

IT equipment is connected to the signal reference grid in an effort to equalize or minimize the potential differences between components. All equipment in the room, including equipment that is mounted to the wall, should be connected to the grid, as should the EGC in each supply branch circuit. **See Figure 10-21.**

The *NEC* does not address this subject, but, as with other bonding or grounding methods used for IT equipment, such bonding cannot substitute for the requirement that the EGC be installed with the branch circuit conductors supplying the IT equipment. The signal reference grid can be thought of as overlaying the required equipment grounding conductors. **See Figure 10-22.**

The signal reference grid serves as a signal reference plane over a broad range of frequencies. Signal reference grids are also commonly referred to as *broadband grounding systems*. The grid structure minimizes potential differences and reduces, eliminates, or controls the conductors connected to computers that tend to resonate at higher frequencies. A grid provides multiple parallel conducting paths between its metal parts. If one path is a high-impedance path because of full or partial resonance, other paths of different lengths will be able to provide a lower-impedance path.

A signal reference grid is commonly constructed of continuous sheet copper, aluminum, or any number of pure or composite metals with good surface conductivity. Listed products are also available for this purpose. **See Figure 10-23.**

Signal reference grids can also be made up of various conductive surfaces, such as the raised floor framing and supports, the suspended ceiling grid system, or a constructed grid under the raised platform or floor of an IT room. Grids of copper or aluminum strips are sometimes installed under a computer-raised floor. This grid provides a constant potential reference network over a broad range of frequencies from direct current (DC) to higher than 30 megahertz, or even in the gigahertz ranges. The grids are typically made in mesh configurations using a minimum of 4 AWG copper or aluminum conductors that have been electrically joined at their intersections, or by thin copper straps that are about two inches wide, also joined at their intersections. **See Figure 10-24.**

These constructed grids are typically placed directly on the subfloor under the raised IT room floor. Cables and conduits under the floor would normally be installed below the raised floor but above the grid. The IT equipment is then connected to the grid using short braided bonding conductors (usually two to four feet in length).

Remember that this common bonding grid is not a substitute for the required EGC supplying the IT equipment; it is in addition to it. Section

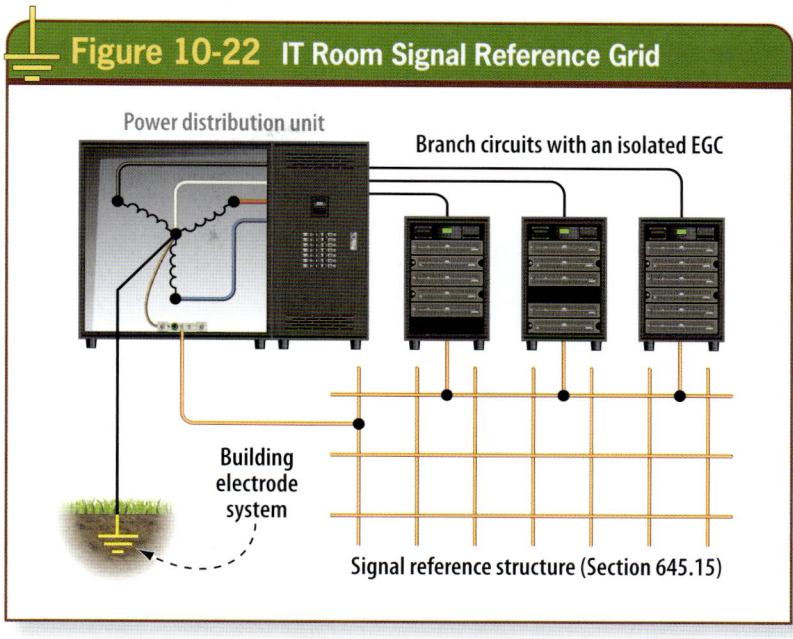

Figure 10-22. *The signal reference structure in IT rooms must overlay the required EGCs of the branch circuits for IT equipment.*

645.15 requires the signal reference grid to be bonded to the EGC provided in the IT equipment. This is accomplished at each separate IT unit in the room and, typically, at each PDU grounding point.

Figure 10-23. *Listed products are available for constructing a signal reference grid in an IT center.*

Courtesy of Harger Lightning and Grounding

Figure 10-24. *Signal reference structures can be in the form of overlapping solid copper straps under the raised platform (floor) of an IT room.*

Courtesy of Harger Lightning and Grounding

SURGE PROTECTION

Surge protection is often desired for IT equipment installations. These devices can provide effective protection against power line surges. The term *Surge-Protective Device (SPD)* is defined in Article 100 of the *NEC*.

> **Surge-Protective Device (SPD).** A protective device for limiting transient voltages by diverting or limiting surge current; it also prevents continued flow of follow current while remaining capable of repeating these functions and is designated as follows:
>
> **Type 1:** Permanently connected SPDs intended for installation between the secondary of the service transformer and the line side of the service disconnect overcurrent device
>
> **Type 2:** Permanently connected SPDs intended for installation on the load side of the service disconnect overcurrent device, including SPDs located at the branch panel
>
> **Type 3:** Point of utilization SPDs
>
> **Type 4:** Component SPDs, including discrete components, as well as assemblies. (CMP-10)

Installations of surge protective devices are covered by the *NEC*. There are general rules to require them in the *NEC*, such as those in Sections 215.18, 225.42, and 230.67 for outside branch circuits, feeders, and services supplying dwelling units. If surge protection is installed by choice or by a specific requirement in the *NEC*, it must comply with the applicable provisions in Part II of Article 242, which is titled *Overvoltage Protection*. **See Figure 10-25.**

Unlike surge arresters rated above 1,000 volts, SPDs are required to be listed. Part III of Article 242 in the *NEC* includes requirements for surge arresters rated over 1,000 volts.

SPDs are not permitted for use on circuits exceeding 1,000 volts and are restricted from use on ungrounded systems, impedance-grounded systems, or corner-grounded systems unless they are specifically listed for use on those systems. This means they have been evaluated to applicable product safety standards. The product certification or listing also drives the requirements for installers to follow the manufacturer's installation instructions, as required in 110.3(B). This requirement resolves many of the questions concerning what is required for SPD installation.

Figure 10-25 Surge Protection Devices (SPDs)

Figure 10-25. SPDs provide a level of equipment protection against line surges.

If an SPD is connected and used with a separately derived system, it must be connected on the load side of the first overcurrent protective device supplied by the separately derived system. This could be a Type 2 or Type 3 SPD because it is located on the load side of a service overcurrent protective device. The supply conductors and grounding conductors for an SPD are not permitted to be smaller than 14 AWG copper or 12 AWG aluminum conductors. The length of conductors for Type 3 SPDs must be at least 30 feet if included in the manufacturer's installation instructions.

Grounding conductor connections to surge protection devices must meet the requirements in Part III of Article 250, and if the grounding electrode conductor is installed in a ferrous metal raceway, it must comply with 250.64(E). This section requires the grounding electrode conductor contained in a ferrous metal raceway to be bonded to both ends of the raceway, thus minimizing impedance and choke effects during surge events.

Surge protective devices (SPDs) must be installed according to the manufacturer's installation instructions.

Courtesy of Bill McGovern, City of Plano, TX

SUMMARY

Electronic and information technology equipment is vulnerable to electromagnetic interference, which can lead to data errors and data loss in some instances. Installations of circuits for electronic equipment such as computers and servers in information technology centers often necessitate installing isolated grounding circuits and receptacles. The minimum requirements for installing isolated grounding circuits and receptacles must be understood.

The *NEC* includes some alternative equipment grounding techniques that are often applied in system designs to address concerns related to normal operation of electronic equipment. Various common methods are applied to achieve optimal performance in the grounding system while simultaneously meeting minimum requirements of the *NEC*. These two must go hand in hand, and safety must never be compromised in an effort to reduce noise in the grounding circuit or system. Article 645 of the *NEC* provides specific rules for wiring information technology rooms, specifically the grounding circuits and systems for such installations. Additional information about constructing IT rooms and protecting IT equipment is contained in *NFPA 75: Standard for the Protection of Information Technology Equipment.*

REVIEW QUESTIONS

1. According to the *NEC*, which of the following is <u>not</u> an acceptable method of reducing objectionable current in the grounding system?
   a. A remedial action approved by the authority having jurisdiction (AHJ)
   b. Disconnection of one or more, but not all, grounding connections or change of the location of the grounding connections
   c. Disconnection of the required EGC of the circuit supplying the equipment
   d. Interruption of the continuity of the conductor or conductive path causing objectionable current

2. The term *ground loop* is used in the IT industry and is best associated with __?__.
   a. a counterpoise system
   b. a ground ring
   c. a grounding electrode system
   d. circulating currents in multiple grounding paths

3. Resonance can occur when the length of a conductor and the __?__ of alternating current are in tune. This effect is similar to the principle of tuning a radio transmitter and antenna for maximum resonance and radiation.
   a. amperage
   b. frequency
   c. voltage
   d. wattage

4. Ground loops are commonly present when computer equipment is connected to multiple circuits supplied from different power sources and those power sources are then interconnected with shielded communication cables.
   a. True   b. False

5. Which of the following is another known solution to the problem of circulating currents in the grounding system?
   a. Installation of a single-point grounding system and use of a single power supply system
   b. Interface devices (surge arresters or surge protective devices)
   c. Use of fiber-optic transmission over completely non-conducting paths or optical isolators
   d. Any of the above

6. Isolated grounding circuits and receptacles are often installed in an effort to reduce electromagnetic interference, which can interfere with data systems and equipment.
   a. True   b. False

7. The insulated EGC connected to an isolated grounding-type receptacle is permitted to pass through one or more panelboards or other enclosures without a grounding connection as long as it terminates at the grounding point at the __?__.
   a. applicable service or separately derived system
   b. building steel
   c. ground rod
   d. surge arrester

8. An isolated grounding-type receptacle is required to be identified by which of the following means?
   a. A green dot on the receptacle
   b. An orange faceplate
   c. An orange triangle on the receptacle
   d. The color orange

9. An EGC must be an effective ground-fault current path, even if it is an isolated/insulated EGC installed with the branch circuit.
   a. True   b. False

10. An isolated/insulated equipment grounding conductor (wire type) installed from an isolated grounding-type receptacle is permitted to be connected only to a ground rod for a true isolated grounding connection.
    a. True   b. False

11. Neutral-to-ground connections downstream of a main bonding jumper in a service or a system bonding jumper in a separately derived system can cause current in the EGC circuit(s) and over other common conductive paths to the source. They are in violation of __?__.
    a. Section 250.24(B)
    b. Section 250.30(A)
    c. both a. and b.
    d. neither a. nor b.

12. When isolated/insulated EGCs are installed with a branch circuit (other than a branch circuit serving a patient care vicinity), there will be __?__ EGC path(s).

    a. 1
    b. 2
    c. 3
    d. 4

13. An auxiliary grounding electrode is permitted to be the only grounding connection for electronic equipment when noise on the equipment grounding circuit is a problem.

    a. True   b. False

14. The isolated/insulated EGC installed and connected to an isolated grounding-type receptacle is required to be identified by which of the following methods?

    a. Bare
    b. Green or green with one or more yellow stripes
    c. Orange
    d. White

15. Although the *NEC* no longer permits the installation of isolated grounding circuits and receptacles in patient care vicinities as restricted by 517.16(A), *NFPA 99: Health Care Facilities Code* still addresses these circuits where installed outside the patient care vicinity. *NFPA 99* also still requires periodic testing of grounding systems, which include installations of isolated grounding receptacles and circuits in health care facilities.

    a. True   b. False

16. To minimize possible differences in potential in the grounding systems for power circuits supplying IT equipment, it is common for IT facilities to be equipped with a single or multiple PDUs.

    a. True   b. False

17. Signal reference grids can be any number of common conductive surfaces in the IT room, including __?__.

    a. a constructed grid under the raised platform or floor of the IT room(s)
    b. the raised floor framing
    c. the suspended ceiling grid system
    d. any of the above

18. A(n) __?__ is a protective device for limiting transient voltages by diverting or limiting surge current; it also prevents continued flow of follow current while remaining capable of repeating these functions.

    a. grounding electrode
    b. SPD
    c. surge arrester
    d. uninterruptable power supply

19. __?__ SPDs are permanently-connected SPDs intended for installation on the load side of the service disconnect overcurrent protective device, including SPDs located at the branch panel.

    a. Type 1
    b. Type 2
    c. Type 3
    d. Type 4

20. If a grounding electrode conductor for an SPD is installed in a ferrous metal raceway, the grounding electrode conductor is required to be bonded to both ends of the raceway, thus minimizing impedance and choke effects during surge events.

    a. True   b. False

21. Article 242 of the *NEC* contains requirements for installations of surge protective devices (SPDs) and surge arresters.

    a. True   b. False

Grounding at Separate Buildings or Structures

At some properties, a single electric utility service supplies multiple buildings or structures. In these installations, the service often directly supplies one of the buildings, and feeders or branch circuits supply the other buildings or structures from the service in that building. Alternatively, the utility service can be freestanding, such as a service mounted on a pole or pad, and feeders or branch circuits supply all buildings or structures on that premises. Specific grounding and bonding rules apply to separate buildings or structures supplied by feeders or branch circuits. The requirements for grounding and bonding at separate buildings or structures are located in *NEC* Section 250.32.

Objectives

» Determine the requirements for a grounding electrode system at buildings or structures supplied by a branch circuit(s) or feeder(s).

» Understand the requirements for equipment grounding conductors to be installed with feeders or branch circuits supplying separate buildings or structures.

» Understand the exception for grounded conductors used for equipment grounding at a separate building or structure supplied by a feeder or branch circuit.

» Determine the bonding requirements for piping systems and structural metal building framing of separate buildings or structures.

» Understand the grounding and bonding rules associated with generators supplying separate buildings or structures.

» Differentiate what constitutes a building or structure from what constitutes equipment by definition.

Chapter 11

Table of Contents

Definitions 254

Supplying Power to Separate Buildings or Structures 254

Purpose of Grounding and Bonding at Separate Buildings or Structures 254

Grounding Electrode Requirement.............. 255

Grounding Electrode Conductor.................. 256

Feeder and Branch Circuit Requirements 258

Ungrounded Systems Supplying Separate Buildings or Structures 260

Supplied by a Separately Derived System.... 260

    Systems with Overcurrent Protection..... 260

    Systems without Overcurrent Protection 260

Building Disconnecting Means Requirements 262

Metal Water Pipe Bonding and Other Bonding 263

Disconnecting Means Remote from Building or Structure 263

Buildings or Structures Supplied by Separately Derived Systems....................... 264

Buildings or Structures Supplied by an Ungrounded System 265

Buildings or Structures Supplied by Generators 266

Summary... 269

Review Questions 270

DEFINITIONS

There are a few key words and terms defined in *NEC* Article 100 that directly relate to grounding and bonding rules for separate buildings and structures.

Branch Circuit (Branch-Circuit). The circuit conductors between the final overcurrent device protecting the circuit and the outlet(s). (CMP-2)

Branch Circuit, Multiwire (Multiwire Branch Circuit). A branch circuit that consists of two or more ungrounded conductors that have a voltage between them, and a neutral conductor that has equal voltage between it and each ungrounded conductor of the circuit and that is connected to the neutral conductor of the system. (CMP-2)

Building. A structure that stands alone or that is separated from adjoining structures by fire walls. (CMP-1)

Feeder. All circuit conductors between the service equipment, the source of a separately derived system, or other power supply source and the final branch-circuit overcurrent device. (CMP-10)

Grounding Electrode. A conducting object through which a direct connection to earth is established. (CMP-5)

Grounding Electrode Conductor (GEC). A conductor used to connect the system grounded conductor or the equipment to a grounding electrode or to a point on the grounding electrode system. (CMP-5)

Structure. That which is built or constructed, other than equipment. (CMP-1)

SUPPLYING POWER TO SEPARATE BUILDINGS OR STRUCTURES

Many buildings or structures are supplied by power from a source other than a utility service. Examples range from a small detached garage supplied by a branch-circuit or feeder from a house (single-family dwelling) to a university campus or industrial facility that has a higher-voltage service supplied by an outdoor substation and 15-kilovolt or 480-volt feeders supplying the buildings.

A service, by definition, is supplied from an electric utility. If the supply source, such as a transformer or generator, is customer-owned, it is not a service; therefore, it is either a feeder or branch circuit. When a building or structure is supplied by a feeder(s) or branch circuit(s), specific rules in the *NEC* must be applied. Section 250.32 of the *NEC* provides the general requirements for grounding and bonding at separate buildings or structures that are supplied by a source other than a service.

PURPOSE OF GROUNDING AND BONDING AT SEPARATE BUILDINGS OR STRUCTURES

Section 250.32 provides the grounding and bonding requirements for buildings and structures supplied by feeders or branch circuits. The purpose of grounding and bonding systems at branch circuit- or feeder-supplied separate buildings or structures is similar to service-supplied buildings or structures. A grounding electrode system (connection to ground) is generally required at separate buildings or structures to establish a reference to ground at that location. This places all normally conductive non–current-carrying metal parts and other conductive materials at or as close as possible to Earth potential at the building or structure served. **See Figure 11-1.**

An important point to emphasize is the differentiation between a *building* and a *structure*. The definitions in Article 100 provide some clarity in that all buildings are structures, while not all structures are buildings. As an example, a school is a building, while a billboard sign along the roadway is a structure.

The other important factor to understand is the difference between *equipment* and a *structure*. Structures are constructed. For example, motor or ground-mounted air-conditioners are equipment covered by Chapter 4 of the *NEC*. When installed outdoors, they are typically anchored to a concrete pad or footing. This does not render the combination of the equipment and footing a structure, and thus predicate the requirements for a grounding electrode. Any grounding electrode at these types of equipment installations is likely an auxiliary grounding electrode, and therefore is optional.

Another example is a parking lot luminaire installed on a pole. The luminaire and the pole are both equipment covered by Article 410 of the *NEC*, and the concrete base is a structure to which the equipment is mounted. The key difference between equipment and a structure is that structures are constructed, as indicated in the definition, while equipment is produced or manufactured, typically in a facility, and delivered to the site for installation.

The definition of the term *structure* is clear that it does not apply to equipment. It is important to establish a clear differentiation between equipment and structures to effectively determine if a grounding electrode is required to be installed by 250.32(A), and if it is permitted as an option in accordance with 250.54. Note that it might be a good design practice to install a grounding electrode for equipment that is installed freestanding and remote from buildings, but it is optional.

The reasons why equipment and systems are grounded at separate buildings or structures are essentially the same reasons why grounding is required if the building or structure is supplied by a utility service. The performance grounding requirements for equipment supplied by a grounded system are provided in 250.4(A)(2). For electrical equipment supplied by an ungrounded system, the performance requirements are in 250.4(B)(1). These performance rules both indicate that normally non–current-carrying conductive materials enclosing electrical conductors or equipment, or forming part of such equipment, must be connected to the Earth (grounded) in a manner that will limit the voltage imposed by lightning or unintentional contact with higher-voltage lines and limit the voltage to ground on these conductive materials. These concepts also apply

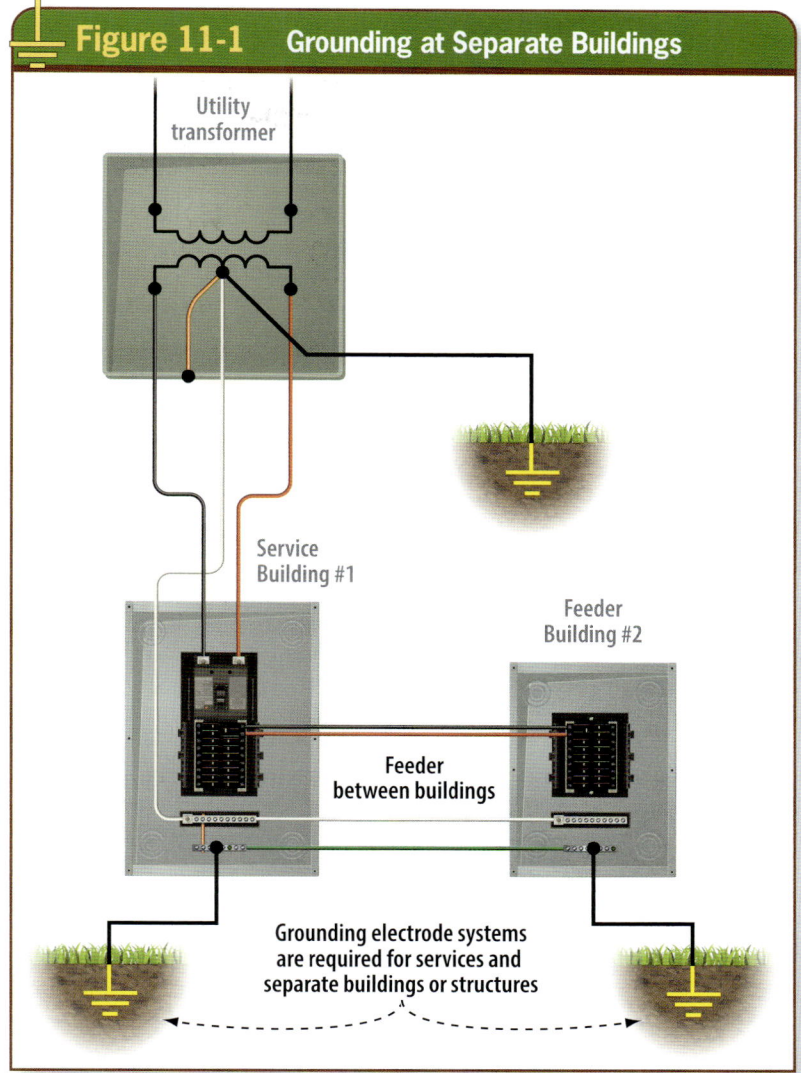

Figure 11-1. *A grounding electrode system and grounding electrode conductor(s) are generally required at buildings supplied by services and buildings supplied by feeders or branch circuits.*

at separate buildings or structures supplied by feeders or branch circuits.

GROUNDING ELECTRODE REQUIREMENT

Section 250.52(A) lists grounding electrodes recognized in the *NEC*, and 250.52(B) lists material and structures not permitted as grounding electrodes. A grounding electrode system is generally required at a separate building or structure supplied by feeders or branch circuits. This condition is relaxed if (1)

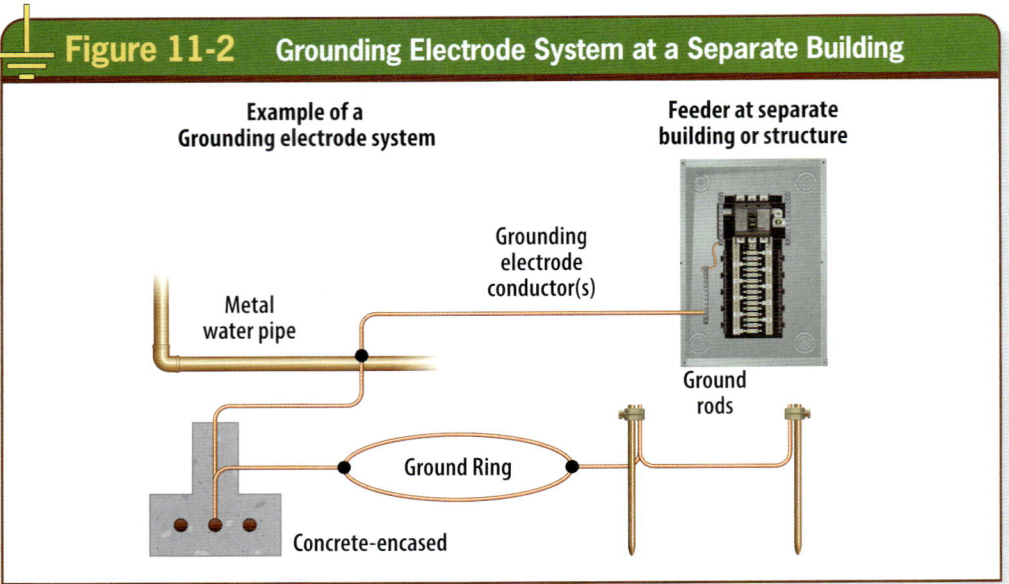

Figure 11-2   **Grounding Electrode System at a Separate Building**

Example of a
Grounding electrode system

Feeder at separate
building or structure

Grounding
electrode
conductor(s)

Metal
water pipe

Ground
rods

Ground Ring

Concrete-encased

Figure 11-2. All grounding electrodes must be bonded together to form a grounding electrode system for the separate building or structure supplied by feeders or branch circuits. Ground clamps for more than one GEC must be identified for the use.

the building or structure is not supplied by electrical power, or (2) the structure or building is supplied by a single branch circuit (including a multiwire branch circuit) that includes an equipment grounding conductor (EGC) for grounding non–current-carrying parts of equipment. Examples of installations meeting this condition are a parking lot light pole or a detached garage supplied by only one branch circuit.

Section 250.32(A) provides the general requirement for a grounding electrode system and grounding electrode conductor(s), in accordance with Part III of Article 250. This means that the same grounding electrode requirements in 250.50 must be applied to separate buildings or structures supplied by feeders or branch circuits. Use the same approach as if the building or structure were supplied by a utility service. If any of the grounding electrodes in 250.52(A) are present at the building or structure served, they must be bonded together to form a grounding electrode system for the separate building or structure. This includes the water pipe electrodes, metal in-ground support structures (metal building frame

electrodes), concrete-encased electrodes, and so forth, as provided in 250.52(A). If none of these electrodes are present to form a grounding electrode system, an electrode must be installed. **See Figure 11-2.**

The exception to the grounding electrode requirement in 250.32 applies to a building or structure that is supplied by a single branch circuit, either an individual or multiwire circuit, that includes an EGC for grounding electrical equipment associated with it. An example of a building supplied this way is a detached garage at a dwelling occupancy, while an example of a structure supplied this way would be a detached carport without doors, walls, or windows at a dwelling occupancy. **See Figure 11-3.**

The terms *multiwire branch circuit* and *structure* are defined in Article 100 and clarify the meaning of the rules in which these terms appear.

GROUNDING ELECTRODE CONDUCTOR

Section 250.32(E) provides requirements for the grounding electrode conductor (GEC) installation at separate buildings or structures supplied by feeders or branch circuits. Part III of Article 250 provides the requirements for grounding electrode conductor installation and sizing. The grounding electrode conductor is generally sized according to Table 250.66, which bases sizing on the largest ungrounded circuit conductor supplying the building or structure. For example, if a building is supplied by a 200-ampere feeder using four 4/0 AWG copper conductors (three phases and a neutral), plus a 6 AWG copper EGC, the minimum size required for the grounding electrode conductor is 2 AWG copper or 1/0 AWG aluminum. **See Figure 11-4.**

If the grounding electrode is a connection to only a rod, pipe, or plate electrode or any combination thereof, per 250.66(A), the maximum size GEC required is a 6 AWG copper or a 4 AWG aluminum conductor. If the grounding electrode conductor is a connection to only a concrete-encased electrode(s), per 250.66(B), the maximum size GEC required is a 4 AWG copper conductor. Section 250.66(C) indicates that if a ring electrode is installed and the grounding electrode conductor is connected only to the ring, it is not required to be larger than the conductor used for the ring. The minimum ground ring conductor size is 2 AWG copper.

The grounding electrode conductors must be installed in accordance with the applicable requirements in 250.64. This includes installing them without a splice or joint, as a general rule, or meeting the requirements for splicing grounding electrode conductors if they are spliced or joined together. Protecting grounding electrode conductors against physical damage is required in accordance with 250.64 as applicable, and connections to grounding electrodes must meet the requirements in 250.70. The grounding electrode conductor requirements for separate building or structure installations supplied by feeders or branch circuits are the same as if the building or structure were supplied by a utility service. Grounding is done the same way,

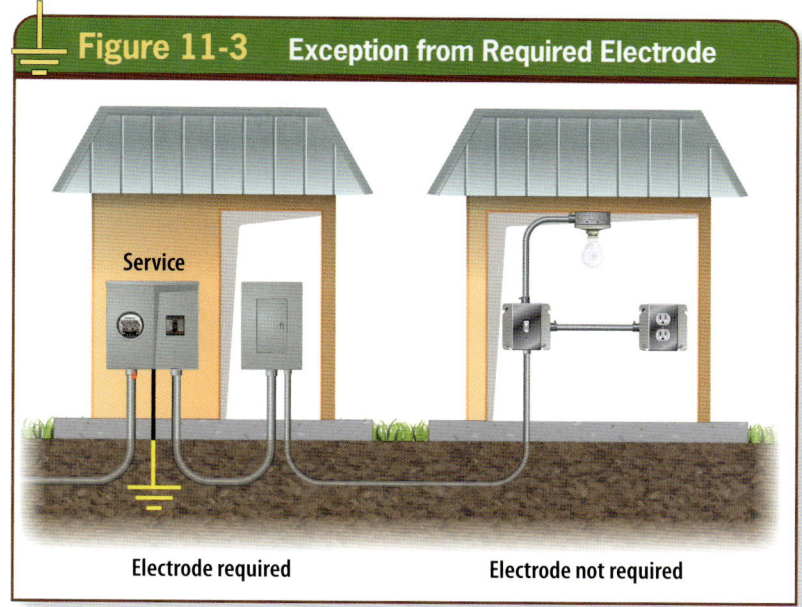

Figure 11-3. A grounding electrode is not required for separate structures supplied by a single branch circuit that includes an EGC. A multiwire branch circuit is also permitted.

except that for new installations, the grounding electrode conductor is connected to the EGC, not the grounded conductor.

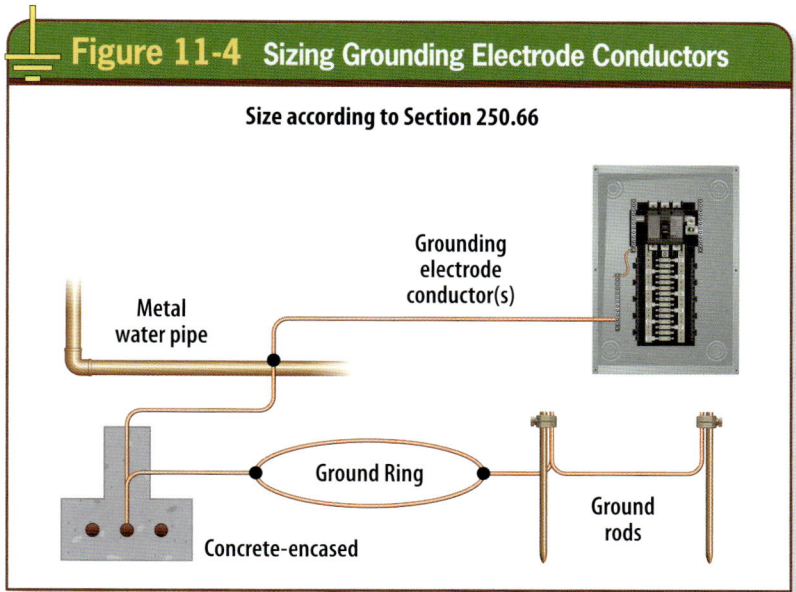

Figure 11-4. A grounding electrode conductor must be installed according to 250.64 and sized according to 250.66. Ground clamps for more than one GEC must be identified for the use.

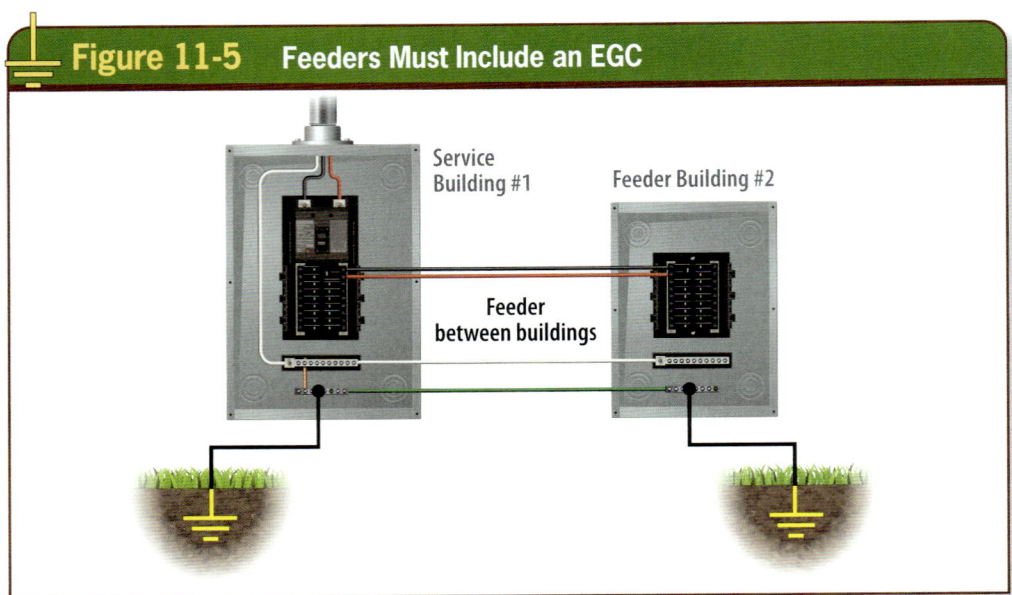

Figure 11-5. *An EGC is generally required with a feeder or branch circuit supplying a separate building or structure.*

FEEDER AND BRANCH CIRCUIT REQUIREMENTS

The *NEC* has a general requirement that feeders include an EGC, as indicated in 215.6. **See Figure 11-5.**

If feeders or branch circuits are installed to supply a separate building or structure, an EGC must also be installed. The EGC can be a wire type, or it can be any wiring method in 250.118 that qualifies as an EGC. If the EGC is a wire type, it must be sized in accordance with 250.122. **See Figure 11-6.**

If a feeder to a separate building or structure is installed using rigid metal conduit (RMC) or intermediate metal

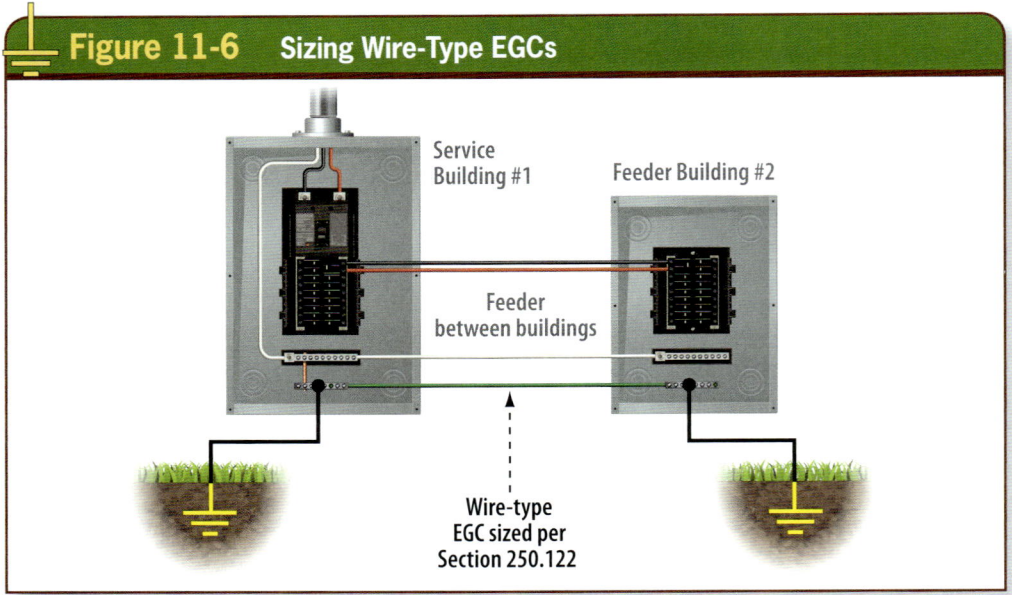

Figure 11-6. *Wire-type EGCs must be sized according to 250.122.*

conduit (IMC), the conduit can serve as the required EGC. Note that any of the metal raceways provided in 250.118 that qualify as EGCs can be used as EGCs. The typical metal raceways installed for feeders supplying separate buildings or structures are rigid metal conduit and intermediate metal conduit.

Corrosion or deterioration and their effect on the life expectancy of wiring methods in contact with the Earth can be an issue in some areas. Corrosion and deterioration protection rules for wiring methods are provided in 300.6. Good design practices include requirements for installing a wire-type EGC to ensure that the safety system is not compromised in the short or long term. Note that the *NEC* falls short of addressing specific life expectancy of grounding electrodes or other materials in contact with the Earth.

In addition, voltage drop considerations must be applied to designs for feeders and branch circuits supplying separate buildings or structures. The EGC must provide an effective ground-fault current path. Section 250.32(B)(1)

restricts a grounded conductor from being connected to a grounding electrode or the EGC for the separate building or structure. **See Figure 11-7.**

There is an exception that allows the grounded (usually the neutral) conductors of feeders or branch circuits to be used for grounding at separate buildings or structures. This exception applies only to existing premises wiring systems and is not applicable to new installations. The conditions of the exception are as follows:

1. An EGC is not included with the supply circuit to the separate building or structure.
2. At the supply and at the served building or structure, there are no common electrically-continuous metallic paths bonded to the grounding systems.
3. Ground-fault protection of equipment is not provided on the supply side of the feeder.

For existing premises wiring systems that use a grounded conductor in this manner, the grounded conductor must not be smaller than the sizes required in 250.122 or 220.61, as applicable.

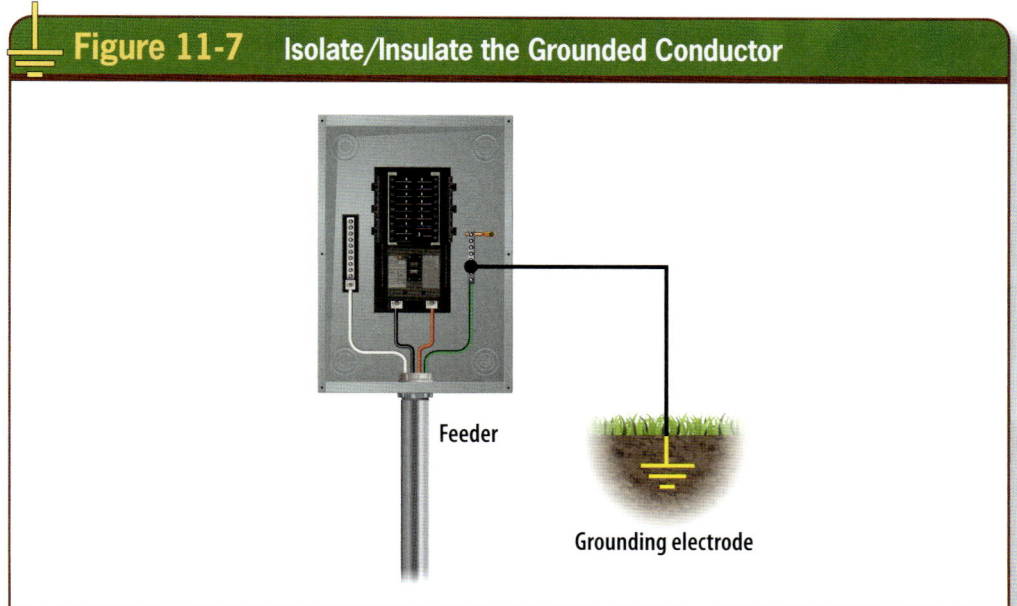

| Figure 11-7 | Isolate/Insulate the Grounded Conductor |

Feeder

Grounding electrode

Figure 11-7. The grounded (usually the neutral) conductor must be isolated from grounded equipment and the EGC [in accordance with 250.32(B)(1)]. Then the EGC and the GEC must be connected to the disconnecting means enclosure.

UNGROUNDED SYSTEMS SUPPLYING SEPARATE BUILDINGS OR STRUCTURES

If an ungrounded source supplies separate buildings or structures by means of feeders or branch circuits, the requirements in 250.32(C) apply. For an ungrounded system, an EGC meeting the requirements in 250.118 must be installed with the supply conductors and must be connected to the building or structure disconnecting means and to the grounding electrode system. The grounding electrode at the separate building or structure served must be connected to the building or structure disconnecting means. **See Figure 11-8.**

SUPPLIED BY A SEPARATELY DERIVED SYSTEM

Section 250.32(B)(2) addresses buildings or structures supplied by separately derived systems, such as generators or transformers installed outside a building or structure. Separately derived systems installed outdoors are required to have a grounding electrode connection at the location of the system. This electrode requirement is for the separately derived system grounding. The separately derived system is not considered a structure; rather, it is equipment anchored to a structure. The electrode is required by 250.30(C), not 250.32(A).

If the disconnecting means is remote from those buildings or structures that are supplied by feeders or branch circuits, the rules in 250.32(D) must be applied.

Systems with Overcurrent Protection

Sometimes overcurrent protection is provided on the secondary side of a separately derived system, as may be the case for unit substation equipment supplying a building or structure. There are also pad-mounted transformers that include overcurrent protection on the secondary side. If overcurrent protection is provided where the conductors originate, the installation of a feeder or branch circuit to a separate building or structure must include an EGC in accordance with 250.32(B)(1). **See Figure 11-9.**

Systems without Overcurrent Protection

Often, separately derived systems, specifically transformer types, do not provide overcurrent protection on the secondary side at the transformer. Section 240.21(C)(4) permits outside feeder conductors from a transformer secondary to supply a building or structure with no secondary (tap) conductor length limitation. However, once the

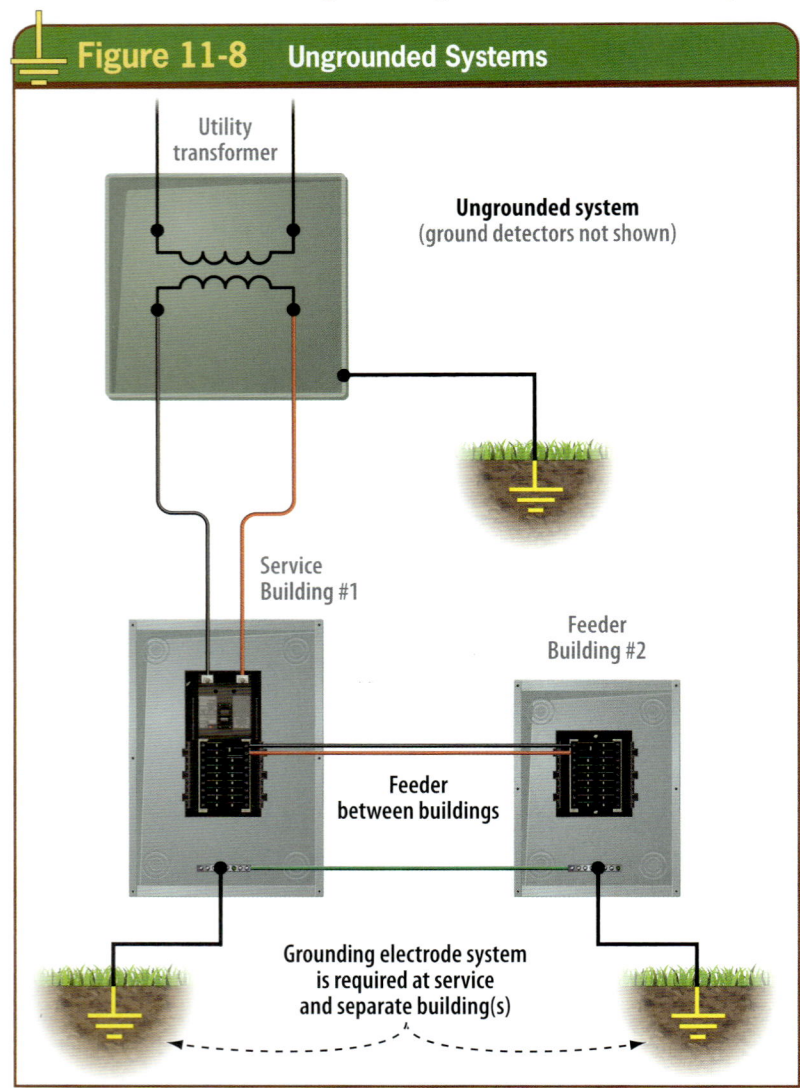

Figure 11-8 Ungrounded Systems

Utility transformer

Ungrounded system
(ground detectors not shown)

Service Building #1

Feeder Building #2

Feeder between buildings

Grounding electrode system is required at service and separate building(s)

Figure 11-8. Grounding electrode systems are required at buildings or structures supplied by services and at buildings supplied by feeders or branch circuits from an ungrounded system.

feeder conductors arrive at the building, the requirements for overcurrent protection apply.

The feeder or branch circuit supplying the separate building or structure is outside, so the rules in Article 225 also apply. The disconnecting means is required by 225.31(A) and must be located in accordance with 225.31(B), either outside or inside the building, nearest to the point of entrance of the feeder conductors. Section 240.21(C)(4)(3) requires the disconnecting means to contain overcurrent protection for the conductors; alternatively, the overcurrent protection must be located immediately adjacent to the disconnecting means.

Note that in accordance with 250.32(A), a grounding electrode system is also required at the building or structure served in this case. If overcurrent protection is not provided where the conductors originate at the separately derived system source, the installation must comply with 250.30(A). If installed with the feeder, the supply-side bonding jumper shall be connected to the building or structure disconnecting means and to the grounding electrode system. This means that the system

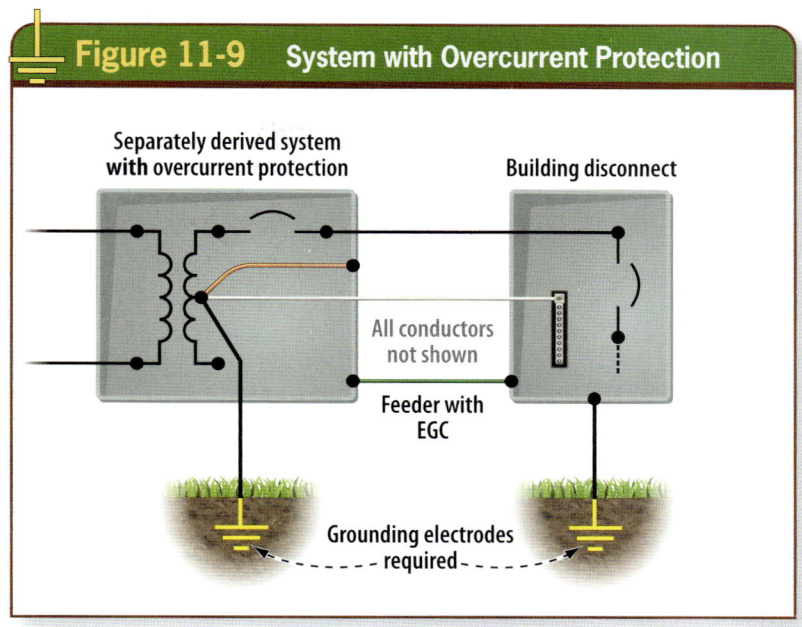

Figure 11-9   **System with Overcurrent Protection**

Separately derived system with overcurrent protection

Building disconnect

All conductors not shown

Feeder with EGC

Grounding electrodes required

Figure 11-9. Where the separately derived system has overcurrent protection at the source, the feeder must include an EGC installed to the separate building or structure disconnecting means.

bonding jumper will be connected in the source enclosure, and a supply-side bonding jumper must be installed from the source enclosure to the disconnecting means. **See Figure 11-10.**

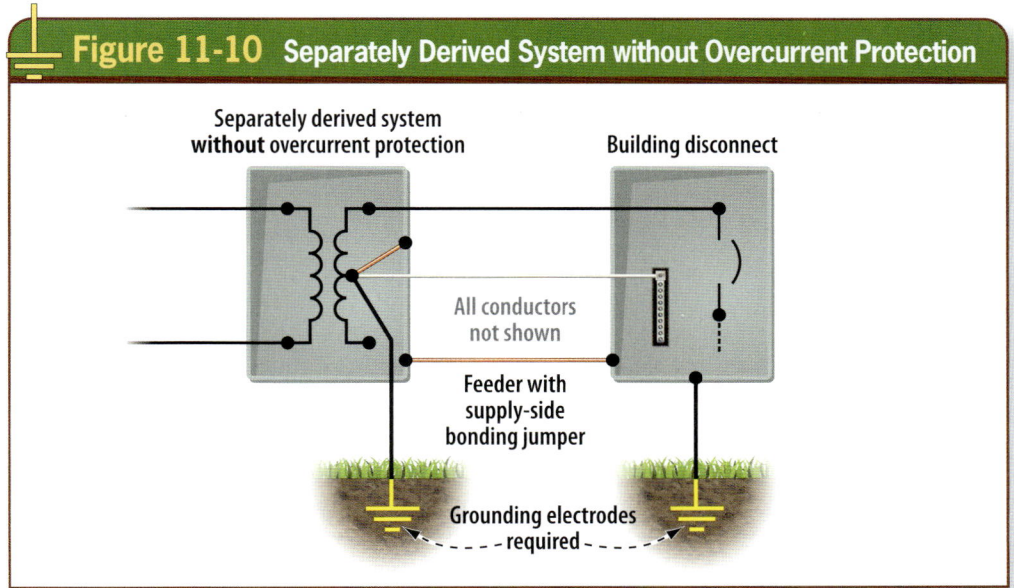

Figure 11-10   **Separately Derived System without Overcurrent Protection**

Separately derived system **without** overcurrent protection

Building disconnect

All conductors not shown

Feeder with supply-side bonding jumper

Grounding electrodes required

Figure 11-10. Where the separately derived system has no overcurrent protection at the source, the feeder must include a supply-side bonding jumper installed to the separate building or structure disconnecting means.

Figure 11-11 Manufacturer Labels in Equipment

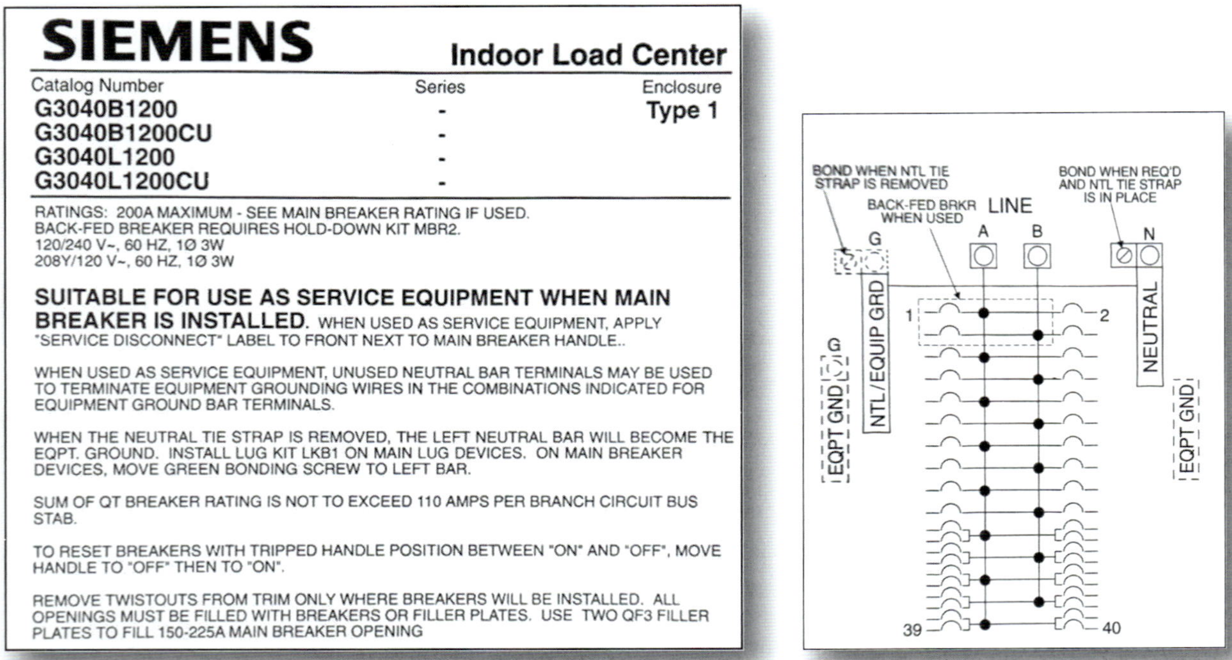

SIEMENS
Indoor Load Center
Catalog Number
Series
Enclosure
G3040B1200 -
Type 1
G3040B1200CU -
G3040L1200 -
G3040L1200CU -
RATINGS: 200A MAXIMUM - SEE MAIN BREAKER RATING IF USED.
BACK-FED BREAKER REQUIRES HOLD-DOWN KIT MBR2.
120/240 V~, 60 HZ, 1Ø 3W
208Y/120 V~, 60 HZ, 1Ø 3W
SUITABLE FOR USE AS SERVICE EQUIPMENT WHEN MAIN BREAKER IS INSTALLED. WHEN USED AS SERVICE EQUIPMENT, APPLY "SERVICE DISCONNECT" LABEL TO FRONT NEXT TO MAIN BREAKER HANDLE..
WHEN USED AS SERVICE EQUIPMENT, UNUSED NEUTRAL BAR TERMINALS MAY BE USED TO TERMINATE EQUIPMENT GROUNDING WIRES IN THE COMBINATIONS INDICATED FOR EQUIPMENT GROUND BAR TERMINALS.
WHEN THE NEUTRAL TIE STRAP IS REMOVED, THE LEFT NEUTRAL BAR WILL BECOME THE EQPT. GROUND. INSTALL LUG KIT LKB1 ON MAIN LUG DEVICES. ON MAIN BREAKER DEVICES, MOVE GREEN BONDING SCREW TO LEFT BAR.
SUM OF QT BREAKER RATING IS NOT TO EXCEED 110 AMPS PER BRANCH CIRCUIT BUS STAB.
TO RESET BREAKERS WITH TRIPPED HANDLE POSITION BETWEEN "ON" AND "OFF", MOVE HANDLE TO "OFF" THEN TO "ON".
REMOVE TWISTOUTS FROM TRIM ONLY WHERE BREAKERS WILL BE INSTALLED. ALL OPENINGS MUST BE FILLED WITH BREAKERS OR FILLER PLATES. USE TWO QF3 FILLER PLATES TO FILL 150-225A MAIN BREAKER OPENING
BOND WHEN NTL TIE STRAP IS REMOVED
BACK-FED BRKR WHEN USED
LINE
BOND WHEN REQ'D AND NTL TIE STRAP IS IN PLACE
G
A
B
N
1
2
NTL/EQUIP GRD
NEUTRAL
EQPT GND
G
EQPT GND
39
40

Figure 11-11. Some panelboards are suitable for use as service equipment and must be so identified. [Reproduction of Siemens panelboard label]

Figure 11-12 Disconnecting Means

Brand X Panelboard
Short Circuit Current Rating 10,000 Amps
SCCR equal to the lowest rated OCPD installed within this panelboard
Suitable for use as service equipment
50QPX2-003
Required grounding electrode

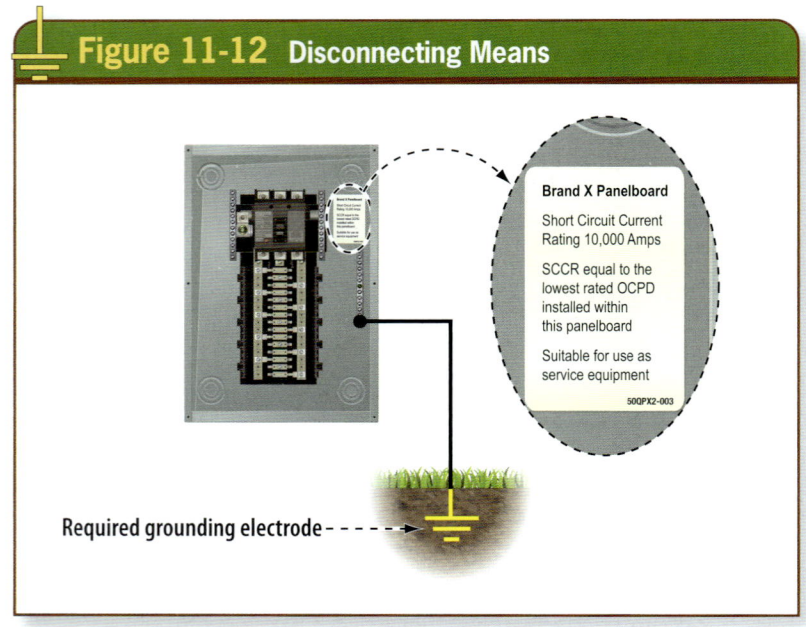

Figure 11-12. A main circuit breaker in a panelboard can be used as the building disconnecting means (see 225.36).

If installed, a grounded (usually a neutral) conductor must be kept separate and isolated from ground and grounded parts at the building or structure disconnecting means, as per 250.30(A).

BUILDING DISCONNECTING MEANS REQUIREMENTS

The disconnecting means installed for a feeder or branch circuit supplying the separate building or structure must be a circuit breaker, molded case switch, general-use switch, snap switch, or other approved means, as stated in 225.36. **See Figure 11-11.**

Where a disconnecting means is installed in accordance with 250.32(B)(1), Exception No.1, it is required to be suitable for use as service equipment. **See Figure 11-12.**

METAL WATER PIPE BONDING AND OTHER BONDING

Bonding metal water piping systems at separate buildings or structures is required, especially for installations where the piping system is not acceptable as a grounding electrode and is not already part of the grounding electrode system. The metal water pipe bonding at separate buildings or structures is treated as if the separate building were supplied by a utility service and not a feeder or branch circuit. Section 250.104(A)(3) requires that metal water piping systems installed in or attached to a building or structure supplied by feeders or branch circuits must be bonded. This includes all exterior metal water piping systems for the building or structure served. The bonding jumper can be attached to the building or structure disconnecting means enclosure, to the EGC run with the supply conductors to the building, or to one or more of the grounding electrodes used at the separate building or structure. **See Figure 11-13.**

The bonding jumper is sized as required by 250.102(D), which requires the bonding jumper to be sized in accordance with 250.122 based on the feeder or branch circuit overcurrent protective device serving the building or structure. The bonding jumper is not required to be larger than the largest ungrounded feeder or branch-circuit conductor supplying the building or structure. For example, if the building were supplied with a 400-ampere feeder, the minimum-size bonding jumper for the water piping system is 3 AWG copper or 1 AWG aluminum.

DISCONNECTING MEANS REMOTE FROM BUILDING OR STRUCTURE

Feeders or branch circuits that supply a separate building or structure are required to have a disconnecting means installed either outside or inside the building or

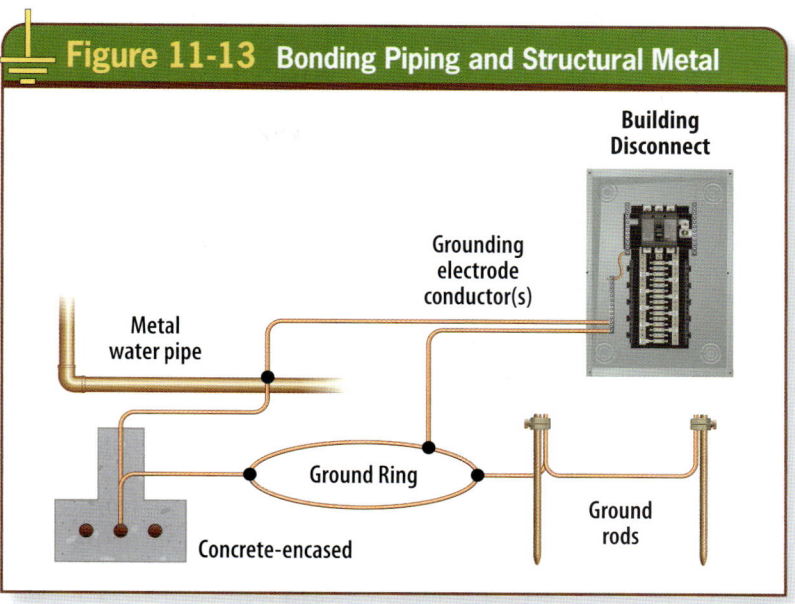

Figure 11-13 Bonding Piping and Structural Metal

Figure 11-13. Bonding of piping systems and building steel is required at separate buildings or structures if they do not qualify as grounding electrodes. Ground clamps for more than one GEC must be identified for the use.

structure nearest to the point of entrance of the supply conductors (see 225.31(B)).

There are four exceptions to this general rule that permit the building or structure disconnecting means to be located elsewhere on the premises. Aside from poles or towers with restrictive conditions, any installation that has safe switching procedures, has qualified persons to service the installation, and is under single management can use the exceptions. See the Exceptions to 225.31(B) in the *NEC* for the full set of qualifying conditions.

There are also other circumstances that permit a feeder or branch circuit disconnect to be remote from a building or structure, as is the case for generators equipped with a disconnecting means and used with emergency systems or legally-required standby systems. The conditions that must be met to be exempted from the building disconnect requirements are as follows:

1. The generator disconnect is readily accessible.
2. The disconnecting means is located within sight from the building it serves.

If these conditions are met, the feeder disconnecting means generally required at the building or structure can be eliminated.

The following conditions and installation requirements must be met when the required building disconnects are located remotely. These are in accordance with 250.32(D); 225.31(B) Exceptions 1 and 2; or, for generators, 700.12(D)(4), 701.12(D)(3), or 702.12.

1. There is no connection made between the grounded conductor of the feeder and the grounding electrode and EGC at the separate building or structure.

2. An EGC is installed to the separate building or structure in accordance with 250.32(B)(1) and is connected to a grounding electrode system at the building or structure, except when an electrode is not required, as provided in 250.32(A) Exception.

3. There are provisions for terminating the grounding electrode conductor and EGC in a box, panelboard, or other enclosure located immediately on the outside or inside of the building or structure. A grounding terminal bar inside the enclosure is a common method for meeting this requirement. **See Figure 11-14.**

BUILDINGS OR STRUCTURES SUPPLIED BY SEPARATELY DERIVED SYSTEMS

Many installations necessitate that buildings be supplied by separately derived systems. The conductors supplied by a separately derived system could be feeders or branch-circuit conductors. Most often, the conductors connected to the secondary of a transformer supply more than a single circuit, so, by definition, they are feeder conductors. Many campus-style electrical systems include a separately derived system supplying separate buildings or structures.

Section 250.32(B)(2) provides specific direction for buildings or structures that are supplied directly from a separately derived system. The first items that should be reviewed are the requirements for grounding electrodes. Each separate building or structure served must have a grounding electrode system. If the separately derived system, such as a transformer, is installed outside the building or structure served, a grounding electrode(s) must be installed and connected at the source location per the requirements in 250.30(C).

Sections 250.32(B)(2)(a) and (B)(2)(b) provide two different installation requirements for separately derived systems supplying separate buildings. The first condition addresses separately derived systems that have overcurrent protection at the source location, such as a unit substation that includes secondary overcurrent protective devices within an assembly enclosure. In these circumstances, a feeder is installed from the load side of an overcurrent protective device at the separately derived system to the building or structure served. The feeder must include an EGC of any type identified in 250.118. If the EGC is a wire type, then the size cannot be smaller than required by 250.122. **See Figure 11-15.**

Figure 11-14   Terminating the GEC and EGC

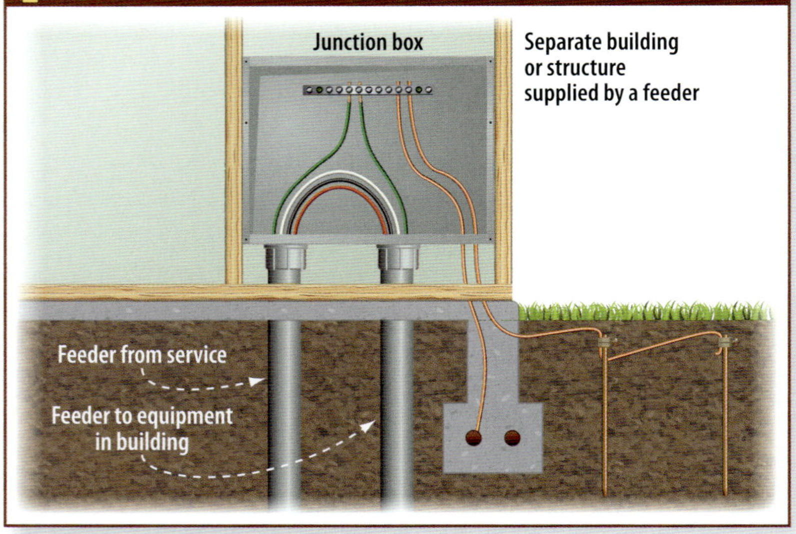

Junction box

Separate building or structure supplied by a feeder

Feeder from service

Feeder to equipment in building

Figure 11-14. A grounding electrode conductor connection to the EGC of building supply circuits is permitted to be made in a junction box. The building disconnect is remote in accordance with 250.32(D).

If there is no overcurrent protective device at the point of origin of the feeder supplied by the separately derived system, then the installation must meet all of the grounding and bonding requirements for separately derived systems set forth in 250.30(A). If the feeder from the separately derived system includes a wire-type supply-side bonding jumper, it must be connected to the building or structure disconnecting means and the required grounding electrode for the building or structure. **See Figure 11-16.**

BUILDINGS OR STRUCTURES SUPPLIED BY AN UNGROUNDED SYSTEM

Although uncommon, when a building or other structure is supplied by an ungrounded separately derived system, the requirements in 250.32(C) must be applied. Remember that an *ungrounded system* is one with no conductor supplied by the system that is intentionally grounded, such as a 3-phase, 3-wire, ungrounded delta system. Section 250.21(B) requires that ground detectors be used for this type of installation. The feeder or branch circuit from an ungrounded separately derived system that supplies a separate building or structure must include either an EGC or a supply-side bonding jumper with the feeder conductors to the building disconnecting means; furthermore, the EGC or supply-side bonding jumper must be connected to the disconnecting means enclosure as well as to the grounding electrode conductor.

The building or structure grounding electrode system must also be connected to the building disconnecting means by a grounding electrode conductor. The size of the grounding electrode conductor is based on requirements in 250.66.

If overcurrent protection is provided where the conductors originate, an EGC in accordance with 250.118 must be installed with the circuit conductors supplying the building or structure. If the EGC is a wire type, it must be sized in accordance with 250.122, which bases

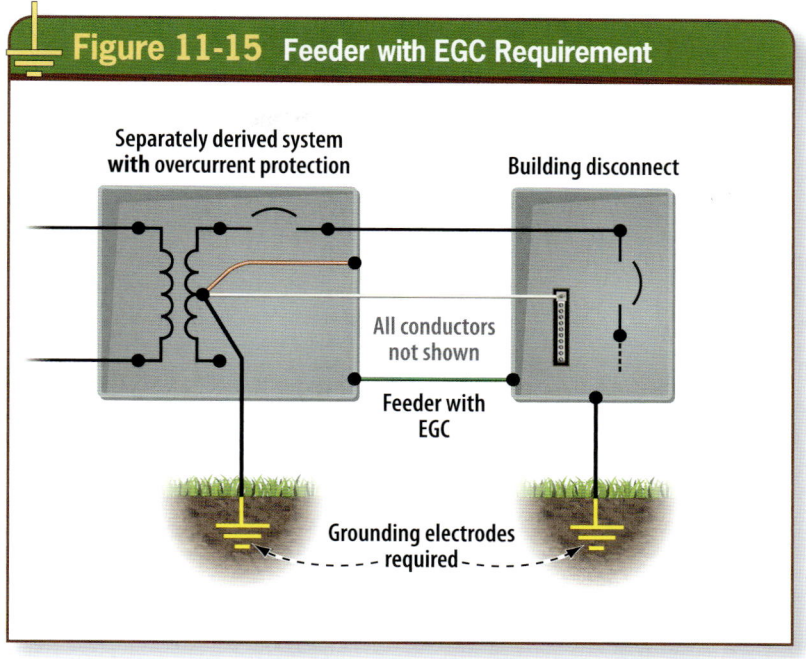

Figure 11-15. Where overcurrent protection is provided at the origin of the feeder from a separately derived system supplying a building or structure, an EGC is required to be installed with the feeder.

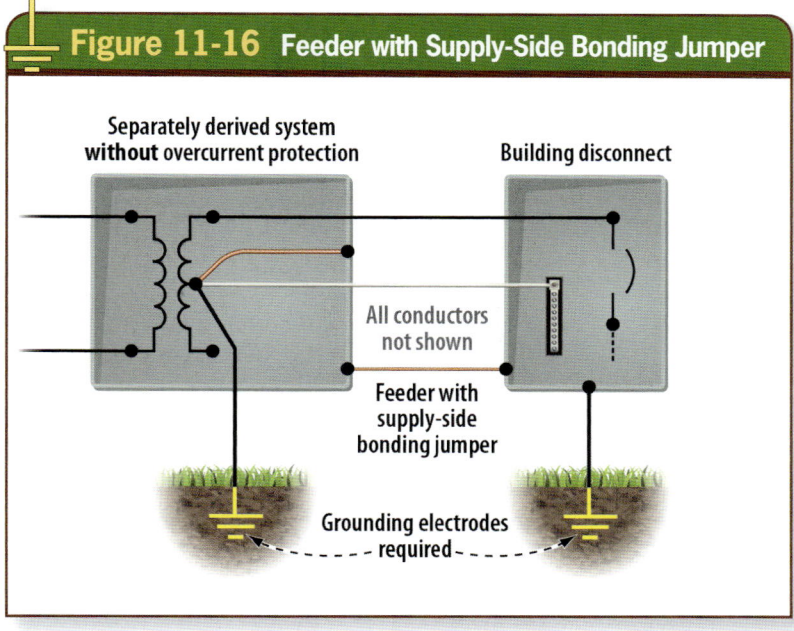

Figure 11-16. Where no overcurrent protection is provided at the origin of the feeder from a separately derived system supplying a building or structure, a supply-side bonding jumper is required to be installed with the feeder.

sizing on the rating of the overcurrent protective device for the feeder or branch circuit supplying the building or structure served. If overcurrent protection is not provided where the conductors originate, the installation must comply with 250.30(B). If installed, the supply-side bonding jumper shall be connected to the building or structure disconnecting means and to the grounding electrode system.

For buildings or structures supplied by ungrounded separately derived systems, a grounding electrode system is required in accordance with 250.32(A). **See Figure 11-17.**

BUILDINGS OR STRUCTURES SUPPLIED BY GENERATORS

Generators are sometimes installed as a part of emergency systems, legally-required standby systems, or optional standby power systems for buildings or structures. **See Figure 11-18.** These installations require feeders from the generators to supply the building or structure; if the generators are located outside the building(s) or structure(s), the requirements in Article 225, specifically 225.31 through 225.39, apply.

Grounding and bonding requirements for permanently-installed generators are found in 250.35. If the transfer equipment for the generator system switches the grounded (neutral) conductor, the generator must be grounded and bonded in accordance with 250.30(A). If the generator system has a grounded conductor and the transfer equipment does not switch the grounded (neutral) conductor, then the grounding and bonding connections for the generator feeder must comply with 250.35(B).

If the overcurrent protective device is located on the generator set, then the feeder must be located on the load side of an overcurrent protective device. In

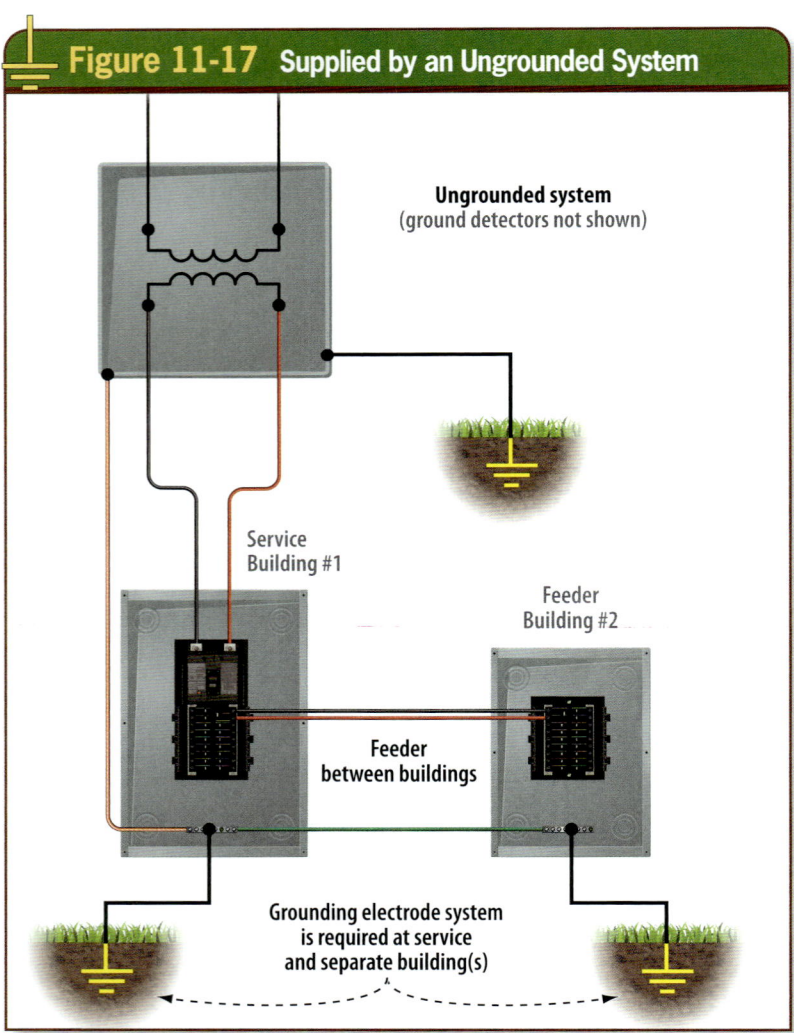

Figure 11-17. Supplied by an Ungrounded System

Ungrounded system
(ground detectors not shown)

Service
Building #1

Feeder
Building #2

Feeder
between buildings

Grounding electrode system
is required at service
and separate building(s)

Figure 11-17. A grounding electrode system is required at buildings supplied by services and buildings supplied by feeders or branch circuits connected to ungrounded systems.

Figure 11-18 Generators

Figure 11-18. Generators often supply emergency or standby systems for separate buildings or structures.

this case, the EGC with the feeder is sized based on the requirements in 250.102(D). This requires sizing based on the rating of the overcurrent protective device at the generator using Table 250.122. **See Figure 11-19.**

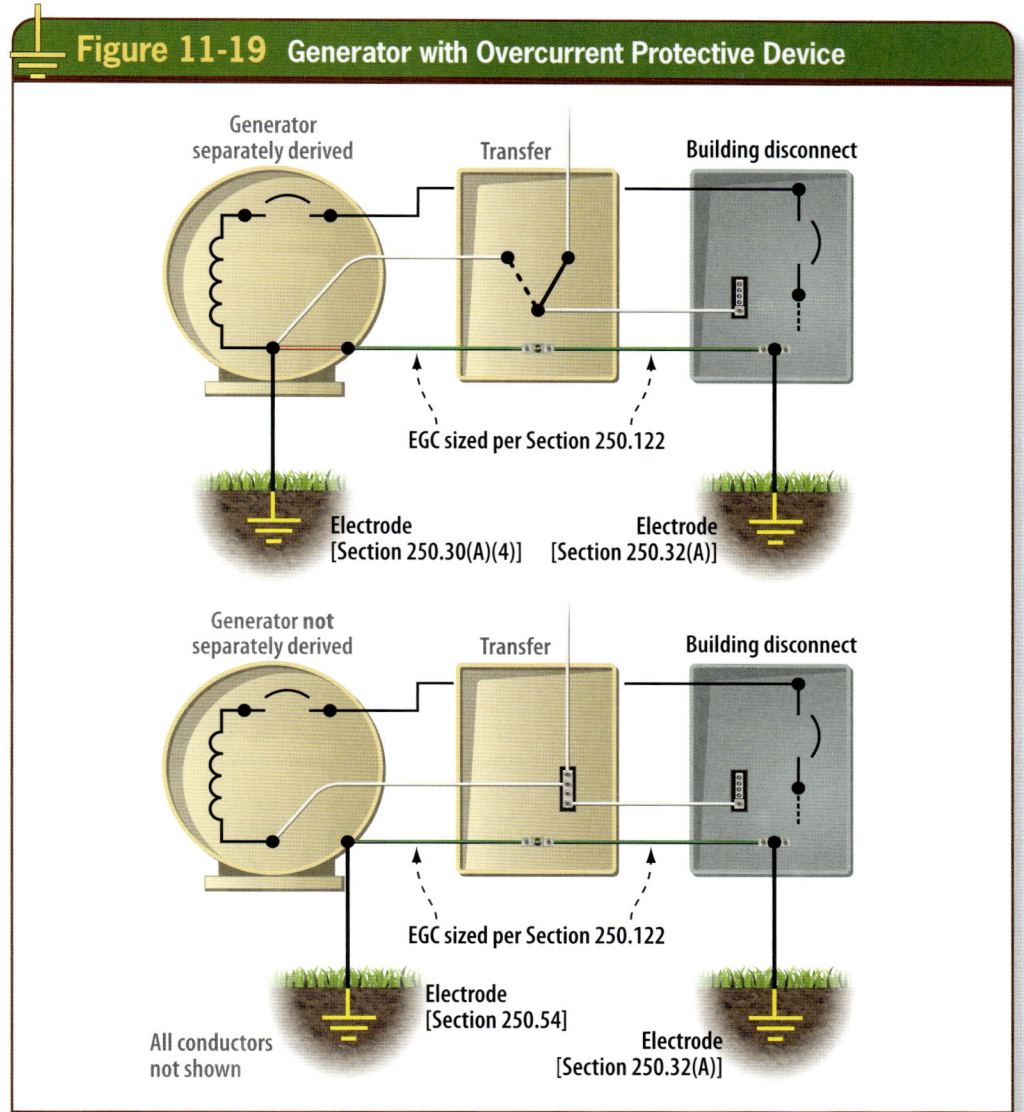

Figure 11-19 **Generator with Overcurrent Protective Device**

Generator separately derived

Transfer

Building disconnect

EGC sized per Section 250.122

Electrode [Section 250.30(A)(4)]

Electrode [Section 250.32(A)]

Generator **not** separately derived

Transfer

Building disconnect

EGC sized per Section 250.122

Electrode [Section 250.54]

All conductors not shown

Electrode [Section 250.32(A)]

Figure 11-19. Where generators with an overcurrent protective device supply a separate building or structure, size the EGC according to 250.122. Note that the ungrounded conductor connections to the transfer switch are not shown.

If the first overcurrent protective device is not located at the generator, a supply-side bonding jumper must be installed with the feeder up to the first system overcurrent protective device enclosure. It must be sized based on the requirements in 250.102(C). The bonding jumper is sized per Table 250.102(C)(1) or by applying 12.5%, which is based on the largest ungrounded phase conductor connected to the generator. **See Figure 11-20.**

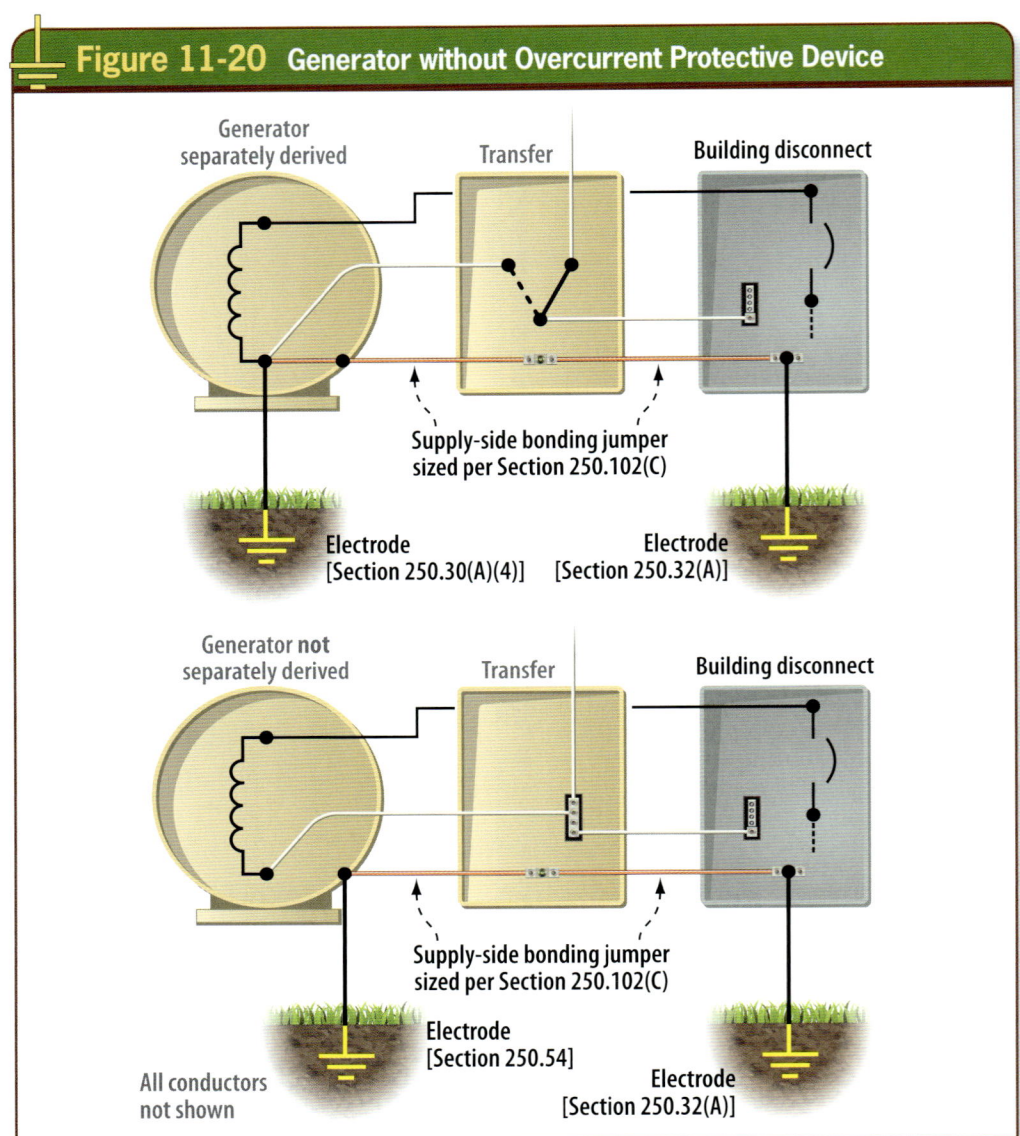

Figure 11-20 **Generator without Overcurrent Protective Device**

Generator separately derived

Transfer

Building disconnect

Supply-side bonding jumper sized per Section 250.102(C)

Electrode [Section 250.30(A)(4)]

Electrode [Section 250.32(A)]

Generator **not** separately derived

Transfer

Building disconnect

Supply-side bonding jumper sized per Section 250.102(C)

Electrode [Section 250.54]

All conductors not shown

Electrode [Section 250.32(A)]

Figure 11-20. Where generators without an overcurrent protective device supply a building or structure, size the supply-side bonding jumper according to 250.102(C)(1) and Table 250.102(C)(1). Note that the ungrounded conductor connections to the transfer switch are not shown.

SUMMARY

Electrical grounding and bonding requirements apply to buildings or structures supplied by feeders or branch circuits. The purpose, performance, and concepts of these grounding and bonding systems are generally the same as if the building or structure were supplied by a utility service.

The requirements for grounding electrodes were reviewed, in addition to the methods for grounding at buildings supplied from grounded systems and ungrounded systems. Alternative restrictive grounding rules for existing buildings or structures were also reviewed, as were grounding rules for buildings that have a disconnecting means remote from the building or structure. Whether a building or structure is supplied by a feeder or branch circuit derived from a grounded system or ungrounded system, grounding and bonding rules apply for those separate buildings or structures.

A structure as defined in the *NEC* is not equipment. The specific grounding and bonding requirements for buildings or structures supplied by generators were also reviewed. In most cases, the grounded conductor must be separated and isolated from ground, the grounding electrode, and the EGC at separate buildings or structures, with the exception of existing premises wiring systems. These rules are not applicable if the building or structure is not supplied with electrical power.

REVIEW QUESTIONS

1. If a building or other structure is supplied by feeders or branch circuits, a grounding electrode system is generally required at the separate building or structure served.
   a. True   b. False

2. A __?__ is a structure that stands alone or that is cut off from adjoining structures by fire walls, with all openings therein protected by approved fire doors.
   a. building
   b. concrete pad
   c. footing
   d. foundation

3. __?__ refers to all circuit conductors between the service equipment, the source of a separately derived system, or another power supply source and the final branch-circuit overcurrent protective device.
   a. Branch circuit
   b. Derived system conductors
   c. Feeder
   d. Service conductors

4. An EGC is required to be installed with a feeder supplying a separate building or structure.
   a. True   b. False

5. Which of the following grounding electrodes, if present, must be used to form a grounding electrode system for a building or structure served by a feeder or branch circuit?
   a. Ground rods
   b. Metal in-ground support structure
   c. Underground metal water pipe
   d. All of the above

6. If polyvinyl chloride (PVC) conduit is installed for a 600-ampere feeder supplying a separate building, a __?__ wire-type copper EGC is required to be installed with the feeder conductors.
   a. 3/0 AWG
   b. 2/0 AWG
   c. 1/0 AWG
   d. 1 AWG

7. According to the *NEC*, a __?__ is that which is built or constructed, other than equipment.
   a. building
   b. footing
   c. foundation
   d. structure

8. The disconnecting means installed in the feeder supplying a separate building always must be suitable for use as service equipment.
   a. True   b. False

9. If no grounding electrodes are present for use at a separate building supplied by a feeder, which of the following applies?
   a. A grounding electrode must be installed.
   b. A grounding electrode is not required.
   c. By exception, a ground ring electrode is required.
   d. The EGC of the feeder can serve as the only grounding means.

10. A grounding electrode is not required for a separate structure that is supplied by only one branch circuit even if the circuit does not include an EGC.,
    a. True   b. False

11. When a single multiwire branch circuit that includes an EGC supplies a separate building or structure and no grounding electrodes are present at the building or structure served, installing a grounding electrode is not required.
    a. True   b. False

12. The grounding electrode conductor installed at a separate building or structure supplied by a feeder must be sized using __?__, which bases sizing on the circular mil area of the largest ungrounded feeder conductor.
    a. Table 8, Chapter 9
    b. Table 250.4
    c. Table 250.66
    d. Table 250.122

13. **A 400-ampere feeder supplies a separate building. If the grounding electrode installed at this separate building is a single ground rod that has less than 25 ohms resistance to ground, the maximum size for the copper grounding electrode conductor required as the sole connection is ___?___.**

    a. 6 AWG

    b. 3 AWG

    c. 2 AWG

    d. 1/0 AWG

14. **Section 250.32(B)(1) generally restricts any installed grounded conductor from being connected to a grounding electrode or the EGC for the separate building or structure.**

    a. True   b. False

15. **The grounded (usually the neutral) conductor of feeders or branch circuits is permitted to be used for grounding at separate buildings or structures only in existing premises wiring systems and in accordance with which of the following conditions?**

    a. An EGC is not included with the supply circuit to the separate building or structure.

    b. Ground-fault protection of equipment is not provided at the service location on the supply side of the feeder.

    c. There are no common electrically-continuous metallic paths between the feeder source and the destination at the building or structure served.

    d. Either a. or b.

16. **Metal water piping systems in separate buildings or structures are required to be bonded to which of the following locations?**

    a. One or more of the grounding electrodes used at the separate building or structure

    b. The building or structure disconnecting means

    c. The EGC run with the feeder conductors to the building

    d. Any of the above

17. **If a generator for standby power supplies a separate building, a separate disconnecting means where the feeder arrives at the building or structure is not required if ___?___.**

    a. the disconnecting means is located within sight from the building it serves

    b. the disconnecting means is suitable for use as service equipment

    c. the generator disconnect is readily accessible

    d. all of the above

18. **When a building is supplied by an ungrounded system, the feeder from the ungrounded system must include an EGC or supply-side bonding jumper with the conductors to the building disconnecting means, and the EGC or supply-side bonding jumper must be connected to the disconnecting means enclosure as well as to the grounding electrode conductor.**

    a. True   b. False

19. **A grounding electrode is required for a lighting pole in a parking lot.**

    a. True   b. False

20. **A building is supplied by a 100-ampere feeder installed to a panelboard located in a separate building. A grounding electrode is required for this installation.**

    a. True   b. False

Grounding Electrical Systems

The *NEC* includes many requirements for electrical system grounding, provisions for systems that are grounded by choice, and restrictions that prohibit certain systems from being grounded. Grounding methods can vary, and include solid grounding, impedance or resistance grounding, grounding through surge arresters, grounding through an inductor, and so forth. Where an electrical system is not grounded, ground detectors are generally required to be installed. Whether a system is grounded by a requirement or by choice, all *NEC* requirements for grounded systems must be applied.

Objectives

» Understand the requirements for grounding electrical systems and differentiate between a system that is grounded and one that is not.

» Determine when a system must be grounded in comparison to when a system is permitted to be grounded by choice.

» Understand when electrical systems are not permitted to be grounded and understand the requirements for ground detectors on ungrounded systems.

» Identify specific installation rules for impedance grounded systems.

» Identify various methods for grounding electrical systems and review common voltages for grounded systems.

Chapter 12

Table of Contents

Definitions 274

System Grounding 274

Methods of System Grounding 276

System Grounding Requirements............... 276

Mandatory System Grounding 277

    Grounded System Voltages.................. 279

    Grounding Using
    Grounding Transformers...................... 280

Optional System Grounding 281

    Grounding Separately Derived Systems ...282

    Requirements for Ground Detection....... 282

    Conductor to be Grounded 283

    Rules for the System Grounded
    Conductor 283

        Sizes 6 AWG or Smaller 284

        Sizes 4 AWG and Larger 284

System Grounding Prohibition 285

Impedance Grounded Systems 286

Ungrounded Systems (Concepts) 287

Summary... 288

Review Questions 288

DEFINITIONS

The *NEC* rules applicable to grounding separately derived systems use several specific terms that must be fully understood for proper application.

Grounding an electrical system means that one system conductor is connected to ground (the Earth) and a reference from the system to ground is established. Once the system is grounded, it will usually have a grounded conductor in addition to one or more ungrounded conductors. Installing and operating an ungrounded system means that no connection or reference to ground (the Earth) from the system conductors is established other than through capacitance. Section 250.30 provides rules for separately derived systems that are grounded and those that are not grounded.

Ground. The earth. (CMP-5)

Ground Fault. An unintentional, electrically conductive connection between an ungrounded conductor of an electrical circuit and the normally non–current-carrying conductors, metal enclosures, metal raceways, metal equipment, or earth. (CMP-5)

Grounded (Grounding). Connected (connecting) to ground or to a conductive body that extends the ground connection. (CMP-5)

Grounded Conductor. A system or circuit conductor that is intentionally grounded. (CMP-5)

Grounded, Solidly. (Solidly Grounded) Connected to ground without inserting any resistor or impedance device. (CMP-5)

Ungrounded. Not connected to ground or to a conductive body that extends the ground connection. (CMP-5)

Voltage to Ground. For grounded circuits, the voltage between the given conductor and that point or conductor of the circuit that is grounded; for ungrounded circuits, the greatest voltage between the given conductor and any other conductor of the circuit. (CMP-1)

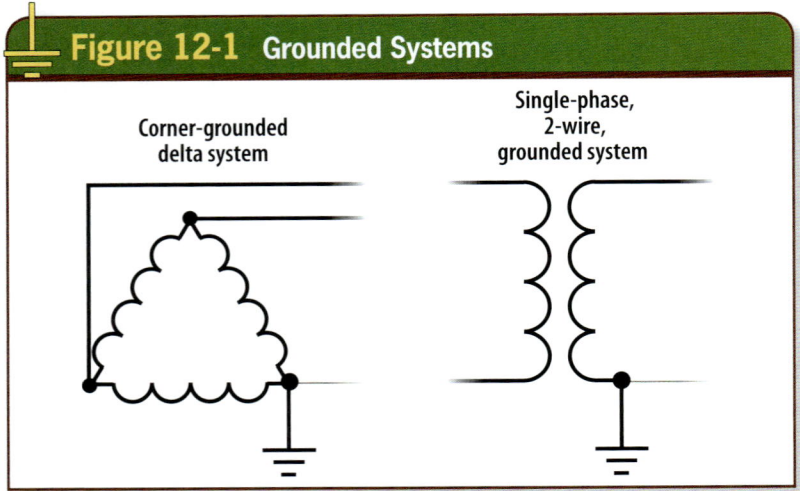

Figure 12-1   Grounded Systems

Corner-grounded delta system

Single-phase, 2-wire, grounded system

Figure 12-1. Grounded systems include a system conductor that is grounded either solidly or through an impedance device.

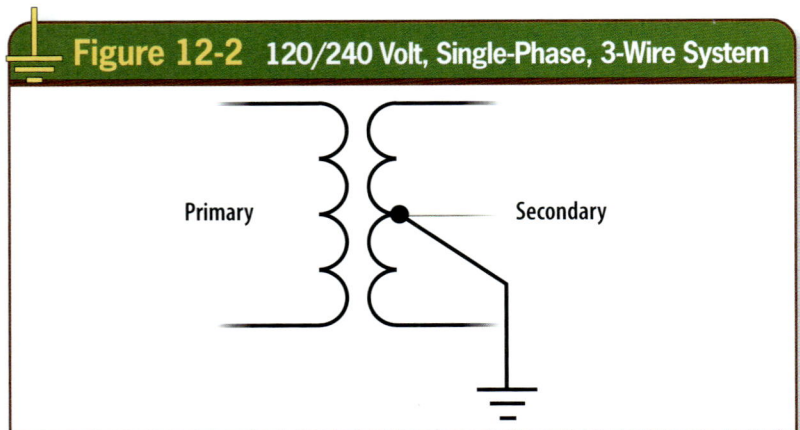

Figure 12-2   120/240 Volt, Single-Phase, 3-Wire System

Primary

Secondary

Figure 12-2. The grounded (neutral) conductor of a 120/240-volt, single-phase, 3-wire system is common to both ungrounded conductors of the system and is connected to the neutral point of the system.

SYSTEM GROUNDING

System grounding is the process of establishing a connection from one conductor of the system to ground (the Earth). Therefore, when a system is grounded, one conductor of the system is connected to ground solidly, whether through an impedance device, resistor, inductor, or some other means. **See Figure 12-1.** Grounded systems usually have a system grounded conductor, although in cases where the system is grounded but there are no line-to-neutral loads, the system grounded conductor is not required.

The system conductors are those supplied from the system, often referred to as the *source of the system*, and include all ungrounded conductors and the conductor that is grounded. For example, in a 120/240-volt, single-phase, 3-wire system, there are three system conductors, two ungrounded conductors, and one grounded (neutral) conductor. **See Figure 12-2.**

In a 208Y/120-volt, 3-phase, 4-wire system, there are four system conductors on the secondary side, three ungrounded phase conductors, and one grounded (neutral) conductor. **See Figure 12-3.**

In a 240/120-volt, 3-phase, 4-wire, delta-connected system, there are four system conductors on the secondary side, three ungrounded phase conductors, and one grounded (neutral) conductor. This type of system is often referred to as a *high-leg delta system.* **See Figure 12-4.**

If an electrical system is not grounded, the system will not have a supply conductor that is connected to the Earth. **See Figure 12-5.**

Webster's New Dictionary defines a *system* as a group of units so combined as to form a whole and to operate in unison. An entire electrical installation that operates in unison is commonly referred to as the *electrical system of the premises.* If a system is grounded, one conductor of the entire system is grounded (usually at the source), creating a grounded system for use on the premises.

Electrical systems on the premises commonly include transformers, batteries, generators, photovoltaic arrays, and/or wind power turbines. The transformer is the most common system installed and used in premises wiring. Often, the utility supply to a commercial or industrial building is at

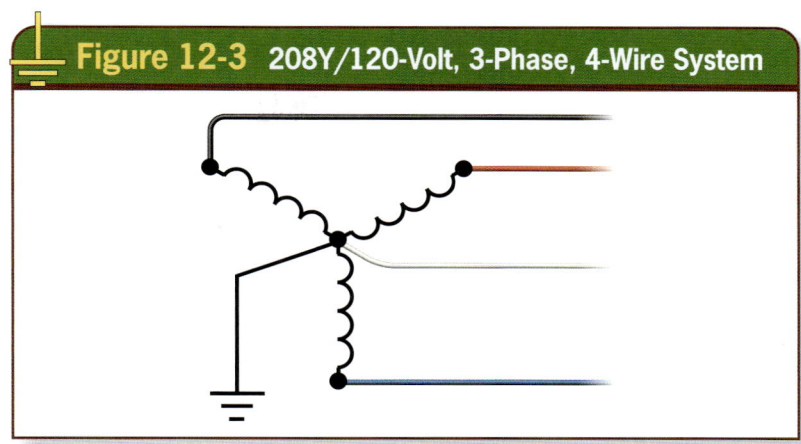

Figure 12-3 208Y/120-Volt, 3-Phase, 4-Wire System

Figure 12-3. The grounded (neutral) conductor of a 208Y/120-volt, 3-phase, 4-wire system is common to all three ungrounded phase conductors of the system and is connected to the neutral point of the system.

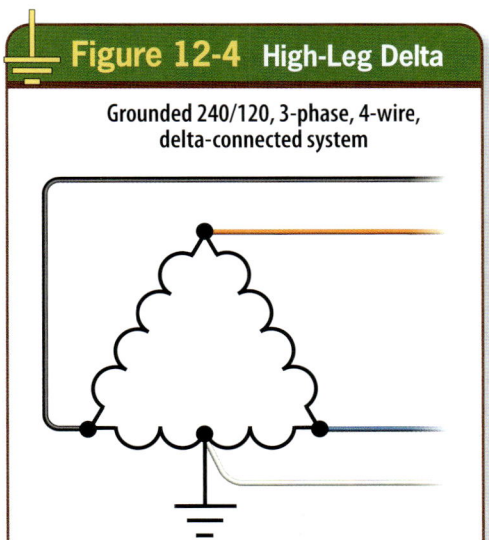

Figure 12-4 High-Leg Delta

Grounded 240/120, 3-phase, 4-wire, delta-connected system

Figure 12-4. This 240/120-volt, 3-phase, 4-wire delta system with a high leg on the B phase has a grounded (neutral) conductor common to both of the ungrounded conductors of one phase winding of the system.

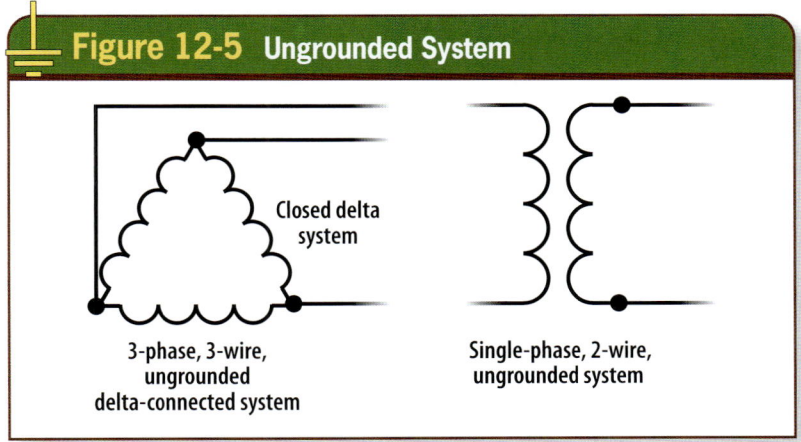

Figure 12-5 Ungrounded System

Closed delta system

3-phase, 3-wire, ungrounded delta-connected system

Single-phase, 2-wire, ungrounded system

Figure 12-5. An ungrounded system has no conductor intentionally connected to ground (Earth).

a voltage and phase configuration such as 480Y/277 volts with three phases and four wires. To create usable voltages for 120-volt loads, transformers are usually installed.

Each transformer that is installed on the premises is a new system (or source). The grounding requirements apply to each of those systems and depend on the voltage produced by the system. In other words, the level of voltage produced by an electrical system is often directly related to the requirements for grounding the system or to whether the system is permitted to operate ungrounded.

METHODS OF SYSTEM GROUNDING

Electrical systems can be grounded in a variety of ways. The grounding method chosen is often based on an *NEC* requirement, an owner specification, or a part of an engineering design, and can vary depending on the application. Common grounding methods include solid grounding (connecting to ground solidly), reactance grounding (grounding through an inductor), resistance grounding (grounding through a low amount of resistance), high-resistance grounding (grounding through the highest permissible resistance), or grounding through surge arresters. **See Figure 12-6.**

Where systems rated 1,000 volts and below are grounded, they are usually solidly grounded or grounded through a high-impedance device. Medium-voltage systems are generally grounded either solidly or through a resistor, and higher-voltage systems are typically grounded through surge arresters.

SYSTEM GROUNDING REQUIREMENTS

From the earliest years of electricity use, there have been many discussions and even heated debates about the benefits of operating systems that are grounded as compared to ungrounded. The *NEC* rules today often make those determinations. Some electrical systems must be grounded, some systems are permitted to be grounded, and some systems are not permitted to be grounded. The *NEC* provisions are broken down for *each* system application. Part II of Article 250 addresses electrical system grounding. The specific mandatory rules for grounding electrical systems are found in 250.20.

Before moving on to the requirements for grounding systems, it is important to review the purpose of grounding an electrical system. Section 250.4(A)(1) describes the purpose of system grounding and what is intended to be accomplished by grounding a

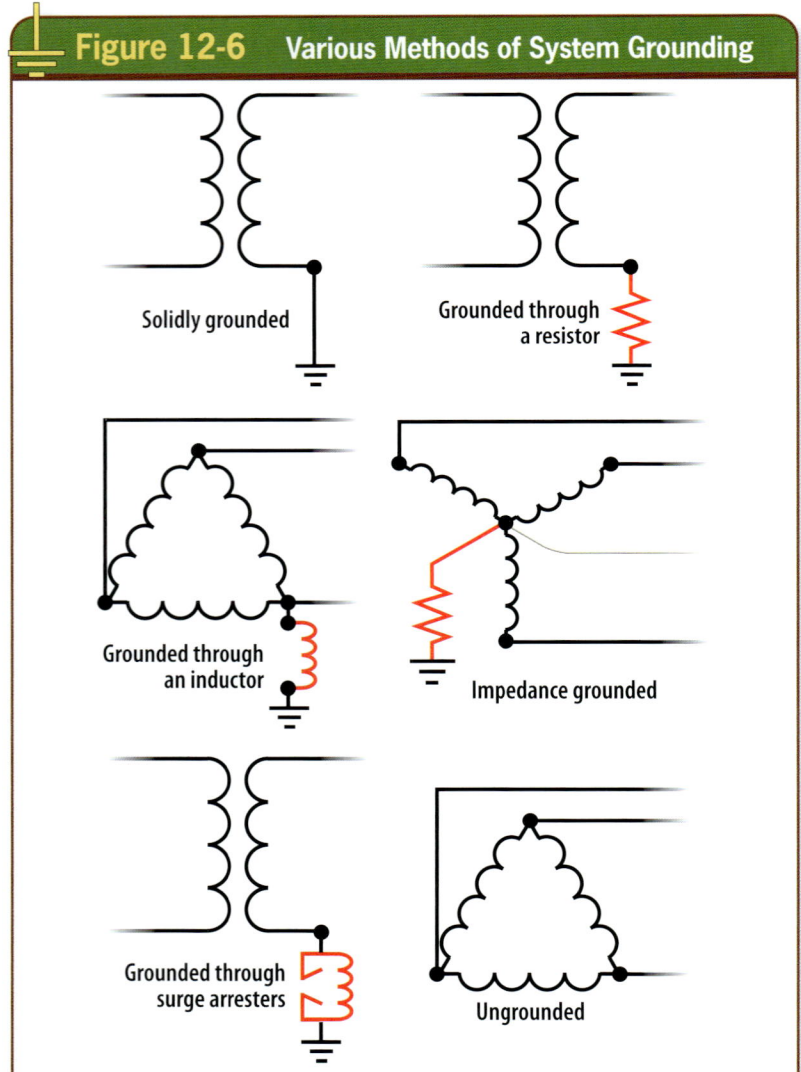

Figure 12-6   Various Methods of System Grounding

Solidly grounded

Grounded through a resistor

Grounded through an inductor

Impedance grounded

Grounded through surge arresters

Ungrounded

Figure 12-6. Systems can be grounded solidly or by other means.

system. Grounded systems are connected to the Earth in a fashion that limits voltage imposed by higher-voltage lines, line surges, lightning events, and so forth. Grounding a system also establishes a reference to the Earth from the system and stabilizes the voltage to ground during normal operation. It should be understood that during abnormal events, such as a line surge or a lightning strike, the system voltage and the voltage on conductive enclosures of the system will attempt to rise for the duration of the event. Grounding helps keep the voltage of the Earth and the connected system and equipment equal during fluctuations caused by the sources listed in 250.4(A)(5).

A ground-fault event attempts to force a rise in voltage on grounded equipment and systems for the duration of the fault condition or until an overcurrent protective device opens the circuit. Grounding helps limit these above-ground voltages during abnormal events such as ground faults, line surges, and lightning events. The effective ground-fault current path provides a means to quickly facilitate overcurrent protective device operation, thus reducing the amount of time the abnormal condition exists.

MANDATORY SYSTEM GROUNDING

Section 250.20 includes the essential text specifying that electrical system grounding is required in accordance with 250.20(A) and (B), depending on the voltage and phase arrangement of the system. If a system is not required to be grounded by the *NEC* but the choice has been made to ground the system anyway, all system grounding rules in the *NEC* apply. An example of a system that is permitted but not required to be grounded is a 480-volt, 3-phase, delta-connected system.

Section 250.20(A) provides the requirements for grounding systems of less than 50 volts. Alternating current (AC) systems of less than 50 volts must

be grounded under any of the following conditions:
1. Where supplied by transformers, if the supply system exceeds 150 volts to ground. For example, if the supply of the transformer is 480 or 277 volts, the low-voltage secondary side must be grounded.
2. Where supplied by transformers, if the supply system is ungrounded. If the supply system for the low-voltage transformer is not grounded (for example, supplied by a 480-volt, ungrounded system), the low-voltage side of the transformer must be grounded.
3. Where installed outside as overhead conductors. For example, a low-voltage system or circuit installed as overhead conductors between buildings on a property should be grounded because the overhead conductors are more vulnerable to lightning.

In these cases, system grounding is necessary for increased safety of the system, conductors, components, and property. **See Figure 12-7.**

Section 250.20(B) addresses the grounding requirements for premises wiring systems of 50 volts to 1,000 volts. Systems in this voltage range

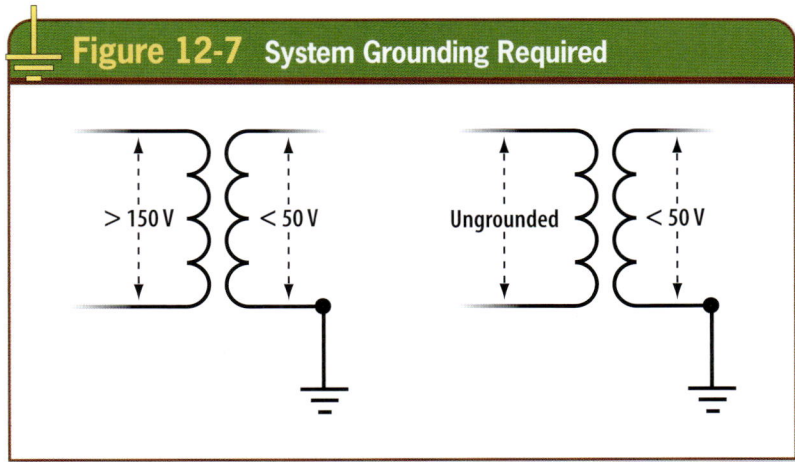

Figure 12-7 System Grounding Required

Figure 12-7. Systems less than 50 volts are required to be grounded in accordance with 250.20(A)(1), (2), or (3). Where conductors of systems less than 50 volts are outside buildings and installed overhead, the systems are required to be grounded.

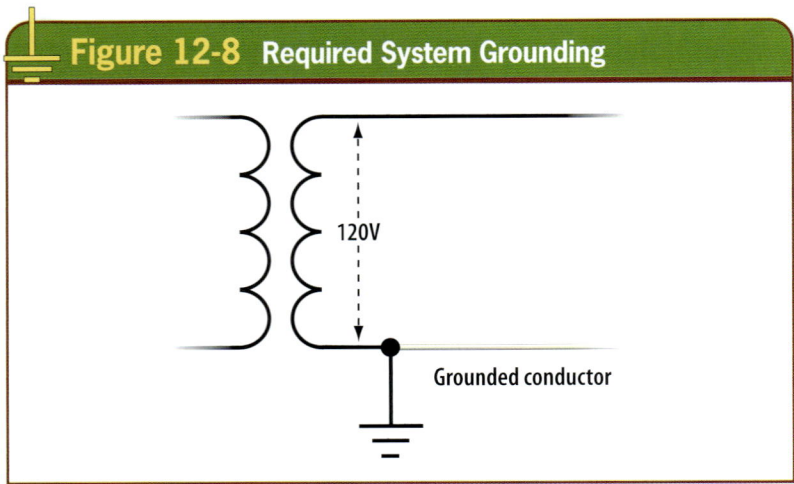

Figure 12-8 **Required System Grounding**

120V

Grounded conductor

Figure 12-8. System grounding is required where the resulting voltage is less than 150 volts to ground from any ungrounded conductors of the system.

must be grounded under any of the following conditions:

1. If the system can be grounded so that the maximum voltage to ground on the ungrounded conductors does not exceed 150 volts
2. If the system is 3-phase, 4-wire, wye-connected and the system neutral conductor is used as a circuit conductor
3. If the system is 3-phase, 4-wire, delta-connected and the midpoint of one phase winding is used as a circuit conductor

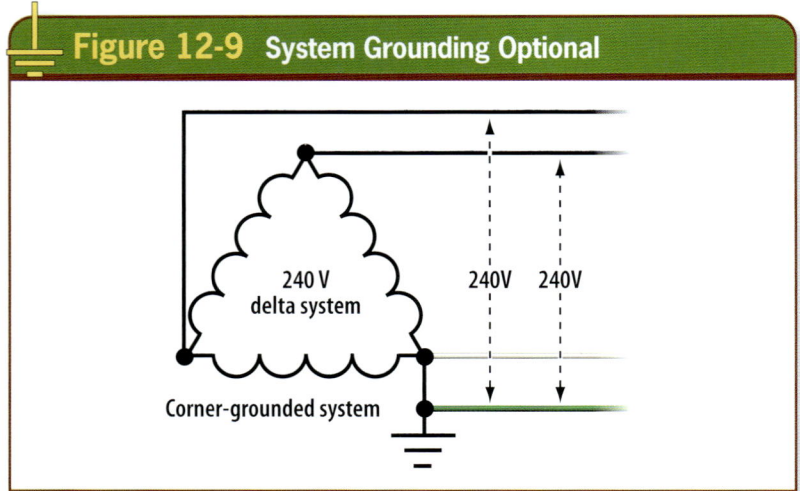

Figure 12-9 **System Grounding Optional**

240 V delta system

240V   240V

Corner-grounded system

Figure 12-9. System grounding is not required where the phase-to-ground voltage is above 150 volts. Examples are 240-volt and 480-volt, 3-phase delta systems.

These grounding requirements apply to many premises wiring systems installed today. In item 1, if the system can be grounded in a way that the phase-to-ground voltage is less than 150 volts, the system must always be grounded. An example of this is a single-phase, 2-wire system with a 120-volt output (secondary). If one conductor or the other is grounded, the phase-to-ground voltage of the system is 120 volts, which is less than 150 volts. **See Figure 12-8.**

Grounding may result in a voltage to ground greater than 150 volts. If a 240-volt, 3-phase, 3-wire, delta-connected transformer is grounded (corner grounded), the voltage to ground is 240 volts, which is greater than 150 volts. Therefore, grounding of this system is optional. **See Figure 12-9.**

Upon reviewing item 2, it becomes clear that this grounding requirement applies to systems such as 480Y/277-volt, 3-phase, 4-wire, wye-connected arrangements if the neutral of such systems is used as a circuit conductor. This is usually the case for these types of systems, so grounding is typically required.

Grounding could be an option for the 480Y/277-volt system because the voltage to ground exceeds 150 volts, but only if the neutral of the system is not used as a circuit conductor, meaning it only supplies phase-to-phase loads. **See Figure 12-10.**

Upon reviewing item 3, it becomes clear that the system addressed in this section is a 3-phase, 4-wire, delta-connected system with a grounded midpoint of one phase winding. In other words, this is a high-leg, delta-connected system. The typical voltages of these systems are 240 volts from phase to phase around the delta and 120 volts from phase to ground on the transformer that has a grounded midpoint. From the B phase of such systems, the phase-to-ground voltage is 208 volts. The high-leg voltage results from taking the square root of three and multiplying it by the phase-to-ground voltage (120 volts). **See Figure 12-11.**

Because the grounded midpoint of the A and C phases creates a 120-volt output, the neutral created is usually used as a circuit conductor to supply 120-volt loads; thus, grounding such systems is required. Section 110.15 requires identifying the high leg of these systems with an orange marking or other effective means. An example of other effective means would be by tagging or using labels acceptable to the authority having jurisdiction. Section 408.3(E) also indicates that the high leg of a system is generally required to be the B phase, except for high-leg connections in metering equipment or multi-section switchboards, switchgear, or panelboards that contain a metering section according to 408.3(E) and (F)(1). Be aware that open delta systems are also used in addition to closed delta systems; however, the kilovolt-ampere (kVA) capacity of open delta systems is approximately 65% of that of closed-delta systems.

Grounded System Voltages

Systems can be solidly grounded in a few ways. How the system is grounded determines the resulting output voltage of the system.

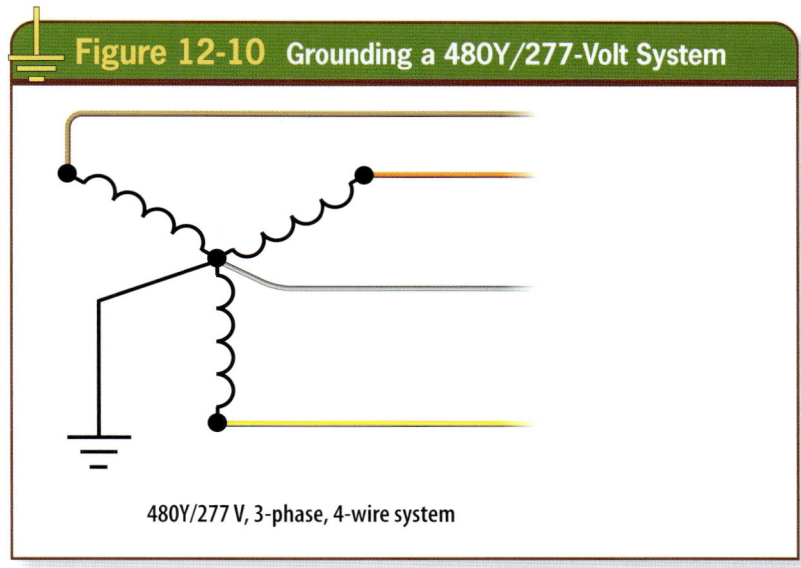

Figure 12-10   Grounding a 480Y/277-Volt System

480Y/277 V, 3-phase, 4-wire system

Figure 12-10. Grounding is optional for a 480Y/277-volt system only where the neutral is not used as a circuit conductor because the phase-to-ground voltage for these systems exceeds 150 volts.

Wye-connected systems are typically grounded at the center or common point of the wye that connects all three of the phase windings. The terminal in a dry-type transformer with a wye-connected secondary is generally identified as the *XO terminal*. This is the center of

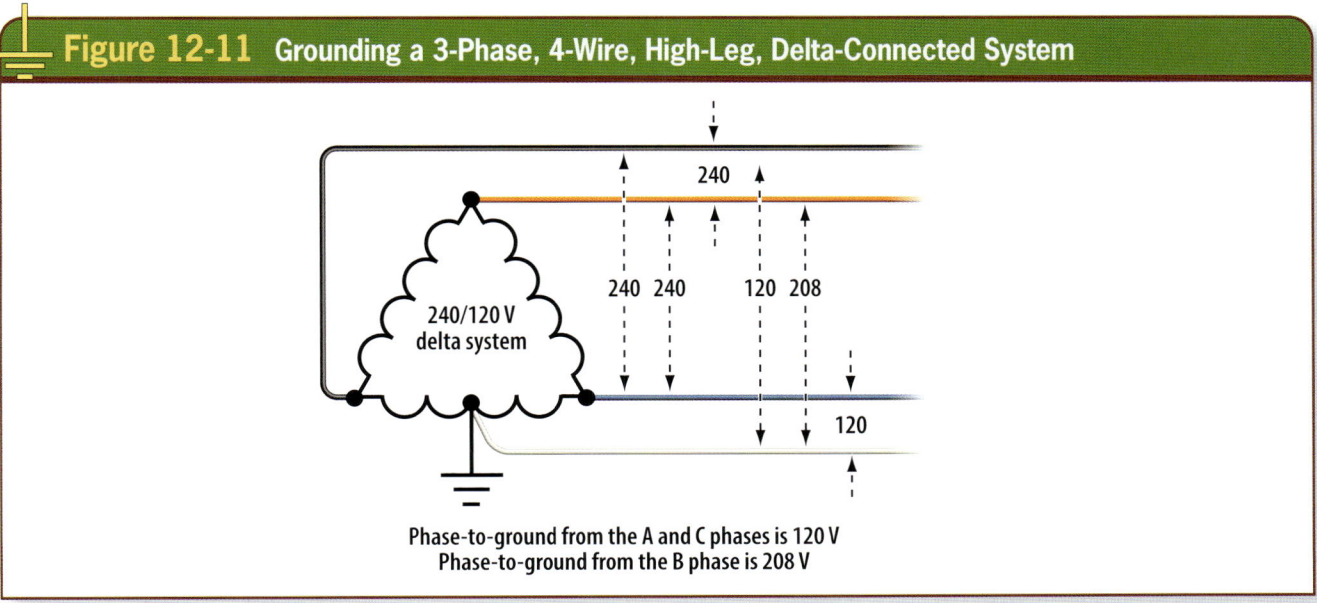

Figure 12-11   Grounding a 3-Phase, 4-Wire, High-Leg, Delta-Connected System

240/120 V delta system

240

240   240   120   208

120

Phase-to-ground from the A and C phases is 120 V
Phase-to-ground from the B phase is 208 V

Figure 12-11. In a 3-phase, 4-wire, high-leg, delta-connected system, a neutral point is established between the ungrounded conductors of one transformer in the closed delta bank. This is the grounded neutral point of the system.

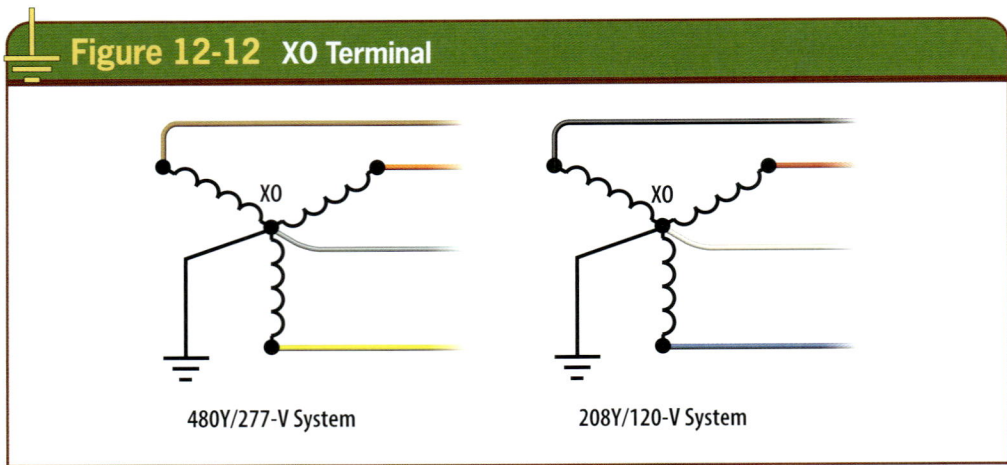

Figure 12-12. *The neutral point of a typical 3-phase, 4-wire, wye-connected system is the point of grounding for such a system and is usually identified as XO.*

the wye and the neutral point of the system. Typical wye-connected systems are 480Y/277-volt and 208Y/120-volt systems. **See Figure 12-12.**

Single-phase, 2-wire systems that are grounded produce a phase-to-ground voltage that is the equivalent of the output voltage of the system. For example, a 120-volt, 2-wire system that is grounded has a voltage of 120 volts between the output conductors and from the ungrounded conductor to ground. **See Figure 12-13.**

Delta-connected systems can be solidly grounded using a few methods. One method of grounding a 3-phase, 3-wire, delta-connected system is to ground one of the phase conductors. This creates a 3-phase, corner-grounded system, sometimes referred to as an end-grounded system. The *NEC* term is *corner-grounded system*. The phase-to-phase voltage of a corner-grounded system is also the phase-to-ground voltage of that system. For example, the phase-to-phase voltage of a 480-volt delta system is 480 volts. If one phase is grounded, the phase-to-ground voltage from the other two phases is 480 volts. **See Figure 12-14.**

Delta-connected systems with the midpoint of one transformer grounded typically are used for 240/120-volt loads, as previously discussed.

Grounding Using Grounding Transformers

Delta systems can also be grounded using zigzag transformers, as covered in 450.5. The term *zigzag* is related to how the transformer is wound around the iron core. This type of transformer is wound in one direction around the core and wound in the opposite direction back around the same core. Zigzag transformers are also commonly referred to as *T-connected transformers*, and they are connected to 3-phase,

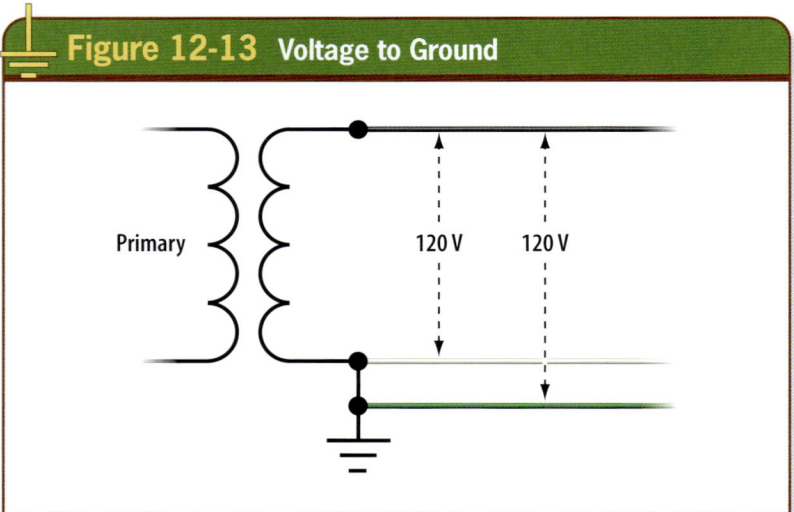

Figure 12-13. *In a single-phase, 2-wire system, the system output voltage (120 volts) is the same as the voltage to ground from the ungrounded conductor of the system.*

3-wire ungrounded systems to create a 3-phase, 4-wire distribution system or to provide a neutral point for grounding purposes. Such transformers must have a continuous per-phase current rating and a continuous neutral current rating. Zigzag-connected transformers are not permitted to be installed on the load side of any system grounding connection. **See Figure 12-15.**

This method of grounding creates a ground reference for delta-connected systems and provides the ability to facilitate overcurrent protective device operation should a ground fault develop on any of the ungrounded conductors of such systems. The phase current in a grounding autotransformer is one-third of the neutral current. These systems should not be used to supply line-to-neutral loads.

Systems with voltages greater than 1,000 volts are covered in Part X of Article 250. Where these systems are mobile or portable, such as on a trailer, they are required to be grounded. If they are permanently installed, they are permitted to be grounded as provided in 250.20(C). If an electrical system is grounded through an impedance device, it must be grounded in accordance with the requirements of either 250.36 or 250.187, depending on the system voltage.

OPTIONAL SYSTEM GROUNDING

Section 250.21(A) provides a list of electrical systems that are permitted, but not required, to be grounded. These systems are as follows:

1. Systems exclusively for industrial electric furnaces for melting, refining, tempering, and so forth
2. Separately derived systems exclusively for rectifiers supplying only adjustable-speed industrial drives
3. Separately derived systems supplied by transformers that have a primary voltage rating of 1,000 volts or less, meeting all of the following conditions:
    a. The system is used exclusively for control circuits.

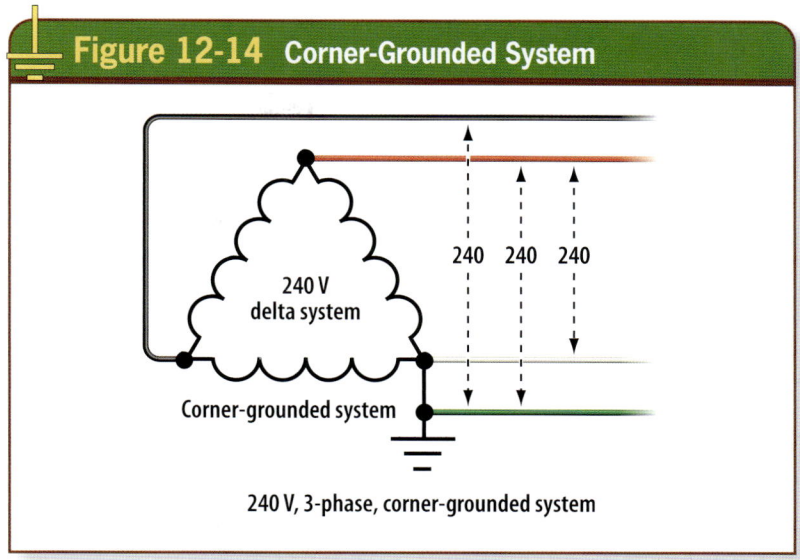

Figure 12-14. *In a corner-grounded system, one phase of the system is grounded. Note that the grounded conductor is a phase conductor, not a neutral conductor.*

    b. Qualified persons service the installation.
    c. Continuity of control power is required.
4. Other systems that are not required to be grounded in accordance with the requirements of 250.20(B)

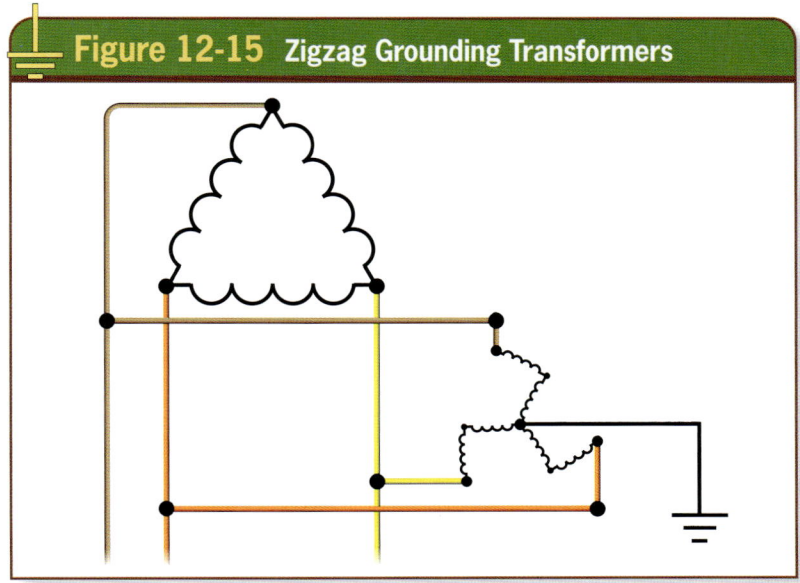

Figure 12-15. *Grounding a delta using zigzag grounding transformers creates a grounded system (no neutral loads) and establishes an effective ground-fault current path.*

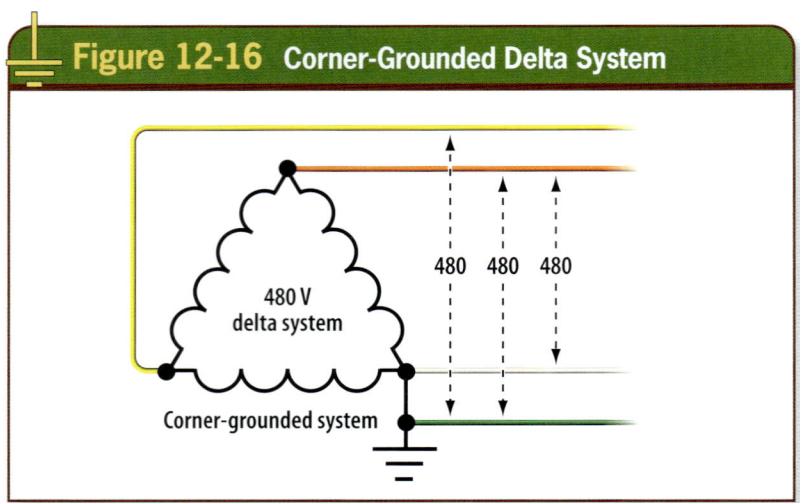

Figure 12-16 **Corner-Grounded Delta System**

Figure 12-16. In a corner-grounded delta system, the phase-to-ground voltage is the same as the phase-to-phase voltage.

Typical systems permitted, but not required, to be grounded include 240-volt, 3-phase, 3-wire and 480-volt, 3-phase, 3-wire delta-connected systems. **See Figure 12-16.**

Grounding Separately Derived Systems

Where a system is separately derived and must be grounded because of the requirements in 250.20 or is grounded as an option in accordance with the

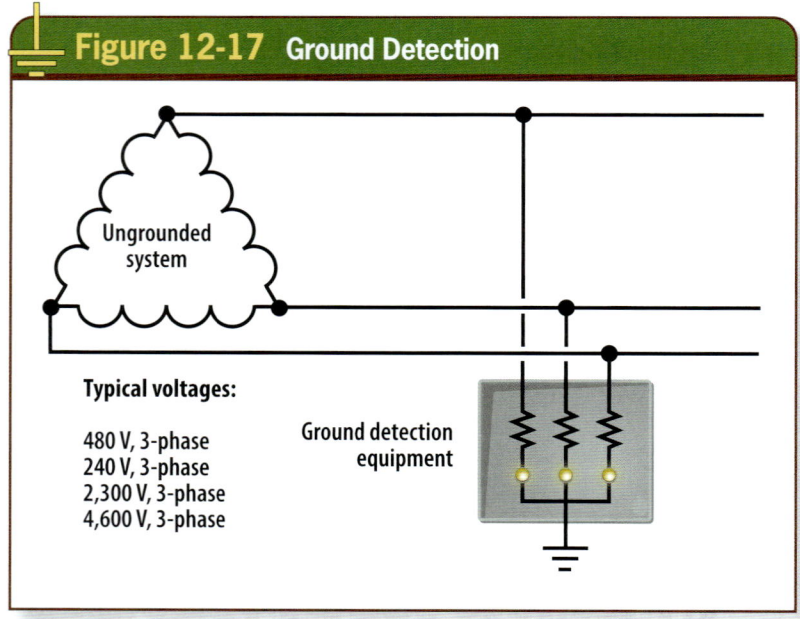

Figure 12-17 **Ground Detection**

Typical voltages:

480 V, 3-phase
240 V, 3-phase
2,300 V, 3-phase
4,600 V, 3-phase

Ground detection equipment

Figure 12-17. Ground detectors are required for ungrounded systems (simplified illustration).

provisions in 250.21, it must be grounded in conformance with all applicable requirements in 250.30(A).

The requirements for grounding a separately derived system are determined by the output voltage, phase arrangement, and use of the system. Another feature for systems supplied by a generator is the switch arrangement provided in the transfer equipment. Informational Note No. 1 in 250.30 provides clear direction on grounding requirements for generators that must be grounded in accordance with 250.30(A) because of the type of transfer switch applied in the design. Direct current (DC) systems are required to be grounded in accordance with Part VIII of Article 250.

Requirements for Ground Detection

For the systems addressed in 250.21(A), system grounding is not required. If a system is not grounded and operates at 120 volts through 1,000 volts, ground detectors are required to be installed. The requirement for ground detection provides the ability to monitor ungrounded systems to detect a first phase-to-ground fault on the system. **See Figure 12-17.**

The first phase-to-ground fault will not cause overcurrent protective device operation, so continued service is achieved. However, it is important that those monitoring the system react to the annunciation, investigate the phase-to-ground condition, and remove it. If the first phase-to-ground condition is not cleared and a second phase-to-ground fault develops on a different phase, the result is a simultaneous phase-to-phase short circuit and phase-to-ground fault event. This type of abnormal event can lead to significant downtime and destruction of equipment.

There are some benefits of operating a system ungrounded if permitted by the *NEC*, but it is important to monitor it and react appropriately if a phase-to-ground condition develops.

Ungrounded systems are often installed and used in industrial facilities where continuity of power is desired for assembly operations and other continuous processes that would be damaged or could cause personnel injury if a first phase-to-ground fault event were to result in interruption of power to the system. The choice to install and operate this type of system is determined by the nature of the process, the operational characteristics of the process, and the operator's or owner's desired method of operation.

Where ground detectors are installed on an ungrounded system, the sensors for such systems must be located as close to the supply source as possible, as indicated in 250.21(B)(2). Listed ground detection equipment is available for use on ungrounded systems.

According to 250.21(C), all enclosures containing equipment and conductors for an ungrounded system must be field-marked "Caution: Ungrounded System Operating–____ Volts" at the source or at the first system disconnecting means. This marking must be durable enough for the environment involved. This marking requirement also applies to switchboards and panelboards that contain ungrounded systems, as provided in 408.3(F)(2). This marking gives qualified people an additional notification of the type of system contained within the enclosure. The phase-to-ground voltage readings of such a system are not as familiar to workers as the grounded system voltage readings.

The marking requirement provides an added level of safety for personnel who must work on ungrounded systems, either to troubleshoot or to add to such systems.

Conductor to be Grounded

As previously discussed, if an AC electrical system is required to be grounded, one conductor of the system must be connected to ground. Section 250.26 provides the information regarding which conductor is the conductor that must be grounded for various systems. For example, for a single-phase, 2-wire system, either conductor supplied by the system can be grounded. This does not create a system neutral conductor; rather, it creates a system grounded conductor. For a single-phase, 3-wire system, the system neutral conductor must be grounded. For multiphase systems having a conductor that is common to all phase conductors, the common (neutral) conductor must be the grounded conductor. In a multiphase system that is corner- or end-grounded, any phase conductor can be grounded. For high-leg, delta-connected systems in which one phase winding is grounded at the midpoint to create a neutral, the neutral is required to be the grounded conductor. **See Figure 12-18.**

Rules for the System Grounded Conductor

The *NEC* provides rules for grounded conductors. Usually, a grounded conductor is used throughout the system. This conductor must be identified according to the requirements in 200.6. It is sometimes referred to in the field as an *identified conductor*, but in *NEC* terms, it is a *grounded conductor that is identified*. The more common term

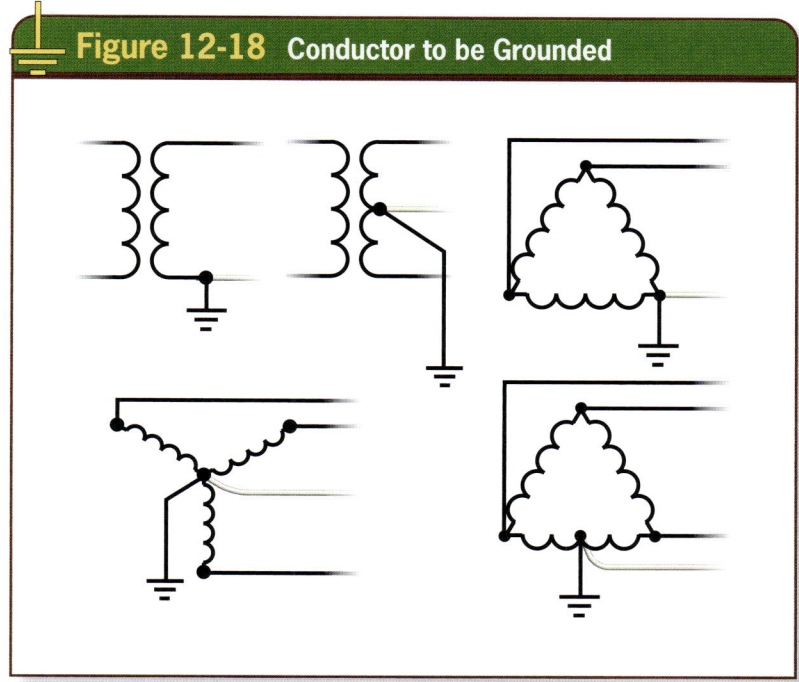

Figure 12-18 Conductor to be Grounded

Figure 12-18. The conductor required to be grounded is specified in 250.26.

used in the field is *neutral*. This is because various other system conductors require identification.

Section 200.6 has two basic identification rules. Section 200.6(A) deals with conductors in sizes 6 AWG and smaller, and 200.6(B) covers sizes 4 AWG and larger. **See Figure 12-19.**

Sizes 6 AWG or Smaller

Grounded (usually neutral) conductors in sizes 6 AWG and smaller must be identified generally by one of the following means:

1. Continuous white or gray insulation along the entire length of the conductor
2. Three continuous white or gray stripes on insulation other than green for the entire length of the conductor

See 200.6(A)(4) through (8) for other specific conditions that require different identification means for grounded conductors.

Sizes 4 AWG and Larger

Grounded (usually neutral) conductors in sizes 4 AWG and larger must be identified generally by one of the following means:

1. Continuous white or gray insulation along the entire length of the conductor
2. Three continuous white or gray stripes on insulation other than green for the entire length of the conductor
3. Identification at the time of installation by a distinctive white or gray marking at the termination of the conductor; the marking must encircle the conductor

There are other important *NEC* rules pertaining to grounded (usually neutral) conductors, such as the general restrictions from overcurrent protection, as provided in 240.22; use with multiwire branch circuits, as provided in 210.4; or installation of a common neutral conductor for use with a feeder, as provided in 215.4(A).

Grounded separately derived systems typically include a neutral point to which a neutral conductor is connected. The terms *neutral point* and *neutral conductor* are defined in Article 100.

Neutral Point. The common point on a wye-connection in a polyphase system or midpoint on a single-phase, 3-wire system, or midpoint of a single-phase portion of a 3-phase delta system, or a midpoint of a 3-wire, direct-current system. (CMP-5)

Neutral Conductor. The conductor connected to the neutral point of a system that is intended to carry current under normal conditions. (CMP-5)

The informational note following the definition of a neutral point in Article 100 clarifies that at the neutral point of the system, the vectorial sum of the nominal voltages from all other phases within the system that use the neutral, with respect to the neutral point, is a zero potential. These terms were added so that the appropriate conductor could be identified whenever this term is used in a requirement, such as in 220.61, 250.26, and 250.36. **See Figure 12-20.**

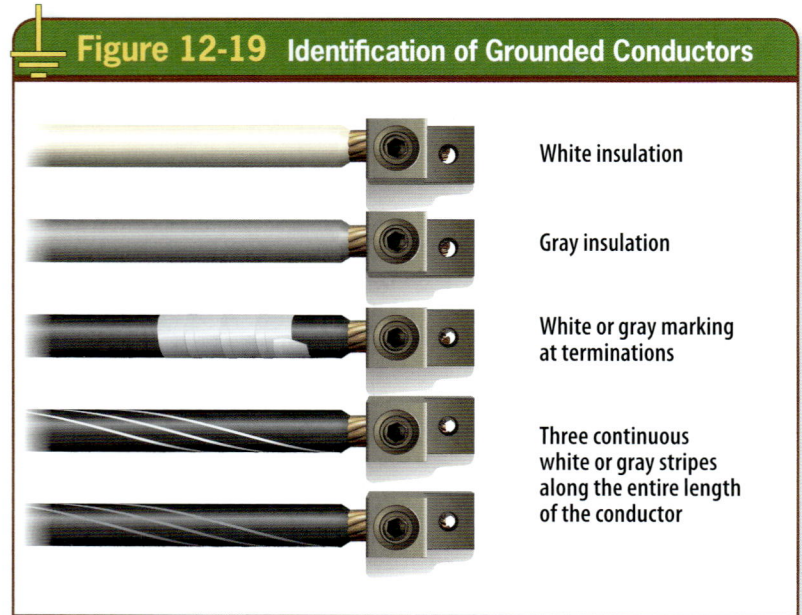

Figure 12-19   Identification of Grounded Conductors

White insulation

Gray insulation

White or gray marking at terminations

Three continuous white or gray stripes along the entire length of the conductor

Figure 12-19. Identification requirements for grounded conductors are provided in 200.6 and 200.7.

It is important to remember that the neutral conductor is usually a current-carrying conductor. Many believe that, because the neutral conductor is a grounded conductor, it is safe to work on a neutral conductor while it is energized. This is a dangerous practice that has led to many serious electrical shocks and electrocutions, specifically with multiwire branch circuits that share a grounded neutral conductor.

The definitions also clarify that a neutral conductor can be present at an outlet box that contains just a single 120-volt circuit supplying a receptacle. The reason is that the grounded conductor is connected to the neutral point of a supply system. Most neutral conductors are grounded conductors, but not all grounded conductors are neutrals. An example of a grounded conductor that is not a neutral is found in a 3-phase, 3-wire, corner-grounded system. The grounded conductor of this system is a grounded phase conductor, not a neutral conductor. All the rules for grounded conductors, including identification, apply to all grounded conductors.

SYSTEM GROUNDING PROHIBITION

The previous sections reviewed applicable rules for systems that are either required or permitted to be grounded. Informational Note No. 2 in 250.20 provides examples of circuits required in the *NEC* that are prohibited from being grounded, but it does not provide an all-inclusive list. For example, 503.155(A) requires circuits for overhead cranes that operate over combustible fibers in Class III hazardous locations to be ungrounded and equipped with a ground detector. Because the system is not grounded, a first phase-to-ground fault will not create a shower of sparks or hot particles that could cause a fire due to the accumulations of fibers on the floor below. This condition is common in Class III locations, such as textile mills, associated with those manufacturing processes.

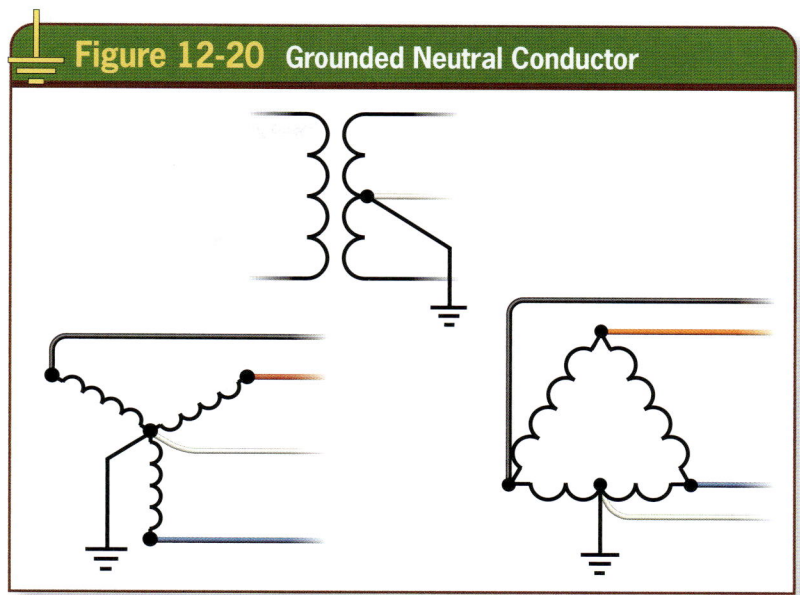

Figure 12-20 Grounded Neutral Conductor

Figure 12-20. Neutral conductors are usually the grounded conductors of systems. The neutral conductor connects to the neutral point of electrical supply systems.

Another type of system that is not permitted to be grounded are the isolated power systems used in health care facilities. The requirements for isolated power systems in health care facilities are provided in 517.61 and 517.160.

As provided in Article 668, circuits for portable equipment within an electrolytic cell working zone are required to be ungrounded. Electrolytic cells are commonly used in the aluminum- and chlorine-processing industries.

Secondary circuits of low-voltage lighting systems are not permitted to be grounded, as indicated in 411.7(A). Also, as provided in 680.23(A)(2), low-voltage lighting systems for underwater pool lighting supplied by isolation transformers are not permitted to be grounded. The listed transformers for these systems are of the isolation type with a grounded metal barrier between the primary and the secondary windings. In previous cycles of the *NEC,* Class 2 load-side circuits for suspended-ceiling low-voltage power grid distribution systems were required to be ungrounded. The 2023 *NEC* now permits, in 393.61, for the Class 2 load-side circuits to be grounded. It is essential for a proper installation to follow the manufacturer's installation requirements when installing suspended-ceiling low-voltage power grid distribution systems for proper grounding and bonding.

Note that these systems or circuits are required to be ungrounded, with the exception of Class 2 load-side circuits for suspended-ceiling low-voltage power grid distribution, but the normally non–current-carrying metal parts of equipment enclosures and raceways that contain these ungrounded systems' conductors and equipment are generally required to be grounded by connection to an equipment grounding conductor (EGC).

IMPEDANCE GROUNDED SYSTEMS

Section 250.20(D) provides the reference to 250.36 and 250.187 for installation of impedance grounded systems depending upon the voltage of the system. Impedance grounded systems have special requirements, all of which need to be followed. This type of installation requires special equipment, and various manufacturers produce the equipment designed specifically for these types of installations.

> **Grounded System, Impedance. (Impedance Grounded System)** An electrical system that is grounded by intentionally connecting the system neutral point to ground through an impedance device. (CMP-5)
>
> **Grounding Conductor, Impedance. (Impedance Grounding Conductor)** A conductor that connects the system neutral point to the impedance device in an impedance grounded system. (CMP-5)

An *impedance grounded system* is a system in which an impedance device, usually a resistor, limits the current in a phase-to-ground fault condition to a low level. This allows the system to remain operational during the ground-fault condition. These types of systems are typically installed in industrial applications. Impedance grounded systems are permitted for use in AC systems with voltages ranging from 480 to 1,000 volts, provided that all of the following restrictions are met:

1. Conditions of maintenance and supervision ensure qualified persons service the installation.
2. Ground detection is provided for the system.
3. Line-to-neutral loads are not supplied by the system.

Sections 250.36(A) through (G) provide the specific installation requirements for impedance grounded systems. The grounding impedance device must be located between the grounding electrode conductor and the impedance grounding conductor connected to the system neutral point. If there is no neutral point, one must be derived from a grounding transformer. More information about grounding transformers is provided in 450.5. The impedance grounding conductor must be fully insulated and have an ampacity of no less than the maximum current rating of the grounding impedance device. It can never be smaller than an 8 AWG copper

Figure 12-21 Impedance Grounded System

Grounding location is between the grounding electrode conductor and the neutral grounding point.

The grounding impedance conductor is used to connect the impedance device with the neutral grounding point of the device.

Grounding impedance conductor must have an ampacity no less than the maximum current rating of the grounding impedance.

Grounding connection can only be made through the grounding impedance device.

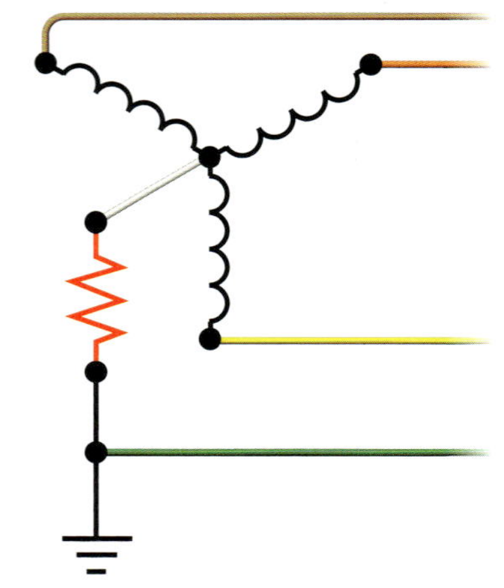

Figure 12-21. *The grounding impedance conductor of an impedance grounding system must be fully insulated, and no impedance bonding jumper connection is permitted on the supply side of the impedance device.*

or 6 AWG aluminum conductor. The system grounding connection must only be made through the impedance device. **See Figure 12-21.**

The impedance grounding conductor is permitted to be installed in a separate raceway from the ungrounded conductors. If an impedance bonding jumper is installed, it must be installed without splices and must be run from the first system disconnect or overcurrent protective device to the grounded side of the grounding impedance device. The grounding electrode conductor must be connected from any point on the grounded side of the grounding impedance device to the equipment grounding connection at the service equipment or first system disconnecting means.

The minimum size required for the impedance bonding jumper of an impedance grounded system must be determined in accordance with either of the following:

1. If the grounding electrode conductor connection is made at the grounding impedance device, the equipment bonding jumper shall be sized in accordance with Table 250.66, based on the size of the service entrance conductors for a service or the derived phase conductors for a separately derived system.
2. If the grounding electrode conductor is connected at the first system disconnecting means or overcurrent protective device, the impedance bonding jumper shall be sized the same as the impedance grounding conductor in 250.36(B).

UNGROUNDED SYSTEMS (CONCEPTS)

The decision to install and operate an ungrounded system is typically a combined effort that includes the design or engineering team, the owner, the operators, and sometimes the authority having jurisdiction. Common reasons for choosing to operate an ungrounded system are providing continuity of electrical operation and minimizing downtime from system outages.

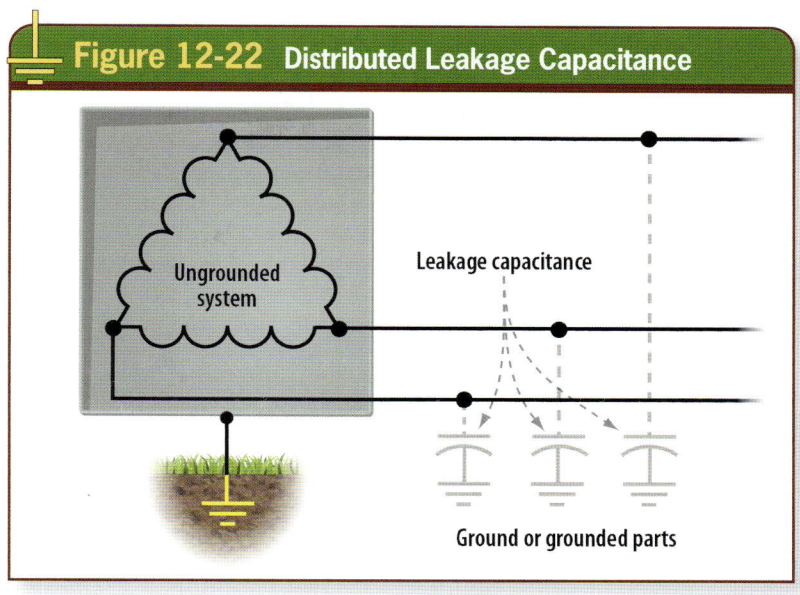

Figure 12-22 Distributed Leakage Capacitance

Ungrounded system

Leakage capacitance

Ground or grounded parts

Figure 12-22. Phase-to-ground voltage in ungrounded systems can result from distributed leakage capacitance.

Common ungrounded delta systems include, but are not limited to:
- 240-volt, 3-phase, 3-wire, delta-connected
- 480-volt, 3-phase, 3-wire, delta-connected
- 2,300-volt, 3-phase, 3-wire, delta-connected
- 4,600-volt, 3-phase, 3-wire, delta-connected
- 13,800 volt, 3-phase, 3-wire, delta-connected

The disadvantage of an ungrounded system is that a first phase-to-ground fault condition can be difficult to find and can take a considerable amount of investigation and time. In theory, the voltage to ground in an ungrounded system is zero volts because there is no ground connection from any system conductor. However, there is distributed leakage capacitance present throughout such systems. Phase-to-ground voltage levels that may appear in a test instrument reading result from capacitance coupling effects from the system circuits. **See Figure 12-22.**

Another important point about voltage to ground levels in ungrounded systems is covered in the definition of the term *voltage to ground*. The definition clarifies that the voltage to ground of a grounded system is the voltage between the given conductor and that point or conductor of the circuit that is grounded. For example, in a 120/240-volt, single-phase system, the voltage is 120 volts from any ungrounded phase conductor to ground. However, for ungrounded systems, the greatest voltage between the given conductor and any other conductor of the circuit is also the phase-to-ground voltage. For example, on a 480-volt, 3-phase, 3-wire, ungrounded delta system, the phase-to-phase voltage is 480 volts, which is also the phase-to-ground voltage for this system, based on the definition.

SUMMARY

Whether the *NEC* requires system grounding or permits system grounding by choice, all of the requirements for grounded systems must be applied whenever systems are grounded; they are not optional. Methods of grounding can vary and include solid grounding, impedance or resistance grounding, grounding through surge arresters, and grounding through an inductor. The method of grounding is not always a matter of choice or a design consideration, but in many cases is required by the *NEC*. The *NEC* indicates the systems required to be grounded, the systems permitted to be grounded, and those systems not permitted to be grounded. Ungrounded systems must generally be provided with ground detection systems to monitor for a first phase-to-ground fault condition. The sensing equipment for ground detection systems must be installed as close as practical to where the system receives its supply.

REVIEW QUESTIONS

1. The *NEC* specifies which electrical systems are required to be grounded, which systems are permitted to be grounded, and which systems are not permitted to be grounded.

   a. True   b. False

2. The term __?__ means connected to ground or to a conductive body that extends the ground connection.

   a. bonded (bonding)
   b. grounded (grounding)
   c. grounded conductor
   d. grounding electrode

3. Grounded conductors __?__ are permitted to be identified either in the same manner as that required for conductors 6 AWG and smaller, or, at the time of installation, by a distinctive white or gray marking at the terminations.

   a. 12 AWG or larger
   b. 4 AWG or larger
   c. of aluminum
   d. of copper

4. A 480-volt, 3-phase, 3-wire, delta-connected system is always required to be grounded.

   a. True   b. False

5. Where 30-volt AC systems are installed outdoors as overhead conductors, they __?__ be grounded.

   a. are not required to
   b. are required to
   c. shall be permitted to
   d. shall not

6. Which of the following conductors is not present when installing an ungrounded system?

   a. Equipment grounding conductor
   b. Grounded conductor
   c. Grounding electrode conductor
   d. Ungrounded conductors

7. Where a system can be grounded so that the maximum voltage to ground from any system ungrounded conductor does not exceed 240 volts, the system must be grounded.

   a. True   b. False

8. The phase-to-ground voltage of a high leg in a 240/120-volt, 3-phase, 4-wire, delta-connected system is __?__.

   a. 120 V   c. 240 V
   b. 208 V   d. 480 V

9. Separately derived systems supplied by transformers having a primary voltage rating of less than 1,000 volts are not required to be grounded if all but which of the following conditions are met?

   a. Continuity of control power is required.
   b. Continuity of line power is required.
   c. Qualified persons service this installation.
   d. This system is used for control circuits only.

10. Where a transformer secondary is 30 volts (AC) and the primary is supplied by an ungrounded source, the secondary must be grounded.

    a. True   b. False

11. Which of the following circuits is required to be grounded?

    a. 120V lighting circuits
    b. Circuits for cranes over Class III locations
    c. Isolated power systems for health care facilities
    d. Lighting systems as provided in 411.7

12. __?__ do/does not have a neutral point.

    a. A 120/240V, single-phase, 3-wire system
    b. 480Y/277V, 3-phase, 4-wire, wye-connected and 208Y/120V, 3-phase, 4-wire, wye-connected systems
    c. A 480V, 3-phase, 3-wire, delta-connected system
    d. A 480Y/277V, 3-phase, 4-wire, wye-connected system

13. A 240-volt, 3-phase, 3-wire, delta-connected system is required to be grounded.

    a. True   b. False

14. Which of the following circuits are permitted to be grounded?

    a. A 480V, 3-phase, 3-wire, delta-connected system
    b. Circuits for portable equipment within electrolytic cell working zones, as provided in Article 668
    c. Isolated power systems in health care facilities covered by 517.61 and 517.160
    d. Secondary circuits of lighting systems for pools, as covered in 680.23(A)(2)

15. A neutral conductor is defined in the *NEC* as the conductor that is connected to the neutral point of a system and that generally carries current under normal conditions.

    a. True   b. False

16. If grounding of a separately derived system is required by the *NEC*, it must be as specified in accordance with __?__.

    a. 250.20(B)   c. 250.30(A)
    b. 250.24   d. 250.30(B)

17. An AC system supplying premises wiring must be grounded where the maximum voltage to ground from any ungrounded conductor does not exceed 150 volts.

    a. True   b. False

18. A phase-to-ground fault in an ungrounded system accidentally grounds the system and activates ground detectors.

    a. True   b. False

19. If a system is not grounded, only one conductor supplied by the system is intentionally grounded.

    a. True   b. False

20. Ground detectors are not required for 240 volt, 3-phase, ungrounded, delta-connected systems.

    a. True   b. False

21. An insulated grounded conductor of a system is required to be identified by any of the following methods except __?__.

    a. gray insulation
    b. green insulation
    c. three white or gray stripes along the entire length of the conductor
    d. white insulation

22. What is the phase-to-ground voltage of the ungrounded conductors supplied by a 480-volt, 3-phase, 3-wire, corner-grounded delta system?

    a. 120 V
    b. 240 V
    c. 277 V
    d. 480 V

23. The voltage readings from the phase conductors to ground of an ungrounded system are due to distributed leakage capacitance in the system.

    a. True   b. False

24. Class 2 load-side circuits for suspended-ceiling low-voltage power grid distribution systems as covered in 393.61 shall be permitted to be grounded.

    a. True   b. False

Grounding and Bonding for Separately Derived Systems

Electrical wiring and power distribution systems for commercial, industrial, institutional, and even some residential occupancies typically include the installation and use of separately derived systems. A separately derived system is generally a separate power source, such as a battery, generator, photovoltaic system, transformer, wind turbine, or other source that produces electrical power. The *NEC* provides an extensive set of rules specifically related to grounding and bonding for separately derived systems. To understand and properly apply these rules to separately derived system installations, the meaning of certain terms must be clear.

Objectives

» Understand what constitutes a separately derived system as it is defined and how to distinguish it from systems that are not separately derived.

» Determine *NEC* grounding and bonding requirements that apply to separately derived systems.

» Understand the application of the common grounding electrode conductor tap concept for multiple separately derived systems.

» Size grounding electrode conductors, system bonding jumpers, supply-side bonding jumpers, and grounded conductors for a separately derived system.

» Determine how to ground and bond a generator-type separately derived system.

» Understand the relationship between the transfer equipment and the system grounding requirements for generator-type separately derived systems.

Chapter 13

Table of Contents

Definitions 292

Determining a Separately Derived System ... 292

Grounding Requirements 293

Grounded Systems 293

    System Bonding Jumper 294

    Supply-Side Bonding Jumper 297

    Grounded Conductor Sizing 298

Grounding Electrodes 299

    Grounding Electrode Conductor for
    a Single System 299

    Grounding Electrode Conductor for
    Multiple Systems 300

Bonding Water Piping and Building Steel 302

Outdoor Source 303

Ungrounded Systems 304

Generators and Transfer Equipment 305

Wind Electrical Systems 309

    Tower Grounding 310

    System Grounding 310

    Equipment Grounding and Bonding 310

    Grounding and
    Bonding Connections 310

Grounding DC Systems 311

    Point of Grounding Connection for
    DC Systems 312

    Size of a DC Grounding
    Electrode Conductor 312

    Size of a DC System Bonding Jumper 313

Ungrounded DC Systems 313

Summary 315

Review Questions 315

DEFINITIONS

The use of *NEC* terminology is important to determine which rules apply to installations and systems. A variety of "*NEC*-defined" terms are used on the subject of grounding and bonding of separately derived systems. A common language of communication must be used when applying *NEC* rules to any system or installation.

Bonded (Bonding). Connected to establish electrical continuity and conductivity. (CMP-5)

Bonding Jumper, Supply-Side. (Supply-Side Bonding Jumper) A conductor installed on the supply side of a service or within a service equipment enclosure(s), or for a separately derived system, that ensures the required electrical conductivity between metal parts required to be electrically connected. (CMP-5)

Bonding Jumper, System. (System Bonding Jumper) The connection between the grounded circuit conductor and the supply-side bonding jumper, or the equipment grounding conductor, or both, at a separately derived system. (CMP-5)

Grounded (Grounding). Connected (connecting) to ground or to a conductive body that extends the ground connection. (CMP-5)

Grounding Electrode. A conducting object through which a direct connection to earth is established. (CMP-5)

Grounding Electrode Conductor (GEC). A conductor used to connect the system grounded conductor or the equipment to a grounding electrode or to a point on the grounding electrode system. (CMP-5)

Separately Derived System. An electrical power supply output, other than a service, having no direct connection(s) to circuit conductors of any other electrical source other than those established by grounding and bonding connections. (CMP-5)

DETERMINING A SEPARATELY DERIVED SYSTEM

One of the keys to determining whether or not a system is separately derived is related to whether or not it has a "direct electrical connection" to another source. Some examples of separately derived systems include generators, batteries, converter windings, transformers, solar photovoltaic systems, and wind turbine generators. Common separately derived systems installed for commercial projects include transformers and generators. **See Figure 13-1.**

Some transformers do not qualify as separately derived systems because one winding in the transformer is common to both the input and the output side of the transformer. These are autotransformers, and they are addressed by requirements in Article 450. Examples of autotransformers include core and coil ballasts in fluorescent luminaires and "auto (buck-boost) transformers" used for raising or lowering voltage levels for particular applications or individual pieces of utilization equipment. **See Figure 13-2.**

Figure 13-1 Separately Derived System

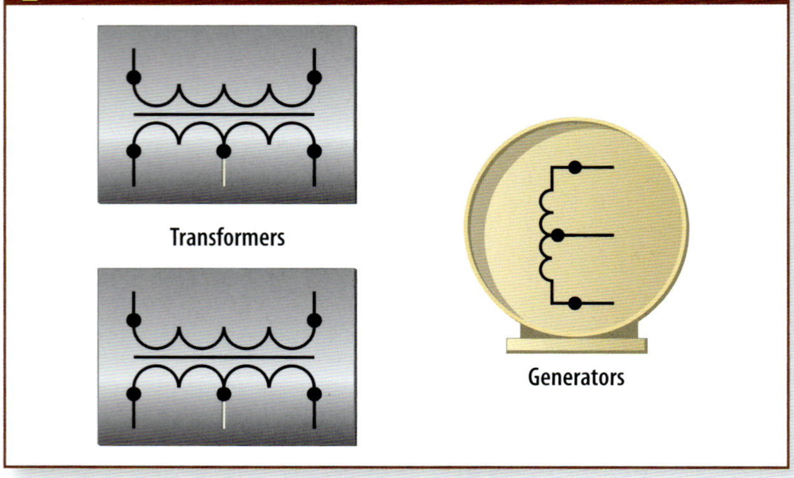

Transformers

Generators

Figure 13-1. Separately derived systems produce electrical power and have no direct electrical connection between the circuit conductors of another supply system.

GROUNDING REQUIREMENTS

Grounding a separately derived system means that the system itself will be connected to the Earth in addition to the enclosure containing the system, if applicable. This grounding connection happens through a grounding electrode. When a system is connected to the Earth, one conductor supplied by the system is intentionally grounded. This creates a grounded conductor of the system to which all *NEC* rules for grounded conductors must apply.

Section 250.30 indicates that the grounding requirements of 250.30(A) and (B) apply to separately derived systems. Sections 250.20, 250.21, and 250.26 are also directly related to the requirement for grounding a separately derived system. If the system is required to be grounded because of provisions in any of these rules, the system must be grounded according to the rules in 250.30(A), which is one of the longest sections in the *NEC*. Multiple separately derived systems installed in parallel must meet the provisions in 250.30. Section 250.30 indicates that multiple separately derived systems of the same type connected in parallel are considered to be a single separately derived system. This is directly related to appropriate locations of a single system bonding jumper, as covered later. An example of this type of system includes multiple generators installed using paralleling equipment.

This chapter breaks the provisions of 250.30(A), (B), and (C) into small pieces to clarify how to ground and bond separately derived systems. The arrangement of this chapter mostly follows the sequence of 250.30.

GROUNDED SYSTEMS

If a separately derived system is required to be grounded, based on the provisions in 250.20 or 250.21, the rules in 250.30(A)(1) through (8) apply.

There are some components that are common to the grounding and bonding scheme of a typical separately derived system (transformer type). **See Figure 13-3.**

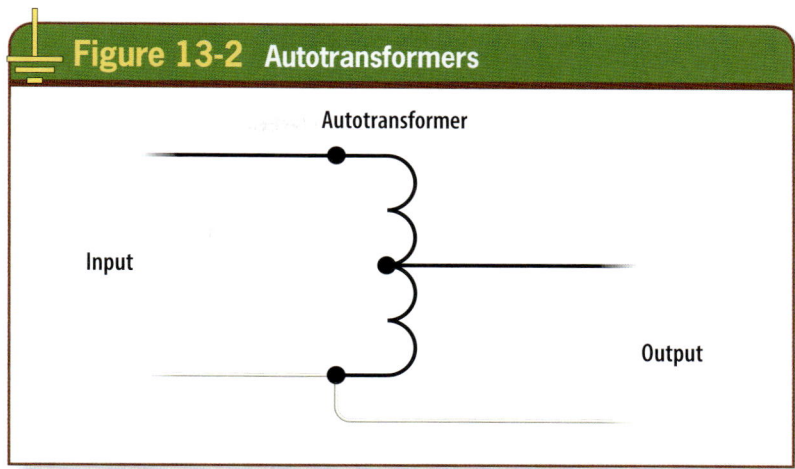

Figure 13-2 Autotransformers

Figure 13-2. Autotransformers are not separately derived systems because one system or circuit conductor is common to both the input and output side of the transformer.

Regardless of the type of separately derived system, the same grounding and bonding components are typically installed, and thus the same *NEC* requirements apply. Because separately derived systems are another source of power, the grounding and bonding system has similar physical and operational characteristics to the grounding and bonding systems used for service equipment.

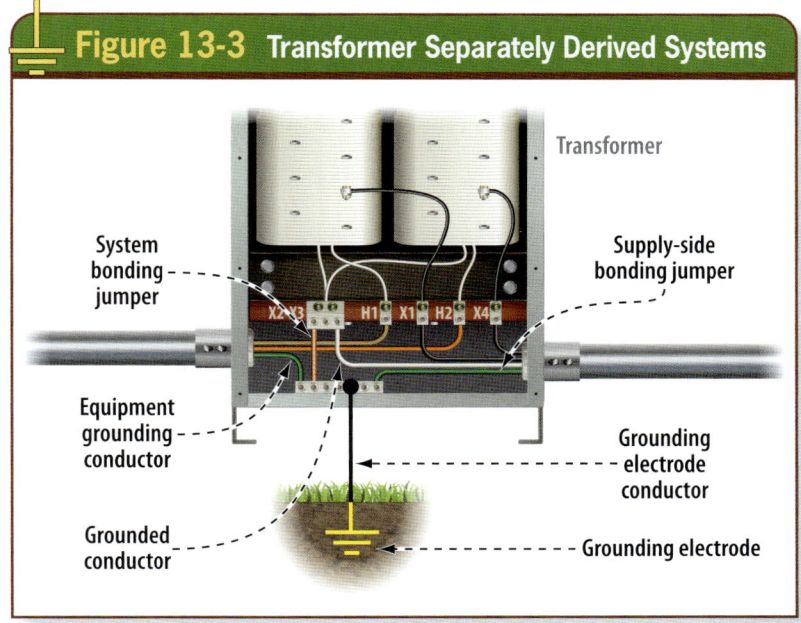

Figure 13-3 Transformer Separately Derived Systems

Figure 13-3. There are multiple grounding and bonding components necessary for transformer separately derived systems.

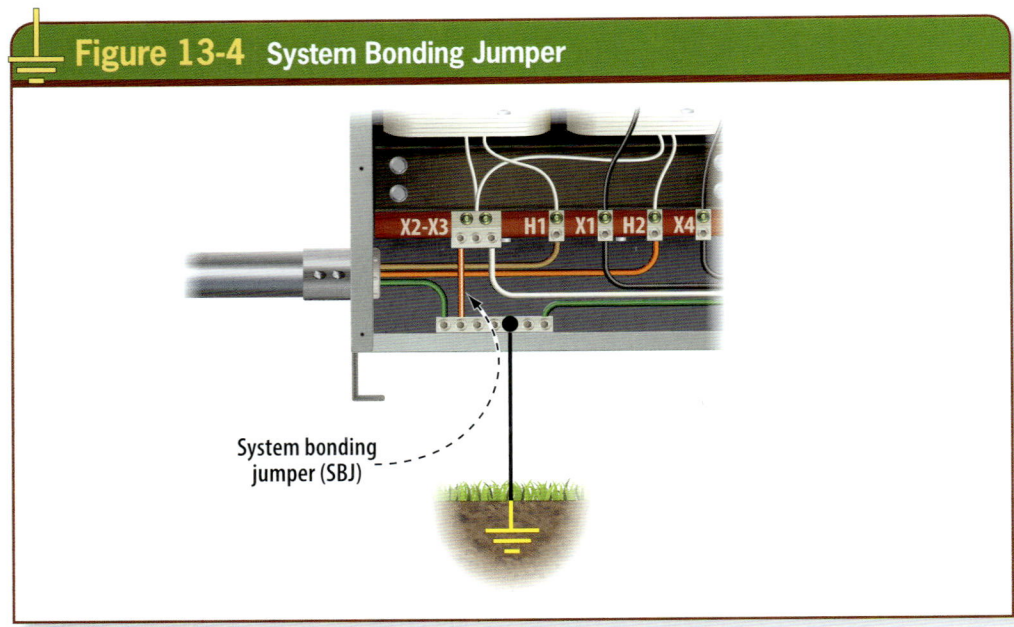

Figure 13-4 System Bonding Jumper

System bonding
jumper (SBJ)

Figure 13-4. A system bonding jumper is permitted to be installed at the dry-type transformer enclosure or the first system overcurrent protective device enclosure supplied by the system.

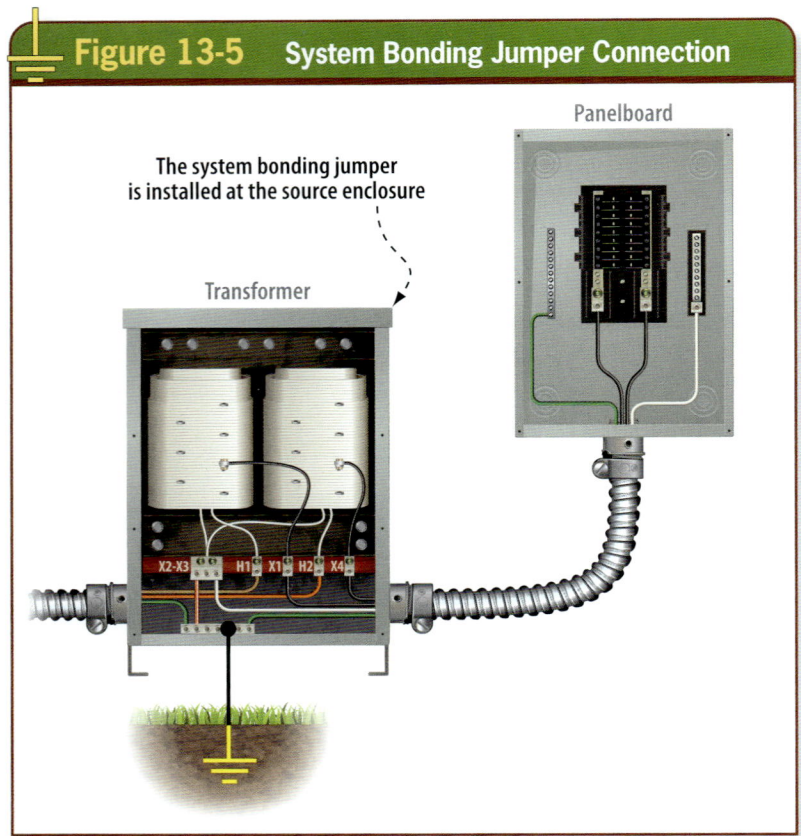

Figure 13-5   **System Bonding Jumper Connection**

Panelboard

The system bonding jumper
is installed at the source enclosure

Transformer

Figure 13-5. The system bonding jumper connects the grounded conductor to the supply-side bonding jumper at the separately derived system source enclosure.

The same concerns about keeping neutral current on its intended conductive path are provided in the rules for separately derived systems. Section 250.30(A) makes it clear that, unless otherwise permitted in Article 250, the grounded (usually a system neutral) conductor is generally not permitted to be reconnected to ground on the load side of the system bonding jumper connection. Impedance grounded systems must meet the requirements for impedance grounding conductor connections, as provided in 250.36 or 250.187.

System Bonding Jumper

The system bonding jumper of a separately derived system connects the grounded conductor of the system to the disconnecting means enclosure, to the supply-side bonding jumper, to the grounding electrode conductor, and to an equipment grounding conductor (EGC) if there is a primary supply to a transformer-type derived system. The system bonding jumper can be located at any point from the source enclosure up to the first system overcurrent protective device or disconnecting means enclosure. **See Figure 13-4.**

If there is no disconnecting means or overcurrent protective device at the load end of the conductors supplied by the system, the system bonding jumper must be installed in and connected to the source enclosure. **See Figure 13-5.**

If the system bonding jumper is installed in the source enclosure, it must connect the grounded conductor to the supply-side bonding jumper and to the metal enclosure. If the system bonding jumper is installed at the first disconnecting means enclosure, it must connect the grounded conductor to the supply-side bonding jumper, the disconnecting means enclosure, and the EGC(s). **See Figure 13-6.**

The *NEC* also requires the system bonding jumper to remain within the enclosure in which it originates. Note that a transformer-type separately derived system usually includes an EGC with the primary feeder connection and a supply-side bonding jumper with the secondary connection. These two are connected to the source enclosure. It is important to understand that this connection does not constitute a direct connection between the system conductors from primary to secondary. **See Figure 13-7.**

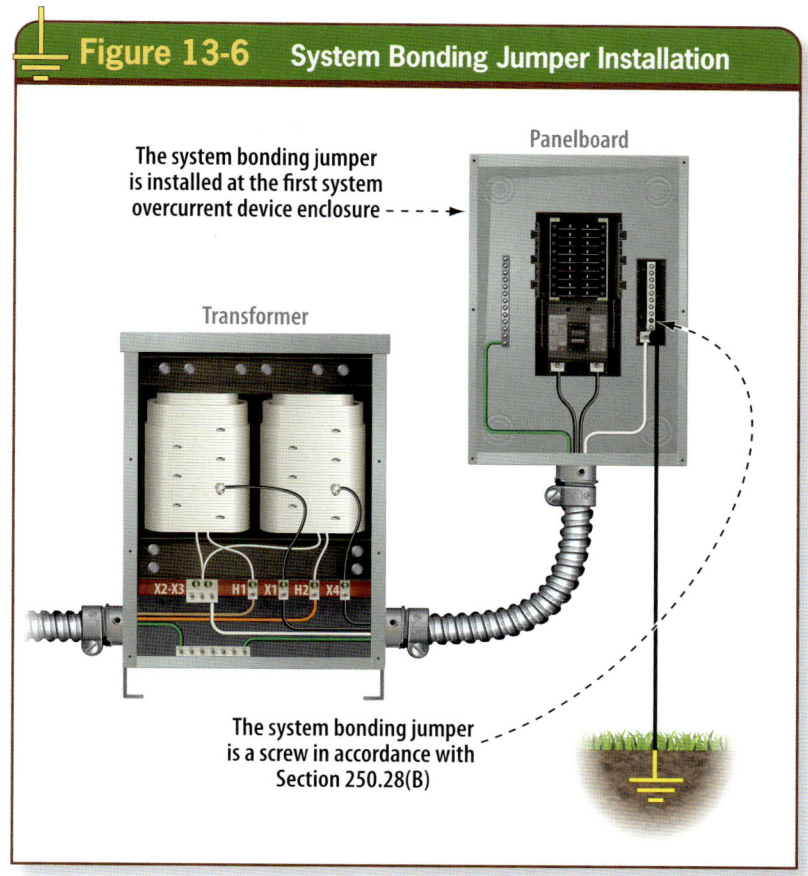

Figure 13-6. The system bonding jumper is permitted to connect the grounded conductor to the supply-side bonding jumper or EGC in the first system disconnecting means enclosure.

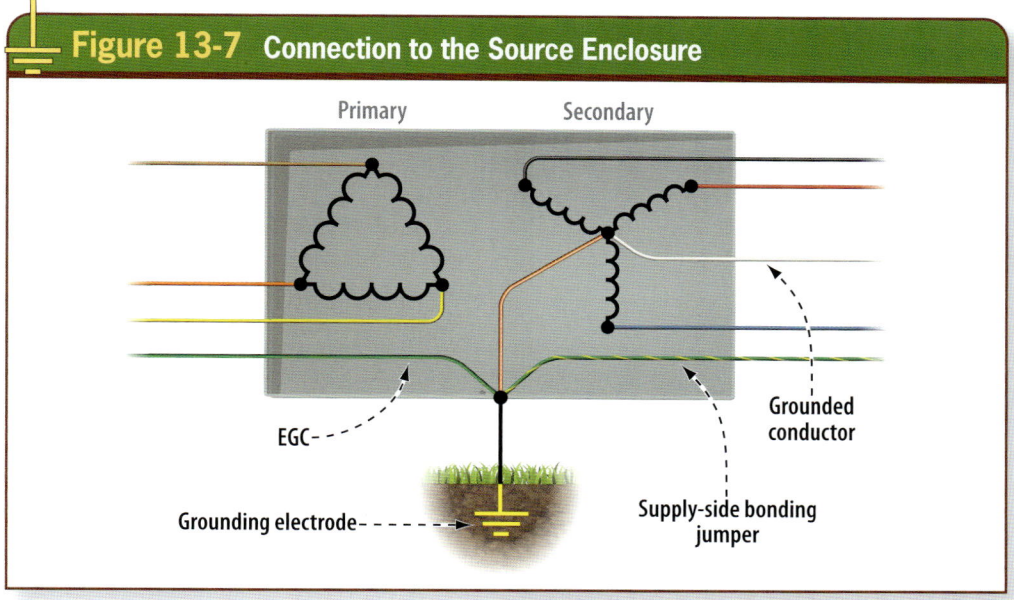

Figure 13-7. The EGC of the transformer primary and the supply-side bonding jumper of the transformer secondary are both connected to the source enclosure.

The system bonding jumper must be installed according to 250.28(A) through (D). If a system bonding jumper is a wire type, it is sized using Table 250.102(C)(1) or 12.5% based on the size of the largest ungrounded derived phase conductor or conductors, or the total circular mil (cm) area of all conductors connected to any one

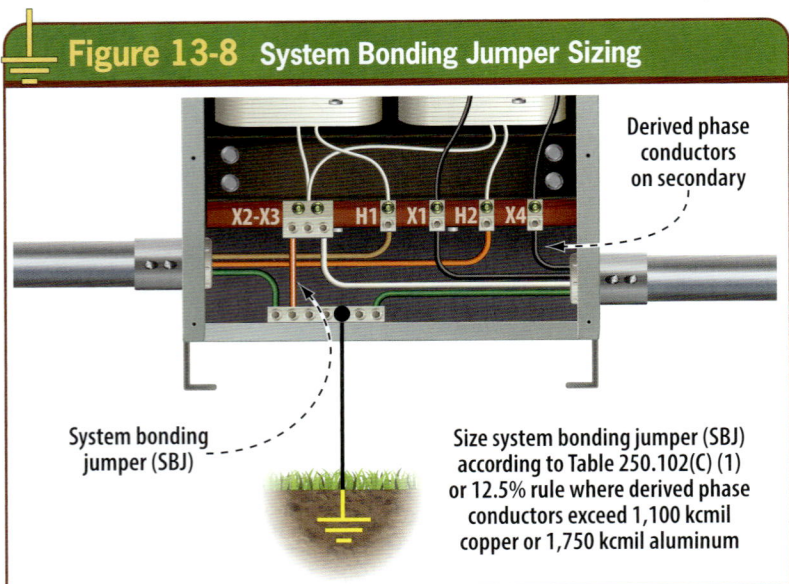

Figure 13-8   System Bonding Jumper Sizing

Derived phase conductors on secondary

X2-X3   H1   X1   H2   X4

System bonding jumper (SBJ)

Size system bonding jumper (SBJ) according to Table 250.102(C)(1) or 12.5% rule where derived phase conductors exceed 1,100 kcmil copper or 1,750 kcmil aluminum

Figure 13-8. *A system bonding jumper is sized based on the circular mil area of the largest ungrounded derived phase conductor and must remain in the enclosure from which it originates.*

Figure 13-9   Section 250.30(A)(1) Exception No. 2

MAIN   ON   OFF
1   ON   OFF
2   ON   OFF
3   ON   OFF

X2-X3   H1   X1   H2   X4

Figure 13-9. *A system bonding jumper is permitted at the source and first disconnecting means as long as it does not create a parallel path for neutral current.*

ungrounded phase at the source, as stated in Note 1 of Table 250.102(C)(1). As an example, if the derived phase conductors connected to a transformer secondary are 750 kcmil copper, the minimum size required for a system bonding jumper is 2/0 AWG copper or 4/0 AWG aluminum or copper-clad aluminum. **See Figure 13-8.**

The term *derived phase conductor* is often used in discussions about separately derived systems. This term relates to the conductors connected to the output side of a power source that is by definition a separately derived system.

The *NEC* refers to these conductors as the ungrounded secondary conductors connected to a transformer-type separately derived system. The circular mil area of these conductors is used for sizing grounding electrode conductors, system bonding jumpers, and supply-side bonding jumpers that are part of a separately derived system. Table 250.102(C)(1) is used for establishing these sizes rather than Table 250.122, which is used for sizing EGCs.

There is normally only one system bonding jumper for a separately derived system that is grounded. By exception, a system bonding jumper is permitted at the source enclosure and the first disconnecting means enclosure, under certain conditions, if there are no parallel paths for current in the grounded conductor of the system.

For example, two system bonding jumpers would be acceptable by 250.30(A)(1) Exception No. 2 for a transformer outside a building or structure that supplies an open-bottom switchboard and has polyvinyl chloride (PVC) conduit installed between the transformer enclosure and the switchboard. In this type of installation, the grounded conductor not only carries the neutral load in normal operation but also serves as an effective ground-fault current path during ground-fault conditions. This is why, in this case, it is important that the minimum size for the grounded conductor be no smaller than the system bonding jumper. Because the Earth does not qualify as an effective current path, this example meets the provisions of the exception. **See Figure 13-9.**

Separately derived systems that consist of multiple sources of the same type are permitted to have a single system bonding jumper installed at the paralleling switchboard, switchgear, or other paralleling equipment connection point instead of at each of the multiple source enclosures.

Supply-Side Bonding Jumper

The requirements for supply-side bonding jumpers installed for separately derived systems are provided in 250.30(A)(2). The conductors supplied from a transformer (separately derived system) are typically installed in accordance with 240.21(C) because they are transformer secondary conductors.

The secondary conductors are routed to the first system overcurrent protective device, usually in a separate enclosure, using a wiring method that provides suitable protection from physical damage. It is also acceptable to use wiring methods such as rigid metal conduit, intermediate metal conduit, or electrical metallic tubing for this type of installation, and these types of wiring methods qualify as the bonding means required between the enclosures. Common wiring methods used for connections to dry-type transformer enclosures are flexible metal conduit or liquidtight flexible metal conduit. An equipment bonding jumper is generally required for these wiring methods. The *NEC* refers to this bonding jumper as a supply-side bonding jumper because it is routed with conductors on the supply side of the overcurrent protective device enclosure where the transformer secondary conductors terminate. **See Figure 13-10.**

The flexible wiring methods are more popular, as they reduce vibration and provide flexibility during installation. A wire-type supply-side bonding jumper is required to be sized in accordance with 250.102(C)(1) based on the size of the largest derived phase conductor connected to the system. **See Figure 13-11.** The supply-side bonding jumper

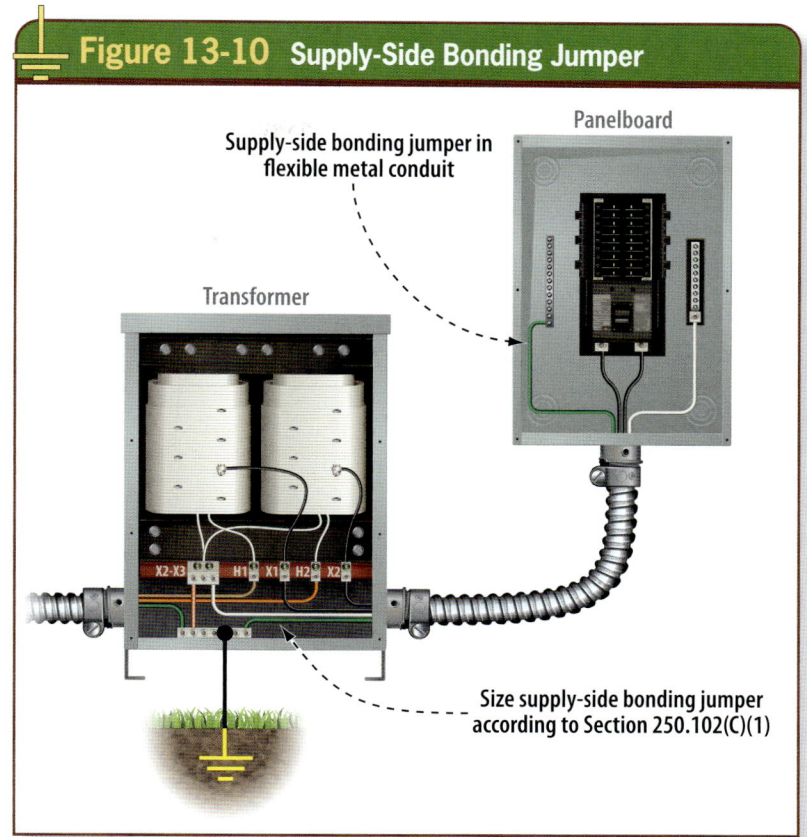

Figure 13-10. A supply-side bonding jumper is typically routed with derived phase conductors and installed with the derived system conductors using flexible wiring methods.

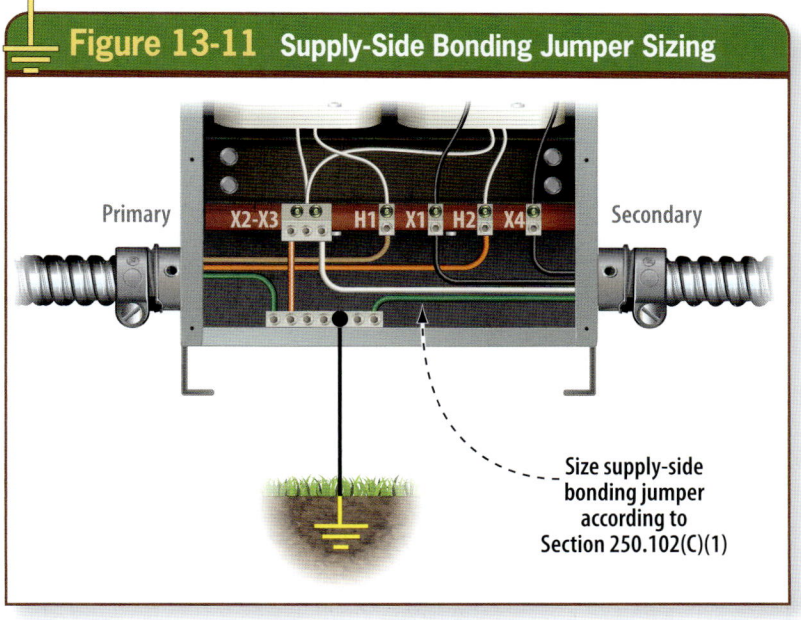

Figure 13-11. The supply-side bonding jumper is sized using 250.102(C)(1) based on the size of the largest ungrounded derived phase conductor.

is not required to be larger than the derived phase conductors supplied by the system. A supply-side bonding jumper is not required between enclosures for installations of two system bonding

Figure 13-12 Grounded Conductor Sizing Requirements

Figure 13-12. Grounded conductors supplied by separately derived systems must be sized according to 250.30(A)(3).

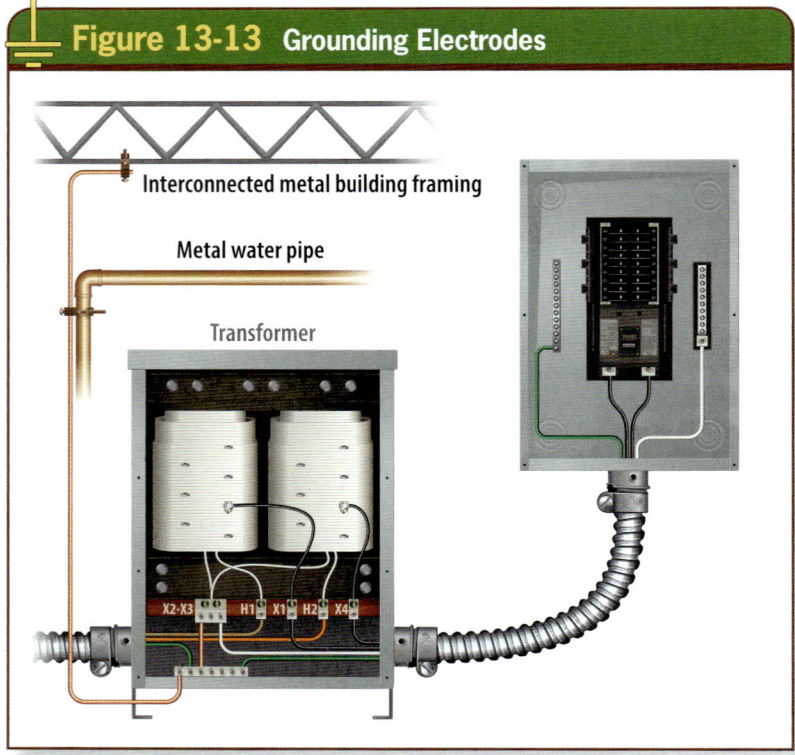

Figure 13-13 Grounding Electrodes

Interconnected metal building framing

Metal water pipe

Transformer

X2-X3   H1 X1 H2 X4

Figure 13-13. The building or structure grounding electrode system is required to be used for separately derived systems.

jumpers meeting the requirements in 250.30(A)(1) Exception No. 2.

Grounded Conductor Sizing

If a separately derived system is grounded, a system connection is intentionally established to ground. Grounded systems include one conductor that is grounded, which could be either a system neutral or a grounded phase conductor. If a grounded conductor is necessary, such as for supplying line-to-neutral loads, then it can be installed. If there is no need for a grounded conductor, one does not have to be installed; however, in that case, a supply-side bonding jumper must be provided for ground-fault current between the separately derived system and the overcurrent protective device.

The grounded conductor sizing requirements are located in 220.61 and 250.30(A)(3). If the system bonding jumper is not located at the source enclosure, the grounded conductor of the separately derived system must meet specific sizing requirements. **See Figure 13-12.**

If installed in a single raceway with the ungrounded phase conductors, the grounded conductor must be sized to carry the load and cannot be smaller than specified in Table 250.102(C)(1), but it is not required to be any larger than the largest ungrounded derived phase conductor of the system. For ungrounded derived phase conductors exceeding the values in Table 250.102(C)(1), the grounded conductor cannot be smaller than 12.5% of the circular mil area of the largest ungrounded conductor or set of ungrounded conductors per phase, as stated in Note 1 of that table.

If the ungrounded derived phase conductors of a system are installed in a parallel arrangement using two or more raceways, the grounded conductor must be installed in parallel in each raceway. The minimum size of the grounded conductor in each raceway is based on the circular mil area of the largest ungrounded derived phase conductor in each raceway, but each cannot be smaller than 1/0 AWG to meet the requirements for parallel conductors in 310.10(G).

For 3-phase, 3-wire, corner-grounded, separately derived systems, the grounded conductor must be no smaller than the ungrounded derived phase conductor supplied by the system. As previously covered, the grounded conductor of an impedance grounded system must meet the sizing requirements in 250.36 or 250.187.

GROUNDING ELECTRODES

Grounding a separately derived system requires a connection to the Earth through a grounding electrode. The *NEC* is specific in 250.30(A)(4) about the electrode that must be used. A building or structure supplied by an electrical service often has grounding electrodes that were inherent to the construction, such as concrete-encased electrodes, in-ground metal structures, and metal underground water pipe electrodes. The *NEC* requires that the electrodes present at the building or structure be used as the electrode for the separately derived system. **See Figure 13-13.**

Grounding Electrode Conductor for a Single System

Grounding separately derived systems requires installing a grounding electrode conductor. For a single separately derived system, the grounding electrode conductor generally must be sized using 250.66, based on the size of the largest ungrounded derived phase conductors. **See Figure 13-14.**

The grounding electrode conductor must connect the grounded conductor of the system to the grounding electrode described in 250.30(A)(5). The connection of the grounding electrode conductor to the system must be made where the system bonding jumper is installed. **See Figure 13-15.**

Exception No. 2 to Section 250.30(A) (5) is applicable to situations in which a separately derived system source is located within equipment that is listed and identified as suitable for use as service equipment and in which a grounding electrode is connected to the

Figure 13-14. The grounding electrode conductor for a separately derived system is generally sized using 250.66, based on the size of the largest ungrounded phase conductor supplied by the system.

Courtesy of Jim Dollard, IBEW Local 98

equipment with a grounding electrode conductor of the minimum size required for the derived system. In cases in which the system bonding jumper is a wire or busbar, the grounding electrode conductor is also permitted to be connected

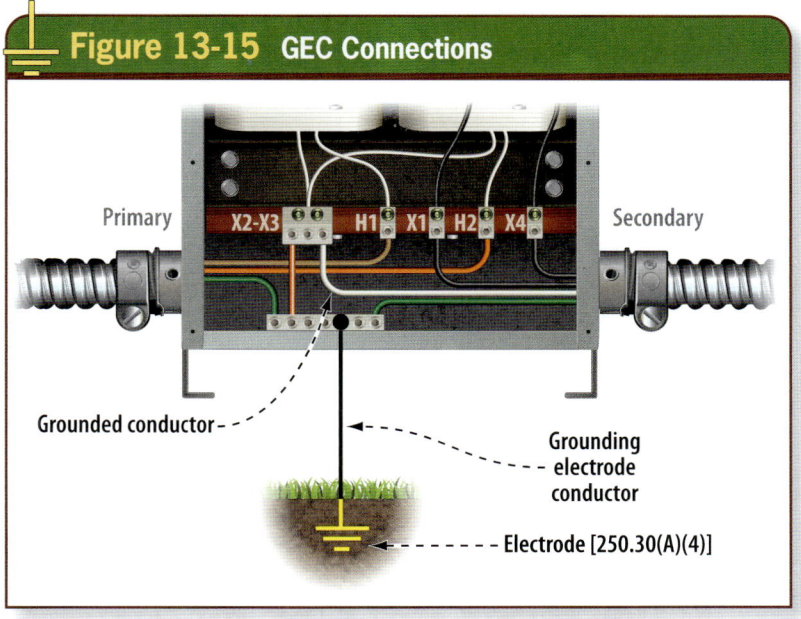

Figure 13-15. A grounding electrode conductor connects the grounding electrode to the grounded conductor of the system.

Figure 13-16   Connected to an Equipment Grounding Terminal Bar

Figure 13-16. By exception, the grounding electrode conductor for a separately derived system is permitted to be connected to an equipment grounding terminal busbar within equipment.

Courtesy of Schneider Electric Square D Company

to an equipment grounding terminal bar within the equipment, provided it is of sufficient size for the separately derived system. **See Figure 13-16.**

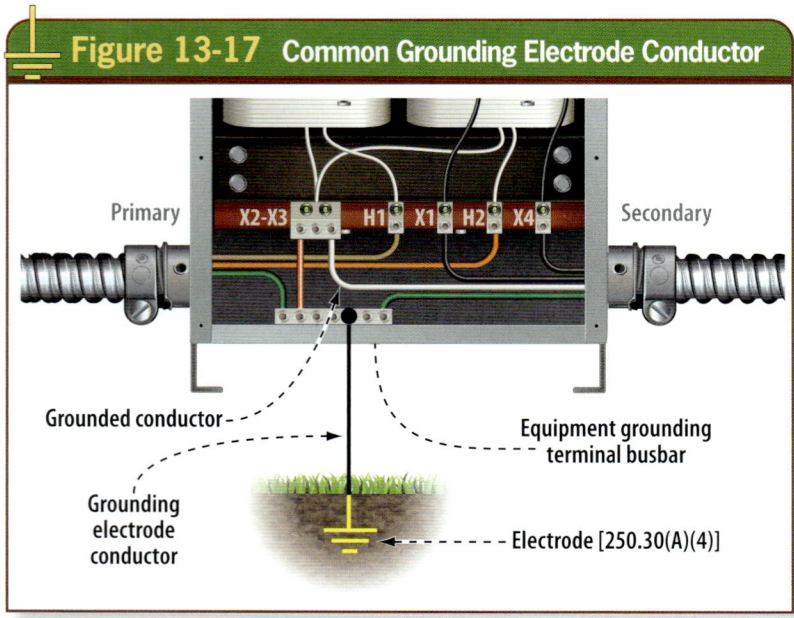

Figure 13-17   Common Grounding Electrode Conductor

Primary

X2-X3   H1   X1   H2   X4   Secondary

Grounded conductor

Equipment grounding terminal busbar

Grounding electrode conductor

Electrode [250.30(A)(4)]

Figure 13-17. The common grounding electrode conductor for multiple separately derived systems is not permitted to be sized smaller than 3/0 AWG copper or 250 kcmil aluminum or copper-clad aluminum.

Grounding Electrode Conductor for Multiple Systems

The phrase *common grounding electrode conductor* is not defined in the *NEC*; however, from the description of its installation requirements in 250.30(A)(6), it can be surmised that this grounding electrode conductor is common to more than one separately derived system. This type of installation requires installing a single common grounding electrode conductor, sized no less than 3/0 AWG copper or 250 kcmil aluminum or copper-clad aluminum. **See Figure 13-17.** A metal water pipe complying with 250.68(C)(1) and a metal frame of a building or structure complying with 250.68(C)(2) can also be used, as indicated in 250.30(A)(6)(a). A wire-type common grounding electrode conductor must be connected to a grounding electrode as specified in 250.30(A)(4).

From each separately derived system, a grounding electrode conductor tap is required to be installed and connected to the common grounding electrode conductor. Each tap conductor must be sized using Table 250.66, based on the largest ungrounded derived phase conductor(s) of the individual derived system it serves.

Figure 13-18   Grounding Electrode Conductor Taps

Figure 13-18. The grounding electrode conductor taps for multiple separately derived systems are permitted to be connected to a busbar to which the common grounding electrode conductor is connected (green marking tape optional).

For example, consider a concrete high-rise structure with separately derived systems installed on each floor. As the closest grounding electrode is located in the basement of the building, an efficient method for grounding all of the separately derived systems is to use this conductor tap concept. The common grounding electrode conductor is typically run vertically through the building core electrical rooms, where connections can be made to it from multiple separately derived systems.

Note that if the only grounding electrode(s) present for connection of the common grounding electrode conductor are those addressed in 250.66(A), (B), or (C), the common grounding electrode conductor is permitted to be sized less than Table 250.66 and in accordance with 250.66(A), (B), or (C) depending on the grounding electrode type.

The connections of the grounding electrode conductor tap to the common grounding electrode conductor must be made at an accessible location. **See Figure 13-18.**

The connections can be made by an exothermic welding process, a connector listed as grounding and bonding equipment, or listed connections to $1/4$-inch thick by two-inch wide copper or aluminum busbar. **See Figure 13-19.**

Figure 13-19   GEC Tap Connections

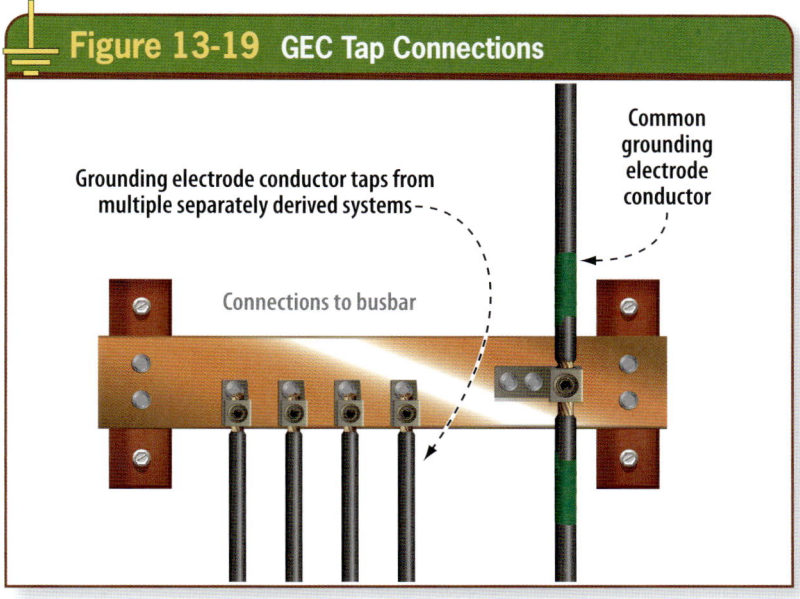

Common grounding electrode conductor

Grounding electrode conductor taps from multiple separately derived systems –

Connections to busbar

Figure 13-19. Connections of grounding electrode conductor taps to a busbar must meet all the requirements in 250.30(A)(6) and be made using a listed connection.

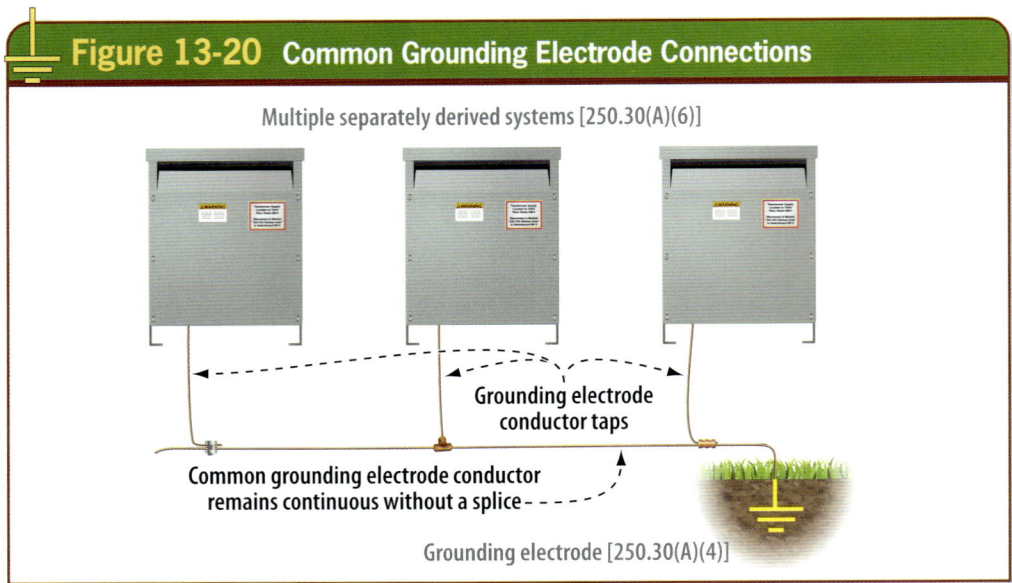

Figure 13-20 Common Grounding Electrode Connections

Multiple separately derived systems [250.30(A)(6)]

Grounding electrode conductor taps

Common grounding electrode conductor remains continuous without a splice

Grounding electrode [250.30(A)(4)]

Figure 13-20. Connections of common grounding electrode conductor taps to the common grounding electrode conductor must be made in a manner such that the latter remains without a splice or joint other than an irreversible compression connection or an exothermic welding connection.

Just as for grounding electrode conductor connections for services, the busbar must be not less than 1/4 inch thick by two inches wide and be long enough to allow for all connections that must be made to it. The connection between the tap and the common grounding electrode conductor must be made in a way that the common grounding electrode conductor remains without a splice or joint. **See Figure 13-20.**

Regardless of whether a grounding electrode conductor is installed for a single separately derived system or whether the common grounding electrode conductor tap concept is used for multiple derived systems, the grounding electrode conductors must be installed in accordance with 250.64(A) through (C) and (E).

BONDING WATER PIPING AND BUILDING STEEL

Section 250.104(D) provides bonding requirements for metal water piping systems and structural steel that are in the same area served by the separately derived system. Installers must be aware that this bonding requirement is typically duplicated for each separately derived system in the building or structure.

For example, in a high-rise building, there may be a transformer-type separately derived system installed on every floor or every other floor. The water piping or structural steel in the same area served by these systems must be bonded to each system. The purpose of this bonding requirement is to put the metal water piping and the structural metal frame at the same potential as the secondary grounding point of the separately derived system. Additionally, this bonding accomplishes another important performance requirement because the bonding jumper provides a direct path for ground-fault current back to the source windings (secondary) and should cause an overcurrent protective device to operate, clearing the faulted condition. Indirect paths back to the source windings would likely have more impedance due to the route of the fault current.

The size of the bonding jumper for both the water piping and the structural steel is, in accordance with Table 250.102(C)(1), based on the largest ungrounded conductor supplied by the separately derived system, but it is not required to be larger than 3/0 AWG copper or 250 kcmil aluminum or copper-clad aluminum. The bonding jumper must be connected to the grounded conductor of the separately derived system, and the connection must be made where the system bonding jumper for the derived system is located. **See Figure 13-21.**

If a common grounding electrode conductor is installed for multiple systems that supply areas with metal water piping systems and exposed structural metal building framing, the metal water piping and structural metal frame must be bonded to the common grounding electrode conductor. The *NEC* does not specifically provide a size for this bonding jumper; however, since the minimum size for the common grounding electrode conductor is 3/0 AWG copper, making this jumper the same size is a conservative approach.

OUTDOOR SOURCE

Some premises wiring installations include separately derived systems such as generators, photovoltaic systems, or transformers installed in outdoor locations. If the source is located outside the building or structure it supplies, a grounding electrode connection to the source is required at the source location outside the building or structure. The grounding electrode used for this connection to the Earth must be in accordance with 250.50. **See Figure 13-22.**

This means that if any of the electrodes in 250.52(A)(1) through (7) are present, they must be used for this grounding connection. In addition to the required grounding connection outside, the installation must meet all applicable requirements in 250.30(A) if the system is grounded or in 250.30(B) if the system is ungrounded. A grounding electrode conductor connection is not required at the outdoor source for an impedance grounded system, which must be installed according to 250.36 or 250.187 for systems greater than 1,000 volts.

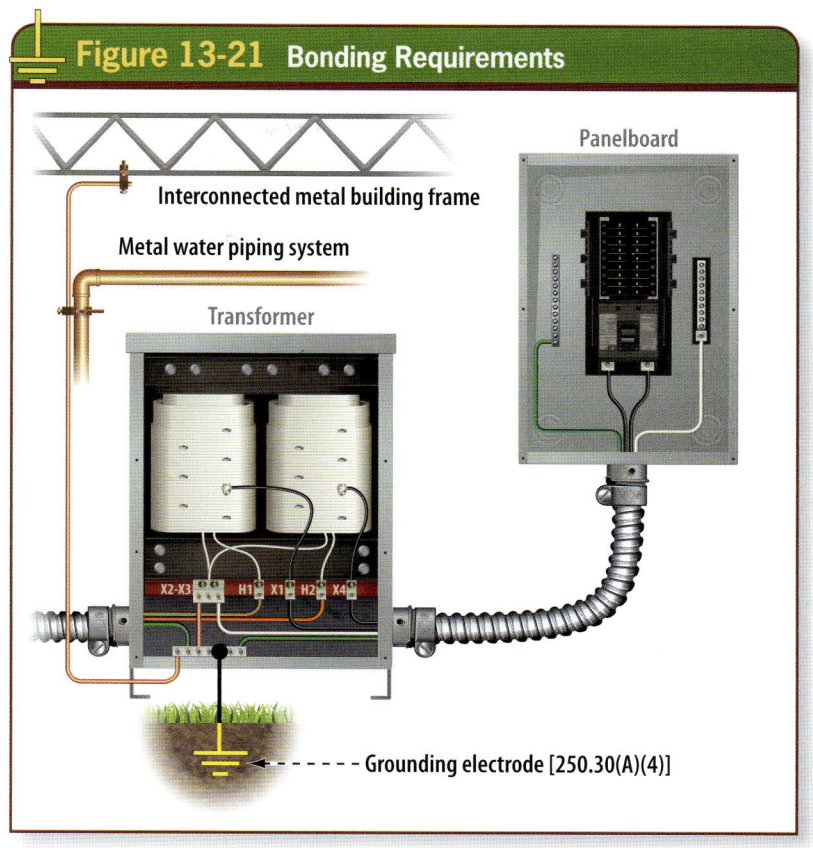

Figure 13-21 Bonding Requirements

Interconnected metal building frame
Metal water piping system
Panelboard
Transformer
Grounding electrode [250.30(A)(4)]

Figure 13-21. *Metal water piping and structural metal building frames are required to be bonded to separately derived systems serving the same area.*

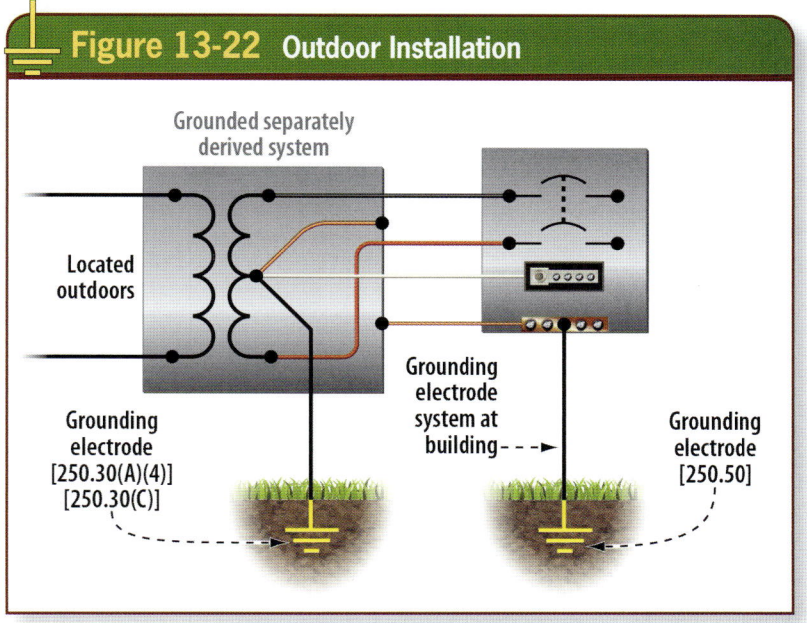

Figure 13-22 Outdoor Installation

Grounded separately derived system
Located outdoors
Grounding electrode [250.30(A)(4)] [250.30(C)]
Grounding electrode system at building
Grounding electrode [250.50]

Figure 13-22. *A grounding electrode connection is required outside for separately derived systems installed outdoors.*

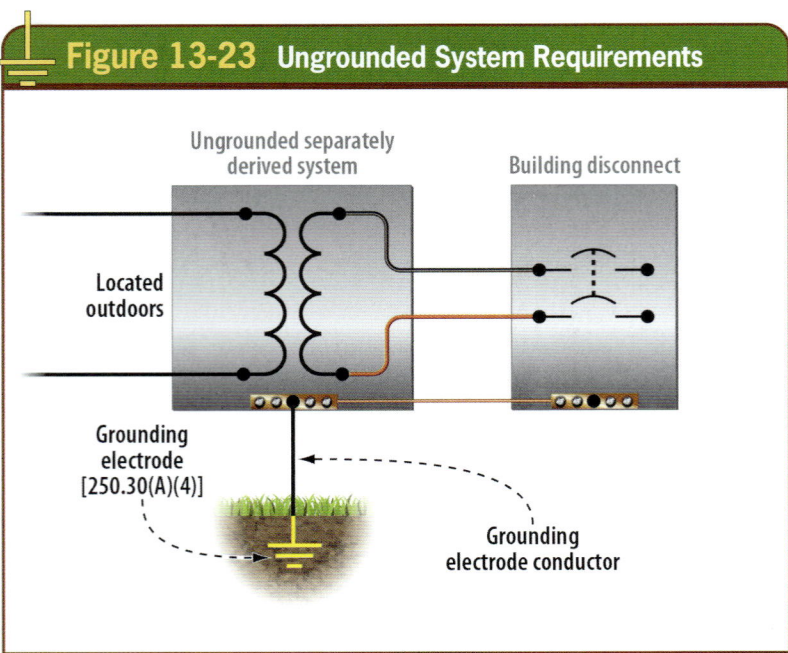

Figure 13-23   Ungrounded System Requirements

Figure 13-23. Grounding requirements apply to enclosures and wiring methods installed for ungrounded separately derived system conductors and equipment.

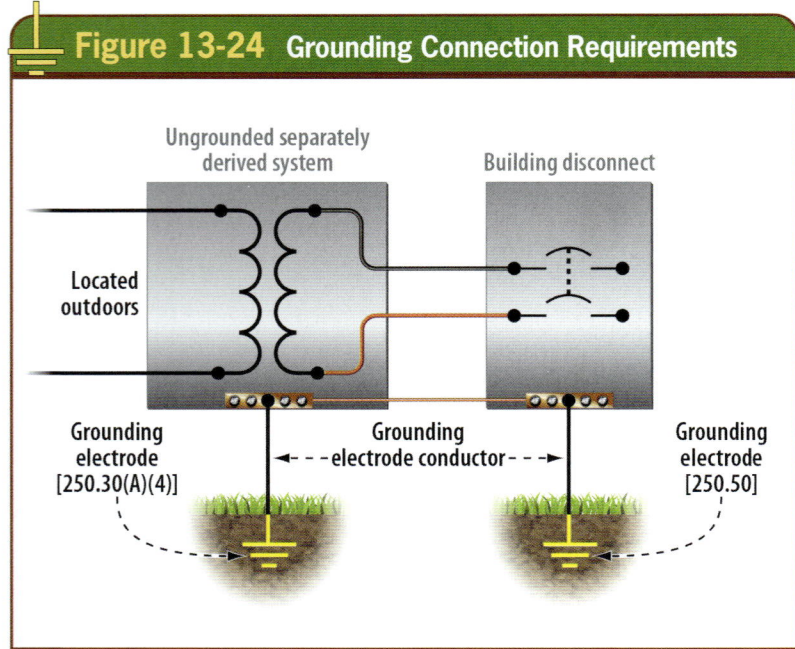

Figure 13-24   Grounding Connection Requirements

Figure 13-24. Grounding electrode conductor connections are required for the conductive wiring methods and equipment installed for ungrounded separately derived systems.

UNGROUNDED SYSTEMS

Ungrounded separately derived systems must meet specific grounding requirements provided in 250.30(B). First, it should be understood that these systems are installed ungrounded, operate ungrounded, and are required to have ground detectors installed in accordance with 250.21(B). No conductor supplied by an ungrounded system is intentionally or solidly grounded, beyond an indirect connection to ground or grounded parts through distributed leakage capacitance. **See Figure 13-23.**

The enclosures and other normally non–current-carrying equipment are required to be connected to ground (the Earth) by a grounding electrode conductor that connects to an electrode that meets the provisions in 250.30(A)(4). The grounding electrode conductor for an ungrounded system is sized the same way the grounding electrode conductor for a grounded system is sized. Use 250.66 based on the size of the largest ungrounded derived phase conductors.

The grounding electrode conductor must connect the metal enclosures to the grounding electrode in one of the methods specified in 250.30(A)(5) or (6), depending on whether it is a single system or the common grounding electrode conductor tap concept for multiple systems is used. The grounding connection can be made at any point on the derived system from the source enclosure to the first system disconnecting means. **See Figure 13-24.**

For ungrounded separately derived systems located outside the building or structure supplied, the grounding electrode conductor connection to the system enclosure must be made outside in accordance with 250.30(C). EGCs or supply-side bonding jumpers are required to be installed with the feeders and branch circuits supplied by an ungrounded system. They still must perform protective functions as described in 250.4(B)(1) through (4). This means that a ground-fault current path must be provided even though the system is installed and operates ungrounded.

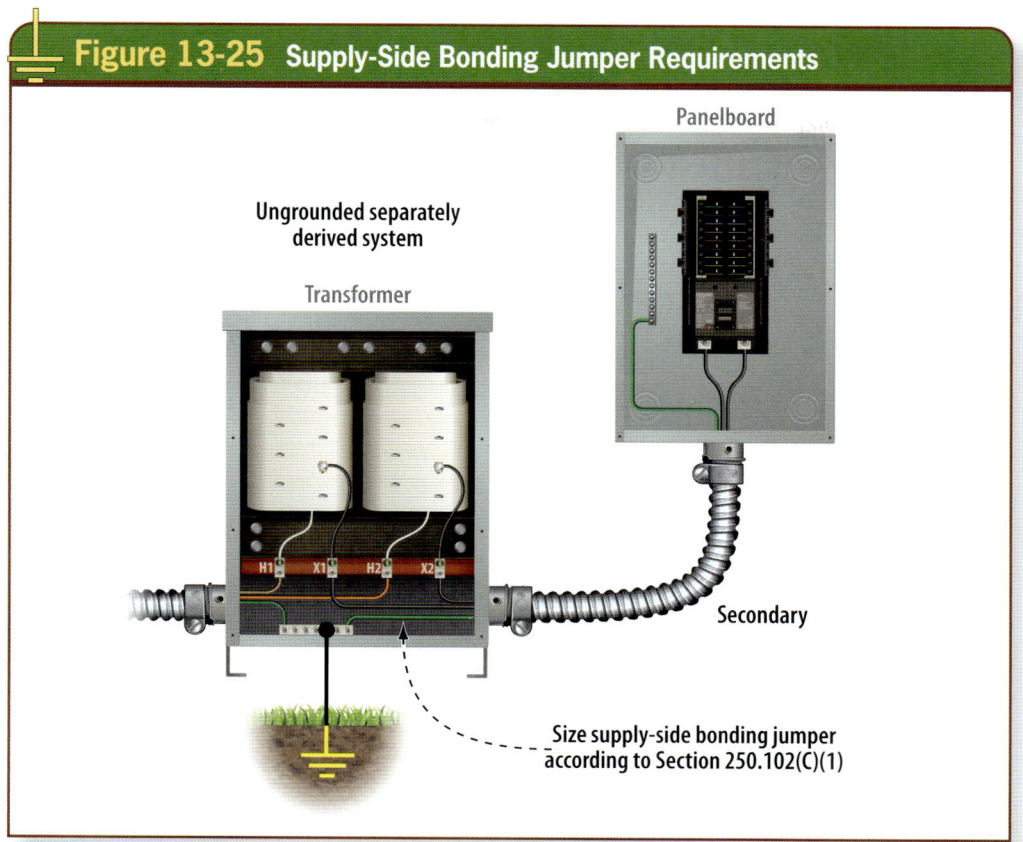

Figure 13-25   Supply-Side Bonding Jumper Requirements

Panelboard

Ungrounded separately derived system

Transformer

H1   X1   H2   X2

Secondary

Size supply-side bonding jumper according to Section 250.102(C)(1)

Figure 13-25. A supply-side bonding jumper is required for ungrounded separately derived systems.

Section 250.30(B)(3) includes requirements to provide a supply-side bonding jumper or path between an ungrounded separately derived system and the first disconnecting means. A wire-type bonding jumper must be sized using 250.102(C) and Table 250.102(C)(1) based on the size of the derived phase conductors. **See Figure 13-25.**

GENERATORS AND TRANSFER EQUIPMENT

Generators can be separately derived systems. How the grounding and bonding connections are made at a generator is usually determined by the type of transfer equipment. There is an important Informational Note following 250.30 that describes the transfer switch and how the grounding connections are made for the generator. **See Figure 13-26.**

Figure 13-26   Generator Requirements

Figure 13-26. If transfer equipment includes a switching action in the grounded conductor, the generator must be grounded according to the requirements in 250.30(A).

Courtesy of Eaton

First, if a transfer switch for a generator includes a switching action in the grounded conductor, then the generator must be grounded as a separately derived system in accordance with all applicable requirements in 250.30(A). This is necessary because in the normal mode the grounded conductor is connected to the service grounding electrode, whereas in the standby mode the grounded conductor is switched over to the generator, which must be grounded as a separately derived system. The result is that in either position of the transfer switch, the system is grounded. **See Figure 13-27.**

If there is no switching action in the grounded conductor by the transfer equipment, then the generator system is grounded with the transfer switch in either position if there is a grounded conductor. A generator system can be grounded (neutral point connected to earth through a grounding electrode conductor), but if no line-to-neutral loads are supplied, there might not be a

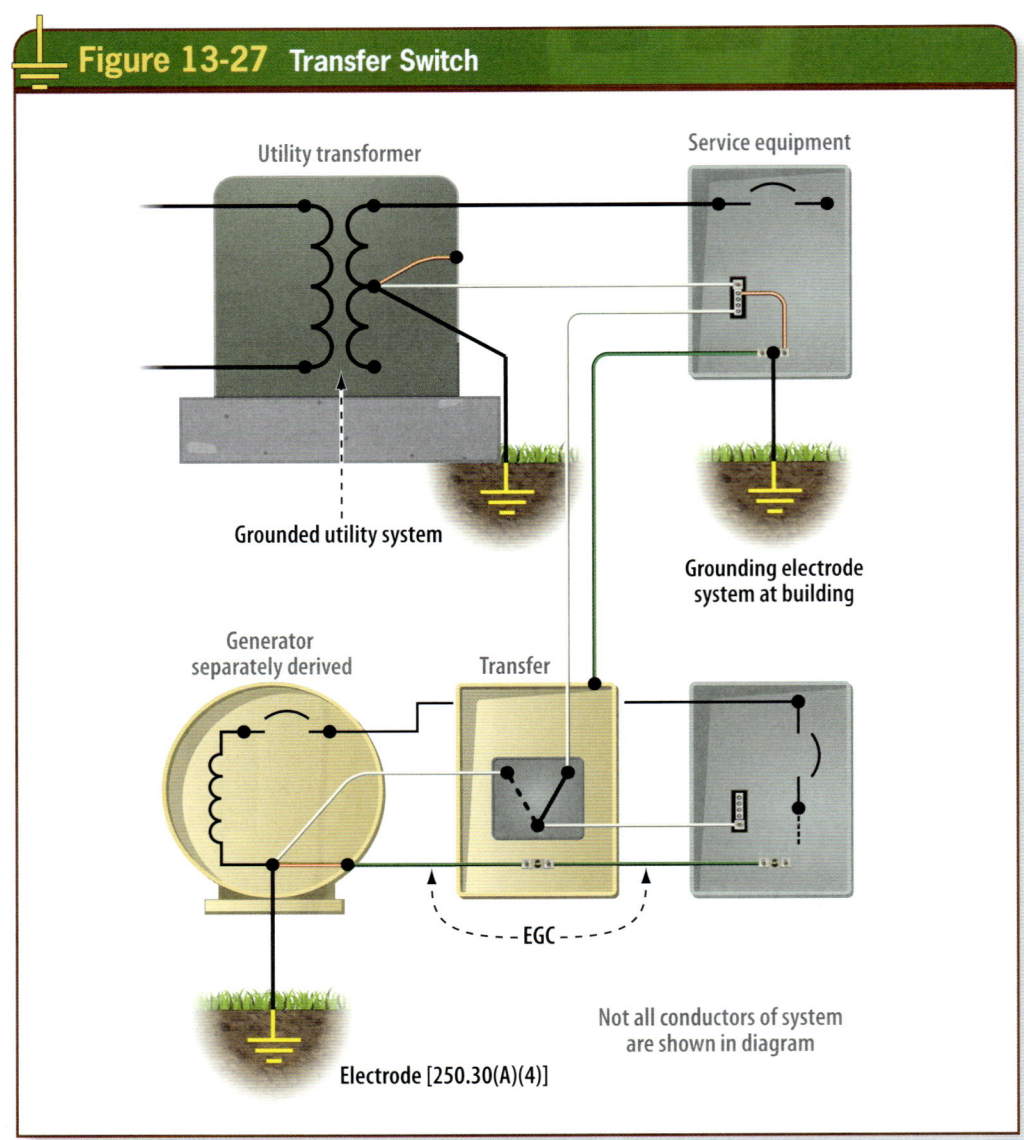

Figure 13-27   Transfer Switch

Figure 13-27. When the generator is grounded as a separately derived system and a transfer switch switches the grounded conductor of the system, the system remains grounded whether the transfer switch is in the normal position or the standby position.

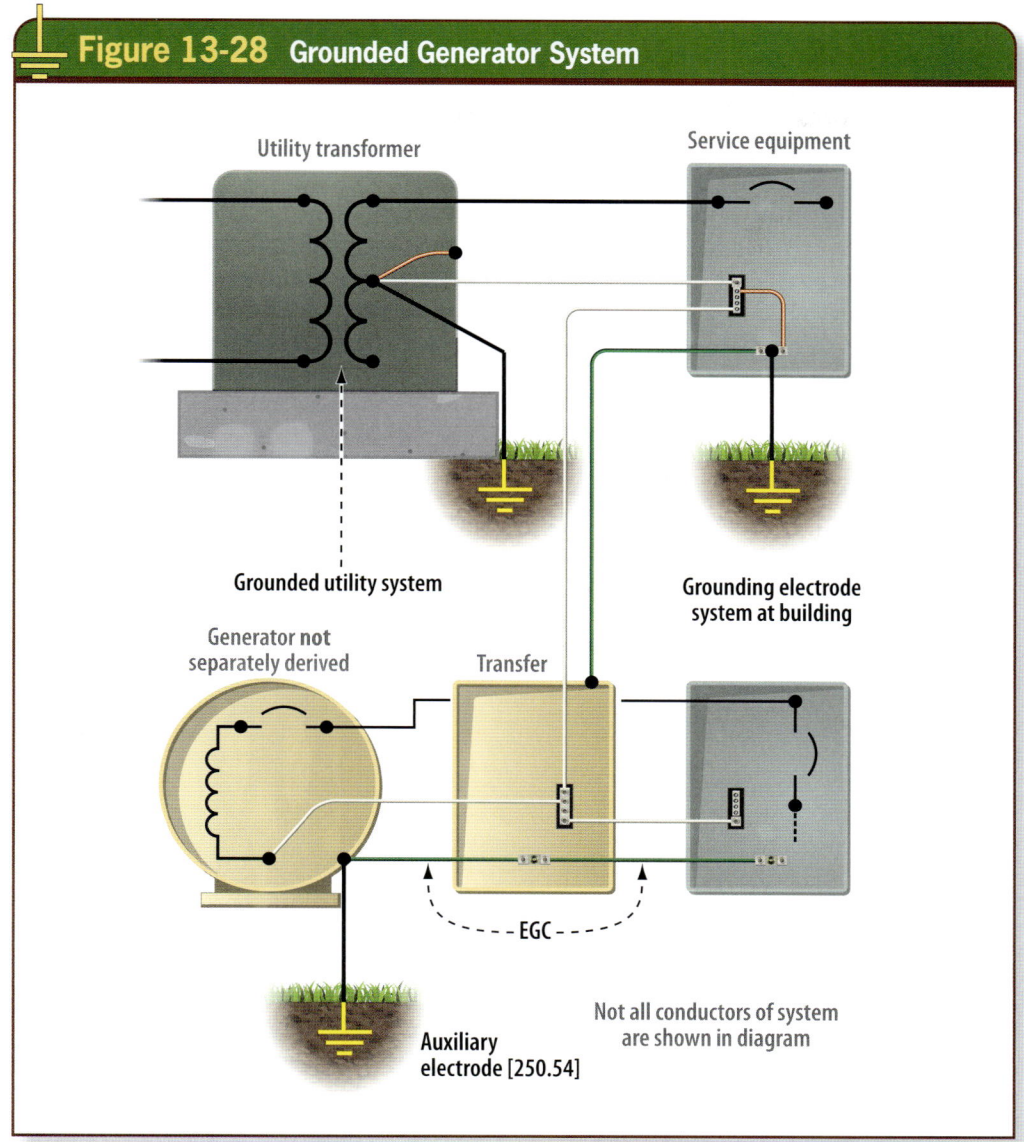

Figure 13-28 Grounded Generator System

- Utility transformer
- Service equipment
- Grounded utility system
- Grounding electrode system at building
- Generator **not** separately derived
- Transfer
- EGC
- Auxiliary electrode [250.54]
- Not all conductors of system are shown in diagram

Figure 13-28. If a transfer switch does not switch the grounded conductor, the separately derived system remains grounded with the switch in either position. The system is grounded by the connection to the service grounded conductor.

grounded conductor. Note that although the generator in this case is not a separately derived system, the grounding and bonding connections must meet the requirements in 250.35 for permanently installed generators. **See Figure 13-28.**

The performance concepts of 250.35 are focused on providing an effective ground-fault current path with the supply conductors of the generator to the first disconnecting means or equipment supplied. Generators grounded as separately derived systems meet this requirement when installed according to the rules in 250.30(A). If the system does not have an overcurrent protective device at the generator, a supply-side bonding jumper must be installed between the generator and the transfer equipment. The sizing requirements for this bonding jumper are related to the location of the first system overcurrent protective device. If the bonding jumper is on the supply side of the first

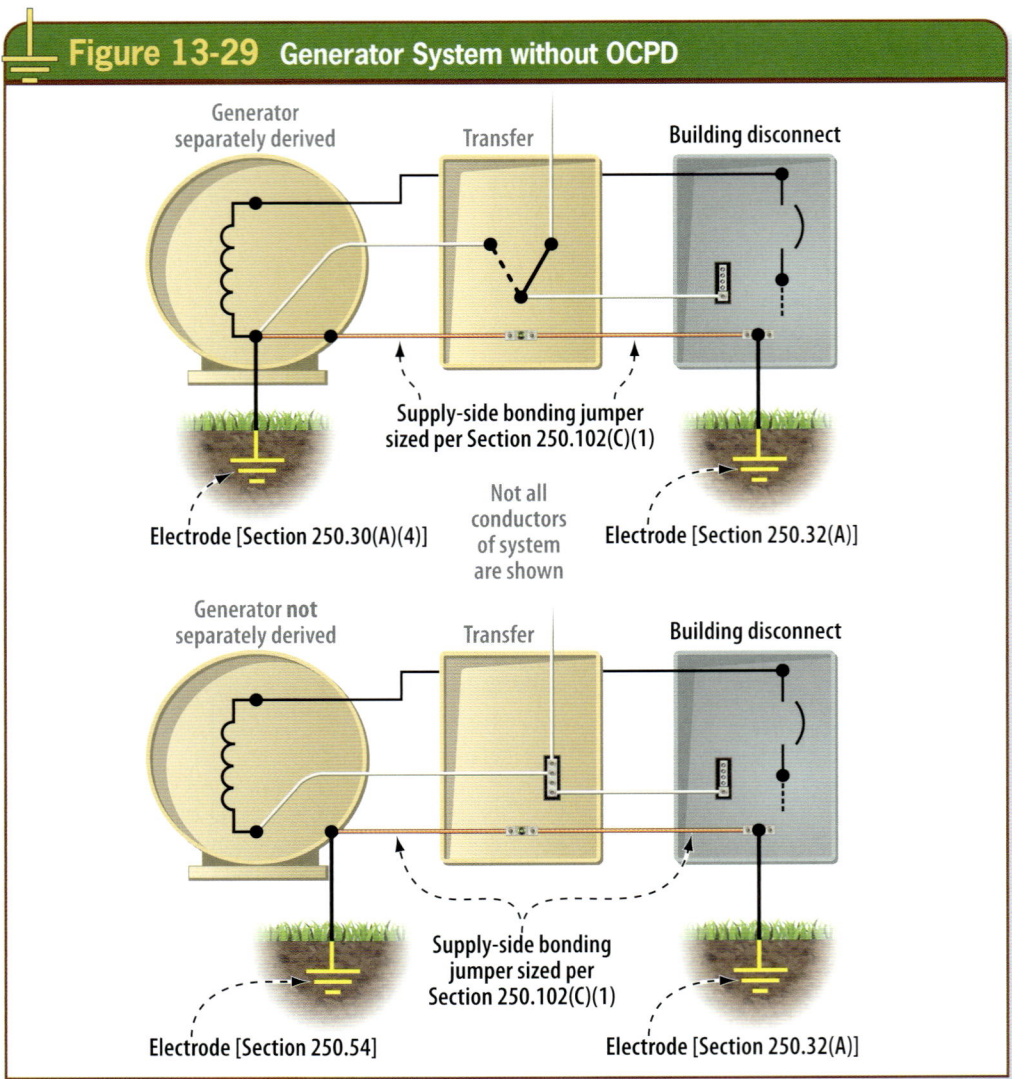

Figure 13-29. *A supply-side bonding jumper for generators without overcurrent protection at the source must be sized according to 250.102(C)(1).*

system overcurrent protective device, it is sized as a supply-side bonding jumper in accordance with 250.102(C) and Table 250.102(C)(1). This means that the size is based on Table 250.102(C)(1) or 12.5%, based on the circular mil area of the ungrounded derived phase conductors supplied from the generator. **See Figure 13-29.**

Generators are often installed outside of buildings or structures with the feeder routed to the transfer equipment, typically inside the building or structure. The conductor installed on the load side of a generator overcurrent protective device is an EGC because it is a load-side installation. This EGC is required to be sized using Table 250.122, based on the rating of the overcurrent protective device supplied. **See Figure 13-30.**

Fact

If an overcurrent protective device is installed at the generator, an EGC is installed with the generator conductors to the first enclosure supplied by the system. This is an EGC and is required to be sized using Table 250.122, based on the rating of the overcurrent protective device.

Generators are often installed outside of buildings or structures with the feeder routed to the transfer equipment, typically inside the building or structure.

Again, the importance of these grounding and bonding connections and sizing rules is to provide an effective path for the ground-fault current back to the generator (source).

WIND ELECTRICAL SYSTEMS

Article 694 of the *NEC* applies to wind electrical systems that consist of one or more wind-driven electric generators. These systems can include generators, alternators, inverters, and controllers. Part V of Article 694 provides the grounding and bonding rules specific to these systems and associated equipment. Exposed non–current-carrying metal parts of

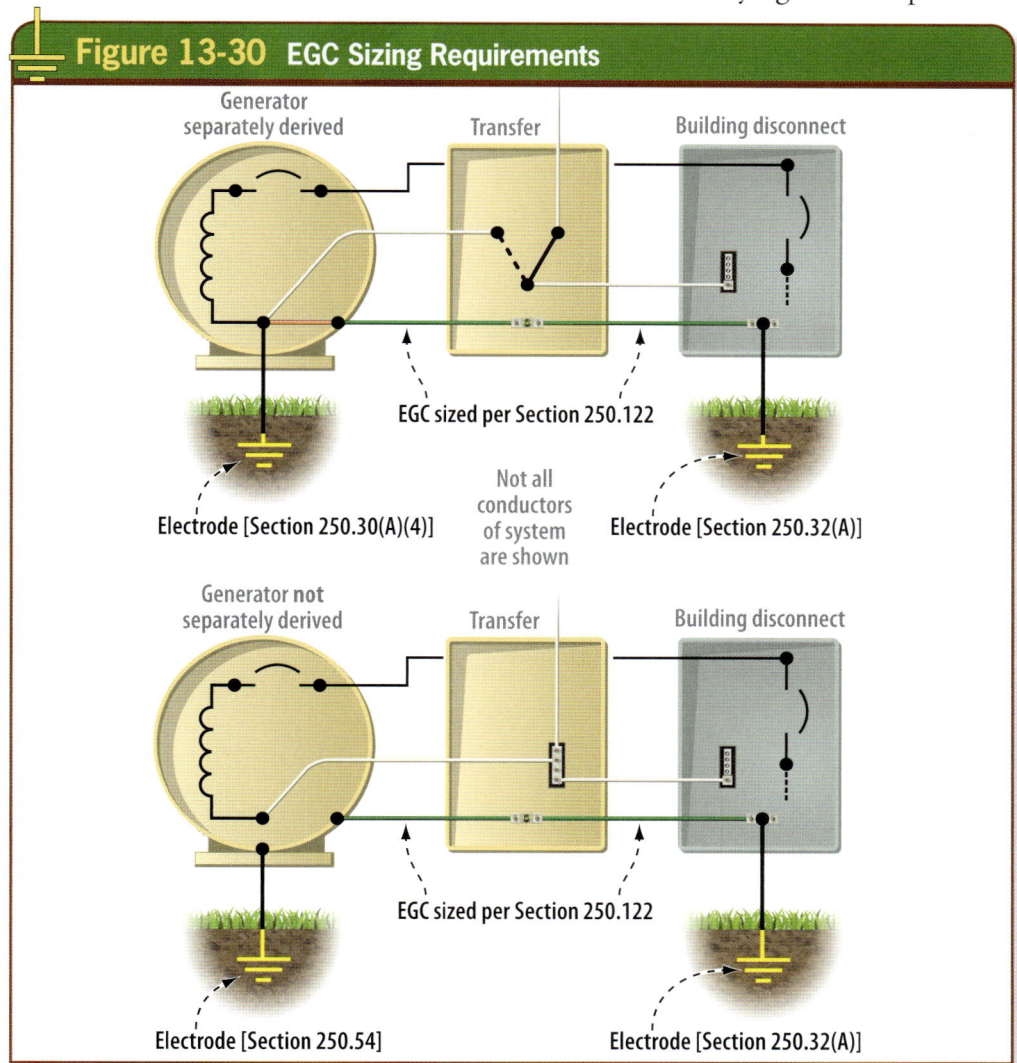

Figure 13-30 EGC Sizing Requirements

Figure 13-30. A load-side EGC for a generator that is provided with overcurrent protection at the source is sized based on 250.122.

towers, turbines, conductor enclosures, and other equipment are required to be grounded in accordance with 250.134 or 250.136, regardless of voltage. Turbine blades and tails that are not likely to become energized are not required to be grounded. **See Figure 13-31.**

Tower Grounding

A wind turbine tower is required to be grounded with one or more auxiliary electrodes to limit voltages imposed by lightning. Auxiliary electrodes must be installed in accordance with 250.54. Electrodes that are part of the tower foundation and that meet the requirements for concrete-encased electrodes are acceptable. A grounded metal tower support that also qualifies as a grounding electrode according to 250.52 is acceptable if it meets the requirements of 250.136. This means the conductive tower can also serve as a grounding means for equipment connected to it in some cases, provided the equipment is connected to an EGC installed with the branch circuit supplying the equipment.

If a tower for a wind generator is installed near any galvanized tower anchors or foundation components, galvanized grounding electrodes must be installed. This is because of the electrolytic corrosion of galvanized foundation and tower anchoring components where copper and copper-clad grounding electrodes are used.

System Grounding

Turbines that drive generators create an electrical power system. Separately derived systems that are required to be grounded in accordance with 250.20 must also be grounded according to the requirements in 250.30(A).

Equipment Grounding and Bonding

An EGC or supply-side bonding jumper is required between a turbine and the premises grounding system. EGC installations must be in accordance with 250.120, and the types of EGCs used are provided in 250.118. Sizing EGCs

Figure 13-31   Wind Electrical

Figure 13-31. Wind electrical systems can be separately derived systems and have to be grounded according to 250.30(A).

Courtesy of NECA ©2010 Rob Colgan

for overcurrent-protected feeders supplied from turbine systems must be done according to 250.122. Supply-side bonding jumper connections must be made in accordance with 250.30(A)(2) and 250.102(C). Guy wires used to support turbine towers are not required to be connected to an equipment grounding conductor.

Grounding and Bonding Connections

The EGC and grounding electrode conductors are required to be connected to a metallic tower by exothermic welding, listed lugs, listed pressure connectors, listed clamps, or other listed means.

Devices such as connectors and lugs shall be suitable for the material of the conductor and the structure to which they connect to ensure compatibility between the two metals. If practicable, contact between dissimilar metals should be avoided to reduce galvanic action and corrosion. All mechanical terminations of EGCs and grounding electrode conductors must be accessible, although buried and concrete-encased electrode connections are generally not required to be accessible.

Auxiliary electrodes and grounding electrode conductors are recognized as lightning protection system components if they meet applicable requirements. If a tower lightning protection system grounding network is installed, the power system electrodes must be bonded to the tower auxiliary grounding electrode system. Guy wire lightning protection system grounding electrodes are not required to be bonded to the tower auxiliary grounding electrode system.

GROUNDING DC SYSTEMS

The current in DC (direct current) systems is unidirectional, meaning the current flows in one direction. This is unlike AC (alternating current), which is bi-directional, meaning AC reverses its direction of flow at regular intervals.

Batteries are a common DC source. Battery systems are used for backup power systems, such as uninterruptible power supplies, and for emergency lighting backup systems. **See Figure 13-32.** Generators and photovoltaic arrays are other sources of DC power. DC generators operate on the principle of magnetic induction.

Grounding requirements for DC systems are related to the type of system and output voltages of the system, similar to the requirements that govern AC system grounding. Grounding for DC systems is covered in Part VIII of Article 250. This part includes grounding requirements specific to DC systems and is required to be applied in addition to the other applicable parts of Article 250.

A DC system is required to be grounded as follows:

1. Two-wire DC systems supplying premises wiring operating at a voltage greater than 60 volts but not exceeding 300 volts are required to be grounded. See the three exceptions to 250.162(A).

2. All 3-wire DC systems are required to be grounded by connecting the neutral conductor of the system to ground.

Figure 13-32 Battery Systems

Figure 13-32. DC systems are often supplied by batteries that are connected to a charging means.

Where a 2-wire DC system is required to be grounded in accordance with 250.162(A), the grounded conductor can be either output conductor, as indicated in 250.26. There are three alternatives that relax this grounding requirement as follows:

1. A DC system supplying only industrial equipment in limited areas and provided with a ground-fault detection system is permitted to be installed and operated ungrounded.
2. If a rectifier-derived DC system is supplied from an AC system in accordance with 250.20, it is permitted to be installed and operated ungrounded.
3. System grounding is not required for the DC source for a fire alarm system if the maximum source circuit current is not more than 0.030 amperes, as specified in Article 760, Part III.

In a 3-wire DC system, the neutral conductor must be grounded. **See Figure 13-33.**

Point of Grounding Connection for DC Systems

The connection of the grounding electrode conductor for a DC system is covered in 250.164. If the DC source is not on the premises, this connection shall be made at one or more supply stations. The grounding electrode conductor connection is not to be made at individual services or at any point on the premises wiring if the source is not on the premises.

When the DC system is on the premises, the grounding electrode conductor connection must be made at one of the following locations:

1. At the source
2. At the first system disconnecting means or overcurrent protective device
3. At another location that provides equivalent protection and uses listed and identified equipment

Size of a DC Grounding Electrode Conductor

The minimum size of a grounding electrode conductor for a DC system is generally required to be no less than the sizes indicated in 250.166(A) and (B).

If the DC system consists of a 3-wire balancer set or a balancer winding with overcurrent protection, as provided in 445.12(D), the grounding electrode conductor shall be no smaller than the neutral conductor and can never be smaller than 8 AWG copper or 6 AWG aluminum. If the DC system is other than a 3-wire balancer set, the grounding electrode conductor cannot be smaller than the largest conductor supplied by the system and can generally not be smaller than 8 AWG copper or 6 AWG aluminum.

Note that if the grounding electrode conductor for a DC system is connected only to a rod, pipe, or plate electrode, the grounding electrode conductor does not have to be sized larger than 6 AWG copper or 4 AWG aluminum. If the grounding electrode conductor is connected only to a concrete-encased electrode, it never has to be larger than 4 AWG copper. If the grounding electrode conductor for a DC system is connected only to a ground ring, it never has to be larger than the size of the ring electrode. These provisions are

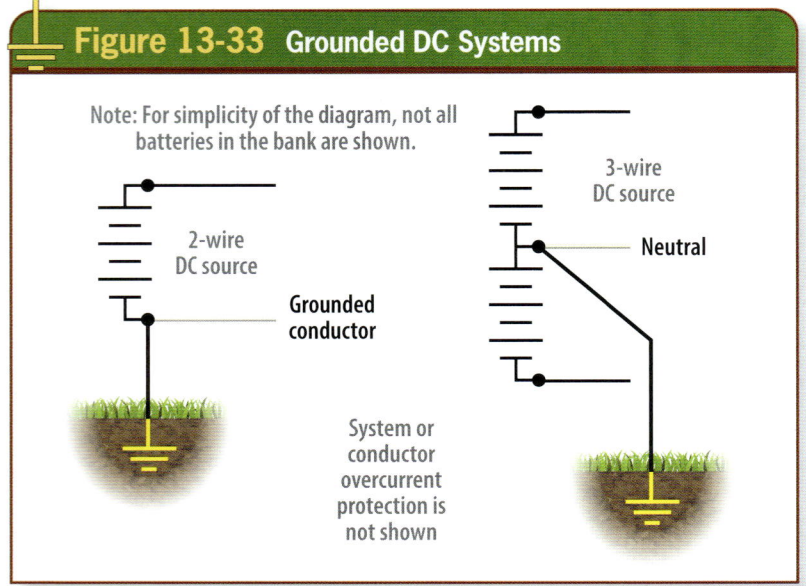

Figure 13-33 **Grounded DC Systems**

Note: For simplicity of the diagram, not all batteries in the bank are shown.

3-wire DC source

2-wire DC source

Neutral

Grounded conductor

System or conductor overcurrent protection is not shown

Figure 13-33. Grounded DC systems must be in accordance with the requirements in Part VIII of Article 250.

for connections of grounding electrode conductors installed for DC systems and are identical to the allowances in 250.66(A) through (C).

Section 250.166 indicates that the minimum size grounding electrode conductor for a DC system must be in accordance with 250.166(A) and (B), unless permitted otherwise by 250.166(C) through (E), but is not required to be larger than 3/0 AWG copper or 250 kcmil aluminum or copper-clad aluminum. The installation requirements in 250.64 apply to grounding electrode conductors for DC systems. This includes, but is not limited to, installing the grounding electrode conductors in a continuous length without a splice or joint and protecting them where they are subject to physical damage.

Size of a DC System Bonding Jumper

Where a DC system is grounded, a system bonding jumper is required to connect the grounded conductor of the system to the EGC(s). This connection must be made at the source or first system disconnecting means where the system is grounded. The size of a DC system bonding jumper must be no less than the system grounding electrode conductor, based on 250.166. **See Figure 13-34.**

The system bonding jumper must be a copper or other corrosion-resistant conductor and can be in the form of a wire, bus, screw, or other suitable conductor. If the system bonding jumper is a screw, the screw must be green. The system bonding jumper(s) must be connected by using any method or combination of methods specified in 250.8.

UNGROUNDED DC SYSTEMS

An ungrounded system has no system conductor that is intentionally grounded. This means that the output conductors will be ungrounded. Where a DC system is a stand-alone power source, such as a battery- or engine-driven generator, the enclosures and other conductive equipment parts are required to be connected to ground (the Earth) by a grounding electrode conductor that connects to an electrode that meets the provisions of Part III of Article 250.

The grounding requirements apply to all metal raceways, cables, and other exposed non–current-carrying metal

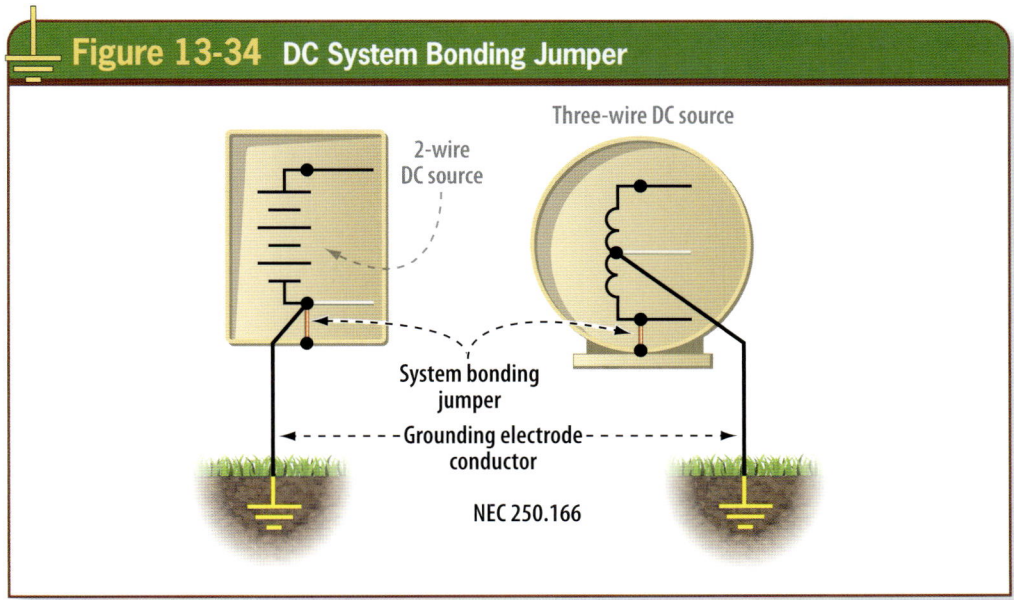

Figure 13-34 DC System Bonding Jumper

Three-wire DC source

2-wire DC source

System bonding jumper

Grounding electrode conductor

NEC 250.166

Figure 13-34. The DC system bonding jumper is typically installed at the source or at the first system disconnecting means enclosure.

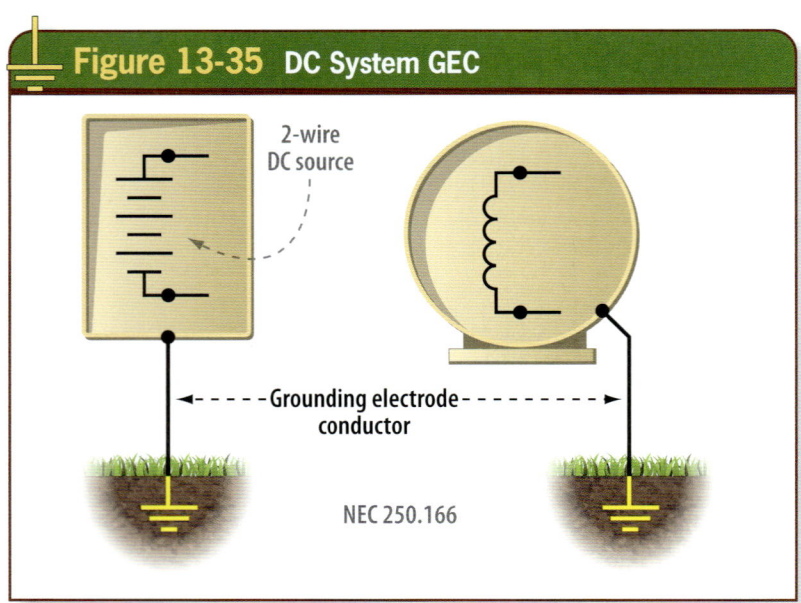

equipment parts. The grounding connection to DC source equipment enclosures, raceways, and so forth is permitted to be made at any point on the system from the source up to the first system disconnecting means enclosure or overcurrent protective device.

The grounding electrode conductor installed for a DC system must be sized based on 250.166. Ground-fault detection systems are required for ungrounded DC systems and are permitted for grounded DC systems. The DC system must be marked at the source or first disconnect to identify the type of DC grounding system applied. **See Figure 13-35.**

Figure 13-35. The grounding electrode conductor for DC power systems must be sized according to 250.166.

Generators are often installed in residential, commercial, and industrial applications. The transfer equipment is key as to whether the generator-produced system is required to be grounded in accordance with Section 250.30(A).

SUMMARY

System grounding requirements are provided in Part II of *NEC* Article 250. Whether or not a system is required to be grounded is specified in 250.20. Separately derived systems can be in the form of a battery system, transformer(s), generator(s), inverter windings, photovoltaic system(s), and so forth. Multiple separately derived systems of the same type that are connected in parallel are considered a single separately derived system. If a system is required to be grounded, the rules in 250.30(A) must be applied. There are specific requirements in 250.30(B) that apply to ungrounded separately derived systems. Part VIII of Article 250 provides specific rules for DC systems that must be grounded.

REVIEW QUESTIONS

1. Which electrical system or source does not qualify as a separately derived system?

   a. Autotransformer
   b. Battery system
   c. Generator
   d. Photovoltaic system

2. A premises wiring system whose power is derived from a source of electric energy or equipment other than a service best describes a __?__.

   a. hydropower facility
   b. separately derived system
   c. utility generating station
   d. utility power network

3. The connection between the grounded circuit conductor and the supply-side bonding jumper, the EGC, or both at a separately derived system best defines a(n) __?__.

   a. EGC
   b. equipment bonding jumper
   c. main bonding jumper
   d. system bonding jumper

4. A conductor installed on the supply side of a service, within a service equipment enclosure, or for a separately derived system that ensures the required electrical conductivity between metal parts required to be electrically connected best defines which component of the grounding and bonding system?

   a. EGC
   b. Equipment bonding jumper
   c. Supply-side bonding jumper
   d. System bonding jumper

5. Common types of separately derived systems installed in commercial and industrial applications are generators and transformers.

   a. True   b. False

6. If there is no disconnecting means or overcurrent protective device at the load end of the conductors supplied by a separately derived system, the system bonding jumper must be installed __?__.

   a. in the first overcurrent protective device enclosure
   b. in the service equipment enclosure
   c. in the source enclosure
   d. outside of the source enclosure on the load side of the system

REVIEW QUESTIONS

7. If the system bonding jumper for a separately derived system is installed at the first disconnecting means enclosure, it must connect the grounded conductor of the system to __?__ .

   a. the disconnecting means enclosure
   b. the EGC(s) and the grounding electrode conductor
   c. the supply-side bonding jumper
   d. all of the above

8. The system bonding jumper for a separately derived system must be sized in accordance with __?__ .

   a. *NEC* Table 8, Chapter 9
   b. Section 250.28(D)
   c. Section 250.122
   d. Section 310.16 and Table 310.16

9. A system bonding jumper for an outdoor transformer separately derived system is not permitted at the source enclosure and the first overcurrent protective device enclosure even if there are no parallel paths for current in the grounded conductor of the system.

   a. True   b. False

10. Where a nonmetallic raceway is installed for the secondary of a separately derived system, a supply-side bonding jumper is required to be installed and sized according to __?__ , based on the largest ungrounded derived system conductor.

    a. *NEC* Table 8, Chapter 9
    b. Table 250.102(C)(1) or 12.5%
    c. Table 250.122
    d. Section 310.16 and Table 310.16

11. The grounded conductor supplied by a separately derived system must be no smaller than required by __?__ if there is no neutral load on the system.

    a. Table 250.2
    b. Table 250.66
    c. Table 250.102(C)(1)
    d. Table 250.122

12. A separately derived system must be grounded by connection to which of the following?

    a. The building or structure grounding electrode system
    b. The metal enclosure of the system
    c. The metal gas piping
    d. The metal water piping system in the area served

13. The grounding electrode conductor for a separately derived system must be connected to the grounded conductor where the __?__ is installed.

    a. equipment bonding jumper
    b. first overcurrent protective device
    c. grounding electrode
    d. system bonding jumper

14. A 400-ampere motor control center is supplied by a separately derived system, and the largest ungrounded conductor is 600 kcmil copper. The minimum size required for the grounding electrode conductor is __?__ if the grounding electrode is a metal water pipe.

    a. 3 AWG copper
    b. 2/0 AWG copper
    c. 1/0 AWG copper
    d. 1 AWG copper

15. Where the common grounding electrode conductor tap concept is used for grounding multiple separately derived systems, the minimum size required for the common grounding electrode conductor is __?__ .

    a. 3/0 AWG copper
    b. 1 AWG copper
    c. 1/0 AWG copper
    d. 2/0 AWG copper

16. Where the grounding electrode tap conductors for multiple separately derived systems are connected to a common grounding electrode conductor, the minimum size required for each tap conductor is __?__ if each system secondary derived ungrounded conductor is sized at 300 kcmil copper.

    a. 3 AWG copper
    b. 1 AWG copper
    c. 3/0 AWG copper
    d. 1/0 AWG copper

17. The connections of grounding electrode conductor taps to a common grounding electrode conductor can be made using all but which of the following methods?

    a. A connection to a 1/4-inch by 2-inch wide copper or aluminum busbar
    b. A connector listed as grounding and bonding equipment
    c. Exothermic welding process
    d. Solder connections

18. If the transfer switch for a generator system switches the grounded (neutral) conductor, the generator must be grounded as a separately derived system.

    a. True   b. False

19. If a generator has a 1,200-ampere overcurrent protective device installed at the generator set, a minimum size of __?__ is required for the EGCs installed in a parallel set of four-inch PVC conduits routed from the generator to the first system overcurrent protective device enclosure.

    a. 1/0 AWG copper
    b. 2/0 AWG copper
    c. 3/0 AWG copper
    d. 4/0 AWG copper

20. If a generator without an overcurrent protective device supplies a building, and a supply-side bonding jumper is installed with 400-kcmil copper feeder conductors, the minimum size required for the supply-side bonding jumper is __?__.

    a. 3 AWG copper
    b. 2 AWG copper
    c. 1 AWG copper
    d. 1/0 AWG copper

21. Metal water piping and structural metal framing that exists in the area served by a separately derived system must be bonded to the grounded conductor of that separately derived system. __?__ is used to size this bonding jumper or conductor.

    a. *NEC* Table 8, Chapter 9
    b. Table 250.102(C)(1) based on the largest ungrounded conductor of the system (but not required to be larger than 3/0 AWG copper or 250 kcmil aluminum or copper-clad aluminum)
    c. Table 250.122
    d. Table 310.16

22. The minimum size required for a copper wire-type system bonding jumper is __?__ if the size of the largest ungrounded derived phase conductor connected to the system is 750 kcmil aluminum.

    a. 4/0 AWG
    b. 3/0 AWG
    c. 2/0 AWG
    d. 1/0 AWG

23. The size of a DC system bonding jumper must be no less than the system grounding electrode conductor based on 250.166.

    a. True   b. False

24. A common grounding electrode conductor installed for multiple separately derived systems is permitted to be which of the following?

    a. 3/0 AWG copper conductor
    b. Interconnected metal building frame
    c. Metal water piping system
    d. Any of the above

25. If the only grounding electrode(s) present for connection of the common grounding electrode conductor are those addressed in 250.66(A), (B), or (C), then the grounding electrode conductor(s) are not permitted to be sized less than Table 250.66.

    a. True   b. False

26. Separately derived systems that consist of multiple sources of the same type are permitted to have a single system bonding jumper installed at the paralleling switchboard, switchgear, or other paralleling equipment connection point instead of at each of the multiple source enclosures.

    a. True   b. False

Special Occupancies and Conditions

Chapter 5 of the *NEC* provides rules for special occupancies that modify or amend the general requirements of Chapters 1 through 7. Electrical installations in special occupancies often require a more restrictive approach in order to address specific concerns, and they can require grounding and bonding rules that exceed general requirements. It is important to establish a clear comprehension of the electrical wiring that exceeds what would normally be required for electrical systems in general locations. Examples include restrictive grounding and bonding requirements for health care facilities and hazardous (classified) locations.

Objectives

» Understand why grounding and bonding requirements for special occupancies are often more restrictive than general *NEC* grounding and bonding requirements.

» Understand the specific grounding and bonding rules for hazardous (classified) locations.

» Understand the special requirements for grounding and bonding in patient care spaces of health care facilities.

» Understand the unique grounding and bonding problems in agricultural facilities and the rules that apply.

» Determine special grounding and bonding requirements for mobile homes and manufactured homes.

» Understand the special requirements for grounding and bonding protection in marinas, boatyards, and docking facilities.

Chapter 14

Table of Contents

Special Rules for Hazardous Locations........ 320

Special Rules for Health Care Facilities........ 324

Grounding of Receptacles and Fixed Equipment in Patient Care Spaces......... 325

Armored Cable (Type AC) 328

Required Wire-Type EGC 329

Isolated Ground Receptacle Installations in Patient Care Spaces... 330

Patient Equipment Grounding Point ... 331

Isolated Power System Grounding.... 332

Grounding and Bonding Requirements for Panelboards 332

Grounding Requirements................. 332

Bonding Requirements 333

Special Rules for Agricultural Installations.... 334

Voltage Gradients 334

Site-Isolating Device............................ 336

Equipotential Planes 337

Mobile and Manufactured Home Grounding and Bonding Rules 338

Mobile Homes 338

Manufactured Homes 339

Special Rules for Marinas, Boatyards, and Docking Facilities 340

Grounding, Bonding, and Equipotential Planes 340

GFPE and GFCI Requirements 341

Summary... 341

Review Questions 342

SPECIAL RULES FOR HAZARDOUS LOCATIONS

Chapter 5 of the *NEC* provides requirements for special occupancies. Articles 500 through 517 contain specific rules for electrical equipment installed and

Figure 14-1   Grounding in Hazardous Locations

Figure 14-1. More specific and strengthened methods of bonding are required for metal raceways and enclosures in hazardous (classified) locations.

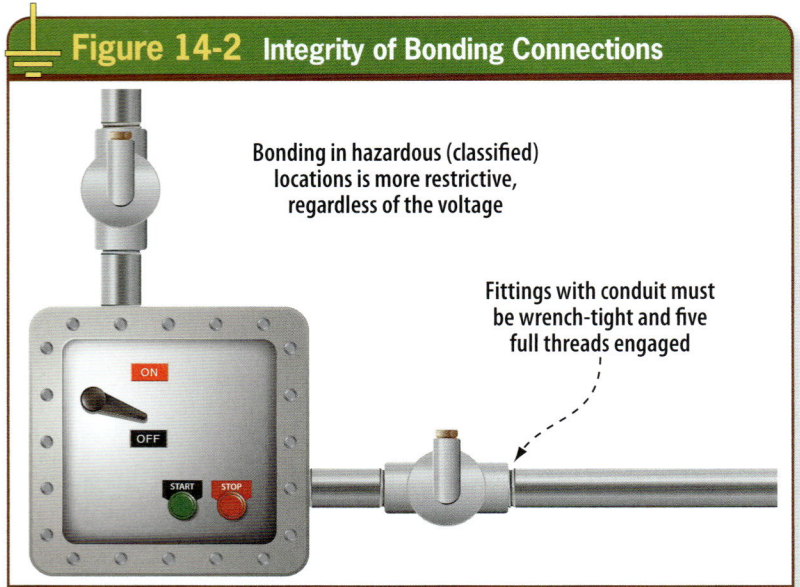

Figure 14-2   Integrity of Bonding Connections

Bonding in hazardous (classified) locations is more restrictive, regardless of the voltage

Fittings with conduit must be wrench-tight and five full threads engaged

ON
OFF
START   STOP

Figure 14-2. The threads of couplings, fittings, boxes, and so forth must have five full threads fully engaged when installed in hazardous (classified) locations.

operated in hazardous (classified) locations, such as motor fuel dispensing facilities, chemical plants, and bulk fuel storage facilities. Chapters 1 through 4 of the *NEC* apply generally to these installations, but special requirements provided in Chapter 5 can and often do significantly modify these general provisions.

For equipment grounding and bonding requirements in Class I, Class II, and Class III hazardous locations, the specific rules are found in 501.30, 502.30, and 503.30, respectively. For hazardous locations using the Zone System, the grounding and bonding rules are provided in Articles 505 and 506, specifically in Sections 505.30 and 506.30. Grounding and bonding requirements that apply to intrinsically safe systems are provided in Sections 504.50 and 504.60, respectively. **See Figure 14-1.**

Special equipment is often required in locations classified as hazardous. This means the equipment needs to be suitable for use in such specific locations. Suitability is often established through product certification (listing) and identification. The *NEC* rules for conductive equipment, raceways, and other electrical wiring installed in these areas are more restrictive in most cases than they are for other nonhazardous locations. In explosive atmospheres, it is essential to control and eliminate all possible sources of ignition, such as an electrical arcing event. A motor controller or switch that has make-and-break arcing contacts is a good example of electrical equipment that must be installed in suitable enclosures that effectively contain arcing under normal conditions.

Part V of Article 250 contains general bonding requirements for electrical installations. Section 250.90 provides a performance requirement that bonding be provided to ensure electrical continuity and adequate capacity for any fault current. Effective bonding is important for safety in general locations, but in hazardous locations there are increased concerns about establishing an effective ground-fault current path that will not create an explosion or fire if a ground fault were to occur.

During a ground-fault event, the amount of fault current is significant enough to the point where special bonding methods are necessary for metal wiring methods. The reason for enhanced bonding is to address concerns about arcing and sparking at any fitting terminations. The strengthened bonding required in these locations ensures an effective path for ground-fault current to facilitate fast operation of the overcurrent protective device protecting the circuit and equipment, which reduces possibilities of "hot spots" developing on enclosures at the point of the ground fault. Otherwise, such hot spots or heated enclosure surfaces could quickly become an ignition source if the arcing event is sustained even for a short period of time.

Section 500.8(E) requires installing conduit with five full threads fully engaged and wrench-tight. The purpose is to maintain the explosion-proof integrity of enclosures and conduit systems, and to prevent arcing or sparking at threaded joints if fault current passes over the conduit during a ground-fault event. **See Figure 14-2.**

Section 250.100 provides requirements for bonding in hazardous locations and applies regardless of the

Figure 14-3 Fittings Must Be Tight

Figure 14-3. Bonding in hazardous locations must be made tight using suitable wiring methods and fittings.

voltage of the system or circuit. This rule indicates that the non–current-carrying metal parts of equipment, raceways, and other enclosures in hazardous locations must be bonded using one of the methods provided in 250.92(B)(2) through (B)(4). **See Figure 14-3.**

This enhanced bonding for metal wiring methods is required even if an equipment grounding conductor of the wire type is installed. **See Figure 14-4.**

Standard locknuts or bushings are not permitted to accomplish the bonding

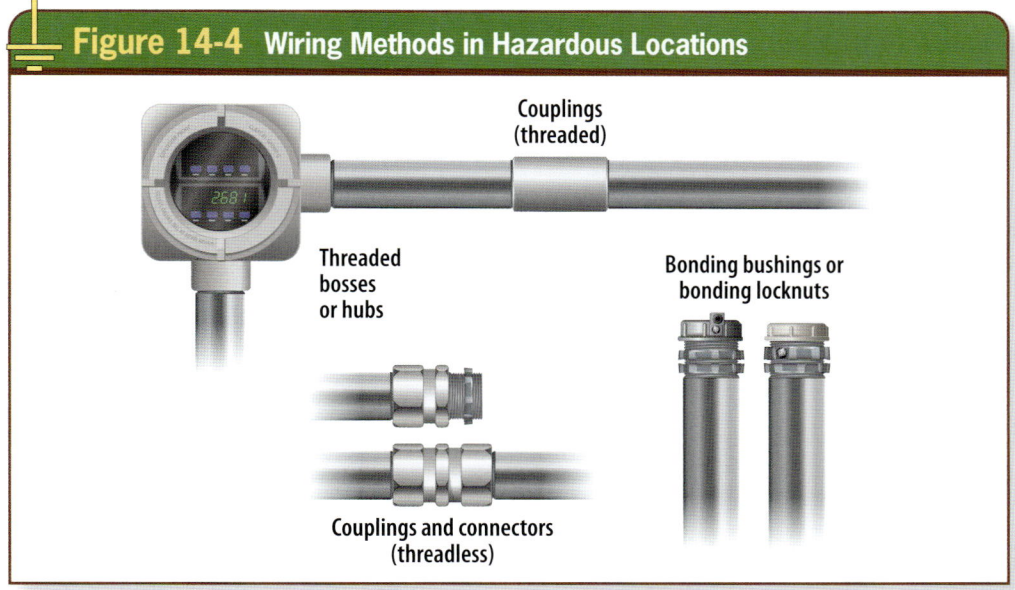

Figure 14-4 Wiring Methods in Hazardous Locations

Couplings (threaded)

Threaded bosses or hubs

Bonding bushings or bonding locknuts

Couplings and connectors (threadless)

Figure 14-4. Any of the methods in 250.92(B)(2) through (4) are acceptable for bonding in hazardous locations.

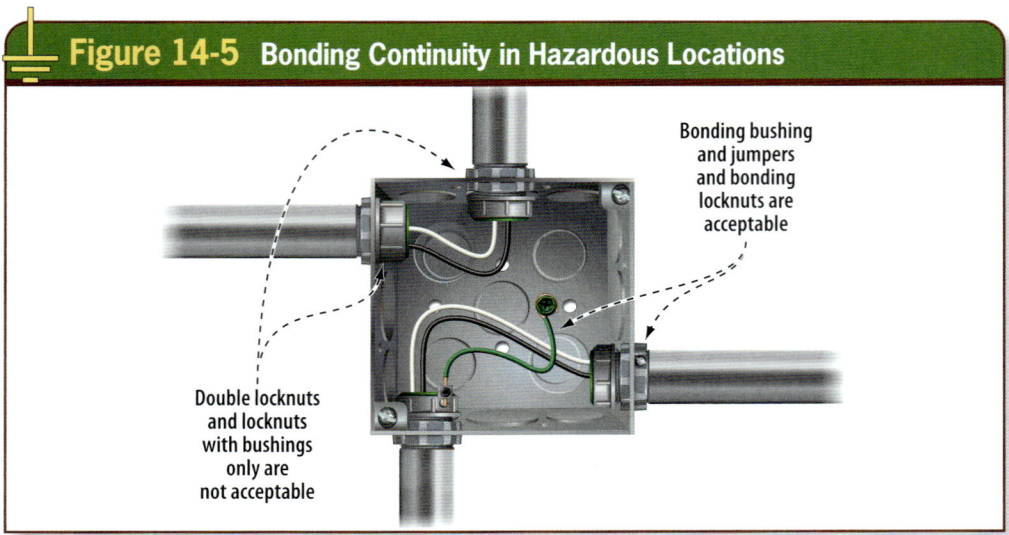

Figure 14-5   **Bonding Continuity in Hazardous Locations**

Bonding bushing
and jumpers
and bonding
locknuts are
acceptable

Double locknuts
and locknuts
with bushings
only are
not acceptable

Figure 14-5. Bonding locknuts and bonding bushings with jumpers are acceptable methods of bonding in hazardous (classified) locations, while standard locknuts or locknut and bushing arrangements are not.

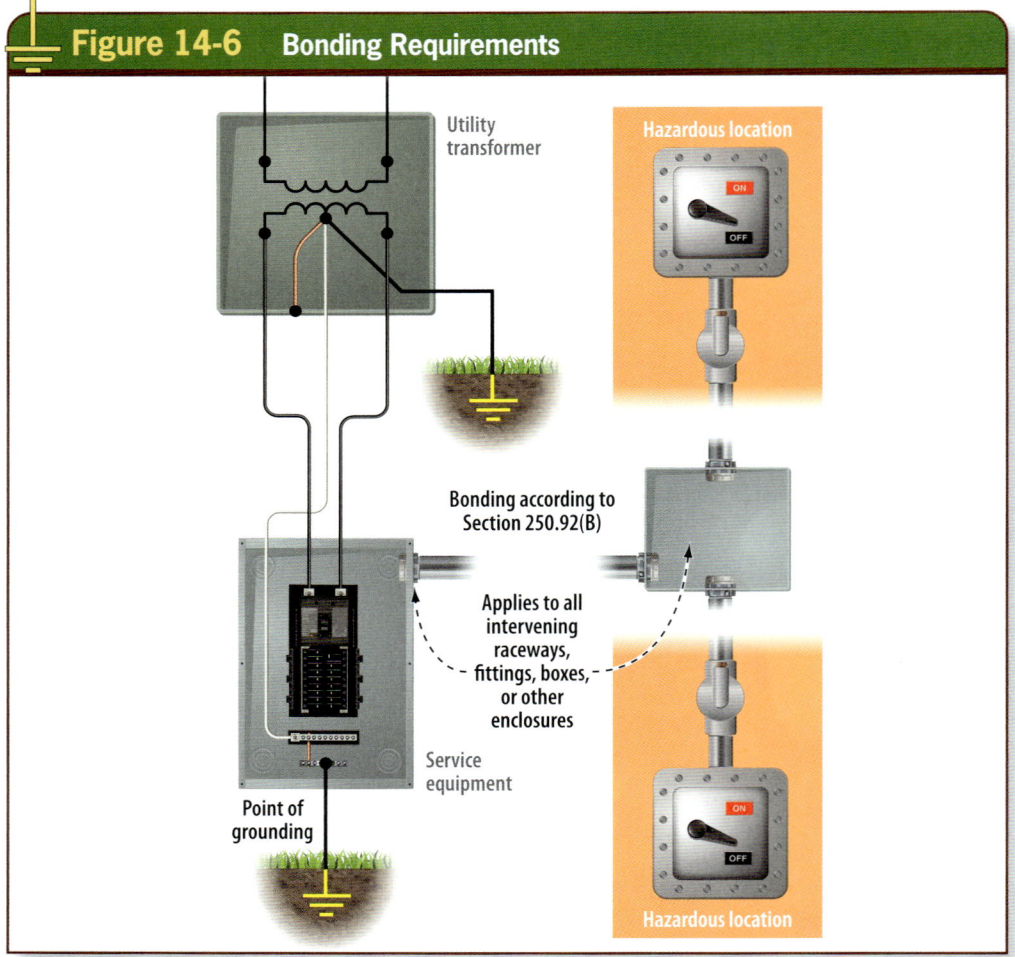

Figure 14-6   **Bonding Requirements**

Utility
transformer

Hazardous location

Bonding according to
Section 250.92(B)

Applies to all
intervening
raceways,
fittings, boxes,
or other
enclosures

Service
equipment

Point of
grounding

Hazardous location

Figure 14-6. Bonding requirements must extend to all intervening metal raceways and back to the applicable service or separately derived system grounding point.

required for wiring in a hazardous location. Where metal wiring connections are made at equipment such as boxes, enclosures, cabinets, and panelboards, effective bonding must be ensured around any joints in the fault current path to prevent sparking and ensure a low-impedance path for any ground-fault current. Standard locknuts and bushings can be used to connect the raceways to the enclosures, but bonding continuity must be ensured around any standard locknut, bushing, or combination, using one of the methods provided in 250.92(B)(2) through (B)(4). **See Figure 14-5.**

During a ground fault, heavy levels of current will be present for the time it takes the overcurrent protective device to open the faulted circuit. Strengthened bonding methods are required for wiring within the hazardous location, and they are also required for the entire metal raceway system, even the sections which are not within the hazardous location. The bonding requirements in Chapter 5 of the *NEC* clarify that the enhanced bonding applies to all metal raceways and enclosures in the hazardous location and to all metal raceways and enclosures of the same circuit(s) run, extending to the grounding point of the applicable service or derived system. **See Figure 14-6.**

The "point of grounding" is typically where the service main bonding jumper is installed or where the system bonding jumper is installed for a separately derived system. This is also typically where the grounding electrode conductor connection is made at the service or system. The *NEC* permits flexible wiring methods such as flexible metal conduit (FMC), liquidtight flexible metal conduit (LFMC), and liquidtight flexible nonmetallic conduit (LFNC) in Division 2 locations under restrictive conditions as provided in 501.10(B) and 502.10(B). LFMC and LFNC must be used with listed fittings. **See Figure 14-7.**

LFMC and LFNC are permitted in Class III locations as indicated in 503.10(A)(3)(2). Where FMC, LFMC, or LFNC is installed for flexibility in hazardous locations, a wire-type equipment bonding jumper is usually required in accordance with 250.102. **See Figure 14-8.**

See the XXX.30(B) sections of Articles 501 through 503, as well as 505.25(B) and 506.25(B). Bonding jumpers are often installed external to the flexible conduit installation using suitable fittings that provide an attachment lug.

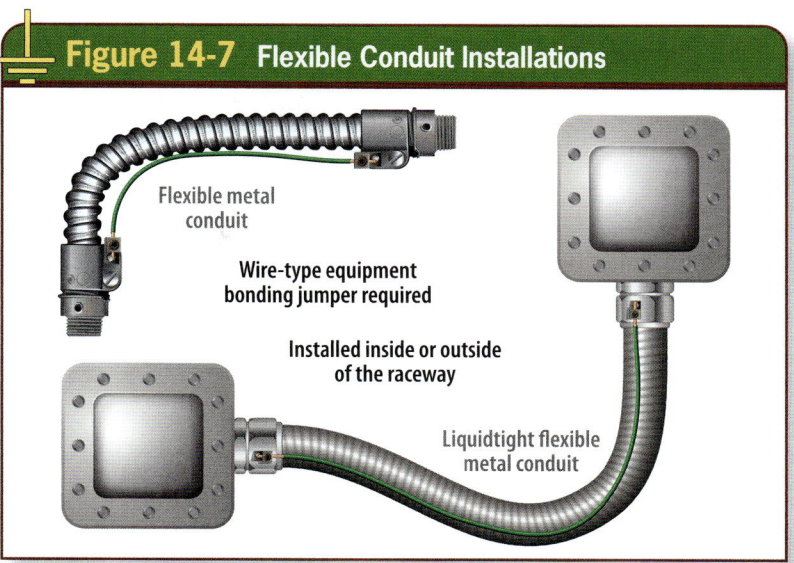

Figure 14-7 Flexible Conduit Installations

Flexible metal conduit

Wire-type equipment bonding jumper required

Installed inside or outside of the raceway

Liquidtight flexible metal conduit

Figure 14-7. A wire-type equipment bonding jumper is required for flexible conduit installations and can be installed inside or outside the conduit.

Figure 14-8 Liquidtight Flexible Metal Conduit

Figure 14-8. A wire-type equipment bonding jumper is required for LFMC installed in hazardous locations.

Figure 14-9   LFMC Installations

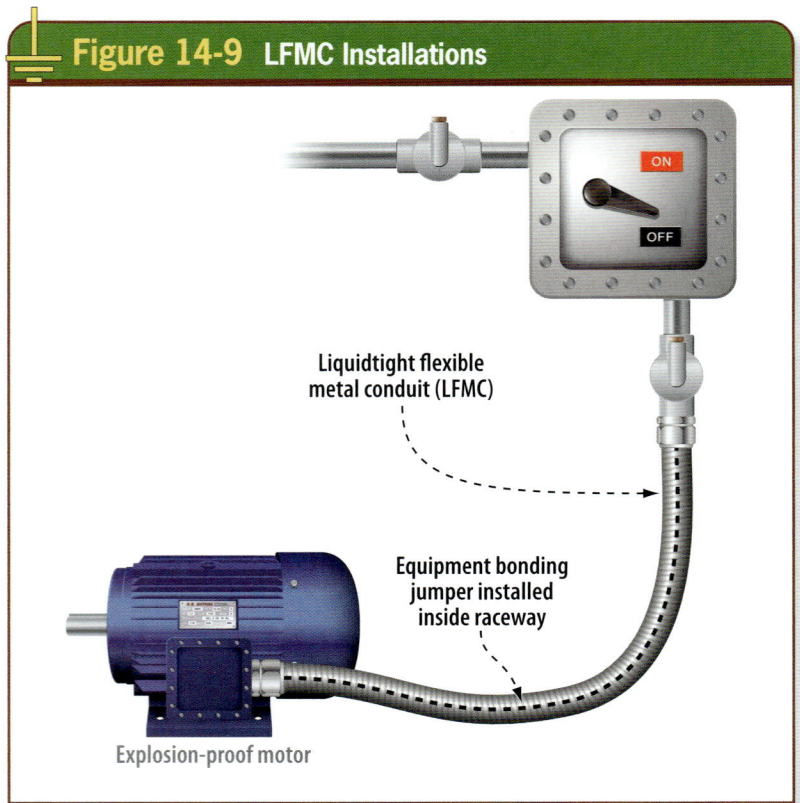

Liquidtight flexible
metal conduit (LFMC)

Equipment bonding
jumper installed
inside raceway

Explosion-proof motor

Figure 14-9. A wire-type equipment bonding jumper is required for LFMC installed in hazardous locations and can be installed inside or outside the raceway.

Figure 14-10   Patient Care Locations

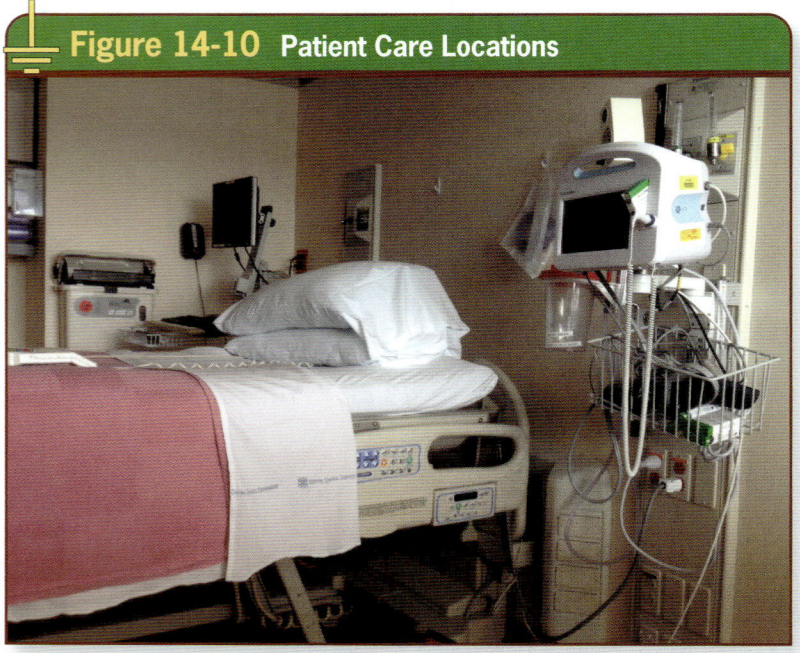

Figure 14-10. Branch circuits serving patient care spaces must meet the requirements in 517.13.

The equipment bonding jumpers can be installed inside or outside the conduit. **See Figure 14-9.**

There are additional concerns about static electricity as an ignition source in hazardous locations. Static electricity problems are typically addressed using equipotential bonding and grounding techniques that reduce potential differences in the hazardous location. In some cases, specifically in indoor applications, raising humidity levels can provide an effective solution, but each static problem requires its own analysis and solution. For information on protection against static electricity and lightning hazards in hazardous locations, see *NFPA 77: Recommended Practice on Static Electricity, NFPA 780: Standard for the Installation of Lightning Protection Systems,* and *API RP 2003-2015, Protection Against Ignitions Arising Out of Static Lightning and Stray Currents.*

SPECIAL RULES FOR HEALTH CARE FACILITIES

Reducing differences of potential between conductive equipment or other objects and the Earth helps minimize shock hazards in normal circuit operation. In patient care spaces, it is even more important that overcurrent protective devices operate quickly during ground-fault conditions to minimize the potential (voltage) on conductive parts. These are the main reasons for the more restrictive requirements on equipment grounding conductors (EGCs) for branch circuits that supply patient care spaces in a health care facility. **See Figure 14-10.**

This requirement applies to all patient care spaces in all health care facilities. It does not apply to the EGCs for feeders that supply the panelboards containing branch circuits for patient care spaces. Electrical feeder EGCs must meet the general requirements in 215.6 that indicate feeders generally must include an equipment grounding conductor. If the feeder is installed in a metal wiring method, such as rigid metal conduit (RMC) or electrical metallic tubing

(EMT), it must comply with the rules in 517.19(E) for Category 1 (Critical Care) spaces. Feeder grounding and bonding requirements are covered in more detail later. For now, this chapter focuses on the rules for the branch circuits serving patient care spaces.

The two separate EGCs for branch circuits serving patient care spaces required by 517.13 emphasize creating redundancy in the safety (grounding and bonding) components of these branch circuits. It is important to remember that many of the requirements for electrical wiring in health care facilities that are included in the *NEC* and in *NFPA 99: Health Care Facilities Code* focus on providing redundancy of systems. In this case, the redundancy is required to be built into the branch circuit safety equipment grounding means.

Grounding of Receptacles and Fixed Equipment in Patient Care Spaces

The requirements for patient care space branch circuit EGCs are provided in 517.13(A) and (B) of the *NEC*. It is important to understand that both subdivisions (A) and (B) must be applied together to each branch circuit serving patient care spaces to meet the minimum requirements in this section. **See Figure 14-11.**

Article 517 of the *NEC* provides specific definitions and rules that assist in identifying patient care spaces and determining where more restrictive grounding and bonding is required. Typically, the engineering design and correspondence with the governing body of the health care facility will have already determined which areas of the facility are patient care spaces, and the design team will have specified suitable wiring methods that meet the redundant equipment grounding requirements.

Branch circuits in patient care spaces of health care facilities are required to provide two independent equipment grounding paths for all non–current-carrying conductive surfaces of fixed electrical equipment likely to become energized that are subject to personnel

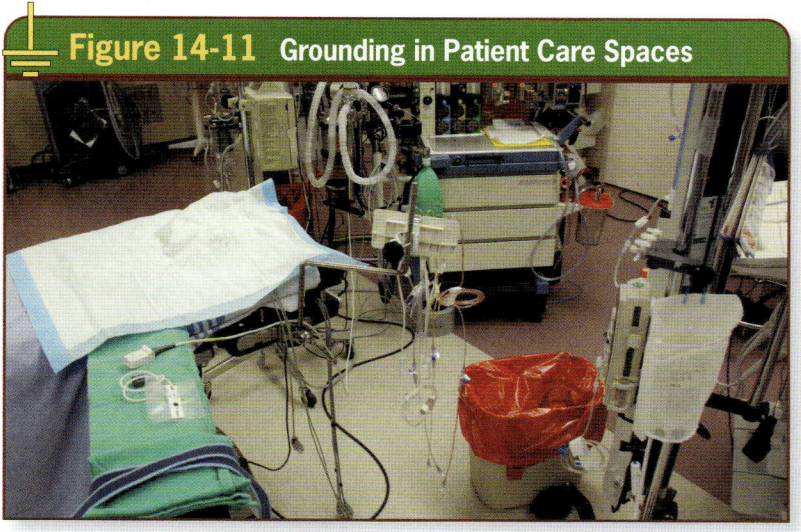

Figure 14-11 Grounding in Patient Care Spaces

Figure 14-11. Branch circuits supplying an operating room in a health care facility must include EGCs that meet the requirements of 517.13.

contact. Section 517.13(A) includes requirements related to the type of wiring method that can be used for branch circuit conductors. Approved wiring methods for patient care spaces include metal raceway systems or cables with a metallic armor or sheath that is acceptable as an EGC in accordance with 250.118. EMT is often specified and installed as a branch circuit wiring method in patient care spaces. **See Figure 14-12.**

TechTip!

Branch circuits in patient care spaces of health care facilities must provide two independent equipment grounding paths.

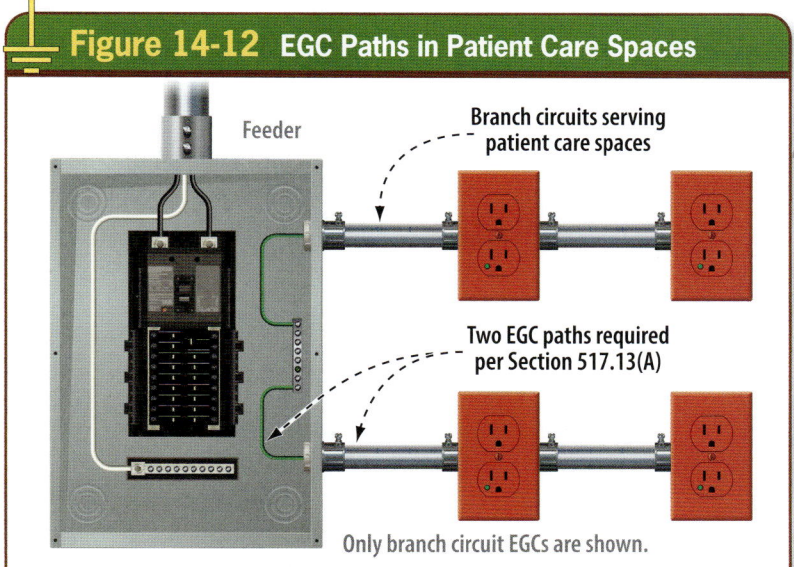

Figure 14-12 EGC Paths in Patient Care Spaces

Feeder

Branch circuits serving patient care spaces

Two EGC paths required per Section 517.13(A)

Only branch circuit EGCs are shown.

Figure 14-12. Two separate EGC paths are required for branch circuits serving patient care spaces.

It is important to recognize the types of conduit, tubing, cables, and other wiring methods that are acceptable as EGCs in accordance with 250.118. A review of this section reveals a fairly extensive list of such qualifying wiring

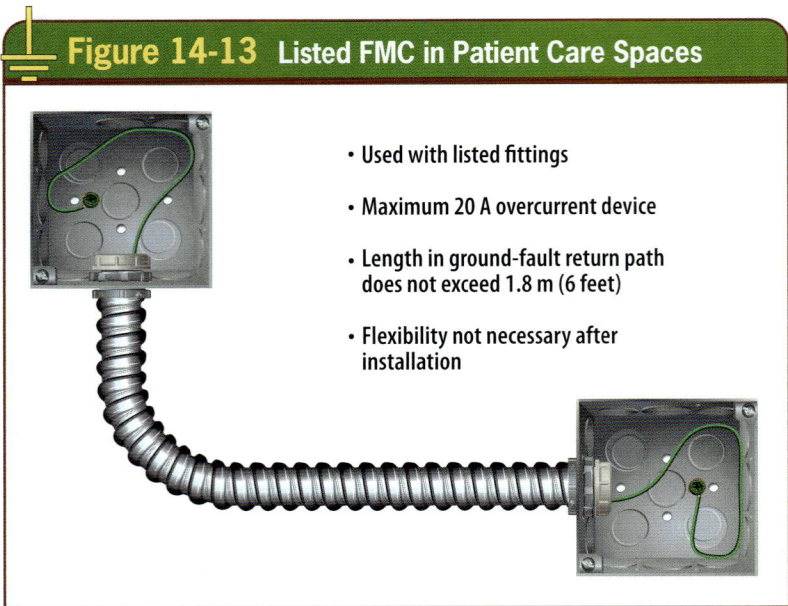

Figure 14-13 | **Listed FMC in Patient Care Spaces**

- Used with listed fittings
- Maximum 20 A overcurrent device
- Length in ground-fault return path does not exceed 1.8 m (6 feet)
- Flexibility not necessary after installation

Figure 14-13. Listed FMC with an insulated copper EGC can be used as a branch circuit wiring method for patient care spaces under restrictive conditions in 250.118(5).

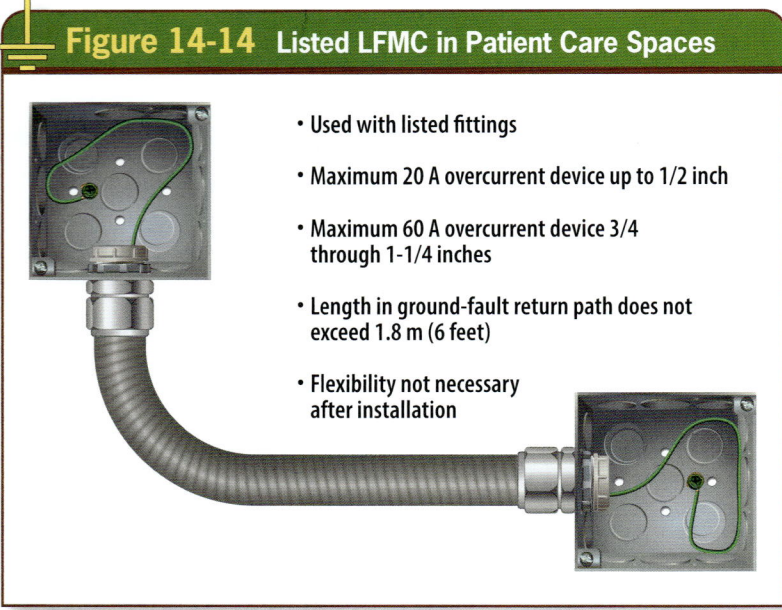

Figure 14-14 | **Listed LFMC in Patient Care Spaces**

- Used with listed fittings
- Maximum 20 A overcurrent device up to 1/2 inch
- Maximum 60 A overcurrent device 3/4 through 1-1/4 inches
- Length in ground-fault return path does not exceed 1.8 m (6 feet)
- Flexibility not necessary after installation

Figure 14-14. Listed LFMC with an insulated copper EGC can be used as a branch circuit wiring method for patient care spaces under restrictive conditions in 250.118(6).

methods: RMC, intermediate metal conduit, EMT, and armored cable (Type AC), among others. These wiring methods inherently provide a path for ground-fault current through the raceway or metallic cable armor itself. The wiring methods listed in 250.118 that meet the minimum requirements in the *NEC* as permitted EGCs are acceptable as wiring methods for branch circuits in patient care spaces.

Remember that the integrity of the EGC has everything to do with the quality of the installation providing suitable supporting and securing means, as well as proper installation of couplings and fittings. The rules addressing securing and supporting requirements for wiring methods approved for patient care spaces are typically located in the XXX.30 section of each applicable wiring method article. For example, the rules for securing and supporting EMT are found in 358.30(A) and (B).

Listed flexible metal conduit (FMC) that meets the minimum requirements of the *NEC* is suitable as an EGC, but under more restrictive conditions. Where listed FMC is installed for branch circuit wiring in a patient care space, it not only must meet the restrictive conditions provided in 250.118(5), but also must include an insulated copper EGC with the branch circuit conductors. This EGC, like all EGCs, must be no smaller than the sizes provided in Table 250.122, based on the rating of the overcurrent protective device ahead of the circuit. **See Figure 14-13.**

Listed FMC might be used in patient care spaces for facilitating flexible connections to equipment such as medical head-wall assemblies in patient care rooms of hospitals. If this wiring method is used in patient care spaces, the wiring method for the branch circuit must meet the requirements of both 250.118(5) and 517.13(A) and (B).

Listed liquidtight flexible metal conduit (LFMC) is, with restrictions, also suitable as an EGC. In order to be suitable for and function as an EGC for branch circuits serving patient care spaces, LFMC must meet the restrictive conditions provided in 250.118(6), and it must include an insulated copper EGC with the branch circuit conductors. Like listed FMC, listed LFMC might be used in a patient care space for facilitating flexible connections to equipment. If this wiring method is used in patient care spaces, it must meet the requirements of both 250.118(6) and 517.13(A) and (B). **See Figure 14-14.**

If cable wiring methods are used for branch circuit wiring in patient care spaces of health care facilities, the conductive armor must provide an effective ground-fault current path and qualify as an EGC. There must also be an insulated copper EGC included in the cable assembly if used for a patient care space. Some Type MC cable is suitable as an EGC, according to 250.118(10).

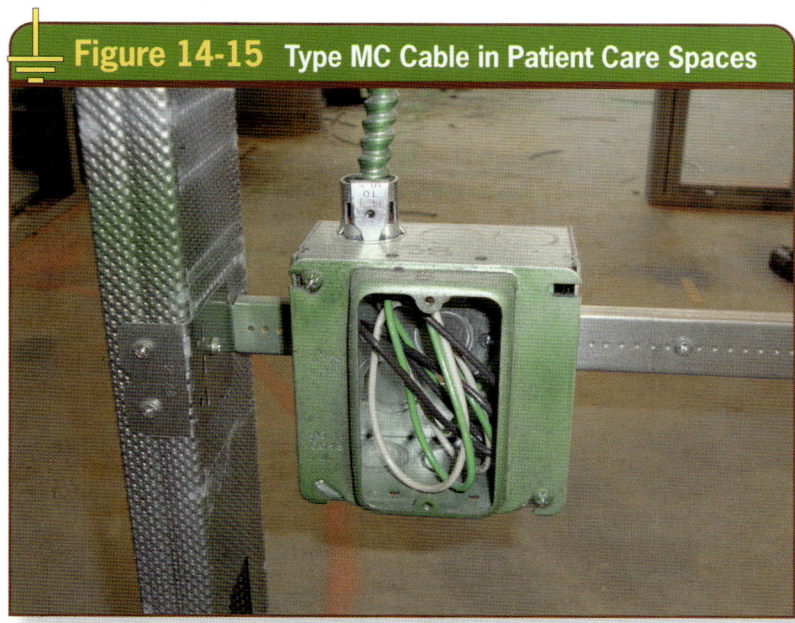

Figure 14-15 Type MC Cable in Patient Care Spaces

Figure 14-15. *Type MC cable is suitable for use in patient care spaces where it includes an insulated copper EGC.*

Courtesy of Southwire Company

Type MC cable that provides an effective ground-fault current path in accordance with one or more of the following:

a. It contains an insulated or uninsulated equipment grounding conductor in compliance with 250.118(1).

b. The combined metallic sheath and uninsulated equipment grounding/bonding conductor of interlocked metal tape–type MC cable that is listed and identified as an equipment grounding conductor

c. The metallic sheath or the combined metallic sheath and equipment grounding conductors of the smooth or corrugated tube-type MC cable that is listed and identified as an equipment grounding conductor

one specific type of MC cable that has interlocking metal-tape–type construction, and the armor is suitable as an EGC. It is similar in construction to standard-type AC cable in that it includes a separate bare conductor within the assembly that is in intimate contact with the cable armor. This type of cable is covered in 250.118(10)(b). **See Figure 14-15.**

This Type MC cable has been listed and evaluated for use as an EGC when installed with suitable listed fittings. Type MC cable is manufactured with and without a contained insulated copper EGC, so it would be acceptable for use in health care facility installations only if the internal insulated EGC is present. As always, to ensure compliance

Only Type MC cables addressed in list items b. and c. are acceptable for use in a patient care space provided they also contain an insulated copper EGC. The UL product categories PJAZ and PJOX provide additional product information about Type MC cable assemblies and fittings. There is at least

For additional information, visit qr.njatcdb.org Item #5333

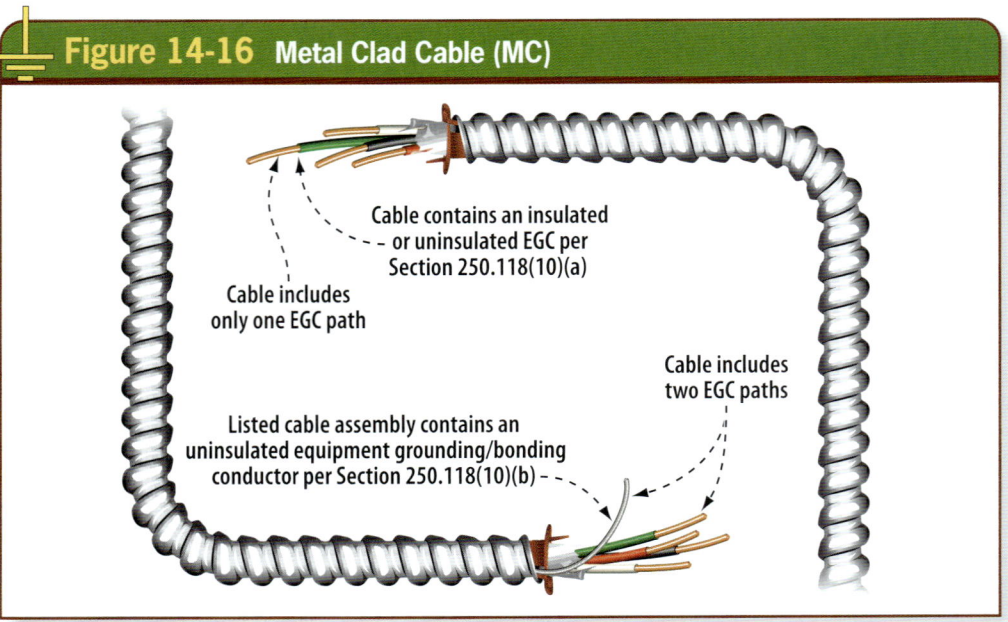

Figure 14-16 Metal Clad Cable (MC)

Cable contains an insulated or uninsulated EGC per Section 250.118(10)(a)

Cable includes only one EGC path

Listed cable assembly contains an uninsulated equipment grounding/bonding conductor per Section 250.118(10)(b)

Cable includes two EGC paths

Figure 14-16. The armor of Type MC cable must qualify as an EGC, and the assembly must contain an insulated copper EGC.

with 110.3(B) of the *NEC*, follow the manufacturer's installation instructions when installing this and any other product. **See Figure 14-16.**

Section 517.13(A) specifies metal raceways or cable assemblies that are acceptable as EGCs in accordance with 250.118. This is one of two EGC paths

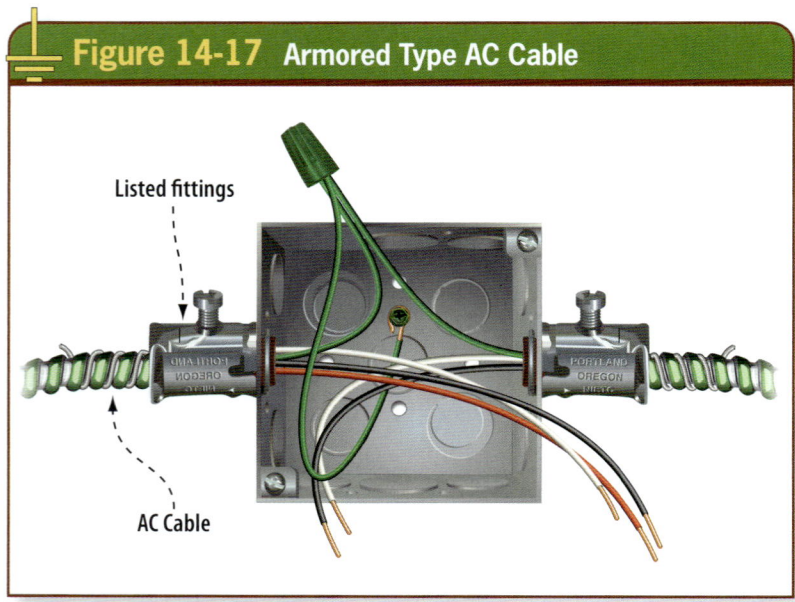

Figure 14-17 Armored Type AC Cable

Listed fittings

AC Cable

Figure 14-17. The armor of Type AC cable is suitable as an EGC.

required for branch circuits serving these spaces. It is important to understand that this effective ground-fault current path in the form of a metal wiring method is characteristically different from the contained insulated copper conductor addressed in 517.13(B).

In review, 517.13(A) indicates that the wiring method selected for the branch circuits in patient care spaces must, on its own, qualify as a suitable EGC in accordance with 250.118, and 517.13(B) requires an insulated copper EGC of the wire type. Nonmetallic types of wiring methods, such as rigid polyvinyl chloride (PVC) conduit, and cables where the outer jacket (armor) of the cable does not qualify as an EGC are not suitable for use as branch circuit wiring methods in patient care spaces.

Armored Cable (Type AC)
The armor sheath of Type AC cable is recognized by 250.118(8) as an EGC. Type AC cable that includes an insulated copper EGC complies with the requirements in 517.13(A) and (B). This cable armor qualifies as an EGC because of the bare internal bonding strip that is in intimate contact with

the armor from fitting to fitting. **See Figure 14-17.** The internal bonding strip in the cable assembly and the interlocking metal-tape–type armor work in combination as an effective ground-fault current path.

The objective of this redundant EGC requirement is to ensure adequate equipment grounding in the patient care space and provide a redundant path for the ground-fault current should a ground-fault condition develop. Quick operation of branch-circuit overcurrent protective devices is essential to safely clear the event. Remember that the rule requires the installation of two separate EGC paths to ensure that one will operate when necessary. A good approach to selecting suitable cable wiring methods that meet this more restrictive requirement is to refer to the UL online Product iQ for the specific product criteria and information.

The fittings used with approved wiring methods are required to be listed. Listed Type AC and MC cable fittings are evaluated for use with these cable assemblies and provide a suitable connection means for establishing continuity and conductivity between the cable armor and the enclosure to which it connects. Each cable must be used with fittings designed and listed for use with that particular cable. Verify with the manufacturer if there is doubt as to compatibility. Cable fittings are required to be identified; this identification can be found on the product itself or on the smallest shipping carton. **See Figure 14-18.**

Required Wire-Type EGC

As previously stated, in addition to the wiring method qualifying as an EGC in accordance with 250.118, each branch circuit and outlet serving patient care spaces must include an insulated copper EGC. This insulated copper conductor must be connected to the grounding terminals of receptacles and the non–current-carrying conductive surfaces of fixed electrical equipment likely to become energized.

This requirement applies to circuits operating at a voltage of more than 100 volts, which generally applies to all line voltage branch circuits serving patient care spaces.

Notice that the equipment grounding requirements in 517.13(B) are more restrictive than what would normally be required in 250.118.

The requirement for an EGC that is both insulated and made of copper exceeds the basic provisions of 250.118. The sizing rules for insulated EGCs required by 517.13(B) are the same as those provided in 250.122 for all branch circuits and feeders. The size is based on the rating of the final overcurrent protective device protecting the circuit in accordance with Table 250.122. The requirements of 250.122 still apply to branch circuit EGCs, including requirements for conditions where the ungrounded (phase) conductors may be increased in size to handle voltage drop conditions.

The requirements of 517.13(A) and (B) apply to the branch circuits serving patient care spaces. Per the definition

For additional information, visit qr.njatcdb.org Item #5333

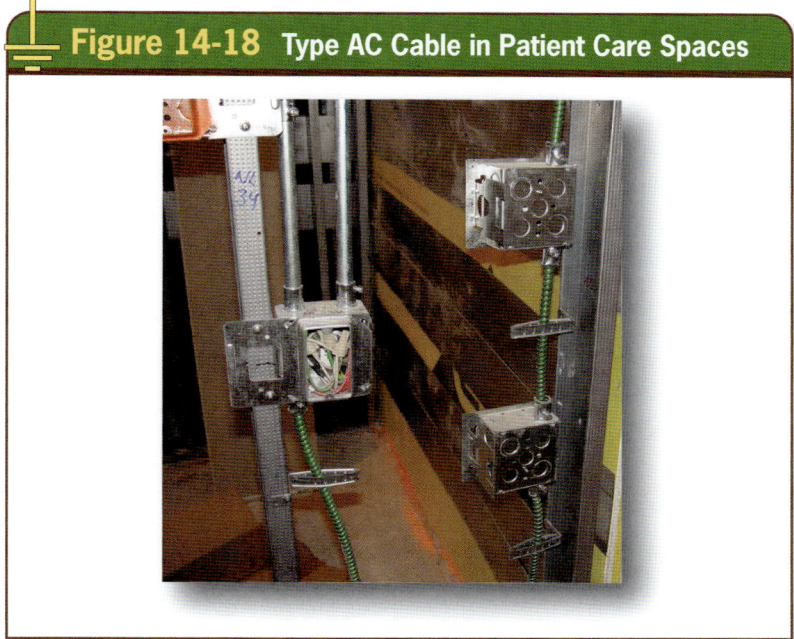

Figure 14-18 Type AC Cable in Patient Care Spaces

Figure 14-18. Type AC cable is suitable for use in patient care spaces if an insulated copper EGC is included in the cable assembly and listed fittings are used.

of *branch circuit*, two separate EGC paths are required with the wiring installed from the branch circuit panelboard (final overcurrent protective device) to the outlet.

> **Branch Circuit (Branch-Circuit).** The circuit conductors between the final overcurrent device protecting the circuit and the outlet(s). (CMP-2)

This means cord- and plug-connected equipment is only required to have a single equipment grounding means. Metal faceplates for receptacles installed in these locations are required to be connected to an effective ground-fault current path and are permitted to be grounded by the 6-32 fastening screw that attaches the plate to the device in accordance with 517.13(B)(1) Exception No 2. **See Figure 14-19.**

Furthermore, the requirements for grounding using an insulated EGC are not necessary for luminaries or switches that are outside the patient care vicinity, as indicated in the Exception to 517.13(B).

The definitions of the terms *branch circuit* and *patient care vicinity* are important in clarifying the minimum requirements of the *NEC*. Even though the term *patient care vicinity* is used in the Exception to 517.13(B), it should be recognized that many governing bodies of health care facilities usually specify the location for patient care as the entire room. Design and engineering firms understand the importance of meeting the minimum requirements, and many times their designs exceed the minimum requirements in the *NEC* and other applicable standards. Be sure to maintain the integrity of the engineered design by following plans and specifications. If changes or modifications are desired or necessary, always work cooperatively with the engineering firms.

The key concept to remember is that the *NEC* requires two grounding paths to be installed for electrical equipment in patient care spaces, assuring that at least one return path for the fault current is always present if one should fail. In the field, this is often referred to as *redundant grounding for patient care spaces*. The concept of this redundancy should be viewed from the standpoint that installation of two suitable EGC return paths is required to ensure at least one path is always present and part of the circuit.

Isolated Ground Receptacle Installations in Patient Care Spaces

Medical equipment manufacturers often specify isolated equipment grounding circuits and isolated grounding receptacles. Reducing the amount of electromagnetic interference (EMI) on the EGC for sensitive electronics is usually the reason for specifying isolated grounding. Isolated grounding receptacles are not permitted in patient care vicinities in accordance with 517.16(A).

Section 517.16(B) provides the requirements for isolated grounding receptacles and circuits installed in health care facilities in areas outside of the patient care vicinity. An equipment

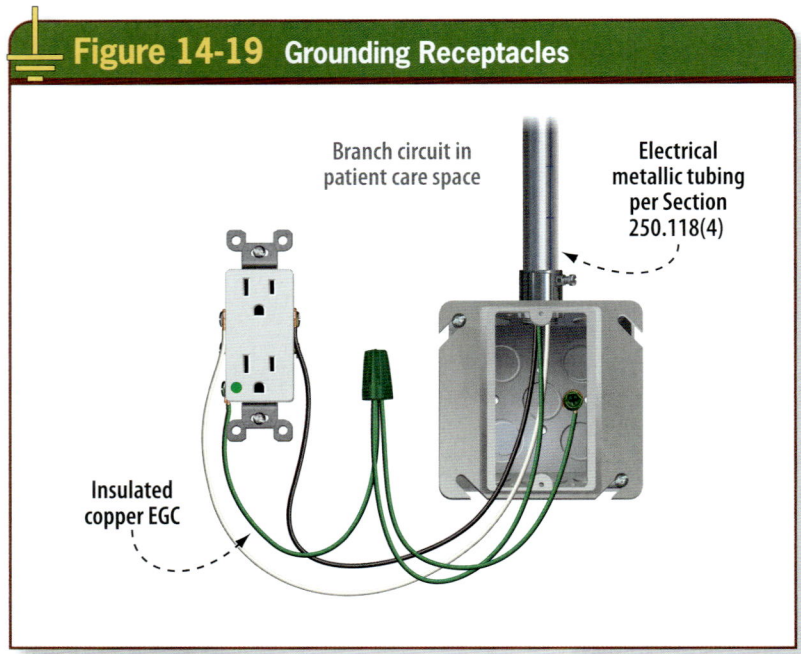

Figure 14-19 **Grounding Receptacles**

Branch circuit in patient care space

Electrical metallic tubing per Section 250.118(4)

Insulated copper EGC

Figure 14-19. An insulated copper EGC is required to be connected to the grounding terminal of receptacles serving patient care spaces.

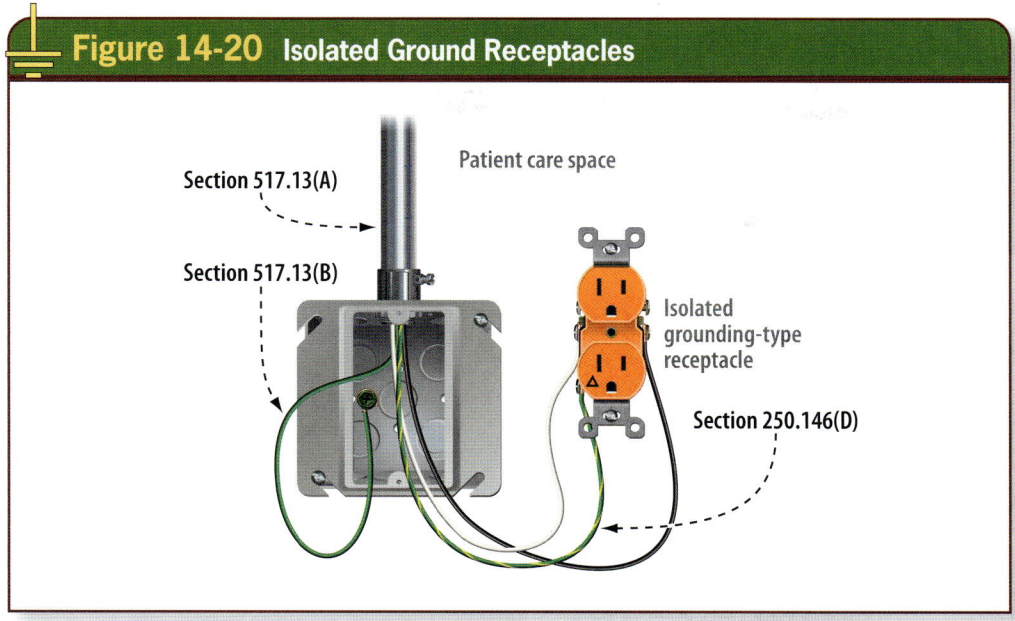

Figure 14-20 Isolated Ground Receptacles

Section 517.13(A)

Section 517.13(B)

Patient care space

Isolated
grounding-type
receptacle

Section 250.146(D)

Figure 14-20. Isolated ground receptacles are identified with an orange triangle on the face of the receptacle and must be connected to an insulated copper EGC installed with the branch circuit in accordance with 250.146(D).

grounding conductor connected to the equipment grounding terminals of isolated grounding receptacles must have green insulation with one or more yellow stripes. An additional insulated equipment grounding conductor, also with green insulation but without any yellow stripes, must be connected to the receptacle enclosure. Both must be installed in a wiring method that by itself is suitable as an EGC in accordance with 517.13(A).

Section 406.3(E) also requires isolated ground receptacles to be identified with an orange triangle on the face of the receptacle. The isolated grounding receptacles installed outside of the patient care vicinity must be installed in accordance with 250.146(D), with an additional green insulated conductor connected to the receptacle enclosure. **See Figure 14-20.**

NFPA 99: Health Care Facilities Code still addresses isolated grounding circuits and receptacles and requires periodic testing of grounding systems, which include installations of isolated grounding receptacles and circuits in health care facilities.

Patient Equipment Grounding Point
An optional method of protecting patients from differences of potential and possible shock hazards is the installation of a patient equipment grounding point. This type of installation usually includes specific devices manufactured for this purpose that incorporate one or more listed grounding and bonding jacks. This type of optional protection is often part of electrical system designs and specifications for operating rooms and other critical procedure locations in health care facilities. At minimum, a 10 AWG equipment bonding jumper must be installed from the grounding terminals of all grounding-type receptacles to the patient equipment grounding point.

Where an isolated power system is used, the patient equipment grounding point is typically connected to the reference grounding bar in the isolated power panel. The bonding conductor used for the equipment grounding point is permitted to be installed separately to each conductive part and receptacle grounding terminal, or it may be looped if this is convenient. See 517.19(D) for the full and complete *NEC* requirements.

Isolated Power System Grounding
Isolated power systems installed in health care facilities must comply with 517.160. **See Figure 14-21.** These systems are required to operate ungrounded. **See Figure 14-22.**

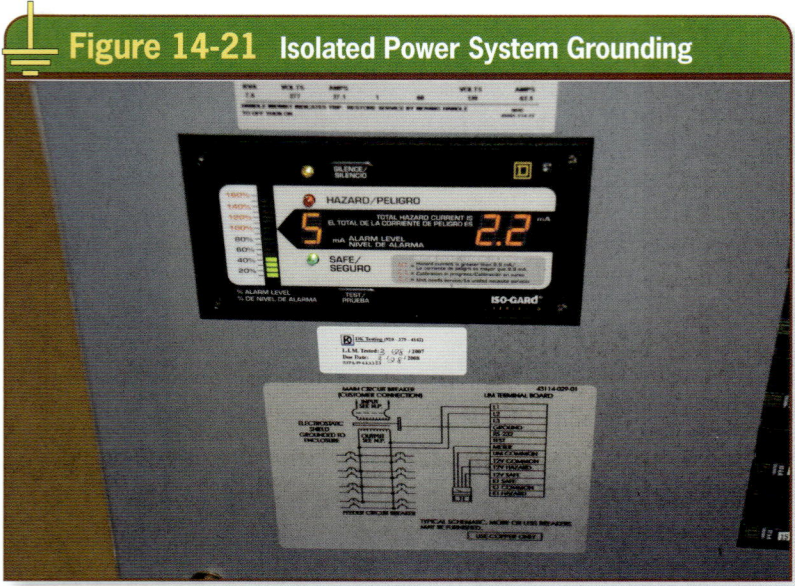

Figure 14-21. *Isolated power systems are an optional method of protection installed in health care facilities, unless required for wet procedure locations.*

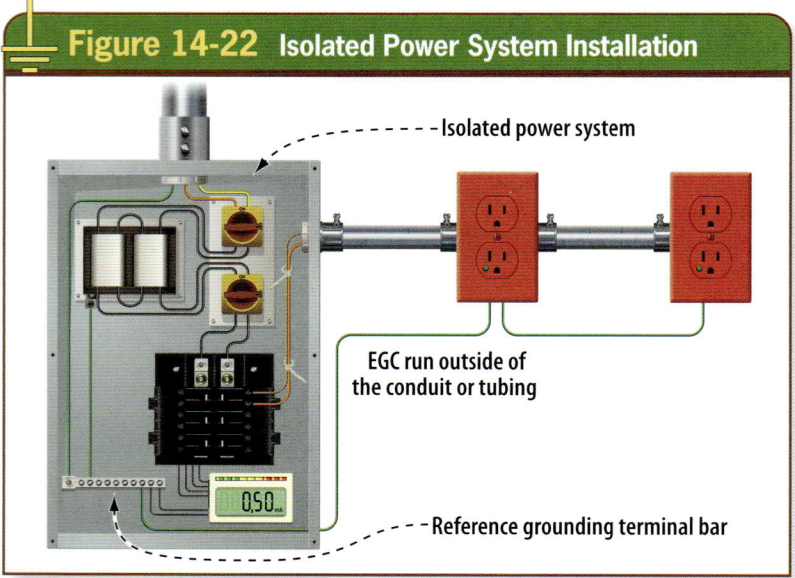

Figure 14-22. *EGCs installed with circuits of isolated power systems must be connected to the reference grounding bus within the isolated power system enclosure.*

EGCs are terminated to the grounding terminal of grounding-type receptacles connected to these systems. Section 517.160(A)(5) requires the orange conductor with a distinct stripe to be connected to the receptacle terminal normally intended for the grounded (neutral) conductor. **See Figure 14-23.**

The most common practice is to install isolated power system EGCs in the same raceway as the circuit conductors. However, because the system is ungrounded, 517.19(G) permits these EGCs to be installed outside of the conduit or enclosure containing the circuit conductors. This is a modification to the general requirements in 300.3(B) and 250.134(2), where the EGC is required to be installed in the same raceway or cable or otherwise run with the ungrounded circuit conductors.

Grounding and Bonding Requirements for Panelboards

Power distribution systems in health care facilities often require multiple levels of panelboards. Specific grounding requirements apply to panelboards serving patient care vicinities.

Grounding Requirements
Section 517.13 applies only to branch circuits serving patient care spaces. As indicated in the definition of the term *branch circuit*, the requirement for two equipment grounding paths is not applicable to feeders. The general requirements for feeder EGCs are found in 215.6. This section requires feeders to provide an EGC where the feeder supplies branch circuits in which EGCs are required. This requirement applies to all feeders regardless of occupancy. Each feeder supplying branch circuits serving patient care spaces must include an EGC in accordance with the provisions of 250.134 for connection of branch circuit EGCs served by the feeder. **See Figure 14-24.**

If a feeder in the critical branch is supplied from a grounded distribution system in a health care facility and is installed using a metal wiring method, the grounding of the switchboard, switchgear, or panelboard is required to be assured by one of the methods in 517.19(E)(1), (2), or (3). This enhanced bonding requirement applies to all junction points (connection points at enclosures, junction and pull boxes, and so on) of the metal raceway, metal Type MC cable armor, or mineral-insulated (Type MI) cable armor.

Enhancing the bonding requirements for feeder equipment grounding means helps to ensure effective operation of branch circuit and feeder overcurrent protective devices in ground-fault conditions. Because only one EGC is required to meet the minimum *NEC* requirements for feeders in health care facilities, enhanced and more restrictive bonding is essential. This bonding at termination points can be accomplished by a grounding (bonding) bushing and a properly sized bonding jumper, threaded bosses or hubs, or other approved devices, such as bonding-type locknuts or bushings.

Bonding Requirements

In addition to the feeder EGC rules in 215.6 and 517.19(E), there are bonding requirements for panelboards that provide branch circuits serving patient care vicinities. Again, this is an effort to minimize potential differences among conductive equipment and parts serving the same patient care vicinity. The *NEC* requires the equipment grounding terminal bars of the normal and essential branch circuit panelboards serving the same individual patient care vicinity to be bonded together.

Section 517.14 requires these components be connected (bonded) together with an insulated copper conductor no smaller than 10 AWG. The bonding conductor installed between the panelboard equipment grounding terminal bars must be continuous, except that it may be broken where it connects to the terminal bar in the panelboard. This applies to all situations in which two or more panelboards supply the same patient care vicinity and are served through separate transfer equipment on the

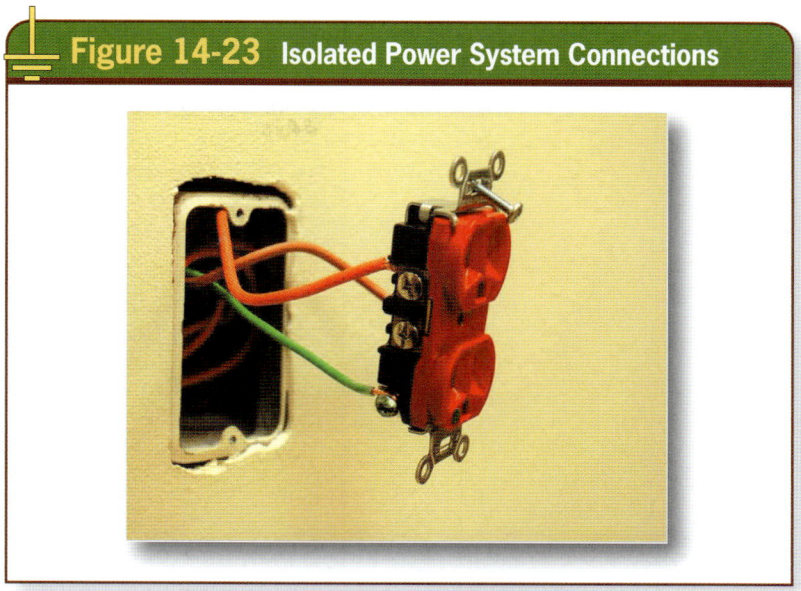

Figure 14-23   Isolated Power System Connections

Figure 14-23. *EGC connections at a receptacle are connected to an isolated power system in an operating room of a hospital. Note there is no grounded conductor connected to the receptacle.*

emergency system. The reason for this additional bonding is that the normal branch circuits and the critical branch circuits serving the same individual patient care vicinity are generally supplied from different separately derived systems. This situation creates the possibility of potential differences between the

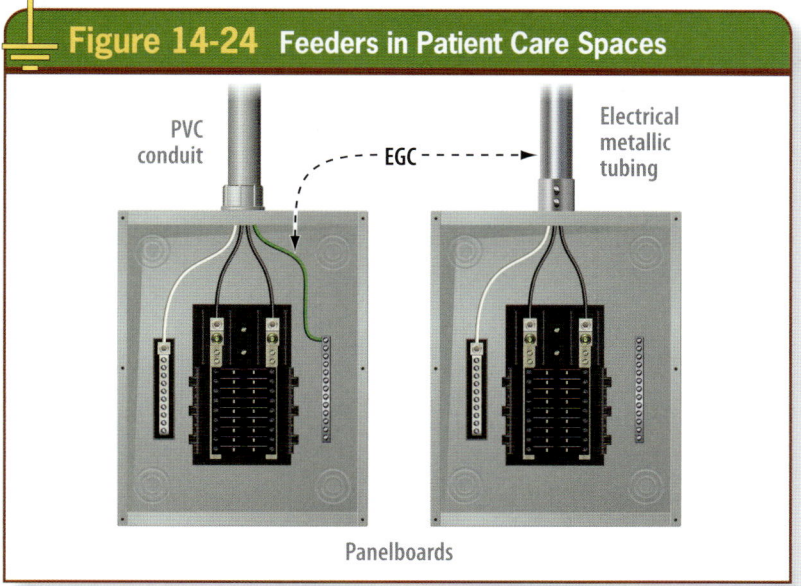

Figure 14-24   Feeders in Patient Care Spaces

PVC conduit

EGC

Electrical metallic tubing

Panelboards

Figure 14-24. *Feeders must include an equipment grounding means, as required in 215.6.*

branch circuit EGCs of the normal and emergency systems. **See Figure 14-25.**

Bonding all equipment grounding terminal bars of all panelboards serving the same patient care vicinity helps minimize the possible potential differences. Note that the *NEC* requires this bonding conductor to be at least a 10 AWG copper conductor and to be insulated. It may be larger depending on the design. For example, if the panelboards serving the same patient care vicinity were located a considerable distance apart, voltage drop concerns could be a problem.

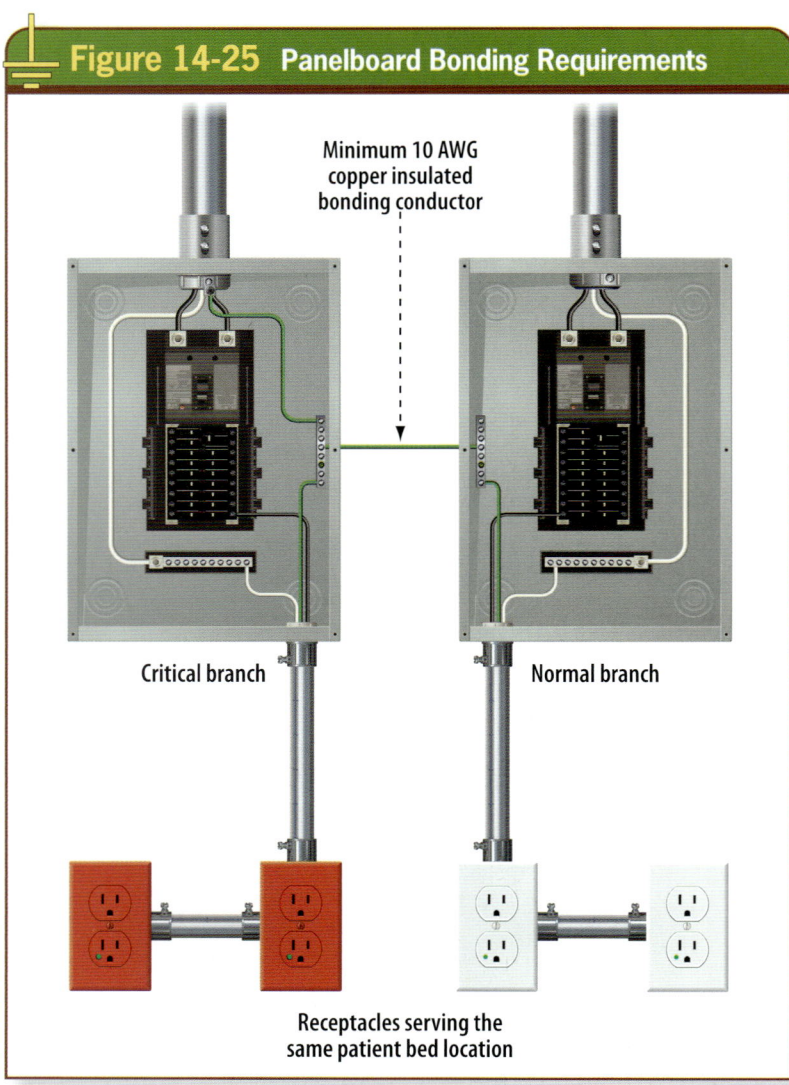

Figure 14-25 **Panelboard Bonding Requirements**

Minimum 10 AWG copper insulated bonding conductor

Critical branch

Normal branch

Receptacles serving the same patient bed location

Figure 14-25. Panelboards serving the same patient care vicinity have to be bonded together with, at minimum, a 10 AWG insulated copper conductor.

SPECIAL RULES FOR AGRICULTURAL INSTALLATIONS

Article 547 provides special requirements for agricultural buildings or any part of a building or structure with the conditions associated with these types of locations. The common adverse conditions encountered in an agricultural building include excessive dust and dust with water accumulations, along with corrosive atmospheres resulting from animal excretion. Corrosion is often accelerated in these areas because of moisture and wet conditions related to periodic washing and sanitizing operations involving the use of water and cleansing agents. Additionally, electrically-heated livestock watering troughs can present an electric shock hazard for livestock and personnel. Methods for safe installation of livestock watering equipment are described in the American Society of Agricultural and Biological Engineers (ASABE) Standard, *ASABE EP342.3-2010 (R2015), Safety for Electrically Heated Livestock Waterers.* See the informational notes following Section 547.44(B) of the *NEC*.

There are three primary concerns for electrical wiring systems in these types of occupancies. The first concern is related to the excessive amount of corrosion and the need to maintain the integrity of the effective ground-fault current path for electrical wiring in these buildings or areas. **See Figure 14-26.** The other two concerns pertain to voltage gradients and the need to maintain the integrity of the effective ground-fault current paths with the feeders and branch circuits installed in these locations. **See Figure 14-27.**

Voltage Gradients

In agricultural facilities, an important concern is that voltage gradients can be present in the Earth. Neutral-to-ground stray current can result in significant behavior problems, loss of production, and even death of farm animals. These stray currents are often referred to as *tingle voltages* and can

Figure 14-26 Agricultural Installations

Figure 14-26. Agricultural buildings and properties have special grounding and bonding requirements.

Courtesy of Tom Garvey

appear between various conductive components: metal piping systems, conductive flooring, metal rails, feeding troughs, and so forth.

Some common causes of stray currents include current being present in primary distribution systems, faulty wiring on or in the farm buildings, and inappropriate neutral-to-ground connections on the load side of the service grounding point or grounding point of a separately derived system.

Ground-fault events can also introduce current into the Earth. Where electrical wiring systems and equipment are effectively grounded and bonded, a phase-to-ground fault condition should operate an overcurrent protective device and clear the abnormal condition quickly.

Fact

Neutral-to-ground stray current can result in significant behavior problems, loss of production, and even death for farm animals.

Figure 14-27 Bonding Agricultural Installations

Figure 14-27. Special bonding requirements apply to animal containment areas in agricultural facilities.

Courtesy of Tom Garvey

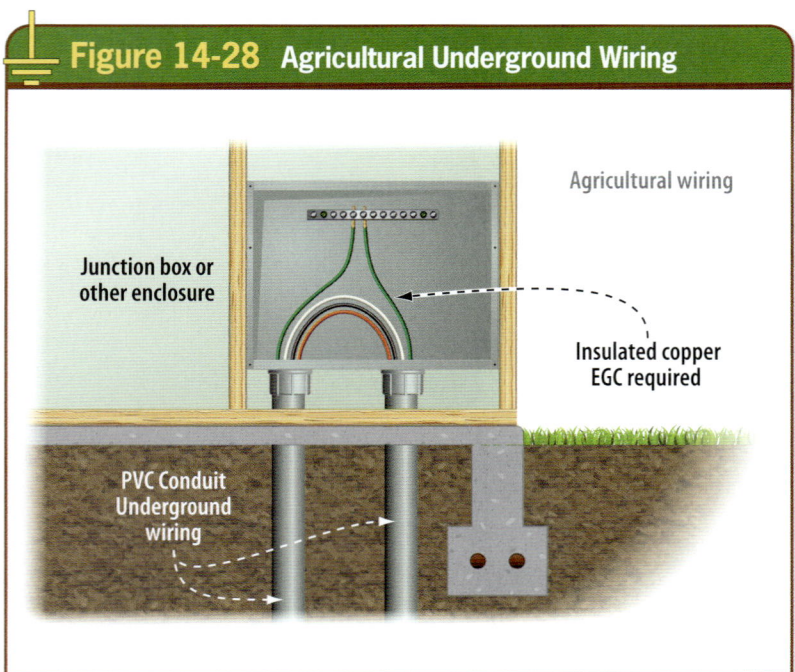

Figure 14-28 Agricultural Underground Wiring

Junction box or other enclosure

Agricultural wiring

Insulated copper EGC required

PVC Conduit Underground wiring

Figure 14-28. Separate EGCs of underground wiring for agricultural installations must be insulated.

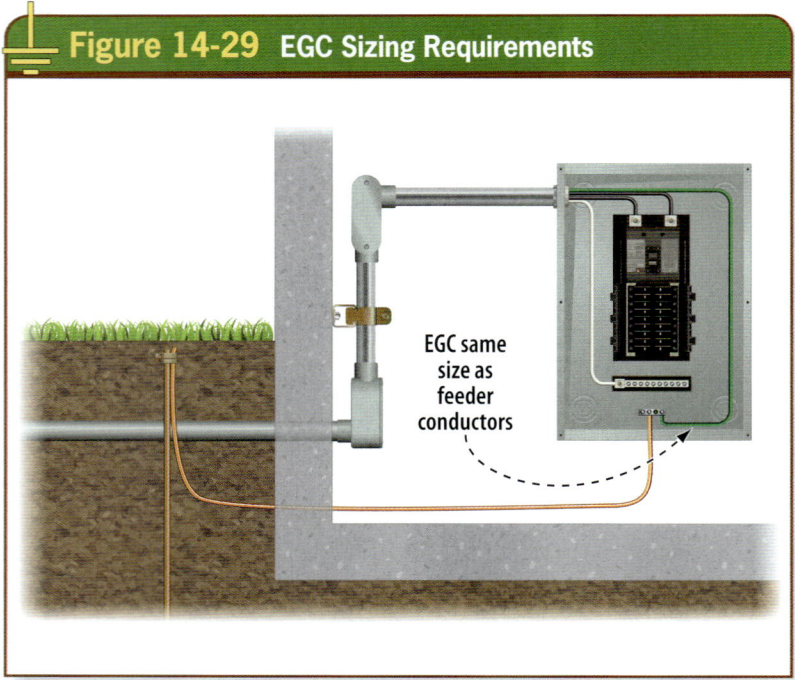

Figure 14-29 EGC Sizing Requirements

EGC same size as feeder conductors

Figure 14-29. The EGC(s) of feeders supplying separate buildings are required to be sized the same as the ungrounded conductors of the circuit.

Site-Isolating Device

Site-Isolating Device. A pole-mounted disconnecting means installed at the distribution point for the purposes of isolation, system maintenance, emergency disconnection, or connection of optional standby systems. (547) (CMP-7)

A separate EGC that is not part of a listed cable assembly, if installed underground, must be insulated as indicated in 547.27. **See Figure 14-28.**

An important requirement for electrical systems in agricultural buildings is to isolate the neutrals, EGCs, and other non–current-carrying parts of equipment for milking parlors, cattle corrals and feeding areas in buildings, and so forth. The *NEC* generally requires isolation of grounded (neutral) conductors at separate buildings or structures in accordance with 250.32(B). These general requirements are supplemented by the provisions in 547.41(B)(3).

Each agricultural building or structure must meet the provisions in 250.32 and must meet the sizing and connection rules in 547.41(B)(3), list items (1) and (2). The minimum size required for the EGC must be no smaller than the largest supply conductor for the building if the same conductor material is used. **See Figure 14-29.** If the EGC is of a different material, it must be adjusted in size in accordance with the equivalent sizes in the appropriate columns of Table 250.122.

Another important requirement deals with the connections between the grounded (neutral) conductor, the EGC, and the non–current-carrying metal parts of equipment. This connection must only be made at the site-isolation device at the distribution point of the premises.

The site-isolating device on an agricultural property is installed to provide a means to disconnect and isolate the agricultural premises' wiring system from the serving utility, typically used under emergency conditions or for maintenance of the load-side wiring system. It also provides an accessible point for the connection of an alternate power source during power outages and prevents back-feeding of the utility grid by the alternate power source. In accordance with 547.41(A)(2), the site-isolating device must be pole mounted, must be an isolating switch, and is not considered to be the service disconnecting means for the agricultural premises. A grounding electrode system or electrode is required to be connected to the grounded conductor at a site-isolating device. **See Figure 14-30.**

When feeders are routed, either overhead or underground, from the site-isolating device to buildings or structures on the premises, the feeder must provide an insulated grounded conductor and a separate EGC, and must meet all other applicable requirements in 250.32.

Equipotential Planes

A common method of protection for farm animals is to establish an equipotential plane in agricultural buildings or structures. The term *equipotential plane* is defined in Article 100.

Equipotential Plane. Conductive elements that are connected together to minimize voltage differences. (CMP-7)

The *NEC* requires an equipotential plane to be installed in livestock areas as prescribed in 547.44(A). The livestock covered by this rule does not include poultry but does include cows, bulls, horses, and other four-legged animals. Animals with four legs present unique problems because of the multiple contact points with the Earth, which provide multiple paths for current to traverse the animal's body. This is the main reason why establishing an equipotential plane is so important.

Areas required to be equipped with an equipotential plane include indoor animal confinement areas with conductive floors where metallic equipment that could become energized is accessible from the livestock area. It is important to remember that concrete floors are considered conductive. An equipotential plane is also required for outdoor animal confinement areas with concrete slabs and where metallic equipment that could become energized is accessible to livestock. The equipotential plane must encompass the entire area where livestock would normally be standing and would be subject to contact with metallic equipment that could become energized.

This type of bonding system is usually installed before the concrete is poured. It can be installed afterward, but that is a far more extensive project and the results are usually less effective.

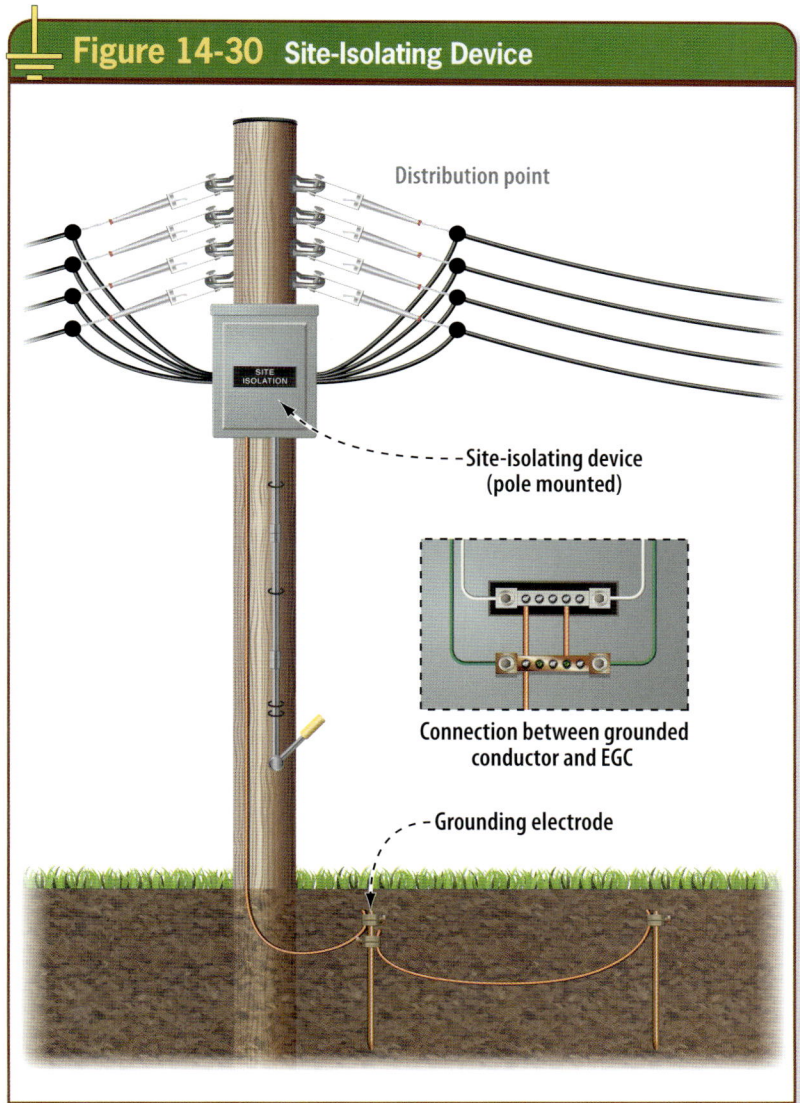

Figure 14-30 Site-Isolating Device

Distribution point

Site-isolating device (pole mounted)

Connection between grounded conductor and EGC

Grounding electrode

Figure 14-30. A site-isolating device must be provided with a grounding electrode system.

The idea is to let the concrete cure and bond simultaneously with the conductive elements contained in it. The equipotential plane can be fabricated using wire mesh, reinforcing bars, and other steel rods and conductive elements. The grid is created when all these conductive elements are bonded together by, at minimum, an 8 AWG solid copper conductor. This conductor must connect all concrete-encased steel and must extend up to metal stanchions and the building grounding electrode system. **See Figure 14-31.**

Listed copper wire mesh products are manufactured and available specifically for this purpose, which can assist installers in establishing an *NEC*-compliant equipotential plane for agricultural buildings and confinement areas where required by Article 547. See UL Product iQ Category KDER for more information about products that are listed as grounding and bonding equipment.

Creating an equipotential plane requires a connection to the building electrical grounding system, which includes the grounding electrodes and the EGCs. The bonding conductor must be a solid copper conductor no smaller than 8 AWG. All connections between the bonding conductor and the conductive elements or wire mesh must be made using pressure connectors or clamps made of brass, copper alloy, or other equivalent and substantial connection means.

An important resource for information about establishing equipotential planes is provided in the American Society of Agricultural and Biological Engineers (ASABE) *EP473.2-2001 (R2015), Equipotential Planes in Animal Containment Areas.*

MOBILE AND MANUFACTURED HOME GROUNDING AND BONDING RULES

Mobile homes and manufactured homes present unique grounding and bonding considerations and require more specific wiring methods.

Mobile Homes

Specific grounding and bonding rules apply to mobile home installations. The main concerns for these installations are that the grounded (usually the neutral) conductors are isolated from grounded equipment and the EGCs. This means that an insulated grounded conductor and a separate EGC must be used to supply a mobile home.

Section 550.32(A) contains requirements for installing a service disconnecting means in a readily accessible outdoor location and within sight of the mobile home. At this location, the grounded conductor is connected to the service disconnecting means enclosure and to the grounding electrode. This is typical of all grounded services, and this serves as the service equipment for the mobile home. This equipment could be a power outlet that is supplied by a feeder from a service in another location.

Where the service equipment is located elsewhere on the property, the required disconnecting means is required to be mounted in a readily accessible outdoor location and within sight from the mobile home. This disconnecting means must also be rated

For additional information, visit qr.njatcdb.org Item #5333

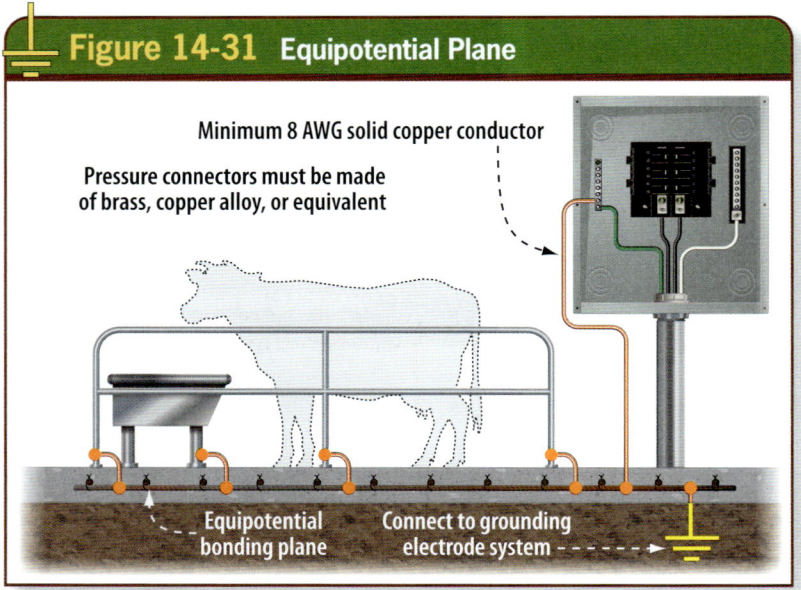

Figure 14-31. *An equipotential bonding plane must be created in concrete slabs of animal confinement areas.*

for no less than what is required for a service disconnecting means for the mobile home. The grounding and bonding at the disconnecting means enclosure must be made according to the requirements in 250.32. The feeder run from the service disconnecting means or the feeder disconnecting means must contain an insulated grounded conductor and a separate EGC.

The feeder conductors are required to be either a listed cord assembly that is factory-installed on the mobile home or a permanently-installed feeder consisting of four insulated, color-coded conductors. Identification of the conductors shall be done by the factory or via field marking to comply with 310.6.

Where the feeder is installed in the field, the requirements in 250.32(B) apply. This means two ungrounded phase conductors—one insulated grounded (neutral) conductor, and a separate EGC—are necessary to supply the panelboard in a mobile home. An exception recognizes grounding alternatives only for existing wiring systems and strictly in compliance with Exception No. 1 to 250.32(B)(1). **See Figure 14-32.**

Manufactured Homes

The service equipment for a manufactured home is permitted to be installed in or mounted on the structure. Multiple conditions must be met for service equipment installed either in or on a manufactured home. First, instructions provided with the home must provide information about securing the home by an anchoring system or to a permanent foundation. The requirements for services in Parts I through VII of Article 230 must be met.

The service disconnecting means must have provisions for connection of a grounding electrode conductor, and it must be routed outside the structure. All grounding and bonding for the service must meet the requirements in Parts I through V of Article 250. The manufacturer's instructions must indicate one

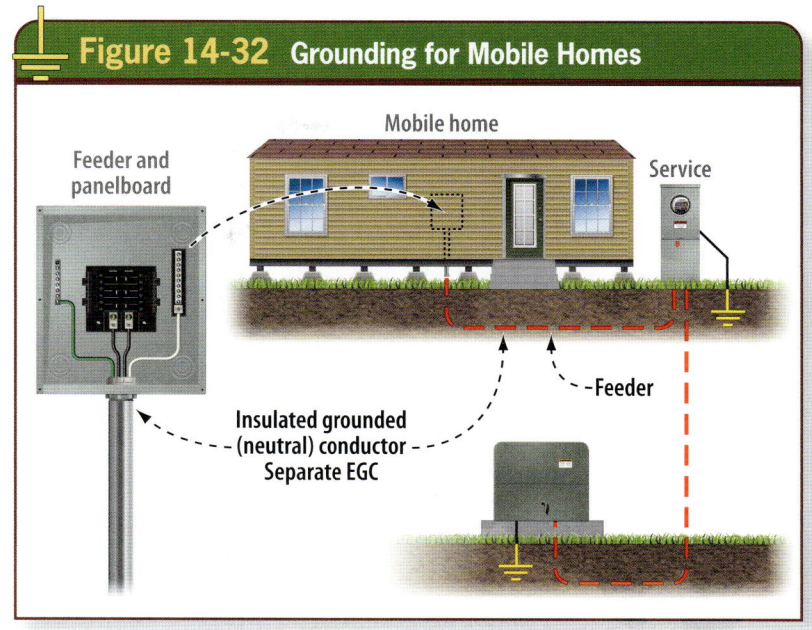

Figure 14-32   Grounding for Mobile Homes

Figure 14-32. Feeders to mobile homes must contain a separate EGC and an insulated grounded conductor.

method of grounding the service equipment, but they must also indicate that the other methods of grounding in Article 250 are permitted, and reference must be made to Article 250 for those alternate methods to achieve the same compliance. Another requirement for manufactured homes, as required by 550.32(B)(7), is that a warning label in compliance with 110.21(B) must be mounted on or adjacent to the service equipment. The warning label shall state the following:

WARNING
DO NOT PROVIDE ELECTRICAL POWER UNTIL THE GROUNDING ELECTRODE(S) IS INSTALLED AND CONNECTED (SEE INSTALLATION INSTRUCTIONS)

If the service equipment for the manufactured home is not installed on or within the structure, the other provisions of 550.32 must be followed. The neutral conductor(s) in mobile home and manufactured home installations must be insulated from ground, grounded metal parts, and the EGC(s).

SPECIAL RULES FOR MARINAS, BOATYARDS, AND DOCKING FACILITIES

Electrical installations for marinas, boatyards, and docking facilities present unique hazards due to their proximity to bodies of water and unique shock and electrocution hazards and require specific grounding and bonding rules. Whenever electricity is near water, the potential for electric shock drowning (ESD) is possible. ESD is the result of low-level currents that enter a person's body while in the water, causing them to be unable to assist themselves and resulting in their drowning. Special grounding and bonding precautions are needed to prevent ESD.

Equipment such as motors and electrical enclosures installed in dusty environments of agricultural facilities must be suitable for the location.

Marinas, boatyards, and docking facilities have unique electrical hazards due to stray currents in the water.

Grounding, Bonding, and Equipotential Planes

Section 555.37 contains the requirements for equipment grounding conductors for marinas, boatyards, and docking facilities. The EGC is required to be sized in accordance with 250.122, to be not less than 12 AWG, and to be insulated due to the corrosive effects that are present in these locations. It is required to be connected to all metal boxes, metal cabinets, metal enclosures, metal frames of utilization equipment, and grounding terminals of grounding-type receptacles.

In addition to the connections required for the EGC, bonding requirements are provided in 555.13. All metal parts in contact with the water, all metal piping, and all non–current-carrying metal parts likely to become energized that are not connected to the EGC are required to be bonded to

Agricultural facilities often include hazardous (classified) locations because of grain dust.

the supply panelboard. The bonding jumper is permitted to be insulated, covered, or bare, but must be a solid copper conductor at least 8 AWG in size. All connections are required to be made in accordance with 250.8.

Requirements for equipotential planes and bonding of equipotential planes are located in 555.14. The purpose of the equipotential plane is to reduce step and touch voltages near electrical equipment due to possible voltage gradients. An equipotential plane is required if the electrical equipment is operating at voltages of more than 250 volts to ground and is located within 10 feet of the body of water. The plane is required to extend from the area around the equipment and directly below it to not less than 36 inches from where a person could contact the equipment.

GFPE and GFCI Requirements

Section 555.35 contains the requirements for ground-fault protection of equipment (GFPE) and ground-fault circuit interrupters (GFCIs). Sources that directly supply docking facilities are required to have GFPE rated at not more than 100 milliamperes. Feeders that supply docking facilities are also required to have GFPE rated at not more than 100 milliamperes. Branch-circuit receptacles, typically installed in a power outlet supplying the watercraft, are required to have GFPE rated at not more than 30 milliamperes. The intent is to keep stray currents in the water low enough to not cause paralysis, while keeping them high enough to not nuisance trip for the watercraft supplied.

Any branch-circuit outlet for other than shore power, not exceeding 150 volts to ground, and either single-phase 60 amperes or less or 3-phase 100 amperes or less is required to be provided with GFCI protection. The requirement applies to all outlets, not just receptacle outlets.

Note the white boxes on the docks, which are the power outlets supplying the boat and are required to have ground-fault protection of equipment (GFPE).

SUMMARY

Electrical installations in hazardous locations require a more comprehensive design than usual, including robust bonding and grounding systems. Health care facilities have many special electrical requirements that require grounding and bonding methods that exceed the basics for general wiring. Specific grounding requirements also exist for agricultural buildings, mobile homes, manufactured homes, and marinas. These subjects should be explored in depth in order to master the grounding and bonding requirements for these special occupancies.

REVIEW QUESTIONS

1. Chapter 5 of the *NEC* provides the requirements for which of the following?

   a. Communications systems
   b. Special conditions
   c. Special equipment
   d. Special occupancies

2. The rules in Chapters 5 through 7 of the *NEC* are in addition to and do not amend or modify the general requirements in Chapters 1 through 7.

   a. True   b. False

3. Regardless of the voltage, the electrical continuity and conductivity of non–current-carrying metal parts of equipment, raceways, and other enclosures in any hazardous (classified) location, as defined in Article 500, are required to be ensured by any of the methods specified for __?__ in 250.92(B)(2) through (4) that are approved for the wiring method used.

   a. load-side equipment
   b. separately derived systems
   c. services
   d. special equipment

4. Specific methods of bonding metal raceways installed in a hazardous (classified) location are required even if there is an EGC or bonding conductor installed inside the raceway.

   a. True   b. False

5. The more restrictive bonding requirements of 250.100 for installations in hazardous locations are necessary to reduce the likelihood that a line-to-ground fault will cause arcing and sparking at connection points of metal raceways and boxes or other enclosures.

   a. True   b. False

6. Which of the following methods of bonding a raceway to a metal enclosure is not acceptable in a hazardous location?

   a. Connections using threaded couplings or listed threaded hubs on enclosures where they are made up wrench-tight
   b. Double locknuts, one on the inside and one on the outside of the enclosure
   c. Other listed devices, such as bonding-type locknuts, bushings, or bushings with bonding jumpers
   d. Threadless couplings and connectors made up tight for metal raceways and metal-clad cables

7. Where flexible metal conduit or liquidtight flexible metal conduit is installed for flexibility as permitted in hazardous (classified) locations, a wire-type __?__ must be installed in accordance with 250.102.

   a. equipment bonding jumper
   b. grounded (neutral) conductor
   c. grounding electrode conductor
   d. system bonding jumper equipment grounding conductor

8. Branch circuits installed in patient care spaces of health care facilities are required to be installed in a metal raceway or using another metal wiring method that qualifies as an EGC in accordance with 250.118, in addition to a(n) __?__ .

   a. bare aluminum conductor
   b. bare copper conductor
   c. insulated aluminum or copper-clad aluminum conductor
   d. insulated copper equipment grounding conductor

9. In a patient care space, the branch circuit wiring can be installed in PVC conduit as long as two insulated copper EGCs are included with the other circuit conductors.

   a. True   b. False

10. The term *patient care vicinity* refers to a distance of __?__ from the patient bed location in a health care facility.

    a. 2'
    b. 3'
    c. 5'
    d. 6'

11. A luminaire installed no less than 7.5 feet above the floor in a patient care vicinity does not have to be provided with two EGC paths as required by the *NEC* in 517.13.

    a. True   b. False

12. Isolated power systems installed in health care facilities are required to be installed and operated as __?__ .

    a. impedance grounded neutral systems
    b. solidly grounded systems
    c. ungrounded systems
    d. any of the above

13. Where an isolated power system is used, the patient equipment grounding point is typically connected to the reference grounding bar in the ___?___.

    a. isolated power panel
    b. operating room
    c. service equipment enclosure
    d. transformer enclosure

14. Where isolated circuit conductors supply 125-volt, single-phase, 15- and 20-ampere receptacles from an isolated power panel, the ___?___ conductor is required to be connected to the terminals on the receptacles that are identified (white or silver) in accordance with 200.10(B) for connection to the grounded circuit conductor.

    a. gray
    b. green
    c. orange
    d. white

15. The redundant EGC requirements of 517.13 apply only to branch circuits serving patient care spaces, not to the feeders.

    a. True   b. False

16. If a critical branch feeder is supplied from a grounded distribution system in a health care facility and is installed using a metal wiring method that qualifies as an EGC according to 250.118, the grounding of the switchboard, switchgear, or panelboard is required to be ensured by any of the following methods except ___?___.

    a. a connection of feeder raceways or Type MC or MI cable to threaded hubs or bosses on terminating enclosures
    b. a grounding bushing and a continuous copper bonding jumper, sized in accordance with Table 250.122, with the bonding jumper connected to the junction enclosure or the ground bus of the panel
    c. other approved devices, such as bonding-type locknuts or bushings
    d. two locknuts, one on the inside and one on the outside of the enclosure

17. The *NEC* requires the equipment grounding terminal bars of the normal and essential branch circuit panelboards serving the same individual patient care vicinity to be bonded together with an insulated continuous copper conductor no smaller than ___?___.

    a. 10 AWG   c. 6 AWG
    b. 8 AWG   d. 4 AWG

18. A(n) ___?___ is an area where wire mesh or other conductive elements are embedded in or placed under concrete, bonded to all metal structures and fixed nonelectrical equipment that may become energized, and connected to the electrical grounding system to prevent a difference in voltage from developing within the plane.

    a. common grounding electrode
    b. equipotential bonding grid
    c. equipotential plane
    d. signal reference grid

19. Which of the following areas of an agricultural facility must have an equipotential plane installed?

    a. Indoor animal confinement areas with conductive floors where metallic equipment that could become energized is located accessible to the livestock area
    b. Outdoor animal confinement areas with concrete slabs where metallic equipment that could become energized is accessible to livestock
    c. Outdoor areas on farms that are not intended to confine animals
    d. Both a. and b.

20. The equipotential plane can be built using wire mesh, reinforcing bars, and other steel rods and conductive elements. The grid is created when all of these conductive elements are bonded together by, at minimum, a(n) ___?___ copper conductor.

    a. 14 AWG stranded
    b. 12 AWG solid
    c. 10 AWG stranded
    d. 8 AWG solid

21. An isolated grounding receptacle and circuit is prohibited in a patient care vicinity.

    a. True   b. False

22. The feeder supplying shore power to watercraft in a marina is required to be supplied with which of the following?

    a. GFPE rated not more than 30 mA
    b. GFCI rated not more than 100 mA
    c. GFPE rated not more than 100 mA
    d. GFPE when rated less than 250 volts to ground

Grounding for Special Equipment

Equipment grounding is an important aspect of electrical safety for people and property. Chapter 6 of the *NEC* provides the requirements for special equipment, such as electric signs, cranes, elevators, audio systems, sensitive electronic equipment, information technology (IT) systems, swimming pools, and solar photovoltaic (PV) systems. Enhanced grounding and bonding rules often apply due to increased shock hazards and the unique construction and operation of special equipment. Grounding special equipment generally requires a connection to any of the equipment grounding conductors specified in *NEC* Section 250.118.

Objectives

- » Understand general grounding and bonding requirements for special equipment.

- » Determine the grounding and bonding rules for special equipment as provided in Chapter 6 of the *NEC*.

- » Determine the more restrictive electrical grounding and bonding requirements for electric signs and outline lighting systems, including light-emitting diode systems.

- » Understand the specific grounding requirements that apply to sensitive electronic equipment.

- » Understand the purpose of enhanced grounding and bonding requirements for swimming pools and similar equipment in aquatic environments.

- » Understand specific grounding and bonding rules for photovoltaic systems.

Chapter 15

Table of Contents

Purpose of Grounding Equipment................ 346

Electric Signs and Outline
Lighting Systems... 346

    Grounding Signs and Outline
    Lighting Equipment 347

    Bonding Metal Parts of Signs and
    Outline Lighting Systems 348

Electric Cranes and Elevators.................... 351

Information Technology Equipment and
Sensitive Electronic Equipment.................. 352

Grounding and Bonding Requirements for
Swimming Pools and Similar Installations 356

    Bonding and Equipment Grounding........ 356

    Pool Pump Motor 357

    Grounding and Bonding
    Underwater Luminaires........................ 358

    Junction Boxes and Other Enclosures 359

    Separate Buildings 359

    Equipotential Bonding Requirements...... 360

        Pool Shells (Conductive) 361

        Perimeter Surfaces......................... 362

        Other Conductive Components........ 363

        Pool Water 364

        Specialized Pool Equipment............ 360

Spas, Hot Tubs, and Permanently
Installed Immersion Pools 365

    Spa and Hot Tub Bonding–Indoor..... 365

    Spa and Hot Tub Bonding–Outdoor ... 366

    Permanently Installed
    Immersion Pools............................. 366

Fountains 366

Therapeutic Pools and Tubs
for Health Care Use.............................. 367

    Bonding 367

    Equipment Grounding 368

    Portable Therapeutic Appliances
    and Equipment 368

    Hydromassage Bathtubs 368

Grounding Requirements for
Solar PV Systems..................................... 369

    System Grounding 369

    Definition.. 370

    Ground-Fault Detector-Interrupter 370

    Equipment Grounding 370

    Grounding Electrode Systems.............. 371

    Electrical Safety 372

Summary.. 373

Review Questions 373

PURPOSE OF GROUNDING EQUIPMENT

Recall that the purpose of the *NEC* is to protect people and property from electrical hazards arising from electricity use. The purpose of grounding and bonding is covered in Section 250.4.

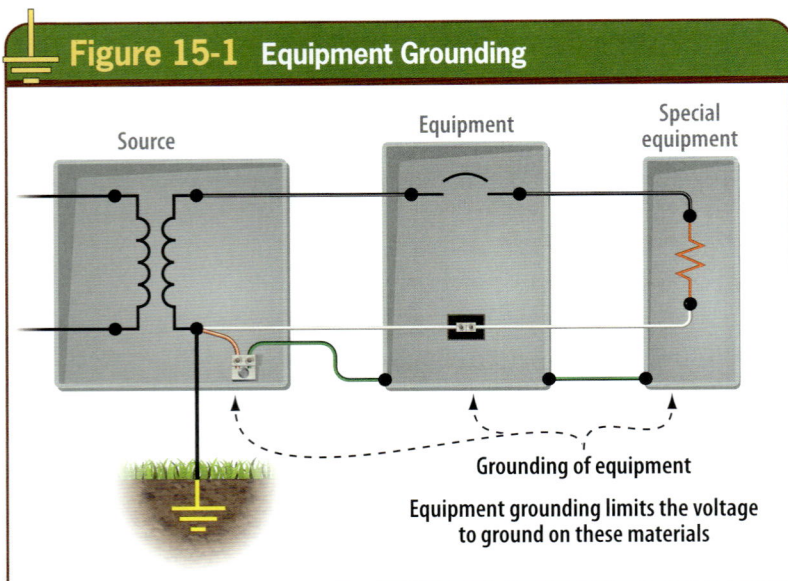

Figure 15-1 Equipment Grounding

Source

Equipment

Special equipment

Grounding of equipment

Equipment grounding limits the voltage to ground on these materials

Figure 15-1. *Grounding places equipment at the same potential as ground (the Earth), thereby limiting above-ground voltages on such equipment.*

Figure 15-2 Electric Signs

Figure 15-2. *Electric signs and outline lighting systems are installed on many buildings or structures.*

Courtesy of Young Electric Sign Company (YESCO)

The reasons for grounding equipment are also outlined in Section 250.4. For grounded systems, equipment grounding and equipment bonding performance requirements are set forth in Sections 250.4(A)(2) and (3). These rules focus specifically on grounding and bonding equipment, not on system grounding, which is covered in Section 250.4(A)(1). Equipment is grounded to limit the voltage to ground on the normally non–current-carrying parts of such equipment. **See Figure 15-1.**

The conductive materials of equipment are bonded together and to the supply source in an arrangement that establishes an effective ground-fault current path. Bonding is an electrical connection that establishes continuity and conductivity between conductive parts. The equipment is required to be grounded and bonded to meet the performance criteria required in Section 250.4. In some cases, additional bonding, such as equipotential bonding, is necessary to address the unique hazards encountered in some special equipment installations, such as pools, hot tubs, electric signs, and outline lighting systems. The main concern with these enhanced bonding requirements is to reduce electrical shock hazards related to these special types of equipment or special occupancies.

ELECTRIC SIGNS AND OUTLINE LIGHTING SYSTEMS

Electric signs and outline lighting systems are common. Nearly every building or structure other than residential occupancies has some type of sign or outline lighting system installed on it. The signs and special lighting systems provide direction to the public and are an essential part of businesses. **See Figure 15-2.** There are a variety of electrical signs and outline lighting systems installed today using technology such as incandescent, fluorescent, neon, cold cathode, fiber optic, and light-emitting diode (LED) systems.

It is easy to see the many technologies used for signage both in this country and around the world. Just visit any major city or attraction; the vast quantity of electric signs and outline lighting used is evident. This equipment is important to society as a whole. **See Figure 15-3.** Article 600 of the *NEC* provides the special requirements for electric signs and outline lighting systems regardless of the voltage. Specific grounding and bonding requirements for signs, outline lighting, and skeleton tubing are provided in Section 600.7. This section is subdivided into two parts. Subdivision (A) includes the grounding requirements, and subdivision (B) provides the special bonding requirements.

Grounding Signs and Outline Lighting Equipment

Article 250 includes equipment grounding rules in Section 250.112, which requires the normally non–current-carrying parts of equipment to be connected to an equipment grounding conductor (EGC), regardless of the voltage.

This is the general grounding rule for signs and outline lighting. Section 250.112(G) recognizes that more specific grounding and bonding must be applied for this type of equipment, and a reference to Section 600.7 is

Figure 15-3 Grounding Electric Signs

Figure 15-3. The electric signs and outline lighting systems installed on buildings or structures convey a variety of messages, provide illumination, and draw the attention of the public.

made. Electric signs and metal equipment of outline lighting systems must be grounded by connection to the EGC of the supplying branch circuit or feeder. **See Figure 15-4.** Any of the types of EGCs referenced in Section 250.118 can be used to accomplish the grounding. If a wire-type EGC is used,

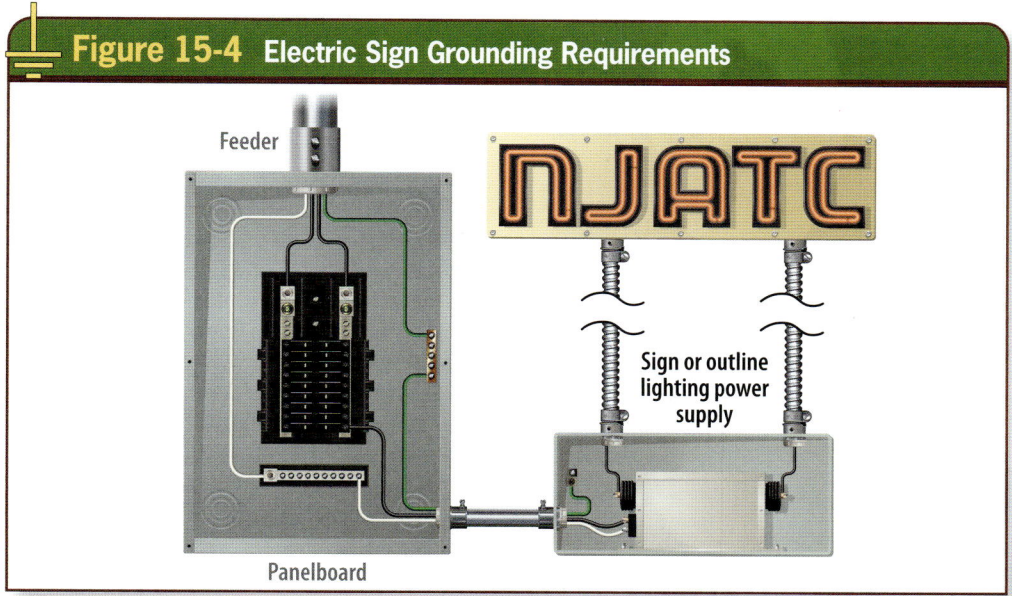

Figure 15-4 Electric Sign Grounding Requirements

Feeder

NJATC

Sign or outline lighting power supply

Panelboard

Figure 15-4. Electric signs and metal parts of outline lighting systems are required to be grounded.

Figure 15-5   Pole-Mounted Electric Signs

Figure 15-5. Pole-mounted electric signs are required to be grounded by connection to an EGC. They are often also grounded by use of an auxiliary grounding electrode installed at the base of the pole.

iStock Photo Courtesy of NECA

Figure 15-6   Grounding Pole-Mounted Signs

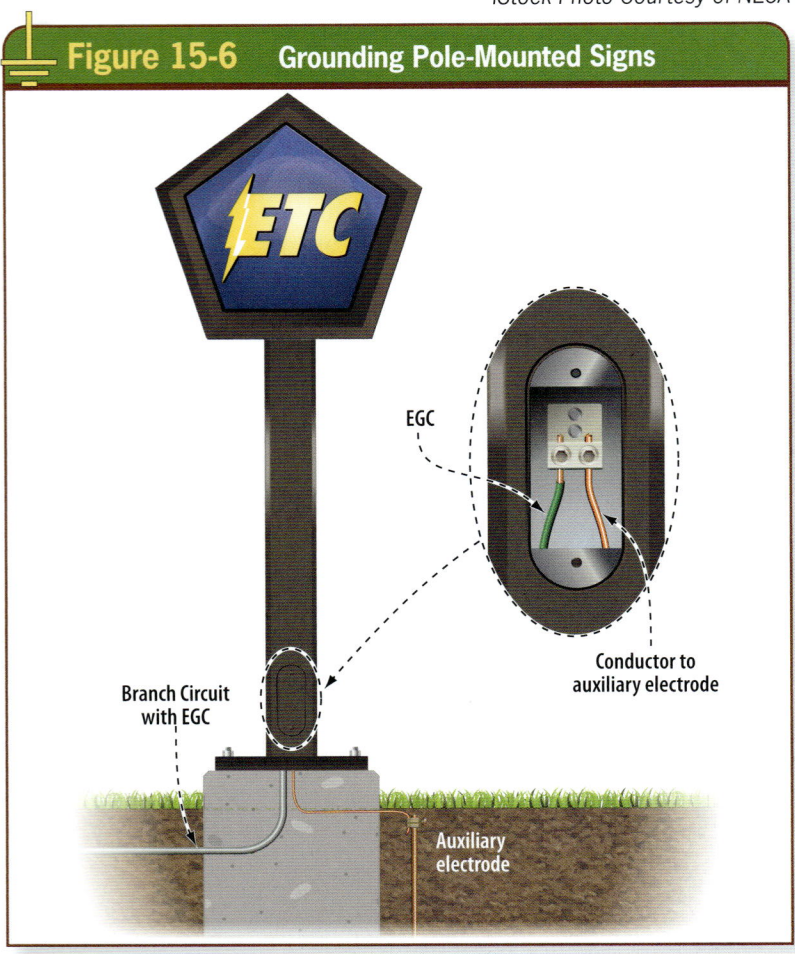

Figure 15-6. Pole signs grounded by connection to an EGC are often connected to an auxiliary grounding electrode that establishes a direct (local) reference to ground at the pole location.

it must be sized in accordance with Section 250.122, which is based on the rating of the circuit breaker or fuse protecting the circuit.

All EGC connections must be made using a method specified in Section 250.8. Grounding and bonding conductor connections are an important aspect of these safety circuits. Recall that sheet metal screws are not acceptable for grounding and bonding connections. The *NEC* relaxes the sign grounding requirement by exception, but only for portable cord-connected signs that are manufactured using a system of double insulation or an equivalent and are so marked.

Sometimes electric signs are pole mounted, as is the case for billboards and pole signs in parking lots. It is often desirable to establish a grounding electrode at these pole locations. **See Figure 15-5.** These electrodes are auxiliary grounding electrodes and are installed as an option. The grounding objectives in this case are to connect the electric sign to the EGC of the supply circuit and to connect it to an auxiliary grounding electrode, usually a ground rod, as covered by Section 250.54. This creates a local ground (Earth) reference for the pole-mounted sign while it is still connected to the EGC of the supply circuit, which is the required effective ground-fault current path. **See Figure 15-6.**

As previously covered, the Earth alone cannot be used as an effective ground-fault current path. Metal parts of a building or structure are also not permitted as an EGC or as a high-voltage secondary return circuit for a neon lighting system.

Bonding Metal Parts of Signs and Outline Lighting Systems

The requirements for bonding metal parts and equipment for electric signs and outline lighting systems are provided in Section 600.7(B). The objectives are to connect all metal parts together and to connect them to the EGC supplying the transformer or power supply. Many electric signs and outline lighting systems have high-voltage transformers that are used to ignite gases in glass tubing, such as the tubing used in neon and cold cathode systems. The branch circuit supplies the primary of the transformer, and the voltage is then raised by a transformer to levels as high as 15,000 volts (7,500 volts line-to-ground). The secondary (high voltage) conductors are routed using suitable wiring methods to the electrodes molded into the ends of the glass tubing. **See Figure 15-7.**

Bonding all metal parts associated with such systems is an important safety requirement. If the metal is not bonded, the high-voltage secondary circuits can induce voltages in the "dead metal," raising potentials and creating shock and fire hazards. **See Figure 15-8.** The *NEC* does relax this bonding requirement for remote metal parts of section signs or outline lighting systems supplied by a Class 2 power supply, such as those used with LED signs. In these cases, the secondary circuit is at Class 2 voltage levels and does not present the same hazards to people and property as the high-voltage secondary circuits commonly used for neon or cold cathode lighting systems.

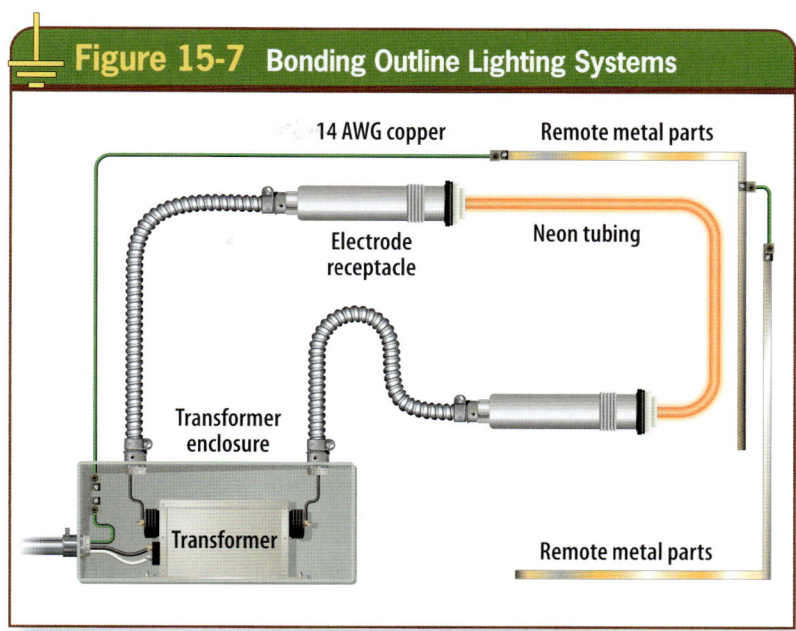

Figure 15-7. *Bonding requirements apply to neon signs and outline lighting systems with remote metal parts.*

All bonding connections to metal parts and equipment associated with signs and outline lighting must be made using a method provided in Section 250.8, and metal parts of buildings or structures cannot be used as the required bonding for signs and outline lighting systems.

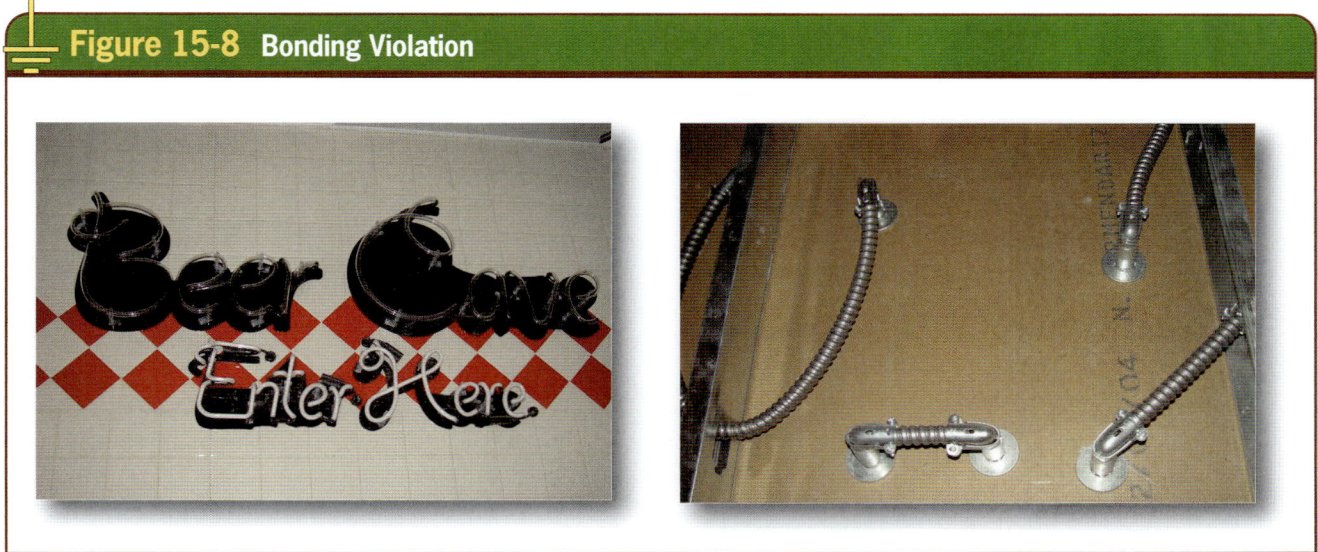

Figure 15-8. *Flexible metal conduit (FMC) enclosing a high-voltage neon secondary circuit is required to be bonded. If even a few sections of FMC are not bonded or grounded, the installation would be in violation of the NEC.*

Bonding conductors for signs and outline lighting systems are required to be no smaller than 14 AWG copper, and where installed external of a sign enclosure or raceway system, they must be protected from physical damage. Note that this external bonding jumper does not have to be sized according to Section 250.102, nor is it limited to six feet in length. The reason for this is that the secondary circuits of these transformers and power supplies are operating at high voltage, and current is in the milliampere range. A ground fault on the transformer or power supply secondary causes operation of the secondary circuit ground-fault protection (SCGFP) device that is integral to the transformer or power supply.

Where high-voltage secondary circuits are installed from transformers or power supplies to the glass tubing of neon lighting or cold cathode systems, the length of high-voltage conductors is limited. These runs are limited by Section 600.32(J)(1) to a maximum of 20 feet in length for metal conduit (measured from each output hub on the transformer to the first tubing

electrode) and to a maximum of 50 feet for nonmetallic conduit. The secondary high-voltage cable must be installed using a suitable wiring method or listed sleeve material. When listed flexible metal conduit (FMC) or listed liquidtight flexible metal conduit (LFMC) is used as the raceway to enclose the high-voltage gas tube oil (GTO) ignition cables, the metal parts of the sign or outline lighting system are permitted to be bonded through the flexible metal raceway wiring method, provided that the total cumulative length of the conduit does not exceed 100 feet. Any small metal parts that do not measure more than two inches in any dimension are not required to be bonded.

If listed nonmetallic conduit is used to enclose high-voltage secondary circuits to neon tubing systems, bonding of metal parts associated with the sign or outline lighting is required. A common example of this is a reverse pan channel letter sign where the individual letters of the sign are all isolated from one another and isolated from the remote transformer or power supply. **See Figure 15-9.** In this case, a bonding conductor is required to be installed separate and remote from the nonmetallic conduit containing the high-voltage secondary circuit. This bonding conductor must be a minimum of 14 AWG copper, and when installed external to the sign or outline lighting system, it must be protected from physical damage.

Many neon signs and outline lighting systems are supplied by electronic power supplies operating above 60 hertz. Section 600.7(B)(6) specifies a distance of at least 1 $1/2$ inches from a nonmetallic conduit containing a high-voltage circuit of 100 hertz or less and no less than 1 $3/4$ inches from a nonmetallic conduit containing a circuit operating above 100 hertz. This is to minimize the effects of unbalanced stress and allow the magnetic lines of flux to remain symmetrical in the high-voltage secondary circuit during normal operation. **See Figure 15-10.**

Figure 15-9   **Bonding Isolated Conductive Parts**

Figure 15-9. Reverse pan channel letter signs with metal parts remote from one another are required to be bonded.

If the bonding conductor or any grounded metal is close to the nonmetallic conduit containing the high-voltage GTO cable, the magnetic lines of force in the AC circuit can become asymmetrical and cause stress and even premature failure of the contained high-voltage GTO cable and the transformer or power supply. For this reason, Section 600.32 also requires spacing from grounded or bonded metal parts for nonmetallic conduits containing high-voltage secondary circuits. The minimum spacing requirements are the same as those provided in Section 600.7(B)(6), which are no less than 1 1/2 inches from circuits of 100 hertz or less and no less than 1 3/4 inches from circuits operating above 100 hertz.

It is important to remember that in higher-frequency circuits, the skin effect is amplified and greater care is needed to ensure that spacing is maintained in order to reduce failures in these types of systems while maintaining effective bonding and grounding of all associated metal parts and equipment for such installations.

ELECTRIC CRANES AND ELEVATORS

Electric cranes and hoists are covered by Article 610 of the *NEC*. Electric cranes are typically supplied by branch circuits that include an EGC. The EGC can be any of the types listed in Section 250.118. If it is a wire type, it must be sized according to Section 250.122. Section 610.61 requires grounding and bonding of all exposed non–current-carrying metal parts of cranes, monorail hoists, and any associated pendant controls. **See Figure 15-11.**

The bonding can be accomplished either by mechanical connections of the conductive parts or by suitable bonding jumpers. The objective is to create a common conductive path for ground-fault current that includes all metal parts associated with the crane or hoist. Moving parts that have metal-to-metal bearing contact are considered bonded

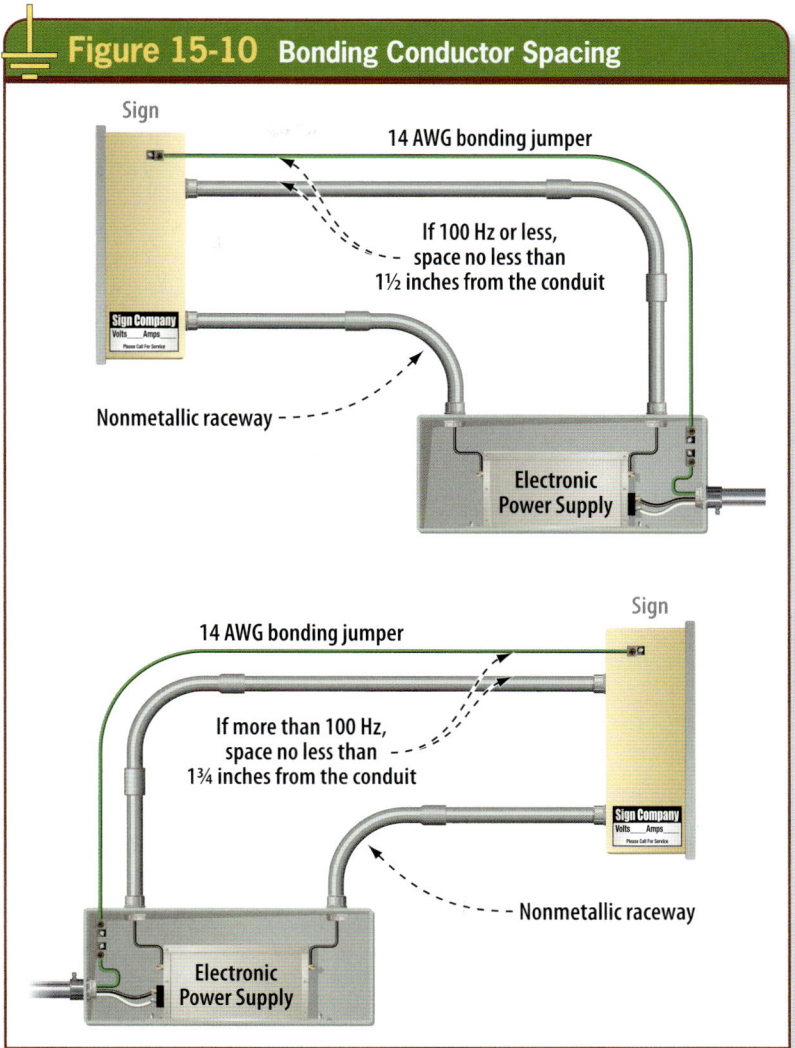

Figure 15-10. *Spacing is required for bonding conductors installed with neon secondary circuits.*

Figure 15-11. *Equipment grounding and bonding are required for electric cranes and hoists.*

together through such contact. On the other hand, a trolley frame and bridge are not considered electrically bonded through the bridge and trolley wheels and tracks; a separate bonding jumper is required around the wheel-to-track contact.

As a reminder, systems supplying crane circuits that operate over Class III locations are not permitted to be grounded due to the restriction in 503.155(A). This is to reduce the probabilities of arcing and sparks that could result from a ground fault in the equipment, which could send hot particles into accumulations of fibers and flyings below the crane, creating a fire hazard. In these locations, the metal parts associated with such equipment must be grounded and bonded, but the system supplying the crane is not permitted to be grounded. Ground detection is required on such ungrounded systems to alert qualified people of the first phase-to-ground fault condition. **See Figure 15-12.**

Elevators and associated equipment are covered by Article 620 of the *NEC*.

Equipment grounding is required for elevators as specified by Section 250.112(E), and specific rules are provided in Part IX of Article 620. **See Figure 15-13.** Metal raceways, mineral-insulated cables, metal tape–type cables (Type MC), or armored cables (Type AC) attached to elevator cars must include an EGC with the supply circuit for elevator equipment.

Elevators and hoists are typically supplied by a circuit or circuits sized according to the applicable requirements in Part II of Article 620. The rating of the overcurrent protective device supplying this equipment determines the minimum size required for wire-type EGCs. Often, more than a single circuit supplies elevator equipment. A larger circuit usually supplies the pump equipment for a hydraulic elevator or the motors for cable elevator equipment, and smaller individual circuits supply car lights and receptacle power. In addition, circuits for servicing the elevator car and pit include EGCs. The frames and metal equipment of electric elevators, including motors, machines, and controllers, and all metal equipment on the elevator car must be grounded and bonded. Nonelectrical elevators that have any electrical circuits in contact with the car must have the frame of the car grounded and bonded in accordance with Parts V and VII of Article 250.

Escalators and moving walks are also covered by the rules in Article 620. Any metal parts associated with these types of equipment must be grounded and bonded.

INFORMATION TECHNOLOGY EQUIPMENT AND SENSITIVE ELECTRONIC EQUIPMENT

Grounding rules for information technology (IT) equipment are provided in Sections 645.14 and 645.15.

Section 645.14 requires separately derived systems for information technology rooms to be installed in accordance with Parts I and II of Article 250. Power systems derived within listed

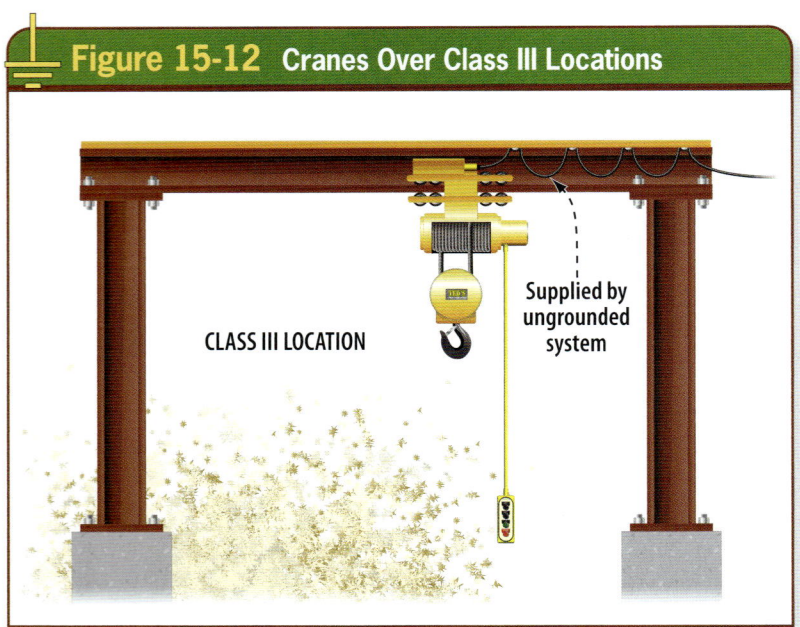

Figure 15-12 Cranes Over Class III Locations

CLASS III LOCATION

Supplied by ungrounded system

Figure 15-12. Cranes installed and operating over Class III locations (where fire hazards are present due to accumulations of fibers and flyings) are required to be supplied by ungrounded systems.

information technology equipment, such as power distribution units (PDUs), and IT systems supplied through receptacles or cable assemblies shall not be considered separately derived for the purposes of applying 250.30. It is important to follow the manufacturer's installation instructions relative to required and permitted grounding and bonding connections for such equipment.

Section 645.15 requires all non–current-carrying metal parts of an information technology system to be grounded and bonded to an EGC in accordance with Parts I, V, VI, VII, and VIII of Article 250. This section also relaxes the equipment grounding requirements for listed IT equipment that is double insulated. Often, information technology equipment is grounded by isolated or insulated equipment grounding circuits in accordance with 250.146(D).

Information technology rooms are usually equipped with power distribution units (PDUs) that supply all equipment associated with the information technology system. **See Figure 15-14.** While listed PDUs typi-

Figure 15-13   Elevator Grounding Requirements

Figure 15-13. *Elevators and associated equipment are required to be grounded by connection to an EGC of the supply circuit.*

cally include transformers, they are usually not treated as separately derived systems for grounding purposes in accordance with Section 250.30(A). They must be grounded according to

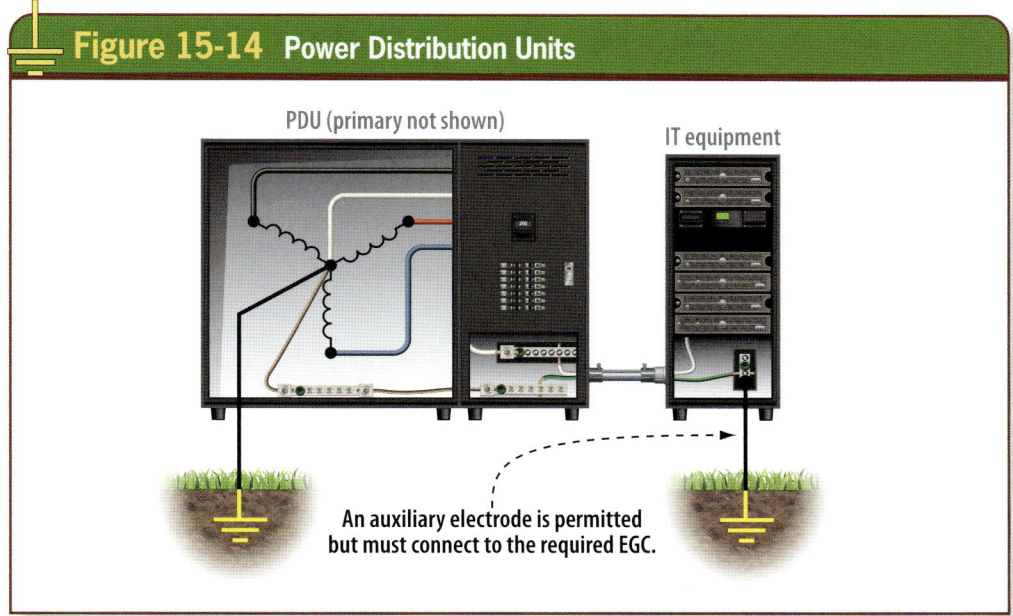

Figure 15-14   Power Distribution Units

PDU (primary not shown)

IT equipment

An auxiliary electrode is permitted but must connect to the required EGC.

Figure 15-14. *PDUs are commonly installed and operated in information technology rooms that are wired according to Article 645 of the NEC.*

specific instructions provided by the manufacturer.

Any signal reference grids installed in information technology rooms must be connected to the EGC provided with the circuits supplying the information technology equipment. Any auxiliary grounding electrodes installed for the IT equipment must meet the requirements in 250.54. Remember that the Earth is not permitted as an effective ground-fault current path.

Article 647 provides specific rules applicable to sensitive electronic equipment installations. Some electronic equipment is sensitive to common voltages such as 120 or 240 volts. The supply systems for sensitive electronic equipment are 3-wire, single-phase systems. The ungrounded conductors have a voltage of 120 volts between them, and from each ungrounded conductor to the neutral of such systems, the voltage is 60 volts. These systems must be grounded in accordance with the requirements in Section 250.30(A) for separately derived systems and must be connected to a grounding electrode as specified in Section 250.30(A)(4). **See Figure 15-15.**

The purpose of such special low-voltage systems is to reduce electromagnetic interference (EMI) in sensitive electronic equipment systems and locations. This type of system is permitted only in commercial and industrial establishments, provided that the system is restricted to areas under supervision of qualified persons. All wiring rules in Sections 647.4 through 647.8 must be applied to such installations.

Sensitive electronic systems are required to be installed using listed equipment for this specific purpose or using standard rated equipment and specific field-applied markings to indicate the voltage and use of the equipment. All junction boxes, equipment, and conductors must be clearly identified to show they are associated with such systems.

Branch circuits and feeders must be provided with a disconnecting means that simultaneously opens all ungrounded conductors. The voltage drop on these systems cannot exceed 1.5% on the branch circuits and 2.5% for the combined voltage drop of the feeder and branch circuit supplied by the system. The voltage is already low on these

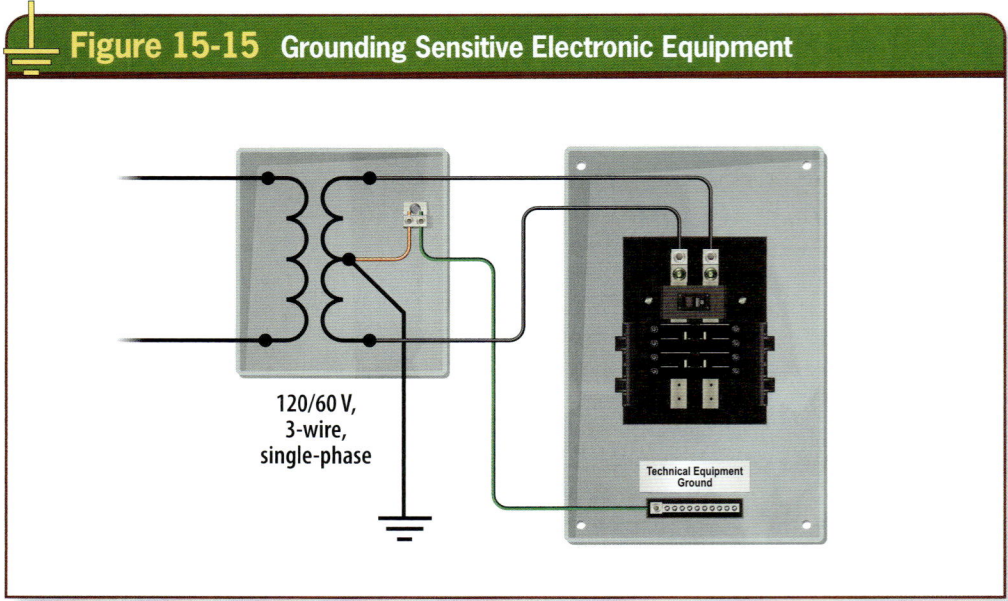

Figure 15-15 Grounding Sensitive Electronic Equipment

120/60 V,
3-wire,
single-phase

Technical Equipment
Ground

Figure 15-15. Special grounding requirements apply to systems supplying sensitive electronic equipment in accordance with Article 647.

systems, and the equipment supplied is vulnerable to damage resulting from lower supply voltages. The *NEC* generally does not address voltage drop as a requirement in Chapters 1 through 4; however, it is addressed as a rule for these systems.

Equipment and receptacles supplied by these systems must be grounded by an EGC run with the supply circuit conductors and terminated on the equipment grounding terminal bus of a panelboard that must be marked "Technical Equipment Ground." This equipment grounding bus is required to be connected to the grounded conductor of the system on the line side of the separately derived system disconnecting means. **See Figure 15-16.** The EGC must be sized according to Table 250.122 and installed with the feeder conductors. The technical equipment grounding bus does not have to be bonded to the equipment enclosure.

Alternative grounding methods recognized in the *NEC* are permitted for these systems if the impedance of the effective ground-fault current return path does not exceed the impedance of the EGC sized according to the requirements in Article 647. These special grounding provisions limit the impedance of a ground-fault path because only 60 volts would be present in a phase-to-ground fault condition, rather than the usual 120 volts.

Section 647.7 includes special rules for receptacles used with this type of system. All 15- and 20-ampere receptacles must be ground-fault circuit interrupter (GFCI) protected. All receptacle outlet strips, adaptors, covers, and faceplates must be marked as provided in the following example:

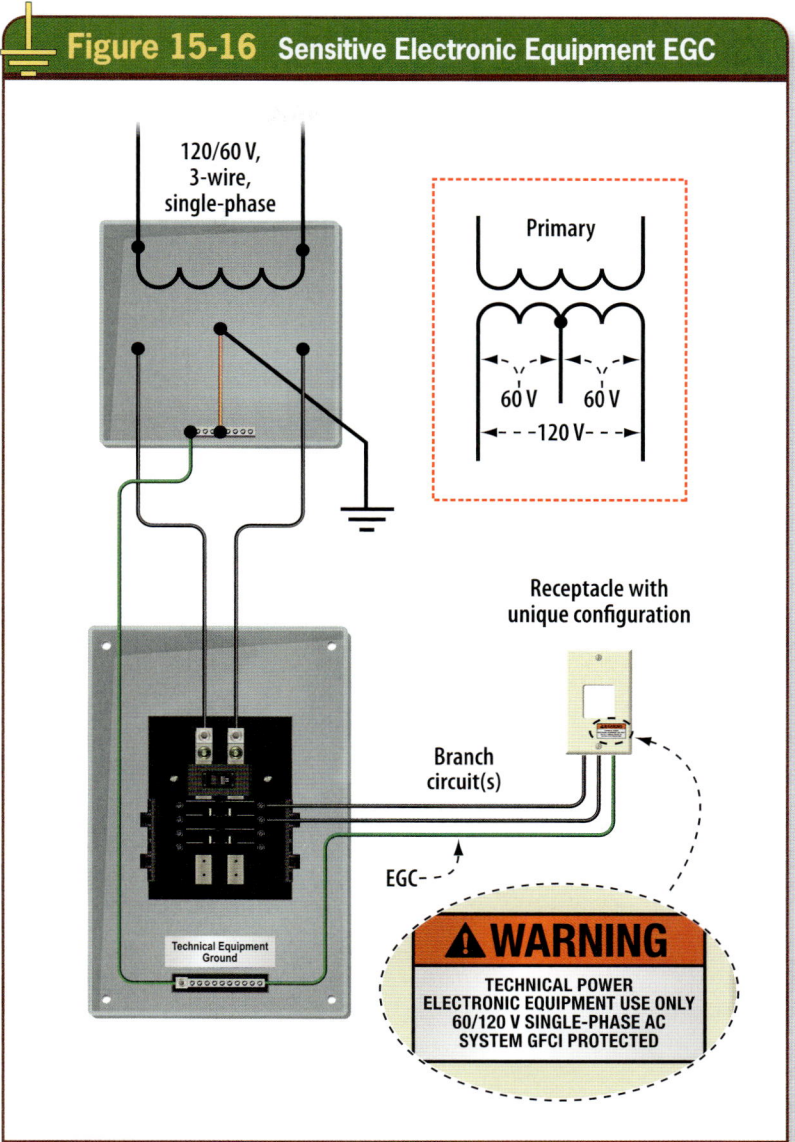

Figure 15-16. *Branch circuits for sensitive electronic equipment must include an EGC.*

> **⚠ WARNING**
>
> **TECHNICAL POWER**
>
> DO NOT CONNECT LIGHTING EQUIPMENT.
>
> FOR ELECTRONIC EQUIPMENT USE ONLY.
> 60/120 V, SINGLE-PHASE AC SYSTEM
> GFCI PROTECTED

All receptacles used with systems covered in Article 647 must have unique configurations and be identified for use with this type of system. See the specific rules for receptacles installed with sensitive electronic equipment installations as provided in Section 647.7. If isolated grounding circuits and receptacles are used with these systems, the installation must be in accordance with Section 250.146(D), except that the EGC of these circuits can be terminated according to Section 647.6(B). There are also specific rules for lighting equipment installed for the purpose of reducing electromagnetic interference originating in lighting equipment. These installations must meet the requirements in Section 647.8. See Article 647 for all rules pertaining to sensitive electronic equipment.

Figure 15-17. *Swimming pools require special grounding and bonding.*

Figure 15-18. *Hot tubs require special grounding and bonding.*

GROUNDING AND BONDING REQUIREMENTS FOR SWIMMING POOLS AND SIMILAR INSTALLATIONS

In wet locations, electric shock hazards are present. Swimming pools and similar installations are aquatic environments and involve applications using special equipment, and as a result more restrictive requirements apply to these installations. Special emphasis is placed on electrical grounding and bonding rules for these types of equipment and installations. **See Figure 15-17.**

Article 680 of the *NEC* provides rules related to receptacle locations, overhead conductor clearances, GFCI protection, and special grounding and bonding. It is important to recognize the arrangement and application of Article 680. Part I applies generally to all other parts of the article. Part II includes special requirements that apply specifically to permanently installed pools. Part III applies to storable pools. Note that Section 680.30 indicates that both Parts I and III apply to storable pools, storable spas, storable hot tubs, and storable immersion pools. This is typical of how the various parts of Article 680 must be applied to swimming pools and similar equipment such as hot tubs. **See Figure 15-18.**

Bonding and Equipment Grounding

Section 680.7 provides a general requirement for bonding and grounding equipment associated with pools or similar installations. The equipment grounding requirement applies to all equipment wired using the methods covered in Chapter 3 of the *NEC*.

Electrical equipment associated with pools and all similar installations are generally required to be grounded. Equipment grounding is required for through-wall lighting assemblies and underwater luminaires. Note that low-voltage lighting equipment listed for use without an EGC is permitted. Any electrical equipment located within five feet of the inside walls of a pool or specified body of water must be grounded. The pool pump motor or equipment associated with the water recirculation system must be grounded. Any junction boxes, transformer enclosures, and GFCIs must be grounded by connection to an EGC. Any panelboards, in addition to service panelboards that supply electrical equipment associated with the pool or any other body of water covered by Article 680, must be grounded by connection to an EGC.

The equipment grounding is accomplished by connection to an EGC of the supply circuit. Equipment associated with pools and similar installations is typically required to be connected to an EGC that is an insulated copper conductor. Insulation on the EGC is required as an extra level of protection against corrosive conditions associated with chemically treated water. **See Figure 15-19.**

Another benefit of installing insulated EGCs is the reduced fault paths during a ground-fault condition in the circuits supplying associated equipment. The main purpose for insulation is related to protection and integrity of the EGC of the circuit.

Insulated copper EGCs are generally required for circuits in corrosive environments installed in rigid metal conduit (RMC), intermediate metal conduit (IMC), electrical metallic tubing (EMT), and liquidtight flexible metal conduit (LFMC). When nonmetallic wiring methods are installed for pools or other equipment covered by Article 680, an EGC is required to be installed. It is typically required to be an insulated copper conductor, sized in accordance with Section 250.122 but no smaller than 12 AWG.

Pool Pump Motors

The branch circuit for a pool pump motor must be installed using the methods specified in Section 680.14, which include RMC, IMC, rigid polyvinyl chloride (PVC) conduit, or reinforced thermosetting resin conduit (Type

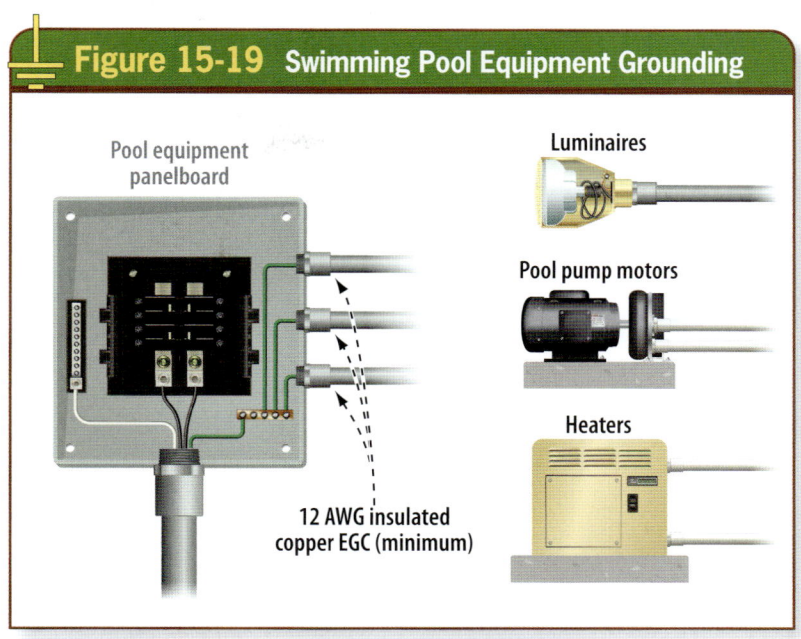

Figure 15-19. *Pool equipment is generally required to be grounded by an insulated copper EGC no smaller than 12 AWG.*

RTRC). Section 680.21(A)(1) specifies permissible wiring methods when flexibility is required, which include liquidtight flexible metal conduit, liquidtight flexible nonmetallic conduit, and metal clad (Type MC) cable listed for the location. In corrosive environments, these wiring methods must include an insulated copper EGC sized according to Section 250.122 but no smaller than 12 AWG. **See Figure 15-20.**

Figure 15-20. *Pool pump motors are required to be grounded by connection to an EGC of the supply circuit.*

Courtesy of Bill McGovern, City of Plano, TX

Figure 15-21   Flexible Conduit Grounding

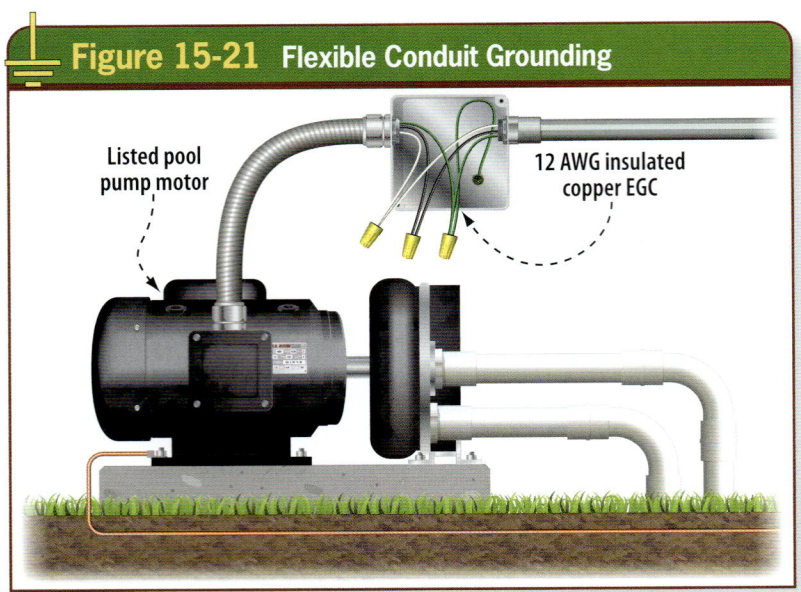

Figure 15-21. *Liquidtight flexible metal conduit or liquidtight flexible nonmetallic conduit for pool pump motors must include an insulated copper EGC no smaller than 12 AWG.*

Note that aluminum EGCs are not permitted for these circuits. Where the wiring is installed on or within a building or structure, EMT is permitted as the wiring method. **See Figure 15-21.**

Where wiring for pools or other equipment covered in Article 680 is installed within a dwelling unit or any accessory building associated with a dwelling unit, any wiring method in Chapter 3 of the *NEC* is permitted. If a cable assembly is installed in these locations, the EGC of the circuit can be bare (uninsulated) but must be covered by the sheath of the assembly. Nonmetallic-sheathed cable or service-entrance (Type SE) cable assemblies are examples of wiring that falls into this category. There are two types of service-entrance cable assemblies: Type SEU and Type SER. The *R* in *Type SER* stands for *round* and indicates the inclusion of an insulated neutral conductor, and the *U* in *Type SEU* indicates that the neutral of the cable is an uninsulated conductor.

Pool pump motors are permitted to be cord-connected if the cord is limited to three feet in length. The cord must include a copper EGC sized according to Section 250.122, and it must be connected to a grounding-type attachment plug.

Grounding and Bonding Underwater Luminaires

Underwater luminaires installed for swimming pools and similar equipment are required to be grounded by connection to the supply circuit EGC. Through-wall lighting assemblies and wet-, dry-, or no-niche luminaires are required to be connected to an insulated copper EGC installed with the circuit conductors. **See Figure 15-22.**

The EGC generally must be installed without joint or splice, except as permitted in 680.23(F)(2)(a) and (b). The EGC must be sized in accordance with Table 250.122, and it must not be smaller than 12 AWG.

Where rigid polyvinyl chloride conduit or liquidtight flexible nonmetallic conduit is installed between a forming shell for a wet-niche luminaire and a junction box or other enclosure, an 8 AWG insulated copper bonding jumper is required to be installed in the conduit to provide electrical continuity between the forming shell and the junction box or other enclosure. The conduit must be sized large enough to

Figure 15-22   Underwater Luminaires

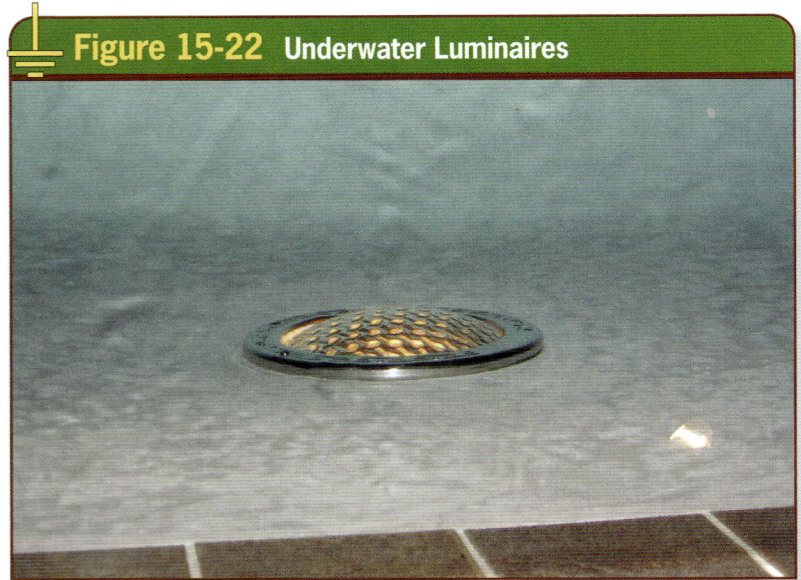

Figure 15-22. *Underwater luminaires are required to be grounded by connection to the EGC of the supply circuit.*

enclose both the 8 AWG insulated copper bonding jumper and the flexible cord that supplies the wet-niche luminaire. The flexible cord assembly provides the EGC connection to the metal luminaire parts. The 8 AWG insulated copper conductor provides the bonding connection for the forming shell. **See Figure 15-23.** The connection must be encapsulated in a listed potting compound to protect it from chemically-treated water and related corrosion.

Listed low-voltage luminaires that do not require grounding, do not exceed the low-voltage contact limit, and are supplied by listed transformers or power supplies are permitted to be installed less than five feet from the inside walls of the pool.

Junction Boxes and Other Enclosures

Junction boxes, transformer enclosures, and enclosures containing GFCIs connected to conduit systems that extend to forming shells or mounting brackets for no-niche luminaires must be provided with a quantity of grounding terminals that exceed the number of conduit entries by at least one. The purpose is to include provisions for any bonding conductor that may also be required for the enclosure.

Boxes for underwater luminaires are required to be listed and equipped with threaded hubs or entries or nonmetallic hub entries. They must be made of copper, brass, plastic, or other corrosion-resistant material. The junction boxes must provide continuity between all metallic conduit entries and grounding terminals inside the box. The EGC terminals of junction boxes for transformers and electrical enclosures for GFCIs must be connected to the equipment grounding terminal of the panelboard supplying any connected dry- or wet-niche luminaire. **See Figure 15-24.**

Separate Buildings

Pool wiring supplied from separate buildings or structures via a feeder or

branch circuit must be in accordance with Section 250.32(B). Where a feeder supplies a separate building or structure, it is also permitted to supply pool equipment if all the grounding requirements of Section 250.32(B) have been satisfied.

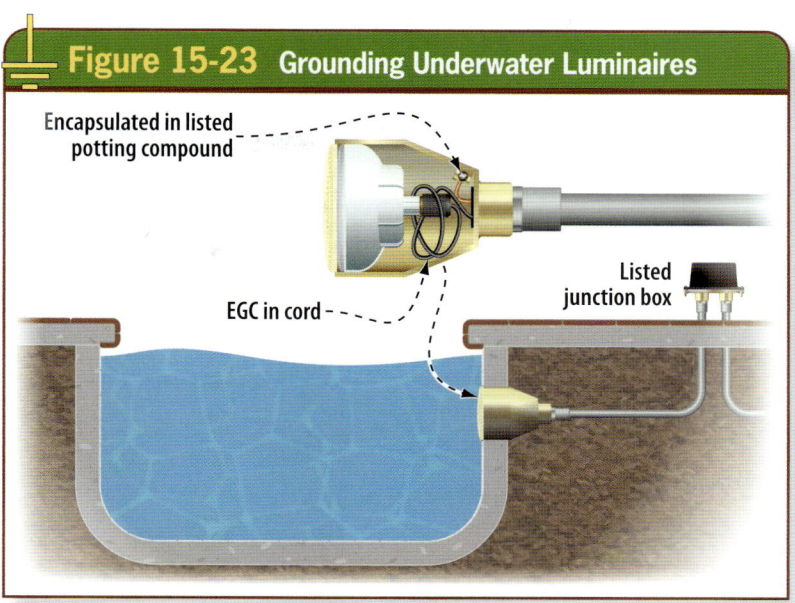

Figure 15-23. Grounding and bonding for underwater luminaires includes an EGC for the luminaire assembly and an 8 AWG copper bonding conductor that connects the forming shell to the equipotential bonding grid for the pool.

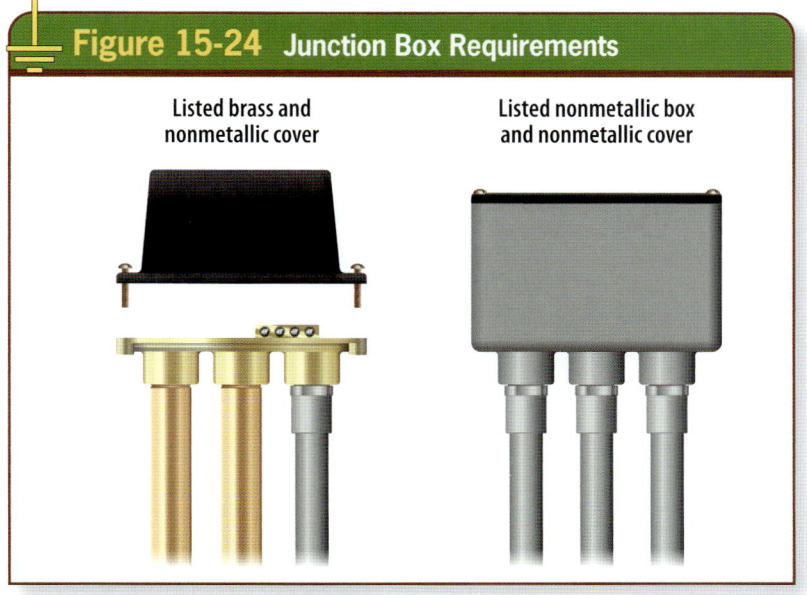

Figure 15-24. Junction boxes for pool equipment are required to be listed for this use.

Figure 15-25 Equipotential Bonding

Figure 15-25. *Equipotential bonding is used to reduce voltage gradients in pool areas.*

Equipotential Bonding Requirements

Section 680.26 provides detailed requirements related to equipotential bonding for pools and similar installations. This equipotential bonding is required to reduce voltage gradients in the pool area. Bonding metal parts together places them at the same potential electrically, reducing the ability for current to be present in any path between them. **See Figure 15-25.** This is the philosophy behind establishing the equipotential bonding requirements for swimming pools and similar installations.

Section 680.26 first looks at what conductive parts and equipment are required to be connected together to form the equipotential bonding structure. Section 680.26(B) indicates that the conductive components provided in 680.26(B)(1) through (7) must be bonded together using solid copper conductors that are 8 AWG or larger. **See Figure 15-26.** The reason for using solid copper wire is that stranded conductors are more vulnerable to corrosive influences associated with pool water chemicals. An 8 AWG solid copper conductor is more resistant to corrosion effects compared to an 8 AWG stranded copper conductor. The bonding conductor can be insulated, covered, or bare.

The grid can be connected by use of RMC made of brass or another corrosion-resistant material. All connections between the 8 AWG solid conductor and the parts required to be bonded must be made using connection methods that comply with Section 250.8. This means they must be suitable for the materials and the location. Installers should understand that no requirement exists for the equipotential bonding conductor to be run to the panelboard or service equipment, nor is the equipotential bonding conductor required to be connected to any grounding electrode. These electrically-conductive components and materials must be bonded together to establish equipotential bonding in the pool area. **See Figure 15-27.**

Pool Shells (Conductive)

Conductive pool shells are typically made of poured concrete, pneumatically-applied concrete, and concrete

Figure 15-26 Equipotential Bonding Requirements

Fence

Bonding grid system (steel, mesh, other)

Pool pump motor

Figure 15-26. *An 8 AWG solid copper conductor is required for the equipotential bonding of metal parts and equipment.*

block with painted or plastered coatings. These surfaces all have varying degrees of water permeability and porosity. Pools with vinyl liners eliminate the contact between the chemically-treated water and the pool shell. Structural reinforcing steel that is not encapsulated must be bonded together by steel tie wires or an equivalent method. This rebar cage is typically the largest conductive component of an in-ground swimming pool and is a key component of the equipotential bonding system.

If the rebar is coated with an encapsulating compound, it is obviously ineffective for use in the equipotential bonding system. Section 680.26(B)(1)(b) provides the requirement for conductive pool shells where the rebar is coated. In this case, a copper grid is required to be constructed for the conductive pool shell. The copper conductor grid must be composed of, at minimum, 8 AWG solid copper wire that is bonded together to form a mesh that conforms to the contour of the pool and pool decking surface. The 8 AWG copper conductors must be arranged in a 12-inch by 12-inch network of conductors uniformly spaced in a perpendicular grid pattern with a four-inch tolerance. The connections at intersecting portions of the 8 AWG conductors must be made using a suitable connection means provided in Section 250.8. The copper grid must be embedded in the concrete no more than six inches from the outer contour of the pool shell. In simple terms, a copper wire mesh basket is created to be used as the grid embedded in the concrete shell of the pool structure.

Listed copper wire mesh products are manufactured and available specifically for this purpose, which can assist installers in establishing *NEC*-compliant equipotential bonding systems for swimming pools and similar installations. Additional information about listed copper wire mesh products is provided in the UL Product iQ category KDER.

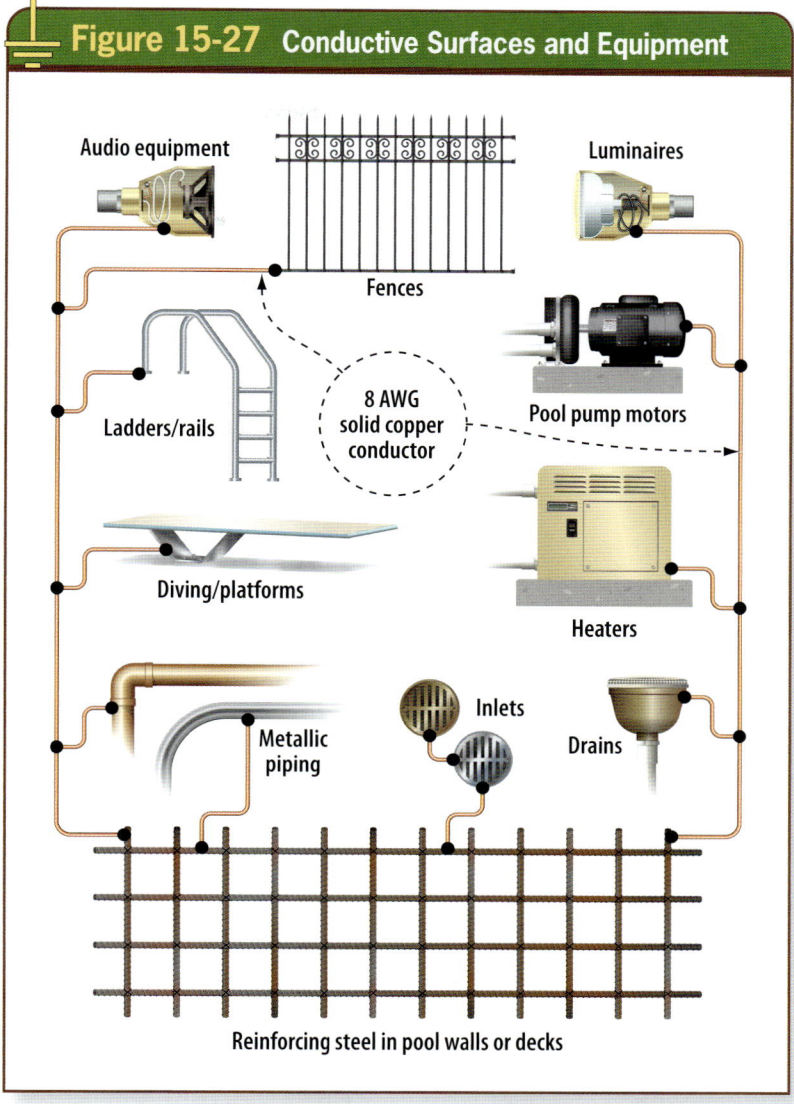

Figure 15-27   Conductive Surfaces and Equipment

Audio equipment

Fences

Luminaires

Ladders/rails

8 AWG solid copper conductor

Pool pump motors

Diving/platforms

Heaters

Metallic piping

Inlets

Drains

Reinforcing steel in pool walls or decks

Figure 15-27. Equipotential bonding requirements for pools apply to multiple conductive surfaces and equipment related to the installation.

Perimeter Surfaces

The perimeter surfaces of a pool are also required to be part of the equipotential bonding system. Perimeter surfaces around a pool can include paved, unpaved, and poured concrete surfaces that surround it. These surfaces are contacted by persons entering and leaving the body of water, and as such the potential between the perimeter surface and the rest of the conductive components of the grid should be the same or as close as

For additional information, visit qr.njatcdb.org Item #5333

possible. The bonding requirement applies to paved and unpaved perimeter surfaces extending a minimum of three feet horizontally from the inside walls of the pool. The three-foot distance is the approximate maximum step distance for a person exiting the pool water.

Bonding of perimeter surfaces is required to be accomplished by the usual reinforcing steel embedded in the concrete perimeter surface or by the installation of a copper wire or wires arranged to form a grid. Where perimeter surfaces are separated by a permanent wall or building, the perimeter surface bonding only must be be applied to the pool side of the wall or building. The conductive bonding for perimeter surfaces must be constructed in accordance with 680.26(B)(2)(a), (b), or (c), using a minimum of one 8 AWG bare solid copper conductor. The perimeter surface bonding must follow the perimeter surface contour of the pool, and any splices must be made using listed connectors

or exothermic welding. This perimeter surface grid conductor must be bonded to the conductive pool shell structure at least four times at uniformly spaced intervals around the contour of the pool. **See Figure 15-28.**

The equipotential bonding grid for perimeter surfaces must be either installed within the poured deck surface or installed under the deck surface at a depth no more than four to six inches from the underside of the deck material. To simplify the requirements for bonding a perimeter surface, it can be done by embedded reinforcing steel, which could be wire mesh (steel or copper) or constructed using a minimum 8 AWG solid copper conductor that forms a grid (mesh) and is arranged to meet the requirements in 680.26(B)(1)(b)(3). The specific criteria for constructing a bonding grid (made up of rebar or a copper ring) for pool perimeter surfaces is provided in Section 680.26(B)(2)(a) or (b).

Section 680.26(B)(2)(c) provides a third option for installing equipotential bonding grids at perimeter surfaces. A copper grid can be installed where structural reinforcing steel is not available or is encapsulated in a nonconductive compound. The copper grid must be installed and meet all of the following conditions:

- It must be constructed of 8 AWG solid bare copper arranged in accordance with 680.26(B)(1)(b)(3).
- It must follow the contour of the perimeter surface extending three feet horizontally beyond the inside walls of the pool.
- Only listed splicing devices or exothermic welding are to be used for connections.
- It must be secured within or under the deck or unpaved surfaces between four to six inches below the subgrade.

Other Conductive Components

All metal parts of the pool structure, including reinforcing rods that are not encapsulated, must be bonded into the equipotential bonding system.

Figure 15-28 Bonding Perimeter Surfaces

Figure 15-28. Equipotential bonding must include the perimeter surfaces around a pool.

Any metal forming shells for underwater luminaires and brackets of no-niche luminaires must be bonded, along with all metal fittings within or attached to the pool structure, with the exception of metal parts no larger than four inches in any dimension and not penetrating the pool structure. All metal equipment associated with the pool circulating system, including pump motors and heaters, and all metal parts and motors for installed pool covers must be bonded. An exception relaxes the bonding requirement for listed equipment with a system of double insulation.

Double-Insulated Pool Pump Motors and Water Heaters

Double-insulated pool pump motors are addressed in 680.26(B)(6)(a) and generally do not have to be bonded to the equipotential bonding system of the pool. However, an 8 AWG solid copper conductor must be connected to the grid and extended to the pool pump motor vicinity to serve as a means of connecting any replacement pump motors that are not double-insulated. If there is no electrical connection between the equipotential bonding grid and the equipment grounding system for the premises, the 8 AWG conductor installed in the vicinity of the double-insulated pool pump motor must be connected to the EGC of the branch circuit supplying the motor. **See Figure 15-29.**

If pool water heaters rated more than 50 amperes have specific instructions about grounding and bonding requirements, only those parts designated must be bonded and grounded.

Fixed Metal Parts

Any fixed metal parts must be bonded to the equipotential bonding system. This includes metal items such as fences, metal sheathing of cables, raceways, piping, awnings, gutters, and door and window frames. Bonding is not required for such metal parts when they are separated from the pool by a permanent barrier or a distance of no less than five feet. If metal parts are located more than 12 feet vertically from the maximum water level or from the top of any observation stands, towers, platforms, or diving structures, they are not required to be bonded to the equipotential bonding grid system.

Pool Water

Section 680.26(C) requires the pool water itself to be bonded. The chemically-treated pool water must be bonded to the equipotential bonding system by contact between the water and the other bonded metal parts.

A minimum of nine square inches of contact is considered sufficient for establishing a bonding connection between the water and the bonding system of conductive parts. This is not an electrical connection but a connection between the water and the electrically-conductive metal parts that are bonded to the equipotential grid of the pool structure. A common example of a pool water bonding connection is a metal handrail or metal ladder in

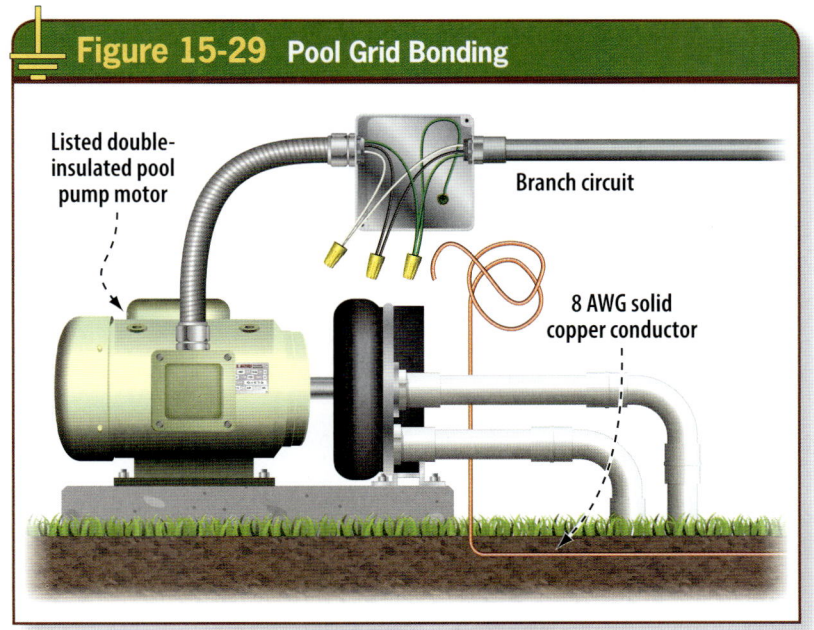

Figure 15-29 Pool Grid Bonding

Listed double-insulated pool pump motor

Branch circuit

8 AWG solid copper conductor

Figure 15-29. The 8 AWG solid copper bonding conductor from the pool bonding grid system must be routed to the double-insulated pool pump motor location and connected to any replacement motor that is not a double-insulated type.

contact with the water and bonded to the grid. **See Figure 15-30.**

If there is no metal in contact with the water, bonding the pool water to the equipotential bonding system for the pool is still a requirement. In these cases, the pool water must be in direct contact with an approved corrosion-resistant conductive surface providing not less than nine square inches of contact with the pool water at all times. This conductive part must be connected to the equipotential bonding system in accordance with 680.26(B). Some manufacturers of pre-formed nonmetallic or fiberglass pool shells make provisions for establishing a bonding connection for the water. This can be accomplished by embedding a small brass or other corrosion-resistant plate in the wall of a nonmetallic pool structure. This should be done by the pool manufacturer.

Specialized Pool Equipment

Section 680.27 includes specific requirements for special underwater equipment such as speaker systems. Such special

Figure 15-30 Bonding Pool Water

Figure 15-30. Pool water is required to be bonded, which can be accomplished by contact with metal parts of handrails, ladders, and so forth.

equipment installed in pools must be identified for the pool installation.

Underwater speakers must be mounted in a metal forming shell connected to a listed junction box as specified in Section 680.24. The forming shell and screen must be of brass or another corrosion-resistant material. The wiring methods that may be used are RMC made of brass or another identified corrosion-resistant metal, LFNC-B, rigid PVC conduit, or Type RTRC. Where any of the preceding nonmetallic wiring methods are used, an 8 AWG insulated solid or stranded bonding conductor must be installed in the conduit to bond the forming shell to the equipotential bonding grid system. The bonding conductor must terminate to the shell and must be coated or encapsulated with a listed potting compound to protect the connection from corrosion. **See Figure 15-31.**

Spas, Hot Tubs, and Permanently Installed Immersion Pools

Spa or hot tub installations must meet the requirements in Parts I and IV of Article 680. This incorporates all general equipment grounding requirements specified in Section 680.7. Spa and hot tub equipment must be grounded. All electrical equipment within five feet of the inside walls of the spa or hot tub must be grounded, including the equipment associated with the circulating system. The grounding is accomplished by connection to the EGC of the supply circuit wiring.

Sections 680.42(B) and 680.43(D) through (F) provide specific bonding and grounding requirements for outdoor and indoor installations respectively. **See Figure 15-32.** The metal frames of spas or hot tubs, specifically the package units, are permitted to accomplish bonding by metal-to-metal contact on a common conductive frame. Any metal strapping that secures wood framing or paneling to the structure does not have to be bonded to the equipotential bonding system. If the spa or hot tub is located outdoors and the wiring is run through a dwelling unit or other associated dwelling unit structure, any of the wiring methods in Chapter 3 of the *NEC* can be installed. The wiring must include a copper EGC no smaller than 12 AWG that is enclosed within the sheath of the cable assembly. This wiring method is permitted only for packaged hot tub or spa assemblies.

Wiring to any underwater luminaire must be installed in accordance with Section 680.23 or 680.33. For a spa or hot tub located indoors, the requirements in Parts I and II of Article 680 apply, except as modified by Section 680.43. These installations must use the wiring methods in Chapter 3 of the *NEC*, with the exception of listed cord- and plug-connected units rated 20 amperes or less.

Spa and Hot Tub Bonding–Indoor

The bonding requirements for indoor spa and hot tub installations are provided in Section 680.43(D). All metal fittings within or attached to the spa or hot tub structure must be bonded. Metal parts of electrical equipment for the spa or hot tub—including the pump motor or motors, the metal raceway and metal piping within five feet of the inside walls of the spa or hot tub if not separated by a permanent barrier, and all metal surfaces within five feet of the inside walls of the spa or hot tub—must be bonded. Any electrical controls or devices that are within five feet of the inside walls of the spa or hot tub and are not associated with the spa or hot tub installation must also be bonded. By exception, small conductive surfaces unlikely to become energized and metal parts of listed package assemblies are not required to be bonded.

Bonding has to be accomplished by any of the following methods:

1. Interconnected fittings or threaded metal piping
2. Metal-to-metal contact on common framing
3. Solid copper conductor sized 8 AWG or larger

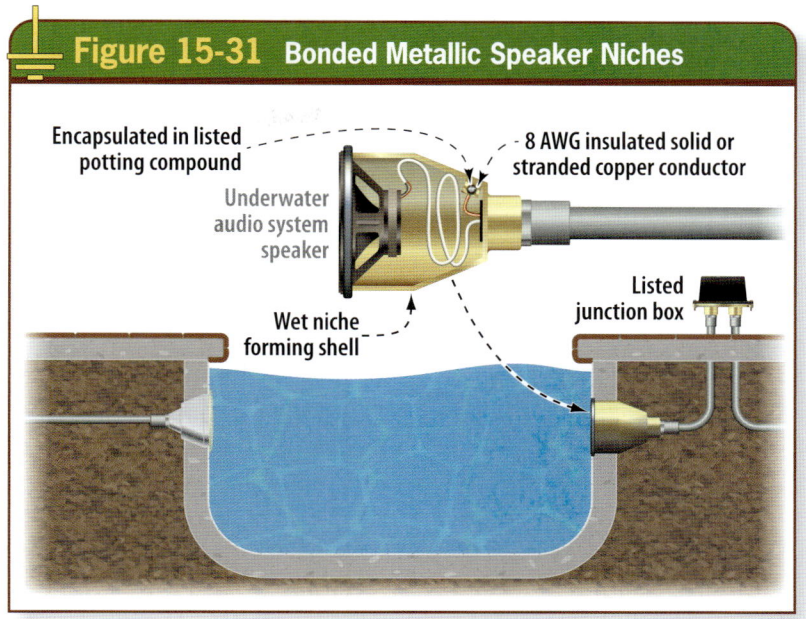

Figure 15-31   Bonded Metallic Speaker Niches

Encapsulated in listed potting compound

8 AWG insulated solid or stranded copper conductor

Underwater audio system speaker

Listed junction box

Wet niche forming shell

Figure 15-31. *Bonding is required for forming shells associated with special equipment such as underwater speakers.*

Spa and Hot Tub Bonding–Outdoor

Section 680.42(B) provides general requirements for bonding outdoor spas and hot tubs. Bonding for spas and hot tubs is permitted when accomplished through metal-to-metal contact to a common conductive frame. Any metal bands or hoops for securing wooden

Figure 15-32   Bonding Spas and Hot Tubs

Figure 15-32. *Special bonding rules apply to both indoor and outdoor spas and hot tubs.*

framing or paneling do not require bonding in accordance with 680.26.

The equipotential bonding requirements in 680.26(B)(2) are not required for spas and hot tubs where the equipment meets all of the following conditions:

1. The spa or hot tub is a listed, self-contained unit for aboveground use.
2. The spa or hot tub is not identified as suitable only for indoor use.
3. The installation is above grade level and in accordance with the manufacturer's instructions.
4. The top rim of the equipment is at least 28 inches above all perimeter surfaces that are within 30 inches horizontally from the spa or hot tub.

Permanently Installed Immersion Pools

The equipotential bonding requirements of 680.26(B) are not required for immersion pools with permanently connected equipment installed on or above finished floors or on nonconductive permitted surfaces.

Audio equipment must not be installed in or on immersion pools, and if audio equipment is installed within six feet of the inside walls of an immersion pool, it must be connected to an EGC and be protected by a GFCI. See 680.45 for more details.

Figure 15-33 Fountains

Figure 15-33. Rules for fountains are provided in Parts I and V of Article 680.

Courtesy of Jim Dollard, IBEW Local 98

Fountains

The specific installation requirements for fountains are located in Part V of Article 680. Section 680.50 indicates that Parts I and V apply to all permanently installed fountains. **See Figure 15-33.** Fountains that have water common to a pool must also comply with the requirements in Part II of Article 680.

Section 680.54 provides specific grounding and bonding rules for fountains covered by Article 680. Essentially, the requirements are similar to those for spas and hot tub installations. A new requirement was added to the 2023 *NEC* in 680.54(C) to require equipotential bonding for splash pads.

Equipment grounding is required for all equipment associated with the fountain, and metal parts such as piping systems associated with the fountain must be bonded to the EGC of the fountain supply circuit or circuits. See these sections for specific requirements.

Portable fountains must meet the requirements in Article 422.

Therapeutic Pools and Tubs for Health Care Use

Part VI of Article 680 includes requirements for pools and tubs intended for therapeutic use in health care facilities, gymnasiums, training centers, and similar areas. Grounding and bonding requirements for these installations are located in Sections 680.62(B) through (D). Any portable therapeutic appliances must meet the requirements in Article 422. **See Figure 15-34.**

Bonding

Special bonding requirements for therapeutic tubs (hydrotherapeutic tanks) are provided in Section 680.62(B) of the *NEC*.

In its simplest form, bonding is the process of establishing continuity and conductivity between conductive parts. The objective is to have them become electrically common to one another by either bonding jumpers or mechanical connections. Hydrotherapeutic tubs and tank assemblies often include a variety of conductive objects as part of the overall assembly or installation. The manufacturer of this type of equipment often provides the necessary bonding connections for all conductive parts of the assembly to meet the requirements of applicable product standards. Where any of the conductive parts associated with the equipment are remote, bonding must be provided in accordance with the requirements in Section 680.62(B).

The following metal parts of therapeutic tubs (hydrotherapeutic tanks) must be bonded together:

- Any metal fittings within or attached to the tub structure
- Any metal parts of electrical equipment associated with the tub water circulating system, including pump motors
- Any metal-sheathed cables and raceways and metal piping that are within 1.5 meters (five feet) of the inside walls of the tub and not separated from the tub by a permanent barrier
- Any metal surfaces that are within 1.5 meters (five feet) of the inside walls of the tub and not separated from the tub area by a permanent barrier
- Any electrical devices and controls that are not associated with the therapeutic tub and located within 1.5 meters (five feet) from the tub unit

Section 680.62(C) includes the various methods permitted for accomplishing the bonding requirements for therapeutic tub installations and associated equipment. Remember the main objectives of bonding as provided in the general definition: bonding establishes continuity and conductivity between conductive parts. All metal or conductive parts identified in this section as required to be bonded must be bonded by one of the following methods:

- Interconnection of threaded metal piping and fittings
- Metal-to-metal mounting on a common frame or base

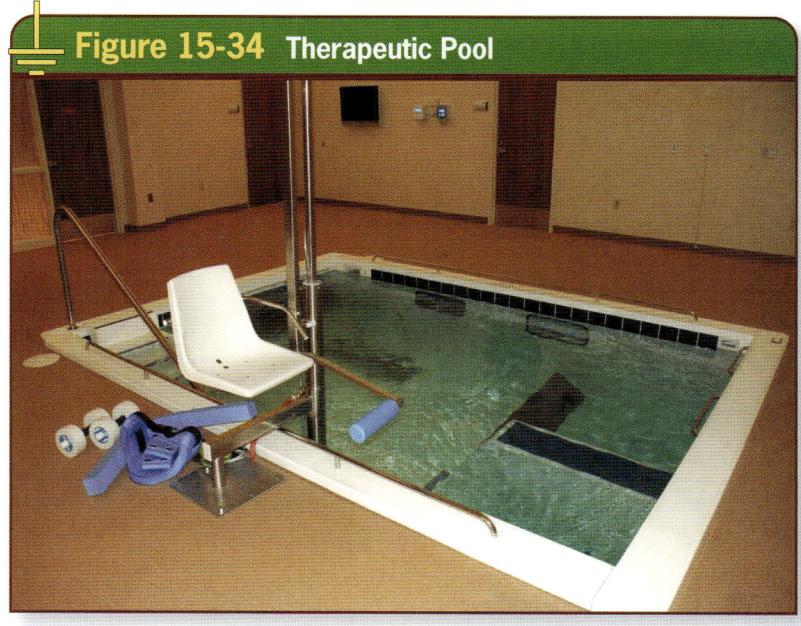

Figure 15-34 Therapeutic Pool

Figure 15-34. Special bonding requirements apply to therapeutic pools and tubs for health care use.

- Connections by suitable metal clamps
- Provision of a solid copper bonding jumper—insulated, covered, or bare—no smaller than 8 AWG

Equipment Grounding

The specific equipment grounding requirements for therapeutic tubs (hydrotherapeutic tanks) are provided in Section 680.62(D) of the *NEC*. These grounding requirements apply to this type of equipment whether it is stationary or fixed. Any equipment located within five feet of the inside walls of the tub or tank must be grounded. This is typically accomplished by connection to the supply circuit EGC. Because Section 680.60 indicates that Parts I and VI apply to this type of equipment, the EGC size rule is provided in Section 680.8(B). The EGC within the supply cord for this type of equipment must not be smaller than 12 AWG copper and must meet the minimum sizing requirements in Section 250.122.

Therapeutic tubs or tanks typically include a water circulation system and can include air pumps and blowers.

Figure 15-35 Hydromassage Tub Requirements

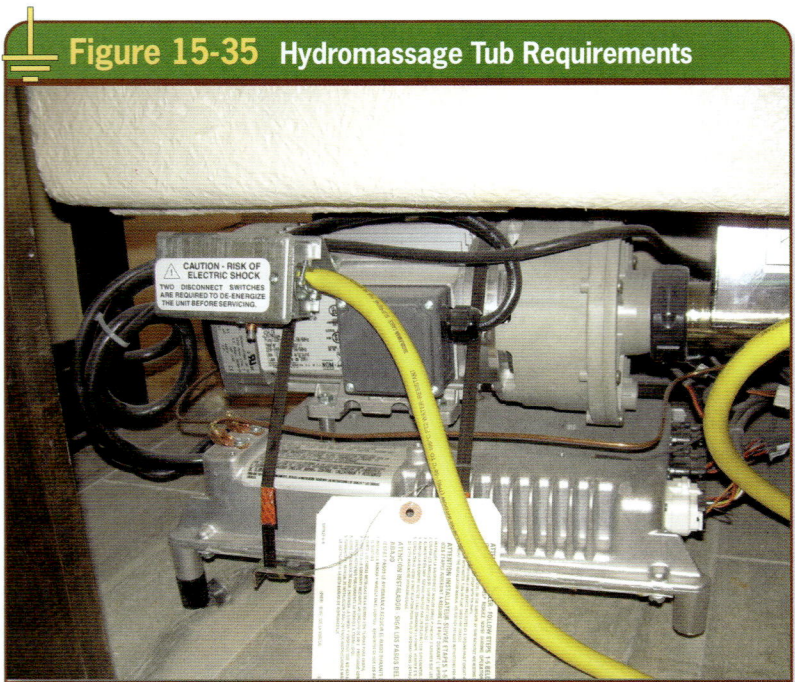

Figure 15-35. *Special bonding and equipment grounding requirements apply to hydromassage bathtubs.*

Any of this type of equipment associated with the therapeutic tub must also be grounded by connection to an EGC. This EGC is typically included within the supply cord of listed package units or assemblies. Once the equipment is

Figure 15-36 Photovoltaic Systems

Figure 15-36. *PV systems have specific grounding and bonding requirements in Article 690.*

I-Stock Photo Courtesy of NECA

plugged into a grounding-type receptacle, the equipment grounding requirements of the *NEC* are satisfied, according to Sections 680.62(D)(1) and (2).

Portable Therapeutic Appliances and Equipment

Any portable therapeutic appliances used in health care facilities are required to meet the grounding requirements in Section 250.114. This section provides general equipment grounding requirements for non–current-carrying metal parts of any cord- and plug-connected equipment that is likely to become energized. Where any of this type of equipment is protected by a system of double insulation (double-insulated), the *NEC* relaxes the grounding requirements, according to the exception to Section 250.114.

Hydromassage Bathtubs

The specific rules for hydromassage bathtubs are provided in Part VII of Article 680. This equipment requires at least one individual branch circuit and may require more depending on the load served. Some of these units are equipped with heaters requiring more than one branch circuit. GFCI protection is required for this equipment, and the GFCI has to be located so as to be readily accessible; any other equipment for the hydromassage bathtub must also be accessible. The GFCI protective device cannot be located under the unit. Equipment grounding is accomplished by connecting the EGC of the branch circuit to the unit. Bonding requirements for hydromassage bathtubs are found in Section 680.74. **See Figure 15-35.**

All metal piping systems and all grounded metal parts in contact with the circulating water must be bonded with, at minimum, an 8 AWG solid copper conductor. This conductor can be insulated, covered, or bare. The bonding jumper must terminate on the circulating pump motor. A terminal lug is typically installed there for this purpose. A bonding jumper connection to a double-insulated motor is not required. The purpose of this bonding requirement is to establish equipotential bonding in the hydromassage bathtub area.

As indicated in Section 680.26(A) for pools, the 8 AWG or larger copper bonding conductor does not have to be run to a panelboard or to the service equipment of the building or structure, nor is it required to be connected to a grounding electrode. This conductor is for equipotential bonding purposes only, not grounding, although interconnection with the EGC is established by connecting the EGC of the branch circuit wiring to the equipment.

GROUNDING REQUIREMENTS FOR SOLAR PV SYSTEMS

Solar photovoltaic (PV) systems and installations are becoming more popular as efforts increase to reduce foreign oil dependency and expand the installation and use of alternative renewable energy sources. Solar PV systems and equipment have specific grounding and bonding requirements provided in Part V of Article 690. **See Figure 15-36.**

System Grounding

The grounding of PV systems is generally accomplished by connection to a grounding electrode or grounding electrode system. General requirements for system grounding are provided in Section 250.20 of the *NEC*. The grounding and bonding rules in Article 250 apply to PV systems and installations except as modified or amended by Article 690. Section 690.41(A) indicates that one of six grounding configurations must be employed for PV systems. This includes 2-wire systems and bipolar systems with one functionally grounded conductor. **See Figure 15-37.**

Photovoltaic system grounding configurations are as follows:

- 2-wire PV arrays with one functionally grounded conductor
- Bipolar PV arrays according to 690.7(C) with a functional ground reference (center tap)

- PV arrays not isolated from the grounded inverter output circuit
- Ungrounded PV arrays
- Solidly grounded PV arrays
- PV systems that use other methods that accomplish equivalent system protection in accordance with 250.4(A) with equipment listed and identified for the use

Definition

A functionally grounded system is often connected to ground through an

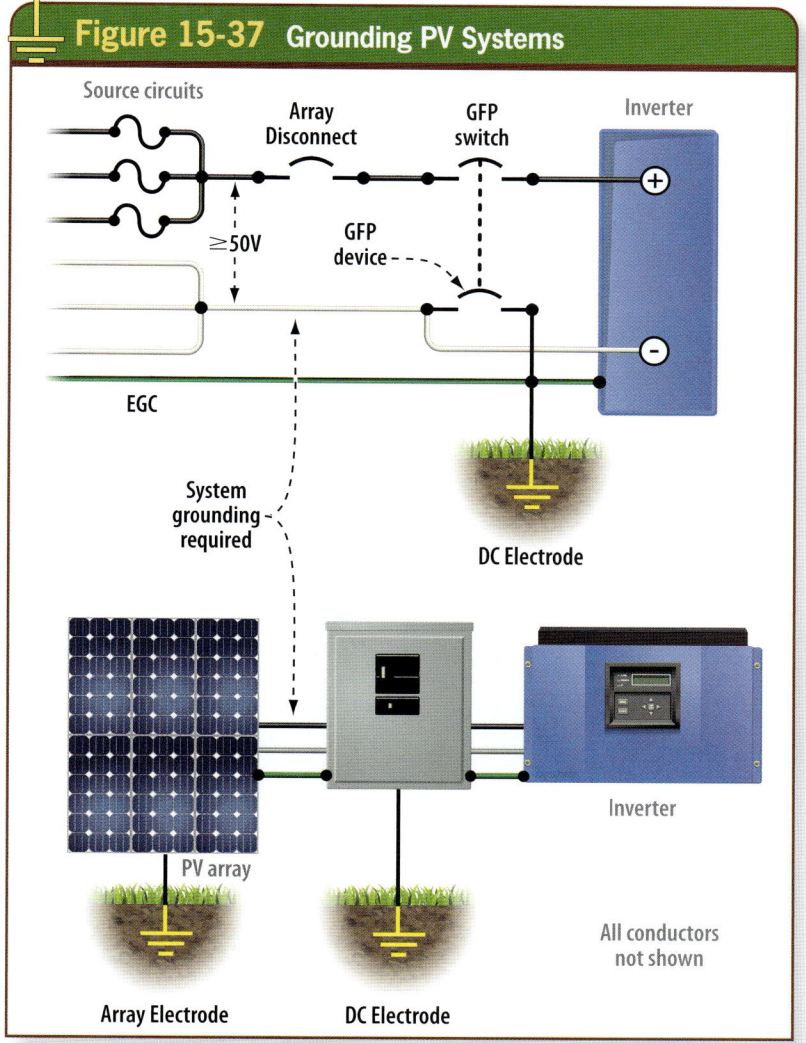

Figure 15-37 Grounding PV Systems

Figure 15-37. *Grounding a PV system can include a direct connection to ground (Earth). (The figure is only a graphic representation of solid grounding and may not include all conductors or details of all grounding configurations mentioned in 690.41(A).)*

electronic means internal to an inverter or charge controller that is equipped with ground-fault protection.

> **Grounded, Functionally. (Functionally Grounded)** A system that has an electrical ground reference for operational purposes that is not solidly grounded. (CMP-4)

A building or structure supporting a PV system shall utilize a grounding electrode system installed in accordance with Part III of Article 250. A system grounding connection for the direct current (DC) circuit can be made at a single point on the output. Systems equipped with ground-fault detector-interrupter (GFDI) protection in accordance with 690.41(B) must have any current-carrying conductor-to-ground connection made at GFDI.

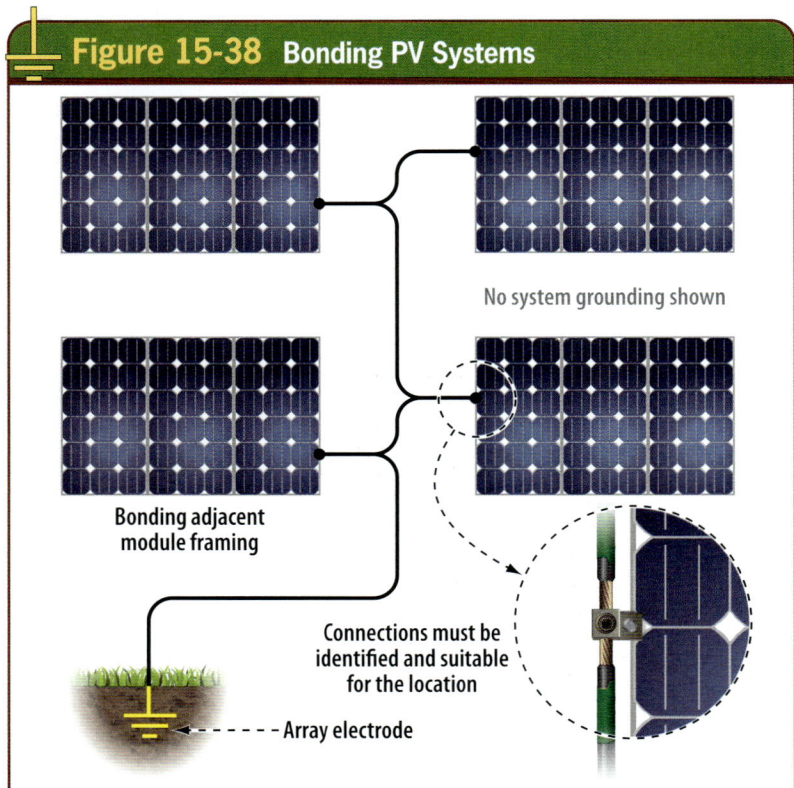

Figure 15-38 Bonding PV Systems

No system grounding shown

Bonding adjacent module framing

Connections must be identified and suitable for the location

Array electrode

Figure 15-38. Exposed metal module framing is required to be bonded to adjacent exposed metal module frames.

Many existing PV systems are functionally grounded systems rather than solidly grounded systems. For functionally grounded PV systems with an interactive inverter output, the AC equipment grounding conductor is connected to associated grounded AC distribution equipment. This connection is often the grounding connection for the GFDI device and equipment grounding of the PV system and array.

Ground-Fault Detector-Interrupter

PV system DC circuits greater than 30 volts or eight amperes must be provided with DC ground-fault detector-interrupter protection in accordance with Sections 690.41(B)(1) through (B)(3). This requirement does not apply where a PV system is solidly grounded, is not installed on or in a building, and includes an array with not more than two modules in parallel.

The GFDI must detect ground fault(s) in the PV array DC conductors and components, including faults in any functionally grounded conductor, and must be listed for GFDI. The faulted circuit(s) must be isolated by one of the means provided in 690.41(B)(2). Indication of ground-fault detection is required by 690.41(B)(3) and must be visible in a readily accessible location.

Not all inverters, charge controllers, or DC-to-DC converters include GFDI. Equipment that does not have GFDI is provided with the following statement in the manual: "Warning: This unit is not provided with a GFDI device."

Equipment Grounding

Regardless of the system voltage, all equipment, metal frames, module frames, and conductor enclosures must be grounded in accordance with Sections 250.134 and 250.136. The exposed metal frames of PV modules must be grounded and bonded with devices listed and identified for such use. Identified devices are also permitted to bond the exposed metal module framing to adjacent exposed metal module frames. **See Figure 15-38.**

The equipment grounding conductor and bonding jumper connections must be made tight and effective, and they must be suitable for the location in which they are installed. **See Figure 15-39.**

EGCs for the PV array must be installed using the same wiring method with the associated circuit conductors, or they must be routed with the circuit conductors as they leave the array vicinity. The size of the EGC for PV source and output circuits must be in accordance with Section 690.45. The general rule is to size the EGC in accordance with 250.122.

If no overcurrent protective device is present to accomplish the sizing requirement, the assumed overcurrent protective device rating shall be in accordance with 690.9(B). Adjustments for voltage drop considerations are not required. EGCs smaller than 10 AWG must be protected according to the provisions in Section 250.120(C). Section 310.3(C) indicates that conductors in sizes 8 AWG and larger are required to be stranded unless specifically permitted to be larger elsewhere in the *NEC*.

Grounding Electrode Systems

Sections 690.47(A) and (B) provide the requirements for grounding electrodes installed for PV systems and equipment. DC systems that are grounded must be connected to a grounding electrode system in accordance with Section 250.166, and for ungrounded DC systems, the grounding electrode requirements of Section 250.169 apply. The grounding electrode conductor for both AC and DC systems must meet the general installation rules in Section 250.64. For AC systems, the grounding electrode system is usually established for the structure or building. The same electrode system must be used for the PV equipment and system grounding. This is a fairly common method of grounding, especially at buildings or structures that are already supplied by a utility service where a grounding electrode system exists. **See Figure 15-40.**

A building or structure supporting a PV array must have a grounding

Figure 15-39. *Bonding jumpers between equipment are required to be effective and suitable for the location in which they are installed.*

Courtesy of IBEW Local 26 Training Center

electrode system in accordance with Part III of Article 250. PV array equipment grounding conductors are required to be connected to the grounding electrode system of the building or structure supporting the PV array in accordance with Part VII of Article 250.6. This connection shall be in addition to any other equipment grounding conductor requirements in 690.43(C). For PV systems that are not solidly

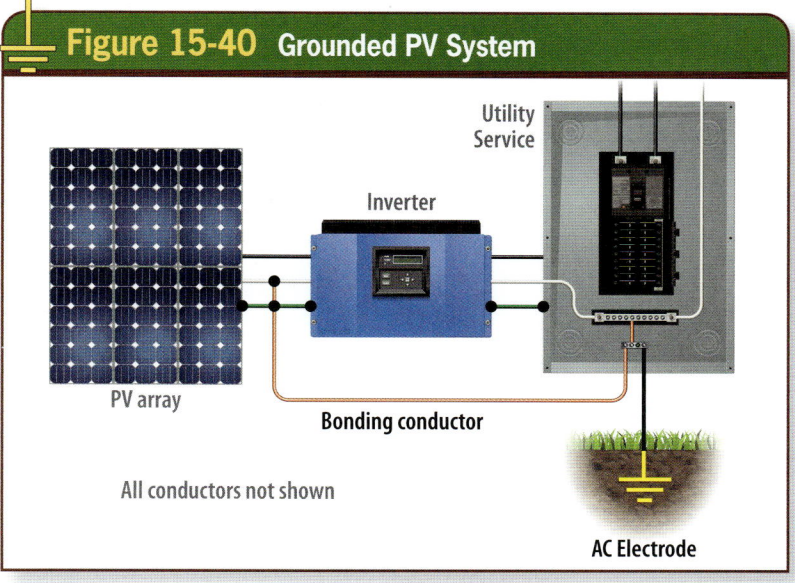

Figure 15-40. *Grounding for PV systems can be accomplished using the same electrode that is used for the building electrical service supplied by the utility.*

grounded, the equipment grounding conductor for the output of the PV system (connected to associated distribution equipment) shall be permitted to be the connection to ground for ground-fault protection and equipment grounding of the PV array. For solidly grounded PV systems, as covered in 690.41(A)(5), the grounded conductor shall be connected to a grounding electrode system by means of a grounding electrode conductor that is sized in accordance with 250.166.

It is common to install additional (auxiliary) grounding electrodes for PV arrays. These arrays are often installed where lightning events are prevalent, such as freestanding ground-mounted installations and installations on building rooftops. Auxiliary grounding electrodes can be installed for this purpose but must meet the requirements in 250.52 and 250.54. The auxiliary electrode(s) are required to be installed at the location of ground- and roof-mounted PV arrays. The electrodes can be connected directly to the array frame(s) or structure. The grounding electrode conductor shall be sized according to 250.66. The structure of a ground-mounted PV array is considered a grounding electrode if it meets the requirements of 250.52(A). Roof-mounted PV arrays shall be permitted to use the metal frame of a building or structure if the requirements of 250.68(C)(2) are met.

Electrical Safety

Note that PV systems will be energized when exposed to the sun, and special electrical safety work practices must be used when working with such systems and equipment. Always conform to OSHA safety regulations and the requirements provided in *NFPA 70E: Standard for Electrical Safety in the Workplace.*

SUMMARY

Chapter 6 of the *NEC* includes some special grounding and bonding rules that are more restrictive than the general rules in Chapters 1 through 4 of the *NEC*. The special equipment described, in many cases, has unique operating characteristics and can present additional electrical hazards to people and property if specific or additional grounding and bonding methods are not applied. Some special equipment bonding and grounding rules in Chapter 6 address shock protection in addition to equipment protection and facilitating overcurrent protective device operation.

REVIEW QUESTIONS

1. Signs and metal equipment of outline lighting systems shall be grounded by connection to the __?__ of the supply branch circuit(s) or the feeder using the types of equipment grounding conductors specified in Section 250.118.

   a. equipment bonding jumper
   b. equipment grounding conductor
   c. grounded conductor
   d. grounding electrode conductor

2. A wire-type equipment grounding conductor for a 30-ampere branch circuit supplying an outline lighting system must be sized no less than __?__.

   a. 12 AWG copper
   b. 10 AWG aluminum
   c. 10 AWG copper
   d. 8 AWG copper

3. All equipment grounding conductor connections for signs and outline lighting systems must be made using a method specified in Section 250.8, which includes all but which of the following?

   a. Listed lugs
   b. Listed pressure connectors
   c. Pressure connectors listed as grounding and bonding equipment
   d. Sheet metal screws with two full threads engaged

4. Metal parts of a building or structure are not permitted as an equipment grounding conductor or as a high-voltage secondary return circuit for a neon lighting system.

   a. True   b. False

5. Copper bonding conductors for metal parts associated with high-voltage neon secondary circuits of outline lighting systems are not permitted to be smaller than __?__.

   a. 14 AWG
   b. 12 AWG
   c. 8 AWG
   d. 6 AWG

6. When listed flexible metal conduit or listed liquidtight flexible metal conduit is used as the raceway to enclose the high-voltage GTO cables, the metal parts of a sign or outline lighting system are permitted to be bonded through the flexible metal raceway wiring method, provided that the total cumulative length of the conduit does not exceed __?__.

   a. 6'
   b. 25'
   c. 50'
   d. 100'

7. Where an external 14 AWG copper bonding conductor for metal parts of outline lighting systems is installed, a distance of at least 1 $1/2$ inches must be maintained from a non-metallic conduit containing a high-voltage circuit operating at __?__ or less, and no less than 1 $3/4$ inches must be maintained from a nonmetallic conduit containing a circuit operating at greater than __?__.

   a. 50 Hz / 50 Hz
   b. 60 Hz / 60 Hz
   c. 100 Hz / 100 Hz
   d. 400 Hz / 400 Hz

REVIEW QUESTIONS

8. The bonding for electric cranes and hoists can be accomplished either by mechanical connections of the conductive parts or by connection of suitable bonding jumpers.

   a. True   b. False

9. The rating of the overcurrent protective device supplying cranes and hoists determines the minimum size required for wire-type equipment grounding conductors. The minimum size of the aluminum equipment grounding conductor for an electric crane supplied by a 125-ampere branch circuit is __?__ .

   a. 10 AWG
   b. 8 AWG
   c. 6 AWG
   d. 4 AWG

10. The frames and metal equipment of electric elevators, including __?__ , are required to be grounded.

   a. all metal equipment on the elevator car
   b. machines
   c. motors and controllers
   d. all of the above

11. Any signal reference grids installed in information technology rooms have to be connected to the __?__ provided with the circuits supplying the information technology equipment.

   a. equipment grounding conductor
   b. grounded conductor
   c. grounding electrode conductor
   d. ungrounded conductor

12. The voltage drop on sensitive electronic equipment covered in Article 647 cannot exceed __?__ on the branch circuits and __?__ for the combined voltage drop of the feeder and branch circuit supplied by the system.

   a. 1.5% / 2.5%
   b. 2% / 4%
   c. 3% / 5%
   d. 5% / 10%

13. Equipment and receptacles supplied by sensitive electronic equipment covered in Article 647 must be grounded by an equipment grounding conductor run with the supply circuit conductors and terminating on the equipment grounding terminal bus of the panel that is marked __?__ .

   a. equipment grounding point
   b. isolated grounding terminal bus
   c. reference grounding bus
   d. technical equipment ground

14. Any electrical equipment located within __?__ of the inside walls of a pool or specified body of water must be grounded.

   a. 5'
   b. 6'
   c. 10'
   d. 15'

15. Pool pump motors must be connected using wiring methods that include an insulated copper equipment grounding conductor sized according to Section 250.122 but no smaller than __?__ .

   a. 14 AWG
   b. 12 AWG
   c. 10 AWG
   d. 8 AWG

16. What is the minimum size of the copper conductor that is installed for an equipotential bonding grid for a swimming pool?

   a. 12 AWG solid
   b. 10 AWG solid
   c. 8 AWG solid
   d. 8 AWG stranded

17. Aluminum equipment grounding conductors are permitted for pool pump motor circuits.

   a. True   b. False

18. Where rigid polyvinyl chloride conduit or liquidtight flexible nonmetallic conduit is installed between a forming shell for a wet-niche luminaire and a junction box or other enclosure, a(n) __?__ insulated copper bonding jumper is required to be installed in the conduit to provide electrical continuity between the forming shell and the junction box or other enclosure.

    a. 14 AWG
    b. 12 AWG
    c. 10 AWG
    d. 8 AWG

19. The pool water is required to be bonded to the equipotential bonding grid.

    a. True   b. False

20. Junction boxes for pool equipment are required to be listed and equipped with threaded hubs or entries or nonmetallic hub entries and must be made of any of the following materials except __?__ .

    a. aluminum
    b. copper or brass
    c. other corrosion-resistant material
    d. plastic

21. Equipotential bonding is required for pools and similar installations in an effort to eliminate voltage gradients in the pool area.

    a. True   b. False

22. Double-insulated pool pump motors are addressed in Section 680.26(B)(6)(b) and generally do not have to be bonded to the equipotential bonding system of the pool. However, a(n) __?__ conductor has to be connected to the grid and extended to the pool pump motor vicinity to serve as a means of connecting any replacement pump motors that are not double-insulated and which would require bonding to the grid.

    a. 10 AWG solid copper
    b. 10 AWG stranded copper
    c. 8 AWG solid copper
    d. 6 AWG stranded copper

23. All electrical equipment within __?__ of the inside walls of the spa or hot tub must be grounded, including the equipment associated with the circulating system.

    a. 3'
    b. 5'
    c. 6'
    d. 10'

24. A solar photovoltaic system that is functionally grounded is one that __?__ .

    a. has an electrical ground reference for operational purposes and is not solidly grounded
    b. is grounded through an impedance device
    c. is resistance grounded
    d. is solidly grounded

25. The exposed metal frames of PV modules must be grounded and bonded with devices __?__ for such use.

    a. approved
    b. classified
    c. engineered
    d. listed and identified

Grounding and Bonding for Communications Systems and Equipment

Communications systems and equipment installed in buildings must comply with the specific rules given in Chapter 8 of the *NEC*. In many cases, it seems that there is complacency about grounding and bonding requirements for communications equipment and systems. Even though these systems typically operate at lower energy levels, improper grounding and bonding can result in severe consequences for equipment and property and present shock hazards. Article 770 and the Chapter 8 articles of the *NEC* provide unique and specific grounding and bonding requirements for communications system installations.

Objectives

» Determine the purpose of grounding and bonding requirements for communications systems and equipment.

» Understand how the grounding and bonding requirements in Article 770 and Chapter 8 of the *NEC* apply to communications systems.

» Determine the requirements for grounding electrodes to be used for intersystem equipment grounding.

» Determine minimum sizes required for grounding electrode conductors and bonding conductors installed for communications systems and equipment.

» Understand the means of connection requirements for grounding electrode conductors installed for communications systems and equipment.

Chapter 16

Table of Contents

Performance and Concepts........................ 378

Definitions 378

Grounding and Bonding Performance.......... 379

Connecting to a Grounding Electrode.......... 380

Grounding Electrode Conductor Installation... 381

Intersystem Grounding and Bonding............ 381

Common Grounding and Bonding Rules
for Communications Systems..................... 384

Grounding and Bonding at Mobile Homes 389

Grounding Requirements 389

Bonding Requirements 390

Radio and Television Equipment
and Antennas... 390

Overvoltages and Lightning Events 392

Summary... 393

Review Questions 393

PERFORMANCE AND CONCEPTS

Article 800 includes general requirements that apply to and are common between Articles 805, 820, 830, and 840. Similar grounding and bonding rules are applicable to each Article and address requirements such as sizing of grounding electrode conductors, installation of bonding jumpers, installation of grounding electrode conductors, and so forth. Each of these articles provides reference to the grounding and bonding requirements set forth in either Section 770.100 or 800.100, as applicable. **See Figure 16-1.**

Grounding, in the simplest form, is the process of connecting an electrically conductive object to ground (the Earth). Bonding is the process of connecting conductive objects together to establish continuity and conductivity. If a system or equipment is grounded, it is connected to the Earth, and if objects are bonded, they are connected together to electrically become one potential—or as close to the same potential as possible. These two processes work in unison to provide safety for communications systems, equipment, and property. Grounding and bonding for communications equipment and systems provide operational grounding and protective grounding functions. **See Figure 16-2.**

DEFINITIONS

The definitions in Article 100 of the *NEC* provide a foundation on which grounding and bonding requirements are built. The meanings of defined terms used in Articles 770, 800, 805, 810, 820, 830, and 840 should be familiar.

Figure 16-1   Communications Systems

Figure 16-1. There are specific and common grounding and bonding requirements that apply to communications systems and equipment covered in Article 770 and Chapter 8 of the NEC.

Ground. The earth. (CMP-5)

Bonded (Bonding). Connected to establish electrical continuity and conductivity. (CMP-5)

Grounded (Grounding). Connected (connecting) to ground or to a conductive body that extends the ground connection. (CMP-5)

Bonding Conductor (Bonding Jumper). A conductor that ensures the required electrical conductivity between metal parts that are required to be electrically connected. (CMP-5)

Grounding Electrode. A conducting object through which a direct connection to earth is established. (CMP-5)

Grounding Electrode Conductor (GEC). A conductor used to connect the system grounded conductor or the equipment to a grounding electrode or to a point on the grounding electrode system. (CMP-5)

Intersystem Bonding Termination (IBT). A device that provides a means for connecting intersystem bonding conductors for communications systems to the grounding electrode system. (CMP-16)

Figure 16-2 Communications Equipment Grounding and Bonding

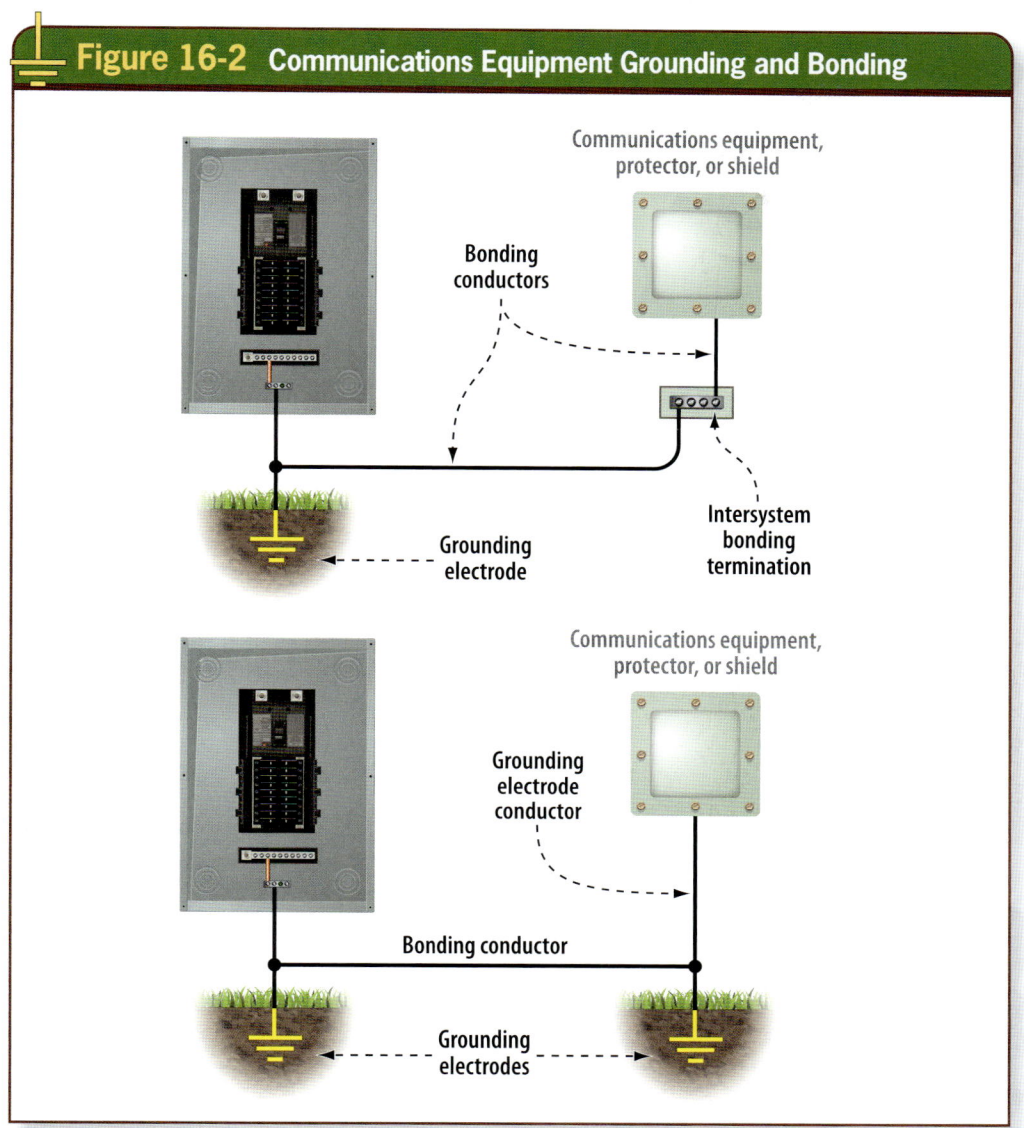

Communications equipment, protector, or shield

Bonding conductors

Grounding electrode

Intersystem bonding termination

Communications equipment, protector, or shield

Grounding electrode conductor

Bonding conductor

Grounding electrodes

Figure 16-2. *Cables and primary protectors of communications systems and equipment must be effectively bonded together and connected to ground (the Earth).*

GROUNDING AND BONDING PERFORMANCE

Grounding is the process of connecting a system or equipment to ground or to a conductive body that extends the ground connection. Grounding requires a connection to the Earth through a grounding electrode. Bonding is the process of connecting objects or entities together. Bonding electrically means that conductive objects are connected to establish continuity and conductivity between them.

Both grounding and bonding are functions necessary for safety when installing communications or other limited-energy systems. The purpose of grounding and bonding for communications and limited-energy systems and equipment is to provide a level of shock

ThinkSafe!

The *NEC* provides the minimum requirements for safe installations of communications system grounding and bonding.

Figure 16-3   Grounding and Bonding Requirements

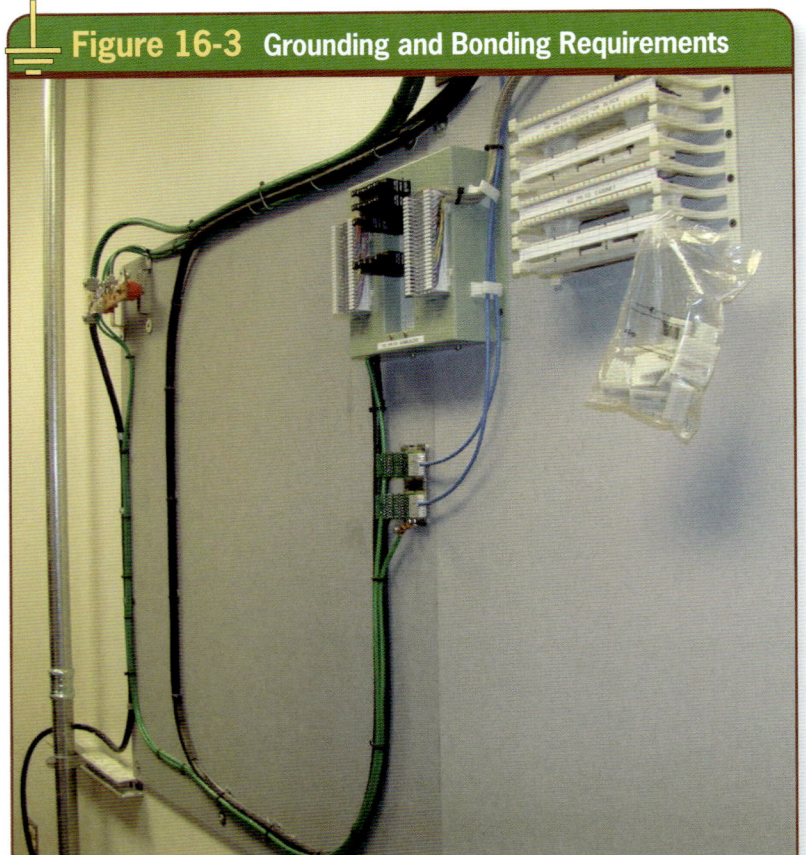

Figure 16-3. Grounding and bonding requirements for communications systems provide a level of protection by providing a path to ground for lightning and surge events.

Figure 16-4   Grounding Communications Systems

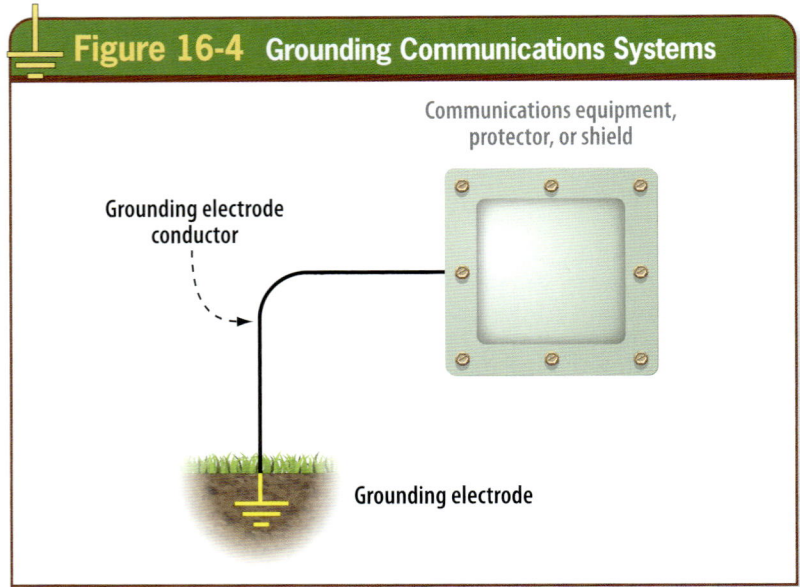

Figure 16-4. Grounding communications systems involves a connection to the Earth through a grounding electrode.

protection and limit damage from voltage surges created by lightning, line surges, or unintentional contact with higher-voltage lines. **See Figure 16-3.**

Grounding protects the equipment and provides a path to the Earth for lightning events; however, the grounding and bonding requirements of Chapter 8 in the *NEC* should not be confused with the requirements for lightning protection systems provided in *NFPA 780: Standard for Installation of Lightning Protection Systems*. Remember, Section 90.2(A) indicates that the purpose of the *NEC* is to protect persons and property from hazards that arise from the use of electricity, whereas lightning is an unpredictable force that is not used by persons. Electrical grounding and bonding requirements in the *NEC* provide varying degrees of protection from lightning events; this is typically not the primary purpose of grounding and bonding as required in the *NEC,* but is one of the functional and performance benefits.

CONNECTING TO A GROUNDING ELECTRODE

System or equipment grounding is accomplished by establishing a connection to the Earth. This connection is made through a grounding electrode. Sections 770.100 and 800.100 require the grounding electrode conductor for optical fiber cable systems and communications systems to be connected to a grounding electrode; more specifically, the same grounding electrode to which the building electrical system is connected. **See Figure 16-4.**

This requirement ensures that both systems and all connected equipment are at the same ground potential. Attempts to install separate grounding electrodes without bonding them to the power system grounding electrode can have hazardous results. This is not permitted by the *NEC* and creates unsafe conditions for persons and property.

The *NEC* requires an *intersystem bonding termination (IBT)* to be installed at the service location or at the disconnecting means for other buildings for connecting these other systems. It is intended specifically for connecting communications system grounding and bonding conductors. In many designs for buildings or structures other than dwelling units, there is often a telephone mounting board designated for all communications equipment and service-point connections. **See Figure 16-5.**

The designation on blueprints is typically "TMGB" and stands for *telecommunications main grounding busbar*. The *NEC* provides the minimum requirements for safe grounding and bonding of communications systems. Engineering designs for communications system grounding and bonding installations may meet or exceed these minimums. In such situations, the engineering designs take precedence if they are more than what the *NEC* requires. **See Figure 16-6.**

GROUNDING ELECTRODE CONDUCTOR INSTALLATION

Communications system grounding electrode conductors must be 14 AWG or larger and be made of copper or another corrosion-resistant material. They can be solid or stranded and must be insulated. Grounding electrode conductors for communications systems should be kept short; more specifically, for one- and two-family dwelling installations, they must not exceed 20 feet. An exception permits a separate grounding electrode to be installed if the grounding conductor length of 20 feet is exceeded. In this case, any separate electrode must be bonded to the power system grounding electrode for the building with a copper or equivalent conductor sized 6 AWG at minimum.

Connections to grounding electrodes for communications circuits must meet the requirements in 250.70.

INTERSYSTEM GROUNDING AND BONDING

General requirements for intersystem grounding and bonding are located in

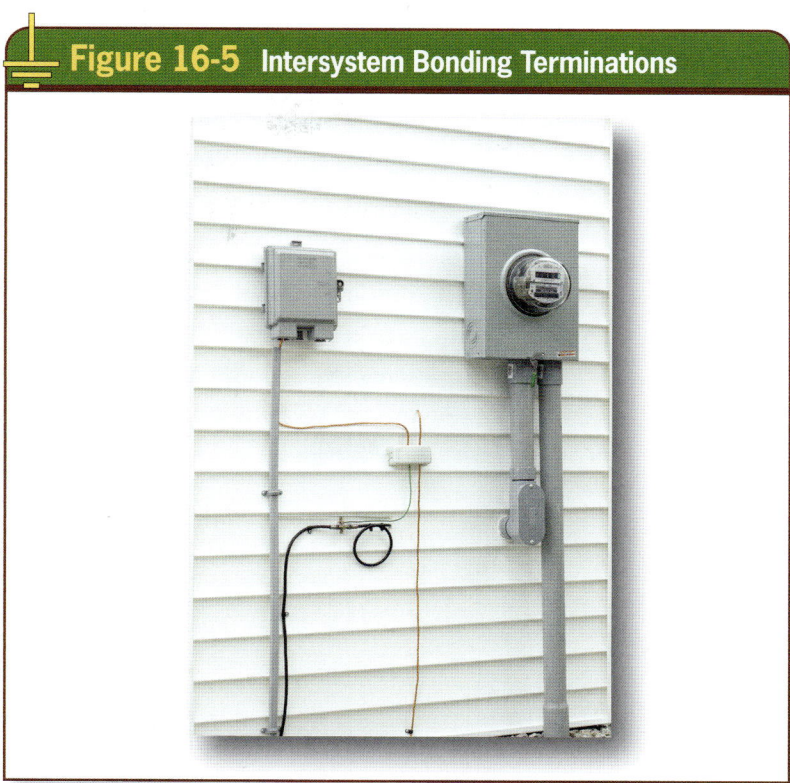

Figure 16-5 Intersystem Bonding Terminations

Figure 16-5. Intersystem bonding terminations are required to provide a connection point for not less than three bonding conductors for communications systems.

Courtesy of ERICO International Corporation

Figure 16-6 Telecom Main Grounding Busbar

Figure 16-6. A telecommunications main grounding busbar (TMGB) for communications grounding is often supplied at the telephone mounting board.

Courtesy of IBEW Local 26 Training Center

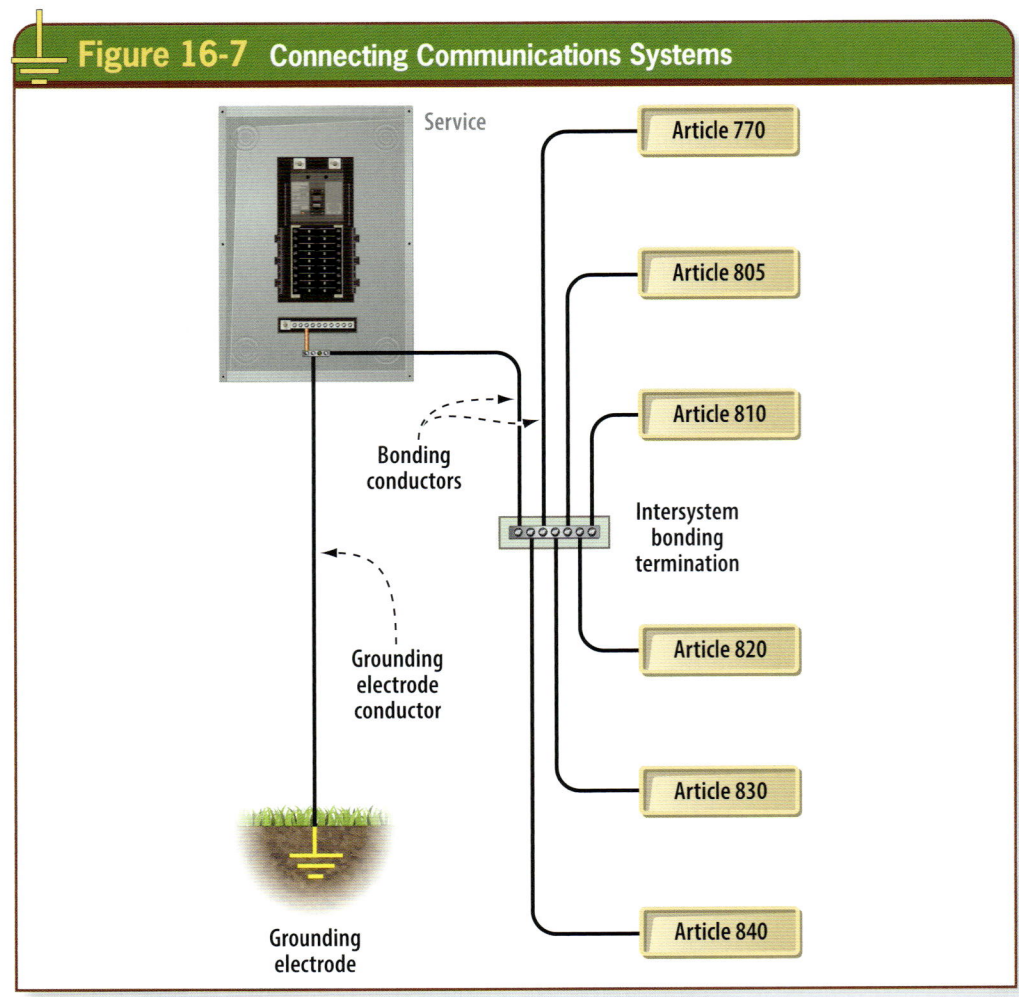

Figure 16-7 Connecting Communications Systems

Service

Article 770

Article 805

Article 810

Bonding conductors

Intersystem bonding termination

Article 820

Grounding electrode conductor

Article 830

Article 840

Grounding electrode

Figure 16-7. An IBT is required for connecting all communications systems.

250.94(A) or (B). Section 250.94(A) contains a general requirement that an intersystem bonding termination (IBT) be installed at the service equipment at the building or structure served. An IBT is also required at each separate building or structure supplied by one or more feeders or branch circuits. The IBT must be installed in a way that leaves it accessible for connection and inspection. **See Figure 16-7.** Section 250.94(B) provides an alternative method for connecting grounding and bonding conductors of communications systems.

Note that not all communications systems are installed when the electrical service is installed; these systems, such as cable TV and other antenna systems,

are usually installed later. The idea of the IBT is that it is in place when a communications system is installed so that the grounding or bonding connection can be made to it. The IBT is connected to ground (the Earth) when it is first installed, and it is connected to the building electrode for the power service of the building. A variety of IBT products are manufactured specifically for this use. The installation of an IBT must not interfere with the opening of a meter enclosure or any electrical equipment enclosure.

If an IBT is used, the terminals shall be listed as grounding and bonding equipment and provide the means for connection of no less than three intersystem bonding conductors. The

IBT itself is permitted to be any of the following:

1. The IBT can be a set of terminals securely mounted and electrically connected to the meter enclosure, the service equipment enclosure, or an exposed nonflexible metal service raceway. **See Figure 16-8.**

2. The IBT can be mounted at the service equipment enclosure, meter enclosure, or service raceway and can be connected to the enclosure or grounding electrode conductor with, at minimum, a 6 AWG copper conductor. **See Figure 16-9.**

3. The IBT can also be a bonding bar located near the grounding electrode conductor for the service. The bonding bar must be connected to the grounding electrode conductor with, at minimum, a 6 AWG copper conductor.

4. The IBT for separate buildings or structures can be mounted at the disconnecting means and be connected to it or to the grounding electrode conductor with, at minimum, a 6 AWG copper conductor.

The installation of an IBT has been a requirement since the 2008 edition of

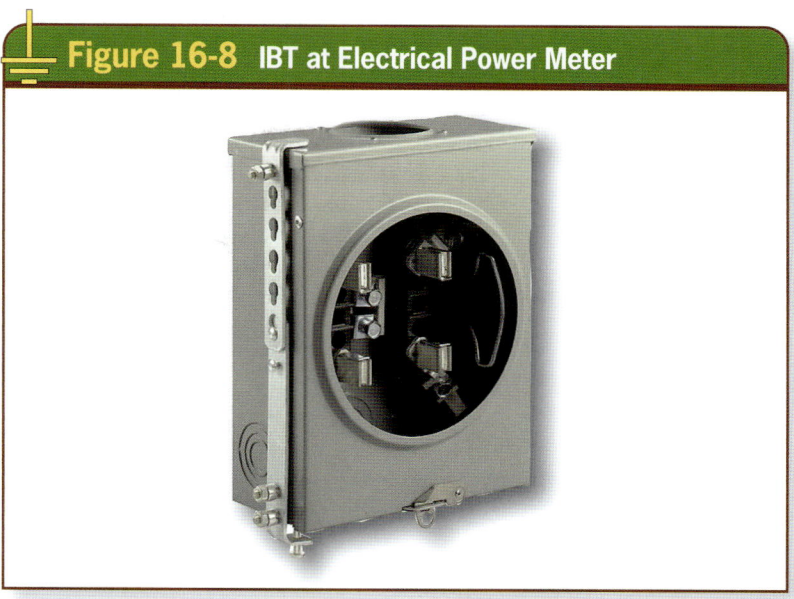

Figure 16-8 **IBT at Electrical Power Meter**

Figure 16-8. *No less than three terminals mounted directly to the electrical power meter can serve as the IBT provision.*

Courtesy of ABB

the *NEC*, but the *NEC* still recognizes that existing buildings or structures may not have an IBT. There are two choices to achieve effective grounding for communications systems in existing buildings: (1) install an IBT in conformance with 250.94, or (2) achieve intersystem grounding and bonding by

Figure 16-9 **Copper Bonding Bar**

Figure 16-9. *An intersystem bonding termination (IBT) in the form of a copper bonding bar is often mounted near the service equipment supplying the building or structure.*

Courtesy of ERICO International Corporation

connecting to an accessible grounding electrode conductor connection point external to the service equipment enclosure or at the disconnecting means at a separate building or structure supplied by one or more feeders or branch circuits. The connection can be made to a nonflexible metal raceway (location 1), at an exposed grounding electrode conductor (location 2), or at another approved means of external connection to a grounded raceway or equipment (location 3). **See Figure 16-10.**

The conductor used to make this connection must be made of copper or

For additional information, visit qr.njatcdb.org Item #5333

another corrosion-resistant material. The terminals must be listed as grounding and bonding equipment. The online UL Product iQ for Electrical Equipment category KDSH provides information about grounding clamps for communications system grounding connections.

Section 250.94(B) recognizes that connections to a copper or aluminum busbar not less than $1/4$ inch thick by two inches wide are permitted. The length of the busbar must be sufficient to accommodate at least three communications grounding or bonding conductor terminations in addition to other grounding and bonding connections.

The requirements for an IBT in 250.94(A) work effectively for dwelling units, but they did not allow for alternative practical solutions utilizing the busbar for connection of multiple electrodes and bonding conductors to be used for communications system grounding in other than dwelling unit installations.

COMMON GROUNDING AND BONDING RULES FOR COMMUNICATIONS SYSTEMS

Articles 770, 805, 810, 820, 830, and 840 of the *NEC* provide specific grounding and bonding requirements for communications and antenna systems covered by the respective articles. All of these articles, with the exception of Article 810, provide similar grounding and bonding requirements. Article 810 covers radio and television equipment and provides specific grounding and bonding rules in 810.20 and 810.21. These rules are covered separately.

Articles 770, 805, 820, 830, and 840 incorporate a parallel numbering sequence to enhance usability of the *NEC*. This training material uses the general grounding and bonding requirements in Article 800 as the basis for covering all common grounding and bonding requirements in these articles and includes some of the

Figure 16-10 GEC Connections

Existing building without an intersystem bonding termination

Install an intersystem bonding termination or connect to one of the following locations:

(1) Exposed nonflexible metallic raceway

(2) Exposed grounding electrode conductor

(3) Approved means for the external connection at the service equipment

Grounding electrode for electrical service

Figure 16-10. *By exception, the grounding electrode conductors of communications systems can be connected to any of three locations.*

differences in tabular form following this section.

Part III of Articles 770, 805, 820, 830, and 840 includes information about protection. Protection is provided by the installation of primary protectors that are typically required to be installed at the communications system point of entrance to the building or structure served. The primary protector can be inherent to communications equipment or landing blocks, or it can be a separate device installed externally. The purpose of primary protectors and discharge units is to provide a level of surge protection and to allow ground discharge of transient overvoltages and unintentional contact with higher-voltage lines. Protectors can be either fused or non-fused types. For specific information about primary protectors, refer to the online UL Product iQ.

For additional information, visit qr.njatcdb.org Item #5333

Part IV of Articles 770, 820, 830, and 840, as well as 820.93, provides the grounding and bonding methods for the communications systems covered by each article. Each of these articles presents the same grounding and bonding requirements and methods for these communications systems, and each provides reference to the general grounding and bonding requirements in Article 800. The requirements for installation for each system are similar except that the minimum size of the grounding electrode conductors may be different in each article.

Section 800.100 is arranged into four subdivisions (A) through (D). These subdivisions provide the minimum requirements for grounding and bonding communications systems.

The first component of the grounding scheme, covered in 800.100(A), is the grounding electrode conductor. This is the conductor installed to connect the primary protector, other metal parts of equipment, and any metallic cable shields to the grounding electrode system. **See Figure 16-11.**

There are six installation requirements for this conductor:

1. The grounding electrode conductor can be insulated, covered, or bare, and it must be listed.
2. The conductor must be solid or stranded copper or another corrosion-resistant material.
3. The conductor must be, at minimum, 14 AWG, and it shall have a current-carrying capacity of no less than the grounded metallic sheath member and protected conductor of the communications cable. This conductor is not required to exceed 6 AWG in size.
4. The length must be kept as short as is practicable, and in dwelling units, it generally cannot exceed 20 feet in length. By exception, the length can exceed 20 feet if an electrode meeting the criteria in 800.100(B)(3)(2) is installed and bonded to the grounding electrode for the power system supplying the building or structure.
5. This conductor must be run as straight as practicable.
6. Where subject to physical damage, it must be protected. If it is installed in a ferrous metal raceway, both ends of the raceway must be bonded to the contained grounding electrode conductor.

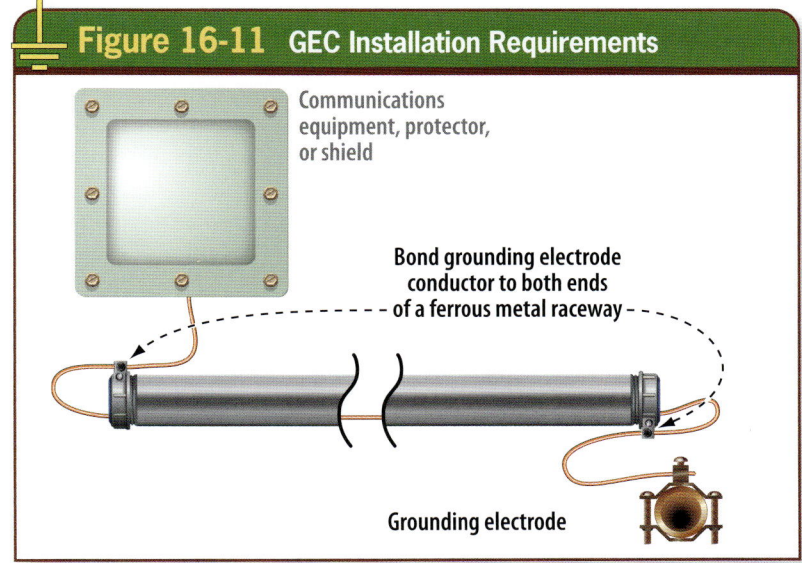

Figure 16-11 **GEC Installation Requirements**

Communications equipment, protector, or shield

Bond grounding electrode conductor to both ends of a ferrous metal raceway

Grounding electrode

Figure 16-11. Grounding electrode conductor installation requirements for communications systems are provided in 800.100(A).

The next component of the grounding scheme is the grounding electrode. By definition, the grounding electrode is a conducting object through which a direct connection to the Earth is established. Section 800.100(B) provides three options for grounding electrodes that must be used. These are provided in somewhat of a hierarchy. The driving language indicates that an electrode, according to 800.100(B)(1) through (3), must be connected to the grounding electrode conductor.

The first option indicates that if an IBT is present for use, the communications system bonding conductor must connect to it.

If an IBT is not present for use, the second option indicates that the grounding electrode connection methods provided in 800.100(B)(2) can be applied.

This grounding electrode conductor connection must be made to the nearest accessible location on any of the following seven points. **See Figure 16-12.**

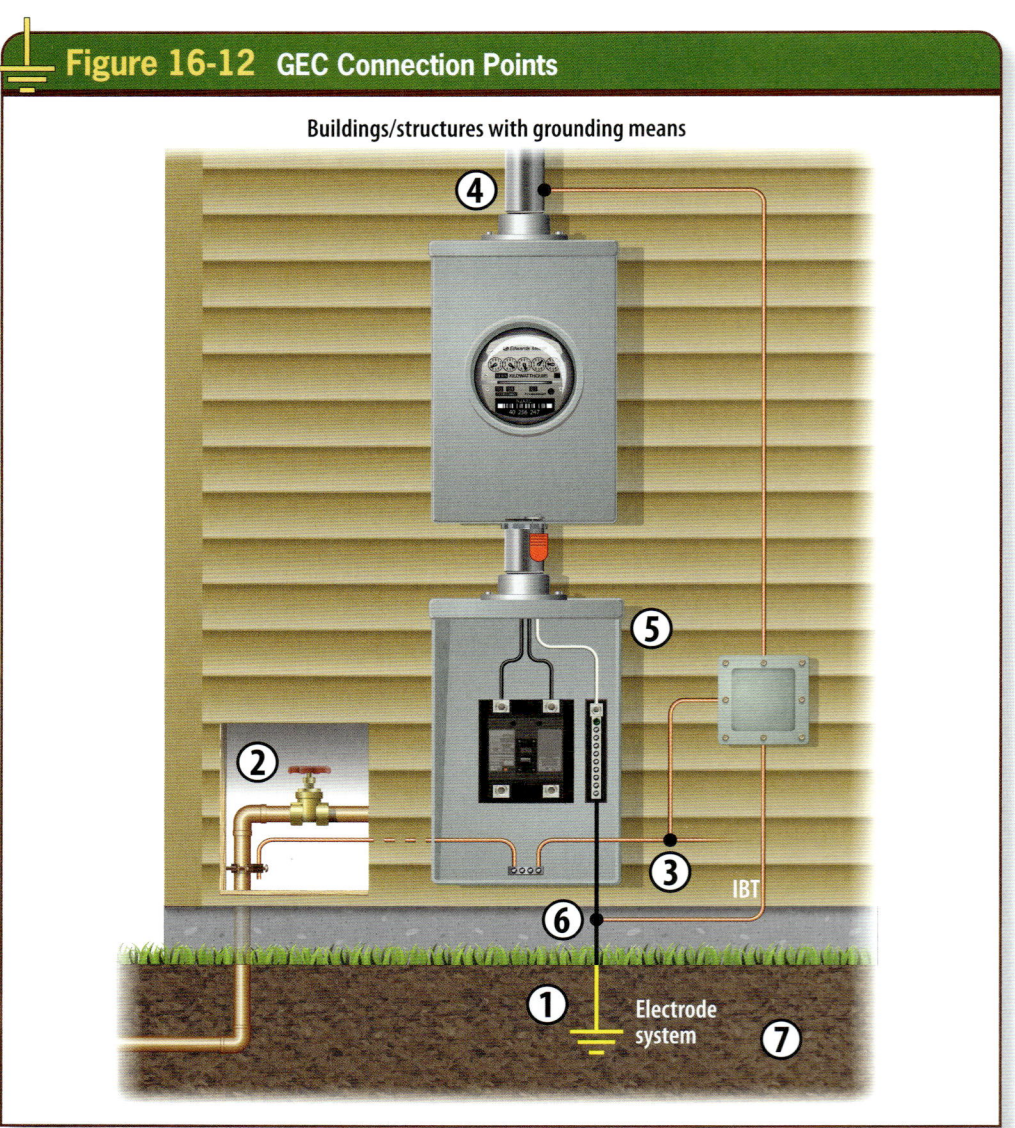

Figure 16-12 **GEC Connection Points**

Buildings/structures with grounding means

Figure 16-12. A grounding electrode conductor can be connected to a grounding electrode or one of the locations identified in the Exception to 250.94(A) if an IBT is not present (existing installations only).

1. The grounding electrode system for the building or structure electrical service
2. The grounded interior metal water piping system within five feet of the point where it enters the building
3. The accessible means external to the service equipment, as provided in 250.94
4. The nonflexible metal power service raceway
5. The service equipment enclosure
6. The grounding electrode conductor or metal enclosure for a grounding electrode conductor
7. The grounding electrode conductor or grounding electrode for a separate building or structure, as required in 250.32

The third electrode option addresses buildings or structures without an IBT or any grounding means. In this case, a grounding electrode that is present must be used; otherwise, one must be installed. Any grounding electrode described in 250.52(A)(1) through (4) can be used as the grounding electrode. If no electrodes previously mentioned are present for use, an electrode meeting the requirements in 250.52(A)(5), (7), and (8) can be installed and used.

Section 800.100(B)(3)(3) also allows a ground rod no less than five feet long and no smaller than ½ inch in diameter to be driven where practicable. The *NEC* indicates that this separate short electrode must be spaced a minimum of six feet from any electrode of another system. This electrode must be bonded to other electrodes that are present. **See Figure 16-13.**

Another component in the communications system grounding scheme is covered in 800.100(C) and is the grounding electrode conductor connection to the electrode. **See Figure 16-14.**

The methods specified in 250.70 must be used to make this connection. This requirement states that grounding electrode conductors be connected to the grounding electrode by exothermic

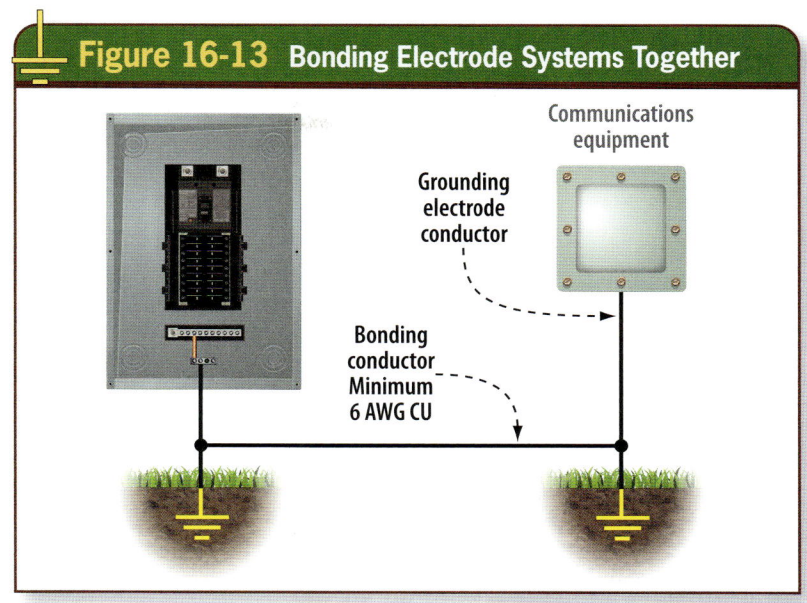

Figure 16-13 **Bonding Electrode Systems Together**

Figure 16-13. If the option of installing a grounding electrode for communications systems is used, this grounding electrode must be bonded to the grounding electrode system for the power service supplying the building or structure.

welding, listed lugs, listed pressure connectors, listed clamps, or other listed means. When making a grounding or bonding connection for a shield, sheath, or non–current-carrying metallic member

Figure 16-14 **GEC Connection Means**

Figure 16-14. The grounding electrode conductor connection means is required to be listed and the rod must be driven flush with the Earth to ensure a minimum eight feet of contact.

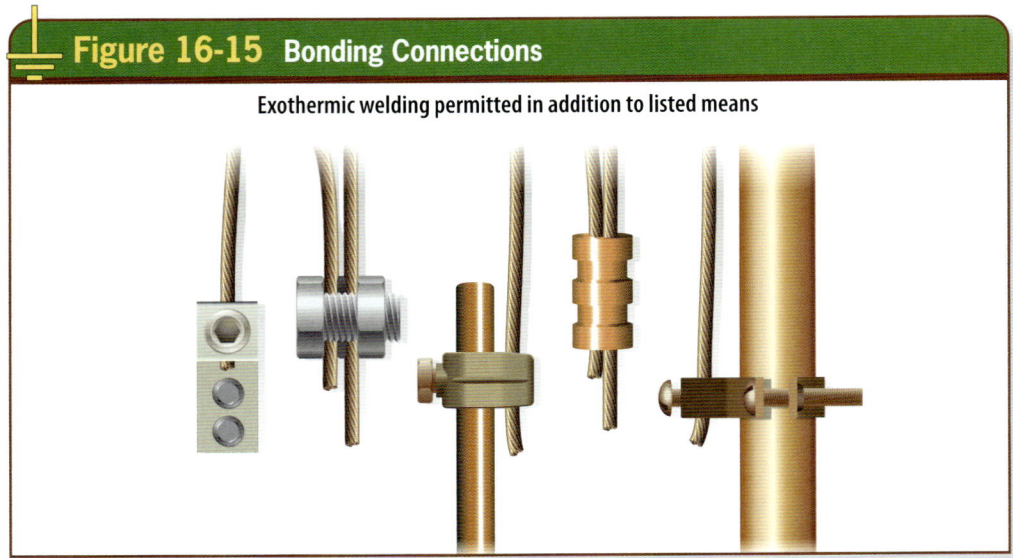

Figure 16-15. *Connections to grounding electrodes are generally required to be listed.*

of a cable to a grounding electrode conductor or bonding conductor, Section 800.180 requires that the device be listed or part of listed equipment. **See Figure 16-15.** The ground clamps must be listed for the materials of the grounding electrode and the grounding electrode conductor. Also, where a grounding electrode conductor is connected to a pipe, rod, or other buried electrode, the connection means must be listed for direct soil burial or concrete encasement.

The fourth common requirement in 800.100(D) addresses bonding of grounding electrodes. This rule indicates that if grounding electrodes for communications system installations are separate from the electrode system for the power supply to the building, the two grounding electrode systems must be bonded together to become one electrically, creating a complete system. The minimum size for the bonding jumper must not be smaller than 6 AWG copper. **See Figure 16-16.**

The grounding and bonding installation requirements in Articles 770 through 840 are similar to those already covered. The differences between the specific requirements in each respective article relate to grounding electrode conductor sizes and conductor materials. The minimum sizes required for grounding electrode conductors are provided generally in Section 800.100 or sized as specifically required in Articles 770, 805, 810, 820, 830, and 840. These articles also provide the permitted material for each conductor and the minimum size required for a bonding

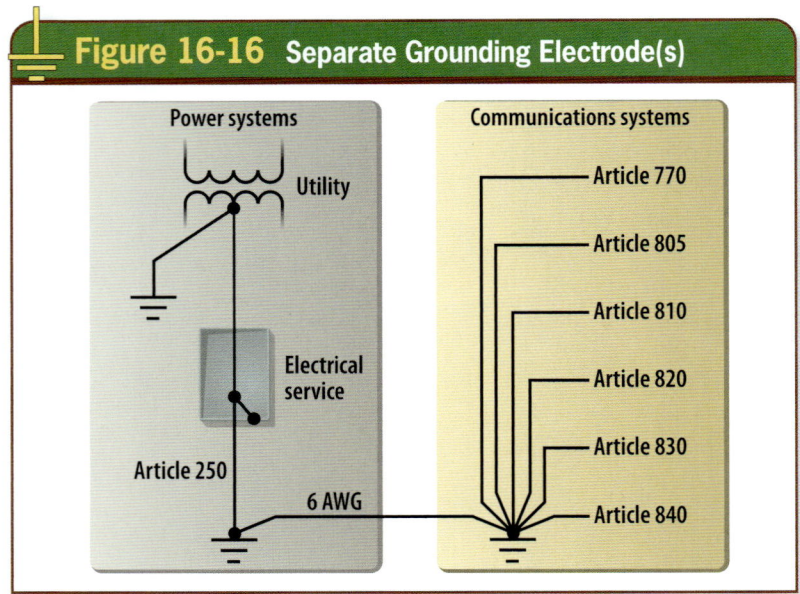

Figure 16-16. *A separate grounding electrode for a communications system must be bonded to the grounding electrode system for the power supply to the building or structure served.*

jumper between separate grounding electrode systems. **See Figure 16-17.**

GROUNDING AND BONDING AT MOBILE HOMES

Section 800.106 provides grounding and bonding requirements for mobile homes. Subdivision (A) includes the grounding rules, and subdivision (B) provides the bonding requirements. It is advisable to refer to each *NEC* communications article for any additional information specific to that communications system that is not provided in the general requirements contained in Article 800.

Grounding Requirements

The grounding requirements for communications systems installed in mobile homes are similar to the requirements for permanent structures. Where the service equipment for a mobile home is located within sight of and no more than 30 feet from the mobile home, the primary protector's grounding terminal must be connected to the grounding electrode conductor or grounding electrode. If the service equipment is located elsewhere on the premises and a feeder disconnect is installed within sight of and no more than 30 feet from the mobile home, the primary protector grounding terminal must be connected to that grounding electrode. Where there is no mobile home service within 30 feet of the mobile home, the primary protector grounding terminal must be connected to a grounding electrode in accordance with the provisions of 800.106(A)(1). If there is no mobile home disconnecting means that is grounded according to 250.32 and located within sight of and no more than 30 feet from the mobile home, the primary protector grounding terminal must be connected to a grounding electrode in accordance with 800.100(A)(2).

These conditions specifically address the need to establish a connection to a grounding electrode from the primary protector at the point of entrance of the communications system to the mobile home structure.

Figure 16-17   GEC Sizing and Material

Article of NEC	Minimum Size of Grounding Electrode Conductor	Grounding Electrode Conductor Material	Section of Article	Electrode System Bonding Jumper
770	14	Copper or equivalent	XXX.100(A)(3)	6 AWG copper
800	14	Copper or equivalent	XXX.100(A)(3)	6 AWG copper
810	10	Copper	XXX.21(H)*	6 AWG copper
	8	Aluminum		
	17	Bronze		
820	14	Copper or equivalent	XXX.100(A)	6 AWG copper
830	14	Copper or equivalent	XXX.93(A) or (B)	6 AWG copper
840	14	Copper or equivalent	XXX.93(A) or (B)	6 AWG copper

*Section 810.58 requires, at minimum, 10 AWG copper, bronze, or copper-clad steel for a protective grounding electrode conductor and, at minimum, 14 AWG copper or equivalent for the operating grounding electrode conductor.

Figure 16-17. *Grounding electrode conductor and bonding jumper sizes and material are provided within Article 770 and each Chapter 8 article.*

Bonding Requirements

The primary protector grounding terminal or grounding electrode must be bonded to the mobile home frame or an available grounding terminal on the mobile home frame. The minimum size required is a 12 AWG copper conductor. This bonding conductor connection must be made if there is no mobile home service equipment or disconnecting means or if the mobile home is supplied by a cord-and-plug connection.

RADIO AND TELEVISION EQUIPMENT AND ANTENNAS

Radio and television equipment is covered by Article 810. Section 810.21 provides the specific grounding and bonding rules for radio and television receiving stations. Grounding electrode conductor installations must be in accordance with 810.21(A) through (K). **See Figure 16-18.**

The conductor material must be copper, aluminum, copper-clad steel, or another corrosion-resistant material. Aluminum or copper-clad aluminum conductors must not be terminated within 18 inches of the Earth due to the corrosion influences. The grounding electrode conductor does not have to be an insulated conductor; it can be bare. It must be secured to the surface on which it is run, but insulated support brackets are not required. If proper support cannot be achieved, the size of the grounding electrode conductor must be increased proportionally to allow for less support by brackets.

If this conductor is subject to physical damage, it must be protected. Where run in a ferrous metal raceway, the contained grounding electrode conductor must be bonded to both ends of the raceway at the points of conductor entrance and emergence. The grounding electrode conductor must be run in a line, as straight as practicable, from the mast or discharge unit to the grounding electrode connection.

The grounding electrode conductor for these systems must connect to an IBT device located as specified in 250.94. **See Figure 16-19.** If there is no IBT device, the conductor must be connected to the nearest of the following:

1. The grounding electrode system for the building or structure electrical service, per 250.50
2. The grounded interior metal water piping system within five feet of the point where it enters the building
3. The accessible means external to the service equipment, as provided in 250.94
4. The nonflexible metal power service raceway
5. The service equipment enclosure
6. The grounding electrode conductor or metal enclosure for a grounding electrode conductor

The other electrode option for radio and television system receiving stations addresses buildings or structures without an IBT or any grounding electrode system. In this case, a grounding electrode that is present must be used; otherwise, one must be installed. Any grounding electrode described in 250.52 can be installed and used. If the

Figure 16-18 Television Equipment Antennas

Figure 16-18. Television equipment antennas must comply with the grounding and bonding rules in Article 810.

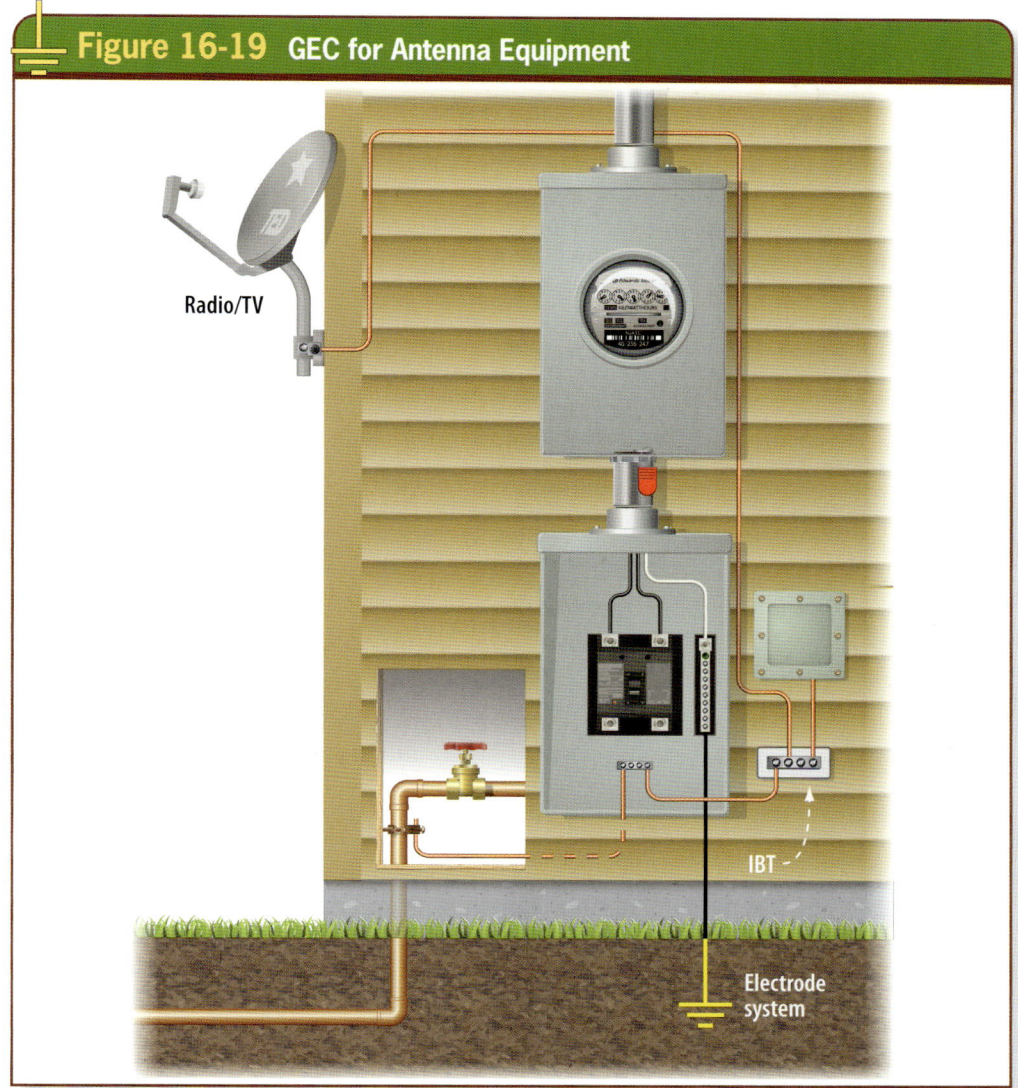

Figure 16-19   **GEC for Antenna Equipment**

Radio/TV

IBT

Electrode
system

Figure 16-19. A grounding electrode conductor for antenna equipment is generally required to be connected to the IBT.

building or structure has no grounding provision as described in 810.21(F)(1) or (2), a connection to an effectively grounded metal structure is permitted.

The grounding electrode conductor for these systems can be run either inside or outside a building. The minimum size of the grounding electrode conductor is based on the type of conductor material: 10 AWG copper, 8 AWG aluminum, or 17 AWG copper-clad steel or bronze. The same grounding electrode conductor is permitted to serve as the operating ground and provide protective functions. It is always the best choice to connect the grounding electrode conductor of these systems to the power system grounding electrode for the building.

If separate electrodes are used, they must be bonded together with, at minimum, a 6 AWG copper conductor. All grounding electrode conductor connections must meet the requirements in 250.70, as previously discussed for the other types of communications systems. Section 810.7 requires listed devices or listed equipment be used to connect grounding or bonding conductors to a shield, sheath, or non–current-carrying metal member of a cable, or to metal equipment or antennas. **See Figure 16-20.**

The grounding requirements for amateur and citizen band transmitting and receiving stations are provided in 810.58. This section indicates that all the requirements in 810.21(A) through (C) are applicable except the minimum size for the protective and operating grounding electrode conductors.

Figure 16-20 Separate Electrodes

Radio/TV

Electrode system

Figure 16-20. If a separate grounding electrode is installed, it must be bonded to the electrode for the power system supplying the building or structure.

The minimum size required for a protective grounding electrode conductor is 10 AWG copper, bronze, or copper-clad steel. The minimum size required for the normal operating grounding electrode conductor is 14 AWG copper or equivalent. Refer to 810.20 for specific requirements pertaining to discharge units required for receiving stations.

OVERVOLTAGES AND LIGHTNING EVENTS

The reasons why communications systems must be connected to the building power system grounding electrode are quite simple, yet such connections are not always made correctly. Using the same grounding electrode as the building electrical service keeps the conductive parts of communications systems and equipment at or close to the same ground (Earth) potential in normal operation. In abnormal events, such as surges related to lightning strikes, the objective is to keep conductive parts of electrical power systems and limited-energy communications systems at the same potential while the potentials rise and fall. This minimizes the possibilities of destructive flashover events within electronic equipment and between electrically-conductive parts and equipment within buildings or structures. If the grounding conductors of a communications system are connected to an electrode separate from the building power service grounding electrode, a lightning event on the building or close to the building can cause conductive parts of equipment in the power system and the communications system to rise at different potentials, creating possible flashovers that can damage equipment or even cause a fire.

SUMMARY

There are important common general grounding and bonding rules for communications systems. The rules are intended to protect persons and property from electrical hazards in normal operation and minimize differences of potential during abnormal events, such as line surges or lightning strikes. Lightning is a powerful and unpredictable force, so meeting the *NEC* requirements is the minimum plan against damage from these natural and erratic events. A lightning protection system in accordance with *NFPA 780* provides another degree of protection above the minimum grounding protection required by the *NEC*.

Communications system grounding electrode conductors must be electrically common to the grounding electrode used for the electrical power system for system and equipment safety and for personnel safety. Article 800 (specifically Section 800.100) provides common rules specific to the grounding and bonding schemes for the communications systems covered in Articles 770, 805, 810, 820, 830, and 840. These articles provide the specific minimum sizing requirements for grounding electrode conductors and bonding conductors installed for these systems. The minimum size conductors should be understood, along with specific rules that address bonding all grounding electrodes together to become one electrically.

REVIEW QUESTIONS

1. Communications systems and circuits in buildings must comply with the applicable rules in Chapter 8 of the *NEC*, in addition to the requirements in *NEC* Chapters 1 through 4 only where referenced from the Chapter 8 articles.

   a. True   b. False

2. The __?__ is a device that provides a means for connecting bonding conductors for communications systems to the grounding electrode system.

   a. grounding electrode
   b. grounding electrode conductor
   c. grounding electrode system
   d. intersystem bonding termination

3. The purpose of grounding and bonding for communications systems and equipment is to provide a level of protection against shock and damage caused by voltage surges created by lightning, line surges, or unintentional contact with higher-voltage lines.

   a. True   b. False

4. Communications system and equipment grounding is accomplished by all but which of the following?

   a. Connection to a conductive body that extends the ground connection
   b. Connection to a conductive body that serves in place of the Earth
   c. Connection to an intersystem bonding termination
   d. Connection to ground through an electrode

5. Communications system grounding electrode conductors must be __?__ or larger and be made of copper or another corrosion-resistant material.

   a. 14 AWG
   b. 12 AWG
   c. 8 AWG
   d. 6 AWG

REVIEW QUESTIONS

6. An intersystem bonding termination device must provide the means for connection of no less than __?__ intersystem bonding conductors.

    a. 2
    b. 3
    c. 4
    d. 5

7. If the intersystem bonding termination is provided through a set of terminals securely mounted to the meter or service equipment enclosure, the terminals have to be __?__ .

    a. approved as grounding and bonding equipment
    b. identified as grounding and bonding equipment
    c. listed as grounding and bonding equipment
    d. marked

8. By definition, a(n) __?__ is a conducting object through which a direct connection to the Earth is established.

    a. equipment grounding conductor
    b. grounding electrode
    c. grounding electrode conductor
    d. intersystem bonding termination

9. The *NEC* specifies that grounding electrode conductors for communications systems are required to be connected to the grounding electrode by all but which of the following means?

    a. Exothermic welding
    b. Listed clamps or other listed means
    c. Listed lugs or listed pressure connectors
    d. Sheet metal screws

10. If a separate grounding electrode is installed for a communications system, this grounding electrode must be bonded to the grounding electrode system for the electrical power supplying the building with, at minimum, a __?__ copper conductor.

    a. 10 AWG
    b. 8 AWG
    c. 6 AWG
    d. 4 AWG

11. The copper conductor required for a protective grounding electrode conductor installed for a television antenna system and equipment must not be smaller than __?__ .

    a. 17 AWG
    b. 14 AWG
    c. 10 AWG
    d. 8 AWG

12. Where the service equipment for a mobile home is located within sight of and no more than 30 feet from the mobile home, the primary protector grounding terminal is required to be connected to the grounding electrode for the service equipment.

    a. True   b. False

13. The purpose of primary protectors and discharge units is to provide a level of surge protection and allow ground discharge of transient overvoltages and unintentional contact with higher-voltage lines.

    a. True   b. False

14. The length of a communications system grounding electrode conductor must be kept as short as practicable in dwelling units, and it generally cannot exceed __?__ in length. By exception, the length can exceed __?__ if an electrode meeting the criteria in 800.100(B)(3)(2) is installed and bonded to the grounding electrode for the power system supplying the building or structure.

    a. 10' / 10'
    b. 20' / 20'
    c. 25' / 25'
    d. 100' / 100'

15. The grounding electrode conductor for a communications system must be insulated, covered, or bare and is required to be __?__ .

    a. approved
    b. identified
    c. listed
    d. provided with green insulation or marked with green marking tape

16. If a communications system grounding electrode conductor is subject to physical damage, it must be protected. If it is installed in a ferrous metal raceway, both ends of the raceway must be bonded to the contained grounding electrode conductor.

    a. True   b. False

17. Where a bonding jumper or conductor is installed for bonding a separate communications system grounding electrode to the power system grounding electrode, the minimum size required is __?__ copper.

    a. 10 AWG
    b. 8 AWG
    c. 6 AWG
    d. 4 AWG

18. The primary protector grounding terminal or grounding electrode must be bonded to a mobile home frame or an available grounding terminal on the mobile home frame, and the minimum size required for the bonding conductor is __?__ copper.

    a. 14 AWG
    b. 12 AWG
    c. 10 AWG
    d. 6 AWG

19. Section 250.94(B) indicates that connections to a copper or aluminum busbar not less than $1/4$ inch thick by two inches wide are permitted. The length of the busbar must be sufficient to accommodate at least three communications grounding or bonding conductor terminations in addition to other grounding and bonding connections.

    a. True   b. False

Ground-Fault Circuit Interrupters and Ground-Fault Protection of Equipment

Grounding and bonding are proven methods to provide protection for electrical systems, but there are other forms of personnel and equipment protection required by the *NEC* that have become popular and have raised the level of protection. Two types of ground-fault protection used in electrical distribution systems are ground-fault circuit interrupter (GFCI) protection and ground-fault protection of equipment (GFPE). Both types of ground-fault protection function similarly during a ground-fault condition, but at different current and time levels. GFCIs protect people while GFPE protects equipment. The *NEC* includes several important requirements for GFCI protection and for protecting large equipment from ground faults that can result in severe damage or even destroy equipment.

Objectives

» Understand the purpose of ground-fault circuit interrupter (GFCI) protection.

» Understand the factors related to the severity of electric shock in humans.

» Understand the operating principles of GFCI protection and ground-fault protection of equipment (GFPE).

» Determine *NEC* requirements for GFCIs and GFPE.

Chapter 17

Table of Contents

Ground-Fault Circuit Interrupters (GFCIs) 398

    Purpose of GFCI Protection................... 398

    Function of GFCIs 399

    NEC Requirements for GFCI Protection.... 400

        GFCIs in Health Care Facilities.......... 402

        GFCI Protection–Receptacle Replacements 403

        Temporary Wiring Installations 404

        Assured EGC Program 405

        Special Purpose Ground-Fault Circuit Interrupters (SPGFCI) 406

Ground-Fault Protection of Equipment (GFPE).................................. 406

    Purpose of GFPE 407

    Types of GFP Equipment...................... 408

        Neutral Ground-Strap System 408

        Zero-Sequence Systems 410

    Selective Coordination......................... 412

    GFPE System Coordination 413

    Applicability in Health Care Facilities....... 413

    Feeder GFPE 415

    Branch Circuit GFPE 415

    Testing of GFPE Systems 415

Summary... 416

Review Questions 416

GROUND-FAULT CIRCUIT INTERRUPTERS (GFCIs)

Few technical advances in the electrical industry have resulted in more lives saved than the invention of and subsequent requirements for ground-fault circuit interrupters installed in dwelling units and in other occupancies.

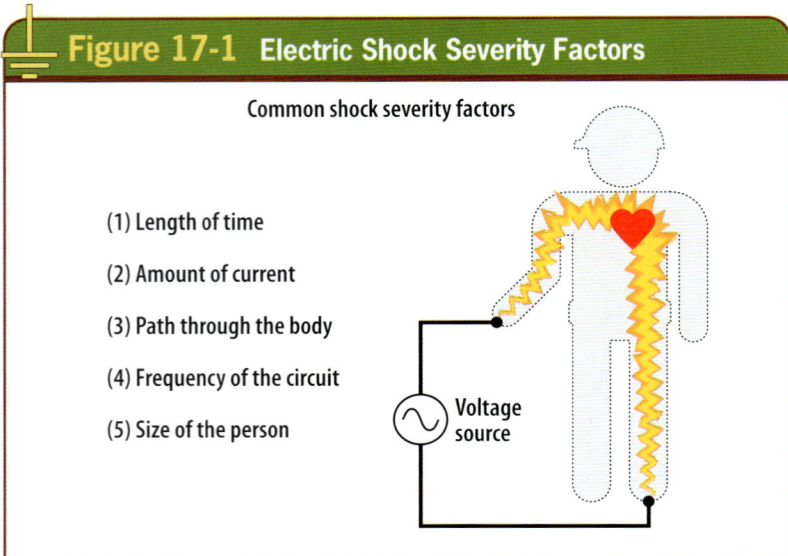

Figure 17-1   **Electric Shock Severity Factors**

Common shock severity factors

(1) Length of time

(2) Amount of current

(3) Path through the body

(4) Frequency of the circuit

(5) Size of the person

Voltage source

Figure 17-1. There are a number of factors that contribute to the severity of electric shock.

Figure 17-2   **GFCI Devices**

GFCI circuit breaker

GFCI receptacle

Figure 17-2. Two types of hardwired GFCI devices (circuit breaker and receptacle) are available for use in locations where GFCI protection is required by the NEC.

Purpose of GFCI Protection

Ground-fault circuit interrupters provide protection against electrocution and minimize the severity of electric shock a person can receive. Although there are no actual statistics available, data collected by the Consumer Product Safety Commission (CPSC) indicates that electrocutions in the United States have been decreasing since the introduction of GFCI protective devices, more specifically Class A GFCIs. Most industry and safety experts agree that GFCI protection has saved numerous lives and significantly reduced the number of shock injuries.

Insulation breakdown or failure is often the cause of electric shock to humans. A person who comes into contact with an energized circuit can become a series path for current, or the person can be in parallel with other fault current paths. If a person is in series contact with the circuit, all current can pass through the body from entry to exit points. Persons in parallel contact with the circuit will become a current path in combination with other current paths over which all the fault current will divide. The severity of shock a person receives is related to the length of time the current is present through the body, the amount of current, the path through the body, the frequency of the current, and the size of the person involved. **See Figure 17-1.**

There are many requirements for GFCI protection provided throughout the *NEC*. These devices operate at a low level of current. The term *ground-fault circuit interrupter* is defined in Article 100 of the *NEC*.

Ground-Fault Circuit Interrupter (GFCI). A device intended for the protection of personnel that functions to de-energize a circuit or portion thereof within an established period of time when a ground-fault current exceeds the values established for a Class A device. (CMP-2)

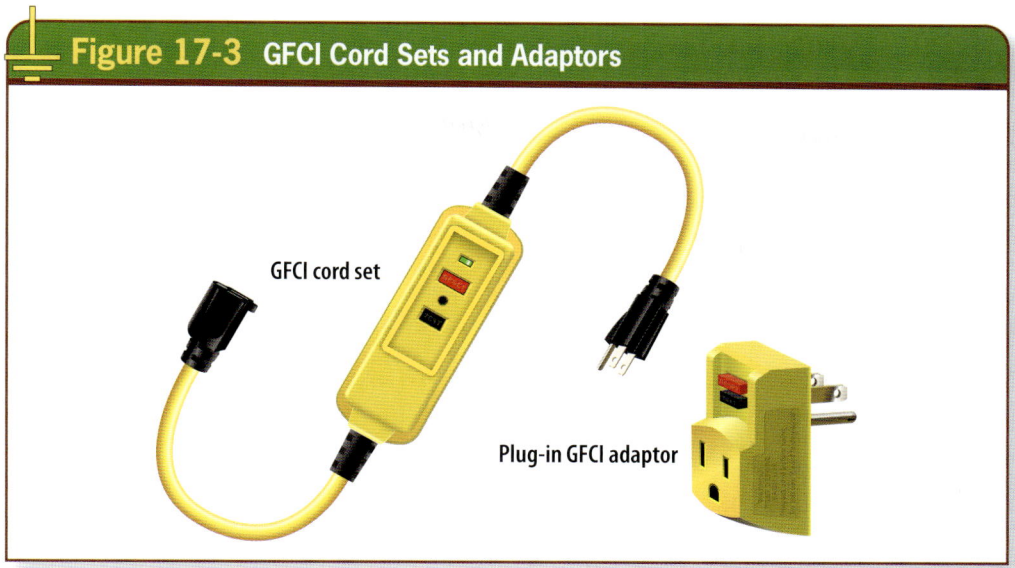

Figure 17-3 GFCI Cord Sets and Adaptors

GFCI cord set

Plug-in GFCI adaptor

Figure 17-3. Portable GFCI devices such as cord sets and adaptors are available for workers to provide protection at the point of use.

Function of GFCIs

The informational note following the definition of GFCI in the *NEC* indicates that Class A GFCIs trip at six milliamperes or greater of ground-fault current and do not trip when the current to ground is less than four milliamperes. Additional information about GFCIs is located in *UL 943, Standard for Ground-Fault Circuit Interrupters*. See the online UL Product iQ for Electrical Equipment category KCXS. **See Figure 17-2.**

For additional information, visit qr.njatcdb.org Item #5333

GFCI protection is available in a few forms. There are GFCI circuit breakers, outlet devices, portable cord sets, portable plug-in devices, and so forth. All GFCI protection devices generally function in the same way. **See Figure 17-3.**

The circuit breaker types of GFCI open the entire circuit if a ground fault occurs, while outlet types only open a portion of the circuit. The principles of operation for a GFCI device are simple. The circuit conductors are monitored by the GFCI-sensing device for equal circuit current in both directions (supply and return). If the return current (on the neutral conductor) of the circuit becomes lower than the current present in the ungrounded conductor, there must be a fault-current path creating an imbalance or differential between them. When this imbalance reaches the level of a four- to six-milliampere difference, the device reacts and opens. **See Figure 17-4.**

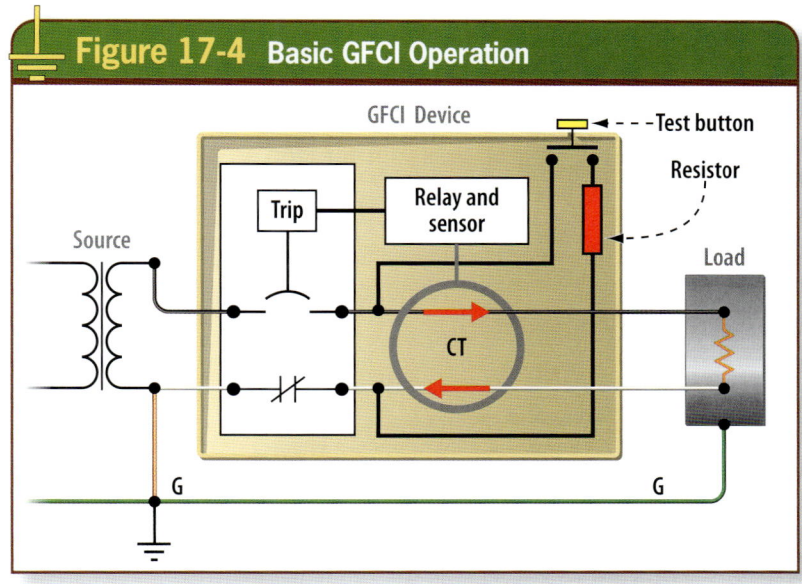

Figure 17-4 Basic GFCI Operation

GFCI Device

Test button

Resistor

Trip

Relay and sensor

Source

Load

CT

G   G

Figure 17-4. GFCIs operate by measuring the current in the supply and return current-carrying conductors of the circuit and by monitoring for an imbalance that exceeds four to six milliamperes.

GFCI devices do not directly limit the level of current through the body during a contact event; instead, they reduce the amount of time the current is present. Extensive research about preventing electrocution of humans resulted in established current and time levels for Class A GFCI devices. GFCI protection monitors current in the complete electrical circuit for an imbalance; therefore, it can be used in grounded systems. An equipment grounding conductor (EGC) is not necessary for a GFCI to operate and provide protection, but the system must be grounded.

NEC Requirements for GFCI Protection

There are several requirements for GFCI protection throughout the *NEC*. The first requirement for GFCI protection included in the *NEC* was for swimming pool installations, where a Class B GFCI device was required for circuits supplying underwater luminaires.

The *NEC* requirements for GFCIs have increased significantly over the years and resulted in reduced shock and electrocution events in the locations in which they are used. GFCIs are required in all locations to be installed in a readily accessible location. GFCI protection is required in dwelling units for all 125-volt through 250-volt receptacles that are installed in locations specified in 210.8(A)(1) through (12) and supplied by single-phase branch circuits rated 150 volts to ground or less.

Section 210.8(A) lists 11 locations where GFCI protection is a requirement for dwelling units. This list does not provide the GFCI protection requirements for special equipment such as pools or hot tubs. Section 210.8(C) requires GFCI protection for lighting outlets not exceeding 120 volts that are installed in crawl spaces. Section 210.8(D) requires the branch circuit or the outlet supplying 12 different types of appliances to have GFCI protection. Section 422.5 requires GFCI protection for seven

Figure 17-5   GFCI Protection in Dwelling Occupancies

All dwelling unit bathrooms	Basements[3]
Garages and accessory buildings[1]	Kitchens
Outdoor receptacles[2]	Sinks (within 6 feet of top inside edge)
Areas with sinks and permanent provisions for food preparation	Boathouses
All crawl spaces at or below grade	Laundry Areas
Receptacles installed within 6 feet of a bathtub or shower stall	Indoor damp and wet locations

[1]GFCI protection requirements apply to all garages and to those accessory buildings that have a floor located at or below grade level not intended as habitable rooms and limited to storage areas, work areas, and areas of similar use.

[2]Receptacles that are not readily accessible and are supplied by a dedicated branch circuit for electric snow-melting or deicing equipment and installed according to 426.28 or 427.22 as applicable, meaning the circuit is equipped with GFPE.

[3]A receptacle supplying a permanently installed premises security system is permitted to omit GFCI protection. Sections 760.41(B) and 760.121(B) prohibit GFCI and arc-fault circuit-interrupter (AFCI) protection for the power circuits supplying fire alarm systems.

Figure 17-5. *GFCI protection is required for all 125-volt through 250-volt receptacles that are in multiple locations of dwelling occupancies as indicated in 210.8(A)(1) through (12) and are supplied by single-phase branch circuits rated at 150 volts or less.*

Figure 17-6   GFCI Protection in Other Than Dwelling Units

All bathrooms[6]

All kitchens[6]

Areas with sinks and permanent provisions for food preparation

Buffet serving areas

Rooftops[1, 6]

At (within 6 feet of) sinks[3, 4, 5, 6]

Outdoor locations[2, 6]

Indoor damp and wet locations

Locker rooms with associated shower facilities

Garages, service bays, similar locations (diagnostic tools, hand tools, portable lighting)[6]

Crawl spaces at or below grade level

Unfinished areas of basements[6]

Within 6 feet of aquariums, bait wells, and similar open aquatic vessels

Laundry areas

Bathtub and shower stalls (within 6 feet of the outside edge of the bathtub or shower stall)

[1] Receptacles installed on rooftops are required to have GFCI protection.

[2] Receptacles that are not readily accessible and are supplied from a dedicated branch circuit for electric snow-melting or deicing equipment are permitted without GFCI protection, by exception.

[3] In industrial establishments only, where the conditions of maintenance and supervision ensure that only qualified personnel are involved, an assured EGC program as specified in 590.6(B)(2) is permitted for only those receptacle outlets that are used to supply equipment that would create a greater hazard if power is interrupted or that have a design that is not compatible with GFCI protection, by exception.

[4] In industrial laboratories, receptacles used to supply equipment where removal of power would introduce a greater hazard are permitted to be installed without GFCI protection, by exception.

[5] For receptacles in patient bed locations of Category 2 (general care) or Category 1 (critical care) spaces of health care facilities shall be permitted to comply with 517.21.

[6] Listed weight supporting ceiling receptacles (WSCR) utilized in combination with compatible weight-supporting attachment fittings (WSAF) installed for the purpose of serving a ceiling luminaire or ceiling-suspended fan shall not be required to be ground-fault circuit-interrupter protected. If a general purpose convenience receptacle is integral to the ceiling luminaire or ceiling-suspended fan, GFCI protection shall be provided.

Figure 17-6. *GFCI protection is required for 125- and 250-volt receptacles supplied by branch circuits rated 150 volts or less to ground, 50 amperes or less, and all receptacles supplied by 3-phase branch circuits rated 150 volts or less to ground, 100 amperes or less, installed in the locations specified in 210.8(B)(1) through (15).*

types of appliances but allows for protection of the appliances with the GFCI device as part of the supply cord or attachment plug or as part of the appliance. **See Figure 17-5.**

Section 210.8(B) requires GFCI protection for all 125-volt through 250-volt receptacles supplied by single-phase branch circuits 150 volts or less to ground, 50 amperes or less, and all receptacles supplied by 3-phase branch circuits rated 150 volts or less, 100 amperes or less if installed in any of fifteen locations provided in 210.8(B). **See Figure 17-6.**

Figure 17-7   GFCI Protection Requirements

Location	Section
Appliances	422.5(A) and (B)
Aircraft hangars	513.12
Audio system equipment	640.10(A)
Marinas, boatyards, docking facilities	555.35
Carnivals, circuses, fairs, and similar events	525.23
Commercial garages	511.12
Electric vehicle charging systems	625.22
Electronic equipment, sensitive	647.7(A)
Elevators, escalators, and moving walkways	620.6
Feeders	215.9
Fountains	680.51(A)
Health care facilities	517.20(A), 517.21
Hydromassage bathtubs	680.71
Mobile and manufactured homes	550.13(B) and (E), 550.32(E)
Natural and artificially made bodies of water	682.15
Park trailers	552.41(C)
Pools, permanently installed	680.21(C), 680.21(D), 680.22(A)(2), 680.22(A)(4), 680.23(A)(3)
Pools, storable	680.32
Sensitive electronic equipment	647.7(A)

Figure 17-7. *The NEC also includes several other requirements for GFCI protection (non-inclusive).*

Section 210.8(B)(10) indicates that GFCI protection is not required for vehicle exhibition halls and showrooms, though it is a requirement for commercial repair garages. GFCI protection installed in locations other than dwelling units must also be in a readily accessible location.

There are other requirements for GFCI protection throughout the *NEC*. **See Figure 17-7.**

GFCIs in Health Care Facilities
Section 517.20 requires special protection from electric shock in wet procedure and patient care spaces in the form of two methods:

1. A power distribution system that inherently limits the possible ground-fault current caused by a first fault to a low value, without interrupting the power supply
2. A power distribution system in which the power supply is interrupted if the ground-fault current exceeds a value of six milliamperes

Section 517.21 addresses GFCI requirements in Category 2 (general care) and Category 1 (critical care) spaces. This *NEC* provision aligns with *NFPA 99*, which indicates that receptacles are not required to be installed in bathrooms or toilet rooms. GFCI protection is required in accordance with 210.8(B)(1) for receptacles installed in bathrooms and toilet rooms of Category 2 (general care) spaces. GFCI protection for receptacles can be omitted in Category 2 (general care) and Category 1 (critical care) spaces where the basin, sink, or other similar plumbing fixture is installed in the patient bed location. **See Figure 17-8.**

In many critical care spaces, there are continuity of power concerns for life support equipment and other essential electrical medical equipment. Patients in such locations are generally incapacitated to the point where they are restricted to beds. In such locations, which include intensive care units and operation recovery units of hospitals, the *NEC* relaxes the GFCI protection requirement.

GFCI Protection–Receptacle Replacements

Receptacles installed on 15- and 20-ampere circuits are generally required to be of the grounding type described in 406.4(A). Section 406.4(D)(3) requires that replacement receptacles have GFCI protection if such protection is now a requirement in the *NEC*. Installers must be aware of the readily accessible locations where GFCI protection is required when servicing and replacing receptacles. Section 406.4(D) provides the alternatives that can be used for receptacle replacements.

If a receptacle is being replaced and a grounding means exists in the outlet box, then a grounding-type receptacle must be installed as the replacement. The grounding terminal of the receptacle replacement must be connected to the EGC of the branch circuit. In many older installations, two-wire branch circuits were installed using knob-and-tube wiring, older nonmetallic-sheathed (NM) cable, or legacy alternating current cable wiring systems that did not provide an EGC in accordance with 250.118. In these installations without EGCs at the outlets, there are three choices for receptacle replacements:

1. A non–grounding-type receptacle can be replaced with another non–grounding-type receptacle or receptacles. **See Figure 17-9.**

2. A non–grounding-type receptacle can be replaced with a GFCI receptacle device that is marked "No Equipment Ground" if no EGC is run to any receptacles on the load

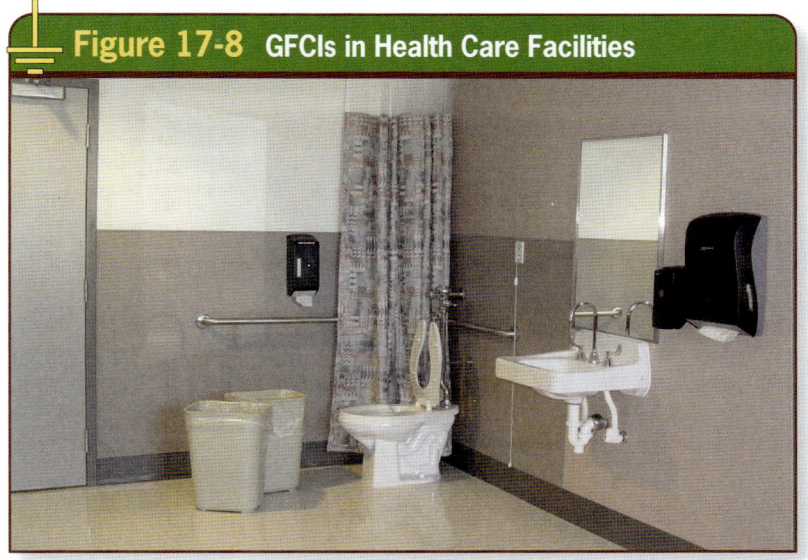

Figure 17-8   GFCIs in Health Care Facilities

Figure 17-8. GFCI protection is not required for receptacles in rooms where the basin, sink, or other similar plumbing fixture is installed in the patient bed location in Category 1 (critical care) and Category 2 (general care) spaces.

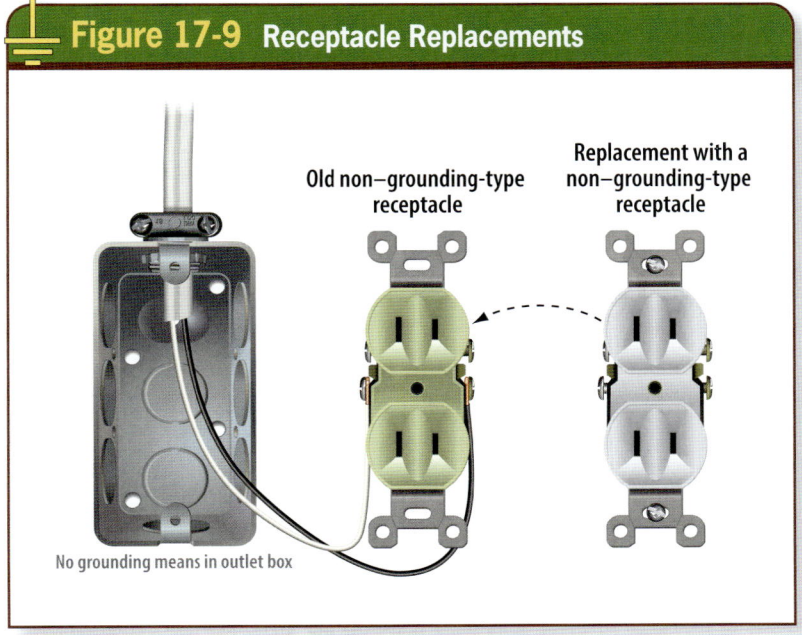

Figure 17-9   Receptacle Replacements

Old non–grounding-type receptacle

Replacement with a non–grounding-type receptacle

No grounding means in outlet box

Figure 17-9. A non–grounding-type receptacle is permitted to be replaced with another non–grounding-type receptacle.

side of the GFCI receptacle device. **See Figure 17-10.**

3. A non–grounding-type receptacle can be replaced with a grounding-type receptacle only where it is

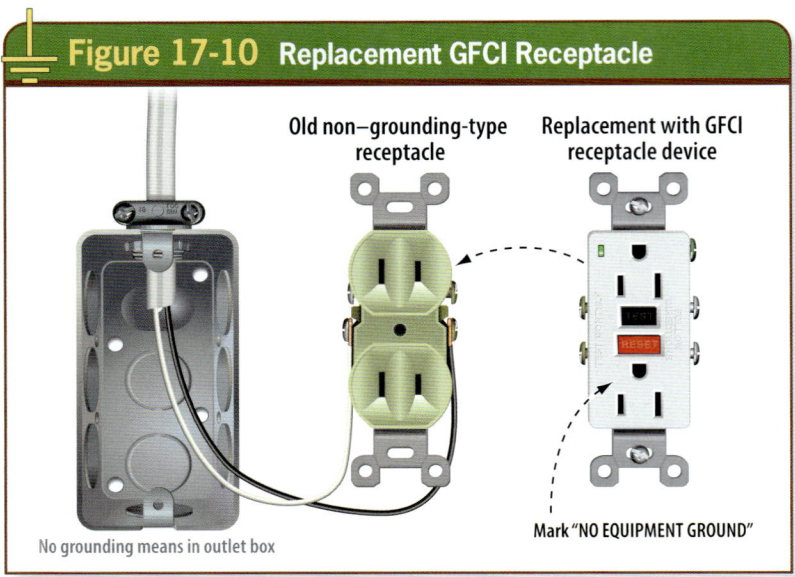

Figure 17-10   Replacement GFCI Receptacle

Old non–grounding-type receptacle

Replacement with GFCI receptacle device

No grounding means in outlet box

Mark "NO EQUIPMENT GROUND"

Figure 17-10. A non–grounding-type receptacle is permitted to be replaced with a GFCI receptacle (GFCI receptacle equipped with pilot light).

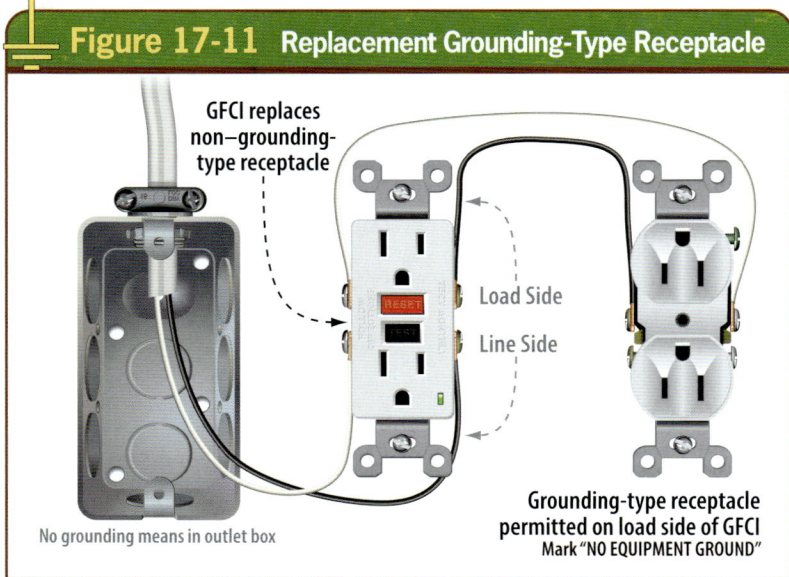

Figure 17-11   Replacement Grounding-Type Receptacle

GFCI replaces non–grounding-type receptacle

Load Side

Line Side

No grounding means in outlet box

Grounding-type receptacle permitted on load side of GFCI
Mark "NO EQUIPMENT GROUND"

Figure 17-11. A non–grounding-type receptacle is permitted to be replaced with a GFCI device that supplies grounding-type receptacles downstream where marked "GFCI Protected" and "No Equipment Ground" (GFCI receptacle equipped with pilot light).

supplied through a GFCI device. Where installed using this alternative, the receptacles supplied through the GFCI device must be marked "GFCI Protected" and "No Equipment Ground." An EGC is not permitted to be installed between the GFCI device and the grounding-type receptacles on the load side of the GFCI device. **See Figure 17-11.**

An exception indicates that in situations where replacement of the receptacle type is impracticable, such as where the outlet box size lacks the capacity to accommodate a GFCI device, the receptacle can be replaced with a new receptacle of the existing type where GFCI protection is provided and the receptacle is marked "GFCI Protected" and "No Equipment Ground," in accordance with 406.4(D)(2).

Temporary Wiring Installations

Temporary wiring installations must have GFCI protection if used by personnel during construction, maintenance, demolition, and similar activities, according to 590.6. Protection is provided by the use of GFCIs or an assured EGC program, under the specific details provided.

Permanent 125-volt; 15-, 20-, and 30-ampere; single-phase receptacle outlets, whether existing or new, must have GFCI protection when used with temporary wiring for the activities previously mentioned. GFCI devices listed and identified for portable use, such as cord sets, are acceptable for providing this protection. **See Figure 17-12.**

An exception relaxes this requirement for industrial establishments where conditions of maintenance and supervision ensure that qualified persons service the installation and interruption by a GFCI would present a greater hazard for persons, and where an assured EGC program is implemented according to 590.6(B)(3).

Temporary 125-volt; 15-, 20-, and 30-ampere; single-phase receptacle outlets must have GFCI protection. Other receptacles (different voltages and ampere ratings) must also have GFCI protection, or the assured EGC program must be used. **See Figure 17-13.**

All 125/250-volt and 125-volt; 15-, 20-, and 30-ampere; single-phase receptacles that are part of portable generators, 15 kilowatts or smaller, must have GFCI protection. All 125- and 250-volt, 15- and 20-ampere receptacles, with or without a generator, used in damp or wet locations must have GFCI protection unless the assured equipment grounding program can be used. GFCI devices listed and identified for portable use, such as cord sets, are acceptable for providing this protection with generators manufactured before January 1, 2011.

Assured EGC Program

The *NEC* allows an assured EGC program as an alternative to the requirement for providing GFCI protection, but only for limited situations. The requirements of an assured EGC program are restrictive. These programs must be in writing and continuously enforced at the site by one or more designated persons. Their responsibilities are to ensure that all cord sets and receptacles not part of the permanent building wiring, in addition to all equipment connected by cord and plug, are installed and maintained according to the provisions of 250.114, 250.138, 406.4(C), and 590.4(D).

The testing requirements for cord sets and receptacles are as follows:

1. Test all EGC continuity of cord sets and cord- and plug-connected equipment.
2. Test all receptacles to verify EGC continuity and that the EGC is connected to the appropriate terminal of the device.
3. Testing must be performed before the first use on site, when there is evidence of damage, before returning equipment to service after a repair, and at intervals not exceeding three months.

The testing must be documented and available to the applicable authority having jurisdiction.

Figure 17-12 Portable GFCI Protection

Figure 17-12. Portable GFCI protection is a common tool that can be carried by persons and used as an added level of shock protection at the point of use.

Courtesy of Jim Dollard, IBEW Local 98

Figure 17-13 Temporary Receptacles

Figure 17-13. GFCI protection is required for temporary receptacles installed and used on construction sites in accordance with NEC 590.6.

Courtesy of Bill McGovern, City of Plano, TX

Special Purpose Ground-Fault Circuit Interrupters (SPGFCI)

A definition for *special purpose ground-fault circuit interrupters* has been added to Article 100 for the 2023 *NEC.*

> **Ground-Fault Circuit Interrupter, Special Purpose (SPGFCI). (Special Purpose Ground-Fault Circuit Interrupter).** A device intended for the detection of ground-fault currents, used in circuits with voltage to ground greater that 150 volts, that functions to de-energize a circuit or portion of a circuit within an established period of time when a ground-fault current exceeds the values established for Class C, D, or E devices.(CMP-2)

There are three different classifications of SPGFCI: Class C, D, or E. Class C SPGFCIs are intended for use in circuits with no conductor operating at more than 300 volts to ground where equipment grounding

or double insulation is provided. Class D SPGFCIs are used in circuits with one or more conductors at more than 300 volts to ground and where a specially sized, reliable, low-impedance grounding path ensures that the voltage across the body during a fault does not exceed 150 volts. Class E SPGFCIs are similar to Class D, with the addition of high-speed tripping that eliminates the need for the oversized ground used for Class D. All three classes of SPGFCIs trip at 20 milliamperes rather than 6 milliamperes for a Class A device.

Section 410.184 allows, by exception, the use of SPGFCIs for protection of horticulture lighting when the circuit exceeds 150 volts to ground. For swimming pools, fountains, and similar locations, 680.5 requires the use of SPGFCIs for circuits of more than 150 volts to ground, and not exceeding 480 volts phase-to-phase. A receptacle of 60 amperes or less, rated 125 volts to 250 volts, is permitted by 680.22(A)(4) to have GFCI or SPGFCI protection when installed within 20 feet of the inside walls of a pool. A parallel requirement for receptacles installed near storable pools is located in 680.32. Luminaires, lighting outlets, and ceiling-suspended paddle fans within a zone of five to ten feet horizontally from the inside walls of a pool are required to have GFCI or SPGFCI protection as required in 680.22(B)(4). The outlet that supplies a self-contained spa or hot tub is required to have either GFCI protection or SPGFCI protection as required in 680.44(A). SPGFCI requirements for fountains are located in 680.58 and 680.59, and they are similar to pools in terms of requirements for receptacles and equipment.

GROUND-FAULT PROTECTION OF EQUIPMENT (GFPE)

Ground-fault protection of equipment differs from ground-fault circuit interrupter protection, as it protects larger equipment.

Figure 17-14   Equipment Damage

Figure 17-14. *Equipment can be destroyed or severely damaged as a result of an arcing fault.*

Courtesy of Bill McGovern, City of Plano, TX

Figure 17-15 Arcing Ground Fault Event Aftermath

Figure 17-15. *An arcing ground fault event can quickly develop into a phase-to-phase short circuit and cause rapid destruction of electrical equipment that is not equipped with ground fault protection.*

Courtesy of Bill McGovern, City of Plano, TX

Purpose of GFPE

Ground-fault protection of equipment (GFPE) protects large equipment from devastating arcing events and destructive burn-downs. **See Figure 17-14.**

GFPE requirements are provided in 210.13, 215.10, 230.95, 240.13, 517.17, and so forth. The *NEC* defines the term *Ground-Fault Protection of Equipment.* **See Figure 17-15.**

> **Ground-Fault Protection of Equipment (GFPE).** A system intended to provide protection of equipment from damaging line-to-ground fault currents by operating to cause a disconnecting means to open all ungrounded conductors of the faulted circuit. This protection is provided at current levels less than those required to protect conductors from damage through the operation of a supply circuit overcurrent device. (CMP-5)

Electric arcs generate significant amounts of heat, and in a circuit of 277 volts to ground, an arcing fault is readily sustained. A ground fault is typically not a solid or "bolted fault" condition, so dynamic arcing impedance is introduced in the circuit. This reduces the fault current seen by a standard overcurrent protective device and increases the time the fault can exist, which allows arcing faults to manifest into destructive events. During an arc event, ionized gas is dispersed, creating a conductive gas or plasma in the atmosphere surrounding the busbars within the equipment. This condition often rapidly escalates from a phase-to-ground fault event to a phase-to-phase short circuit condition. This is why 230.95 requires ground-fault protection of equipment.

GFPE is generally required for solidly-grounded wye services and feeders of more than 150 volts to ground but not exceeding 1,000 volts phase-to-phase for each disconnect rated at or above 1,000 amperes. GFPE is required for nominal 480Y/277-volt, 3-phase, 4-wire, wye-connected systems. The maximum settings are 1,200 amperes and not longer than one

ThinkSafe!

During an arcing ground-fault event, the air becomes ionized, creating a conductive plasma effect.

Figure 17-16   GFPE Requirements

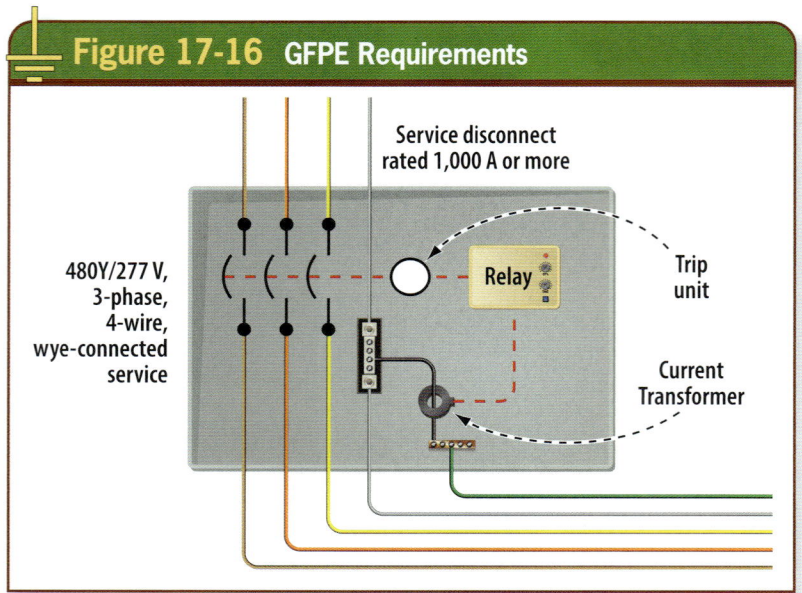

480Y/277 V, 3-phase, 4-wire, wye-connected service

Service disconnect rated 1,000 A or more

Relay

Trip unit

Current Transformer

Figure 17-16. GFPE is required for service disconnects rated 1,000 amperes or more that have an applied nominal voltage of more than 150 volts to ground not exceeding 600 volts phase-to-phase.

second for fault currents 3,000 amperes or more. As indicated in 210.13, GFPE is also required for large branch circuits of more than 150 volts to ground but not exceeding 600 volts phase-to-phase for each disconnect

Figure 17-17   GFPE Systems

Figure 17-17. A GFPE system can be in the form of a circuit breaker that includes internal GFPE sensing and relaying devices.

rated at or above 1,000 amperes. It is important to note that GFPE is not permitted for fire pumps or in systems where a non-orderly shutdown or interruption would introduce additional hazards. **See Figure 17-16.**

Types of GFP Equipment

Two popular types of ground-fault protection (GFP) equipment are *ground-strap type* GFP equipment and *zero-sequence* (residual) GFP equipment. The main bonding jumper is implied by the term *ground strap* or *neutral ground strap*. Both types provide protection from load-side ground faults. Ground-fault protection installed in service equipment does not provide protection on the line side of the GFP system in the equipment. A line-side ground-fault event is not detected by the GFP sensors, and equipment can be severely damaged or destroyed by the ground-fault event.

Neutral Ground-Strap System

Ground-fault protection of equipment typically includes a current transformer (CT), a control relay, and a shunt-trip circuit breaker or fused switch. In this type of system, the main bonding jumper passes through a sensing window in the CT. **See Figure 17-17.** This type of protection is permitted for use in service equipment and typically would not be installed on the load side of the service due to the restrictions of load-side neutral-to-ground connections as provided in 250.24(B) and 250.142(B).

The ground-strap type system functions by sensing a high amount of fault current passing through the main bonding jumper during a ground fault. At predetermined values of current and time, the ground fault protection system sends a signal through a relay that activates the shunt trip mechanism to rapidly open the disconnect or circuit breaker. The pickup ampere setting cannot exceed 1,200 amperes, and the maximum time permitted is not to exceed one second. Though it sounds brief, this is a significant amount of time (60 cycles).

A diagram of this type of GFPE clearly shows the complete circuit from the utility transformer to the service equipment. **See Figure 17-18.** Follow the neutral from the utility transformer to the service disconnect and locate the main bonding jumper. This is the effective ground-fault current path from the service to the utility source. Any ground-fault current will be present in the neutral conductor for the duration of time it takes the GFPE to open the circuit. All of the fault current will pass through the current transformer and around the main bonding jumper in the service equipment.

As required, grounding electrodes are installed and connected at the utility transformer and the service equipment. Even though the Earth is included in this circuit, only a minimum amount of current will be present in this path because of the high impedance in the Earth. During a ground-fault event, current will divide over any and all paths that are present and common to the power

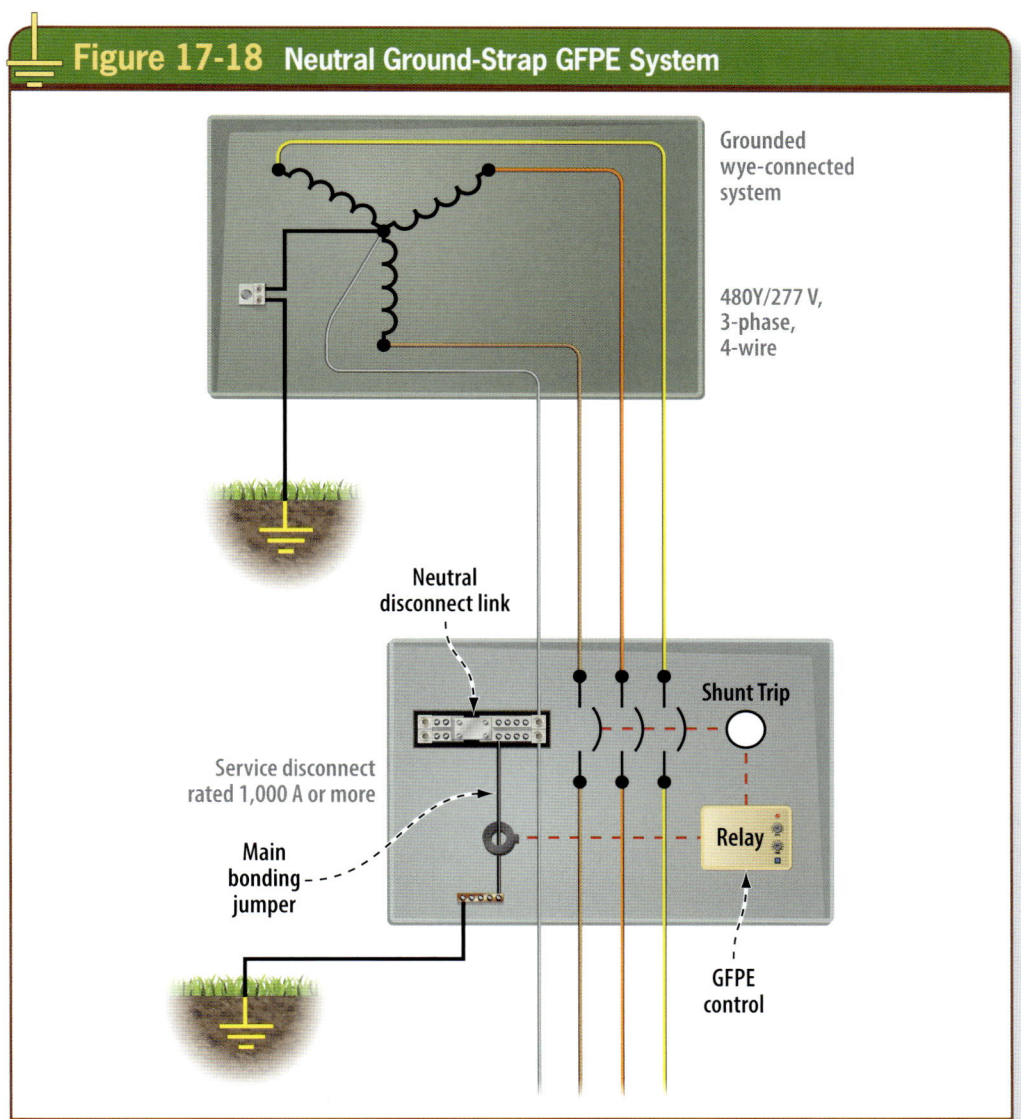

Figure 17-18 Neutral Ground-Strap GFPE System

Figure 17-18. A neutral ground-strap GFPE system includes a current transformer (CT) around the main bonding jumper.

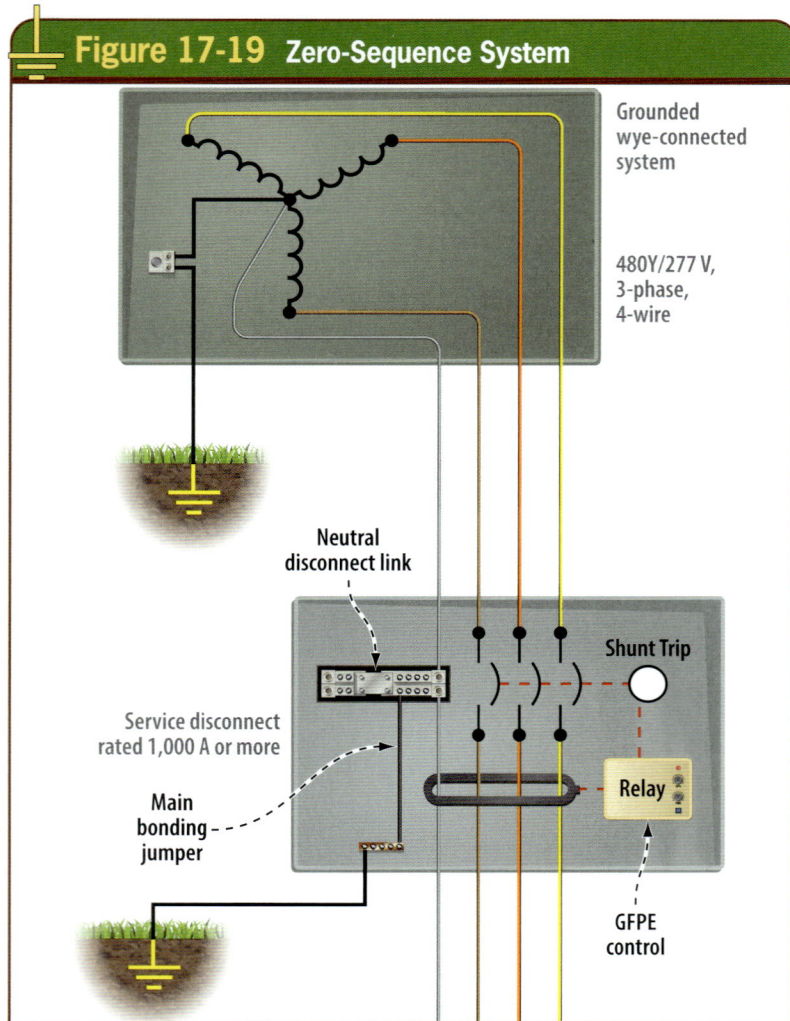

Figure 17-19 Zero-Sequence System

Grounded
wye-connected
system

480Y/277 V,
3-phase,
4-wire

Neutral
disconnect link

Shunt Trip

Service disconnect
rated 1,000 A or more

Relay

Main
bonding
jumper

GFPE
control

Figure 17-19. GFPE can be provided using a basic zero-sequence system.

Figure 17-20 Current Transformer

Figure 17-20. In a zero-sequence GFPE system, all ungrounded phase conductors and the neutral conductor pass through the zero-sequence-sensing current transformer.

source. The amount of current in each path is limited by the impedance in that particular path. The inductive reactance and high impedance of those other fault current paths results in most of the fault current being carried through the main bonding jumper at amounts far greater than the current on other paths that are separate from the service-entrance conductors.

Zero-Sequence Systems

Another type of GFPE system is the zero-sequence type, sometimes referred to as a residual GFPE system. The term *residual* is used because this type of equipment sums up all current in the phases and grounded conductor, and any excess current is residual or left over. This type of system is equipped with a control relay, sensing current transformer(s), and shunt-trip circuit breaker or disconnecting means. **See Figure 17-19.**

In this type of GFPE system, all ungrounded phase conductors and the neutral conductor are routed through a single current transformer. **See Figure 17-20.**

The main bonding jumper and equipment grounding conductor do not pass through the CT. **See Figure 17-21.**

Zero-sequence types are also available with a separately-installed CT around the neutral bus. The difference between the two types is that in the latter, the neutral CT is optional. **See Figure 17-22.**

This type of GFPE can be used on 3-phase systems or on 3-phase, 4-wire, wye-connected systems. Many zero-sequence GFPE systems have the CTs built into the circuit breaker, and the external CT is installed in the equipment around the neutral bus. In the zero-sequence system, the vector sum of normal current in the system ungrounded conductors and neutral conductor is around zero due to the canceling effect between the conductors of a 3-phase, 4-wire, wye-connected system. When a ground-fault

Figure 17-21 Zero-Sequence System Installation

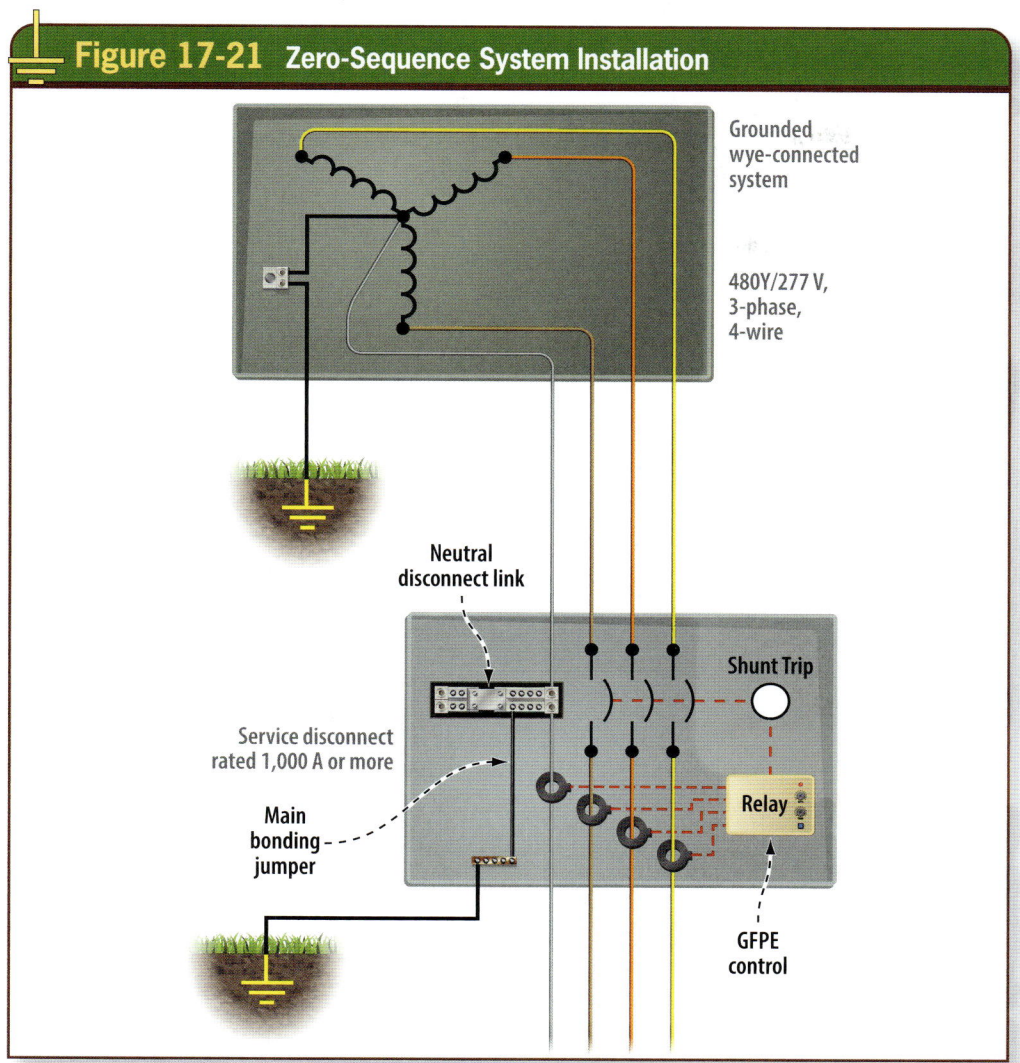

Figure 17-21. *GFPE can be provided using a basic zero-sequence system with an optional neutral sensing window.*

event occurs, the fault current path is through the main bonding jumper and other fault current paths outside of the CT. This imbalance activates the GFPE relay and shunt-trip mechanism to open the circuit breaker or switch. The output of the current transformer is proportional to the level of fault current passing through, and the ground-fault relay is field-adjustable to a maximum of 1,200 amperes. The time-delay settings of GFPE are usually at durations less than one second (typically from 2 to 35 cycles). Keep in mind that the maximum ampere setting permitted in the *NEC* is 1,200 amperes and the

Figure 17-22 Zero-Sequence GFPE

Figure 17-22. *GFPE is commonly provided using a zero-sequence GFPE (breaker type).*

Courtesy of IBEW Local 26 Training Center

Figure 17-23 Circuit Breaker-Type GFP

Figure 17-23. GFPE can include a zero-sequence current transformer for circuit breaker-type GFP.

Courtesy of IBEW Local 26 Training Center

Figure 17-24 Selective Coordination

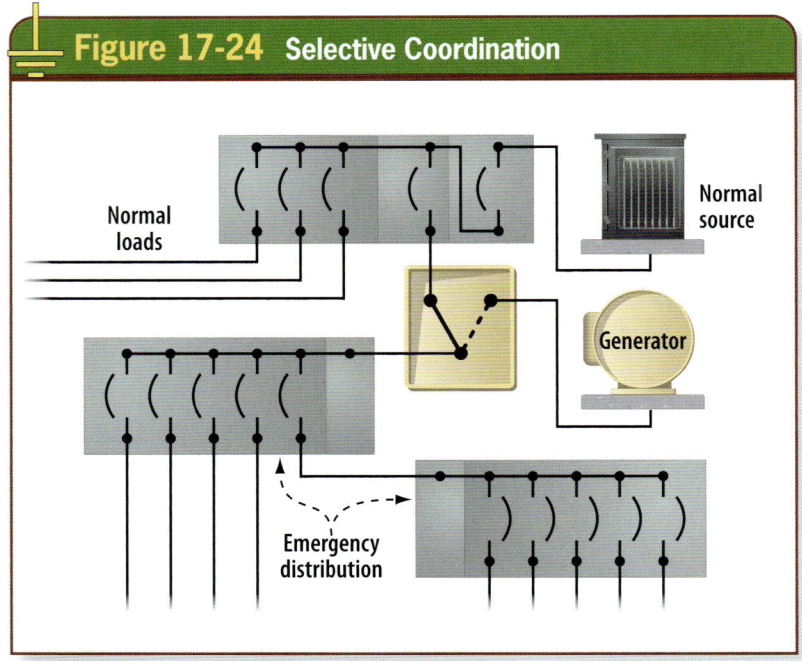

Normal loads

Normal source

Generator

Emergency distribution

Figure 17-24. Selective coordination is required for overcurrent protective devices used in emergency systems in accordance with NEC 700.27.

maximum time is one second. **See Figure 17-23.**

Selective Coordination

Coordinating the trip sequence of overcurrent and GFPE devices in power distribution systems is necessary where selective coordination is required by the *NEC*. Selective coordination can be accomplished by various combinations of overcurrent and GFPE devices that are carefully applied and engineered into the power distribution system. Selective coordination can be accomplished using circuit breakers, fuses, or combinations of fuses and circuit breakers. The last of these three methods mentioned is probably the most common in today's electrical systems. Where GFPE systems are installed, they should also be designed such that they are selective to the point where the offending ground-fault event opens only the closest upstream device from the fault event.

A definition of the term *selective coordination* is included in Article 100 of the *NEC* and is applicable for selective coordination of overcurrent protective devices for elevators in 620.62, emergency systems in 700.32, legally required standby systems in 701.32, and critical operations power systems in 708.54.

Coordination, Selective (Selective Coordination). Localization of an overcurrent condition to restrict outages to the circuit or equipment affected, accomplished by the selection and installation of overcurrent protective devices and their ratings or settings for the full range of available overcurrent, from overload to the maximum available fault current, and for the full range of overcurrent protective device opening times associated with those overcurrents. (CMP-10)

Where overcurrent protective devices and GFPE are selectively coordinated, they provide the benefits of restricting outages to the circuit or equipment closest to the ground-fault or short-circuit event by operating the local overcurrent or GFPE device, rather than causing the entire system to suffer a failure. **See Figure 17-24.**

The requirements for selective coordination in 700.32 apply to all emergency systems. Article 517 includes specific rules for essential electrical systems in health care facilities. It is important to remember that the essential electrical system in a hospital includes the life safety branch, the critical branch, and the equipment branch.

Section 517.26 establishes a correlation between Article 517 and Article 700, and specifically omits requirements for selective coordination such as 700.32. Section 517.31(G) requires overcurrent protective devices serving the essential electrical system to be coordinated for a period of time that a fault exceeds 0.01 seconds. Note that this type of overcurrent protective device coordination is not considered selective coordination. Additional information about selective coordination is provided in **Annex C**.

GFPE System Coordination

System coordination is often desired and necessary to avoid power outages and interruption of service. Coordination of power in distribution systems can be easily accomplished using selectively coordinated standard overcurrent protective devices or ground fault protection of equipment.

The point of a ground-fault in any system is never known, so system coordination is all about anticipating a ground-fault event and strategically placing GFPE at each level desired in the system. Applying GFP equipment in multiple levels of feeders on the load side of the service GFP device can effectively localize ground faults and simultaneously provide power continuity. A coordinated system of ground fault protection in cascading feeder levels affords the ability to isolate an offending fault to one location or feeder level. Coordination often incorporates zone-selective interlocks or differential relay settings. Zone-selective interlocking involves installing signaling circuit wiring between each level of GFPE.

GFPE operates well below the normal ratings of the overcurrent protection feature in the same device. GFPE systems are affected only by ground-fault currents; they do not interfere with normal operation of the overcurrent protective device. Where necessary or desired, the tripping function of the service GFPE device can be delayed while the feeder GFPE closest to the offending fault is activated. A magnetic current relay initiates the required time delay relay, providing effective system coordination and continuity of service.

GFPE coordination involves analyzing the system characteristics and applying appropriately selected GFP equipment in a way that effectively provides a deliberate separation of the tripping ranges from the lowest-rated to the highest-rated GFPE system device. This type of system provides excellent protection of property for large low-voltage electrical distribution equipment. **See Figure 17-25.**

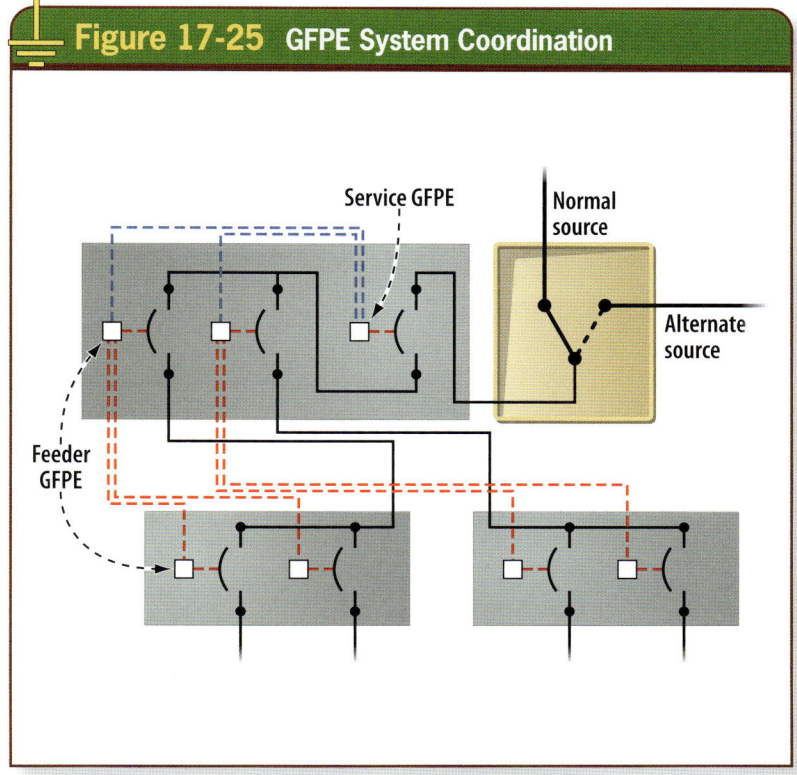

Figure 17-25 GFPE System Coordination

Service GFPE

Normal source

Alternate source

Feeder GFPE

Figure 17-25. *System coordination can be accomplished using zone-interlocking of the GFPE devices at multiple levels.*

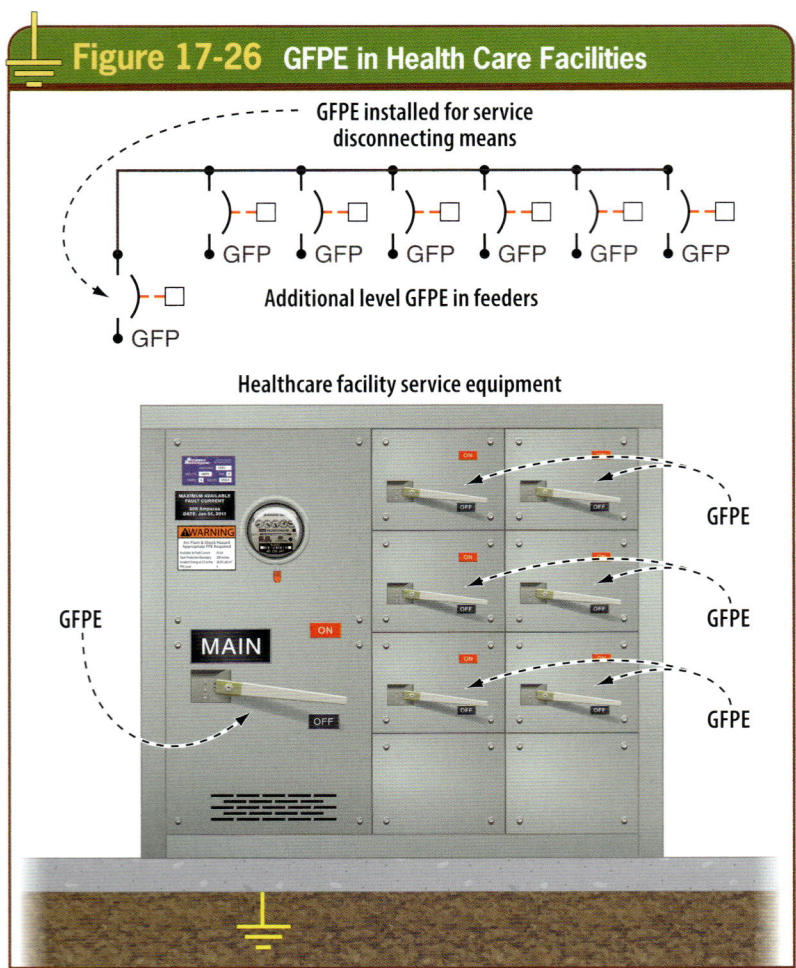

Figure 17-26. GFPE in Health Care Facilities

GFPE installed for service disconnecting means

GFP GFP GFP GFP GFP GFP

Additional level GFPE in feeders

GFP

Healthcare facility service equipment

GFPE

GFPE

GFPE

GFPE

MAIN

Figure 17-26. An additional level of GFPE is required in health care facilities according to 517.17.

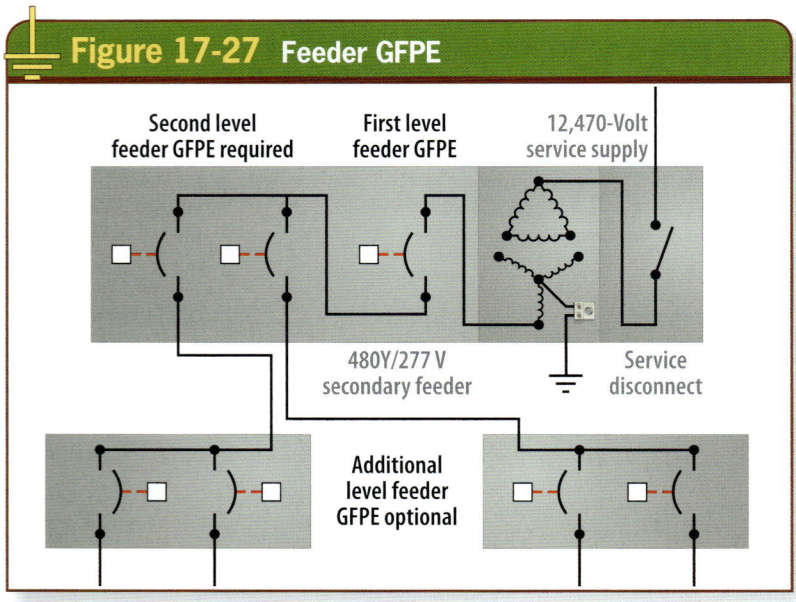

Figure 17-27 Feeder GFPE

Second level feeder GFPE required

First level feeder GFPE

12,470-Volt service supply

480Y/277 V secondary feeder

Service disconnect

Additional level feeder GFPE optional

Figure 17-27. GFPE is required for feeders in accordance with 215.10.

Applicability in Health Care Facilities

Special rules for GFPE of electrical systems apply to health care facilities. Section 517.17(B) indicates that when GFPE is provided at the service or feeder disconnecting means, as specified by 230.95 or 215.10, an additional level of GFPE must be installed in the next level of feeder disconnecting means downstream toward the load, and they must be selectively coordinated. This is logical and consistent with the objectives of *NFPA 99: Health Care Facilities Code*, which are concerned with continuity of electrical power in health care facilities, especially where critical care and life support are essential for patients. **See Figure 17-26.**

The primary purpose for this is to ensure that a ground-fault event in the electrical system does not open the service GFPE disconnecting means but instead opens the device closest to the fault, thus isolating the offending circuit while maintaining continuity of power to the rest of the facility.

According to 517.17, additional levels of GFPE are not permitted to be installed on the load side of an essential electrical system transfer switch. This feeder protection is required to be 100% selective so that if a ground fault occurs downstream from the feeder overcurrent protective device, only the closest overcurrent protective device will open and the upstream devices will remain closed.

Each level of the GFP system must be performance tested when first installed. The GFPE performance testing must be conducted by a qualified person(s) and must incorporate primary current injection during this process. It is critical to GFPE system functionality that the grounded (neutral) conductor be completely isolated from any grounding connections downstream from the service. This is a general requirement of the *NEC* in 250.24(B) and is a specific requirement of manufacturer's installation and testing instructions for GFPE.

This is especially important where transfer switches and alternate power sources are installed in the premises wiring system. In most cases, 4-pole transfer switches are installed on 3-phase, 4-wire systems to allow the system to function properly without grounding connections to the neutral or grounded conductor downstream from the service disconnecting means, which includes GFPE for that particular separately derived system.

Feeder GFPE
Section 215.10 requires ground-fault protection of equipment for feeders of the same voltage and current ratings that qualify services for GFPE. An example is where the service to a property is provided at 12,470 volts. In this case, the electrical system is delivered at more than 600 volts, and the service disconnecting means is at that voltage level. A transformer steps the voltage down to 480Y/277, creating the need for ground-fault protection of equipment in the feeder position.

The exceptions that are provided in 230.95 for services are also provided in 215.10 for feeders. Each separately derived system with feeders at voltages and configurations requiring GFPE must be protected if it qualifies because of the ampacity level of the feeder and equipment supplied. Remember that ground-fault protection of equipment is generally required for each disconnecting means rated at 1,000 amperes or greater installed on a 480Y/277-volt system. **See Figure 17-27.**

Branch Circuit GFPE
Section 210.13 requires equipment ground-fault protection for branch circuits of the same voltage and current ratings that qualify services and feeders for GFP equipment. GFPE is also required for large branch circuits of more than 150 volts to ground but not exceeding 600 volts phase-to-phase for each disconnecting means rated at or above 1,000 amperes. An example of a large branch circuit is a single unit of

industrial utilization equipment such as a furnace or motor.

Testing of GFPE Systems
GFPE devices must be performance tested when first installed on site to ensure proper operation. Testing GFPE verifies that the system will interrupt a ground-fault event at selected current pick-up and time settings. The performance testing must be performed by qualified persons and in accordance with the manufacturer's instructions and must include primary current injection.

Section 230.75 provides the requirement for a means to disconnect the grounded conductor (usually a neutral) within the service equipment for testing purposes. Although the term *neutral disconnect link* is not used in the *NEC*, this term is used by manufacturers of equipment that is suitable for use as service equipment. The link can be in the form of a busbar or, in smaller equipment, can be the terminal to which the neutral conductor is connected. Once the link is removed, a test can verify that the neutral is isolated from grounding connections on the load side of the service disconnect. The test records must be made available to the authority having jurisdiction. **See Figure 17-28.**

Figure 17-28 Neutral Disconnect

Figure 17-28. A neutral disconnect means (link) is required by 230.75 and is typically used for testing associated with GFPE.
Courtesy of IBEW Local 26 Training Center

Fact
Branch circuits are the conductors from the final overcurrent protective device to the outlet where the equipment is connected.

SUMMARY

Two types of ground fault protection addressed in the *NEC* are ground-fault circuit interrupter (GFCI) protection and ground-fault protection of equipment (GFPE). Common traits of these two types of ground-fault protection are that they require a ground fault to operate, they interrupt current when they operate, and they open the faulted circuit within a prescribed time frame. GFCIs protect people from serious shock and electrocution, while GFPE protects electrical equipment from arcing burn-downs and destruction. The principles of operation of GFCI and GFPE are similar except for a difference in current and time pickup levels. There are several *NEC* rules that require GFCI protection for persons and require installing GFPE to protect property. There are specific performance testing requirements for GFPE when it is first installed at the site. The test must be performed by qualified person(s), and it must include primary current injection.

REVIEW QUESTIONS

1. **GFCIs and GFPE provide ground-fault protection. What are the common operating characteristics for both of these types of GFP?**

   a. They interrupt current when they operate.
   b. They open the faulted circuit within a prescribed time frame.
   c. They require a ground fault to operate.
   d. All of the above.

2. **A GFCI operates within which of the following current ranges?**

   a. 1 to 5 A
   b. 4 to 6 mA
   c. 15 to 20 A
   d. 1,000 to 1,200 A

3. **Ground-fault circuit interrupters for dwelling units and structures other than dwelling units are required to be readily accessible.**

   a. True
   b. False

4. **GFCIs provide protection for people from electrocution and minimize shock hazards.**

   a. True
   b. False

5. **The __?__ is a device intended for the protection of personnel that functions to deenergize a circuit or portion thereof within an established period when a current to ground exceeds the values established for a Class A device.**

   a. arc-fault circuit interrupter protective device
   b. GFCI
   c. GFPE
   d. overload protective device

6. **GFCI devices do not limit the magnitude of current; they limit the time the current is present.**

   a. True
   b. False

7. **Which of the following locations in dwelling units does not require GFCI protection for 125-volt, 15- and 20-ampere receptacles?**

   a. Bathrooms
   b. Garages
   c. Kitchens
   d. Living areas

8. **Which of the following locations in structures other than dwelling units does not require GFCI protection for 125-volt, 15- and 20-ampere receptacles?**

   a. Bathrooms

   b. Indoor locations in general use areas

   c. Outdoor locations

   d. Within 6' of a sink

9. **In industrial establishments only, where the conditions of maintenance and supervision ensure that only qualified personnel are involved, an assured EGC program as specified in 590.6(B)(3) is not permitted for receptacle outlets used to supply equipment that would create a greater hazard if power is interrupted or that has a design that is not compatible with GFCI protection, by exception.**

   a. True   b. False

10. **In wet procedure locations of health care facilities where interruption by a GFCI cannot be tolerated, GFCI protection is not permitted to be installed. In these instances, which type of protective equipment is required?**

    a. Ground detectors only

    b. Ground-fault protection of equipment (GFPE)

    c. Isolated power systems in accordance with 517.160

    d. Overload protection

11. **In older installations without an equipment grounding conductor(s) at the outlets, which of the following is acceptable?**

    a. A non–grounding-type receptacle can be replaced with a GFCI receptacle device as long as it is marked "No Equipment Ground" and no EGC is run to any receptacles on the load side of the GFCI receptacle device.

    b. A non–grounding-type receptacle can be replaced with a grounding-type receptacle only where it is supplied through a GFCI device. Where installed using this alternative, the receptacles supplied through the GFCI device have to be marked "GFCI Protected" and "No Equipment Ground." An EGC is not permitted to be installed between the GFCI device and the grounding-type receptacles on the load side of the GFCI device.

    c. A non–grounding-type receptacle can be replaced with another non–grounding-type receptacle or receptacles.

    d. Any of the above is acceptable.

12. **On construction sites, GFCI protection is required for 125-volt, single-phase receptacles that are not part of the permanent wiring of a building or structure. This GFCI protection is not applicable to __?__ receptacle outlets.**

    a. 15A

    b. 20A

    c. 30A

    d. 50A

REVIEW QUESTIONS

13. Ground-fault protection of equipment is required to protect large equipment from arcing burn-downs and destruction caused by phase-to-ground faults.

    a. True   b. False

14. GFPE is generally required for solidly grounded wye electrical services and feeders of more than 150 volts to ground but not exceeding 1,000 volts phase-to-phase for each service disconnect rated __?__ or more.

    a. 400 A
    b. 800 A
    c. 1,000 A
    d. 1,200 A

15. GFPE is required for fire pumps or continuous industrial processes where a non-orderly shutdown would introduce additional or increased hazards, as indicated in the exceptions to 230.95 in the *NEC*.

    a. True   b. False

16. The maximum current setting for service disconnecting means ground-fault protection of equipment is __?__, and the maximum time delay for operation is __?__ for ground-fault currents equal to or greater than 3,000 amperes.

    a. 600 A / 2 seconds
    b. 1,000 A / 5 seconds
    c. 1,200 A / 1 second
    d. 3,000 A / 1 second

17. __?__ is a system intended to provide protection of equipment from damaging line-to-ground fault currents by operating to cause a disconnecting means to open all ungrounded conductors of the faulted circuit. This protection is provided at current levels less than those required to protect conductors from damage through the operation of a supply circuit overcurrent protective device.

    a. Arc-fault circuit interrupter protection
    b. GFCI protection
    c. GFPE
    d. Short-circuit protection

18. The ground-strap type of GFPE consists of a current transformer, a control relay, and usually a shunt-trip circuit breaker. In this type of GFPE system, the __?__ passes through the current transformer for sensing ground-fault current.

    a. EGC
    b. grounded conductor
    c. grounding electrode conductor
    d. main bonding jumper or system bonding jumper

19. Zero-sequence GFPE typically consists of a control relay, a shunt trip circuit breaker, and a current transformer that is placed around all of the circuit conductors, including the grounded (neutral) conductor.

    a. True   b. False

20. __?__ is localization of an overcurrent condition to restrict outages to the circuit or equipment affected, accomplished by the selection and installation of overcurrent protective devices and their ratings or settings for the full range of available overcurrent, from overload to the maximum available fault current, and for the full range of overcurrent protective device opening times associated with those overcurrents.

    a. GFPE
    b. Overload protection
    c. Selective coordination
    d. Short-circuit protection

21. Which of the following systems does not require selective coordination of the overcurrent protective devices?

    a. Emergency systems
    b. Legally required standby systems
    c. Life safety systems in hospitals
    d. Overcurrent protective devices for elevators where more than one driving machine disconnecting means is supplied by the same source

22. Section 517.17(B) indicates when GFPE is provided at the service or feeder disconnecting means, as specified by 230.95 or 215.10, an additional level of GFP is not required to be installed in the next level of feeder disconnecting means downstream toward the load.

    a. True   b. False

23. The additional level of ground-fault protection of equipment required by 517.17(B) is not permitted to be installed in which of the following locations or applications?

    a. Between the on-site generating unit or units described in 517.35(B) and the essential system transfer switch or switches
    b. On electrical systems that are not solidly grounded wye systems with greater than 150 volts to ground but not exceeding 1,000 volts phase-to-phase
    c. On the load side of an essential electrical system transfer switch
    d. All of the above

24. Sections 230.95(C) and 517.17(D) of the *NEC* require GFPE systems to be performance-tested when first installed on site to ensure proper operation.

    a. True   b. False

25. Ground-fault circuit interrupters for other than dwelling units are not required to be readily accessible.

    a. True   b. False

26. GFCI protection is required in dwelling units for all 125-volt through 250-volt receptacles installed in locations specified in 210.8(A)(1) through (12) and supplied by single-phase branch circuits rated 150 volts to ground or less.

    a. True   b. False

Grounding Rules for Medium- and High-Voltage Systems

The system grounding requirements and methods discussed so far have generally been limited to systems of 1,000 volts and less. The grounding requirements for systems and circuits of greater than 1,000 volts are provided in Part X of *NEC* Article 250. There are various methods to accomplish the grounding required for medium- and high-voltage systems. Equipment grounding and specific rules for grounding cable shielding are necessary for medium- and high-voltage installations and systems. There are important requirements for providing an effective ground-fault current path for services supplied by grounded electrical systems operating at over 1,000 volts.

Objectives

» Understand the grounding methods used for medium- and high-voltage systems.

» Determine the requirements for alternating current (AC) substation grounding electrode systems.

» Identify the grounding rules in the *NEC* that apply to medium- and high-voltage systems.

» Understand the purpose of grounding cable shielding.

» Understand the purpose of grounding through surge arresters.

Chapter 18

Table of Contents

Requirements for Grounding Systems 422

Grounding Methods for Systems
Over 1,000 Volts 422

Solidly Grounded Systems 423

    Single-Point Grounding 423

    Multi-Point Grounding 424

Grounding Service-Supplied Alternating
Current Systems 425

Impedance Grounding.............................. 426

Portable or Mobile Equipment Grounding..... 427

Grounding Equipment 428

Substation Grounding Requirements 430

    Equipotential Grounding Grid................. 431

    Metal Structures and Fences 432

Conductor Shielding and Stress Reduction... 433

Grounding Through Surge Arresters............ 435

Engineered Grounding System Designs 437

    NEC Compliance 438

    Engineered Plans and Specifications...... 438

    Site Power Distribution Network............ 438

    Site Grounding and Bonding Plan 439

    Other Connected Sources 440

    Impedance Grounded Systems............. 440

Common Grounding and
Bonding Components 440

Summary.. 441

Review Questions 442

REQUIREMENTS FOR GROUNDING SYSTEMS

Grounding methods and requirements for systems operating at more than 1,000 volts, such as 5- and 15-kilovolt (kV) systems, differ slightly from those for systems of 1,000 volts or less. A typical 5-kilovolt system is one with a phase-to-phase voltage of 4,160 volts. The phase-to-ground voltage is approximately 2,400 volts. An example of a 15-kilovolt system is one with a phase-to-phase voltage of 12,470 volts. The phase-to-ground voltage is approximately 7,200 volts.

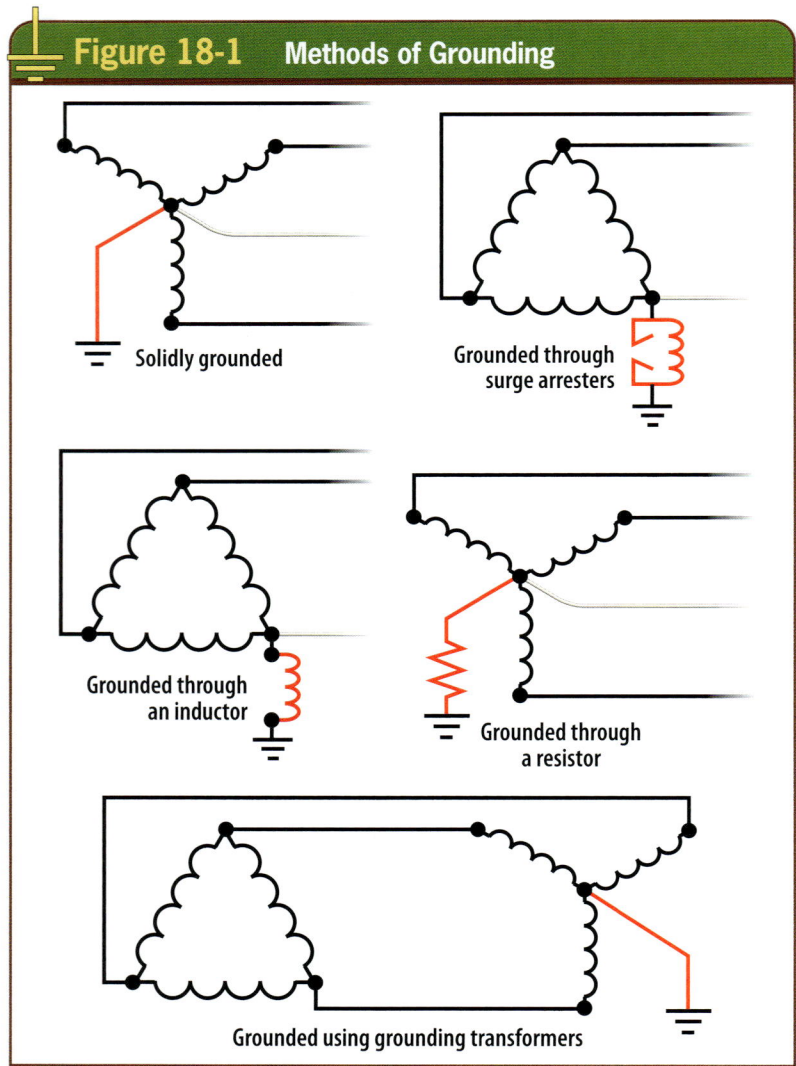

Figure 18-1   Methods of Grounding

Solidly grounded

Grounded through surge arresters

Grounded through an inductor

Grounded through a resistor

Grounded using grounding transformers

Figure 18-1. There are various grounding methods for systems of more than 1,000 volts.

According to ANSI C84.1-2020, low-voltage systems are those 1,000 volts or less. Medium-voltage systems are classed as voltages greater than 1,000 volts and less than 100,000 volts. Electrical systems greater than 100 kV are classed as high-voltage systems. The *NEC* addresses voltages as either 1,000 volts or less and over 1,000 volts. The *NEC* provides several rules related to the grounding of these systems and associated equipment. Part X of Article 250 provides the rules for grounding and bonding systems of more than 1,000 volts. Section 250.180 indicates that if systems over 1,000 volts are grounded, they shall comply with all applicable requirements of Sections 250.1 through 250.178 and with 250.180 through 250.194, which supplement and modify the preceding sections.

The reasons for grounding systems of more than 1,000 volts are the same as the reasons for grounding systems of 1,000 volts or less. These systems are grounded to limit voltages imposed by lightning events, line surges, or unintentional contact with higher-voltage lines and to provide voltage stabilization during normal operation of the system. If systems of more than 1,000 volts are grounded, the requirements in 250.180 through 250.194 must apply accordingly, depending on the type of grounding used for the system. It is important to realize that grounding and bonding provisions in Parts I through IX are only modified or supplemented by Part X of Article 250. Also, remember that 90.3 indicates that Chapters 5, 6, and 7 of the *NEC* can modify or amend any of the requirements within Article 250.

GROUNDING METHODS FOR SYSTEMS OVER 1,000 VOLTS

A few grounding methods are permitted to create a ground reference for such systems. **See Figure 18-1.** Electrical systems greater than 1,000 volts are usually grounded in one of three ways. They can be solidly grounded, grounded through an impedance device, or grounded through a set of grounding transformers that create a reference to ground. Part X of Article 250 provides specific rules for systems grounded at a single point and for systems that are grounded at multiple locations.

SOLIDLY GROUNDED SYSTEMS

A solidly grounded electrical system is one that has a direct electrical connection to ground with no intentional impedance installed between the Earth connection and the system. A commonly grounded system operating at greater than 1,000 volts is a 4,160-volt, 3-phase, 4-wire, wye-connected system. In this system, there is a derived neutral that is the grounded conductor. **See Figure 18-2.**

The requirements for grounding such systems are found in 250.184(A). There, the neutral of such systems is generally required to be an insulated conductor with 600-volt–rated insulation. Bare neutral conductors of such systems are only permitted if the following specific conditions can be satisfied:

1. The bare neutral is installed with the service entrance conductors.
2. The bare neutral is installed with a service lateral or underground service conductor.
3. The bare neutral is installed with the direct-buried portion of a feeder.

The neutral conductor of solidly grounded neutral systems can also be bare when installed as an outdoor overhead conductor. In this case, only the portion installed overhead is permitted to be bare.

Exception No. 3 to 250.184 also permits a bare neutral conductor for solidly grounded neutral systems if the neutral is isolated from the phase conductors and protected from physical damage. See Exceptions 1 through 3 to 250.184.

The neutral conductor of a solidly grounded system must be of sufficient current-carrying capacity for the load served and generally must not be smaller than one-third of the ampacity

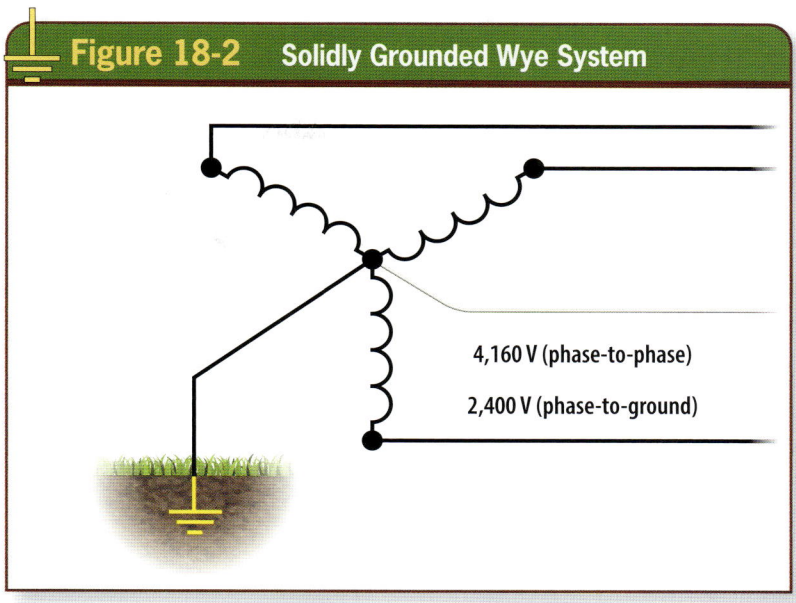

Figure 18-2. Solid grounding is often applied for a 5,000 (4,160/2,400)-volt system.

of the ungrounded phase conductors supplied by the system. By exception, the *NEC* does permit the neutral for these systems to be sized no smaller than 20% of the ungrounded phase conductor ampacity only in commercial and industrial establishments where conditions of engineering supervision are in place. **See Figure 18-3.**

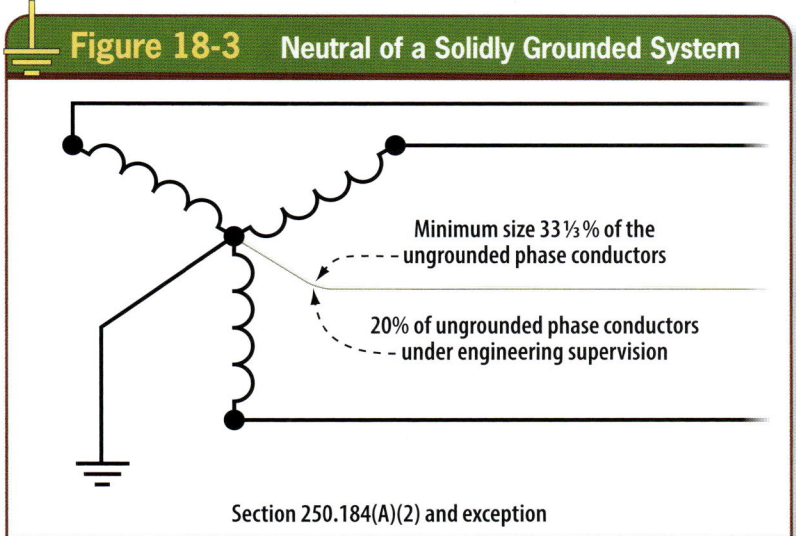

Figure 18-3. The neutral of solidly grounded systems is generally not permitted to be smaller than 33 1/3% of the ungrounded conductor supplied by the system.

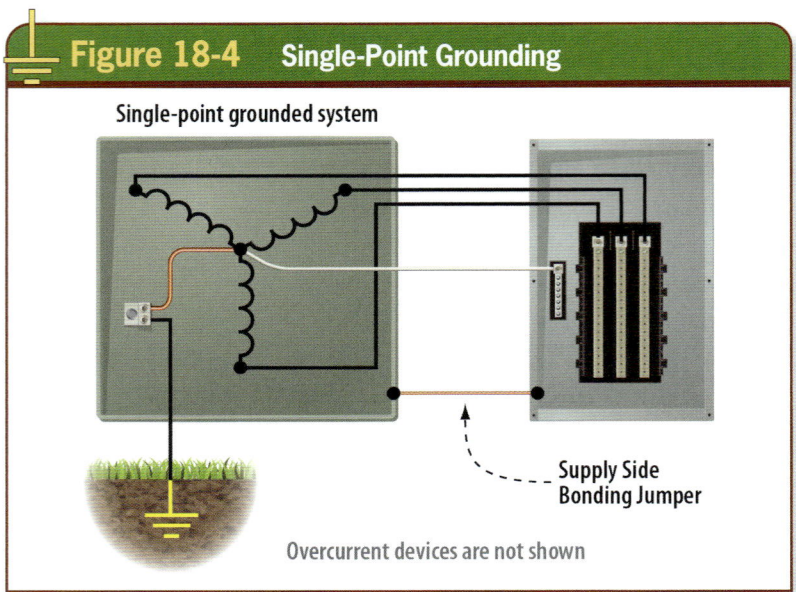

Figure 18-4 Single-Point Grounding

Single-point grounded system

Supply Side
Bonding Jumper

Overcurrent devices are not shown

Figure 18-4. Single-point grounded systems are connected to a grounding electrode, and an EGC must be installed with the circuit supplying equipment.

Single-Point Grounding

Single-point grounding of a system means the system is grounded at only one point, and no neutral-to-ground connections can be made downstream of that initial grounding location. A common aspect of each of the grounded systems at lower voltages is that they are all grounded at one point, unless grounding the neutral downstream is permitted by exception. In a single-point grounded neutral system, the neutral is grounded typically at the source. Then an EGC is run from the single point of grounding of the system, with allowable grounding connections to the Earth from that EGC. **See Figure 18-4.**

Section 250.184(B) indicates that single-point grounded neutral systems can be supplied from a separately derived system or from a source of a multi-grounded neutral system with an equipment grounding conductor connected to the multi-grounded neutral conductor at the source of the single-point grounded neutral system. The connection to the Earth for a single-point grounded neutral system is made through a grounding electrode meeting the applicable requirements in Part III

of Article 250. A grounding electrode conductor is required from the neutral conductor of such systems to the grounding electrode.

The EGC for single-point grounded neutral systems connects the neutral of the system to the grounding electrode, just as it does for other separately derived systems. This bonding jumper is essentially a system bonding jumper, but note the difference in terminology in Part X of Article 250. Functionally, these components of the grounding and bonding system perform in the same way. If feeders are routed to each load, such as equipment or separate buildings or structures, an EGC must be installed. The EGC must be routed with the ungrounded phase conductors of the system and cannot carry any continuous load current. This EGC can be insulated or bare and must have sufficient current-carrying capacity for the maximum fault likely to be imposed on it.

Note that ribbon shielding or metal tape shielding on medium- and high-voltage cables is usually not of sufficient size to serve as an EGC. The shielding serves a different purpose. A neutral conductor is not required to be run with the feeder conductors unless there is a load requiring it. Section 250.184(B) requires the neutral of single-point grounded neutral systems to be isolated from ground except at the grounding point, which is the source location in most cases.

Multi-Point Grounding

The *NEC* also recognizes multi-point grounded (or multi-grounded) neutral systems. The rules for multi-grounded neutral systems are provided in 250.184(C). As the term implies, there are multiple grounding points to the neutral of such systems. In multi-grounded neutral systems, the system neutral is typically derived and grounded at the source and then distributed for long distances. Grounding is required from the neutral at multiple points along its route. **See Figure 18-5.**

Multi-grounded neutral systems are commonly used in installations where the system supplies several buildings or structures, such as in a campus distribution system. Multi-grounded neutral systems are also permitted for use in underground systems where the neutral conductor is exposed, and in systems where the multi-grounded neutral system is installed overhead between poles. The grounding of such a system must be accomplished at each transformer and at additional locations by connection to the Earth through a grounding electrode. Connection to an EGC of the circuit is not permitted.

The neutral conductor of a multi-grounded neutral system must be connected to ground (the Earth) at intervals not exceeding 1,300 feet. This maximum distance between grounding points on these systems is often referred to as the *four grounds per mile* method of grounding.

Where a multi-grounded neutral system employs shielded cables, the cable shielding must be grounded at each cable connection point where the shields are exposed and subject to contact by persons. An exception to Section 250.184(C)(5) indicates that for a multi-point grounded system, a grounding electrode is not required to bond the neutral conductor in an uninterrupted conductor exceeding 400 meters (1,300 feet) if the only purpose for removing the cable jacket is for bonding the neutral conductor to a grounding electrode.

GROUNDING SERVICE SUPPLIED ALTERNATING CURRENT SYSTEMS

Requirements for providing an effective ground-fault current path from the supply source to the service equipment were added to Section 250.186 in the 2014 edition of the *NEC*. Section 250.24(D) requires a grounded conductor to be brought to the service for grounded systems 1,000 volts or below. The main purpose for this conductor is to ensure that a low-impedance path is provided for ground-fault current to return to the utility

supply transformer or source and facilitate overcurrent protective device operation during a ground-fault event. The same need exists for services and systems over 1,000 volts, but before 2014, there were no specific *NEC* requirements.

Some jurisdictions previously relied on the performance requirements of 250.4(A) to ensure that a suitable ground-fault return path was provided for these systems. This section establishes the requirement for a suitable, effective ground-fault current path to be provided for services and systems over 1,000 volts and also accounts for instances where the serving utility may or may not provide a neutral (grounded conductor) with their distribution system. Where a utility does not provide a neutral conductor, there is generally a static line or other ground-fault return path to which the supply-side bonding jumper can be connected, completing the return circuit. This new provision enhances consistency with essential performance requirements in 250.4.

For grounded systems, 250.186(A) requires that a grounded conductor be installed and routed with the ungrounded conductors to each service disconnecting means. This grounded

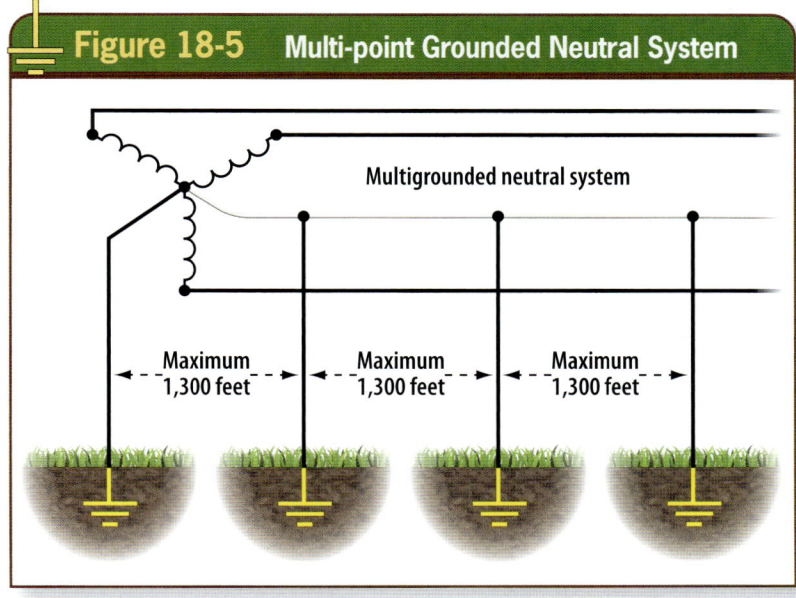

Figure 18-5 Multi-point Grounded Neutral System

Multigrounded neutral system

Maximum 1,300 feet   Maximum 1,300 feet   Maximum 1,300 feet

Figure 18-5. Multigrounded neutral systems of more than 1,000 volts include a system neutral that is grounded in more than one location.

conductor must be connected to the service disconnecting means using a main bonding jumper. The minimum size required for the grounded conductor must not be less than the size of the required grounding electrode conductor and must not be smaller than the sizes specified in Table 250.102(C)(1), but it does not need to be larger than the largest ungrounded service-entrance conductor. If the size of the ungrounded service conductors exceeds the values in Table 250.102(C)(1), use 12.5% for establishing the minimum size, as stated in Note 1 of that Table. Sections 250.186(A)(1) and (2) provide minimum sizing requirements for the grounded conductor when routed in a single raceway or enclosure or when routed in a parallel arrangement to the service equipment.

For ungrounded systems, there is no grounded conductor. Section 250.186(B) requires a supply-side bonding jumper be installed and routed with the ungrounded conductors to each service disconnecting means. This supply-side bonding jumper must be connected to the equipment grounding conductor terminal bus in each service disconnecting means. The minimum size required

for the supply-side bonding jumper must not be less than the size of the required grounding electrode conductor and must not be smaller than the sizes specified in Table 250.102(C)(1), but it shall not be required to be larger than the largest ungrounded service-entrance conductor. If the size of the ungrounded service conductors exceeds the values in Table 250.102(C)(1), use 12.5% for establishing the minimum size, as stated in Note 1 of that Table. Sections 250.186(B)(1) and (2) provide minimum sizing requirements for the supply-side bonding jumper when routed in a single raceway or enclosure or when routed in a parallel arrangement to the service equipment.

IMPEDANCE GROUNDING

Another method of grounding systems of more than 1,000 volts is through an impedance device. The impedance device is typically a resistor or impedance coil inserted in the AC circuit. Impedance grounding means that there is intentional opposition to current inserted between the grounded (neutral) conductor of the system and the grounding electrode conductor. These systems are referred to in the *NEC* as *impedance grounded systems*. The impedance intentionally limits the amount of current that will return to the source during ground-fault conditions, so the overcurrent protective device does not operate, therefore maintaining continuity of service. Impedance grounded systems are installed in an effort to limit current to the source in ground-fault conditions, thereby providing a measure of protection for equipment. By limiting the current through an intentional impedance device, any arcing condition is kept to a lower magnitude.

Special equipment is manufactured for use on these types of installations. Impedance grounded systems are only permitted under controlled conditions. First, certain conditions of maintenance and supervision must be met, ensuring that only qualified persons service these installations. This is an

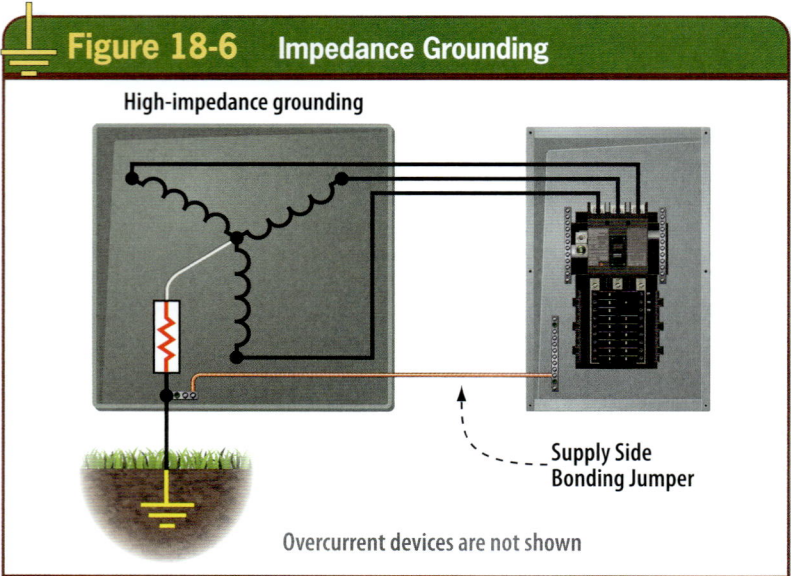

Figure 18-6   Impedance Grounding

High-impedance grounding

Supply Side Bonding Jumper

Overcurrent devices are not shown

Figure 18-6. Impedance grounded systems of more than 1,000 volts are connected to ground (the Earth) through an impedance device.

important requirement, because in a ground-fault condition, qualified persons must understand the proper response and course of action to remove the faulted condition. Second, ground-fault detection systems are installed to notify qualified persons of a first phase-to-ground fault event. The third condition that must be met to allow the use of this type of system is that no line-to-neutral loads are served. See 250.186(1) through (3).

The grounding impedance device for an impedance grounded system must be installed in series with the grounding electrode conductor and the neutral point of the system source, which could be located at a transformer or a generator. The system neutral grounding connection is only permitted to be made through the impedance device. The grounded conductor must be insulated for the maximum neutral voltage. An Informational Note to 250.187(B) indicates that for a 3-phase wye system, the maximum neutral voltage is 57.7% of the phase-to-phase voltage.

The impedance grounding conductor of the system extends from the center point of the wye connection to the line side of the impedance device, and a grounding electrode conductor extends from the load side of the impedance device to the grounding electrode. The grounded conductor of an impedance grounding system does not carry current during normal conditions because impedance grounded systems are not allowed to supply line-to-neutral loads. Therefore, it does not comply with the defined term "neutral conductor," and is now defined instead as an "impedance grounding conductor." It is the conductor that connects the impedance device to the system neutral point, which is the point of grounding for this type of system. Just as with other supply circuits from grounded systems, an EGC is required to be installed. The EGC connection at the source must be made on the load side of the impedance device. **See Figure 18-6.**

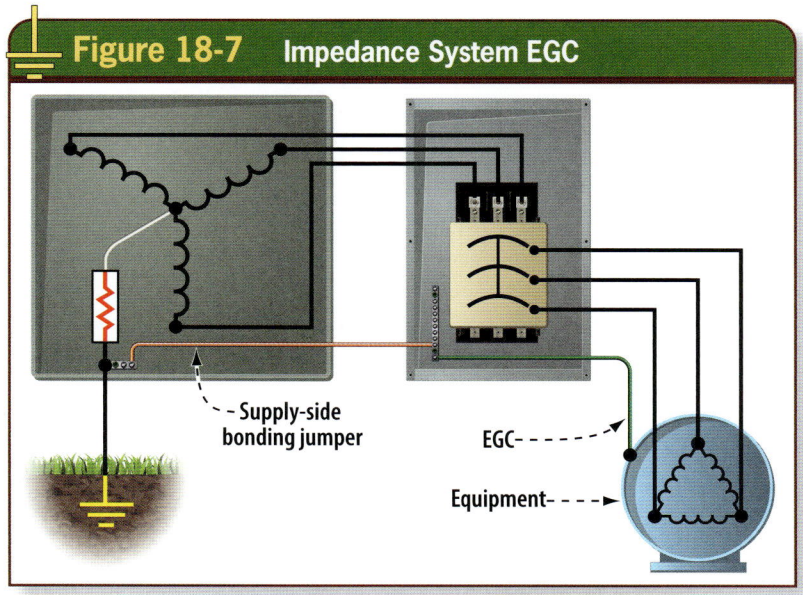

Figure 18-7   Impedance System EGC

- - Supply-side bonding jumper

EGC- - - -

Equipment- - - -

Figure 18-7. *EGCs are required to be installed for impedance grounded systems.*

The EGC can be a bare conductor, or it can be insulated. It must connect to the grounding electrode conductor and to the equipment grounding terminal bus of the source equipment enclosure. The EGC is installed for grounding equipment supplied by the system and serves three important functions: it grounds equipment, it performs bonding, and it serves as an effective ground-fault current path in ground-fault conditions. **See Figure 18-7.**

PORTABLE OR MOBILE EQUIPMENT GROUNDING

An important factor related to grounding a system rated more than 1,000 volts is whether the system supplies equipment that is portable or mobile. Section 250.20(C) indicates that if portable or mobile equipment is supplied by a medium- or high-voltage system (more than 1,000 volts, or 1 kV), the system must be grounded in accordance with 250.188. Sections 250.188(A) through (F) provide the requirements for grounding systems that supply portable or mobile equipment. Portable or mobile equipment must be supplied by a system with a neutral ground through an impedance device unless the system is delta-connected, in which case a grounded neutral must be derived.

This section of the *NEC* requires both system and equipment grounding. Exposed non–current-carrying metal parts of portable or mobile equipment must be grounded by connection to an EGC that is connected to the point at which the system neutral is impedance grounded.

Ground-fault detection and relaying must be provided to automatically deenergize any high-voltage system component that has developed a ground-fault condition. The continuity of the EGC must be continuously monitored, and the supply system must be automatically deenergized upon loss of continuity of the EGC. This rule emphasizes the importance placed on the EGCs of such equipment. The EGC is a safety circuit and must be effective at all times while the system is energized. If the continuity of the EGC is not established, the system supplying the mobile or portable equipment cannot be energized.

The impedance device must be connected to a grounding electrode to establish the ground connection. The grounding electrode for systems supplying portable or mobile equipment must be isolated and separated by at least six meters (20 feet) from any other system or equipment grounding electrode, and there can be no direct connection between the grounding electrodes, such as buried pipe and fence. High-voltage trailing cable and couplers for interconnection of portable or mobile equipment are required to meet the specific requirements of Part III of Article 400 for cables. The cable couplers must also meet the requirements in 490.65.

GROUNDING EQUIPMENT

Grounding of equipment associated with medium- and high-voltage systems is required for fences, enclosures, housings, support structures, and so forth, and for all non–current-carrying metal parts of fixed, portable, or mobile equipment. Note that equipment that is isolated from ground and cannot be contacted by persons in contact with the ground is not required to be grounded. This exception applies to pole-mounted equipment such as transformer and capacitor cases that are elevated.

Grounding is accomplished through a grounding electrode conductor. Section 250.190(B) provides installation requirements for grounding electrode conductors for systems of more than 1,000 volts. The sizing requirements for grounding electrode conductors is based on the use of Table 250.66, using the largest ungrounded service, feeder, or branch circuit conductor supplying the equipment. **See Figure 18-8.** The minimum size required for the grounding electrode conductor is 6 AWG copper or 4 AWG aluminum.

Feeders and branch circuits of more than 1,000 volts often include EGCs. The EGCs must be of sufficient capacity. As previously reviewed, EGCs installed with circuits of more than 1,000 volts cannot be smaller than 6 AWG copper or 4 AWG aluminum unless they are an integral part of a cable assembly. If a cable assembly shield is a concentric type and is suitable for ground-fault current performance, the

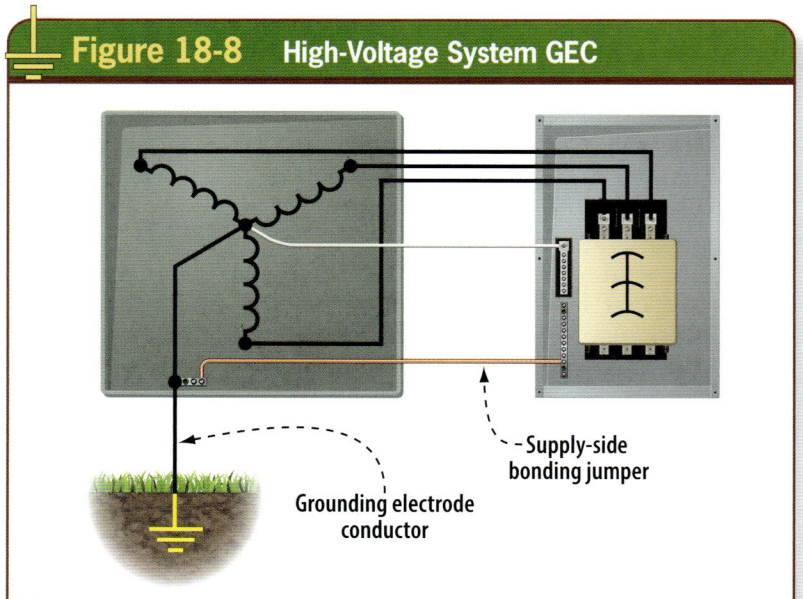

Figure 18-8 **High-Voltage System GEC**

Supply-side bonding jumper

Grounding electrode conductor

Figure 18-8. The grounding electrode conductor for a system over 1,000 volts must generally be sized using Table 250.66.

shield can serve as the required EGC. For solidly grounded systems, a cable ribbon shield or tape shield of the cable assembly is not permitted as an EGC because of its inadequate size. This shielding material is typically under the minimum capacity necessary to perform during ground-fault conditions, as EGCs must be able to provide an effective ground-fault current path to facilitate overcurrent protective device operation. The EGC is sized using Table 250.122 based on the rating of the overcurrent protective device protecting the feeder circuit. **See Figure 18-9.**

As an example, if a pad-mounted transformer is single-point grounded and includes overcurrent protection at the bushing on the output side of the transformer, the rating of the overcurrent protection integral with the bushing establishes the minimum size of the EGC. A 150-ampere bushing results in a 4 AWG copper EGC for this feeder circuit.

The overcurrent rating for a circuit breaker in these types of systems is typically a combination of the current transformer and current pickup setting of a protective relay system in the circuit breaker assembly. Remember that the minimum size of the EGC for systems of 1,000 volts and higher is 6 AWG copper or 4 AWG aluminum if

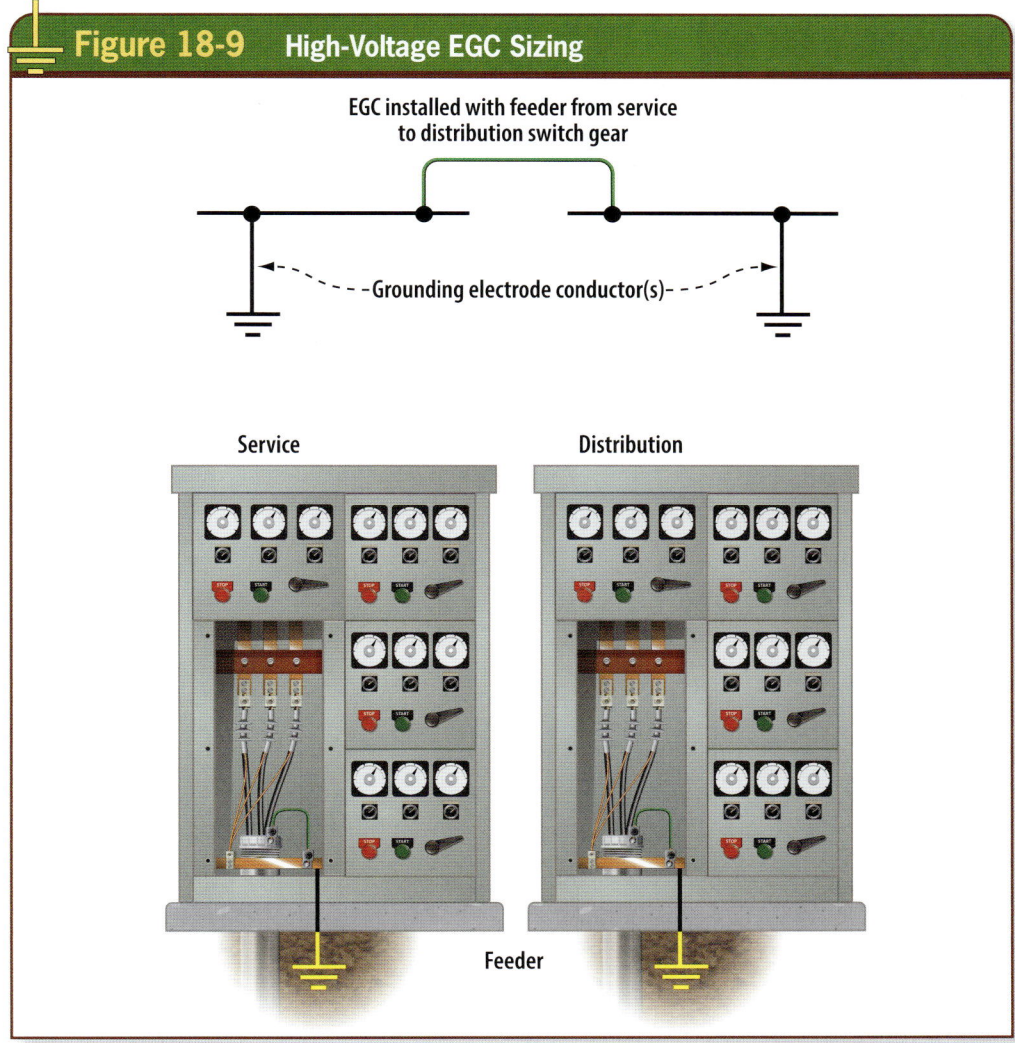

Figure 18-9 High-Voltage EGC Sizing

EGC installed with feeder from service to distribution switch gear

-Grounding electrode conductor(s)-

Service

Distribution

Feeder

Figure 18-9. EGCs with feeders of more than 1,000 volts must generally be sized using Table 250.122.

an EGC is not an integral part of a cable assembly.

Article 495 of the *NEC* provides grounding requirements for equipment rated more than 1,000 volts. The frames of switchgear and control assembly enclosures are generally required to be grounded. This type of equipment usually includes an equipment grounding terminal bar for landing all grounding electrode conductors and EGCs supplied for the equipment.

Equipment rated over 1,000 volts often includes surge protection as an integral part of the assembly; it could also be provided as an accessory feature. Medium- and high-voltage systems are vulnerable to line surges and events that are high in magnitude and can be destructive. Surge arresters are often installed on these systems and provide a level of protection against such events. **See Figure 18-10.**

Figure 18-10 Surge Arresters

Figure 18-10. *Surge arresters are often included in metal-clad switchgear for medium-voltage systems.*

Courtesy of Jim Dollard, IBEW Local 98

Figure 18-11 Substation Grounding

Figure 18-11. *Grounding is required for substation installations.*

SUBSTATION GROUNDING REQUIREMENTS

The *NEC* does not provide many specific details and requirements for substation grounding. Section 250.191 requires a grounding electrode system in accordance with the applicable requirements for grounding electrodes contained in Part III of Article 250. Common grounding electrodes installed for alternating current (AC) outdoor substation grounding are concrete-encased electrodes, ground rings, and ground rods. *IEEE 80, Guide for Safety in AC Substation Grounding*, provides specific information about outdoor AC substation grounding. **See Figure 18-11.**

Article 100, Part II, Over 1000 Volts, Nominal, defines the term *substation* as follows:

Substation. An assemblage of equipment (e.g. switches, interrupting devices, circuit breakers, buses and transformers) through which electric energy is passed for the purpose of distribution, switching, or modifying its characteristics. (CMP-9)

When dealing with substation grounding, both system and equipment grounding are accomplished. Outdoor AC substations typically include a variety of conductive parts and equipment that must be grounded. **See Figure 18-12.**

Equipotential Grounding Grid

One of the primary objectives in the grounding of metal parts at an outdoor substation is to establish an equipotential grid to which all conductive parts can be connected. This grid is installed underground outside of a fenced enclosure, typically about three feet from the fence, and completely encircling the enclosure. There are usually several grounding electrodes (typically ground rods) driven and connected to this ring system. Any underground metal structures, such as piping and framing, should be bonded to this grid. The minimum size of the grid conductor should be no less than 4/0 AWG copper and no less than 25% of the capacity of the system. **See Figure 18-13.**

Figure 18-12   Grounding Substation Installations

Figure 18-12. Grounding for substation installations is typically designed using IEEE 80, Guide for Safety in AC Substation Grounding.

Courtesy of Donald R. Cook, Shelby County, AL

Figure 18-13   Example of Substation Grounding Grid

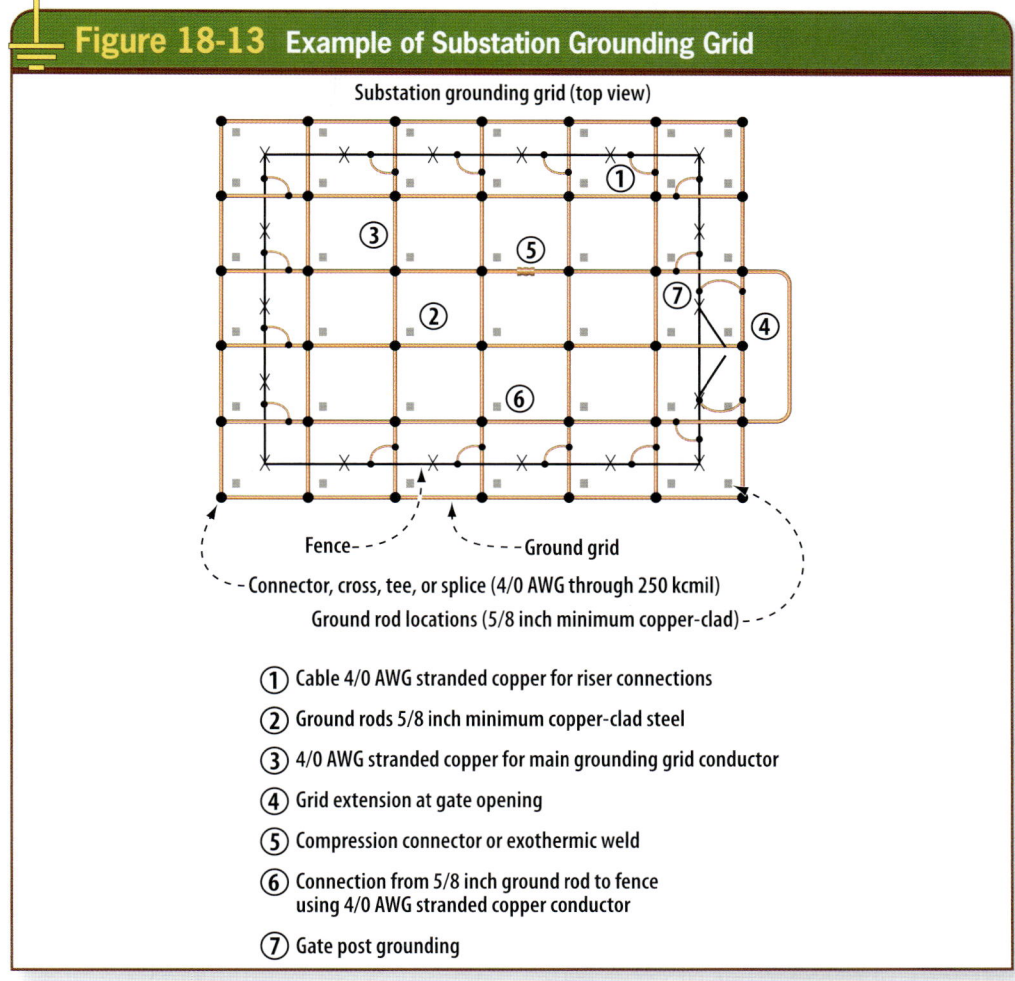

Substation grounding grid (top view)

Fence - - -   - - - Ground grid
- - Connector, cross, tee, or splice (4/0 AWG through 250 kcmil)
Ground rod locations (5/8 inch minimum copper-clad) - -

① Cable 4/0 AWG stranded copper for riser connections

② Ground rods 5/8 inch minimum copper-clad steel

③ 4/0 AWG stranded copper for main grounding grid conductor

④ Grid extension at gate opening

⑤ Compression connector or exothermic weld

⑥ Connection from 5/8 inch ground rod to fence using 4/0 AWG stranded copper conductor

⑦ Gate post grounding

Figure 18-13. Substation fences and other conductive parts must be grounded and bonded to the substation grounding grid system.

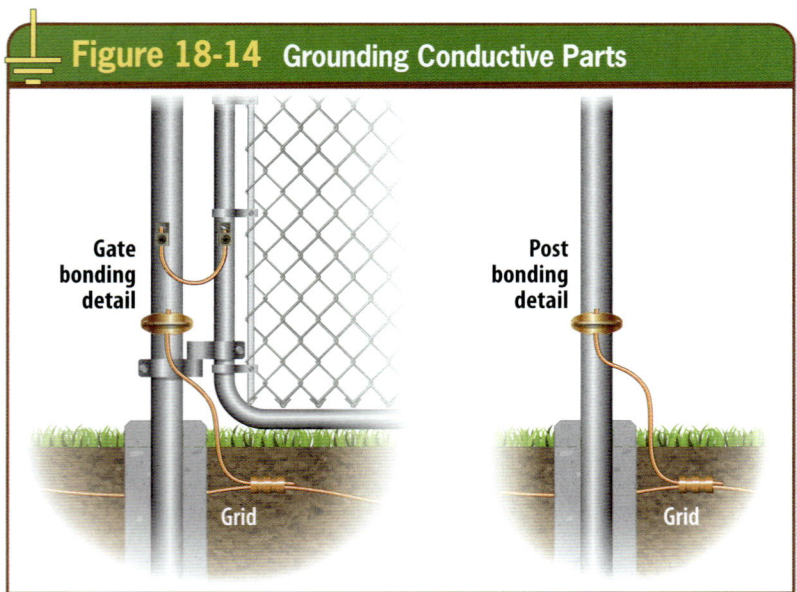

Figure 18-14 Grounding Conductive Parts

Gate bonding detail

Post bonding detail

Grid

Grid

Figure 18-14. Substation grounding of fences and other conductive parts is accomplished using suitable bonding connection devices and bonding jumpers.

Connections from metal enclosures and structures within the fenced area and connection from the fence should be made using conductors sized 4/0 AWG copper at a minimum and no less than 25% of the output conductors of the system. **See Figure 18-14.**

There are different philosophies on fence grounding at outdoor substations. Most designs require the fence to be commonly connected to the grounding grid. Another approach is to isolate the fence from the grid system in the event of a phase-to-ground fault that elevates the potential of all connected metal parts. In this case, the potential of the fence is raised, which can present a shock hazard for persons coming in contact with the fence during the event. Most designs require the fence to be grounded common to the grid grounding system.

Sometimes the grid system for a substation is created by laying copper conductors in a checkerboard arrangement at three-foot intersections to create a mat that is buried in the ground upon which the substation is constructed. This establishes a convenient point of connection for any conductive metal in the substation. The length of bonding jumper

connections is reduced when this method of grid construction is used.

The connections for substation grounding grid systems must be effective and strong. The common methods used for these connections are exothermic welding processes and irreversible compression connectors. It is important to use compression connectors that are listed as grounding and bonding equipment. These connectors are evaluated to endure the stresses of rising and falling potentials caused by various events occurring on these systems during both normal operation and abnormal events.

One form of protection provided by the grid system for outdoor substations is the ability to dissipate lightning strikes effectively to minimize possibilities of damage.

Metal Structures and Fences

Section 250.194 provides specific requirements for grounding and bonding metal fences and metal structures enclosing or surrounding electrical substations.

There are many exposed live medium- and high-voltage parts in substations. For multiple reasons, including ordinances, safety, security, and economics, metal fences are often installed around substations.

Since fences are often accessible to the general public and other personnel, they must be grounded and bonded to limit the rise of hazardous potential on the fence. This set of provisions in the *NEC* establishes basic prescriptive requirements for grounding and bonding of metal fences built in and around electrical substations. Although many substation fences are required by and typically covered by utility regulations and the *National Electrical Safety Code (NESC)*, these provisions in Part X of Article 250 provide a basis for installations covered by the *NEC*.

For situations where step and touch potential considerations indicate that additional grounding and bonding is required, alternate designs performed under engineering supervision are allowed. Designers are also referred to the industry standard on the grounding of fences in and around substations, which is *IEEE 80, Guide for Safety in AC Substation Grounding.*

Where metal fences are located within 16 feet of exposed electrical conductors or equipment, they must be bonded to the grounding electrode system with wire-type bonding jumper(s). Bonding jumpers are required at each fence corner and at 160-foot maximum intervals along the fence. At substations where bare overhead conductors cross the fence, bonding jumpers must be installed on each side of the crossing. Any fence gates are required to be bonded to the gate support post, and each gate support post must be bonded to the grounding grid and/or electrode system. Any gates or openings in the fence must have a bonding connection installed across the opening by a buried bonding jumper. The grounding grid or grounding electrode system must be located so as to cover the swing of all gates. Barbed wire strands above the fence also must be bonded to the grounding grid.

Where subject to contact by persons, all exposed conductive metal structures, including guy wires within eight feet vertically or 16 feet horizontally of exposed conductors or equipment, must be bonded to the grounding grid and/or electrode systems in the area.

CONDUCTOR SHIELDING AND STRESS REDUCTION

Cables installed for medium- and high-voltage systems are generally required to be of a shielded type. The cable shields can be in a concentric strand or conductive tape arrangement. **See Figure 18-15.**

Cable shielding provides a method to evenly distribute voltage stress and drain it off to ground at the termination points of the cable. **See Figure 18-16.**

Section 315.44 provides *NEC* requirements for cable shields. The primary purposes of shielding are to confine the voltage stresses to the insulation, dissipate insulation leakage current, drain off

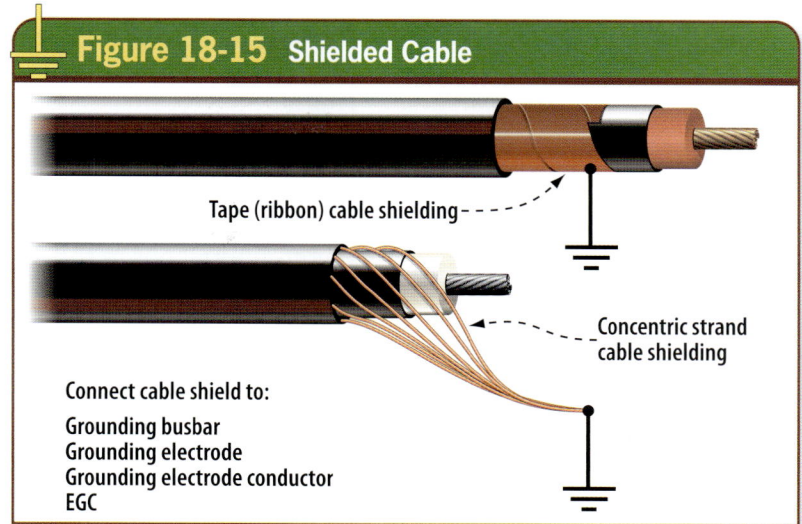

Figure 18-15 Shielded Cable

Tape (ribbon) cable shielding

Concentric strand cable shielding

Connect cable shield to:
Grounding busbar
Grounding electrode
Grounding electrode conductor
EGC

Figure 18-15. Cable shielding can be concentric stranding or ribbon (tape).

the capacitive charging current, and carry the ground-fault current to facilitate the operation of ground-fault protective devices in the event of an electrical cable fault.

Solid dielectric insulated conductors operating above 2,000 volts in permanent installations are required to have ozone-resistant insulation and must be

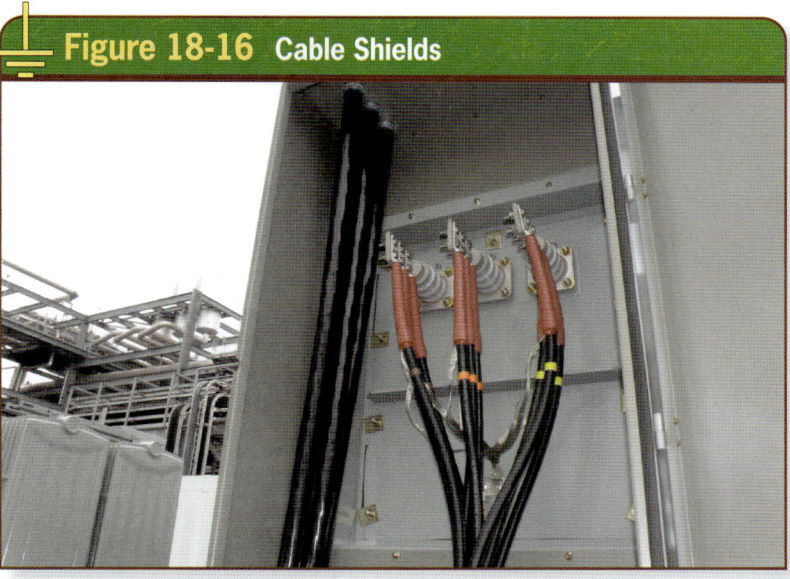

Figure 18-16 Cable Shields

Figure 18-16. Cable shields are required to be connected to ground through a grounding electrode conductor, grounding busbar, EGC, or grounding electrode.

Courtesy of Cogburn Brothers, Inc.

of a shielded type. All metallic shields (whether ribbon tape or concentric stranding) must be connected to a grounding electrode conductor, grounding busbar, EGC, or grounding electrode. In enclosed equipment, the cable shield connections are typically made to the equipment grounding bus in the enclosure. **See Figure 18-17.**

At a pole installation, the cable shields are typically connected to a grounding electrode conductor or directly to a grounding electrode. There are stress reduction kits that include the provisions for connecting cable shields to ground. Load-break elbow assemblies also include such provisions to bleed off stress at termination points.

Non-shielded cables are permitted for circuits up to 2,400 volts, but only under the following restrictive conditions:

1. They must be listed by a qualified electrical testing laboratory.
2. The insulation must be resistant to electric discharge and surface tracking or must be covered with a material that is ozone-resistant in addition to being resistant to surface tracking and electric discharge.
3. Where installed in wet locations, the conductor must have an overall nonmetallic jacket or metallic sheath.
4. The conductor insulation thickness must be no less than the values in Table 315.10(B).

Non-shielded cables are also permitted for circuits up to 5,000 volts in replacement applications only under the following restrictive conditions:

1. They must be listed by a qualified electrical testing laboratory.
2. Conditions of maintenance and supervision must ensure that only qualified persons service the system.
3. The insulation must be resistant to electric discharge and surface tracking or must be covered with a material that is ozone-resistant in addition to being resistant to surface tracking and electric discharge.
4. Where installed in wet locations, the conductor must have an overall nonmetallic jacket or metallic sheath.
5. The conductor insulation thickness must be no less than the values in Table 315.10(B).

This allowance of non-shielded cables can be applied in situations where existing non-shielded cables need to be replaced and connected to existing equipment. This allowance is not applicable to situations in which existing equipment is relocated or reinstalled on the same property and new wiring is installed to the old equipment.

Directly buried cables are required to be of a type suitable for the location and are generally required to be shielded if rated above 2,000 volts. Cables rated 2,001 to 2,400 volts are permitted to be non-shielded types if that cable has an overall metallic sheath or armor. The metallic sheath or armor must be connected to a grounding electrode conductor or grounding terminal bar in the equipment, EGC, or grounding electrode.

Figure 18-17 **Grounded Cable Shields**

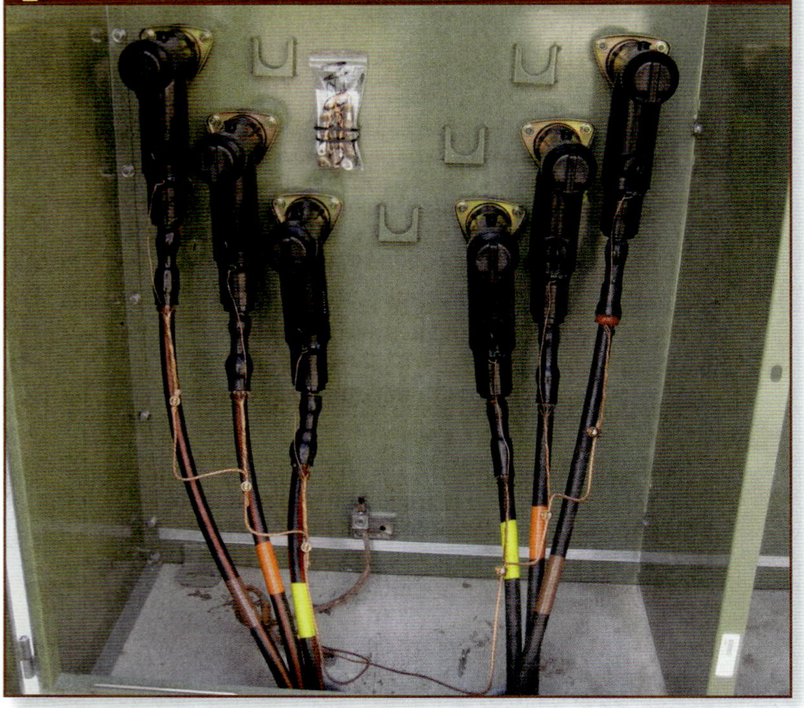

Figure 18-17. Cable shields are required to be connected to ground using suitable conductors and connectors.

There is an exception that relaxes the cable shielding requirement for airfield lighting equipment and cable installations. **See Figure 18-18.** In these types of installations, series circuits rated up to 5,000 volts that are controlled by regulators are permitted to be installed using non-shielded cable as long as they meet the criteria set forth in the Federal Aviation Administration Advisory Circulars, which include guidelines and practices for installing airport lighting systems. **See Figure 18-19.**

GROUNDING THROUGH SURGE ARRESTERS

Surge arresters are devices that protect electrical equipment and installations from transients (voltage spikes). This is accomplished by either limiting or shunting to ground all voltage events above a set threshold. Grounding of medium- and high-voltage systems can be accomplished using surge arresters.

Surge Arrester. A protective device for limiting surge voltages by discharging or bypassing surge current; it also prevents continued flow of follow current while remaining capable of repeating these functions. (CMP-10)

Figure 18-18 Airfield Lighting Exception

Figure 18-18. Regulators may supply airport runway lighting with 5,000-volt cable that is unshielded.

The term *clamping voltage* is often used and associated with surge arresters. The clamping voltage is the voltage at which the surge arrester initiates its transient protection operation.

Surge arresters are an effective means for protection against overvoltages

Figure 18-19 Airfield Lighting Exception

Figure 18-19. Regulators for airport runway lighting can be wired using 5,000-volt cable that is not a shielded type.

Figure 18-20 Surge Arresters

Figure 18-20. Surge arresters are usually installed for overhead lines for systems operating at more than 1,000 volts.

Courtesy of Bill McGovern, City of Plano, TX

Figure 18-21 Overhead Surge Arresters

Figure 18-21. Surge arresters installed for overhead lines of more than 1,000 volts are connected to ground through a grounding electrode conductor.

Courtesy of Bill McGovern, City of Plano, TX

and events such as lightning strikes. Surge arresters are often installed for overhead lines because of their vulnerability to lightning strikes. **See Figure 18-20.**

A common application of surge arresters is found on the tops of utility poles where medium- and high-voltage lines are installed. The surge arresters are connected directly to the medium- or high-voltage line conductors. They are connected to ground by a grounding electrode conductor. **See Figure 18-21.**

Surge arresters operate during overvoltage events, such as lightning strikes or significant line surges. This device uses a spark gap configuration that allows the contacts of the device to close, draining excessive overvoltages to the Earth during the duration of the line surge or even unintentional contact with higher-voltage lines. Surge arresters typically protect the primary of higher-voltage systems.

Surge arrester grounding conductors must be connected to any of the following:

1. A grounding electrode
2. A grounding electrode conductor
3. The grounded conductor at the service
4. The equipment grounding terminal bus within medium- or high-voltage equipment enclosures

Surge arresters are not required by the *NEC*, but Article 242, Part III, provides requirements for surge arresters installed on electrical systems that are covered by the *NEC*. For example, if surge arresters are installed on a premises system, a surge arrester must be installed in each of the ungrounded supply conductors.

Surge arresters are not permitted where the phase-to-ground power frequency voltage exceeds the rating of the surge arrester. The surge arresters applied in an electrical system must have a rating equal to or greater than the maximum continuous operating voltage available at the point it is connected to the system. Surge arresters can be installed inside or outside buildings or structures, but they must be inaccessible to unqualified persons.

The conductor for connecting surge arresters must be as short as practicable and must not have any unnecessary bends. Where surge arresters are connected to line conductors or equipment, the connecting conductor cannot be smaller than 6 AWG. Copper or aluminum is permitted for this connection, but the rules for aluminum conductors must be followed.

Where a surge arrester is installed in an ungrounded primary system, the spark gap or listed device must have a 60-hertz breakdown voltage of at least twice the primary circuit voltage, but not necessarily more than 10 kV (10,000 volts). In addition, there must be at least one other grounding connection on the grounded conductor of the secondary that is located no less than 20 feet from the surge arrester grounding electrode.

Where surge arresters are connected to multi-grounded neutral primary systems, the spark gap or listed device must have a 60-hertz breakdown of no more than 3 kV (3,000 volts), and there must be at least one other grounding connection on the grounded conductor of the secondary that is no less than 20 feet from the surge arrester grounding electrode.

Where the grounding conductor for a surge arrester is installed in a ferrous metallic raceway or enclosure, the grounding conductor must be bonded to the enclosure at points of entrance and emergence from the raceway or enclosure. This type of installation of physical protection for surge arrester grounding conductors must meet the requirements in 250.64(E). This section requires grounding electrode conductors installed in ferrous metal raceways to be bonded at both ends of the raceway where the grounding electrode conductor enters or emerges from the raceway. This minimizes the choke effect on the grounding electrode conductor. **See Figure 18-22.**

ENGINEERED GROUNDING SYSTEM DESIGNS

In larger commercial and industrial complexes and facilities, the grounding and bonding system designs often exceed the minimum requirements of the *NEC.* It is important that those grounding and bonding systems meet the minimum requirements of the *Code,* even though the design may be much more substantial than a typical grounding and bonding scheme for a smaller building or facility. It is important to

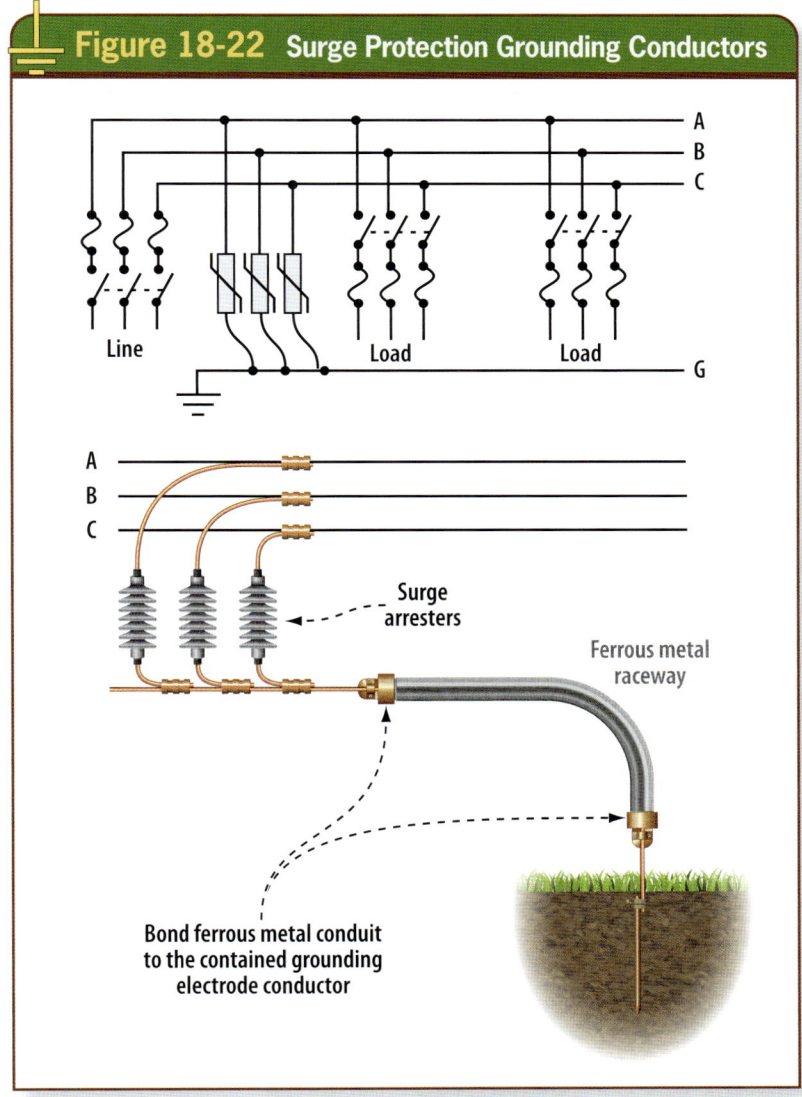

Figure 18-22 Surge Protection Grounding Conductors

Figure 18-22. Ferrous metal sleeves for surge arrester grounding conductors must be bonded to the contained grounding electrode conductor at both ends of the sleeve.

become familiar with typical engineered grounding and bonding schemes that are often installed in larger commercial or industrial buildings or multi-building complexes.

NEC Compliance

Previous discussions placed a strong emphasis on the importance of understanding the basics of grounding and bonding requirements and what each function is intended to accomplish. Whether an electrical service or system is small or large, the fundamental grounding and bonding principles for these must still be applied. The grounding and bonding system's performance is the same. It is important to remember that large and complex grounding and bonding designs can be in addition to, and often overlay, the minimum requirements. The minimum requirements of the *NEC* must be followed relative to the connection to earth through a grounding electrode or system of electrodes, grounding and bonding of equipment, branch circuit and feeder equipment grounding conductors, and so forth.

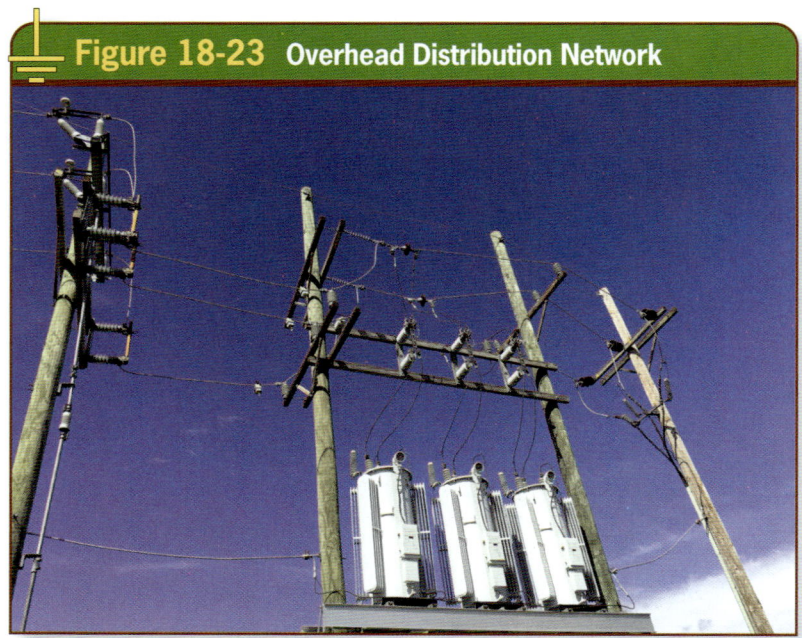

Figure 18-23 Overhead Distribution Network

Figure 18-23. Older commercial and industrial facilities often utilize poles and overhead lines for their power distribution.

Engineered Plans and Specifications

Many large industrial facilities have their own engineering teams on site, and others hire engineering firms to develop electrical system designs for their facilities. These designs must also be in compliance with the applicable electrical code requirements.

Many authorities having jurisdiction require plans and specifications to be submitted when applying for construction and renovation building permits. Most of these jurisdictions have an administrative provision that requires installers and owners to complete the installation in accordance with the *NEC* and in accordance with the engineered plans and specifications. This is also the law in many states.

Any deviation from the engineered design must be under the authorization of the registered professional or engineering firm responsible. Always check with the local authority having jurisdiction for the *Code* requirements and any other applicable regulations within that jurisdiction.

Site Power Distribution Network

Many large commercial and industrial facilities and complexes are served by medium- and high-voltage electric utility services or other energy source(s), and they require heavy switching and transformers or substations as an integral part of the electrical distribution network on the property. The power is typically delivered at voltage levels not suitable for the inside of buildings or structures. A network of heavy feeder duct banks or tunnel systems and strategically-located substations are common essential components of these designs. Some older properties installed overhead poles for their power distribution systems. **See Figure 18-23.**

Most of these systems installed today are underground. These power networks are often supplied from the same source and double-ended, and they use tie-breaker(s) in equipment for isolation of single-source systems or isolation of separate sources arriving from two directions. Double-ended power systems provide convenience for preventive maintenance operations and can allow isolation of a portion of the loop that may have failed or is otherwise in need of repair or servicing.

Grounding and bonding systems are integral to all parts of these power distribution networks. An example of such a distribution network would be a utility service on a property supplied by medium or high voltage directly from a switching yard and substation and metered at that substation and switching yard. Sometimes this equipment is installed within buildings or structures on the property in properly constructed and rated rooms. From the substation(s), medium voltages such as 13,800 volts, 12,470 volts, 4,160 volts, and 2,300 volts are typically produced from a transformer(s). **See Figure 18-24.**

Switching and overcurrent protection (protective relay systems) are provided for both the transformers and the conductors. Feeders are routed and distributed to locations on the property using a series of effectively-routed underground tunnels or duct banks for the feeder conductors. Duct banks for medium- and high-voltage feeders are often concrete-encased and identified with a red die on the top before final grading processes. There are usually several manholes installed in the duct bank network to allow for access and ease of initial installation as well as to provide for more effective maintenance and testing of the distribution system. As the feeders arrive at the building or structure served, there is typically a unit substation or transformer installed to step the voltage down to levels suitable for use in supplying loads within the buildings or structures. **See Figure 18-25.**

Site Grounding and Bonding Plan

It is common for engineered designs to include an overhead plan view of a property showing the grounding and bonding plan and layout (site plan) for the power distribution network. This

Figure 18-24 Outdoor Substations

Figure 18-24. Many commercial and industrial properties own and maintain outdoor substations.

Photo from iStock

Figure 18-25 Substation Assemblies

Figure 18-25. Substations are also available as unit assemblies for indoor or outdoor applications.

Courtesy of Schneider Electric Square D Company

plan could include grounding details such as triads, duct bank cross-sections, and manhole grounding arrangements. **See Figure 18-26.**

Other Connected Sources

In some more elaborate designs, energy storage systems and renewable energy sources such as fuel cells, photovoltaic arrays, and wind turbines can be connected to the distribution network through interactive utility inverters and other suitable transfer equipment. Where there are multiple interconnected power sources connected to the same network grid, the requirements in Article 705 apply.

These types of power distribution networks are commonly referred to as *micro-grids.* A micro-grid is defined as "a local energy grid with control capability, which means it can disconnect from the traditional grid and operate autonomously." A utility source is usually connected to a micro-grid, but not always. There are some micro-grids that operate completely independently from the utility grid, and these are referred to as *islanded power systems.*

For additional information, visit qr.njatcdb.org Item #2608

All systems connected to the grid or micro-grid must meet the minimum requirements for system and equipment grounding.

Impedance Grounded Systems

Impedance grounded systems are sometimes used for these types of facilities to limit ground-fault current and to provide an enhanced level of reliability and safety. The more specialized equipment used with these systems necessitates that only qualified persons service and maintain these systems.

COMMON GROUNDING AND BONDING COMPONENTS

There are grounding and bonding elements in the grounding and bonding scheme that are common for low-voltage, medium-voltage, and high-voltage systems. The common elements are grounding electrodes, grounding electrode conductors, equipment grounding conductors, bonding jumpers, and system bonding jumpers. Additionally, there are often static grounding and bonding components and systems as well as lightning protection systems.

Typically, there is a grounding electrode system at each switching yard and substation. In the engineering community, a grounding electrode commonly known as a triad grounding system, or *grounding triad,* is often installed at each manhole, transformer, or substation. This creates a local reference to the Earth

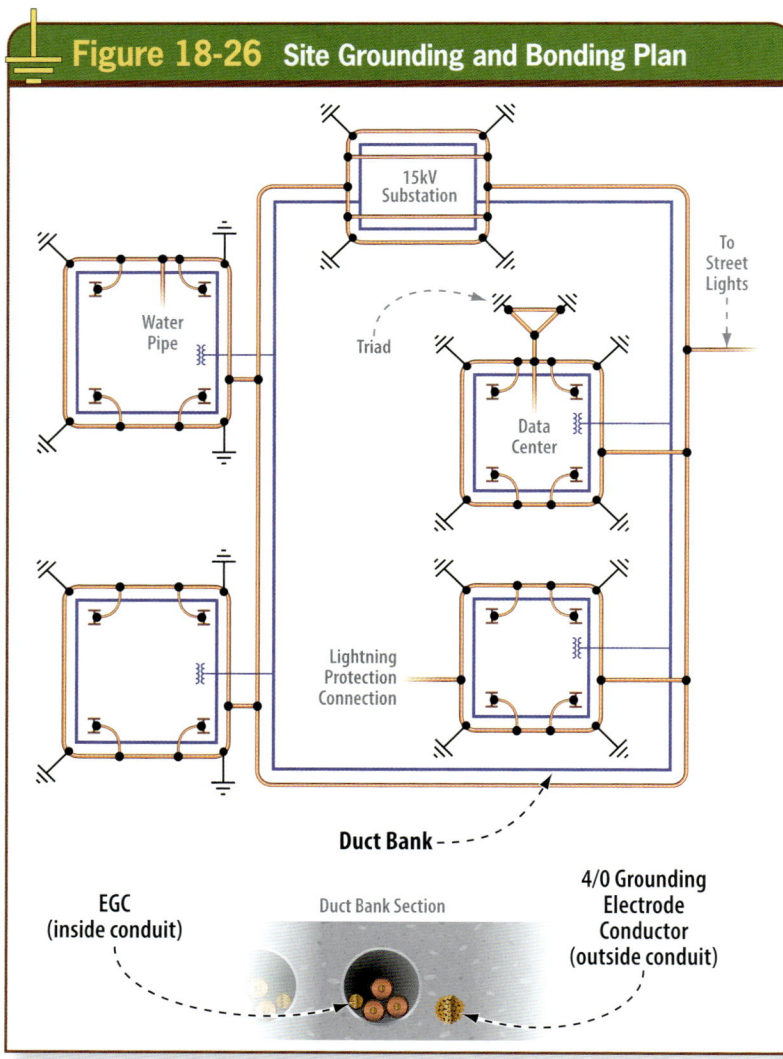

Figure 18-26 Site Grounding and Bonding Plan

15kV Substation

Water Pipe

Triad

To Street Lights

Data Center

Lightning Protection Connection

Duct Bank

EGC (inside conduit)

Duct Bank Section

4/0 Grounding Electrode Conductor (outside conduit)

Figure 18-26. *A site-grounding plan shows specific grounding details as necessary.*

at that point on the system. Equipment grounding conductors are installed with the feeders of the power distribution network. The sizing and installation requirements for grounding electrode conductors, grounded conductors, equipment grounding conductors, system and main bonding jumpers, and equipment bonding jumpers are provided in Article 250 of the *NEC*.

As indicated in the definition of the term *grounding electrode*, a direct connection to Earth is accomplished. In a triad ground rod arrangement, and in any other ground rod installation where two or more rods are installed, the distance between the rods should be twice the depth of the rod if the same size (length) rods are used, based on *IEEE 3003.2-2014, Recommended Practice for Equipment Grounding and Bonding in Industrial and Commercial Power Systems.* The following are samples of installations that could be specified to form a grounding triad type of electrode:

- Example: two eight-foot rods
  - Eight feet plus eight feet = 16 feet apart
- Example: two eight-foot rods and one ten-foot rod
  - Eight feet plus ten feet = 18 feet apart
  - Eight feet plus eight feet = 16 feet apart

SUMMARY

Medium- and high-voltage systems are required to be grounded under certain conditions; more specifically, where they supply portable or mobile equipment. There are a few different grounding methods and requirements for systems operating at more than 1,000 volts, such as 5- and 15-kilovolt systems. A typical 5-kilovolt system is one with a phase-to-phase voltage of 4,160 volts. The phase-to-ground voltage of this system is approximately 2,400 volts. An example of a 15,000-volt system is one with a phase-to-phase voltage of 12,470 volts. The phase-to-ground voltage in this system is approximately 7,200 volts. These systems are properly referred to as medium-voltage systems.

The *NEC* provides some methods and several rules related to the grounding of these systems and associated equipment. Three methods of grounding for medium-voltage systems are solid grounding, impedance grounding, and grounding accomplished through grounding transformers. Sizing requirements for grounding electrode conductors and equipment grounding conductors used with systems over 1,000 volts are similar to those rules for systems operating at 1,000 volts or less. In commercial and industrial complexes and facilities, grounding and bonding system designs often exceed the minimum requirements of the *NEC*. It is important that those grounding and bonding systems meet the minimum requirements of the *Code*, even though such designs may be much more substantial than a typical grounding and bonding scheme for smaller buildings or facilities. The basic principles of grounding and bonding and how the system performs is related to selected and applied engineered designs and those installations meeting the minimum requirements of the *NEC*.

REVIEW QUESTIONS

1. Which part(s) of Article 250 provides requirements for grounding systems of more than 1,000 volts?

   a. Parts I and X only

   b. Parts I, III, and X only

   c. Part X and all other parts of Article 250 as modified by Part X

   d. Part X only

2. Ungrounded systems greater than 1,000 volts are permitted to be __?__.

   a. grounded through a set of grounding transformers

   b. grounded through an impedance device

   c. solidly grounded

   d. grounded by any of the above methods

3. A solidly grounded electrical system of more than 1,000 volts is one that has a direct electrical connection to ground with a grounding impedance device installed between the Earth connection and the system.

   a. True   b. False

4. The grounded conductor of a solidly grounded system of more than 1,000 volts must be of sufficient current-carrying capacity for the load served and generally must not be smaller than __?__ of the ampacity of the ungrounded phase conductors supplied by the system.

   a. 10%

   b. 25%

   c. 33 1/3%

   d. 50%

5. Single-point grounding of a system of more than 1,000 volts means the system is grounded at only one point and no neutral-to-ground connections can be made downstream of that initial grounding location.

   a. True   b. False

6. Where a multi-grounded neutral system of more than 1,000 volts is installed, the connections to ground from the neutral conductor must be made at intervals not exceeding __?__.

   a. 100'

   b. 1,300'

   c. 2,500'

   d. 5,280'

7. The connection to the Earth for a single-point grounded neutral system is made through a grounding electrode meeting the applicable requirements in Part __?__ of Article 250.

   a. I

   b. II

   c. III

   d. X

8. The grounding impedance device for an impedance grounded system must be installed in series with the grounding electrode conductor and the neutral point of the system source, which could be at a transformer or a generator. Any equipment grounding conductor connections at the source have to be made on __?__ of the impedance device.

   a. either side

   b. the line side

   c. the load side

   d. any of the above

9. For systems of more than 1,000 volts, the sizing requirements for grounding electrode conductors are based on __?__, using the largest ungrounded service, feeder, or branch circuit conductor supplying the equipment.

   a. Table 8, Chapter 9

   b. Table 250.4

   c. Table 250.66

   d. Table 250.122

10. Equipment grounding conductors installed with circuits of more than 1,000 volts cannot be smaller than 6 AWG copper or 4 AWG aluminum, unless they are an integral part of a cable assembly.

    a. True   b. False

11. Where a separate equipment grounding conductor is installed with feeder conductors supplied from a 4,160-volt system, the minimum size shall not be smaller than the sizes provided in __?__.

    a. Table 1, Chapter 9

    b. Table 250.66

    c. Table 250.122

    d. Table 310.16

12. **If a 4,160-volt feeder is protected by a 100-ampere fuse, the minimum size wire-type copper equipment grounding conductor for the circuit is ___?___ .**

    a. 8 AWG

    b. 6 AWG

    c. 4 AWG

    d. 2 AWG

13. **Section 250.191 requires a grounding electrode system in accordance with the applicable provisions of Part III of Article 250.**

    a. True   b. False

14. **A common size for grounding grids installed at substations is no less than ___?___ .**

    a. 4/0 AWG copper

    b. 3/0 AWG copper

    c. 2/0 AWG copper

    d. 1/0 AWG copper

15. **The primary purposes of shielding are to confine the voltage stresses to the insulation, dissipate insulation leakage current, drain off the capacitive charging current, and carry the ground-fault current to facilitate operation of ground-fault protective devices in the event of an electrical cable fault.**

    a. True   b. False

16. **Non-shielded cables are permitted for circuits up to 2,400 volts but only under which of the following conditions?**

    a. The insulation must be resistant to electric discharge and surface tracking or covered with a material that is ozone resistant and resistant to surface tracking and electric discharging.

    b. They must be listed by a qualified electrical testing laboratory, and the conductor insulation thickness must be no less than the values in Table 311.10(B).

    c. Where installed in wet locations, the conductor must have an overall nonmetallic jacket or metallic sheath.

    d. All of the above.

17. **A ___?___ is a protective device for limiting surge voltages by discharging or bypassing the surge current; it also prevents continued flow of follow current while remaining capable of repeating these functions.**

    a. primary protector

    b. surge arrester

    c. surge protective device

    d. transient voltage surge suppressor

18. **A ___?___ is a local energy grid with control capability, which means it can disconnect from the traditional grid and operate autonomously.**

    a. microgrid

    b. nanogrid

    c. network grid

    d. signal reference grid

19. **When a grounding triad is installed, ___?___ or more ground rods are driven in a particular configuration with specific separations between them.**

    a. 1

    b. 2

    c. 3

    d. 4

20. **Article 315 of the *NEC* covers medium-voltage conductors and cables, including installation requirements, construction specifications, and ampacities.**

    a. True   b. False

Annexes

Table of Contents

Annex A – Investigation and Testing of Footing-Type Grounding Electrodes for Electrical Installations 445

Annex B – Steel Conduit and EMT for Equipment Grounding 455

Annex C – Selecting Protective Devices - Short Circuit Current Calculations 467

Annex D – Chapter 9, Table 8 Conductor Properties 479

Annex E – Lightning Protection Systems 481

Investigation and Testing of Footing-Type Grounding Electrodes for Electrical Installations

H. G. Ufer, Associate Member IEEE

Summary: Footing-type grounding electrodes installed in the concrete foundations of residential and small commercial buildings designed to meet a maximum ground resistance value of 5 ohms are described. Bare solid copper electrode wires of various lengths, steel reinforcing rods, and 10-foot lengths of hot galvanized rigid steel conduit were used to determine the resistance values.

City Electrical Inspector members reported to the Southwestern Section of the International Association of Electrical Inspectors that a continuous metallic water piping system, as recommended by the *National Electrical Code*, is not always available. A committee was appointed to investigate this condition and report its findings at subsequent meetings.

The *National Electrical Code*, which is prepared by a committee of the National Fire Protection Association, is the recognized American standard for the safe installation of electrical equipment in the United States. It is based on the combined experience of all groups in the electrical industry and all factual information that is available at the time of each edition's preparation. One of the oldest basic requirements of the code is the protection of electric installations by grounding; it recommends that a metallic underground water piping system, either local or supplying a community, should always be used as the grounding electrode when such a piping system is available.

Paper 63-105, recommended by the IEEE Safety Committee and approved by the IEEE Technical Operations Committee for presentation at the IEEE Western Appliance Technical Conference, Los Angeles, Calif., November 4, 1963. Manuscript submitted August 1, 1963; made available for printing May 5, 1964.

H. G. Ufer is with Underwriters' Laboratories, Inc., Santa Clara Calif.

There is increasing concern over the fact that a continuous metallic water piping system is not presently available in some areas for electrical grounding purposes. This is partially a result of the use of nonmetallic water pipe mains and laterals to bring water into a building, in addition, it has become standard practice in the installation of cast-iron water pipe mains and laterals to use neoprene gaskets to join the pipe sections and discontinue the poured-lead joint formerly used.

Inspection authorities have also reported that the ground is interrupted by the installation of insulating joints on the plumbing system. When copper pipe is employed for the hot water side and galvanized steel pipe for the cold water side, such joints are used to prevent decomposition of the copper pipe resulting from electrolysis. These insulating joints are also used for water-softening tanks. Bonding jumpers are not always used, and often they are removed.

The driven grounds required by the code in lieu of connection to a water system are not always dependable. In the paper "Grounding of Electric Services to Water Piping Systems" presented to the American Water Works Association on October 26, 1960, A. G. Clark of Los Angeles reported that the use of driven grounds had not proven satisfactory. The minimum resistance obtained on representative ground rods was 10 ohms, and most of the driven representative rods tested were several times the minimum value of 10 ohms.

In view of these conditions and failures, it seemed that an adequate means could be provided for grounding, one which is not likely to be disturbed, which requires very little maintenance, and which does not require connection to the water pipe systems to provide an adequate low resistance ground.

The purpose of this paper is to report the development and testing of such a grounding method and to record the results of tests on installations in residential and small commercial buildings since 1961.

BACKGROUND

A number of military installations of ammunition and pyrotechnic storage facilities were built during World War II and provided with lightning protection systems which required a permanent ground connection. The author conducted the field inspection of the lightning protection for Underwriters' Laboratories, Inc., and had first-hand knowledge of the construction details. Installations in Arizona have been selected for this paper because the climate is normally hot and dry during most of the year.

Tucson Installation

The installation at the Davis-Monthan Air Force Base, Tucson, Ariz., consists of six bomb-storage vaults and four pyrotechnic storage sheds. The bomb storage vaults are steel-reinforced concrete, while the storage sheds are steel angle frame construction covered with corrugated galvanized sheet steel.

The soil at this location is sand and gravel, and the average rainfall is 10.91 inches. Low resistance grounding had to be provided for each of these buildings in order to discharge to ground any static charge of electricity caused by wind and sandstorms or by exposure to lightning during a rainstorm or thundershower.

In the areas where these bomb storage vaults and storage sheds were located, an underground water piping system was usually not available. Driven ground rods or a bare copper counterpoise could be used, but at the time these installations were being rushed to completion, strategic materials were being conserved. It was decided to provide grounding by the use of the steel angle iron or the reinforcing rods of the building structure.

For the bomb storage vaults (igloos as designated by the Armed Forces) with the dimensions 10 by 30 or 40 feet, the two parallel exterior walls were provided with footings. The front and rear walls of these vaults were not provided with footings.

The footings were dug to a depth of 2 feet. In these footings, ½-inch vertical reinforcing rods approximately 30 inches long were spaced about 12 inches apart. They were pushed a few inches into the earth to maintain their position and spacing while the concrete was poured. Across the upper ends of these vertical rods, ½-inch steel reinforcing bars were laid horizontally and welded to every alternate vertical rod embedded in the concrete of the footings.

From this base, an umbrella or igloo-shaped structure was formed of ½-inch steel reinforcing rods welded to the rod at the top of the footing. The lightning rod terminals were erected on the top of each igloo by welding them to the reinforcing rod assembly. The umbrella was then covered with concrete.

For the 10- by 4-foot storage sheds, grounding was obtained through the angle iron framework, which had the ends of the vertical angle iron at each corner of the building embedded in concrete to a depth of 1 foot.

Complete foundations were not provided for these small storage sheds. As a bond between the various metal parts of the frame and metal covering for these buildings, welding or bolts were used; the adjacent edges of the corrugated steel sheets were overlapped and secured by rivets or sheet metal screws.

Flagstaff Installation

The installation at the Navajo Ordnance Depot is located adjacent to U.S. Highway 66, approximately 15 miles west of Flagstaff, Ariz. This ordnance depot has an area of about 56 square miles. The soil is clay, shale, gumbo, and loam, with small-area stratas of soft limestone. The average rainfall is 20.27 inches.

The original contract in 1942 called for the construction of 800 bomb storage vaults. These were of the same construction and dimensions as those erected at the Davis-Monthan Air Force Base at Tucson.

A water piping system either local or serving a community was not available for grounding at this location; accordingly, the method used at Tucson was utilized.

GROUND RESISTANCE TESTS

Each bomb storage vault and storage shed erected in Tucson was inspected, and ground resistance tests were made on each of these structures. The 800 bomb storage vaults erected at the Navajo Ordnance Depot were inspected during the construction, and then individual ground resistance tests were performed.

Ground resistance readings were taken between each lightning protection terminal and ground on all igloos and storage sheds. Tests were also made between all exposed conductive material and ground on each installation described in this paper.

Readings were taken by using a heavy-duty megger ground tester with a scale from 0 to 50 ohms graduated in increments of 1 ohm for each scale division. Each vault and shed was required to have a ground resistance reading of not more than 5 ohms.

The method of measuring the ground resistance values was a 3-point method.[1] The sequence of ground resistance tests for the two lengths of conduit recorded in Table I were as follows: Two lengths of rigid-steel ¾-inch galvanized conduit were laid 5 inches apart in a trench 36 inches deep, 11 feet long, and 15 inches wide. One length of conduit was enclosed in 2 inches of concrete (electrode 1), while the second length was not enclosed in concrete (electrode 2); the second length was then enclosed in 2 inches of concrete and placed back in the trench beside the other length of conduit, and the dirt was replaced in the trench. The resistance measurements were repeated with the results recorded in Table I.

Check Tests

In order to determine the adequacy of the ground resistance values which were made in 1942, it was decided to conduct check tests. Permission was obtained from the Office of Chief of

Operations in Washington, D.C., to check the grounds at Flagstaff Army Base and Tucson Air Force Base.

Check tests were made at the Navaho Ordnance Depot on July 12, 1960, and at the Davis-Monthan Air Force Base on August 23, 1960. Additional check tests were made at the Navajo Ordnance Depot on October 20, 1960.

At both bases, the readings taken measured from 2 to 5 ohms in all instances. All readings taken in 1942 when the original installations were made did not exceed 5 ohms to ground.

The Surveillance Officer at the Navajo Ordnance Depot must regularly check and record the ground resistance for each igloo, shell loading, and storage building. These tests are made at approximately 30–60-day intervals. During the 20 years since these ground systems were installed and accepted by the author, the maintenance department at the Navajo Ordnance Depot

Table I	Beverly Hills Tests		
Type of Ground Electrode	Installation Date	Testing Date	Resistance (Ohms)
Electrode 1	11/1/60	11/1/60	5.1
	11/1/60	11/2/60	9
	11/2/60	11/10/60	9
	11/2/60	11/30/60	9
Electrode 2	11/1/60	11/1/60	90
	11/1/60	11/2/60	100
		11/10/60	
		11/30/60	
Bare pipe and pipe in concrete	11/1/60	11/1/60	14
		11/2/60	
		11/10/60	
		11/30/60	
Electrode 2 in concrete	1/9/61	1/9/61	8.7
	1/9/61	1/11/61	9
Electrodes 1 and 2 in concrete (series)	1/9/61	1/9/61	5
		1/11/61	
Cast-iron sewer and gas pipes-bonded		1/11/61	30
Sewer, gas, and water pipes-bonded		1/11/61	1
Sewer, gas, and water pipes connected to electrode 1		1/11/61	2

has not been requested to replace or repair any grounding components on the 800 bomb storage vaults or igloos.

As far as could be determined, there was no report of repairs or maintenance on grounding systems which had been installed on igloos and sheds at the Davis-Monthan Air Force Base.

PRELIMINARY INVESTIGATION

Beverly Hills Tests

Test A

In view of the previous findings, an installation was made using conduit electrodes laid horizontally in the ground and enclosed in 2 inches of concrete at a depth of 3 feet.

Two 10-foot lengths of ¾-inch galvanized steel rigid conduit were laid side by side horizontally in a single trench 15 inches wide, 11 feet long, and 36 inches deep. The installation was made in Beverly Hills, Calif., on property which has not been cultivated during the past 30 years.

The resistivity of the soil in the area was 3,830 ohms per cm³ (cubic centimeter). A ground rod driven to a depth of 20 feet was calculated to have a resistance of 5 ohms based on this resistivity test. (Table II shows the soil resistivity of the various sites mentioned in the paper.)

Table II	Resistivity of Soils	
Installation	**Soil**	**Resistivity (Ohms/Meter²)**
Beverly Hills	Loam and sand	38.30
Bishop	Gravel and sand	120
Riverside	Red adobe	50
Livermore	Loam and Sand	35
Hayward	Large rocks and gravel	150
Portland, Ore.	Loam and sand	40
Long Beach	Heavy loam and sand	25* / 20†
Twenty-nine Palms	Decomposed granite and sand	70
Burbank	Sandy loam	40

* Metal sign on 12-inch steel I-beams.
† Concrete floor slab in garage.

In the first series of the conduit electrode tests, one length of conduit (electrode 1) was enclosed in 2 inches of concrete; the other (electrode 2) was not enclosed in concrete. These two lengths of conduit were then covered with the soil removed to prepare the test trench.

In the second series of tests, the soil in the trench was removed and the bare conduit electrode 2 was enclosed in 2 inches of concrete.

The ground resistance of the water, gas, and sewer drainage piping systems bonded together was 1 ohm. The resistance was also 2 ohms when connected to electrode 1. This low resistance value is apparently due to the fact that the local water pipe system is galvanized steel pipe for the mains and the lateral. The lateral extends underground 40 feet from the water meter to the supply connection in the building.

The overhead electric service is a 3-wire a-c 110–240-volt system. The neutral of this service is grounded to the water pipe in the building. The utility ground for this service is provided at the utility pole which is 25 feet from the water meter.

The results of the ground resistance tests on the two conduit electrodes are recorded in Table I.

Test B

On January 16, 1960, one 10-foot length of 3½-inch galvanized rigid steel conduct was installed vertically to a depth of 5 feet and enclosed in 2 inches of concrete.

The ground resistance was 10 ohms; this value has remained constant during the 2-year period of these tests, and the readings are apparently stable.

In 1961, during a discussion of these supplemental tests, it was suggested that a test should be made on the property where the supplemental tests were conducted to be sure that the ground did not contain any foreign items which would favor these tests. In September 1961, the same Biddle ground megger, electrodes, and wire leads were used for this test as were used to measure the ground resistance of electrodes 1 and 2 (Table I).

The megger was placed adjacent to the ground in which pipe electrodes 1 and 2 were placed for the supplemental tests, and it was connected to these electrodes for a ground resistance test. The two 25-foot *No. 14* American wire gage (AWG) stranded wire leads were connected to the proper terminals on the megger. The probes were 36-inch ⅜-inch copper weld rods, and two were now placed in the ground 50 feet apart. On a 30-foot arc, ground resistance readings were taken at 3-foot intervals until probe 1 was adjacent to probe 2.

The probes were returned to their original positions. The readings were repeated with the wire leads reversed on the terminals of the megger. Six readings measured 11 ohms, two measured 12 ohms, and two measured 13 ohms; the average reading was 11.1 ohms. The measurements were repeated with the wire leads reversed on the megger, with no change recorded in these measurements.

The megger was then connected to electrodes 1 and 2 with the probes separated 50 feet and attached to the proper terminals on the ground megger. The ground resistance of electrodes 1 and 2 measured 5 ohms, as has been reported in Table I.

BUILDING INSTALLATIONS

The military installation tests demonstrate that a low-resistance ground electrode can be obtained in an area where conditions are unfavorable and that such a ground can be continuously effective over many years without maintenance. Encasement of electrodes in concrete and compaction provided by building weight may be the controlling factors in these installations. The supplementary tests on a limited-size electrode indicates that a concrete encasement contributes to the reduction of ground resistance.

A number of tests on footing-type electrode installations in building constructions were made during 1961, 1962, and 1963. Arrangements were made to have this type of electrode installed in conventional residential and commercial buildings. Locations selected were widely separated and these buildings were provided with a footing-type concrete building foundation. No. 4 AWG solid bare copper wires were embedded in the center of the concrete footing approximately 2 inches above the base of the footing.

The concrete building foundations are T-shaped. The base is about 12 inches wide, while the top of the footing where the foundation still is attached is 6 inches wide. The depth of these footings varies from 2 to 4 feet below the grade level of the building plot.

Two separate lengths of wire were installed in each foundation with one wire approximately 30 feet and the other 100 feet in length. The ends of each wire extended out of the foundation walls to provide ready access for testing and for connection to the neutral terminal in the service switch if and when authorized by the local electrical department.

The area of these buildings on which these installations were made varied from 500 to 8,000 square feet. They are 1-story residential- and commercial-type buildings.

An installation was made with five 20-feet lengths of No. 4 AWG bare solid copper wire embedded in a concrete building foundation. The ends of each 20-foot length of wire extended from the walls of the concrete footing to be available for measurements. This special installation was made to determine the length of the grounding wire required to provide 5-ohm maximum resistance values.

Bishop Installation

An experienced electrical contractor was contacted in Bishop, Calif., and requested to assist in this program of installing footing-type grounding electrodes in a residential or small commercial building. He had previously received a contract to wire a new residence at Bishop, and the owner agreed to have the footing-type grounding electrode installed.

The California Electric Power Company is the serving public utility for this area. They were notified that this installation was to be made, and their engineer conducted the ground resistance tests when the installation was completed.

The electrical contractor installed two No. 4 AWG bare copper wires in the concrete forms for the building foundation, one 76 feet long, the other 82 feet. The ends of both wires were brought out of the side of the concrete foundation below the location for the installation of the main switch. The grounding wires were run in opposite directions, and the ends were brought out of the side of the foundation at a location diametrically opposite the wall on which the main switch was installed.

The contractor was requested to install one long wire in the foundation about 100 feet long and one short wire about 25 feet long. This was suggested so that test results would show how much wire was required to provide a ground resistance of not more than 5 ohms.

This installation was made during September 1961. The building is a one-story masonry residence with an area of approximately 1,500 square feet. The concrete footing is T-shaped. The base is 14 inches wide, while the top of the footing where the wood sills are installed is 8 inches wide. The footing is approximately 12 inches below grade level. The No. 4 AWG bare copper grounding conductor was installed in the center of the concrete foundation about 2 inches above the base of the footing.

Table III

Location	Installation Date	Testing Date	Ground-Wire Resistance (Ohms)	Ground-Wire Length (Feet)	Soil Condition	Average Rainfall (Inches)
Bishop	9/61	11/1/61	4.0, 4.6	76, 82	Gravel, sand	11
		12/2/62	1.2	76 + 82		
Riverside	10/61	11/30/61	2.8, 2.8	100, 100	Red adobe	8
		11/16/62	0.3, 0.3			
Livermore	3/62	4/62	1.3, 1.3	25, 75	Loam, sand	14
		9/62	1.55	25 + 75		
Hayward	9/62	9/62	4.2, 2.8	22, 73	Gravel, large rocks	14
		10/62	3.2, 1.8	22, 22 + 73		
Hayward	3/26/62	4/3/62	1.29, 1.25	25, 75	Adobe, topsoil	14
Portland	2/62	7/62	1.1, 0.8	78, 90	Loam, sand	42.67
		12/62	1.1, 0.8			
Long Beach*	12/61	12/62	0.6		Heavy loam, sand	13.32
		3/25/63†	0.8			
Long Beach‡	8/61	8/62	4.5	30	Heavy loam, sand	13.32
		3/25/63†	5.4			
Twenty-nine Palms	2/8/63	3/22/63	4.3, 2.2§	50, 100	Decomposed granite, sand	0.5
			6.5, 4.4‖			
Palm Springs, Calif.¶	9/20/63	11/21/63**	1	100	Loam, sand	6.74

* Billboard of two 12-inch steel I-beams 8 feet in ground.
† Four days after rain.
‡ No. 6 wire embedded in garage floor.
§ Biddle megger.
‖ Siemens and Halske megger.
¶ More recent data.
** Heavy rain 10/20/63.

On November 1, 1961, 60 days after the concrete had been poured for the foundation, the test engineer for the California Electric Power Company measured the ground resistance with a Siemens & Halske ground megger and recorded a ground resistance of 4.6 and 4.0 ohms (see Table III).

The serving public utility is the authority responsible for the inspection and test of all electrical installations in the area. Their engineer authorized the electrical contractor to use this footing-type grounding electrode for grounding the electrical installation of this building and permitted the contractor to connect the free ends of both grounding conductors together to provide a grounding conductor with a total length of approximately 158 feet in the concrete footing. The ground resistance for the 158 feet measured 4.6 ohms.

During a regular trip to Bishop in December 1962, the engineer for the California Electric Power Company made the second test on the ground resistance of the footing-type grounding electrode and recorded a reading of 1.2 ohms.

Riverside Installation

The Chief Electrical Inspector of Riverside, Calif., arranged for the installation of footing-type grounding electrodes in the concrete footings of a residence erected in Riverside. Two No. 4 AWG bare copper wire electrodes, each 100 feet in length, were installed in the concrete footings of this residence. In all locations, the electrode wire was placed in the concrete forms to be spaced about 2 inches above the base of the footing. The author was at this location during construction to see that the electrode wires were properly placed above the base of the foundation footing before the concrete was poured for the foundation.

The Chief Electrical Inspector of Riverside supervised these installations. The first ground resistance readings were taken November 30, 1961, about 60 days after the concrete was poured, and the second readings were taken 1

year later; all readings are recorded as 2.8 ohms, for the years 1961 and 1962.

Livermore Installation

The City Engineer for Alameda County, Calif., selected the locations for the installation of footing-type grounding electrodes. Two residential buildings were provided with these grounding electrodes. One building was in Livermore, and two electrode wires, 25 and 71 feet long, were installed in the concrete foundation.

Ground resistance measurements were made by the Testing Engineer for the Pacific Gas and Electric Company. The first measurements were made April 3, 1962, 1 week after the concrete was poured and 12 days after rain. The second measurements were made September 20, 1962, 6 months after the concrete was poured. The engineer reported that the adobe and topsoil was very dry and cracked, since the last rain in the area consisted of 0.22 inch in April.

All readings were recorded as less than 2 ohms with very slight difference between the long and short wires in the first readings. For the second measurements, the resistance for the short and long wires were measured separately and when connected together were recorded as 1.5 ohms. Fig. 1(A) shows the position of the wires, while Fig. 1(B) gives the results of the measurements.

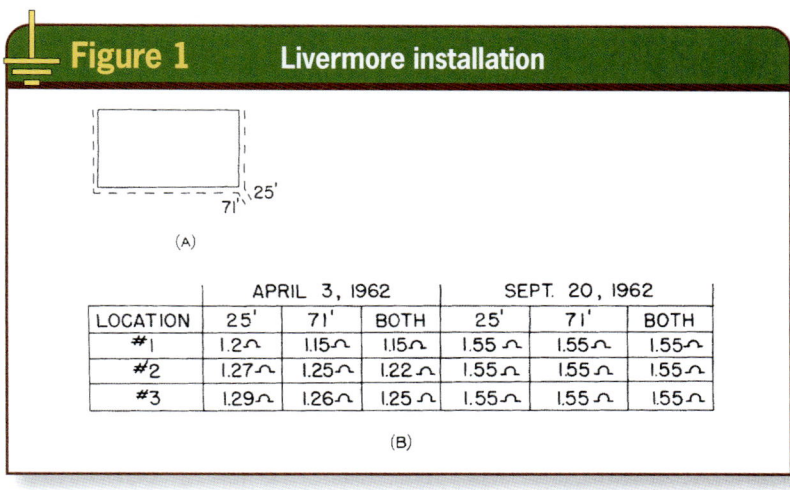

Figure 1   Livermore installation

(A)

	APRIL 3, 1962			SEPT. 20, 1962		
LOCATION	25'	71'	BOTH	25'	71'	BOTH
#1	1.2 Ω	1.15 Ω	1.15 Ω	1.55 Ω	1.55 Ω	1.55 Ω
#2	1.27 Ω	1.25 Ω	1.22 Ω	1.55 Ω	1.55 Ω	1.55 Ω
#3	1.29 Ω	1.26 Ω	1.25 Ω	1.55 Ω	1.55 Ω	1.55 Ω

(B)

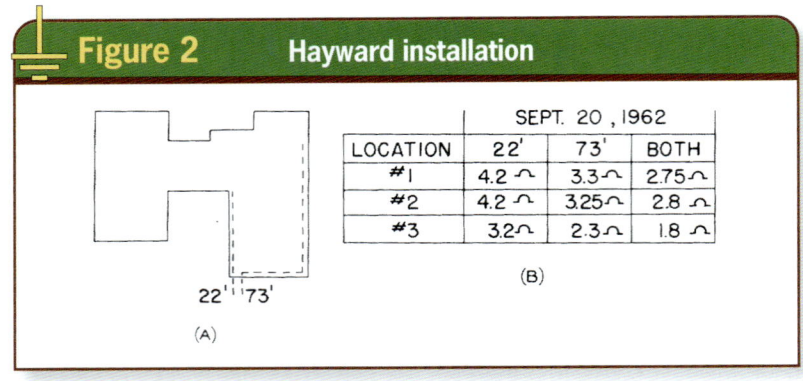

Figure 2	Hayward installation

SEPT. 20, 1962

LOCATION	22'	73'	BOTH
#1	4.2 Ω	3.3 Ω	2.75 Ω
#2	4.2 Ω	3.25 Ω	2.8 Ω
#3	3.2 Ω	2.3 Ω	1.8 Ω

(B)

22' 73'

(A)

Hayward Installations

The City engineer of Alameda County selected these locations in Hayward for a footing-type grounding electrode, since the soil was mostly rock and not dependable for a good ground. The last rain in this area was in April 1962.

Two No. 4 AWG bare copper wires, 22 and 73 feet long, were placed in the concrete foundation for this residence, as shown in Fig. 2(A). The foundation was poured on September 1, 1962.

Ground resistance readings were measured on September 20, 1962, by a test engineer of the Pacific Gas and Electric Company, the serving public utility in this area. As shown in Fig. 2(B), the 22-foot wire measured 4.2 ohms; the 73-foot wire, 3.3 ohms; both wires (95 feet), 2.75 ohms. Measurements taken October 1962 were 3.2 ohms for the 22-foot wire, 1.8 for the 73-foot wire, and 1.8 ohms for both wires.

At a second location, where the earth was a mixture of adobe and topsoil, two wires were placed in the foundation wall

as shown in Fig. 3. Readings were made by an engineer from the Pacific Gas and Electric Company on April 3, 1962, 1 week after the footing was poured; the weather was dry, since the last rain had been on March 22, 1962. See Fig. 3 and Table III for the results. A 25-foot ground wire is apparently adequate in this location as the resistance measured less than 2 ohms for both the long and short wires.

Portland Installation

The Chief Electrical Inspector of Portland, Ore., selected the location for a footing-type grounding electrode. Two No. 4 AWG bare copper wires, approximately 78 and 100 feet long, were installed in the concrete foundation of a 1-story commercial warehouse with dimensions of 80 by 100 feet. The building foundation, the floor slab and the 6-inch precast walls were reinforced concrete, and a wood truss roof on wood columns was provided.

The ground resistance tests recorded 0.8 ohm. The power company then tested the water pipe ground and recorded 1.1 ohms. They decided that the probable cause of the 0.8-ohm reading was that the footing ground wire might possibly have been touching the water pipe. The author visited this installation to check the test results but found that the ground wires had been cut off where they extended beyond the faces of the concrete foundation.

The No. 4 AWG wires installed in the concrete foundation contacted the steel reinforcing rods, but apparently this did not affect the values of the ground resistance readings.

One of the engineers of the serving public utility cut away the concrete where the wire had been cut and was able to attach the testing conductor to this wire; he recorded a ground resistance of about 1 ohm. Because of the location of this building and the availability of an adequate water pipe system for grounding the electrical installation, the water pipe is being used for the ground since the ends of the No. 4 grounding wires are not accessible.

Figure 3	Second Hayward installation

WELL CASING

#4 STRANDED BARE COPPER

#4 STEEL REINFORCING BARS

X Y

{ x−1=1.15, x−2=1.25, x−3=1.26 ohms
y−1=1.2, y−2=1.27, y−3=1.29 ohms
(x+y)−1=1.15, (x+y)−2=1.22, (x+y)−3=1.25 ohms }

Long Beach Installations

The first installation consists of a steel frame sign mounted on two 12-inch steel I-beams spaced 25 feet apart. The ends of the I-beams stand 8 feet in the ground. The ground resistance measured 0.6 ohms.

The second Long Beach installation consists of 30 feet of No. 6 AWG bare copper wire embedded in the floor slab of a 2-car garage. The wire is in the shape of a circular loop with one end extending beyond the face of the floor slab for connection to the neutral terminal of the service switch for this installation.

Ground resistance readings were made by the Chief Electrical Inspector of Long Beach with a *Model 263A* Vibroground. The resistance measured 4.5 ohms.

The second resistance tests for both Long Beach installations were made 4 days after rain.

Twenty-Nine Palms Installation

The Chief Electrical Inspector for San Bernardino County arranged to have the footing-type ground electrode installed in the concrete foundation of the Methodist Church of Twenty-nine Palms. This is a desert location, and ground resistance tests recorded for this area by other engineers indicated that an adequate ground is not available.

The church is a 1-story building approximately 35 by 70 feet with a wood-frame stucco construction on a poured concrete foundation. The concrete footing is T-shaped 12 inches below grade; the base is 14 inches wide, while the upper end is 6 inches wide with ½-inch bolts embedded in the concrete for attaching the foundation sills.

Two No. 4 AWG bare copper wire electrodes were used, one 50 and the other 100 feet long. One end of the 50-foot wire extended beyond the face of the foundation wall on the north elevation of the building (point C), while the other end of this 50-foot conductor extended beyond the face of the foundation on the west elevation (point A). At point A, one end of the 100-foot electrode wire extends from the face of the foundation wall, while the other

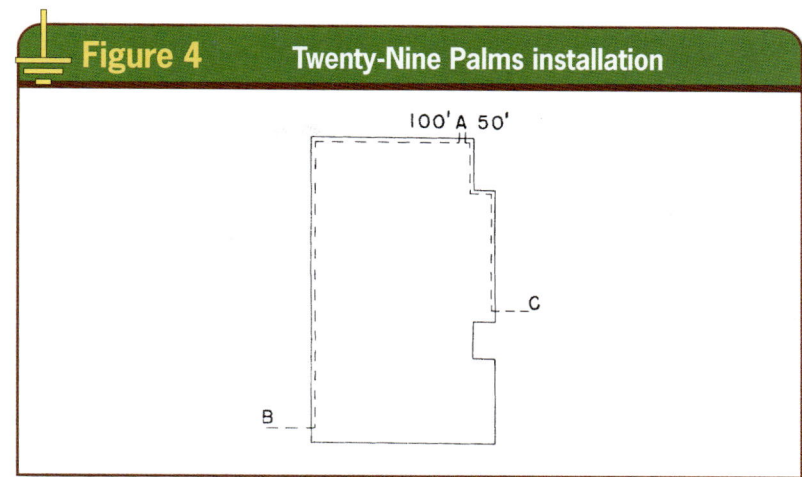

Figure 4   **Twenty-Nine Palms installation**

end of this 100-foot wire extends itself from the face of the east elevation (point B) of the concrete foundation wall of this building (refer to Figure 4).

Ground resistance measurements were taken by the Test Engineer for the California Electric Power Company using a Siemens & Halske ground megger *No. 2585269* with a range of 0–25 ohms and 0–2,500 ohms. These readings were checked with a Biddle ground megger *No. 167273* having three ranges, 0–3, 0–30, and 0–300 ohms.

As bare wire leads were used to record the readings taken with the Siemens & Halske megger, it was decided to use bare and insulated wire for connections from the megger to each of two reference grounds and to connect the third lead to the ground being tested. The results are recorded in Table IV.

Table IV

Location of Ground Lead	Type of Lead	Length of Grounding Conductor (Feet)	Resistance (Ohms) Siemens and Halske Megger	Biddle Megger
Point C	Insulated wire	50	6.0	4.3
		150*	4.4	2.2
		50	6.5	4.3
Point A	Bare wire	50	3.1	5.5
		100	2.0	3.2
		150*	2.0	3.2

*Short and long wires.

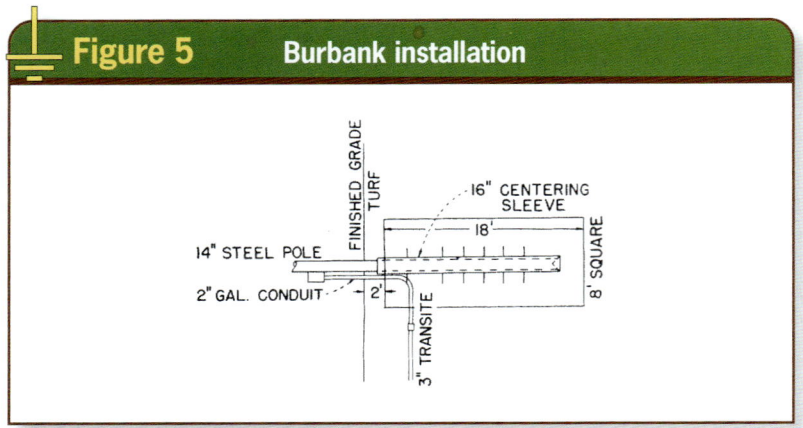

Figure 5 Burbank installation

or less using a Biddle megger and a recognized test method.[2]

Burbank Installation

During 1952, a steel floodlight pole was installed in the athletic field of John Burroughs High School, Burbank, Calif. The pole is 14 inches in diameter at the base and tapers to a diameter of 8 inches at the top; it is 100 feet long over-all and is set in a centering sleeve from the base to a height of 16 feet. This assembly is set in a reinforced concrete footing 8 by 8 by 18 feet. (See Fig. 5.)

Since installation, the resistance to ground has been checked at least once each year by the Burbank city engineers. On July 2, 1963, the ground resistance was checked with their ground megger and measured 1.8 ohms. The soil is sandy loam, the soil resistivity in this area is 4,000 ohms per cm^3 (Table II), and the average rainfall is about 10 inches.

An engineer for the California Electric Power Company reported a grounding installation in a residence in San Bernardino County where a 4-inch concrete floor slab was poured over a membrane. The hot and cold water pipes enclosed in concrete were the grounding conductors. An insulating joint was installed in the water supply pipe lateral. The engineer for the California Electric Power Company measured the ground resistance of the copper water pipes enclosed in the concrete floor slab as over 10 ohms. He then measured the ground resistance of the underground water pipe lateral a short distance from the insulating joint as 1.3 ohms.

LENGTH OF ELECTRODES

Installations were made at Alameda and San Bernardino Counties where the soil conditions were not favorable for an adequate ground. At Hayward and Livermore, the soil was poor and filled with large rocks. In these locations, electrode wires 22 and 25 feet long were installed (Table III).

An installation was also made in the desert at Twenty-nine Palms where the soil conditions are not favorable for an adequate ground. The ground resistance readings for these installations, including the 50-foot electrode at Twenty-nine Palms, record measurements of 5 ohms

CONCLUSION

The 20-year history of the military installation in Arizona, together with the shorter records of installations in conventional residential and small commercial buildings, apparently establishes the adequacy and advisability of footing-type grounding electrodes where a continuous underground water system is not available or dependable; in addition, it indicates that the grounding conductor ought to be placed about 2 inches above the base of the concrete foundation footing.

REFERENCES

1. Guide for Measuring Ground Resistance and Potential Gradients in the Earth. *AIEE Report No. 81,* July 1960.
2. Master Test Code for Resistance Measurement. *AIEE Standard No. 550,* May 1949.

Steel Conduit and EMT for Equipment Grounding

In addition to *system* grounding, electrical systems require *equipment* grounding and bonding to safeguard personnel and protect equipment. Properly sized and bonded equipment grounding conductors (EGCs) ensure that all metal parts of electrical equipment are at the same electrical potential as the earth to prevent electric shock and provide a low-impedance path to facilitate the operation of the circuit overcurrent protective devices.

Section 250.118 of the *National Electrical Code (NEC)* recognizes several types of conductors that are permitted to be used as EGCs. In other words, they qualify as EGCs because they provide an effective path for ground-fault current. Rigid metal conduit (RMC), intermediate metal conduit (IMC), and electrical metallic tubing (EMT) are included in 250.118 as list items (2), (3), and (4) respectively, followed by other types of metal raceways, metal cable trays, and metal cables that qualify as EGCs.

Steel conduit and EMT are widely used in secondary power distribution systems both indoors and outdoors. Electrical systems on the load side of the service point are required to be designed in such a way that the steel conduit or EMT does not carry any appreciable electric current under normal operating conditions but performs grounding and bonding functions. During ground-fault conditions, the metal conduit or EMT, acting as an EGC, will carry most of the return fault current, or, in some cases, will be the only return path of the fault current to the source. This is due to the conductive path of the conduit or tubing, which is usually larger in surface area compared to any wire-type EGC contained within the conduit or tubing run. The ground-fault current divides over all paths available as it returns to the source. The metal conduit around a contained EGC provides impedance in circuit, which will limit or choke the conductor as it is subjected to the heavier levels of ground-fault current. As a result, most of the fault current will be present in the metallic raceway containing the wire-type EGC.

In reality, the conduit is only one of the fault current return paths. **See Figure B-1.**

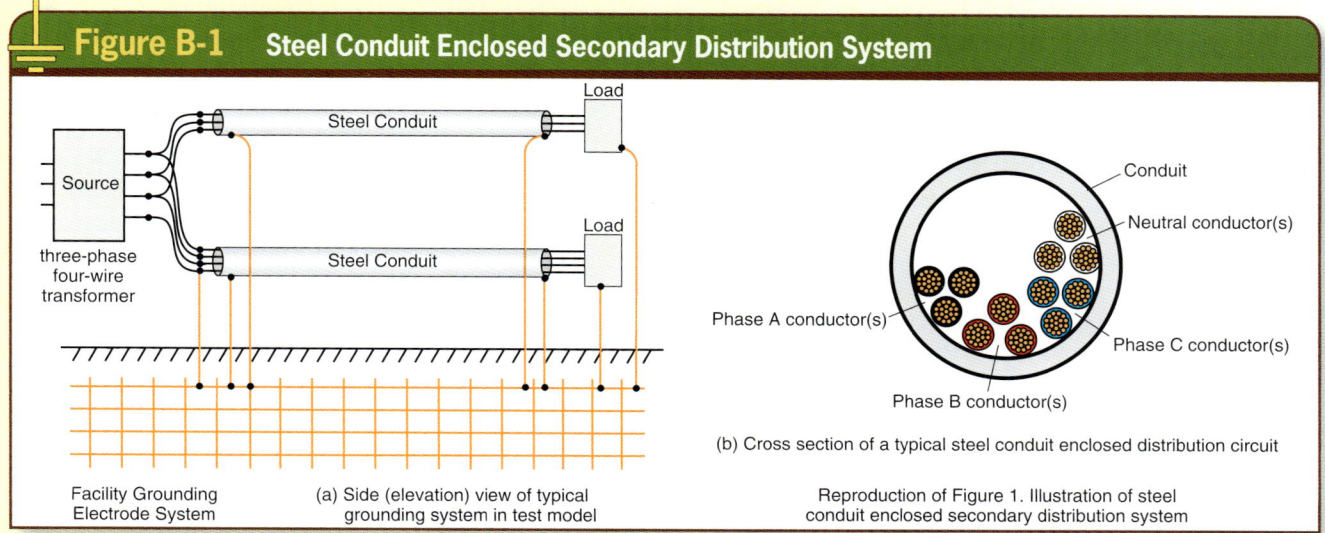

Figure B-1   Steel Conduit Enclosed Secondary Distribution System

Load

Steel Conduit

Source

Load

Steel Conduit

three-phase
four-wire
transformer

Facility Grounding
Electrode System

(a) Side (elevation) view of typical
grounding system in test model

Conduit

Neutral conductor(s)

Phase A conductor(s)

Phase C conductor(s)

Phase B conductor(s)

(b) Cross section of a typical steel conduit enclosed distribution circuit

Reproduction of Figure 1. Illustration of steel
conduit enclosed secondary distribution system

Figure B-1. Diagram of Conduit Run Model

Specifically, in a practical system, the fault current will split among several parallel paths when returning to the source. For example, for a phase-to-neutral-to-ground fault in the bottom conduit, the fault current will return to the source through the conduit (path BA), neutral (path DC), and facility ground (path FE).

For a phase-to-conduit fault, the fault current will return to the source through the conduit and facility ground. It is possible that for a phase-to-conduit fault, the only return path for the fault current is the conduit. For a single phase-to-neutral conductor fault, the fault-current return path is through the neutral conductor; the neutral conductor may have higher impedance than the previously noted fault-current paths.

Therefore, the type of fault becomes the limiting factor. This report is focused on the performance of steel conduit as the EGC, so the situation where the only fault path is the steel conduit was used as the worst-case condition. It is important to discuss other fault conditions that represent worst-case scenarios and determine the design procedure of electrical installations. For a single phase-to-neutral conductor fault, the fault-current return path is through the neutral conductor only. The steel conduit does not participate in the fault circuit. For neutral conductor sizes, as recommended in standards, the impedance of the fault-current path for a neutral conductor is higher than the impedance for a fault to the steel conduit. Consequently, for these cases, the maximum allowable length of the circuit is dictated by the phase-to-neutral conductor fault. **See Figure B-2.**

NEC REQUIREMENTS

Article 250 of the *NEC* contains the general requirements for grounding and bonding of electrical installations, as well as other specific requirements. Sections 250.4(A) and (B), titled "General Requirements for Grounding and Bonding," set forth performance requirements that detail what must be accomplished by the grounding and bonding of non–current-carrying parts of the electrical system. The conductive parts must form an effective low-impedance path to the source to safely conduct any fault current and facilitate the operation of overcurrent protective devices protecting the enclosed circuit conductors and any equipment in the circuit.

Part VI of Article 250 specifically covers equipment grounding. This part of the article includes the list of acceptable EGCs in Section 250.118.

For metallic conduit and EMT to perform effectively as EGCs, it is crucial that they are installed properly with

Figure B-2   Current Paths

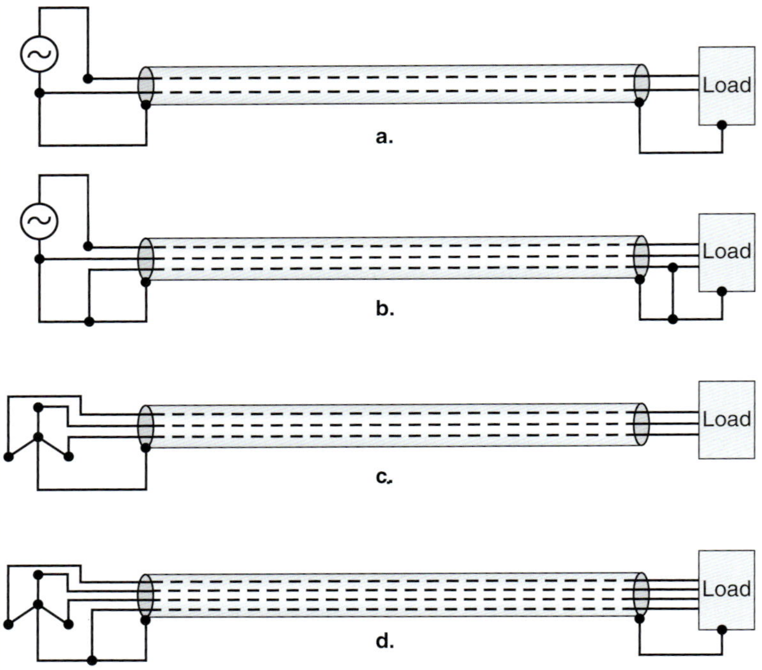

a.

b.

c.

d.

Reproduction of Figure 2.1 a. through d. Illustration of simplified installations

a. Single-phase circuit without equipment grounding conductor
b. Single-phase circuit with equipment grounding conductor
c. Three-phase circuit without neutral conductor
d. Three-phase circuit with neutral conductor

Figure B-2. Current, be it normal current or ground-fault current, will take any and all paths to return to the power source. The amount of current in a particular path is related to the amount of impedance in that path.

tight joints at every fitting. Skillful and professional installation in electrical construction is extremely important for the integrity of the safety system. If a fault occurs, this helps ensure a continuous, low-impedance path back to the overcurrent protective device, which will open the circuit.

If the joints are not made up tight and there is a break in the circuit under fault conditions, there is a possibility of electric shock for anyone who comes in contact with the conduit. *NEC* Sections 300.10 (Electrical Continuity) and 300.12 (Mechanical Continuity) state that metal raceways, cable armor, and other metal enclosures for conductors shall be metallically joined together into a continuous electrical conductor and shall be connected to all boxes, fittings, and cabinets so as to provide effective electrical continuity. Section 250.120 of the *NEC* requires that all connections, joints, and fittings "shall be made tight using suitable tools."

Section 250.122 covers the sizing of wire-type EGCs and includes Table 250.122 titled "Minimum Size Equipment Grounding Conductors for Grounding Raceway and Equipment." An important mandatory note at the bottom of Table 250.122 makes it clear that the conductor sizes given in Table 250.122 may not be adequate to comply with 250.4(A)(5), *Effective Ground-Fault Current Path* (Grounded Systems) and 250.4(B)(4), *Path for Fault Current* (Ungrounded Systems). These installations may have to be evaluated to ensure that they can provide the required effective ground-fault current path.

CONDUIT AS EFFECTIVE GROUND-FAULT CURRENT PATH

The *NEC* does not dictate any particular size of conduit or tubing to serve as the EGC for an upstream overcurrent protective device. A metal raceway sized properly for the conductor fill will provide an adequate equipment ground-fault return path.

In 1966, Eustace C. Soares, a renowned expert in the area of grounding electrical systems, published the first edition of his book, *Grounding Electrical Distribution Systems for Safety*. It included tables showing acceptable lengths of steel conduit and EMT (metal raceways) for equipment grounding based on his calculations. The work of Soares did not include any field testing of conduit systems to measure performance as effective paths for fault current.

In the early 1990s, U.S. steel conduit producers decided to undertake a research project in order to confirm the work done by Soares to provide scientific proof that steel conduit and EMT do provide an adequate equipment ground-fault return path and to develop software that would assist engineers and others in determining the appropriate maximum lengths of steel conduit and EMT serving as EGCs.

GEORGIA INSTITUTE OF TECHNOLOGY GROUNDING RESEARCH

One of the top experts in the field of grounding, Dr. Sakis Meliopoulos, professor of Electrical and Computer Engineering at Georgia Tech, facilitated the EGC research project, which was completed in 1994. This research represented the first substantial update on the impedance and permeability of steel conduit in over forty years.

The first phase of the grounding research at Georgia Tech consisted of resistance and permeability testing of various steel raceways that were purchased from local distributors. Based on this information, Dr. Meliopoulos and his team developed a computer model and validated the results through actual field testing. The next step was to develop a computer software program that allowed the user to calculate the appropriate length of steel conduit or EMT runs necessary to meet *NEC* requirements.

It is important to discuss other fault conditions that represent worst-case conditions and determine the design procedure of electrical installations. For a single phase-to-neutral conductor fault in the systems of Figures B-2(a), B-2(b), and B-2(d), the fault-current return path is through the neutral conductor only. The steel conduit does not participate in the fault circuit.

For neutral conductor sizes, as recommended in standards, the impedance of the fault-current path for these cases is higher than the impedance of the steel conduit. Consequently, for these cases, the maximum allowable length of the circuit is dictated by the phase-to-neutral conductor fault.

In recent years, fault-current levels have increased. For this reason, it is important to examine existing parameters for EGCs. A program was initiated to evaluate the performance of steel EMT, IMC, or RMC during faults in secondary distribution systems.

A relevant issue is that of the grounding of steel conduit. For many technical and safety reasons, electric power installations must be grounded. Two main considerations are: (a) protection in the event of faults and (b) avoidance of electric shocks.

As the capacity of secondary distribution systems increases, so does the short-circuit capacity (available fault-current levels) and associated protection and safety concerns. Performance of EGCs can be best determined by exact modeling and testing of the system under various excitation and fault conditions. This research project did exactly that.

The above issues have been investigated by performing the following tasks: (1) modeling of steel-conduit-enclosed multi-conductor systems, (2) laboratory testing of several representative steel conduit types under low currents for the purpose of characterizing the magnetic material, (3) computer simulation of steel-conduit-enclosed secondary distribution systems, (4) full-scale testing of steel-conduit-enclosed power systems under high fault current, (5) full-scale measurement of arc voltage for various fault-current levels, and (6) analysis of full-scale test results and conclusions.

Details of these tasks are included in the Georgia Institute report.

A summary of the results of the investigation is as follows:

- A model of steel-conduit-enclosed power distribution systems has been developed and validated with laboratory and full-scale measurements 256 feet in length.
- Comparably sized steel EMT, IMC, and RMC will allow the flow of higher fault current than an EGC as listed in *NEC* Table 250.122.
- Steel EMT, IMC, and RMC that does not exceed the maximum allowable length meets the performance requirements of Section 250.4 of the *NEC*. As a matter of fact, the performance of steel conduit sized in accordance with Chapter 9, Table 1 of the *NEC*, when compared to the minimum size EGCs in Table 250.122, allows the flow of higher fault current. This is due to the lower impedance of the steel conduit.
- Steel EMT, IMC, and RMC are of sufficiently low impedance to limit the voltage to ground and facilitate the operation of the circuit protective devices in runs not exceeding the maximum allowable lengths detailed in this report. In most cases, the maximum allowable lengths exceed those permitted by the IAEI Soares *Book on Grounding* using the same arc voltage and ground-fault current.
- The arc voltage of 50 volts and the ground-fault current of 500% of the overcurrent protective device rating, as stated in the IAEI Soares *Book on Grounding* 1993 edition, is overly conservative as a design guide. Testing for this project confirmed the actual voltage across a fault arc to be up to 30 volts for an arc length of 75 mils and fault current ranging from 400 to 2,500 amperes.
- A validated computer model has been developed that computes the

maximum allowable lengths for specific conductor size enclosed by specific sizes of steel EMT, IMC, or RMC conduit under fault conditions.

- Where lengths do not exceed the maximum allowable length computed by the method, supplemental grounding conductors in secondary power systems enclosed in steel EMT, IMC, or RMC are not necessary. The supplemental conductor is sometimes required by the *NEC* in critical installations such as healthcare areas where dual protection for patients is considered prudent.

- The recommended maximum allowable lengths listed in the IAEI Soares *Book on Grounding* were compared to those computed with the validated model for the same arc voltage and ground-fault current. In general, it was found that the IAEI Soares *Book on Grounding* numbers are in reasonable agreement, except in some cases where errors were found. The detailed comparison and explanation is given in Section 5 of the report.

- The maximum allowable length for a specific system depends on conductor size, steel conduit size, and fault type. In many cases, the maximum allowable length for a phase-to-neutral fault is shorter than the maximum allowable length for a phase-to-steel conduit fault. Thus, in most cases, the steel conduit is *not* the limiting factor. In these cases, use of a supplemental grounding conductor will not increase the maximum allowable length. The use of the validated computer model to compute the maximum allowable length in specific systems is recommended.

FINDINGS IN GEORGIA TECH GROUNDING RESEARCH

A few years later, Georgia Tech conducted research on steel conduit to show how steel conduit reduces electromagnetic fields. This research was added to the grounding research and ultimately rolled into the GEMI (Grounding and Electro Magnetic Interference) software analysis program, available for free download at www.steelconduit.org.

The GEMI research project and resulting software analysis program proved that listed steel conduit and EMT clearly exceed the minimum equipment grounding requirements of the *NEC*. In addition, the GEMI research on grounding verified the following:

- Comparably-sized steel RMC, IMC, and EMT allow the flow of higher fault current than an aluminum or copper EGC as provided in *NEC* 250.118.

- RMC, IMC, and EMT provide a low-impedance path to ground and facilitate the operation of the overcurrent protective devices in runs not exceeding the maximum allowable lengths detailed in the research report.

- Where lengths do not exceed the maximum allowable length computed by the GEMI software, supplemental grounding conductors in secondary power systems enclosed in steel EMT, IMC, or RMC do not add to safety in a phase-to-neutral fault. The use of supplementary EGCs, when participating in the fault circuit, reduce the overall impedance and may or may not increase the allowable length of the run, depending on the size and system design.

- The maximum allowable length for a specific system depends on conductor size, steel conduit size, and fault type. In many cases, the maximum allowable length for a phase-to-neutral fault is shorter than the maximum allowable length for a phase-to-steel conduit fault. Thus, in most cases, the steel conduit is not the limiting factor in a conductor-to-neutral fault.

For additional information, visit qr.njatcdb.org Item #2610

SUMMARY

A computer model has been developed that computes the impedance of steel conduit with enclosed power conductors. The computer model has been validated with full-scale tests. The model can predict the effect of temperature on the total impedance. The measured impedances during the full-scale tests are within the values predicted with the model in the temperature range 25°C to 55°C. It is important to note that the full-scale tests consisted of repeated short circuit tests on samples of steel conduit systems, and therefore the pre-fault temperature was different for each test, but it was generally in the above stated range of temperatures.

Tests of arc voltage were performed. These tests consisted of generating an arc between two electrodes. The current through the arc was controlled by current-limiting impedance. The separation distance between the two electrodes was measured before and after the test. This separation distance was always longer than the thickness of power conductor insulation and, therefore, represents worst-case (or conservative) results.

Full-scale tests were performed with and without supplementary grounding conductors. Supplementary grounding conductors, when participating in the fault circuit, reduce the overall impedance. However, it is important to note that the limiting factor in the capability of the system to interrupt a fault is the size of the phase conductor and neutral. Specifically, for systems designed with present standards, the fault circuit for a phase-to-neutral fault (where a steel conduit or supplementary ground conductor is not involved in the fault) presents the maximum impedance and, therefore, will draw the minimum fault current as compared to other faults at the same location. Use of supplementary EGCs does not add to the safety of the systems in these instances. The maximum allowable length of a steel conduit run can be increased to some degree by the addition of a supplemental grounding conductor. This varies by size and system design. An additional project has been undertaken to expand the model to compute the maximum allowable length in these cases. **See Figure B-3.**

The power source utilized for the tests had the following short circuit characteristics:
- Available fault current at the 120 V tap: 85k A
- Available fault current at the 277 V tap: 83k A

The waveforms of the electric current and voltage were recorded by means of a strip chart recorder, as well as by means of a digitizing scope and subsequent transfer of the data to a personal computer for further analysis. Georgia

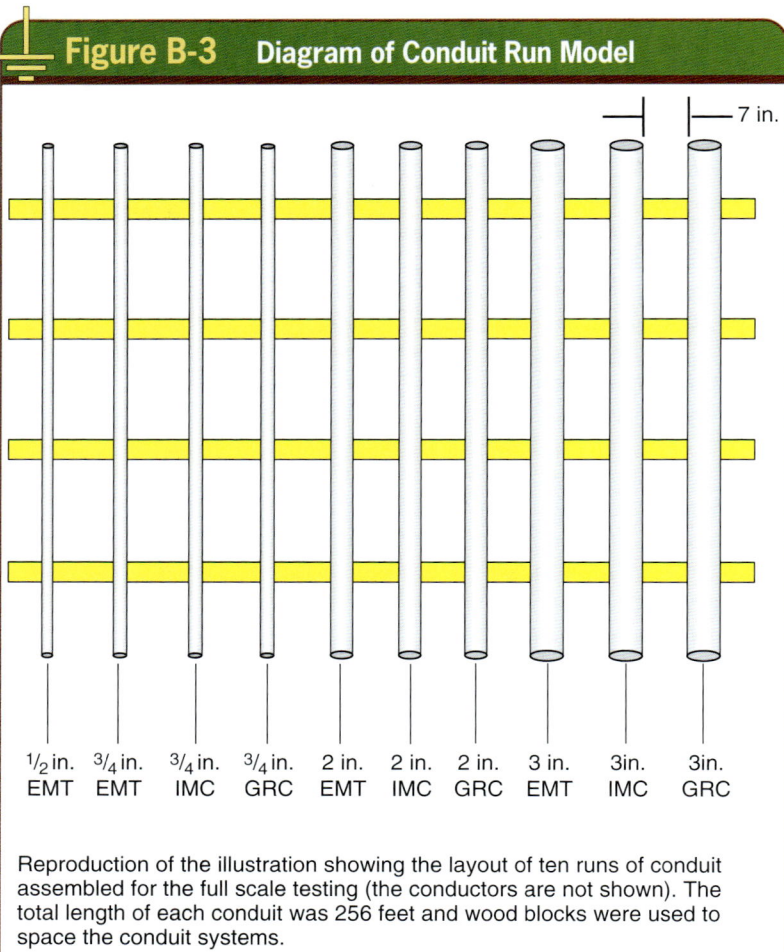

Figure B-3 **Diagram of Conduit Run Model**

7 in.

| ½ in. EMT | ¾ in. EMT | ¾ in. IMC | ¾ in. GRC | 2 in. EMT | 2 in. IMC | 2 in. GRC | 3 in. EMT | 3in. IMC | 3in. GRC |

Reproduction of the illustration showing the layout of ten runs of conduit assembled for the full scale testing (the conductors are not shown). The total length of each conduit was 256 feet and wood blocks were used to space the conduit systems.

Figure B-3. These conduit runs, installed for testing purposes, are 256 feet in length and spaced seven inches apart.

Tech provided the digitizing oscilloscope and personal computer.

In this report, the Georgia Institute also examined the practice as provided by the original Soares *Book on Grounding*. The findings suggested to use 500% of the overcurrent protective device rating as the ground-fault current for interruption of the fault to be overly conservative. Present industry experience with protective devices indicates that a fault current of 300% of the overcurrent protective device rating will result in a reliable operation of the overcurrent protective device. Current practice of overcurrent protective device testing is to test the overcurrent protective device for an electric current up to 300% of its rating. Based on these observations, it could be concluded that even the value of 400% of the overcurrent protective device rating for ground-fault current would be conservative.

All computations of maximum run and all conclusions are based on the impedance of the steel-conduit-enclosed power distribution system. For this reason, it is expedient to base the comparison of the experimental results and the modeling results on the impedance of the steel-conduit-enclosed distribution system. This comparison is summarized in Figures B-4 and B-5 for systems with 120- and 277-volt nominal voltage, respectively. In addition, Figures B-6 and B-7 summarize the measured data for all cases. The tables represent the summary of all tests done at Kearney Laboratories in McCook, IL (the May 1993 and August 1993 tests). For the complete NEMA Report of the Georgia Institute of Technology research project on Modeling and Testing of Steel EMT, IMC, and RIGID (GRC) Conduit, visit www.steelconduit.org.

Workers pulled wire and prepared the conduit models for ground-fault testing.

The conduit testing model was prepared to simulate a variety of ground-fault current injection tests.

For additional
information, visit
qr.njatcdb.org
Item #2610

TABLES REFLECTING MAXIMUM LENGTH OF METALLIC CONDUIT AS EFFECTIVE GROUND-FAULT CURRENT PATH

Figure B-4 Summary of Test Results and Model Comparison for 120-Volt Systems			
Steel Conduit Type and Conductor Size	**Electric Current Level (Amperes)**	**Measured Impedance R, X**	**Computed Impedance R,X**
EMT - 1/2Z CU-#10	247	187.9, 30.55	188.71, 24.72
EMT - 1/2S CU-#10	228	199.3, 32.15	189.05, 26.65
EMT - 3/4 CU-#8	345	130.1, 25.33	121.11, 21.59
IMC - 3/4 CU-#8	437	102.0, 29.63	105.29, 27.55
GRC - 3/4 CU-#8	417	106.8, 30.13	105.31, 31.65
EMT - 2 CU-3/0	1,822	21.56, 10.64	22.37, 8.96
IMC - 2 CU-3/0	1,849	19.78, 12.89	18.83, 12.56
GRC - 2 CU-3/0	1,758	21.22, 13.04	18.82, 13.41
EMT - 3 CU-350	2,578	13.15, 9.86	13.02, 10.44
IMC - 3 CU-350	2,609	12.69, 10.21	11.56, 10.43
GRC - 3 CU-350	2,433	13.98, 10.86	11.81, 10.41

Z – zinc diecast coupling
S – steel coupling
CU – copper conductor

Figure B-4. *Steel conduit is the only return path. Impedance values are given in milliohms per 100 feet.*

Steel Conduit Type and Conductor Size	Electric Current Level (Amperes)	Measured Impedance R, X	Computed Impedance R,X
EMT - 1/2Z CU-#10	537	193.1, 17.76	187.49, 14.84
EMT - 1/2S CU-#10	546	193.7, 18.6	187.47, 14.66
EMT - 3/4 CU-#8	851	123.9, 15.5	119.64, 12.20
IMC - 3/4 CU-#8	1,066	95.1, 18.98	100.79, 115.57
GRC - 3/4 CU-#8	997	95.9, 20.25	97.63, 18.36
EMT - 2 CU-3/0	4,628	20.39, 7.28	21.95, 6.17
IMC - 2 CU-3/0	5,422	15.78, 8.40	16.64, 7.53
GRC - 2 CU-3/0	5,271	15.71, 9.59	15.12, 8.15
EMT - 3 CU-350	7,092	11.33, 6.94	11.72, 6.84
IMC - 3 CU-350	6,775	9.71, 8.35	8.69, 7.52
GRC - 3 CU-350	6,920	10.18, 8.68	8.40, 7.87

Figure B-5 Summary of Test Results and Model Comparison for 277-Volt Systems

Z – zinc diecast coupling
S – steel coupling
CU – copper conductor

Figure B-5. Steel conduit is the only return path. Impedance values are given in milliohms per 100 feet.

Figure B-6   Summary of Test Results and Model Comparison for 120-Volt Systems

Steel Conduit Type and Conductor Size	Electric Current Level (Amperes)	Measured Impedance R, X	Measured Impedance with a Supplementary Conductor	Measured Impedance with a Supplementary Conductor and Earth Return
EMT - 1/2Z CU-#10	247	187.9, 30.55	157.2, 17.53	137.77, 62.59
EMT - 1/2S CU-#10	228	199.3, 32.15	163.8, 17.53	137.8, 62.6
EMT - 3/4 CU-#8	345	130.1, 25.33	105.6, 15.56	99.2, 60.05
IMC - 3/4 CU-#8	437	102.0, 29.63	89.84, 16.53	90.43, 16.33
GRC - 3/4 CU-#8	417	106.8, 30.13	91.19, 15.87	91.69, 16.08
EMT - 2 CU-3/C	1,822	21.56, 10.64	11.34, 10.34	11.38, 10.35
IMC - 2 CU-3/0	1,849	19.78, 12.89	11.29, 10.57	11.30, 10.58
GRC - 2 CU-3/0	1,758	21.22, 13.04	11.42, 10.46	11.44, 10.43
EMT - 3 CU-350	2,578	13.15, 9.86	6.89, 10.15	6.93, 10.20
IMC - 3 CU-350	2,609	12.69, 10.21	6.95, 9.96	7.03, 10.00
GRC - 3 CU-350	2,433	13.98, 10.86	6.89, 10.15	6.94, 10.13

Z – zinc diecast coupling
S – steel coupling
CU – copper conductor

Figure B-6. *Effect of other parallel fault current return paths. Impedance values are given in milliohms per 100 feet.*

Figure B-7 Summary of Test Results and Model Comparison for 277-Volt Systems

Steel Conduit Type and Conductor Size	Electric Current Level (Amperes)	Measured Impedance R, X	Measured Impedance with a Supplementary Conductor	Measured Impedance with a Supplementary Conductor and Earth Return
EMT - 1/2Z CU-#10	537	193.1, 17.76	161.92, 13.77	135.88, 48.66
EMT - 1/2S CU-#10	546	193.7, 18.6	161.4, 13.86	134.6, 48.62
EMT - 3/4 CU-#8	851	123.9, 15.5	102.44, 12.51	89.70, 47.00
IMC - 3/4 CU-#8	1,066	95.1, 18.98	88.48, 15.14	88.54, 15.55
GRC - 3/4 CU-#8	997	95.9, 20.25	88.16, 15.91	90.06, 15.23
EMT - 2 CU-3/0	4,628	20.39, 7.28	11.87, 8.87	11.90, 8.87
IMC - 2 CU-3/0	5,422	15.78, 8.40	11.23, 8.97	11.18, 8.95
GRC - 2 CU-3/0	5,271	15.71, 9.59	11.23, 9.23	11.28, 9.24
EMT - 3 CU-350	7,092	11.33, 6.94	7.15, 8.63	7.11, 8.59
IMC - 3 CU-350	6,775	9.71, 8.35	6.74, 8.32	6.81, 8.36
GRC - 3 CU-350	6,920	10.18, 8.68	7.14, 8.70	6.91, 8.69

Z – zinc diecast coupling
S – steel coupling
CU – copper conductor

Figure B-7. *Effect of other parallel fault current return paths. Impedance values are given in milliohms per 100 feet.*

COMPLETE REPORT AND OTHER INFORMATION

The GEMI software is available for free download at www.steelconduit.org. Click on "Resources/GEMI Analysis Software." This is where you can also find copies of the two complete Georgia Institute of Technology research reports on grounding and on shielding against electromagnetic fields.

This Annex provided simplified information related to testing the effectiveness of steel conduit systems as ground-fault current paths. Full information is provided in the GEMI analysis reports developed by the Georgia Institute of Technology.

All information contained in Annex B is courtesy of Steel Tube Institute.

For additional information, visit qr.njatcdb.org
Item #2610

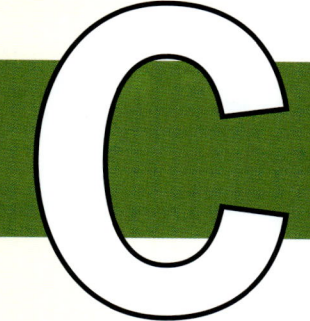
Selecting Protective Devices - Short-Circuit Current Calculations

A key factor in protecting electrical equipment and persons from electrical hazards is the proper application of overcurrent protective devices such as circuit breakers and fuses. Section 110.10 of the *NEC* requires the total circuit impedance, overcurrent protective devices, component short-circuit current ratings, and other circuit characteristics to be selected and coordinated so that the circuit protective devices can effectively respond and operate in a fault event. Overcurrent protective devices such as circuit breakers and fuses must be appropriately applied to ensure that the short-circuit current rating of any system component is not exceeded should a short circuit or high-level ground fault occur.

The term *available fault current* (sometimes referred to as *short-circuit current*) of a system is the largest current that a system can deliver at any point on a wiring system. The level of available fault current in a system is typically highest at the source. The amount of available fault current can usually be obtained from the serving utility company, or it can be calculated conservatively by using an infinite bus value. This value must be known to properly calculate the amount of fault current at any point on the electrical system from the source to the furthest outlet.

This Annex provides significant useful information about applying overcurrent protection in electrical systems. Multiple examples are provided to show how to accomplish this, including methods to accomplish selective coordination.

A handy fault current calculator is available as a mobile application for smart devices and can assist in determining fault-current levels when these values are in question.

For additional information, visit qr.njatcdb.org Item #5349

Selecting protective devices

4 Power system analysis

Contents

Contents	Section page
4.1 Fault current calculations	1
4.2 Selective coordination	11
4.3 Arc Flash	30

4.1 Fault current calculations

Contents

Contents	Section page
4.1.1 Overview	1
4.1.2 Code requirements	1
4.1.3 One- line diagram	1
4.1.4 Procedures and methods	3
4.1.5 Point-to-point calculation method	4
4.1.6 Equipment evaluations	8
4.1.7 Tables	8

4.1.1 Overview

The fault current calculation is the most basic calculation performed on a power distribution system and vital for the proper electrical equipment application. There are several NEC sections with requirements directly pertaining to the proper electrical product application and available fault current. Safe and reliable electrical equipment application, including OCPDs, relies on such power systems analysis study information obtained from fault current and selective coordination studies.

4.1.2 Code requirements

Knowing available fault current throughout the power distribution system is important for proper product application. The NEC recognizes the importance of fault currents in many areas within its requirements, including these important topics and sections:

Available fault current markings

- 110.24 — Service Entrance Equipment
- 409.22(B) — Industrial Control Panels
- 440.10(B) — Air Conditioning & Refrigeration Equipment
- 620.51(D)(2) — Elevator Control Panels
- 670.5(B) — Industrial Machinery

Applying solutions within their ratings

- 110.9 — Interrupting Rating
- 110.10 — Component Protection
- 110.24 — Available Fault Current
- 240.4 — Conductor Protection
- 250.122 — Equipment Grounding Conductor Protection
- 409.22(A) — Industrial Control Panels
- 440.10(A) — Air Conditioning & Refrigeration Equipment
- 620.16(B) — Elevator Control Panels
- 690.5(A) — Industrial Machinery

Marking Short-Circuit Current Ratings (SCCR)

- 230.82(3) — Meter Disconnect
- 409.110(4) — Industrial Control Panels
- 430.8 — Motor Controllers
- 430.98 — Motor Control Centers
- 440.4(B) — Air Conditioning & Refrigeration Equipment
- 620.16(A) — Elevator Control Panel
- 670.3(A)(4) — Industrial Machinery
- 700.5(F) — Transfer Equipment for Emergency Systems
- 701.5(D) — Transfer Equipment for Legally Required and Standby Systems
- 702.5(C) — Transfer Equipment for Optional Standby Systems
- 708.24(F) — Transfer Equipment for Critical Operations Power Systems

Selective coordination

- 620.62 — Selective Coordination for Elevator Circuits
- 645.27 — Critical Operations Data Systems
- 695.3(C) — Multi-building Campus-Style Complexes
- 700.32 — Emergency Systems
- 701.32 — Legally Required Standby Systems
- 708.54 — Critical Operations Power Systems

4.1.3 One-line diagram

The one-line diagram, often referred to as a single-line, plays an important role in many aspects of power distribution system design, maintenance and construction. A one-line diagram graphically represents the power distribution system. Developing this diagram is the first step in making fault current, selective coordination and incident energy studies. This diagram should show all fault current sources and significant circuit elements. Significant circuit element reactance and resistance values should be included in the diagram. The one-line diagram should be updated any time the power distribution system changes. Changes must be reviewed with attention paid to the impact upon the studies that are based on this diagram's contents.

4.1.3.1 Fault current contributors

Fault current sources in a power distribution system include:

Utility

Utilities provide power through a transformer or series of transformers depending upon where in the distribution system the facility obtains its power. Most rural locations have a transformer dedicated to a facility or multiple facilities. In some urban areas, for reliability sake, power is derived from utility secondary networks where utility transformers are operated in parallel. Available fault currents on these secondary network systems are very high, in a range greater than 100 kA and upwards of 200 kA.

The fault current that's typically provided from the utility is an infinite bus calculation based upon the supply transformer's kVA size and minimum impedance. For applications on a secondary network, consulting the utility is the only way to obtain the available fault current for any given installation.

Generators

On-site generation for backup power must be a consideration for the power distribution system equipment. In most cases, the local generation will not provide fault currents greater than what can be seen from a utility. When large systems have multiple generators installed in parallel, such as hospitals or other similar applications, it's conceivable that available fault currents are greater than that available from utility sources.

4

Section 4 — Power system analysis

The available fault current will depend upon the kVA of the generator and sub-transient reactance. NEC Section 445.11 specifies the information to be included on the generator's nameplate consisting of:

- Sub-transient, transient, synchronous, and zero sequence reactances

- Power rating category

- Insulation system class

- Indication if the generator is protected against overload by inherent design, overcurrent protective relay, circuit breaker or fuse

- Maximum fault current for inverter-based generators, in lieu of the synchronous, sub-transient, and transient reactances

The sub-transient reactance is an impedance value used in determining generator fault contribution during the first cycle after a fault occurs. In approximately 0.1 second, the reactance increases to the transient reactance which is typically used to determine the fault current contribution after several cycles. In approximately 1/2 to 2 seconds, the generator's reactance increases to the synchronous reactance, which is the value that determines current flow after a steady-state condition is reached by the system, should fault currents be permitted to flow this long.

Motors

Voltages collapse during a fault, and when this happens to operating motors, their rotors will continue to turn and convert this rotating motion from a load to a fault current source. It's not practical to consider the contribution of each small motor in a system. IEEE rules of thumb deal with small motors by combining motors ≤ 50 Hp at each bus to which they are attached, and modeling them as one motor with an assumed sub-transient reactance, with 25% the typical assumed value.

The basic equation to determine motor contribution I_{sc} is:

$$I_{sc} \text{ motor} = (\text{Motor FLA} \times 100) \div \%Xd'$$

Motor sub-transient reactances range from 15% to 25%, with 25% being the more popular value used.

The motor's sub-transient reactance is an impedance value used to determine the motor's fault current contribution during the first cycle after a fault occurs. In approximately 0.1 second, the reactance increases to the transient reactance, which is typically used to determine the fault current contribution after several cycles. In approximately 1/2 to 2 seconds, the motor's reactance increases to the synchronous reactance, which is the value that determines current flow after a steady-state condition is reached by the system, should fault currents be permitted to flow this long.

Alternate power sources

Alternative energy sources are becoming more and more common in power distribution systems. In addition to an inverter that's collecting energy from solar or wind power, batteries are also fault current contributors and should be considered when appropriate.

4.1.3.2 Fault current reducers (impedances)

Impedance components considered in fault calculations, and shown on one-line diagrams, include:

Conductors

Unlike rotating machinery and transformers, conductors have a resistance and reactance mix to add to the power distribution system. Impedance values can be obtained from Table 9 of the NEC. As an example, a 500 kcmil copper conductor in a metallic raceway has impedance values as:

- Resistance = 0.029 ohms per 1000 ft.

- Reactance = 0.048 ohms per 1000 ft.

This is typically expressed in a rectangular format as:

$$Z_{Conductor} = 0.029 + j0.048 \text{ ohms per 1000 ft.*}$$

* "j" is a 90 degree operator signifying a vector at a 90 degree angle. Each impedance is comprised of real and reactive components. Real components are a magnitude at a 0 angle and inductive reactive components are a magnitude at a 90 degree angle.

This conductor's impedance is represented graphically in Figure 4.1.3.2.a.

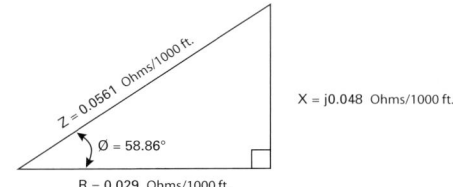

Figure 4.1.3.2.a

Conductor length is important when determining the impedance of any branch, feeder or service circuit. The impedance for 200 ft. of the 500 kcmil conductor referenced above is calculated as:

$$Z_{(200 ft.)} = 200 \text{ ft.} \times \frac{0.029 + j0.048 \text{ Ohms}}{1000 \text{ ft.}}$$

$$Z_{(200 ft.)} = 0.0006 + j0.0096 \text{ Ohms}$$

Impedance values will be different depending upon the conductor size and material (copper or aluminum) as well as what raceway in which it is installed and its length. One-line diagrams must have enough information to determine the correct conductor ampacity as well as determining the correct impedance for calculations.

Transformers

Transformers add considerable impedance to a power distribution system. The transformer nameplate will include its %Z value, which is based on the transformer's secondary.

When actual information is not available, the rule of thumb for typical transformer impedance ranges is shown in the following table.

kVA 3-phase	X/R	Range of %Z
112.5	3.0	1.6 — 2 Min — 6.2
150	3.5	1.5 — 2 Min — 6.4
225	4.0	2.0 — 2 Min — 6.6
300	4.5	2.0 — 4.5 Min — 6.0
500	5.0	2.1 — 4.5 Min — 6.1
750	6.0	3.2 — 5.75 — 6.75 — 6.8
1000	7.0	3.2 — 5.75 — 6.75 — 8.0
1500	7.0	3.5 — 5.75 — 6.75 — 6.8
2000	8.0	3.5 — 5.75 — 6.75 — 6.8
2500	9.0	3.5 — 5.75 — 6.75 — 6.8

Table notes:

1. Underlined values are from ANSI C57.12.10-1977[1], ANSI C57.12.22-1980 [2] and NEMA 210-1976 [10].
2. Network transformers with three-position switches have 5.0%Z for 500-750 kVA. See ANSI C57.12-40-1982 [3].
3. Three-phase banks with three single-phase transformers may have values as low as 1.2%.

The infinite bus calculation for the maximum fault current that can be possibly seen on the transformer's secondary is calculated with the following equation:

$$I_{Infinite\ bus} = \frac{FLA \times 100}{\% Z}$$

Selecting protective devices

A transformer manufacturer determines impedance through what's called a short-circuit test conducted as follows:

- The transformer secondary is short-circuited
- Voltage on the primary is increased until full load current flows in the secondary
- The applied voltage divided by the rated primary voltage (times 100) is the transformer's impedance

Example: For a 480 volt rated primary, if 9.6 volts cause secondary full load current to flow through the shorted secondary, the transformer impedance is 9.6 ÷ 480 = 0.02 = 2%Z.

Busway

Busway presents a flexible method to distribute power in a facility. Busway originated in Detroit's automotive industry during late 1920's in a response to a need for overhead wiring systems that would simplify electric motor-driven machine connections and permit a convenient arrangement for these production line machines. Busway has grown in popularity for many applications beyond manufacturing.

Busway presents a very low impedance to the distribution system, making it a very efficient power distribution means with attractive low voltage drop qualities. Busway also presents a low impedance during fault events.

The following table presents typical busway impedance values for use in voltage drop and fault calculations.

Typical busway parameters, line-to-neutral, in mΩ/100 ft, 25°C

Current rating (amps)	Aluminum		Copper	
	R	X	R	X
600	2.982	1.28	2.33	1.57
800	2.00	0.80	1.63	1.25
1000	1.60	0.64	1.27	0.92
1200	1.29	0.55	0.97	0.69
1350	1.03	0.44	0.86	0.63
1600	0.89	0.38	0.72	0.55
2000	0.70	0.32	0.58	0.46
2500	0.57	0.26	0.41	0.32
3000	0.46	0.21	0.37	0.29
4000	0.34	0.16	0.28	0.21
5000	—	—	0.20	0.16

Reactors

There are various reasons that reactors are used in a power distribution system. One reason is to limit fault current. Current-limiting reactors, connected in series, are primarily used to reduce fault currents and to match the impedance of parallel feeders. For example, to reduce the available fault and arcing current at the equipment, low voltage motor control centers can be supplied with three single-phase reactors that limit available fault current.

Reactors are also used in grounding neutrals of generators directly connected to the distribution system bus to limit the line-to-ground fault to somewhat less than the three-phase fault at the generator terminals. If the reactor is so sized, in all probability, the system will remain effectively grounded.

4.1.4 Procedures and methods

To determine the fault current at any point in the system, first secure an up to date one-line diagram. The one-line diagram must include all major fault current sources and impedances to fault currents. Next, an impedance diagram is created that includes all major power system components represented as impedances.

The impedance tables in Section 4.1.7 include three-phase and single-phase transformers, cables and busway. Use these tables if information from the manufacturer is not readily available.

Fault current calculations are performed without current-limiting devices in the system. To determine the maximum "available" fault current, calculations are made as though these devices are replaced with copper bars. This is necessary to project how the system and the current-limiting devices will perform.

Also, multiple current-limiting devices do not operate in series to "compound" a current-limiting effect. The downstream, loadside fuse will operate alone under a fault condition if properly coordinated.

The application of the point-to-point method permits determining available fault currents with a reasonable degree of accuracy at various points for either three-phase or single-phase electrical distribution systems. This method can assume unlimited primary fault current (infinite bus) or it can be used with limited available primary fault current.

4.1.4.1 Maximum and minimum fault currents

Fault current calculations should be performed at all critical points in the system including:

- Service entrance equipment
- Transfer switches
- Panelboards
- Load centers
- Motor control centers
- Disconnects
- Motor starters

Normally, fault studies involve calculating a bolted three-phase fault condition. This can be characterized as all three phases "bolted" together to create a zero impedance connection. This establishes a "worst case" (highest current) condition that results in maximum three-phase thermal and mechanical stress in the system.

From this calculation, other fault condition types can be approximated. This "worst case" condition should be used for interrupting rating, component protection, "Table" method for determining PPE per NFPA 70E and selective coordination.

Arc flash hazard analysis calculations should consider both maximum and minimum fault current calculations. Incident energy depends upon current and time. For lower arcing current values, clearing times could be longer than those for higher arcing current values, which could result in higher incident energy values. Therefore, an arc flash analysis must consider both spectrum of available fault current for calculating arcing currents, which are then compared with OCPD TCC curves to determine clearing times.

There are several variables in a distribution system affecting calculated bolted three-phase fault currents. Variable values applicable for the specific application analysis must be selected. The point-to-point method presented in this section includes several adjustment factors given in notes and footnotes that can be applied, and that will affect results. Some of the parameters that must be considered include utility source fault current, motor contribution, transformer percent impedance tolerance and voltage variance.

In most situations, the utility source(s) or on-site energy sources (such as generators) are the major fault current contributors. The point-to-point method includes steps and examples that assume an infinite available fault current from the utility source. Generally, this is a good assumption for highest, worst case conditions since the property owner has no control over the utility's system and future utility changes. In many cases, a large increase in the utility available fault current does not increase the building system's fault current a great deal on the secondary of the service transformer. However, there are cases where the actual utility medium voltage available fault current provides a more accurate fault current assessment (minimum bolted fault current conditions) that may be needed to assess arc flash hazards.

When motors are in the system, motor fault current contribution is also a very important factor to include in any fault current analysis. When a fault occurs, motor contribution adds to the fault current magnitude, with running motors contributing four to six times their normal full load current. Series rated combinations can't be used in specific situations due to motor fault current contributions (see the section on Series Ratings in this book).

4

Section 4 — Power system analysis

EA•TON BUSSMANN SERIES

For short time duration capacitor discharge currents, certain IEEE (Institute of Electrical and Electronic Engineers) publications detail how to calculate these currents if they are substantial.

4.1.5 Point-to-point calculation method

The application of the point-to-point method permits determining available fault currents with a reasonable degree of accuracy at various points for either 3 Ø or 1 Ø electrical distribution systems.

4.1.5.1 Basic pint-to-point calculation

The following are the basic steps to employ in the point-to-point method of calculating fault current.

Step 1: Determine the transformer full load amps (FLA) from either the nameplate, the following formulas or Table 4.1.7.1.a:

$$3 \text{ Ø Transformer } I_{FLA} = \frac{kVA \times 1000}{E_{L-L} \times \sqrt{3}}$$

$$1 \text{ Ø Transformer } I_{FLA} = \frac{kVA \times 1000}{E_{L-L}}$$

Step 2: Find the transformer multiplier (see Notes 1 and 2).

$$\text{Multiplier} = \frac{100}{\%Z_{transformer}}$$

Note 1. Get %Z from nameplate or Table 4.1.7.1.a. Transformer impedance (Z) is used to determine what the fault current will be on the transformer secondary.

Note 2. 25 kVA and larger UL 1561 listed transformers have a ± 10% impedance tolerance. Fault current levels can be affected by this tolerance. Therefore, for high end ,worst case, multiply %Z by 0.9. For low end of worst case, multiply %Z by 1.1. Transformers constructed to ANSI standards have a ± 7.5% impedance tolerance (two-winding construction).

Step 3: Determine by formula or Table 4.1.7.1.a the transformer let-through fault current. See Notes 3 and 4.

$$I_{sc} = \text{Transformer FLA} \times \text{Multiplier}$$

Note 3. Utility voltages may vary ± 10% for power and ± 5.8% for 120 volt lighting services. Therefore, for highest fault current conditions, multiply values as calculated in Step 3 by 1.1 or 1.058 respectively. To find the lower end worst case, multiply results in Step 3 by 0.9 or 0.942 respectively.

Note 4. Motor fault current contribution, if significant, may be added at all fault locations throughout the system. A practical motor fault current contribution estimate is to multiply the total motor current in amps by 4. Values of 4 to 6 are commonly accepted.

Step 4: Calculate the "f" factor.

$$3 \text{ Ø Faults } f = \frac{\sqrt{3} \times L \times I_{3\emptyset}}{C \times n \times E_{L-L}}$$

For the next two equations, see note 5

$$1 \text{ Ø Line-line faults } f = \frac{2 \times L \times I_{L-L}}{C \times n \times E_{L-L}}$$

$$1 \text{ Ø Line-neutral faults } f = \frac{2 \times L \times I_{L-N}}{C \times n \times E_{L-N}}$$

Where:

L = Conductor length to the fault in feet
C = Constant from Table 4.1.7.6.a of "C" values for conductors and Table 4.1.7.7.a of "C" values for busway
n = Number of conductors per phase (adjust C value for parallel runs)
I = Available fault current in amps at circuit's beginning
E = Circuit voltage

Note 5. On a single-phase center-tapped transformer, the L-N fault current is higher than the L-L fault current at the secondary terminals. The fault current available (I) for this case in Step 4 should be adjusted at the transformer terminals as follows: At L-N center tapped transformer terminals, $I_{L-N} = 1.5 \times I_{L-L}$ at transformer terminals.

Depending upon wire size, at some distance from the terminals, the L-N fault current is lower than the L-L fault current. The 1.5 multiplier is an approximation and will theoretically vary from 1.33 to 1.67. These figures are based on a change in turns ratio between primary and secondary, infinite source available, zero feet from transformer terminals and 1.2 x %X and 1.5 x %R for L-N versus L-L resistance and reactance values. Begin L-N calculations at the transformer's secondary terminals, then proceed point-to-point.

Step 5: Calculate "M" (multiplier) or take from Table 4.1.7.2.

$$M = \frac{1}{1+f}$$

Step 6: Calculate the available short-circuit symmetrical RMS fault current at the point of fault. Add motor contribution, if applicable.

$$I_{SC RMS Sym.} = I_{SC} \times M$$

Step 6A: Significant motor fault current contribution may be added at all fault locations throughout the system. A practical motor fault current contribution estimate is to multiply the total motor current in amps by 4. Values of 4 to 6 are commonly accepted.

4.1.5.2 Point-to-point calculation when available primary fault current is known

The following procedure can be used to calculate the fault current level at a downstream transformer's secondary in a system when the fault current level at the transformer primary is known.

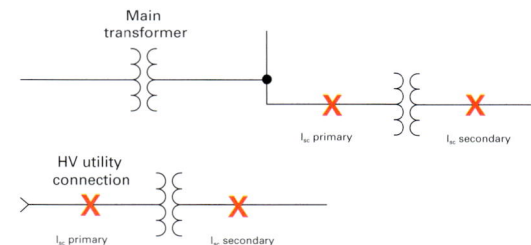

Figure 4.1.5.2.a

Step A: Calculate the "f" factor (I_{SC} primary known)

3 Ø Transformer (I_{SC} primary and I_{SC} secondary are 3 Ø fault values);

$$f = \frac{I_{SC} \text{ primary} \times V_{Primary} \times \sqrt{3} \times \%Z}{100,000 \times kVA_{transformer}}$$

1 Ø Transformer (I_{SC} primary and I_{SC} secondary are 1 Ø fault values: I_{SC} secondary is L-L);

$$f = \frac{I_{SC} \text{ primary} \times V_{primary} \times \%Z}{100,000 \times kVA_{transformer}}$$

Step B: Calculate "M" (multiplier)

$$M = \frac{1}{1+f}$$

Step C: Calculate the fault current at the transformer secondary (see Note under Step 3 of "Basic point-to-point calculation procedure")

$$I_{SC secondary} = \frac{V_{primary}}{V_{secondary}} \times M \times I_{SC primary}$$

E·T·N **BUSSMANN** SERIES

Selecting protective devices

4.1.5.3 Point-to-point calculation

Example 1

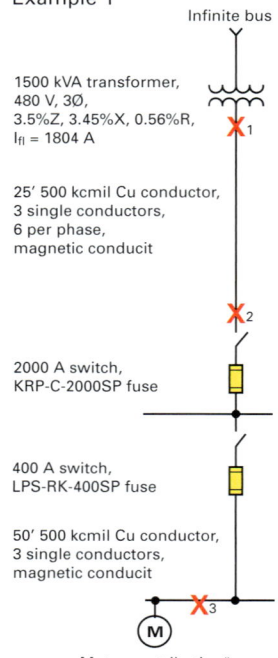

1500 kVA transformer, 480 V, 3Ø, 3.5%Z, 3.45%X, 0.56%R, I_{fl} = 1804 A

25' 500 kcmil Cu conductor, 3 single conductors, 6 per phase, magnetic conduit

2000 A switch, KRP-C-2000SP fuse

400 A switch, LPS-RK-400SP fuse

50' 500 kcmil Cu conductor, 3 single conductors, magnetic conduit

Motor contribution*

Note: The following "Step" numbers pertain to the steps described in "4.1.5.1. Point-to-point calculation basic."

Fault X1

Step 1: $I_{fL} = \dfrac{1500 \times 1000}{480 \times \sqrt{3}}$ = 1804 A

Step 2: Multiplier = $\dfrac{100}{3.5 \times 0.9^{\dagger}}$ = 31.746

Step 3: I_{SC} = 1804 A x 31.746 = 57,279 A
I_{SC} motor contribution* = 4 x 1804 A = 7217 A
I_{SC} total = 57,279 A + 7217 A = 64,496 A

Fault X2

Step 4: $\dfrac{\sqrt{3} \times 25 \times 57,279}{22,185 \times 6 \times 480}$ = 0.0388

Step 5: M = $\dfrac{1}{1+ 0.0388}$ = 0.9626

Step 6: I_{SC} = 57,279 A x 0.9626 = 55,137 A
I_{SC} motor contribution* = 4 x 1804 A = 7217 A
I_{SC} total = 55,137 A + 7217 A = 62,354 A

* See Note 4 on page 4-4. Assumes 100% motor load. If 50% of this load is from motors, I_{SC} motor contribution = 4 x 1804 A x 0.5 = 3608 A.
† See Note 2 on page 4-4.

Fault X3

Step 4: $\dfrac{\sqrt{3} \times 50 \times 55,137}{22,185 \times 1 \times 480}$ = 0.4484

Step 5: M = $\dfrac{1}{1+ 0.4483}$ = 0.6904

Step 6: I_{SC} = 55,137 A x 0.6904 = 38,067 A
I_{SC} motor contribution = 4 x 1804 A = 7217 A
I_{SC} total = 38,067 A + 7217 A = 45,284 A

4

Example 2

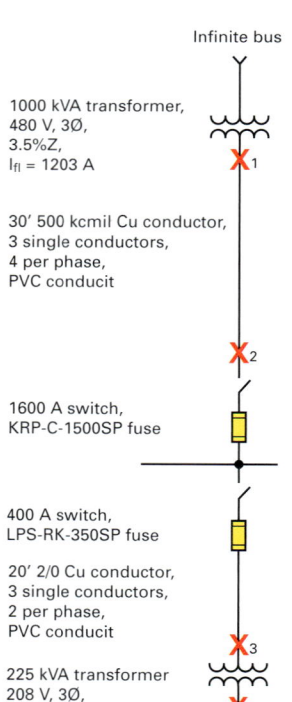

1000 kVA transformer, 480 V, 3Ø, 3.5%Z, I_{fl} = 1203 A

30' 500 kcmil Cu conductor, 3 single conductors, 4 per phase, PVC conduit

1600 A switch, KRP-C-1500SP fuse

400 A switch, LPS-RK-350SP fuse

20' 2/0 Cu conductor, 3 single conductors, 2 per phase, PVC conduit

225 kVA transformer 208 V, 3Ø, 1.2%Z

Note: The following Step numbers pertain to the steps described in "4.1.5.1 Point-to-point calculation basic."

Fault X1

Step 1: $\dfrac{1000 \times 1000}{480 \times \sqrt{3}}$ = 1202.8 A

Step 2: Multiplier= $\dfrac{100}{3.5 \times 0.9}$ = 31.746

Step 3: I_{SC} = 1202.8 A x 31.746 = 38,184 A

Fault X2

Step 4: $\dfrac{\sqrt{3} \times 30 \times 38,184}{26,706 \times 4 \times 480}$ = 0.0387

Step 5: M = $\dfrac{1}{1+ 0.0387}$ = 0.9627

Step 6: I_{SC} = 38,184 A x 0.9627 = 36,761 A

Fault X3

Step 4: $\dfrac{\sqrt{3} \times 20 \times 36,761}{2 \times 11,424 \times 480}$ = 0.1161

Step 5: M= $\dfrac{1}{1+ 0.1161}$ = 0.8960

Step 6: I_{SC} = 36,761 A x 0.8960 = 32,937 A

Fault X4

Step A: $\dfrac{32,937 \times 480 \times \sqrt{3} \times 1.2 \times 0.9}{100,000 \times 225}$ = 1.3144

Step B: M = $\dfrac{1}{1+ 1.3144}$ = 0.4321

Step C: I_{SC} = $\dfrac{480 \times 0.4321 \times 32,937 \text{ A}}{208}$ =32,842 A

Section 4 — Power system analysis

4.1.5.4. Single-phase system fault currents

Fault current calculations on a single-phase center tapped transformer system require a slightly different procedure than 3 Ø faults on 3 Ø systems.

Primary available fault current

It is necessary that the proper impedance be used to represent the primary system. For 3 Ø fault calculations, a single primary conductor impedance is used from the source to the transformer connection. This is compensated for in the 3 Ø fault current formula by multiplying the single conductor or single-phase impedance by 1.73 (√3).

However, for single-phase faults, a primary conductor impedance is considered from the source to the transformer, and back to the source. This is compensated for in the calculations by multiplying the 3 Ø primary source impedance by two.

Center-tapped transformer impedance

The center-tapped transformer impedance must be adjusted for the half-winding (generally line-to-neutral) fault condition.

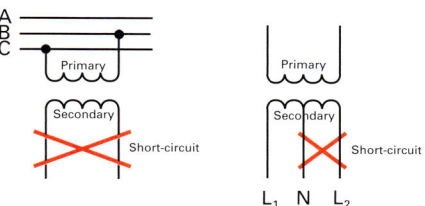

| **Figure 4.1.5.4.a** | **Figure 4.1.5.4.b** |

Figure 4.1.5.4.b illustrates that during line-to-neutral faults, the full primary winding is involved, but only the half-winding on the secondary is involved. Therefore, the actual transformer reactance and resistance of the half-winding condition is different than the actual transformer reactance and resistance of the full winding condition. Thus, adjustment to the %X and %R must be made when considering line-to-neutral faults. The adjustment multipliers generally used for this condition are:

- 1.5 times full winding %**R** on full winding basis
- 1.2 times full winding %**X** on full winding basis

Note: %R and %X multipliers given in "Impedance Data for Single-Phase Transformers" Table may be used. However, calculations must be adjusted to indicate transformer kVA ÷ 2.

Cable and two-pole switch impedances

The cable and two-pole switch impedance on the system must be considered "both-ways" since the current flows to the fault and then returns to the source. E.g., if a line-to-line fault occurs 50 feet from a transformer, then 100 feet of cable impedance must be included in the calculation. (Figure 4.1.5.5.c.)

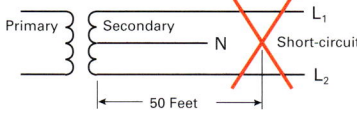

Figure 4.1.5.4.c

The following calculations illustrate 1 Ø fault calculations on a single-phase transformer system. Both line-to-line and line-to-neutral faults are considered.

Note in these examples:

- The multiplier is 2 for some electrical components to account for the single-phase fault current flow
- The half-winding transformer %X and %R multipliers for the line-to-neutral fault situation along with impedance and reactance data

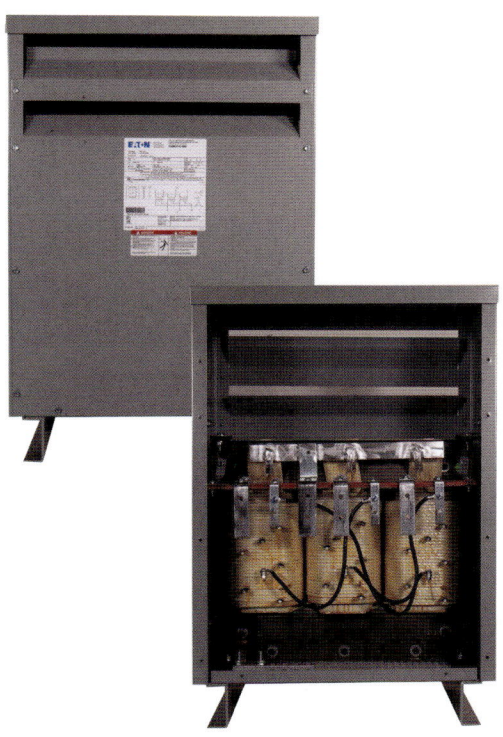

Selecting protective devices

Single-phase system fault current calculation

Example 1 — line-to-line

Note: The following Step numbers pertain to the steps described in "4.1.5.1. Point-to-point calculation basic."

Fault X1

Step 1: $\dfrac{75 \times 1000}{240} = 312.5$ A

Step 2: Multiplier = $\dfrac{100}{1.4 \times 0.9^{*}} = 79.37$

Step 3: I_{SC} L-L = 312.5 A x 79.37 = 24,802 A

Fault X2

Step 4: $\dfrac{2 \times 25 \times 24,802}{22,185 \times 1 \times 240} = 0.2329$

Step 5: M = $\dfrac{1}{1 + 0.2329} = 0.8111$

Step 6: I_{SC} L-L = 24,802 A x 0.8111 = 20,116 A

Fault X3

Step 4: $\dfrac{2 \times 50 \times 20,116}{4774 \times 1 \times 240} = 1.7557$

Step 5: M = $\dfrac{1}{1 + 1.7557} = 0.3629$

Step 6: I_{SC} = 20,116 A x 0.3629 = 7300 A

* In addition, 25 kVA and larger UL 1561 listed transformers have a ±10% impedance tolerance that can affect fault current. Therefore, for high end worst case, multiply %Z by 0.9. For low end of worst case, multiply %Z by 1.1. Transformers constructed to ANSI standards have a ±7.5% impedance tolerance (two-winding construction).

Example 2 — line-to-neutral

Note: The following Step numbers pertain to the steps described in "4.1.5.1. Point-to-point calculation basic."

Fault X 1

Step 1: $\dfrac{75 \times 1000}{240} = 312.5$ A

Step 2: Multiplier = $\dfrac{100}{1.4 \times 0.9} = 79.37$

Step 3*: I_{SC} L-L = 312.5 A x 79.37 = 24,802 A
I_{SC} L-N = 24,802 x 1.5 = 37,202 A

Fault X2

Step 4: $\dfrac{2 \times 25 \times 37,202}{22,185 \times 1 \times 120} = 0.6987$

Step 5: M = $\dfrac{1}{1 + 0.6987} = 0.5887$

Step 6*: I_{SC} L-L = 37,202 A x 0.5887 = 21,900 A

Fault X3

Step 4: $\dfrac{2 \times 50 \times 21,900^{**}}{4774 \times 1 \times 120} = 3.8323$

Step 5: M = $\dfrac{1}{1 + 3.8323} = 0.2073$

Step 6*: I_{SC} = 21,900 A x 0.2073 = 4540 A

* The L-N fault current is higher than the L-L fault current at the single phase center-tapped transformer's secondary terminals. The available fault current (I) for this case in Step 4 should be adjusted at the transformer terminals as follows: At L-N center-tapped transformer terminals, I_{L-N} = 1.5 X I_{L-L} at transformer terminals.
** Assumes same size neutral and line conductors.

Infinite bus

75 kVA transformer,
120/240 V, 1Ø,
1.40%Z, 31.22%X, 0.68%R,
I_{fl} = 312.5 A

X_1

25' 500 kcmil Cu conductor,
3 single conductors,
magnetic conduit

X_2

400 A switch,
LPS-RK-400SP fuse

50' 3 AWG Cu conductor,
3 single conductors,
magnetic conduit

X_3

4

Section 4 — Power system analysis

EAT•N BUSSMANN SERIES

4.1.6 Equipment evaluations

The first step to properly applying electrical solutions, as well as complying with Code requirements, is the fault current study. Once the fault current levels are determined, equipment has to be evaluated for proper application including:

- OCPD interrupting ratings
- System selective coordination
- Component protection (SCCR)
- Incident energy analysis

See the various sections in this handbook for further information on these topics.

Low voltage fuses have their interrupting rating expressed as symmetrical component of fault current. They are given an RMS symmetrical interrupting rating at a specific power factor. This means that the fuse can also interrupt the asymmetrical current associated with this rating. Thus, only the symmetrical component of fault current need be considered to determine the necessary low voltage fuse interrupting rating.

The NEC includes requirements for marking and/or documenting, available fault current for various locations throughout the power distribution system. NEC 110.24 requires field marking service equipment (other than dwelling units and certain industrial facilities) with the maximum available fault current. Additionally, other requirements include either marking the available fault current on the equipment, or documenting the available fault current covering industrial control panels, HVAC equipment, elevator control panels and industrial machinery.

In addition to OCPD interrupting ratings and equipment SCCR ratings, the available fault current is used for other purposes, including determining selective coordination and arc flash boundary, along with the proper arc rated PPE per NFPA 70E. Whether determined by the incident energy method or arc flash PPE category method (70E 130.5), the available fault current is required.

4.1.7 Tables

4.1.7.1 Fault currents available from transformers

Table 4.1.7.1 includes values based on actual field nameplate data or from utility transformer worst case impedance.

Voltage and phase	kVA	Full load amps	% impedance[†] (nameplate)	Fault current amps[††]
120/240 single-phase*	25	104	1.50	12,175
	37.5	156	1.50	18,018
	50	208	1.50	23,706
	75	313	1.50	34,639
	100	417	1.60	42,472
	167	696	1.60	66,644
120/208 three-phase**	45	125	1.00	13,879
	75	208	1.00	23,132
	112.5	312	1.11	31,259
	150	416	1.07	43,237
	225	625	1.12	61,960
	300	833	1.11	83,357
	500	1388	1.24	124,364
	750	2082	3.50	66,091
	1000	2776	3.50	88,121
	1500	4164	3.50	132,181
	2000	5552	4.00	154,211
	2500	6940	4.00	192,764
277/480 three-phase**	75	90	1.00	10,035
	112.5	135	1.00	15,053
	150	181	1.20	16,726
	225	271	1.20	25,088
	300	361	1.20	33,451
	500	602	1.30	51,463
	750	903	3.50	28,672
	1000	1204	3.50	38,230
	1500	1806	3.50	57,345
	2000	2408	4.00	66,902
	2500	3011	4.00	83,628

Table 4.1.7.1.a

* Single-phase values are L-N at transformer terminals. These figures are based on change in turns ratio between primary and secondary, 100,000 kVA primary, zero feet from transformer terminals, 1.2 %X and 1.5 %R multipliers for L-N vs. L-L reactance and resistance values, and transformer X/R ratio = 3.

** Three-phase fault currents based on "infinite" primary.

† 25 kVA or greater UL Listed transformers have a ± 10% impedance tolerance. Fault current shown in Table 4.1.7.1.a reflect -10% condition. Transformers constructed to ANSI standards have a ± 7.5% impedance tolerance (two-winding construction)

†† System voltage fluctuations will affect the available fault current. For example, a 10% increase in system voltage will result in a 10% greater available fault current than as shown in Table 4.1.7.1.a.

Available Fault Current Calculator

FC[2] is an online or downloadable application (for both Apple and Android devices) that utilizes the point-to-point method for calculating and documenting available fault current levels in single- and three-phase systems.

It's capable of producing equipment labels in English, Spanish or French for local language needs. Scan the QR code or visit the Bussmann division website at Eaton.com/bussmannseries.

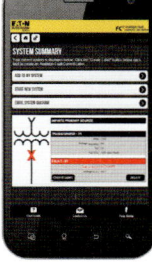

E:T•N BUSSMANN SERIES

Selecting protective devices

4.1.7.2 "M" multiplier

The "M" multiplier is used in the point-to-point calculations. The basic equation used to derive the values in this table is:

$$M = \frac{1}{1+f}$$

The "f" value is based upon a calculation as described in Step 4 of the point-to-point method for calculating fault currents.

f	M	f	M	f	M
0.01	0.99	0.50	0.67	7.00	0.13
0.02	0.98	0.60	0.63	8.00	0.11
0.03	0.97	0.70	0.59	9.00	0.10
0.04	0.96	0.80	0.55	10.00	0.09
0.05	0.95	0.90	0.53	15.00	0.06
0.06	0.94	1.00	0.50	20.00	0.05
0.07	0.93	1.20	0.45	30.00	0.03
0.08	0.93	1.50	0.40	40.00	0.02
0.09	0.92	1.75	0.36	50.00	0.02
0.10	0.91	2.00	0.33	60.00	0.02
0.15	0.87	2.50	0.29	70.00	0.01
0.20	0.83	3.00	0.25	80.00	0.01
0.25	0.80	3.50	0.22	90.00	0.01
0.30	0.77	4.00	0.20	100.00	0.01
0.35	0.74	5.00	0.17	—	—
0.40	0.71	6.00	0.14	—	—

Table 4.1.7.2 "M" multiplier

4.1.7.3 Transformer single-phase impedance data

Table 4.1.7.3 is reprinted from IEEE Std. 242-1986 (R1991), IEEE Recommended Practice for Protection and Coordination of Industrial and Commercial Power Systems. Copyright 1986 by the Institute of Electrical and Electronics Engineers, Inc. with the permission of the IEEE Standards Department.

25 kVA and greater UL Listed transformers have a ± 10% tolerance on their impedance nameplate.

kVA 1 Ø	Suggested X/R ratio for calculation	Normal range of percent impedance*	Impedance multipliers — line-neutral faults**	
			for %X	for %R
25.0	1.1	1.2–6.0	0.6	0.75
37.5	1.4	1.2–6.5	0.6	0.75
50.0	1.6	1.2–6.4	0.6	0.75
75.0	1.8	1.2–6.6	0.6	0.75
100.0	2.0	1.3–5.7	0.6	0.75
167.0	2.5	1.4–6.1	1.0	0.75
250.0	3.6	1.9–6.8	1.0	0.75
333.0	4.7	2.4–6.0	1.0	0.75
500.0	5.5	2.2–5.4	1.0	0.75

Table 4.1.7.3

* National standards do not specify %Z for single-phase transformers. Consult manufacturer for values to use in calculation.

**Based on winding rated current (one-half nameplate kVA divided by secondary line-to-line voltage).

This table has been reprinted from IEEE Std. 242-1986 (R1991), IEEE Recommended Practice for Protection and Coordination of Industrial and Commercial Power Systems, Copyright 1986 by the Institute of Electrical and Electronics Engineers, Inc. with the permission of the IEEE Standards Department.

4.1.7.4. Impedance data for sing e-phase and three-phase transformers supplement

The data included in Table 4.1.7.4 provides actual transformer nameplate ratings taken from field installations. 25 kVA and greater UL Listed transformers have a ± 10% tolerance on their nameplate impedance.

kVA		%Z	Suggested X/R ratio for calculation
1 Ø	3 Ø		
10	—	1.2	1.1
15	—	1.3	1.1
—	75	1.11	1.5
—	150	1.07	1.5
—	225	1.12	1.5
—	300	1.11	1.5
333	—	1.9	4.7
—	500	1.24	1.5
500	—	2.1	5.5

Table 4.1.7.4

4.1.7.5 Various fault current types as a percent of three-phase bolted faults (typical)

This table provides some general information on various fault current types as a percentage of three-phase bolted fault currents. These are general rules of thumb that should nct replace actual calculations that can be provided by software applications.

Fault type	Percentage
Three-phase bolted	100%
Line-to-line bolted	87%
Line-to-ground bolted	25-125%* (Use 100% near transformer, 50% otherwise)
Line-to-neutral bolted	25-125% (Use 100% near transformer, 50% otherwise)
Three-phase arcing	89% maximum
Line-to-line arcing	74% maximum
Line-to-ground arcing minimum	38% minimum

Table 4.1.7.5

* Typically much lower, but can actually exceed the three-phase bolted fault if it is near the transformer terminals. Will normally be between 25% to 125% of three phase bolted fault value.

4

Section 4 — Power system analysis

4.1.7.6 "C" values for conductors

Table 4.1.7.6 data is used as part of the point-to-point fault current calculation when determining the "f" factor as part of Step 4.

AWG or kcmil	Three single conductors						Three-conductor cable					
	Steel conduit			Non-magnetic conduit			Steel conduit			Non-magnetic conduit		
	600 V	5 kV	15 kV	600 V	5 kV	15 kV	600 V	5 kV	15 kV	600 V	5 kV	15 kV
Copper												
14	389	—	—	389	—	—	389	—	—	389	—	—
12	617	—	—	617	—	—	617	—	—	617	—	—
10	981	—	—	982	—	—	982	—	—	982	—	—
8	1557	1551	—	1559	1555	—	1559	1557	—	1560	1558	—
6	2425	2406	2389	2430	2418	2407	2431	2425	2415	2433	2428	2421
4	3806	3751	3696	3826	3789	3753	3830	3812	3779	3838	3823	3798
3	4774	4674	4577	4811	4745	4679	4820	4785	4726	4833	4803	4762
2	5907	5736	5574	6044	5926	5809	5989	5930	5828	6087	6023	5958
1	7293	7029	6759	7493	7307	7109	7454	7365	7189	7579	7507	7364
1/0	8925	8544	7973	9317	9034	8590	9210	9086	8708	9473	9373	9053
2/0	10,755	10,062	9390	11,424	10,878	10,319	11,245	11,045	10,500	11,703	11,529	11,053
3/0	12,844	11,804	11,022	13,923	13,048	12,360	13,656	13,333	12,613	14,410	14,119	13,462
4/0	15,082	13,606	12,543	16,673	15,351	14,347	16,392	15,890	14,813	17,483	17,020	16,013
250	16,483	14,925	13,644	18,594	17,121	15,866	18,311	17,851	16,466	19,779	19,352	18,001
300	18,177	16,293	14,769	20,868	18,975	17,409	20,617	20,052	18,319	22,525	21,938	20,163
350	19,704	17,385	15,678	22,737	20,526	18,672	22,646	21,914	19,821	24,904	24,126	21,982
400	20,566	18,235	16,366	24,297	21,786	19,731	24,253	23,372	21,042	26,916	26,044	23,518
500	22,185	19,172	17,492	26,706	23,277	21,330	26,980	25,449	23,126	30,096	28,712	25,916
600	22,965	20,567	17,962	28,033	25,204	22,097	28,752	27,975	24,897	32,154	31,258	27,766
750	24,137	21,387	18,889	29,735	26,453	23,408	31,051	30,024	26,933	34,605	33,315	29,735
1000	25,278	22,539	19,923	31,491	28,083	24,887	33,864	32,689	29,320	37,197	35,749	31,959
Aluminum												
14	237	—	—	237	—	—	237	—	—	237	—	—
12	376	—	—	376	—	—	376	—	—	376	—	—
10	599	—	—	599	—	—	599	—	—	599	—	—
8	951	950	—	952	951	—	952	951	—	952	952	—
6	1481	1476	1472	1482	1479	1476	1482	1480	1478	1482	1481	1479
4	2346	2333	2319	2350	2342	2333	2351	2347	2339	2353	2350	2344
3	2952	2928	2904	2961	2945	2929	2963	2955	2941	2966	2959	2949
2	3713	3670	3626	3730	3702	3673	3734	3719	3693	3740	3725	3709
1	4645	4575	4498	4678	4632	4580	4686	4664	4618	4699	4682	4646
1/0	5777	5670	5493	5838	5766	5646	5852	5820	5717	5876	5852	5771
2/0	7187	6968	6733	7301	7153	6986	7327	7271	7109	7373	7329	7202
3/0	8826	8467	8163	9110	8851	8627	9077	8981	8751	9243	9164	8977
4/0	10,741	10,167	9700	11,174	10,749	10,387	11,185	11,022	10,642	11,409	11,277	10,969
250	12,122	11,460	10,849	12,862	12,343	11,847	12,797	12,636	12,115	13,236	13,106	12,661
300	13,910	13,009	12,193	14,923	14,183	13,492	14,917	14,698	13,973	15,495	15,300	14,659
350	15,484	14,280	13,288	16,813	15,858	14,955	16,795	16,490	15,541	17,635	17,352	16,501
400	16,671	15,355	14,188	18,506	17,321	16,234	18,462	18,064	16,921	19,588	19,244	18,154
500	18,756	16,828	15,657	21,391	19,503	18,315	21,395	20,607	19,314	23,018	22,381	20,978
600	20,093	18,428	16,484	23,451	21,718	19,635	23,633	23,196	21,349	25,708	25,244	23,295
750	21,766	19,685	17,686	25,976	23,702	21,437	26,432	25,790	23,750	29,036	28,262	25,976
1000	23,478	21,235	19,006	28,779	26,109	23,482	29,865	29,049	26,608	32,938	31,920	29,135

Table 4.1.7.6.a "C" values for conductors.

Note: These values are equal to one over the impedance per foot, and based upon resistance and reactance values found in IEEE Std. 241-1990 (Gray Book), IEEE Recommended Practice for Electric Power Systems in Commercial Buildings & IEEE Std. 242-1986 (Buff Book), IEEE Recommended Practice for Protection and Coordination of Industrial and Commercial Power Systems. Where resistance and reactance values differ or are not available, the Buff Book values have been used. The values for reactance in determining the C Value at 5 kV & 15 kV are from the Gray Book only (Values for 14-10 AWG at 5 kV and 14-8 AWG at 15 kV are not available and values for 3 AWG have been approximated).

Selecting protective devices

4.1.7.7 "C" values for busway

Table 4.1.7.7 data is used as part of the point-to-point fault current calculation when determining the "f" factor as part of Step 4.

| | Busway | | | | |
| | Plug-in | Feeder | | High impedance | |
Ampacity	Copper	Aluminum	Copper	Aluminum	Copper
225	28700	23000	18700	12000	—
400	38900	34700	23900	21300	—
600	41000	38300	36500	31300	—
800	46100	57500	49300	44100	—
1000	69400	89300	62900	56200	15600
1200	94300	97100	76900	69900	16100
1350	119000	104200	90100	84000	17500
1600	129900	120500	101000	90900	19200
2000	142900	135100	134200	125000	20400
2500	143800	156300	180500	166700	21700
3000	144900	175400	204100	188700	23800
4000	—	—	277800	256400	—

Table 4.1.7.7.a "C" values for busway.

Note: These values are based on a survey of industry and equal to one over the impedance per foot for busway impedance. Busway manufacture information should be consulted for specific applications

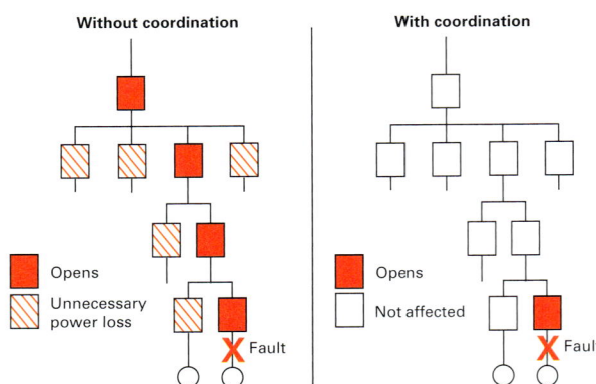

4

Chapter 9, Table 8 Conductor Properties

Table 8 Conductor Properties

Size (AWG or kcmil)	Area mm²	Area Circular mils	Stranding Quantity	Stranding Diameter mm	Stranding Diameter in.	Overall Diameter mm	Overall Diameter in.	Overall Area mm²	Overall Area in.²	DC Copper Uncoated ohm/km	DC Copper Uncoated ohm/kFT	DC Copper Coated ohm/km	DC Copper Coated ohm/kFT	Aluminum ohm/km	Aluminum ohm/kFT
18	0.823	1620	1	—	—	1.02	0.040	0.823	0.001	25.5	7.77	26.5	8.08	42.0	12.8
18	0.823	1620	7	0.39	0.015	1.16	0.046	1.06	0.002	26.1	7.95	27.7	8.45	42.8	13.1
16	1.31	2580	1	—	—	1.29	0.051	1.31	0.002	16.0	4.89	16.7	5.08	26.4	8.05
16	1.31	2580	7	0.49	0.019	1.46	0.058	1.68	0.003	16.4	4.99	17.3	5.29	26.9	8.21
14	2.08	4110	1	—	—	1.63	0.064	2.08	0.003	10.1	3.07	10.4	3.19	16.6	5.06
14	2.08	4110	7	0.62	0.024	1.85	0.073	2.68	0.004	10.3	3.14	10.7	3.26	16.9	5.17
12	3.31	6530	1	—	—	2.05	0.081	3.31	0.005	6.34	1.93	6.57	2.01	10.45	3.18
12	3.31	6530	7	0.78	0.030	2.32	0.092	4.25	0.006	6.50	1.98	6.73	2.05	10.69	3.25
10	5.261	10380	1	—	—	2.588	0.102	5.26	0.008	3.984	1.21	4.148	1.26	6.561	2.00
10	5.261	10380	7	0.98	0.038	2.95	0.116	6.76	0.011	4.070	1.24	4.226	1.29	6.679	2.04
8	8.367	16510	1	—	—	3.264	0.128	8.37	0.013	2.506	0.764	2.579	0.786	4.125	1.26
8	8.367	16510	7	1.23	0.049	3.71	0.146	10.76	0.017	2.551	0.778	2.653	0.809	4.204	1.28
6	13.30	26240	7	1.56	0.061	4.67	0.184	17.09	0.027	1.608	0.491	1.671	0.510	2.652	0.808
4	21.15	41740	7	1.96	0.077	5.89	0.232	27.19	0.042	1.010	0.308	1.053	0.321	1.666	0.508
3	26.67	52620	7	2.20	0.087	6.60	0.260	34.28	0.053	0.802	0.245	0.833	0.254	1.320	0.403
2	33.62	66360	7	2.47	0.097	7.42	0.292	43.23	0.067	0.634	0.194	0.661	0.201	1.045	0.319
1	42.41	83690	19	1.69	0.066	8.43	0.332	55.80	0.087	0.505	0.154	0.524	0.160	0.829	0.253
1/0	53.49	105600	19	1.89	0.074	9.45	0.372	70.41	0.109	0.399	0.122	0.415	0.127	0.660	0.201
2/0	67.43	133100	19	2.13	0.084	10.62	0.418	88.74	0.137	0.3170	0.0967	0.329	0.101	0.523	0.159
3/0	85.01	167800	19	2.39	0.094	11.94	0.470	111.9	0.173	0.2512	0.0766	0.2610	0.0797	0.413	0.126
4/0	107.2	211600	19	2.68	0.106	13.41	0.528	141.1	0.219	0.1996	0.0608	0.2050	0.0626	0.328	0.100
250	127	—	37	2.09	0.082	14.61	0.575	168	0.260	0.1687	0.0515	0.1753	0.0535	0.2778	0.0847
300	152	—	37	2.29	0.090	16.00	0.630	201	0.312	0.1409	0.0429	0.1463	0.0446	0.2318	0.0707
350	177	—	37	2.47	0.097	17.30	0.681	235	0.364	0.1205	0.0367	0.1252	0.0382	0.1984	0.0605
400	203	—	37	2.64	0.104	18.49	0.728	268	0.416	0.1053	0.0321	0.1084	0.0331	0.1737	0.0529
500	253	—	37	2.95	0.116	20.65	0.813	336	0.519	0.0845	0.0258	0.0869	0.0265	0.1391	0.0424
600	304	—	61	2.52	0.099	22.68	0.893	404	0.626	0.0704	0.0214	0.0732	0.0223	0.1159	0.0353
700	355	—	61	2.72	0.107	24.49	0.964	471	0.730	0.0603	0.0184	0.0622	0.0189	0.0994	0.0303
750	380	—	61	2.82	0.111	25.35	0.998	505	0.782	0.0563	0.0171	0.0579	0.0176	0.0927	0.0282
800	405	—	61	2.91	0.114	26.16	1.030	538	0.834	0.0528	0.0161	0.0544	0.0166	0.0868	0.0265
900	456	—	61	3.09	0.122	27.79	1.094	606	0.940	0.0470	0.0143	0.0481	0.0147	0.0770	0.0235
1000	507	—	61	3.25	0.128	29.26	1.152	673	1.042	0.0423	0.0129	0.0434	0.0132	0.0695	0.0212
1250	633	—	91	2.98	0.117	32.74	1.289	842	1.305	0.0338	0.0103	0.0347	0.0106	0.0554	0.0169
1500	760	—	91	3.26	0.128	35.86	1.412	1011	1.566	0.02814	0.00858	0.02814	0.00883	0.0464	0.0141
1750	887	—	127	2.98	0.117	38.76	1.526	1180	1.829	0.02410	0.00735	0.02410	0.00756	0.0397	0.0121
2000	1013	—	127	3.19	0.126	41.45	1.632	1349	2.092	0.02109	0.00643	0.02109	0.00662	0.0348	0.0106

Notes:
1. These resistance values are valid **only** for the parameters as given. Using conductors having coated strands, different stranding type, and, especially, other temperatures changes the resistance.
2. Equation for temperature change: $R_2 = R_1 [1 + \alpha (T_2 - 75)]$ where $\alpha_{cu} = 0.00323$, $\alpha_{AL} = 0.00330$ at 75°C.
3. Conductors with compact and compressed stranding have smaller bare conductor diameters than those shown. See Table 5A for actual compact cable dimensions.
4. The IACS conductivities used: bare copper = 100%, aluminum = 61%.
5. Class B stranding is listed as well as solid for some sizes. Its overall diameter and area are those of its circumscribing circle.

Lightning Protection Systems

LIGHTNING PROTECTION FUNDAMENTALS

Lightning strikes can cause damage to buildings or structures and to the electrical wiring systems and equipment installed within those buildings. A lightning protection system consists of a low-impedance network of strike termination devices that are suitably connected to a special grounding electrode system installed to dissipate lightning into the Earth. Lightning protection systems are an effort to divert direct or indirect lightning strikes around the building or structure and equipment so as to safely dissipate the lightning into the Earth. **See Figure E-1.**

Purpose of a Lightning Protection System

Lightning protection systems provide a deliberate pathway to ground for lightning. The lightning protection system must be capable of dissipating the high energy to ground (Earth) as effectively and directly as possible.

The installation of a lightning protection system is no guarantee that equipment inside a building or structure or the building itself will not be damaged by a lightning strike. The system provides a plan to provide a reasonable degree of protection from these events. Some people have false beliefs that installing a lightning protection system on a building protects everything inside from lightning. A lightning protection system is the best-known form of protection from these natural events, but there is no guarantee. Lightning is an unpredictable force that is continuously being studied.

The purpose of *NFPA 780: Standard for the Installation of Lightning Protection*

Systems is to provide safeguarding of people and property from hazards arising from lightning exposure. Lightning protection systems do not prevent lightning strikes, nor do they attract lightning from distances greater than the conventional attractive area of a building or structure.

> **Lightning Protection System.** A complete system of strike termination devices, conductors (which could include conductive structural members), grounding electrodes, interconnecting conductors, surge protection, and other connectors and fittings required to complete the system.

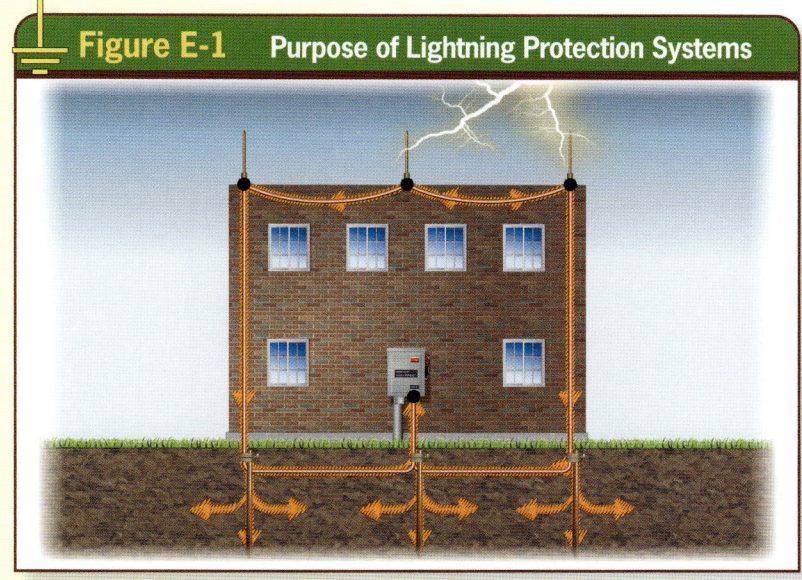

Figure E-1 Purpose of Lightning Protection Systems

Figure E-1. Lightning protection systems provide a path to Earth for lightning strikes.

As seen in this definition, there are several parts that are necessary to construct a lightning protection system. **See Figure E-2.**

A closer look at this definition reveals several concepts related to how the system is intended to perform. First, the system intercepts a lightning strike through the strike termination device. Then the force is diverted down to the Earth. This process is accomplished by

Figure E-2. Basic lightning protection system parts are strike termination devices, down conductors, grounding electrodes, interconnecting conductors, and surge protective devices.

Courtesy of Harger Lightning and Grounding

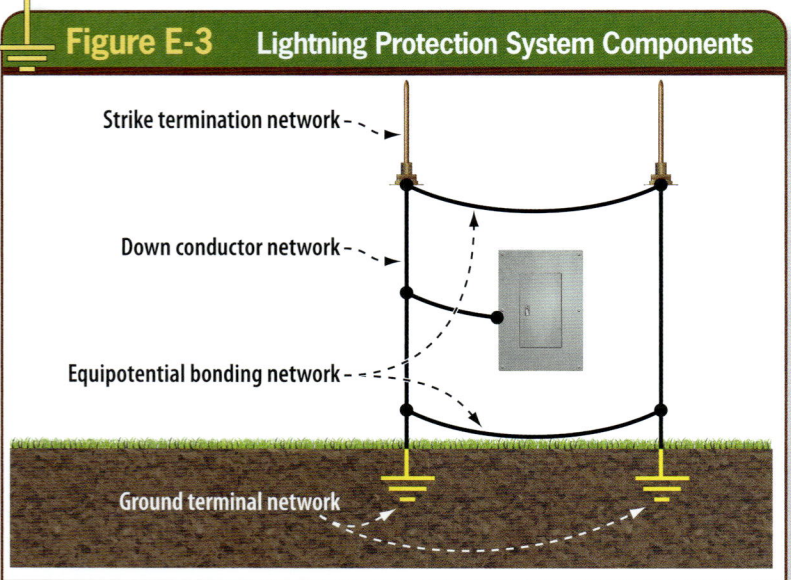

Figure E-3. A lightning protection system consists of several networks.

the specific down conductors or structural components of the building that serve as down conductors. It is important that bonding is provided from conductive parts on or within the building or structure to reduce flashover possibilities during the event. The grounding electrodes of a lightning protection system provide the dissipation safely into the ground (Earth). Lastly, effectively-applied surge protective devices handle any unwanted transient surges attempting to enter the building via the electrical supply system.

System Components

Lightning protection systems consist of several components that make up the entire system. Starting from the top of a structure and working towards the ground, the system includes a strike termination network (usually air terminals or lightning rods), a down conductor network, a grounding terminal or grounding electrode network, an equipotential bonding network, and appropriate surge protection. **See Figure E-3.**

There are two classes of materials and components used in lightning protection systems. Class I materials are used on structures 75 feet tall or less, and Class II materials and components must be used in buildings taller than 75 feet. Class I materials are typically smaller and lighter than Class II materials. Copper or aluminum conductors can be used, but attention must be given to areas subject to corrosive influences. The conductors used for lightning protection systems have a more finely-woven stranding characteristic than those typically used for electrical wiring systems.

Installation Methods and Criteria

In general, each strike termination device must be provided with two separate paths (down conductors) to the ground. A main conductor must be used between strike termination devices and for the down conductors. *NFPA 780* provides the minimum sizes required for roof conductors and down conductors. **See Figure E-4.**

A low-impedance path is necessary in the down conductor network to reduce opposition in this path to Earth. Care should be taken to minimize the number of bends and to ensure that any necessary bends are long radius (as gradual as possible). A bending radius must never be less than eight inches. Sharp bends invite flashover possibilities. If the voltage of a strike exceeds the breakdown voltage of air space between a down conductor and another conductive object, a side flash can occur during a lightning strike. *NFPA 780* provides a formula that simplifies the calculation required to determine the probability of side flash.

Network conductors on the roof, in addition to the down conductors of the system, must be securely fastened at appropriate intervals. A minimum of two down conductors is required for any application. For structures exceeding 250 feet in perimeter, additional down conductors must be installed. Down conductors should be placed at the corners of a building or structure and be separated as widely as possible in between. Good designs locate down conductors away from public areas or provide suitable protection around the down conductor in these areas. The ground network of a lightning protection system provides a low-impedance connection to the Earth. This low-impedance connection through multiple ground terminals helps minimize peak voltages that would be present on the system during an event.

The ground network must be designed and installed to reduce the possibility of step and touch potentials. The grounding electrode network of the system is the largest contributor to the level of overvoltage that will appear on the system if the ground connection is not effective. An ineffective grounding network will create opposition to the quick and safe dissipation of a lightning strike.

Each down conductor must be terminated to a grounding electrode dedicated to the lightning protection system. The grounding electrodes can be copper, copper-clad steel, or stainless-steel types. Electrodes of the ground network can be

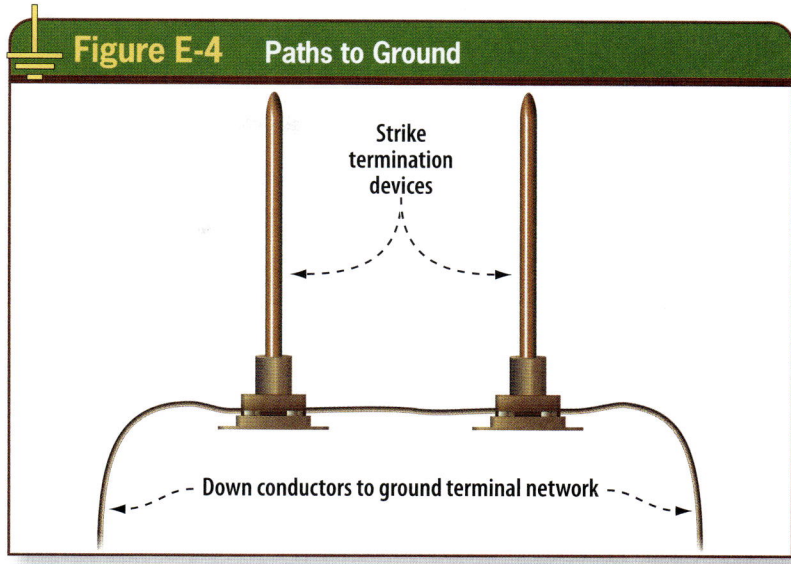

Figure E-4. *Two paths are required from air strike termination devices.*

rods, rings, plates, radials, and concrete-encased electrodes. Connections to the electrodes must be made by exothermic welding, bolting, brazing, or high compression connections listed for the application. Grounding clamps listed for direct burial application are permitted. **See Figure E-5.**

When possible, the electrodes should be installed below the frost line. As previously discussed, the ground network of the lightning protection system must be bonded to the grounding electrode

Figure E-5. *Lightning protection conductors use a variety of connection methods.*

Courtesy of Harger Lightning and Grounding

system for the power service supplying the building or structure.

Surge Protective Devices

Line surges can occur for a number of reasons. Among these are direct lightning strikes to incoming lines or strikes in close proximity to the building that will cause a rise in potential on the incoming lines. This includes the electrical service lines and any limited-energy lines such as telephone systems, antenna systems, or broadband communications systems.

Other means may need to be employed for protecting internal electrical and electronic systems against lightning electromagnetic pulses. Potential equalization can be achieved at electrical services by installing suitable listed surge protective devices (SPDs) and spatial shielding. All lightning protection systems require protection against incoming surges. The degree of surge protection is related to the size of the equipment and system being protected. **See Figure E-6.** Installation requirements for surge protective devices are provided in Part II of Article 242 of the *NEC*.

Quality Control for Lightning Protection Systems

The UL Master Label program and the Lightning Protection Institute (LPI) Certified System program provide quality assurances for installed lightning protection systems. Each program requires using components that are certified to *UL 96*. UL issues a master label for systems and components inspected on completion of the installation. The components of the system must be listed to *UL 96*, and the installation must meet the requirements in *UL 96A*. Master Label certificates must be renewed every five years or if the building changes structurally. **See Figure E-7.**

LPI Certified Systems offer similar quality control assurances. LPI Certified Systems must be inspected to verify conformance with LPI 175, *NFPA 780*, and *UL 96A* as applicable. There are lightning protection installers who have been trained to evaluate and apply appropriately-designed systems to any building or structure. Electrical contractors often offer this service as part of their contracting business, or they may subcontract to organizations that specialize in lightning protection system installation and certification.

More information about lightning protection system installations and the requirements for such systems are available in *NFPA 780: Standard for the Installation of Lightning Protection Systems*; *UL 96A, Installation Requirements for Lightning Protection Systems*; and LPI 175, *Lightning Protection Institute Standard of Practice for the Design –Installation – Inspection of Lightning Protection Systems*. For information about listed products suitable for use in lightning protection systems, refer to *UL 96, Lightning Protection Components*.

Figure E-6. *Type 2 surge protection is installed at the service on the load side of an overcurrent protective device.*

Courtesy of Eaton

Figure E-7. *A UL Master Label provides standards compliance and quality assurance for lightning protection systems.*

Index

A

Accessibility, 53, 53f, 144–145
  water piping system, 174–175, 175f
Agricultural installations
  equipotential planes in, 337–338, 338f
  site-isolating devices in, 336–337, 336–337f
  voltage gradients in, 334–335, 335f
Air piping, 177, 177f
Airfield lighting, 435, 435f
Alternating current (AC) systems
  Earth in the circuit of, 2, 3
  grounding electrodes and utility services, 93–94, 94f
  grounding service supplied, 425–426
  opposition to, 28–30, 29f
  Table 250.102(C)(1)-Sizing the Grounded Conductor, Main Bonding Jumper, System Bonding Jumper, and Supply-Side Bonding Jumper for Alternating-Current Systems, 67–68, 67f
  Table 250.66-Sizing AC Grounding Electrode Conductors, 66–67, 66f
  ungrounded systems, 16–17, 17f
Aluminum conductors, 130–131, 131f
Amperes (current). *See* Current (amperes)
Antennas, 390–392, 390–392f
APIRP 2003-2015, Protection Against Ignitions Arising Out of Static Lightning and Stray Currents, 324
Appliances, grounding of, 224–225, 225f
Armored-clad cable, 191–192, 191–192f, 328–329, 328–329f
Article 100-Definitions of Grounding and Bonding Terms, 56–64
Article 200-Use and Identification of Grounded Conductors, 64, 64f
Article 250-Arrangement and Use, 65, 65f
  methods of grounding equipment, 213–216, 214–216f
  Part I-General, 65–66

Table 250.102(C)(1)-Sizing the Grounded Conductor, Main Bonding Jumper, System Bonding Jumper, and Supply-Side Bonding Jumper for Alternating-Current Systems, 67–68, 67f
Table 250.122-Sizing Equipment Grounding Conductors, 68–69, 68f
Table 250.66-Sizing AC Grounding Electrode Conductors, 66–67, 66f
Article 517, 69–70, 69f
Article 600, 70
Article 770, 71
ASABE EP342.3-2010 (R2015), Safety for Electrically Heated Livestock Waterers, 334
ASABE EP473.2-2001 (R2015), Equipotential Planes in Animal Containment Areas, 338
Assured EGC program, 405
Authority having jurisdiction (AHJ), 54, 54f
Autotransformers, 292, 293f
Auxiliary grounding electrodes, 91–92, 92f, 225–226, 226f, 244, 244f
  in isolated/insulated circuits, 244

B

Boatyards, 340–341, 340–341f
Bonding
  around expansion fittings and loose-joined metal raceways, 173–174, 173–174f
  boxes with concentric and eccentric knockouts, 169, 169f
  bushings, 164, 165f
  cleaning coated surfaces, 163, 163f
  communications systems and equipment (*See* Communications systems and equipment)
  concepts of, 3–4f, 3–5
  connections, 162f, 163
  definition of, 56–57
  definition of terms, 158

to ferrous metal conduits, 151–153, 151–153f
function of, 13, 13f
grounding and, 8, 8–9f
lightning protection systems, 179, 179f
load side rules, 170–171, 170f
maintaining continuity in, 160–161, 161f
metal parts of signs and outline lighting systems, 348–351, 349–351f
metal water pipe systems, 174–177f, 174–178, 263, 263f, 302–303, 303f
multiple occupancy buildings, 175–176, 176f
other metal parts, 174–179, 174–179f
performance criteria, 158–160, 159–160f
reducing washers, 168–169, 168f
requirements for hazardous locations, 320–324, 320–324f
for safety, 2
at separate structures (*See* Separate buildings or structures)
service rules, 166–168, 167f
sizing requirements for supply side and load side applications, 164–165f, 165–166
structural metal building frames, 178–179, 178f
swimming pools and similar installations (*See* Swimming pools and similar installations)
Bonding jumpers, 56–58, 56–58f, 161, 161f
function and purpose of, 164, 164–165f
grounded conductor routing and connections, 104–105, 104–105f
installation and size of equipment, 171–172f, 171–173
length of, 164
main, in service equipment, 105–107, 105–107f
supply-side, 63–64, 169–170, 169–170f, 297–298, 297f
system, 62, 67–68, 67f, 294–296f, 294–297
Table 250.102(C)(1)-Sizing the Grounded Conductor, Main Bonding Jumper, System Bonding Jumper, and Supply-Side Bonding Jumper for Alternating-Current Systems, 67–68, 67f
water pipe electrode supplement and, 90–91, 91f

Bracketed text, *NEC,* 54–55
Branch circuits, 69, 69f, 213–214, 214f, 254
ground-fault protection of equipment (GFPE), 415
requirements for, 258–259, 258–259f
Busbars, 138, 138f
telecommunications main grounding, 381, 381f
Bushings, bonding, 164, 165f

C
Cable assemblies in parallel, 201–202, 202–203f
Cable shielding, 433–435, 433–435f
Chapter 8, *NEC,* 71
Choke effect, 150–151, 151f
Circuits
branch, 69, 69f, 213–214, 214f, 254, 258–259, 258–259f, 415
Earth in, 2, 3
equipment grounding conductors (EGC) for motor, 202, 202f
fundamentals of, 24, 24f
isolated/insulated (*See* Isolated/insulated grounding circuits and receptacles)
multiple, in single raceway or cable tray, 199, 199f
normal and fault current in, 30–31, 30f
Ohm's Law and, 24–28, 25–28f
opposition to current in, 28–30, 29–30f
path for current through the body, 6–8, 6–8f
short, 32–33, 32f, 467–478
Clamp-on electrode resistance meters, 88f, 89
Clamping voltage, 435
Cleaning of coatings, 163, 163f
Coatings, cleaning of, 163, 163f
Communications systems and equipment
common grounding and bonding rules for, 384–389, 385–389f
connecting to grounding electrode, 380–381, 381f
definitions, 378
grounding electrode conductor installation, 381
intersystem grounding and bonding, 381–384, 382–384f
at mobile homes, 389–390
overvoltages and lightning events, 392
performance and concepts, 378, 378–379f

radio and television equipment and antennas, 390–392, 390–392f
Compressed air systems, 177, 177f
Concentric and eccentric knockouts, 169, 169f
Concrete-encased electrodes, 83–84, 83f
Conductive materials, 13, 13f
Conductive pool shells, 360–361, 361f
Conductor insulation, 41–42f, 41–43
Conductor properties, 479
Conductor shielding, 433–435, 433–435f
Conduit fill requirements, 152, 152f
Conduit hubs and clamps, 152, 153f
Connections
  bonding, 162f, 163
  equipment grounding conductors (EGC), 194, 195f, 216, 216f, 222, 222f
  grounded conductor, 104–105, 104–105f
  grounding electrode conductors (GEC), 139–140, 139–140f, 148–150, 148–150f
  receptacle grounding, 216–219, 217–219f
  separable, 222
Consumer Product Safety Commission (CPSC), 8, 398
Continuity
  between attachment plugs and receptacles, 222
  bonding, 160–161, 161f, 173–174, 173–174f
  equipment grounding conductor (EGC), 219–220, 219–220f
Copper-clad conductors, 130–131
Corner-grounded system, 280, 281f, 282f
Cranes, electric, 211, 211f, 351–352, 351–352f
Current (amperes), 26–28, 26–28f
  in circuits, normal and fault, 30–31, 30f
  in equipment grounding conductors (EGC), 202–203
  in grounding electrode conductor (GEC), 130, 130f
  normal, 33
  objectionable, 234, 237–238, 238f
  operating overcurrent protective devices, 33–34, 33–34f
  opposition to, 28–30, 29–30f
  overcurrent protection, 31–33, 31f
  time and, 35, 36–40f, 41
  Watt's Wheel and, 28, 28f
Current paths, 30–31, 30f

effective ground-fault, 62–63, 63f
ground fault, 34, 35f
grounded systems, 14–15, 14f, 15
through the body, 6–8, 6–8f
ungrounded systems, 18
Current-limiting overcurrent protection device, 41

D
Daisy-chained installation, 169, 169f
Definitions, *NEC*, 55, 55f
Direct current (DC) systems
  Earth in the circuit of, 2, 3
  grounded, 311–313, 311–313f
  grounding electrode conductors (GEC) for, 134–135, 135f
  opposition to, 28–30, 29f
  ungrounded, 313–314, 314f
Disconnection, service
  grounding electrode conductors (GEC), 140–144, 140–144f
  grounding electrodes, 119–120, 119–120f
  at separate buildings or structures, 174–177f, 174–178, 262, 262f, 263, 263f
Docking facilities, 340–341, 340–341f
Double-insulated pool pump motors, 363, 363f
Dual-fed service equipment, 107, 107f

E
Earth in the circuit, 2–3, 3f
Earth, grounding requiring connection to, 58, 58f, 128–130, 129f
Effective ground-fault current path, 62–63, 63f
Effectiveness, grounding path, 144–145, 144–145f
Electric cranes, 211, 211f, 351–352, 351–352f
Electric shock drowning (ESD), 340
Electric shock severity, 398, 398f
Electric signs, 211, 211f, 346–351, 346–351f
Electrical metallic tubing (EMT), 160, 160f
Electrical systems, 10–19
  grounded, 10–15, 11–15f, 158, 159f
  ungrounded, 15–19, 15–19f, 120–122, 121f, 158, 159f
Electromagnetic interference (EMI), 234–235f, 234–236
Elevators, 211, 211f, 351–352, 353f

Enclosures. *See* Raceways and enclosures
Engineered grounding system designs,
 437–440, 438–440f
Equipment grounding conductors (EGC),
 30, 31, 33, 34, 60, 61f
 annealed or opened, 42
 assured EGC program, 405
 auxiliary grounding electrodes, 91–92,
  92f, 225–226, 226f
 basics of, 184–192, 185–192f
 bonding with, 170–171, 171f
 cable assemblies in parallel, 201–202,
  202–203f
 capacity of, 43–44, 43f
 Code recognized, 186, 186f
 connections, 194, 195f, 216, 216f, 222,
  222f
 continuity, 219–220, 219–220f
 current in, 202–203
 defined, 184, 184f
 with feeder conductors, 213–214, 214f
 with feeder taps, 202, 203f
 identification, 195–196f, 195–197
 installation of, 192–194, 192–194f
 installation of circuit conductors,
  193–194, 194f
 isolated, 242–243, 243f
 metallic cable assemblies, 189–192,
  190–192f
 for mobile homes, 338–339, 339f
 for motor circuits, 202, 202f
 multiple circuits in single raceway or
  cable tray, 199, 199f
 for parallel runs, 199–201, 199–201f
 in patient care spaces, 329–330, 330f
 performance of, 184–186, 185–186f
 protection of, 43, 193
 rules for grounding, 208–209, 208–209f
 site-isolating device, 336–337, 336f
 sizing, 197–202, 197–202f
 sizing criteria, 196–197, 196f
 swimming pools and similar
  installations, 356–357, 357f
 Table 250.122-Sizing Equipment
  Grounding Conductors, 68–69, 68f
 types of, 186–192
Equipotential bonding, 360–364,
 360–364f
Equipotential grounding grid, substation
 installations, 431–432, 431–432f
Equipotential planes, 337–338, 338f,
 340–341
Exothermic welding, 89, 89f

Expansion fittings, bonding around,
 173–174, 173–174f
Explanatory information, *NEC*, 54, 54f

F
Feeder conductors, 213–214, 214f
 at separate buildings or structures,
  258–259, 258–259f
Feeder ground-fault protection of
 equipment (GFPE), 414f, 415
Feeder taps, equipment grounding
 conductors (EGC) with, 202, 203f
Fences, metal, 432–433
Ferrous metal enclosures, 151, 151f
Fixed metal parts, pool, 363
Flexible conduit installations, 323, 323f
Flexible metal conduit (FMC), 349–350,
 349f
Floor boxes, 218, 218f
Footing-type grounding electrodes,
 445–446
 background on, 446–447
 building installations, 449–454
 conclusion, 454
 ground resistance tests, 447–448
 length of electrodes, 454
 preliminary investigation, 448–449
Foundations, 3, 3f
Fountains, 366, 366f

G
Gas piping systems, 92–93, 93f
 bonding, 176–177, 177f
Generators
 separate buildings or structures supplied
  by, 266–268, 266–268f
 separately derived systems, 305–309,
  305–309f
Grids, 246–247, 246–247f
Ground detection, 282–283, 282f
Ground fault, 12, 58–59, 274
 current path, 14–15, 14f, 15, 34, 35f,
  62–63, 63f
Ground loop, 234, 234f
Ground resistance testing, 87–89, 87–89f
Ground ring electrodes, 84–85, 84f
Ground-fault circuit interrupters (GFCI),
 8, 341
 assured EGC program, 405
 in dwelling occupancies, 400f, 401
 function of, 398–399f, 399–400
 in health care facilities, 402–403, 403f
 NEC requirements for, 400–405f,
  400–406

in other than dwelling units, 401, 401f
protection-receptacle replacements,
    403–404, 403–404f
purpose of, 398, 398f
special purpose, 406
temporary wiring installations,
    404–405, 405f
Ground-fault detector-interrupter, solar
    PV systems, 370
Ground-fault events, 5, 5f
in agricultural facilities, 335
Ground-fault protection of equipment
    (GFPE), 341, 406
applicability in health care facilities,
    414–415, 414f
branch circuit, 415
feeder, 414f, 415
neutral ground-strap system, 408–409f,
    408–410
purpose of, 407–408, 407–408f
selective coordination, 412–413, 412f
system coordination, 413, 413f
testing of, 415, 415f
types of, 408–411, 408–412f
zero-sequence systems, 410–411,
    410–412f
Grounded conductors, 59–60, 59–60f
disconnect requirements for services,
    119–120, 119–120f
dual-fed service equipment, 107, 107f
as first line of defense, 101–102,
    101–102f
functions (purposes) of, 109–110, 109f
grounding scheme for services,
    102–104, 103f
identification, 115–116, 115–116f,
    283–285, 283–285f
isolated from ground, 215–216, 215f
load-side use, 113–115, 114f
main bonding jumpers in service
    equipment, 105–107, 105–107f
marking equipment for ungrounded
    systems, 121–122, 121f
minimizing impedance in service,
    107–109, 108f
neutral, 110–116, 110–116f
requirements for service equipment
    (listing), 116–118, 117–119f
routing and connections, 104–105,
    104–105f
separately derived systems, 298–299,
    298f
service equipment, 100–101, 100f

for service raceways and enclosures, 122,
    122f
sizing for parallel installations, 112–113,
    113f
sizing requirements, 111–112, 111–112f
supply-side grounding at other than a
    service, 122–123
Table 250.102(C)(1)-Sizing the
    Grounded Conductor, Main Bonding
    Jumper, System Bonding Jumper, and
    Supply-Side Bonding Jumper for
    Alternating-Current Systems, 67–68,
    67f
used for grounding, 228, 228f
Grounded electrical systems, 10–15,
    11–15f, 158, 159f
Grounded equipment, 12, 12f, 59–60,
    59–60f
Grounded utility supply systems,
    100–101, 100f
Grounding concepts, 3–4f, 3–5
bonding and, 8, 8–9f
definitions in, 2, 2f, 56–57, 59
Earth connection, 58, 58f, 128–130,
    129f
minimization of shock hazards, 5, 5f
performance code language and, 9–10,
    9f
for safety, 2
Grounding conductors, surge protection,
    437, 437f
*Grounding Electrical Distribution Systems
    for Safety,* 44
Grounding electrode conductors (GEC),
    61–62, 62f
for antennas, 390–392, 390–392f
basics of, 128, 128f
communications systems and
    equipment, 381, 384–389, 384–389f
connection locations, 146–147, 147f
connection point, 139–140, 139–140f
connection to busbar, 138, 138f
connections, 148–150, 148–150f
current in, 130, 130f
for DC systems, 134–135, 135f
effectiveness (integrity) of grounding
    path, 144–145, 144–145f
installation of, 135–144, 136–144f
magnetic field concerns, 150–153,
    151–153f
materials for, 130–131, 131f
methods of connection at service
    disconnecting, 140–144, 140–144f

for multiple systems, 300–302,
300–302f
protection from physical damage, 136,
136–137f
purpose of, 128–130, 129f
routing from individual electrodes, 137,
137f
at separate buildings, 256–257, 257f
Table 250.66-Sizing AC Grounding
Electrode Conductors, 66–67, 66f
using Table 250.66 and, 133–134, 133f
Grounding electrodes, 2, 2f, 3, 50, 51f,
60–61, 60f
accessibility, 53, 53f
auxiliary, 91–92, 92f, 225–226, 226f,
244, 244f
bonding of lightning protection systems
to service electrode systems, 94, 95f
concrete-encased, 83–84, 83f
connecting communications systems
and equipment to, 380–381, 381f
defined, 76–77, 77f
establishing system of, 79–80, 80f
footing-type (*See* Footing-type
grounding electrodes)
ground ring electrodes, 84–85, 84f
installation requirements, 85–94,
86–95f
introduction to, 76, 76f
items not permitted as, 92–93, 93f
mandatory, 80–81
metal in-ground support structures,
82–83, 82f
metal underground water pipe, 81–82,
81–82f
other local metal underground systems
or structures, 85
plate electrodes, 85, 85f, 89, 89f
purpose and performance of, 77–78f,
77–79
rod and pipe electrodes, 85, 86–87, 86f
at separate buildings (*See* Separate
buildings or structures)
separately derived systems, 299–302,
299–302f
soil resistivity and ground resistance
testing, 87–89, 87–89f
solar PV systems, 371–372, 371f
spacing requirements, 89–90, 90f
system requirements, 79, 79f
types of, 81–85, 81–85f
utility services and, 93–94, 94f
water pipe electrode supplement and
bonding jumpers, 90–91, 91f

Grounding transformers, 280–281, 281f
Grounding-type receptacles, 223, 224f
Grounding, electrical equipment
appliances, 224–225, 225f
auxiliary electrode requirements,
225–226, 226f
communications systems and equipment
(*See* Communications systems and
equipment)
conductor enclosure and raceway
grounding requirements, 212–213,
212–213f
connections of equipment grounding
conductors, 216, 216f
continuity between attachment plugs
and receptacles, 222
feeders and branch circuits, 213–214,
214f
general rules for, 208–212, 208–212f
identification of wiring device terminals,
221, 221f
isolating the grounded conductor from
ground, 215–216, 215f
methods, 213–216, 214–216f
panelboard terminal bars, 214–215,
214–215f
in patient care spaces, 324–332,
324–332f
purpose of, 208, 208f
receptacle grounding connections,
216–219, 217–219f
receptacle replacements, 223–224,
223–224f
by secure metal supports, 227–228, 227f
at separate structures (*See* Separate
buildings or structures)
snap switches, 220–221, 221f
use of grounded conductor for, 228,
228f
Grounding, electrical systems
definitions in, 274
grounded system voltages, 279–280,
280f
grounding separately derived systems,
282
impedance grounded systems, 286–287,
286–287f
mandatory system grounding, 277–281,
277–281f
optional system grounding, 281–285,
282–285f
requirements for, 276–277
requirements for ground detection,
282–283, 282f

system grounding, 274–275f, 274–276
system grounding prohibition, 285–286
ungrounded systems (concepts),
   287–288, 287f
using grounding transformers, 280–281,
   281f
various methods of, 276, 276f
Grounding, nonelectrical equipment, 227
Grounding, special equipment
   electric cranes and elevators, 211, 211f,
      351–352, 351–352f
   electric signs and outline lighting
      systems, 346–351, 346–351f
   information technology equipment and
      sensitive electronic equipment,
      352–356, 353–355f
   purpose of, 346, 346f
   solar PV systems, 368–371f, 369–372
   swimming pools and similar
      installations, 356–368f, 356–369

H
Hazardous locations, 320–324, 320–324f
Health care facilities
   ground-fault circuit interrupters (GFCI)
      in, 402–403, 403f
   ground-fault protection of equipment
      (GFPE) in, 414–415, 414f
   grounding and bonding requirements
      for panelboards in, 332–334,
      332–334f
   isolated grounding circuits in, 243–244,
      243f
   patient care spaces, 324–332, 324–332f
   special rules for, 324–334, 324–334f
   therapeutic pools and tubs in, 366–369,
      367–368f
High-leg delta system, 275, 275f
Hot tubs, 364–366, 365f
Hydromassage bathtubs, 368–369, 368f

I
Identification
   equipment grounding conductors
      (EGC), 195–196f, 195–197
   grounded conductor, 115–116,
      115–116f
   of wiring device terminals, 221, 221f
IEEE 3003.2-2014, Recommended Practice
   for Equipment Grounding and Bonding
   in Industrial and Commercial Power
   Systems, 441
IEEE 80, Guide for Safety in AC Substation
   Grounding, 430, 433

Immersion pools, 364–366, 365f
Impedance, 3, 28–29
   grounded systems, 286–287, 286–287f
   minimization in service grounded
      conductors, 107–109, 108f
Impedance grounding, 426–427,
   426–427f, 440
In-ground support structures, 82–83, 82f
Incident energy, 41
Individual EBJ installation, 170, 170f
Information technology centers,
   grounding and bonding in, 244–245,
   245f, 352–356, 353–355f
Installation
   circuit conductors, 193–194, 194f
   equipment bonding jumpers, 171–172f,
      171–173
   equipment grounding conductors
      (EGC), 192–194, 192–194f, 224,
      224f
   grounding electrode, 85–94, 86–95f
   grounding electrode conductors (GEC),
      135–144, 136–144f
   lightning protection systems, 482–484
Insulated Cable Engineers Association
   (ICEA), 42, 42f
Interrupting rating, 32
Intersystem bonding termination (IBT),
   380–381, 381f
   grounding and bonding, 381–384,
      382–384f
Islanded power systems, 440
Isolated/insulated grounding circuits and
   receptacles
   electromagnetic interference (EMI) in
      grounding circuits and, 234–235f,
      234–236
   equipment grounding circuits for
      equipment, 242–243, 243f
   in health care facilities, 243–244, 243f,
      330–331, 331f, 332, 332–333f
   in information technology centers,
      244–245, 245f
   isolated ground-type receptacles,
      218–219, 218–219f
   objectionable currents in grounding
      paths and, 237–238, 238f
   panelboards and isolated grounding
      circuits, 241–242, 242f
   power quality system grounding analysis
      and, 238–239
   purpose of, 236–237, 236–237f
   signal reference structures (grids),
      246–247, 246–247f

surge protection, 248–249, 248–249f
use of auxiliary grounding electrodes in, 244, 244f
wiring rules for, 239–241, 239–241f

J
Junction boxes, 359, 359f

L
Let-Go thresholds, 8
Lightning events, communications systems and equipment and, 392
Lightning protection systems, 94, 95f
bonding, 179, 179f
components of, 482
installation methods and criteria, 482–484
purpose of, 481–482
quality control, 484
surge protective devices, 484
Liquidtight flexible metal conduits, 323–324, 323–324f, 327–328, 327f
Load-side use
bonding jumpers, 164–165f, 165–166, 170–171, 170f
grounded conductors, 113–115, 114f
Loose-joined metal raceways, bonding around, 173–174, 173–174f
Luminaires, 211, 211f

M
Magnetic field concerns with grounding electrode conductors (GEC), 150–153, 151–153f
Main bonding jumpers, 105–107, 105–107f
Mandatory grounding electrodes, 80–81
Mandatory system grounding, 277–281, 277–281f
Manufactured homes, 339
Marinas, 340–341, 340–341f
Medium- and high-voltage systems
common grounding and bonding components, 440–441
conductor shielding and stress reduction, 433–435, 433–435f
engineered grounded system designs, 437–440, 438–440f
equipotential grounding grid, 431–432, 431–432f
grounding methods for systems over 1,000 volts, 422–423, 422f
grounding of equipment, 428–430, 428–430f

grounding service supplied alternating current systems, 425–426
grounding through surge arresters, 430, 430f, 435–437, 436–437f
impedance grounding, 426–427, 426–427f
metal structures and fences, 432–433
multi-point grounding, 424–425, 425f
portable or mobile equipment grounding, 427–428
requirements for grounding systems, 422, 422f
single-point grounding, 424, 424f
solidly grounded systems, 423–425, 423–425f
substation grounding requirements, 430–432f, 430–433
Megohmmeters, 41f
Metal piping systems, 13, 13f
bonding, 174–177f, 174–178, 263, 263f
gas, 92–93, 93f
in-ground support structures, 82–83, 82f
local metal underground, 85
at separate buildings or structures, 263, 263f
separately derived systems, 302–303, 303f
underground water pipe electrodes, 81–82, 81–82f
Metal supports, grounding by, 227–228, 227f
Metal-clad cable, 190–191, 190–191f, 328–329, 328–329f
Metallic cable assemblies, equipment grounding conductors (EGC), 189–192, 190–192f
Mobile homes, 338–339, 339f
communications systems and equipment at, 389–390
Motor circuits, equipment grounding conductors (EGC) for, 202, 202f
Motor controllers, 210f, 211
Motor frames, 210, 210f
Motors, pool pump, 357–358, 357–358f
double-insulated, 363, 363f
Multi-conductor cables, 196
Multi-point grounding, 424–425, 425f
Multiple buildings or structures supplied by feeder or branch circuits, 176, 176f
Multiple occupancy buildings, metal water piping in, 175–176, 176f

Multiple service disconnects, 104–105, 104–105f

N

National Electrical Code® (NEC), 2
  arrangement and application of, 50–52, 51–52f
  Article 100-Definitions of Grounding and Bonding Terms, 56–64
  Article 200-Use and Identification of Grounded Conductors, 64, 64f
  Article 250-Arrangement and Use, 65–68f, 65–69
  bracketed text, 54–55
  Chapter 8, 71
  Code-making panel responsibilities, 55–56
  defined terms, 55, 55f
  enforcement and approvals, 52
  explanatory information, 54, 54f
  performance language in, 9–10, 9f
  permissive language in, 54, 54f
  purpose of, 50, 50f
  requirements, exceptions, alternatives, and information, 53, 53f
  special occupancies, equipment, and conditions, 69–70f, 69–71
National Electrical Contractors Association *(NECA),* 8
National Electrical Installation Standards (NEIS), 8
Neon signs, 349, 349f, 350, 350f
Neutral conductors, 110–116, 110–116f
Neutral disconnect requirement for services, 119–120, 119–120f
Neutral ground-strap system, 408–409f, 408–410
Neutral points, 110–116, 110–116f
NFPA 54: National Fuel Gas Code, 178
NFPA 70E: Standards for Electrical Safety in the Workplace, 372
NFPA 77: Recommended Practice on Static Electricity, 324
NFPA 780: Standard for The Installation of Lightning Protection Systems, 12, 16, 94, 178, 324, 380, 484
NFPA 99: Health Care Facilities Code, 243–244, 331, 414
Non-grounding type receptacles, 223–224, 223f
Normal current, 33

O

Objectionable currents, 234, 237–238, 238f
Ohm's Law, 24–28, 25–28f, 88
Ohms (resistance), 3, 25–26, 26f
Outdoor source, 303, 303f
Outline lighting systems, 346–350f, 346–351
Overcurrent protection, 31–33, 31f, 41
  amperes operating, 33–34, 33–34f
  at separate buildings or structures, 260, 261f
Overhead distribution networks, 438, 438f
Overhead surge arresters, 436, 436f
Overvoltages, communications systems and equipment, 392

P

Pad-mounted utility transformers, 101, 102f
Panelboards, 262f
  equipment grounding terminal bars, 214–215, 214–215f
  grounding and bonding in health care facilities, 332–334, 332–334f
  and isolated grounding circuits, 241–242, 242f
Parallel cable assemblies, 201–202, 202–203f
Parallel circuit through the body, 7–8
Parallel installations, grounded conductor sizing for, 112–113, 113f
Parallel runs, equipment grounding conductors (EGC) for, 199–201, 199–201f
Performance criteria, bonding, 158–160, 159–160f, 379–380, 380f
Performance language, 9–10, 9f
Performance, equipment grounding conductors (EGC), 184–186, 185–186f
Perimeter surfaces, pool, 361–362, 362f
Permissive language, 54, 54f
Photovoltaic systems. *See* Solar PV systems
Pipe organs, 210, 210f
Plate electrodes, 85, 85f
  installation, 89, 89f
Pole-mounted electric signs, 348, 348f
Pole-mounted utility transformers, 101, 101f
  grounding not required for, 209, 209f
Pool pump motors, 357–358, 357–358f
  double-insulated, 363, 363f

Pool shells, 360–361, 361f
Pool water bonding, 363–364, 364f
Power (watts), 28, 28f
Power distribution units (PDUs), 234–235, 235f, 353, 353f
Power quality system grounding analysis, 238–239
Practical Safeguarding, 2
Protection
  equipment grounding conductors (EGC), 43, 193
  ground-fault circuit interrupters (GFCI) (*See* Ground-fault circuit interrupters (GFCI))
  grounding electrode conductors (GEC), 136, 136–137f
  lightning, 94, 95f, 179, 179f
  overcurrent, 31–34, 31f, 33–34f, 41, 260, 261f
  surge, 248–249, 248–249f
Pump motors, pool, 357–358, 357–358f

R

Raceways and enclosures
  bonding around expansion fittings and loose-joined metal, 173–174, 173–174f
  conductor enclosure and grounding requirements for, 212–213, 212–213f
  ferrous metal, 151, 151f
  grounding of service, 122, 122f
  individual bonding jumpers, 170, 170f
  multiple circuits in single, 199, 199f
  underwater luminaires, 359, 359f
Radio and television equipment and antennas, 390–392, 390–392f
Receptacles
  ground-fault circuit interrupters (GFCI) protection-receptacle replacements, 403–404, 403–404f
  grounding connections, 216–219, 217–219f
  grounding in patient care spaces, 324–332, 324–332f
  isolated/insulated (*See* Isolated/insulated grounding circuits and receptacles)
  replacements, 223–224, 223–224f
Reducing washers, 168–169, 168f
Resistance (Ohm's), 3, 25–26, 26f
  soil resistivity and ground resistance testing, 87–89, 87–89f
  Watt's Wheel and, 28, 28f

Rigid metal conduits (RMC), 57, 160, 214, 214f
Rod and pipe electrodes, 85
  installation, 86–87, 86f
Routing, grounded conductor, 104–105, 104–105f

S

Safety
  grounding and bonding for, 2
  protecting conductor insulation, 41–42f, 41–43
  protecting wire-type equipment grounding conductors, 43
  solar PV systems, 372
  by system design, 41–42f, 41–43
Selective coordination, ground-fault protection of equipment (GFPE), 412–413, 412f
Self-grounding, 218, 218f
Sensitive electronic equipment, 352–356, 353–355f
Separable connections, 222
Separate buildings or structures
  bonding at, 176, 176f
  definitions in, 254
  disconnecting means requirements, 262, 262f, 263–264, 264f
  feeder and branch circuit requirements, 258–259, 258–259f
  grounding electrode conductor, 256–257, 257f
  grounding electrode requirement, 255–256, 256f
  metal water pipe bonding, 263, 263f
  pool wiring, 359
  purpose of grounding and bonding at, 254–255, 255f
  supplied by generators, 266–268, 266–268f
  supplied by separately derived system, 260–262, 261f, 264–265, 265f
  supplied by ungrounded system, 265–266, 266f
  supplying power to, 254
  ungrounded systems supplying, 260, 260f
Separately derived systems
  bonding water piping and building steel, 302–303, 303f
  DC systems, 311–313, 311–313f
  definitions in, 292
  determining, 292, 292–293f

generators and transfer equipment, 305–309, 305–309f
grounded conductor sizing, 298–299, 298f
grounding electrodes, 299–302, 299–302f
grounding of, 282, 293–298f, 293–299
grounding requirements for, 293
outdoor source, 303, 303f
separate buildings or structures supplied by, 260–262, 261f, 264–265, 265f
supply-side bonding jumper, 297–298, 297f
system bonding jumper, 294–296f, 294–297
ungrounded systems, 304–305, 304–305f
wind electrical systems, 309–310f, 309–311
Series circuit through the body, 6–7, 7f
Service conductors, 100–101, 100f
Service disconnection
grounding electrode conductors (GEC), 140–144, 140–144f
grounding electrodes, 119–120, 119–120f
Service equipment requirements, 116–118, 117–119f
bonding, 166–168, 167f
disconnect, 119–120, 119–120f
Shielding, conductor and cable, 433–435, 433–435f
Shock hazards, minimization of, 5, 5f
Short-circuits, 32–33, 32f, 42–43
current calculations, 467–478
current withstand chart, 42, 42f
Signal reference structures (grids), 246–247, 246–247f
Single-point grounding, 424, 424f
Site grounding and bonding plans, 439–440, 440f
Site power distribution network, 438–439, 438–439f
Site-isolating devices, 336–337, 336–337f
Sizing
bonding jumpers supply side and load side applications, 164–165f, 165–166, 169–170, 169–170f
equipment grounding conductors (EGC), 196–197, 196f
grounded conductors, 111–112, 111–112f
grounded conductors parallel installations, 112–113, 113f

grounding electrode conductors (GEC), 131–133, 131–133f
Table 250.102(C)(1)-Sizing the Grounded Conductor, Main Bonding Jumper, System Bonding Jumper, and Supply-Side Bonding Jumper for Alternating-Current Systems, 67–68, 67f
Table 250.122-Sizing Equipment Grounding Conductors, 68–69, 68f
Table 250.66-Sizing AC Grounding Electrode Conductors, 66–67, 66f
Snap switches, 220–221, 221f
Soil resistivity, 87–89, 87–89f
Solar PV systems, 368f, 369
definition, 369–370
electrical safety, 372
equipment grounding, 370–371, 370–371f
ground-fault detector-interrupter, 370
grounding electrode systems, 371–372, 371f
system grounding, 369, 369f
Solidly grounded systems, 423–425, 423–425f
Source of the system, 274f, 275
Spacing requirements, 89–90, 90f
Spas, 364–366, 365f
Speakers, underwater, 364, 365f
Special occupancies and conditions
agricultural installations, 334–338, 335–338f
health care facilities, 324–334, 324–334f
marinas, boatyards, and docking facilities, 340–341, 340–341f
mobile and manufactured home groundings and bonding rules, 338–339, 339f
National Electrical Code® (NEC), 69–70f, 69–71
special rules for hazardous locations, 320–324, 320–324f
Steel conduit and EMT, 455–456
complete report and other information, 465
as effective ground-fault current path, 457
Georgia Institute of Technology grounding research and, 457–459
NEC requirements, 456–457
summary, 460–461
tables reflecting maximum length of, 462–465

Structural steel, 13, 13f
  bonding, 178–179, 178f, 302–303,
    303f
Substation grounding, 430–432f,
  430–433
Supply-side bonding jumpers, 63–64
  separately derived systems, 297–298,
    297f
  service rules, 166–168, 167f
  sizing requirements, 164–165f,
    165–166, 169–170, 169–170f
  Table 250.102(C)(1)-Sizing the
    Grounded Conductor, Main Bonding
    Jumper, System Bonding Jumper, and
    Supply-Side Bonding Jumper for
    Alternating-Current Systems, 67–68,
    67f
Supply-side grounding at other than a
  service, 122–123
Surface covers, 218, 218f
Surge arresters, 430, 430f, 435–437,
  436–437f
Surge-protective devices (SPD), 248–249,
  248–249f
  lightning protection systems, 484
Swimming pools and similar installations,
  356, 356f
  bonding and equipment grounding,
    356–357, 357f
  equipotential bonding requirements,
    360–364, 360–364f
  fountains, 366, 366f
  grounding requirements for solar PV
    systems, 368–371f, 369–372
  junction boxes and other enclosures,
    359, 359f
  pool pump motors, 357–358, 357–358f
  separate buildings, 359
  spas, hot tubs, and permanently installed
    immersion pools, 364–366, 365f
  therapeutic pools and tubs for health
    care use, 366–369, 367–368f
  underwater luminaires, 358–359, 359f
System and equipment grounding, 9–10,
  9f, 274–275f, 274–276
System bonding jumpers, 62
  separately derived systems, 294–296f,
    294–297
  Table 250.102(C)(1)-Sizing the Grounded
    Conductor, Main Bonding Jumper,
    System Bonding Jumper, and Supply-
    Side Bonding Jumper for Alternating-
    Current Systems, 67–68, 67f

System coordination, ground-fault
  protection of equipment (GFPE), 413,
  413f
System design, safety by, 41–42f, 41–43

T

T-connected transformers, 280–281, 281f
Table 250.102(C)(1)-Sizing the Grounded
  Conductor, Main Bonding Jumper,
  System Bonding Jumper, and Supply-
  Side Bonding Jumper for Alternating-
  Current Systems, 67–68, 67f
Table 250.122-Sizing Equipment
  Grounding Conductors, 68–69, 68f
Table 250.66-Sizing AC Grounding
  Electrode Conductors, 66–67, 66f,
  133–134, 133f, 138, 138f
Telecommunications main grounding
  busbar, 381, 381f
Temporary wiring installations, ground-
  fault circuit interrupters (GFCI) for,
  404–405, 405f
Terminal bars, 214–215, 214–215f
Therapeutic pools and tubs, 366–369,
  367–368f
Time-current curves, 35, 36–40f, 41
Tingle voltages, 334–335
Tower grounding, 310
Transfer equipment, separately derived
  systems, 305–309, 305–309f
Transformers, grounding, 280–281, 281f
Turbines, 310–311

U

*UL 467, Grounding and Bonding
  Equipment*, 136
UL 514A, Metallic Outlet Boxes, 169
Underground water pipe electrodes,
  81–82, 81–82f, 85
  bonding jumpers and, 90–91, 91f
Underwater luminaires, 358–359, 359f
Underwater speakers, 364, 365f
Ungrounded electrical systems, 15–19,
  15–19f, 158, 159f, 287–288, 287f
  DC systems, 313–314, 314f
  marking equipment for, 121–122, 121f
  requirements for services supplied by,
    120–121, 121f
  at separate buildings or structures, 260,
    260f, 265–266, 266f
  separately derived systems, 304–305,
    304–305f

Utility services grounded systems, 100–101, 101f. *See also* Grounded conductors

V

Voltage, 24–25, 25f
    agricultural facilities, 334–335, 335f
    grounded system, 11, 11f, 279–280, 280f
    tingle, 334–335
    Watt's Wheel and, 28, 28f

W

Water heaters, pool, 363, 363f
Water pipe electrodes, underground, 81–82, 81–82f, 85
    bonding jumpers and, 90–91, 91f
Water piping systems bonding, 174–177f, 174–178
    at separate buildings or structures, 263, 263f
    separately derived systems, 302–303, 303f
Water pumps, 212, 212f
Watt's Wheel, 28, 28f
Wind electrical systems, 309–310f, 309–311
Wire-type conductors, 162f, 163, 196–197, 196f, 329–330, 330f
Wiring device terminals, identification of, 221, 221f

Z

Zero-sequence systems, 410–411, 410–412f
Zigzag-connected transformers, 280–281, 281f